UNIVERSITY OF ST. THOMAS LIBRARIES

WITHDRAWN
UST
Libraries

Active Margin Basins

Edited by

Kevin T. Biddle
Exxon Company, International
Houston, Texas, U.S.A.

AAPG Memoir 52

Chief Editor of World Petroleum Basin Memoirs:
Anny B. Coury

Published by
The American Association of Petroleum Geologists
Tulsa, Oklahoma 74101, U.S.A.

Copyright © 1991
The American Association of Petroleum Geologists
All Rights Reserved
Published 1991

ISBN: 0-89181-330-6

AAPG grants permission for a single photocopy of any article herein for research or noncommercial educational purposes. Other photocopying not covered by the copyright law as Fair Use is prohibited. AAPG participates in the Copyright Clearance Center. For permission to photocopy more than one copy of any article, or parts thereof, contact: Permissions Editor, AAPG, P.O. Box 970, Tulsa, Oklahoma 74101.

Association Editor: Susan Longacre
Science Director: Gary D. Howell
Publications Manager: Cathleen P. Williams
Special Projects Editor: Anne H. Thomas
Production: Custom Editorial Productions, Inc., Cincinnati, Ohio

Dust jacket photo by Kevin Biddle. Mixed gravity map and digital topography map, southern California.

Basin Memoir Editors:

Interior Cratonic Basins
Morris W. Leighton
Dennis R. Kolata
Donald F. Oltz
J. James Eidel

Interior Rift Basins
John C. Harms
Susan M. Landon
Michael Steckler

Divergent/Passive Margin Basins
J. D. (Jack) Edwards
P. A. Santogrossi

Active Margin Basins
Kevin T. Biddle

Foreland Basins and Foldbelts
Roger W. Macqueen
Dale A. Leckie

About the Editor

Kevin Biddle currently is a Geological Advisor with Exxon Company, International, in Houston, Texas, where his primary responsibility is new ventures exploration in Latin America. Dr. Biddle joined Exxon in 1978 and spent the next ten years at Exxon Production Research Company (EPR). His last assignment with EPR was as Research Supervisor of the Basin Framework section. He holds a Bachelor of Science degree in Earth Sciences from the University of California at Santa Cruz (1973) and a Master of Arts (1976) degree and a Ph.D. (1979) in Geology from Rice University.

Dr. Biddle's interests range from stratigraphy and basin analysis to tectonic controls on sedimentation with an emphasis on how basins form and deform with time. He is particularly interested in deformation associated with large-scale strike-slip faulting. This interest stems perhaps from having grown up a few miles to the west of the San Andreas fault. In 1985, he co-edited (along with Nicholas Christie-Blick) a collection of papers on strike-slip deformation, basin formation, and sedimentation, which was issued as SEPM Memoir 37. Participating in the AAPG's Petroleum Basin Memoir series volume on Active Margins provided him an opportunity to continue this line of work.

Foreword

The World Petroleum Basins project was conceived in order to aid explorationists in deciding what basins, or parts of basins, show the greatest promise for the most profitable return on investment. One way of achieving this goal is to supply these explorationists with a broad comparable base of data and concepts to improve their forecasts through analog techniques. Several volumes have been compiled to emphasize the geological factors that combine to make a basin productive and therefore assist in evaluating the geological risks of a new venture. Ultimate rankings, risk assessments, and business decisions are usually tempered, and often overridden, by nongeological factors, such as economic climate, geography, and political stability, which lie beyond the scope of this project

Based on such consideration, the Publications Committee of the American Association of Petroleum Geologists resolved in 1983 to publish a series of volumes on world petroleum basins, with each volume being devoted to one major class of basins. An ad-hoc committee,[1] headed by Anny B. Coury, U.S.G.S., Denver, was charged with the overall planning of the project, with deciding the optimum number of volumes, and with the general organizational of each volume. It was further decided that in each volume (or major basin class) a significant portion should be devoted to the detailed description of one maturely explored "type" or "model" basin, while additional summary papers would discuss other basins or provinces of similar type.

Studies comparing frontier provinces, basins, or undrilled prospects with productive counterparts of similar tectonostratigraphic setting have always played an important role in the evaluation of the hydrocarbon potential of new ventures. Such "analogs," when properly documented and of the same general class as the new target, can assist in establishing plausible limits to as-yet-unknown geological factors, such as reservoir thickness and distribution, trap size and kind, and maturation history, within the framework of the target that is to be assessed. Statistically, such exercises in comparative geology raise the confidence level of the assessment. They also serve to provide concepts to be evaluated and tested in similar areas.

To use analogs in basin assessment without bias or danger of miscomparison or misapplication, a geologist needs to have access to a broad-based worldwide data base that must be integrated with innovative geological concepts. The assemblage of such concepts and a data base, is, however, a time-consuming and costly task that most smaller organizations cannot afford. The entry of many smaller oil companies and independents into the international exploration theater during recent decades has therefore increased the demand for a readily available catalog of analogs.

The committee in charge of the petroleum basin series wrested long and hard over the issues of classification, choice of analogs, and number of volumes needed to do justice to the overall objective. Several recently published classification schemes of petroleum basins and provinces (e.g., Bally, 1975; Bally and Snelson, 1980; Kingston et al., 1983a, 1983b; Klemme, 1975, 1980, 1986) were discussed to arrive at a compromise for volume topics and principal analogs. The ad-hoc committee proposed and the full Publications Committee of the Association voted in 1984 to proceed with the preparation and publication of five volumes, one for each of the following classes:

Divergent/Passive Margin Basins
Interior Cratonic Basins
Active Margin Basins
Foreland Basins and Foldbelts
Interior Rift Basins

For each volume one or more editors were selected, who then proceeded to solicit contributions for that volume. Each editor was given a generalized table of contents and a suggested list of illustrations so that the series would have internal coherence and so that the basin descriptions would be more easily comparable. We hope that these volumes will become a welcome and valuable collection of analogs assisting explorationists for many years to come.

REFERENCES CITED

Bally, A. W., 1975, A geodynamic scenario for hydrocarbon occurrences: Proceedings of the 9th World Petroleum Congress, Tokyo, v. 2: Essex, England, Applied Science Publication, p. 33-34.
Bally, A. W., and S. Snelson, 1980, Realms of subsidence, in D. A. Miall, ed., Facts and principles of world petroleum occurrences: Canadian Society of Petroleum Geologists Memoir 6, p. 9-94.
Kingston, D. R., C. P. Dishroon, and P. A. Williams, 1983a, Global basin classification system: AAPG Bulletin, v. 67, p. 2175-2193.
Kingston, D. R., C. P. Dishroon, and P. A. Williams, 1983b, Hydrocarbon plays and global basin classification: AAPG Bulletin, v. 67, 2194-2198.
Klemme, H. D., 1975, Giant oil fields related to their geologic setting—a possible guide to exploration: Bulletin of Canadian Petroleum Geology, v. 23, p. 30-66.
Klemme, H. D., 1980, Petroleum basins—classifications and characteristics: Journal of Petroleum Geology, v. 3, p. 187-207.
Klemme, H. D., 1986, Field size distribution related to basin characteristics, in D. D. Rice, ed., Oil and gas assessment—methods and applications: AAPG Studies in Geology 21, p. 85-89.

---

[1]Kaspar Arbenz, Anny B. Coury, Michael A. Fisher, James A. Helwig, David R. Kingston, H. Douglas Klemme

Table of Contents

	Introduction *Kevin T. Biddle* 1
Chapter 1	The Los Angeles Basin: An Overview *Kevin T. Biddle* 5
Chapter 2	The Los Angeles Basin: Oil in an Urban Setting *William Rintoul* 25
Chapter 3	Structural Geology and Tectonic Evolution of the Los Angeles Basin, California *Thomas L. Wright* 35
Chapter 4	Review of the Neogene Biostratigraphy and Stratigraphy of the Los Angeles Basin and Implications for Basin Evolution *Gregg H. Blake* 135
Chapter 5	Central Los Angeles Basin: Subsidence and Thermal Implications for Tectonic Evolution *Larry Mayer* 185
Chapter 6	Geochemistry of Los Angeles Basin Oil and Gas Systems *Alan W. A. Jeffrey, Hossein M. Alimi, and Peter D. Jenden* 197
Chapter 7	Stratigraphic Controls of Oil Fields in the Los Angeles Basin *Robert S. Yeats and John M. Beall* 221
Chapter 8	Oil and Gas Production from Submarine Fans of the Los Angeles Basin *Tom Redin* 239
Chapter 9	Taranaki Basin, New Zealand *Julie Palmer and Geoff Bulte* 261
Chapter 10	Middle and Upper Magdalena Basins, Colombia *Steven Schamel* 283
Chapter 11	A New Geologic Model Related to the Distribution of Hydrocarbon Source Rocks in the Falcón Basin, Northwestern Venezuela *Tito Boesi and Donald Goddard* 303

Introduction

Kevin T. Biddle

Exxon Company, International
Houston, Texas, U.S.A.

"The most distinctive characteristic of the Los Angeles basin is its structural relief and complexity in relation to its age and size" (Yerkes et al., 1965, p. A16); however, its very complexity caused no small amount of discussion in designing and naming this volume of the AAPG World Petroleum Basin Memoirs. (See the Foreword for a discussion of the scope of these memoirs.) The series coordinators decided early that the Los Angeles basin should be included in the World Petroleum Basins project because of its interesting geology and importance as a hydrocarbon producer. Initially, the Los Angeles basin was considered for a convergent-margin volume, presumably in recognition of the late-stage shortening that has taken place in the Los Angeles region of southern California. There is little doubt, however, that the Los Angeles basin has formed and deformed within the evolving San Andreas transform system (Atwater, 1970, 1989; Campbell and Yerkes, 1976; Blake et al., 1978; Engebretson et al., 1985; Wright, this volume). There is also little doubt among those who have worked in the area that the initial subsidence of the Neogene Los Angeles basin was caused by extension (Yeats, 1968; Crowell, 1974, 1976, 1987; Wright, this volume). The series coordinators decided, therefore, that to portray the Los Angeles basin as a model for basins formed in convergent-margin settings would be misleading.

The title of this volume, *Active Margin Basins,* is a compromise, but, like many compromises, this title falls short of completely describing its subject. An active margin can be dominantly convergent, divergent, or transform. The California margin was (and is) a transform margin during the key episodes of Los Angeles basin evolution. Although both extension and shortening have taken place along this transform margin in both time and space, the dominant motion has been strike displacement between the North American and Pacific plates. The margin is certainly active, but it would be better to view the Los Angeles basin as an example of the possible complexities associated with transform-margin basins.

One might ask why the Los Angeles basin was chosen to be one of the model basins described in this series of volumes on petroleum basins. A number of reasons make the Los Angeles basin not only a good choice, but an interesting one as well. First, a transform-margin basin provides an excellent complement for other basin types treated in this series (see the Foreword by Arbenz). The San Andreas transform system is the best studied transform margin in the world, and the Los Angeles basin is perhaps the most studied basin within that system. As such, the Los Angeles basin qualifies as a reasonable model for the transform-margin class of basins. Second, our understanding of deformation associated with strike-slip tectonics is evolving. Increasing recognition of mixed mode faulting in strike-slip settings (Nur and Boccaletti, 1989), ideas on the importance of crustal rotation during deformation (Luyendyk and Hornafius, 1987), and new work on the strength of major strike-slip faults (Zobeck et al., 1987) may change the way we view strain fields in areas of strike-slip tectonics. An updated description of the Los Angeles basin provides documentation and tests for some of these evolving ideas. Third, several contributors to this volume supply information collected during the last twenty-five years of oil exploration in the area. Much of this information has not been available before, and it adds greatly to our knowledge of the basin. This new information should be useful not only to oil explorationists but to those interested in California tectonics in general. Finally, the large oil and gas reserves of the basin make it the world's most productive basin in terms of hydrocarbons per volume of sedimentary fill (see Biddle, this volume). Understanding why this basin is so rich in fossil fuels should be of interest to all explorers.

The last point raises a caution. In his foreword to this volume, Arbenz notes that one of the goals of this series on world petroleum basins was to provide a well-described model that might serve as an analogue for similar, but less explored, basins. It would be unwise to choose an endmember property of any population as a basis on which to draw an analogy, and it would also be unwise to use the amount of hydrocarbons in the Los Angeles basin in that fashion. Other aspects of the Los Angeles basin's history provide good bases for comparison with basins formed in similar settings: its complex, polyphase evolution being one such basis (Biddle, this volume).

Each volume of the World Petroleum Basin Memoirs was designed to be similar in scope, but the content of this volume was constrained by the number of volunteered and completed contributions. This volume includes introductory papers on hydrocarbon occurrence and basin origin (Biddle) and a historical view of the area's exploration history (Rintoul). These papers are followed by major contributions on the structural geology and tectonic history of the basin (Wright) and on biostratigraphy and stratigraphy (Blake). Wright's paper is lengthy and contains a wealth of information. A table of contents is given in the introduction to his study to help the reader locate topics of interest. Mayer discusses subsidence history and thermal implications in the basin. His paper is followed by a treatment of the geochemistry of both source rocks and hydrocarbons (Jeffrey et al.). The final two papers on the Los Angeles basin cover stratigraphic controls on oil fields (Yeats and Beall) and production from submarine fans (Redin). Although the most important aspects of the Los Angeles basin's history are covered, a notable absence is a treatment of reservoir properties in the basin.

The second part of this text is composed of shorter contributions on basins thought to share similar origins with the Los Angeles basin. Identifying the basins to include in this section also caused considerable discussion. The basins that are included are shown in Figure 1. Of these, the Taranaki and Falcon basins bear the greatest resemblance to the Los Angeles basin. Both formed within transform settings and were initiated by extension followed by superimposed deformation. The Middle and Upper Magdalena basins of Colombia did not form along a transform margin, but along a complicated margin with a protracted history of convergence. These two basins did have an early history of extension followed by shortening accompanied by a certain amount of strike slip, however. Contributions on other basins, such as the Sahkalin basin of the eastern Soviet Union, were scheduled for inclusion in the volume but were not available when the volume went to press.

All of the contributions in this volume were reviewed by at least two outside reviewers. Their comments proved very useful, and their efforts are gratefully acknowledged.

K. Balshaw-Biddle
G. H. Blake
N. H. Bostick
K. M. Campion
M. G. Fitzgerald
A. R. Green
T. P. Harding
J. C. Ingle
C. M. Isaacs
T. S. Loutit
E. Lozano
T. H. McCulloh
D. W. Phelps
C. Rossen
C. Schubert
J. D. Shane
P. D. Snavely III
K. O. Stanley
R. G. Stanley
C. R. Tapscott
F. F. Weber, Jr.
R. C. Wright
T. L. Wright
R. S. Yeats

REFERENCES

Atwater, T. M., 1970, Implications of plate tectonics for the Cenozoic tectonic evolution of western North America: GSA Bulletin, v. 81, p. 2005–2018.

Atwater, T. M., 1989, Plate tectonic history of the northeast Pacific and western North America, in E. L. Winterer, D. M. Hussong, and R. W. Decker, eds., The eastern Pacific Ocean and Hawaii: GSA, The Geology of North America, v. N, p. 21–72.

Blake, M. C., Jr., R. H. Campbell, T. W. Dibblee, Jr., D. G. Howell, T. H. Nilsen, W. R. Normark, J. C. Vedder, and E. A. Silver, 1978, Neogene basin formation in relation to plate-tectonic evolution of the San Andreas fault system, California: AAPG Bulletin, v. 62, p. 344–372.

Campbell, R. H., and R. F. Yerkes, 1976, Cenozoic evolution of the Los Angeles basin area—relation to plate tectonics, in D. G. Howell, ed., Aspects of the geologic history of the California continental borderland: Pacific Section, AAPG Miscellaneous Publication 24, p. 541–558.

Crowell, J. C., 1974, Origin of late Cenozoic basins in southern California, in W. R. Dickinson, ed., Tectonics and sedimentation: SEPM Special Publication 22, p. 190–204.

Crowell, J. C., 1976, Implications for crustal stretching and shortening of coastal Ventura basin, California, in D. G. Howell, ed., Aspects of the geologic history of the California continental borderland: Pacific Section, AAPG Miscellaneous Publication 24, p. 365–382.

Crowell, J. C., 1987, Late Cenozoic basins of onshore southern California: complexity is the hallmark of their tectonic history, in R. V. Ingersoll and W. G. Ernst, eds., Cenozoic basin development of coastal California: Englewood Cliffs, New Jersey, Prentice-Hall, Inc., Rubey v. VI, p. 207–241.

Engebretson, D. C., A. Cox, and R. G. Gordon, 1985, Relative motions between oceanic and continental plates in the Pacific basin: GSA Special Paper 206, 59 p.

Luyendyk, B. P., and J. S. Hornafius, 1987, Neogene crustal rotations, fault slip, and basin development in southern California, in R. V. Ingersoll and W. G. Ernst, eds., Cenozoic basin development of coastal California: Englewood Cliffs, New Jersey, Prentice-Hall, Inc., Rubey v. VI, p. 259–283.

Nur, A., and M. Boccaletti, 1989, Active and recent strike-slip tectonics: EOS, v. 70, n. 35, p. 806.

Wright, T. L. 1991, Structural geology and tectonic evolution of the Los Angeles basin, California, in K. T. Biddle, ed., Active margin basins: AAPG Memoir 52, p.35–134.

Yeats, R. S., 1968, Rifting and rafting in the southern California borderland, in W. R. Dickinson and A. Grantz, eds., Proceedings of conference on geologic problems of San Andreas fault system: Stanford, California, Stanford University Publications, p. 307–322.

Yerkes, R. F., T. H. McCulloh, J. E. Schoellhamer, and J. G. Vedder, 1965, Geology of the Los Angeles basin, California—an introduction: USGS Professional Paper 420-A, p. A1–A57.

Zobeck, M. D., et al., 1987, New evidence on the state of stress of the San Andreas fault system: Science, v. 238, p. 1105–1111.

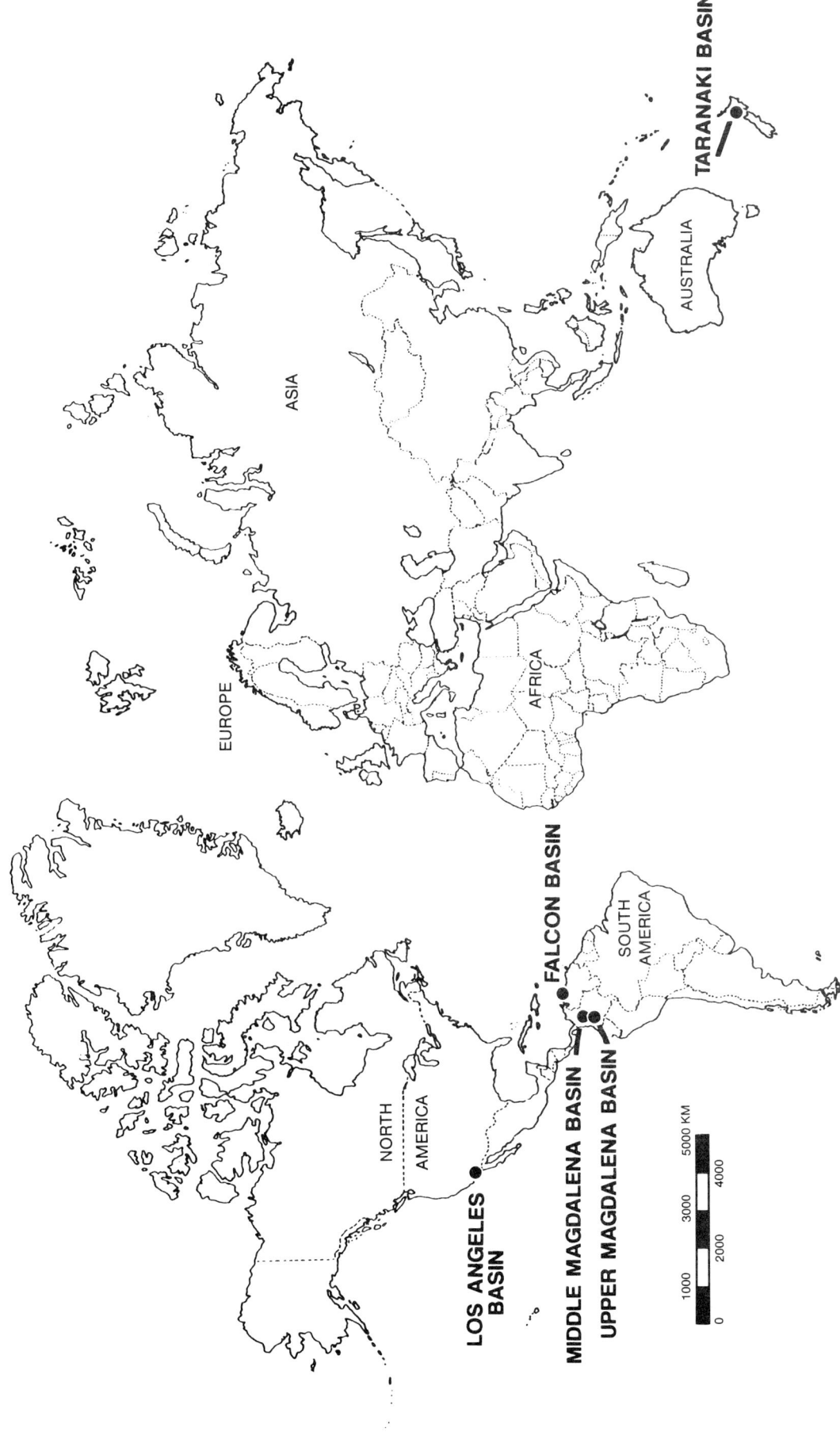

Figure 1. Location of basins discussed in this volume.

Chapter 1

The Los Angeles Basin
An Overview

Kevin T. Biddle

Exxon Company, International
Houston, Texas, U.S.A.

ABSTRACT

The Los Angeles basin is a polyphase Neogene basin within the San Andreas transform system between the Pacific and North American plates. The basin was initiated in the mid-Miocene by widespread extension associated with significant strike slip and rotation of the Transverse Ranges of southern California. Late Miocene to early Pliocene extension, which accompanied the opening of the Gulf of California, led to the principal phase of basin opening. The early Pliocene to Recent deformational history of the basin is characterized by shortening associated with the active North Los Angeles fold and thrust system.

The Los Angeles basin is the richest basin in the world in terms of hydrocarbons per volume of sedimentary fill. Each phase of basin evolution has contributed to the basin's productivity. Some aspects of the basin's history that have affected the occurrence of oil and gas are related to basin-forming mechanisms and can be used to guide thinking in similar settings. Other first-order controls on hydrocarbon occurrence stem from processes that operated on a much larger scale and are not related to basin type. Deposition of thick, high-quality source rock within the Los Angeles basin is the result of such regional controls.

The polyphase history of the Los Angeles basin demonstrates the complexity that can occur along active transform margins. Such complexity can be expected in basins that have formed in similar settings.

INTRODUCTION

The Los Angeles basin is one of many Neogene basins along the western margin of California (Figure 1). While it shares aspects of its evolutionary history with a number of these basins, the Los Angeles basin can be distinguished from most of them by its large reserves of oil and gas. This basin has been called one of the richest sedimentary basins in the world (Barbat, 1958; Yerkes et al., 1965; Gardett, 1971; Wright, 1987). Current estimates place its recoverable reserves at just under 10.5 billion oil-equivalent barrels (California Department of Conservation, 1985, 1987). The small areal extent of the basin, its thick sedimentary succession, and its large discovered reserves make the Los Angeles basin the world's most productive basin in terms of hydrocarbons per volume of sedimentary fill (Barbat, 1958; Yerkes et al., 1965; Wright, 1987) (Figure 2).

Because of its productivity, the Los Angeles basin and surrounding area have received a great deal of attention over the past century. Discussions on the oil potential of the area are documented back to 1865 (Whitney, 1865). The most complete work on the basin to date is the study by Yerkes et al. (1965). In addition to covering the evolution of the basin, these authors summarized previous literature on the area in an annotated, chronological format through 1962. Numerous studies from 1965 to the present are discussed in the papers included in this volume (see particularly Wright, *Structural Geology and Tectonic Evolution of the Los Angeles Basin,* this volume) and need not be discussed further here. The volume of research on this basin qualifies it as one of the best studied basins along the most studied transform margin in the world. This level of documentation supports its selection as one of the model basins in the AAPG Petroleum Basin Memoirs.

The present-day Los Angeles basin is a topographic and structural basin located along the intersection of the Peninsular Ranges, the Transverse Ranges, and the continental borderlands of southern California (Yerkes et al., 1965; Wright, this volume) (Figure 3). It is a small basin (Figure 4), but it contains very thick, dominantly Neogene sedimentary fill (Figure 5). In the past, the depositional basin was considerably larger than the present-day structural basin (Yerkes et al., 1965; Wright, this volume; Yeats and Beall, *Stratigraphic Controls of Oil Fields in the Los Angeles Basin,* this volume). Like many basins, the Los Angeles basin is a polyhistory basin (Kingston et al., 1983), and each significant phase of its development has contributed to the hydrocarbon productivity of the area.

The initial phase (or phases) of basin development involved extension associated with significant strike slip (Wright, this volume). These extensional events created the container into which potential reservoir and source rocks were deposited and in which early traps formed. This extension took place within the expanding Pacific–North American transform margin (Atwater, 1970, 1989; Crowell, 1974b; Campbell and Yerkes, 1976; Blake et al., 1978; Engebretson et al., 1985).

A later phase of evolution, associated with deformation in the Transverse Ranges and the North Los Angeles fold and thrust belt (Davis et al., 1989; Wright, this volume), involves north-south shortening with limited strike slip. Shortening continues today and also occurs within the San Andreas transform system. During this phase, source rocks continued to mature, depressed into the oil window by sedimentary and structural loading. Migration occurred and additional traps were formed. Continuing deformation has modified or destroyed traps in some parts of the basin (Wright, this volume; Yeats and Beall, this volume). With time, this phase of deformation may lead to the destruction of the Los Angeles basin.

The papers in this volume provide up-to-date summaries and release new information on the history of the Los Angeles basin. In particular, information collected during exploration for hydrocarbons in the 1960s and 1970s improves our understanding of the basin. The goal of this overview is to set the stage for the contributions that follow. Here, the exploration history of the basin (including the volumes of hydrocarbons discovered to date) is summarized, proposed mechanisms of basin formation and deformation are discussed, and speculation on factors that may have controlled the occurrence of oil and gas in the area is made. This chapter concludes with a discussion of which of these factors are basin-specific and which are controlled by events far removed from the Los Angeles basin.

OIL AND GAS IN THE LOS ANGELES BASIN

The Los Angeles basin area has a century-long history of exploration for hydrocarbons. (For an historical view of exploration in the area, see Rintoul, *The Los Angeles Basin: Oil in an Urban Setting,* this volume). The Brea Olinda field on the eastern side of the basin was established in 1880 and is credited as the first discovery (Figure 6). Since that time, an additional 67 oil fields have been found in the basin. These fields are listed in Table 1 along with their trap types, variety of hydrocarbons, year of discovery, recoverable oil and gas reserves, and combined recoverable reserves in oil-equivalent barrels. The location of each field is shown in Figure 7.

The exploration history of the basin is summarized in Figure 6 in terms of reserves added per year (Figure 6A) and reserves added through individual discoveries (Figure 6B). Until about 1918, during the early phase of exploration, the discovery rate was slow and reserve additions were modest. In the next fifteen years, almost all of the large fields in the basin were discovered, including the supergiant Wilmington field—the largest field in the basin. The discovery well for Wilmington, the last giant found in the basin, was drilled in 1932, although it was many years before the areal extent and volume of this field were fully appreciated. Improving technology, increasing understanding of hydrocarbon occurrence, and fluctuating prices all played a role in the pace of exploration during this important period of the basin's development (see Rintoul, this volume).

Thirty-eight additional fields have been discovered since Wilmington. These thirty-eight fields have a combined estimated ultimate recovery (EUR) of 600 million oil-equivalent barrels (MOEB) (Table 1), constituting about 6%

Date Submitted: 4/15/90
Date Accepted: 5/31/90

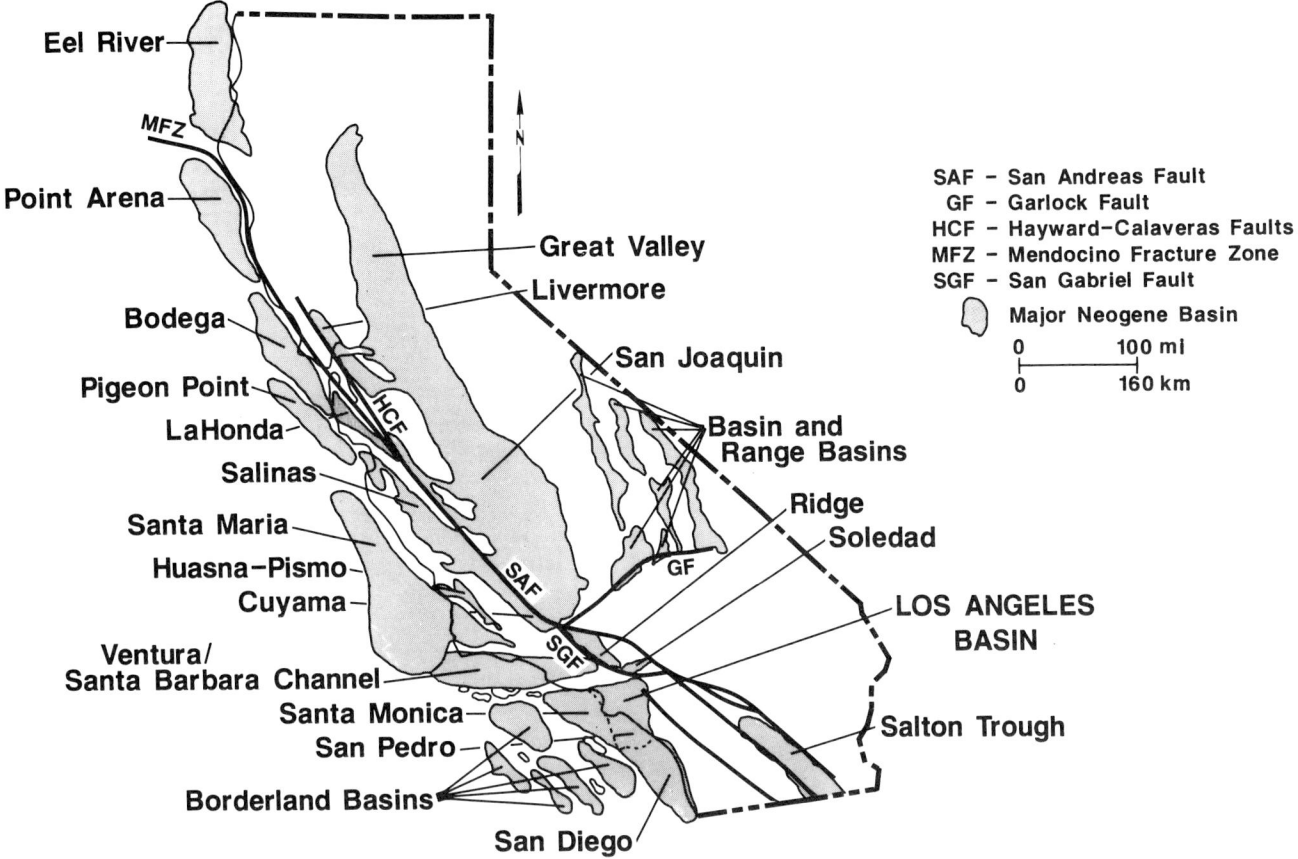

Figure 1. Location of major Neogene basins, California.

of the basin's total EUR. Nevertheless, the last field discovered in the basin (the Beta field, discovered in 1976) has reserves of 218 MOEB and is one of the largest oil fields discovered in the United States in the last twenty years.

A field-size distribution curve (Figure 8) shows that the basin contains three supergiant fields (Wilmington, Huntington Beach, and Long Beach), a number of intermediate size fields, and numerous small fields (see also Table 1). In considering the forty-eight fields with reserves greater than 1 MOEB, the mean field size for this population is 216 MOEB and the median is about 32 MOEB. If all sixty-eight fields are considered, the mean field size drops to 154 MOEB, with half the fields containing less than about 8 MOEB.

The Wilmington field alone, with reserves of 3 billion oil-equivalent barrels (GOEB), contains about 30% of the basin's total EUR. The three supergiants combined hold about 52% (5.4 GOEB) of the recoverable hydrocarbons found to date in the basin.

Exploration has stagnated since the 1976 discovery of the Beta field, representing the longest period of nondiscovery in the basin since its first field was found in 1880. The lack of new discoveries in the basin is due only in part to its mature exploration state. It also results from significant restraints placed on exploration by environmental concerns, regulatory limitations, and the difficulties associated with exploration in an urban setting. Reserve estimates have continued to rise without new field discoveries because of pool extension and improved recovery techniques. These types of reserve additions should continue for several decades (Beyer, 1988).

IDEAS ON BASIN ORIGIN

It has long been recognized that the present Los Angeles basin is the product of several events, or phases, of evolution. Before the advent of plate tectonics, early workers discussed various phases of basin evolution in terms of descriptions of structures, sedimentary fill, and timing of events. Most early authors who studied the basin did not speculate on mechanisms or causes of basin formation.

Driver (1948, p. 104) endeavored to "present a concept of the accumulation, distribution, and distortion of the various rocks within the Los Angeles Basin." He subdivided the rocks into the following categories: basement complex, Cretaceous and Eocene, Oligocene and Miocene, upper Miocene, lower Pliocene, and post–lower Pliocene. He recognized that the present basin physiography resulted from late-stage deformation, which he considered to be the result of post–middle Pleistocene activity. Driver (1948) attributed the associated uplift, subsidence, and deformation to the general processes of diastrophism.

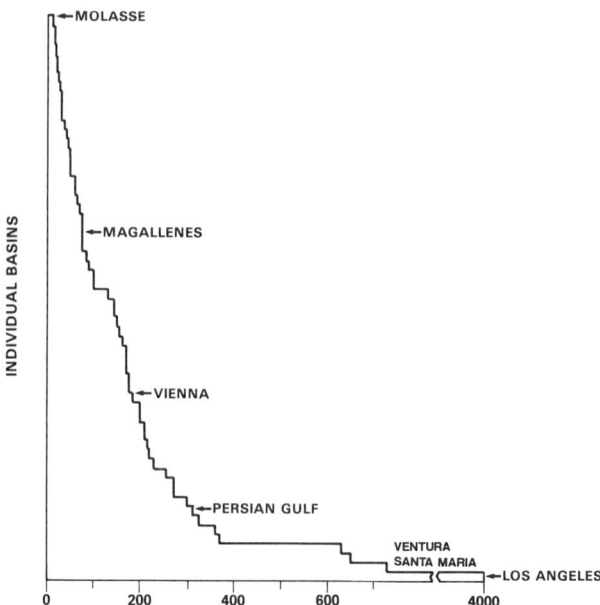

Figure 2. Basin productivity, measured in hydrocarbons per volume of sedimentary fill. The vertical axis represents individual basins, from least prolific at the top to most prolific at the bottom. Although not inclusive of all producing basins, all major basin types are represented. The Molasse basin of Germany, the Magallanes basin of Argentina and Chile, the Vienna basin of Austria, the Persian Gulf area, and the Ventura and Santa Monica basins are identified for reference (D. White, unpublished data).

In 1958, Barbat pointed out that geomorphic features to the west of the Los Angeles basin had been compared with the basins and ranges of Nevada and other adjacent states. He stressed comparisons between the offshore continental borderlands and an unfilled Los Angeles basin. This combination of comparisons tacitly implied some form of extensional origin for both areas.

Barbat divided the tectonic history of the basin into two phases (Barbat, 1958). His first phase culminated with a mid-Miocene period of deformation that, he suggested, formed the basic pattern of alternating basins and ranges in southern California. He then divided the sedimentary succession of the basin into a prebasinal record (Upper Cretaceous–lower Miocene) and a basinal record (middle Miocene–Pleistocene).

Barbat's second tectonic phase included crustal movements that took place after mid-Miocene tectonism and culminated in the so-called mid-Pleistocene Pasadenan orogeny of Stille (1936). During this second tectonic phase, the basinal sediments were folded and faulted, the basin margins were uplifted, and a major syncline was formed in the central basin area. Barbat suggested that a number of preexisting structures were reactivated—some with a different sense of displacement—during this second tectonic phase.

In 1965, Yerkes et al. published their major synthesis of the geology of the Los Angeles basin. These authors divided the evolution of the basin into five phases. They included two prebasinal phases that described the basement and Upper Cretaceous to lower Miocene sedimentary rocks, a middle Miocene basin-inception phase, a late Miocene to early Pleistocene principal phase of subsidence and deposition, and a post–mid-Pleistocene basin-disruption phase.

Yerkes et al. (1965) recognized that the depositional Los Angeles basin began to form in the middle Miocene and that late Miocene and Pliocene subsidence established much of its present form and structural relief. These authors also emphasized the effects of the late-stage deformation in modifying the earlier formed basin. Yerkes et al. described the sedimentary section deposited during each stage and the structural geology of the area, but they did not speculate on the causes or mechanisms of basin formation and deformation. In general, the phases of basin evolution described by Yerkes et al. have withstood the test of time (see Wright, this volume).

With the growing acceptance of plate tectonics and an increasing understanding of the geology of the Pacific Ocean floor, workers began to address the causes of basin formation in southern California. Yeats (1968) adopted the hypothesis that the Los Angeles basin was a filled Miocene rift whose origin was concurrent with extension in the Gulf of California and with sea-floor spreading in that region. He also proposed that granitic rocks of the area and their sedimentary veneer were detached from underlying Franciscan and correlative rocks and had rafted apart above major detachment surfaces. The proposed detachments were suggested to be rejuvenated Mesozoic thrust planes.

Although Yeat's rafting idea is now out-of-date, his concept of detached blocks and associated extensional deformation presaged current discussion about the role of crustal detachments in middle Miocene extension and rotation in southern California (Luyendyk et al., 1980, 1985; Luyendyk and Hornafius, 1987; Nicholson and Seeber, 1989). In an update of his earlier work, Yeats (1987) stated that one central concept—that the mid-Miocene volcanic and extensional event in southern California is different from preceding subduction tectonics and succeeding diffuse transform tectonics—is still valid.

In a series of papers in the mid 1970s, John Crowell presented his ideas on basin formation and sedimentation within the San Andreas transform system (Crowell, 1974a, b, 1976). His ideas still guide thinking today. In 1974, Crowell (1974a) stated that basins and sites of depositions along transform faults are related mainly to bends and irregularities in fault-zone traces. He also noted that all of southern California could be considered a Cenozoic sliced-and-splayed transform margin between the North American and Pacific plates.

In a second paper later that year, Crowell (1974b) developed a general model of formation for pull-apart basins and concluded that the Los Angeles basin began as an irregular pull-apart basin in the early mid-Miocene. Crowell's sketch of an idealized pull-apart basin is

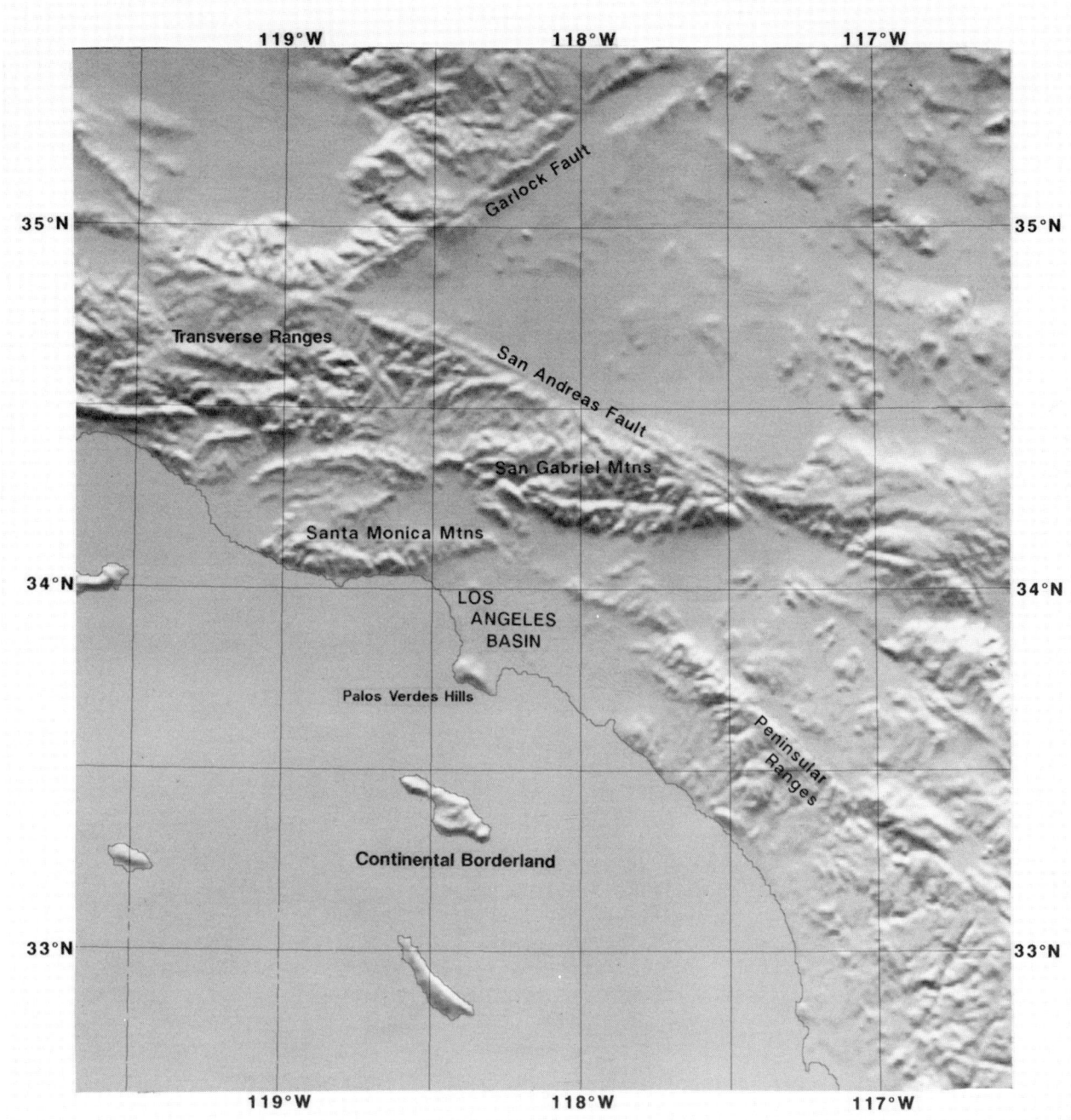

Figure 3. Location of the present-day Los Angeles basin and surrounding mountain ranges. Shaded topography is artificially illuminated from the north and was produced from a Defense Mapping Agency 30-second digital terrain dataset (Godsen, 1981).

reproduced in Figure 9. In this figure, the pull-apart basin forms in extension between two overlapping zones of strike slip. Extension is accommodated by dip- and oblique-slip faults between the bounding strike-slip faults and by combined strike slip and dip slip along segments of the bounding strike-slip faults. If sufficient extension occurs, the crust may be so severely attenuated that the center of the basin may be floored by a volcanic-sedimentary complex. Crowell contended (1974b) that the Newport-Inglewood fault zones formed the western strike-slip margin of the Los Angeles basin in the early mid-Miocene. Faults in the southern part of the basin, such as the Cristianitos fault (Figure 4) in the southern part of the basin, may correspond to the oblique-slip faults of Crowell's model (Figure 9). Other faults that delineated the original Neogene Los Angeles basin have been obscured by later

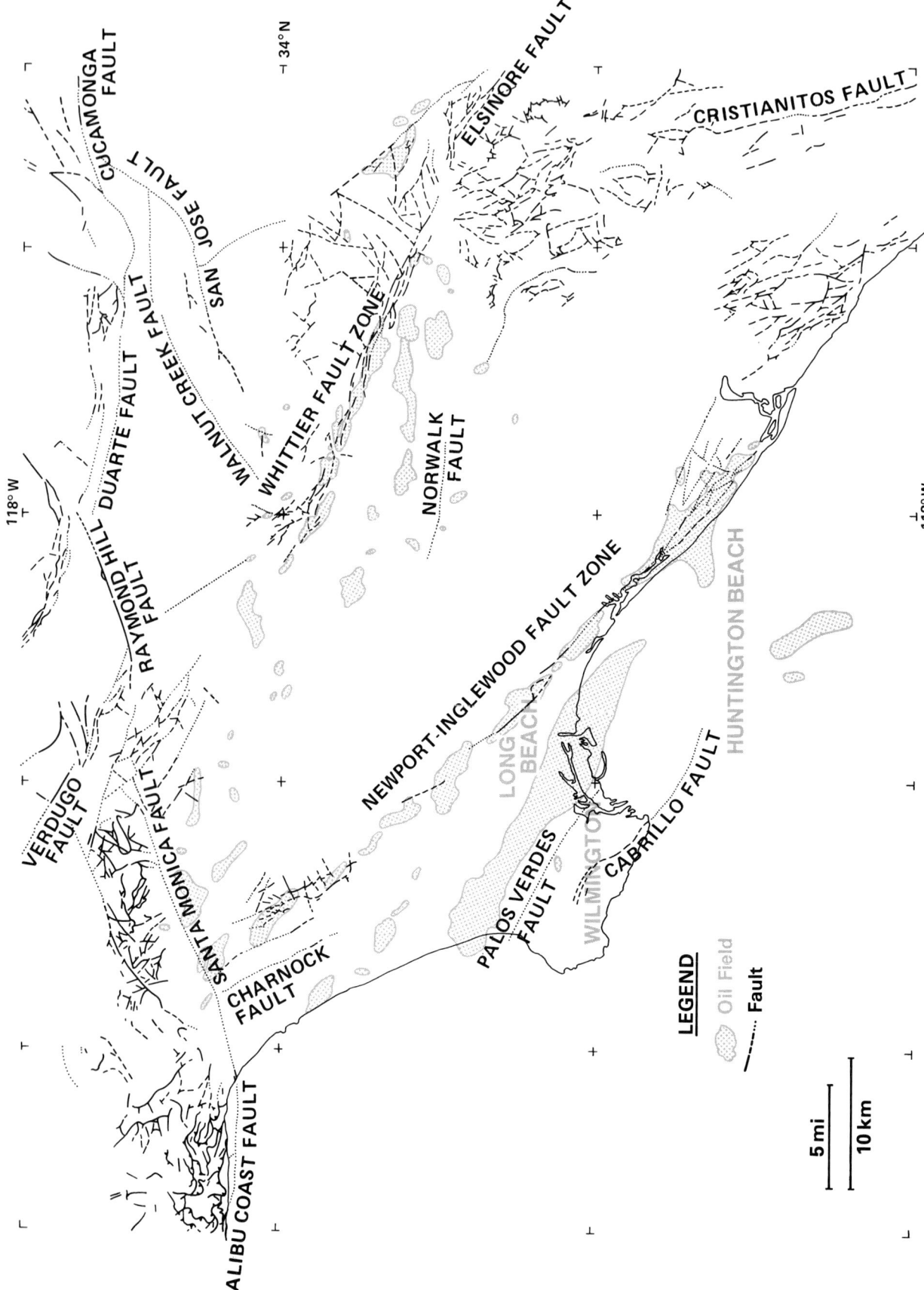

Figure 4. Location of surface faults and oil fields, Los Angeles basin. Major faults, fault zones, and the three largest oil fields in the basin are identified for reference. (Sources: Jennings, 1962; Rogers, 1965, 1967; Jennings and Strand, 1969; California Department of Conservation, 1986).

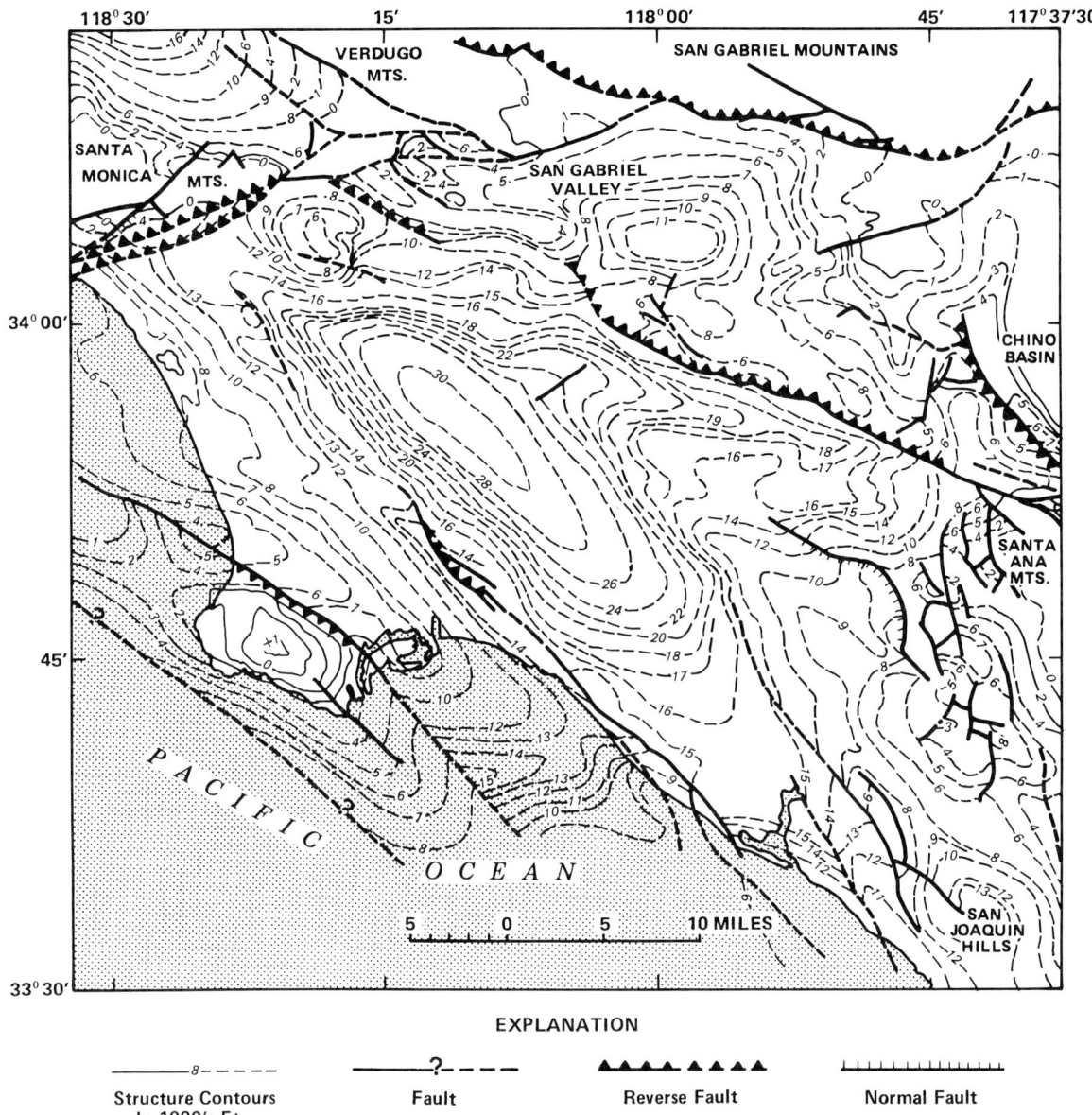

Figure 5. Major structural features and structural contours on the inferred top of basement, Los Angeles basin (from Yerkes et al., 1965, based on McCulloh, 1960). Most of the sedimentary section above the contoured top-of-basement surface is thought to be Neogene. Interpretation of the deepest part of the basin is based on gravity data and other indirect evidence; the deepest penetration to date is the American Petrofina Corehole No. 1, drilled to a vertical depth of 20,736 ft (6320 m).

deformation. Crowell (1974b) speculated that the original basin began as a pull-apart basin over a "hot spot" associated with the passage of the East Pacific Rise along the coast of California. He also recognized that the later history of the Los Angeles basin is not so clearly related to a pull-apart origin.

In 1987, Crowell expanded and updated his evaluation of basin evolution in southern California. According to his scenario for the evolution of late Cenozoic basins, basin formation was initiated by rifting about 22 Ma followed by later thermal subsidence. He also discussed the role of crustal rotation indicated by paleomagnetic data (see below) in the formation of the Los Angeles basin. Crowell concluded that the Los Angeles basin and related southern California basins formed as irregular pull-apart basins at a time of irregular rifting within the simple-shear system of the San Andreas transform boundary (Crowell, 1987).

Paleomagnetic data from the Transverse Ranges, which began to appear in publications in the late 1970s and early 1980s (Kamerling and Luyendyk, 1979; Luyendyk et al., 1980; Hornafius, 1985; Hornafius et al., 1986), indicate that large clockwise rotations of crustal blocks are an integral part

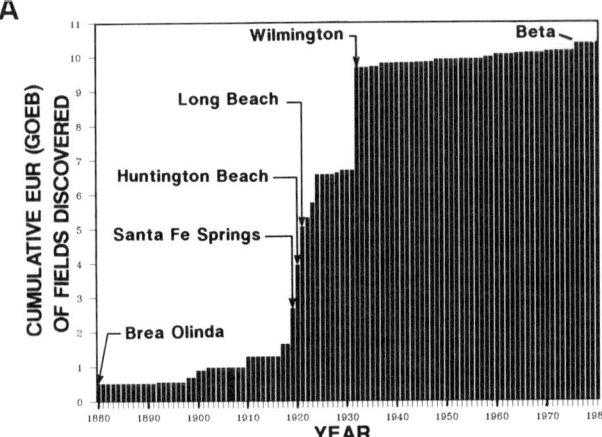

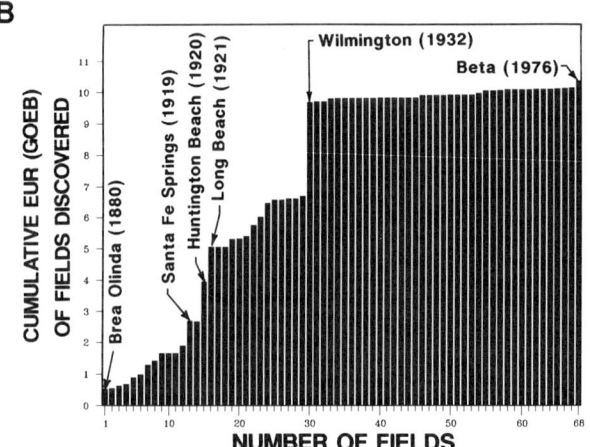

Figure 6. Discovery histories for the Los Angeles basin. (A) Reserve additions by year. (B) Reserve additions by individual field discovery in sequential order in billions of oil-equivalent barrels (GOEB).

of the Neogene deformation in southern California. These data have played a central role in recent discussions of deformation within the San Andreas transform system and have led to modifications of ideas on the formation of late Cenozoic basins in the region. Bruce Luyendyk and colleagues developed a geometric model that incorporates this rotation into the deformational interpretation of the area. Their model includes specific implications for the opening of the Los Angeles basin during the mid-Miocene (Luyendyk et al., 1980; Luyendyk and Hornafius, 1987). These authors proposed the model to explain rotation of blocks bounded by east-trending faults and nested between blocks limited by northwest-trending faults.

Paleomagnetic data indicate that as much as 25% of the surface area of southern California has undergone clockwise rotation about vertical axes and that some areas have rotated as much as ninety degrees (Luyendyk and Hornafius, 1987). The timing of this rotation in southern California is reasonably well constrained as beginning shortly before the middle Miocene and producing fifty to sixty degrees of clockwise rotation before the beginning of the late Miocene. Since that time, an additional thirty to forty degrees of rotation has occurred (Luyendyk and Hornafius, 1987). Cox (1980) reviewed possible mechanisms for rotation in such settings. All mechanisms discussed involved simple-shear boundary conditions.

The rotational model of Luyendyk et al. (1980) predicts that right slip, left slip, and rotation occur simultaneously and that during rotation triangular basins open by crustal dilation at the junctures between rotated and nonrotated blocks (see Figure 12). Luyendyk and Hornafius (1987) suggested that as much as 24,000 km² of post–mid-Miocene dilation may have occurred between the northern edge of the Transverse Ranges and the Mexican border. This dilation represents about 15% of that total area.

The geometric model of Luyendyk et al. (1980) begins with a fracture pattern that resulted from the first contact of the Pacific and North American plates. The key aspects of this initial configuration are the northerly trending fault systems of the future Transverse Ranges and the northwesterly trending fault systems to the northwest and southeast (Figures 10, 11). In the mid-Miocene, deformation began by simple shear driven by the dextral shear couple between the Pacific and North American plates. The Transverse Ranges and similar blocks rotated clockwise within this shear couple, while the northwest-trending blocks deformed by right-lateral simple shear (Luyendyk et al., 1980) (Figure 12).

The Los Angeles basin is located at the boundary between the rotated blocks of the Transverse Ranges and the nonrotated blocks to the south (Figure 12). In this model, the northwest-trending fault systems south of the Transverse Ranges terminate at the southern boundary of those ranges and bound a series of triangular basins. Luyendyk and Hornafius (1987) suggested that the Los Angeles basin is a particularly good example of one of these triangular basins, which, borrowing a term from S. W. Carey (1958), they called a *sphenocasm*.

The geometric model (Luyendyk et al., 1980) predicts that the triangular basins south of the Transverse Ranges would open by extension and be bounded on the southwest by northwest-striking right-oblique faults, to the east by north-striking normal faults, and to the north by east-striking oblique faults. In the Los Angeles basin, the Newport-Inglewood fault system would form the southwestern basin boundary, the Cristianitos fault (see Figure 4) and similar faults would constitute the eastern basin edge, and the Santa Monica–Raymond fault system would form the northern edge of the basin.

Luyendyk and Hornafius (1987, p. 279) concluded that the Los Angeles basin has the correct shape, age, subsidence history, and style of boundary faults to fit the criteria for a basin formed by rotational tectonics. They also noted that their model predicts basin geometry, but it does not specify the style of crustal dilation within the basin. These authors indicated that patterns of basin fill in the triangular basins should be similar to those predicted by Crowell (1974b) (Figure 9) for pull-apart basins.

Wright (this volume) provides a detailed treatment of the tectonic history of the Los Angeles basin. He interprets basin formation to be the result of Miocene extensional events that were caused by plate interaction along the

Text continues on p. 14.

Table 1. Los Angeles basin oil and gas fields.

	Field	Trap[1]	Type[2]	Discovery year[3]	Oil[4] (EUR in kbbl)	Gas[4] (EUR in mcf)	OEB[5] (in kbbl)
1	Alondra	S	A	1946	2,154	1,408	2,406
2	Anaheim (abd)	S	A	1951	4	—	4
3	Bandini	C	A	1953	5,969	15,469	8,738
4	Belmont Offshore	S	A,B	1948	68,500	41,931	76,006
5	Beta	S	A	1976	214,272	21,866	218,186
6	Beverly Hills	C	A	1900	164,131	215,163	202,645
7	Boyle Heights (abd)	S	A	1955	273	113	293
8	Brea-Olinda	C	A	1880	438,691	481,986	524,967
9	Buena Park, East (abd)	C	A	1942	197	20	201
10	Buena Park, West (abd)	C	A	1944	50	17	53
11	Cheviot Hills	S	A	1958	26,180	142,492	51,686
12	Chino-Soquel	S	A	1950	324	349	387
13	Coyote, East	S	A	1909	121,829	60,804	132,713
14	Coyote, West	S	A	1909	257,522	271,005	306,032
15	Dominguez	S	A	1923	276,846	387,394	346,190
16	El Segundo	C	A,B	1935	14,744	34,725	20,960
17	Esperanza	S	A	1956	1,331	699	1,456
18	Gaffey (abd)	S	A	1955	10	—	10
19	Howard Townsite	S	A	1947	6,162	27,810	11,140
20	Huntington Beach	S	A	1920	1,138,034	861,117	1,291,805
21	Hyperion	C	A	1944	798	209	835
22	Inglewood	S	A	1924	400,048	285,002	451,063
23	Kraemer	S	A	1918	3,925	1,078	4,118
24	Kraemer, Northeast (abd)	S	A	1953	?	—	—
25	Kraemer, West (abd)	S	A	1956	10	—	10
26	La Mirada (abd)	C	A	1946	25	10	27
27	Lapworth (abd)	C	A	1935	55	—	55
28	Las Cienegas	S	A	1960	65,349	55,550	75,293
29	Lawndale	S	A	1928	3,747	6,729	4,958
30	Leffingwell (abd)	S	A	1946	763	2,460	1,203
31	Long Beach	S	A	1921	927,428	1,087,440	1,121,773
32	Long Beach Airport	S	A	1954	11,572	35,003	17,838
33	Los Angeles City	C	A	1892	23,575	—	23,575
34	Los Angeles Downtown	S	A,B	1964	15,233	22,922	19,336
35	Los Angeles, East	C	A	1946	6,936	12,401	9,156
36	Mahala	S	A	1920	4,077	1,586	4,361
37	Montebello	S	A	1917	202,004	234,712	243,917
38	Newgate	S	A	1956	296	370	362
39	Newport	S	A	1922	187	259	233
40	Newport, West	S	A,B	1923	77,647	8,371	79,145
41	Olive	C	A	1953	3,020	1,209	3,236

Table continues on p. 14.

Table 1. Los Angeles basin oil and gas fields (continued).

	Field	Trap[1]	Type[2]	Discovery year[3]	Oil[4] (EUR in kbbl)	Gas[4] (EUR in mcf)	OEB[5] (in kbbl)
42	Playa del Rey	C	A	1929	63,008	62,061	74,118
43	Potrero	S	A	1928	15,672	72,967	28,733
44	Prado-Corona	S	A,B	1966	1,632	5,192	2,561
45	Richfield	S	A	1919	217,340	173,067	248,319
46	Rosecrans	S	A	1924	83,339	166,330	113,112
47	Rosecrans, East	S	A	1959	202	234	243
48	Rosecrans, South	C	A	1940	8,835	20,661	12,533
49	Rowland (abd)	S	A	1931	2	—	2
50	Salt Lake	S	A	1902	53,683	211,894	91,612
51	Salt Lake, South	C	A	1970	10,091	4,503	10,897
52	Sansinena	C	A	1898	60,840	74,661	74,204
53	San Vicente	C	A	1968	21,043	19,433	24,522
54	Santa Fe Springs	S	A,B	1919	622,254	836,512	771,990
55	Sawtelle	S	A	1965	15,274	13,100	17,619
56	Seal Beach	S	A,B	1924	217,236	219,786	256,484
57	Sherman (abd)	S	A	1965	93	50	102
58	Sunset Beach	S	A	1954	6,910	9,591	8,627
59	Talbert (abd)	C	A	1947	126	4	127
60	Torrance	C	A	1922	247,562	162,573	276,593
61	Turnbell (abd)	S	A	1941	766	582	870
62	Union Station	S	A	1967	1,895	5,298	2,843
63	Venice Beach	C	A	1966	4,030	2,678	4,508
64	Walnut	S	A	1948	131	25	135
65	Whittier	C	A	1898	55,731	52,193	65,074
66	Whittier Heights, North (abd)	S	A	1944	85	84	235
67	Wilmington	S	A,B	1932	2,788,158	1,192,802	3,001,670
68	Yorba Linda	ST	A	1930	94,781	2,174	95,170
	Totals				9,074,637	7,628,134	10,439,275

[1]S = structural; C = combination; ST = stratigraphic
[2]A = Oil and associated gas; B = Nonassociated gas (California Department of Conservation, 1985)
[3]Department of Energy (1985); California Department of Conservation (1985, 1987); California Division of Oil and Gas (1974)
[4]EUR = estimated ultimate recovery; kbbl = thousand barrels; mcf = million cubic feet. Source: California Department of Conservation (1985)
[5]Oil-equivalent barrels calculated by using 5600 cubic feet of gas equals one oil-equivalent barrel
(abd) = abandoned

margin of North America (Wright, this volume, his Figure 1). Wright incorporated rotation of the Transverse Ranges into two detailed reconstructions of the Los Angeles basin: a middle Miocene reconstruction (circa 14 Ma) and an early Miocene one (circa 20 Ma) (Wright, this volume, his Figures 36, 37). These reconstructions provide the best available view of how the basin may have opened.

Wright suggests that the Neogene evolution of the Los Angeles basin can be used to test the hypothesis that the basin opened as a consequence of the rotation of crustal blocks within the Transverse Ranges. It is interesting to note that he concludes that the data used in his reconstructions are compatible with the rotational model, but they do not provide proof that rotation has taken place.

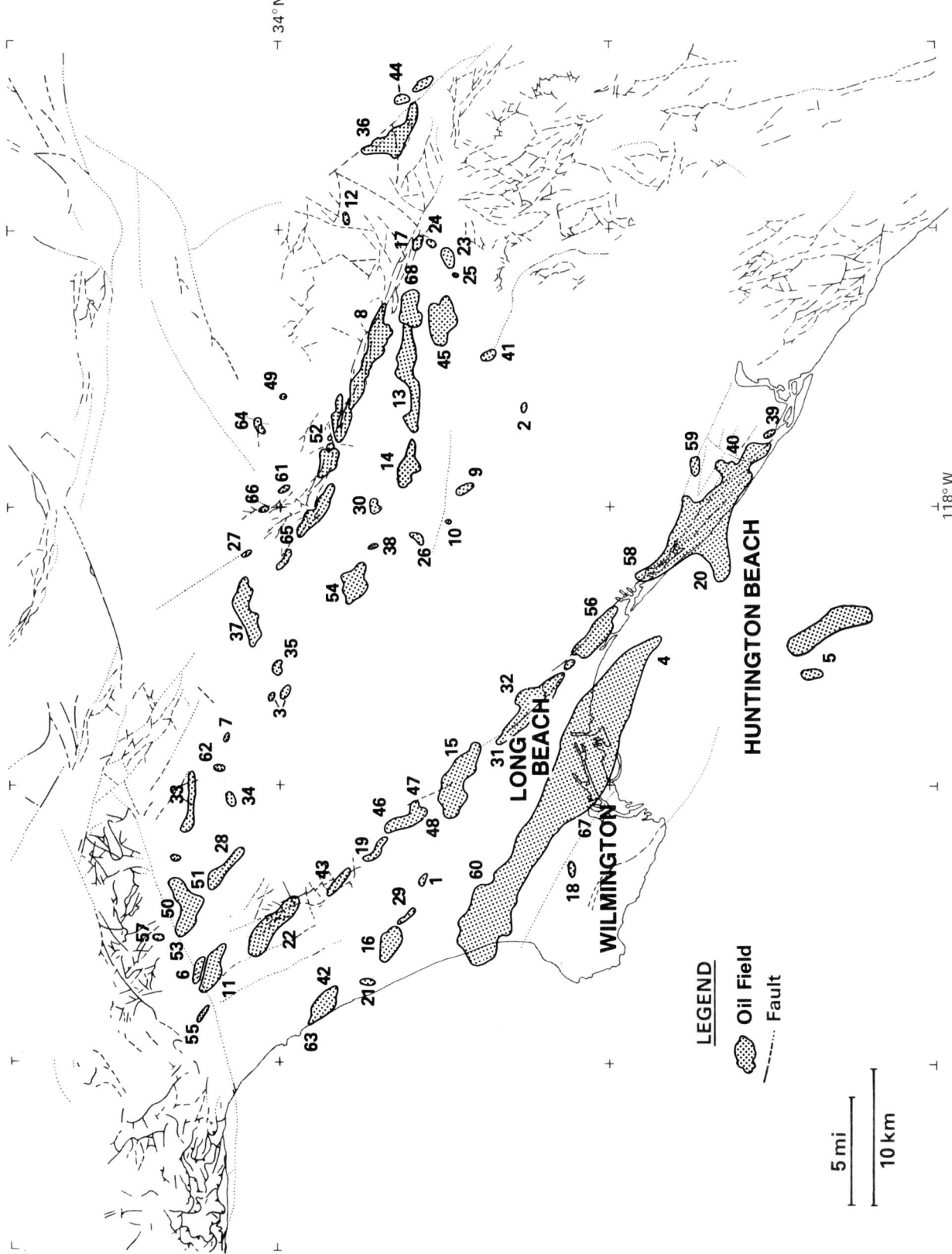

Figure 7. Location of oil fields, Los Angeles basin. Numbers represent individual fields. Field names are listed in Table 1.

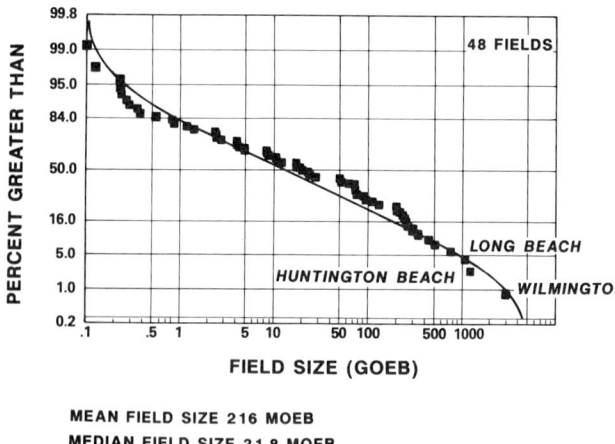

Figure 8. Field-size distribution, Los Angeles basin. Only fields with recoverable reserves greater than 1 MOEB are plotted. See text for discussion.

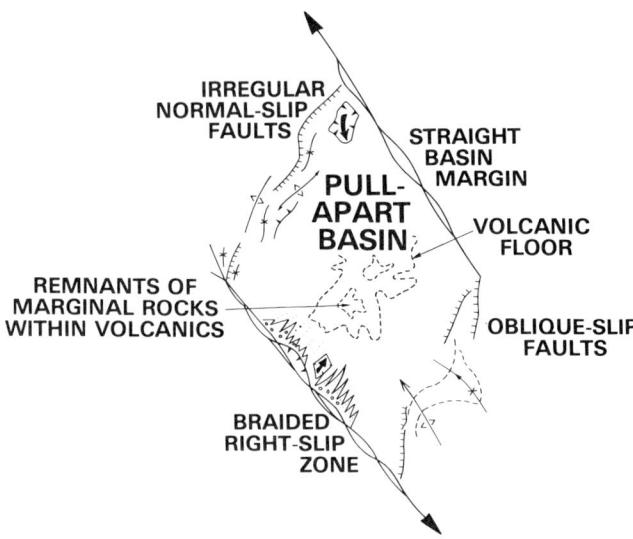

Figure 9. Sketch map of an idealized pull-apart basin. The basin forms by extension between two overstepping strike-slip faults. The extension is accommodated by dip- and oblique-slip faults (after Crowell, 1974b).

In conclusion, there is general agreement among those who have studied the Los Angeles basin that the basin formed in the middle Miocene and late Miocene–early Pliocene by extension associated with strike slip within the evolving transform system between the North American and Pacific plates. The kinematics of basin formation are only just starting to become clear. As more data become available on Miocene and Pliocene displacement histories for individual faults, our understanding of this critical episode of basin formation in southern California will improve. The information contained in this volume should help further that improvement.

LATE-STAGE SHORTENING

Many authors have recognized that the present-day shape and extent of the Los Angeles basin is the result of deformation that has taken place late in the basin's history (Barbat, 1958; Yerkes et al., 1965; Crowell, 1974b, 1987; Davis et al., 1989; Wright, this volume, among many others). It is also recognized that this deformation is dominantly compressive in the Los Angeles area and is expressed as north-south shortening throughout much of the basin (Davis et al., 1989; Wright, this volume). Barbat (1958) called this deformation the *second phase of the basin's tectonic history.* Yerkes et al. (1965) termed it the *basin-disruption phase of evolution.*

Recent work has provided a new view of the style of deformation that is associated with this late shortening. Davis et al. (1989), influenced by fold and fault geometry and recent seismicity, produced a series of retrodeformable cross sections that span the Los Angeles basin (Figure 13). These authors believe that the Transverse Ranges and the southern Coast Ranges are part of active, basement-involved fold and thrust belts.

They have noted that the structural geology of the Los Angeles area is dominated by three major compressional uplifts: the Santa Monica Mountains anticlinorium, the Palos Verdes anticlinorium and western shelf, and the Verdugo Hills–San Rafael Hills and the San Gabriel Mountains. Davis et al. (1989) stated that these features are best explained as structures associated with thrust ramps off major detachments.

In constructing their cross sections, Davis et al. (1989) used techniques that allow the cross sections to be kinematically and geometrically restored to a reasonable predeformational state. They assumed that the deformation is dominated by basement-involved shortening and that major anticlines or anticlinoria are best explained by fault-bend or fault-propagation folds. They also assumed that no significant amount of Pliocene and Quaternary strike slip (i.e., no more than 3 km) has occurred on any of the faults that intersect their cross sections. Davis et al. (1989) inferred that the thrust faults that control structural geometry in the Los Angeles basin merge with regional detachments beneath the Transverse Ranges and northern Peninsular Ranges. The depth to the master detachment(s) is thought to be about 10 to 15 km. This depth is indicated by both the fold analysis of Davis et al. (1989) and by the base of regional seismicity (Hauksson, 1987).

Two simplified cross sections from Davis et al. (1989) are reproduced in Figure 13 (also see Figure 45 *in* Wright, this volume). These cross sections run from the Palos Verdes Hills in the south, across the Torrance, Alondra, and Howard Townsite oil fields in the central part of the basin, to the Santa Monica Mountains in the north (see Figures 3, 4, 7). Each of these cross sections can be restored to a reasonable preshortening state and is therefore a permissible solution. Both cross sections restore the Santa Monica Mountains anticlinorium as a south-vergent feature formed by blind thrusts that emanate from a master

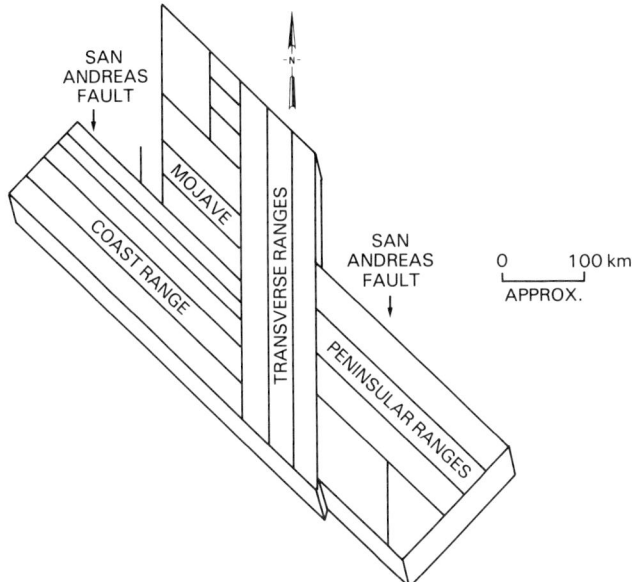

Figure 10. Inferred initial fracture pattern and geometry of southern California before rotation of crustal blocks in Miocene (from Luyendyk et al., 1980).

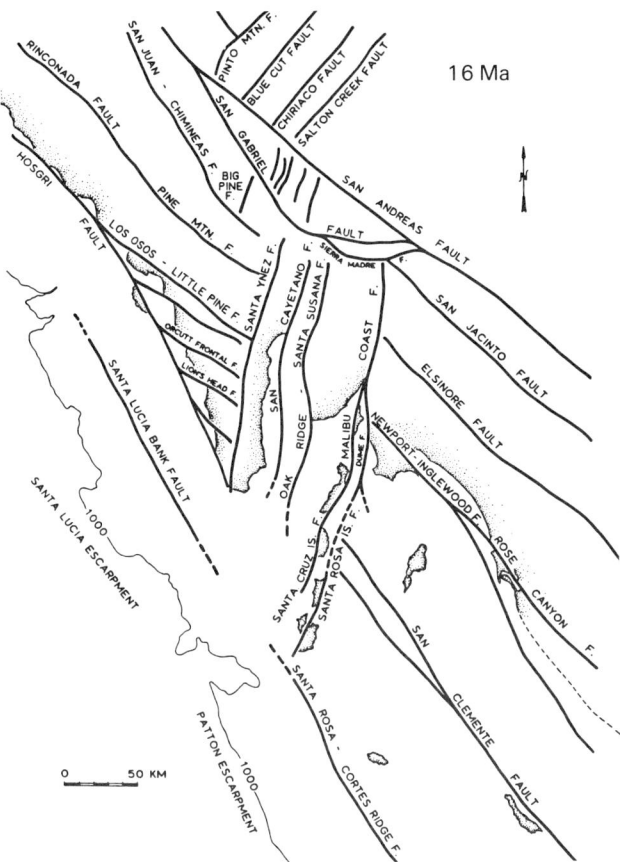

Figure 11. Palinspastic reconstruction of southern California at 16 Ma. Major block-bounding faults are identified. Compare with Figures 10 and 12 (from Hornafius et al., 1986).

detachment surface at depth. The Palos Verdes Hills and adjacent western shelf and the central part of the basin are treated differently in the two cross sections.

In the first cross section (Figure 13A), the Palos Verdes Hills and the western shelf are shown as anticlinal structures associated with north-vergent blind thrusts. An alternate solution (Figure 13B) relates these structures to north-vergent back thrusts off a south-vergent detachment. This cross section (Figure 13B) also shows the entire Los Angeles basin as detached at depth.

Based on regional geologic relationships, Davis et al. (1989) suggested that late-stage convergence began between 4.0 and 2.2 Ma and continues today. Their cross sections show that since shortening began, 15 km of convergence has taken place between the Palos Verdes Hills and the San Andreas fault. As much as 30 km of convergence may have occurred between the continental margin and the San Andreas fault (Davis et al., 1989).

While the occurrence of late-stage shortening in the Los Angeles basin area has been appreciated for a long time, the approach of Davis et al. (1989) provides a new view of the consequences of that shortening. Their conclusions seem most applicable to the northern margin of the basin. There is no consensus yet whether or not their cross sections adequately portray the structure of the rest of the basin (see Wright, this volume, for a more detailed discussion).

An intriguing benefit of the retrodeformable cross sections of Davis et al. (1989) is the view that the restored section provides of the middle Miocene–early Pliocene depositional Los Angeles basin (Figure 14). After removing the effects of late-stage shortening, the basin appears as a narrow, deep basin bounded by high-angle normal-separation faults in cross sectional view. The restored cross section does not provide information about strike separation along these faults. Comparison of the present-day cross sections (Figure 13) and the restored cross section (Figure 14) reemphasizes the distinct polyphase history of the Los Angeles basin.

DISCUSSION

It has been stated that the Los Angeles basin represents optimum conditions for the generation and entrapment of hydrocarbons (Barbat, 1958; Gardett, 1971; Wright, 1987). Several authors have speculated on the combination of factors that produced these optimum conditions (Barbat, 1958; Yerkes et al., 1965; Wright, 1987).

Barbat (1958) suggested that eight factors governed the hydrocarbon productivity of the basin. These eight factors include aspects of source-rock formation, migration, reservoir continuity, trap, and preservation of trapped fluids. Barbat concluded that choosing a few basic controls on productivity was not an easy task. He ended by writing that "no matter how the Los Angeles Basin may differ from other producing areas, the differences generally favor the

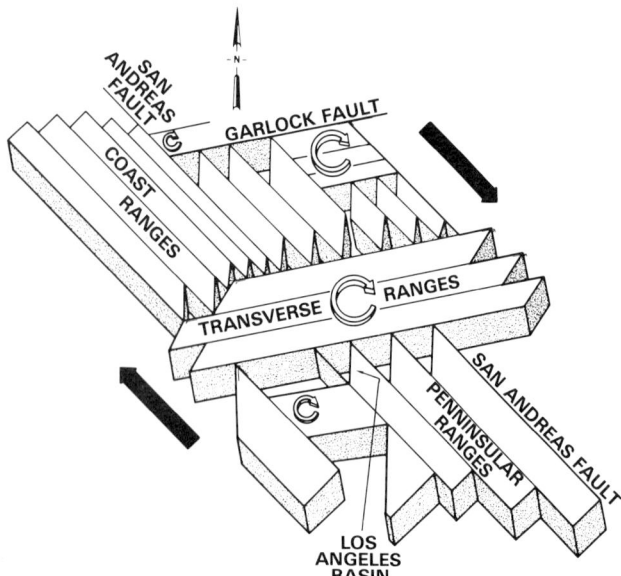

Figure 12. Geometric model of southern California after Miocene block rotation. Note position and shape of Los Angeles basin just to the south of the rotated blocks of the Transverse Ranges (from Luyendyk et al., 1980).

Los Angeles Basin" (Barbat, 1958, p. 76). Yerkes et al. (1965) also concluded that a nearly unique combination of factors and serendipitous timing of events account for the extreme productivity of the area.

Despite the unique aspects of the Los Angeles basin, it is instructive to speculate on which controls of hydrocarbon occurrence are basin-specific (i.e., somehow associated with the mechanisms of basin formation) and which result from events far removed from the basin.

One of the first-order controls on hydrocarbon occurrence is the deposition of source rocks within a basin. The sources of hydrocarbons in the Los Angeles basin are organic-rich rocks of the Modelo, Puente, and Monterey formations (see Blake, *Review of the Neogene Biostratigraphy and Stratigraphy*, this volume; Jeffrey et al., *Geochemistry of the Los Angeles Basin Oil and Gas Systems*, this volume). Intervals within these formations can be exceedingly good source rocks, reaching total organic carbon contents as high as 18% (Jeffrey et al., this volume). The organic matter is dominated by sapropelic material with excellent capacity for generating liquids upon maturity.

The source rocks within the Los Angeles basin are part of an extensive succession of Miocene and lower Pliocene organic-rich and diatomaceous rocks, of which the Monterey Formation is characteristic. The Monterey Formation and Monterey-like rocks are widespread in coastal California (Garrison et al., 1981). Similar diatomaceous rocks of roughly the same age are common around the rim of the North Pacific Ocean (Ingle, 1981). It is clear that controls on the deposition of these rocks operated over a very large area and are not related to the formation of any single basin or basin type.

Barron (1986) summarized the paleo-oceanographic and tectonic controls on deposition of the Monterey Formation and related siliceous rocks in California. He concluded that diatomaceous sediments appeared abruptly in the North Pacific Ocean at the same time (17.5 to 16.5 Ma) that they were disappearing from the Caribbean and low-latitude North Atlantic. Barron (1986) interpreted this switch in the sites of silica deposition as a result of increased production of silica-corrosive North Atlantic bottom water in the late early Miocene.

Surface productivity along coastlines rimming the North Pacific was enhanced about two million years later by major polar cooling that lead to intensified upwelling along the continental margins. Rising sea level in the late early Miocene, coupled with subsidence along the evolving transform margin of California, isolated many areas from significant terrigenous sediment input. All of these factors led to the deposition of widespread, relatively pure biogenic and organic-rich sediments along the California margin at a time when Neogene basins, such as the Los Angeles basin, were forming.

Preservation of organic material in the water column and within near-surface sediments after deposition is an important aspect of source-rock formation. Yerkes et al. (1965) commented that the high initial organic content of the sediment was preserved, in part, because of poor circulation in the constricted Los Angeles basin. Restricted circulation is thought to have been caused by variations in bathymetry and the presence of sills within the basin or near the basin margins.

It is commonly thought that the abundant organic material in the Los Angeles basin was deposited in anoxic bottom water during periods of high surface productivity. Isaacs (1987) has pointed out, however, that controls on the distribution and preservation of organic matter in the Monterey Formation are more complicated than this generalization and that in some areas organic matter is most abundant in rocks deposited in intermediate oxygen conditions at times of lower surface productivity.

It is likely that restricted circulation and oxygen-deficient bottom waters did contribute significantly to the preservation of organic matter in the Modelo, Puente, and Monterey formations of the Los Angeles basin. It is also likely that the mechanism of formation of the Los Angeles basin contributed to restricted circulation. Narrow, deep depocenters are characteristic of basins formed by extension associated with strike slip (see Ballance and Reading, 1980; Biddle and Christie-Blick, 1985). Rapid subsidence (leading to deep-water conditions) over the limited area of the Los Angeles basin is a good example of this (Mayer, *Central Los Angeles Basin: Subsidence and Thermal Implications for Tectonic Evolution*, this volume).

The basin geometry led to restricted circulation, which, in turn, contributed to oxygen depletion in the deepest parts of the basin. Summerhayes (1981) pointed out, however, that the oxygen-minimum zone along the California margin was strongly developed because of regional oceanographic controls and that this condition also affected preservation of organic matter in Monterey-like sediments.

The thermal evolution of the Los Angeles basin is a function of how the basin formed. In extensional settings, such as the one in which the Los Angeles basin formed,

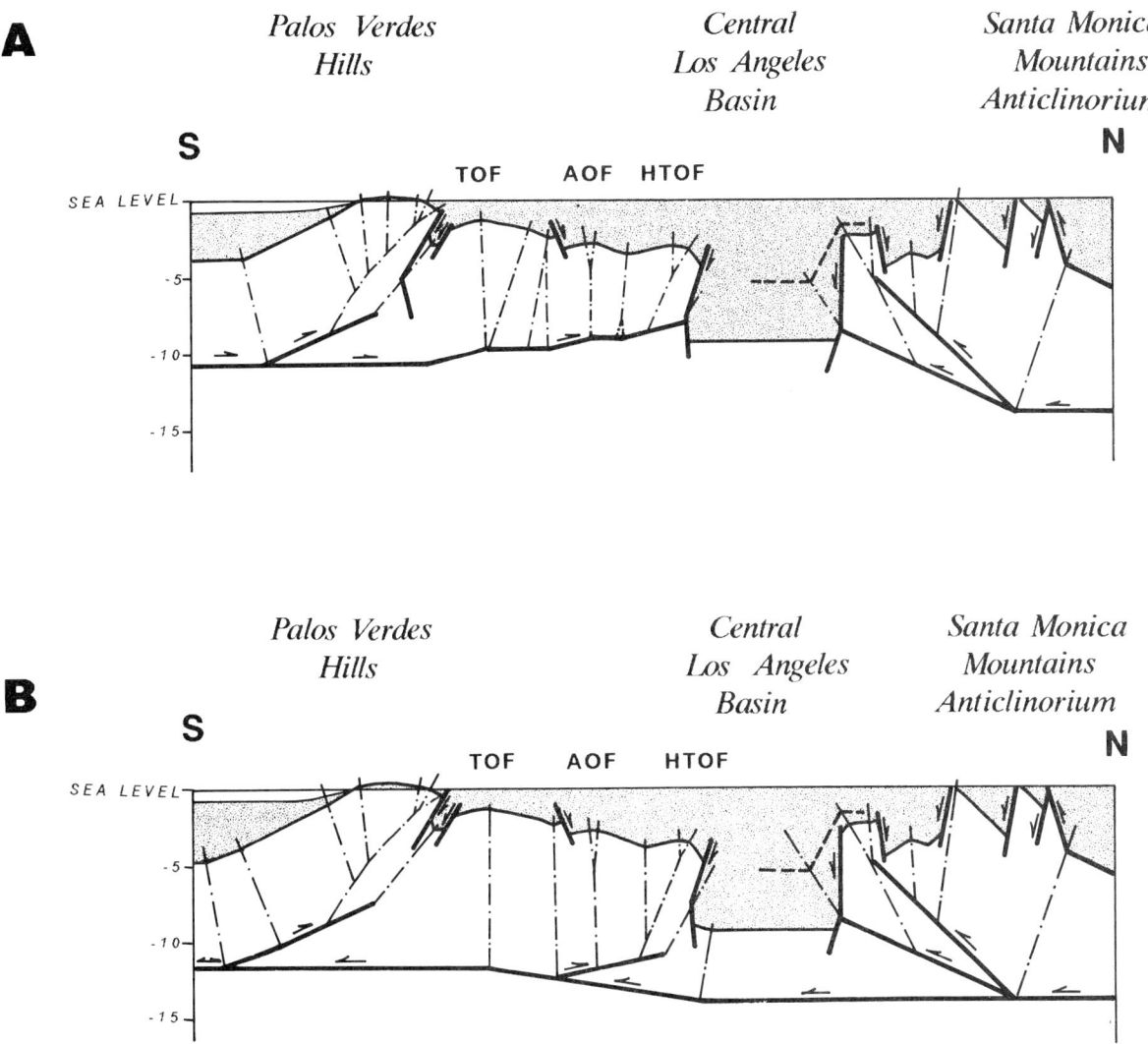

Figure 13. Simplified retrodeformable cross sections across the Los Angeles basin (Davis et al., 1989). Stippled pattern represents the sedimentary fill of the basin. (A) North- and south-vergent solution with no detachment beneath the center of the basin. (B) South-vergent solution with master detachment at depth beneath the Los Angeles basin. Abbreviations used: AOF—Alondra oil field, HTOF—Howard Townsite oil field, TOF—Torrance oil field.

the amount of extension is the primary control on lithospheric and crustal thermal structure (Sleep, 1971; McKenzie, 1978; Royden et al., 1980). Extension elevates isotherms, and, once extension has ended, cooling of the associated thermal anomaly produces additional subsidence above that anomaly. Both elevation of isotherms during extension and decay of the associated thermal anomaly help establish the heat flux through basins formed by extension. By analyzing the subsidence history of a basin and combining this analysis with present-day temperature data, the temperature history of the basin can be reconstructed and used to predict when and where source rocks matured.

The subsidence history of the Los Angeles basin has been examined by Turcotte and McAdoo (1979), Sawyer et al. (1987), and Mayer (this volume). Turcotte and McAdoo concluded that the subsidence history of the southwestern block of the Los Angeles basin is consistent with thermal cooling and local isostatic adjustment mechanisms. They also concluded that the one-dimensional thermal subsidence model that they applied to the southwestern block is not directly applicable to the central part of the basin. Sawyer et al. (1987) and Mayer (this volume) both note that simple one-dimensional models are inadequate to explain the rapid subsidence of the Los Angeles basin. One-dimensional models ignore the effects of lateral heat flow, which have been shown to be an important mechanism of heat loss in small, narrow basins (Pitman and Andrews, 1985).

Pitman and Andrews (1985) examined the thermal history of small pull-apart basins from a theoretical perspective. They showed that such basins lose most of their anomalous

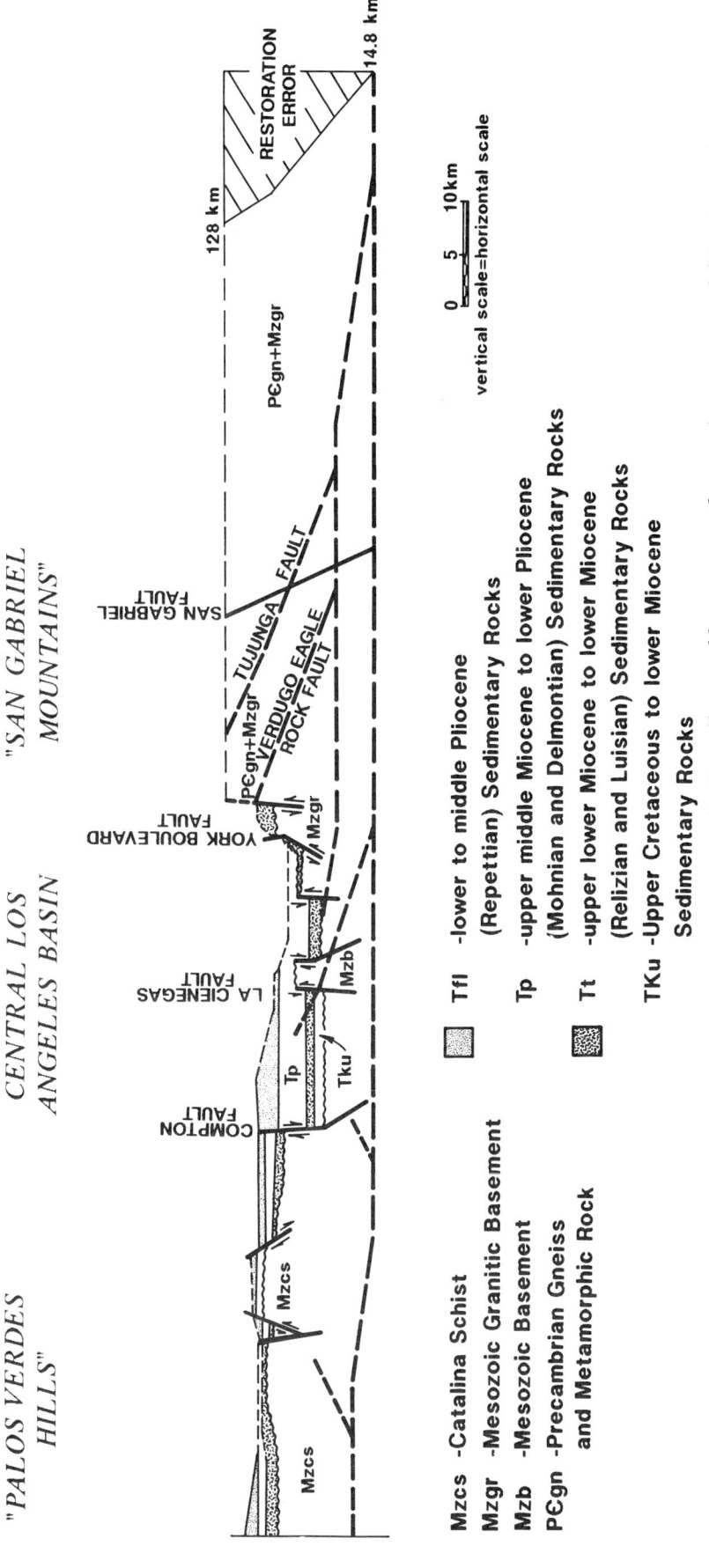

Figure 14. Restored cross section of the Los Angeles basin with the effects of late-stage shortening removed (Davis et al., 1989). This restored section provides a view of the middle Miocene–early Pliocene depositional basin.

heat during the stretching phase that is associated with very rapid synrift subsidence. Because of rapid synrift cooling, geothermal gradients in the basin remain fairly stable during rifting and cooling. This has important ramifications for maturation of source rocks.

Jeffrey et al. (this volume) show that most of the oil produced in the Los Angeles basin is fairly low gravity (less than 25° API) and that limited higher gravity oils, condensate, and nonassociated gas are present. The central synclinal area is thought to have generated most of these hydrocarbons. Philippe (1965) concluded that the bulk of generation in the basin is occurring deeper than 8000 ft (2438 m). All of this suggests that source rocks are at low-to-moderate levels of maturity even where fairly deeply buried in the central part of the basin. This level of maturity results from the moderate temperature structure of the basin, which is a consequence of the style of basin formation.

The distribution of migration carrier beds, potential reservoirs, and seals are partly controlled by the style of basin formation and partly by broader, regional paleogeographic considerations. Migration pathways and reservoirs in the Los Angeles basin are dominated by Miocene and Pliocene turbidites and associated deep-sea rocks (see Redin, this volume; Yeats and Beall, this volume). These directly overlie and are interbedded with potential source rocks in many parts of the basin. These relationships allow excellent communication from mature source to traps.

Because of rapid syn-extension subsidence in narrow pull-apart basins, such basins may develop deep-water conditions early in their history (Pitman and Andrews, 1985). Mid-Miocene and Pliocene extension and subsidence quickly established a deep-water setting in the Los Angeles basin (Wright, this volume; Blake, this volume; Mayer, this volume; Redin, this volume), providing the environment for deposition of both source and reservoir rocks.

The paleogeographic setting of the Los Angeles basin in an inboard position along the California margin allowed the basin to trap large volumes of sand and gravel derived from the hinterlands. As a result, thick successions of Miocene and Pliocene sand-rich strata are common in the basin. These form the extensive reservoir columns of the major fields in the basin (Redin, this volume). Basins outboard of the Los Angeles basin received smaller amounts of age-equivalent coarse-grained sediment (cf. Gorsline and Emery, 1959; Gorsline and Douglas, 1987).

Trap styles are another feature of the basin that are closely related to basin type. Several trap types exist, and trap geometry varies from simple to complex (Wright, this volume). Viable traps were formed during the early phases of extension associated with strike slip and during the later phase of shortening with limited strike slip.

Transform-margin settings respond quickly to changes in motion between the bounding plates. As a result, transform-margin basins tend to exhibit polyphase histories, and the Los Angeles basin is a good example. One consequence of this is that traps that formed early may be modified or destroyed by ongoing or superimposed deformation in this type of basin. This has occurred in the Los Angeles basin, where late-stage shortening has breached some traps (Wright, this volume; Yeats and Beall, this volume).

Each phase of Neogene evolution has contributed to the hydrocarbon productivity of the Los Angeles basin. The earlier, pre-Neogene history of subduction and terrane accretion (Howell et al., 1987; Wright, this volume) played only a limited role in affecting the occurrence of oil and gas. Pre-Neogene events may have influenced where some major features in the basin, such as the Newport-Inglewood fault zone, formed. The Miocene and Pliocene extensional events formed the container into which source rock and most of the reservoir rocks were deposited. These events also blocked out the thermal structure of the basin and formed a variety of traps. The late-stage shortening event formed additional traps and modified or destroyed some early structures that contained oil. Sedimentation associated with late-stage shortening depressed earlier deposited source rocks into the oil window over large areas of the basins and defined late-stage migration pathways. Continuing shortening may eventually lead to the destruction of the Los Angeles basin. Just as the conditions for the formation and trapping of hydrocarbons in the Los Angeles basin are exceptional, the timing of exploration and production with respect to the life cycle of the basin may be nearly ideal.

CONCLUSIONS

The Los Angeles basin is a member of the general class of strike-slip basins (Mann et al., 1983) that form in transform-margin settings. It has experienced a polyphase history as have many other basins that formed in similar settings. The Los Angeles basin originated as a Neogene depositional basin in the mid-Miocene to early Pliocene by two episodes of extension associated with strike-slip deformation (Wright, this volume). The Pliocene-to-Recent history of the basin is dominated by shortening associated with limited strike slip (Davis et al., 1989; Wright, this volume).

Each phase of basin evolution has affected the hydrocarbon productivity of the area. The Neogene extensional phases created the container into which source and reservoir rocks were deposited. These events also defined the thermal structure of the basin and formed early traps. Burial of upper Miocene organic-rich rocks may have forced hydrocarbon generation in the deeper parts of the basin during the later extensional episode. Late-stage shortening has changed the basin outline, modified (and in some cases destroyed) earlier formed traps, and created new traps. Sedimentation associated with late-stage deformation has forced more potential source rock into the oil window and has defined modern migration pathways.

Some of the aspects of basin evolution that make the Los Angeles basin so productive are inherent in the mechanisms of basin formation. Other aspects controlled by events removed from the basin are not a direct reflection of basin type. The incredible resources of the Los Angeles basin are the result of a serendipitous combination of basin-forming mechanisms and events far removed from the basin itself. Source-rock quality and volume are two critical factors that were controlled by conditions affecting large areas of the North Pacific rim and are not directly related to the formation of the Los Angeles basin. However, the rapidly subsiding, deep Los Angeles basin formed at the right time, in the right place, with an appropriate geometry to serve as a significant container for organic-rich material deposited during mid-Miocene to early Pliocene time. Aspects of basin evolution that are directly related to mechanisms of basin formation include style and rate of subsidence, thermal evolution, and structural styles. Aspects that are more indirectly related to basin formation are styles of deposition and the deep-water to nonmarine stacking pattern of the sedimentary succession.

One characteristic of the Los Angeles basin that appears to be applicable to other active transform-margin basins is the complex polyphase history. This class of basins forms in a setting that is quick to respond to changes in motion between the major bounding plates. These changes are common along intercontinental transform zones and result in superimposed deformation. It should be kept in mind that although the Los Angeles basin is a good example of the complexities associated with this class of basins, each individual basin has its own history. Knowledge gained from studying the basins discussed in this volume can guide thinking on other similar basins, but it cannot replace detailed local work.

ACKNOWLEDGMENTS

I would like to thank Exxon Production Research Company and Exxon Company, International for allowing me to participate in the World Petroleum Basins project and for approving this overview for publication. Discussions with Tom Wright, Tod Harding, Steve May, and Ken Stanley were helpful while preparing this paper. Bob Brovey and Ben Merembeck of Exxon Production Research Company provided the artificially illuminated topographic map (Figure 3). An earlier version of the manuscript was reviewed by K. Balshaw-Biddle and K. O. Stanley. Their comments are gratefully acknowledged.

REFERENCES

Atwater, T. M., 1970, Implications of plate tectonics for the Cenozoic tectonic evolution of western North America: GSA Bulletin, v. 81, p. 2005-2018.

Atwater, T. M., 1989, Plate tectonic history of the northeast Pacific and western North America, in E. L. Winterer, D. M. Hussong, and R. W. Decker, eds., The eastern Pacific Ocean and Hawaii: GSA, The Geology of North America, v. N, p. 21-72.

Ballance, P. F., and H. G. Reading, eds., 1980, Sedimentation in oblique-slip mobile zones: International Association of Sedimentologists Special Publication, n. 4, 265 p.

Barbat, W. F., 1958, The Los Angeles basin area, California, in L. G. Weeks, ed., Habitat of oil: AAPG, p. 62-77.

Barron, J. A., 1986, Paleo-oceanographic and tectonic controls on deposition of the Monterey Formation and related siliceous rocks in California: Palaeogeography, Palaeoclimatology, Palaeoecology, v. 53, p. 27-45.

Beyer, L. A., 1988, Summary of geology and petroleum plays used to assess undiscovered recoverable petroleum resources of Los Angeles basin province, California: USGS Open-File Report 88-450L, 62 p.

Biddle, K. T., and N. Christie-Blick, eds., 1985, Strike-slip deformation, basin formation, and sedimentation: SEPM Special Publication, n. 37, 386 p.

Blake, G. H., 1991, Review of the Neogene biostratigraphy and stratigraphy of the Los Angeles basin and implications for basin evolution, in K. T. Biddle, ed., Active margin basins: AAPG Memoir 52, p. 135-184.

Blake, M. C., Jr., R. H. Campbell, T. W. Dibblee, Jr., D. G. Howell, T. H. Nilsen, W. R. Normack, J. C. Vedder, and E. A. Silver, 1978, Neogene basin formation in relation to plate-tectonic evolution of San Andreas fault system, California: AAPG Bulletin, v. 62, p. 344-372.

California Department of Conservation, 1985, 71st annual report of the state oil and gas supervisor: California Department of Conservation, Division of Oil and Gas, Publication PR06, 157 p.

California Department of Conservation, 1986, Field boundaries, western portion District 1, Los Angeles and Orange counties: California Department of Conservation, Division of Oil and Gas, Map 100.

California Department of Conservation, 1987, 73rd annual report of the state oil and gas supervisor: California Department of Conservation, Division of Oil and Gas, Publication PR06, p. 53-84.

California Division of Oil and Gas, 1974, California oil and gas fields, v. 11, south, central, coastal and offshore California: California Division of Oil and Gas, Report TR12.

Campbell, R. H., and R. F. Yerkes, 1976, Cenozoic evolution of the Los Angeles basin area—relation to plate tectonics, in D. G. Howell, ed., Aspects of the geologic history of the California continental borderland: Pacific Section, AAPG Miscellaneous Publication 24, p. 541-558.

Carey, S. W., 1958, A tectonic approach to continental drift, in S. W. Carey, convenor, Continental drift: a symposium: Hobart, University of Tasmania, p. 177-355.

Cox, A., 1980, Rotation of microplates in western North America, in D. W. Strangeway, ed., The continental crust and its mineral deposits: Geological Association of Canada Special Paper 20, p. 305-321.

Crowell, J. C., 1974a, Sedimentation along the San Andreas fault, in R. H. Dott, Jr., and R. H. Shaver, eds., Modern and ancient geosynclinal sedimentation: SEPM Special Publication 19, p. 292-303.

Crowell, J. C., 1974b, Origin of late Cenozoic basins in southern California, in W. R. Dickinson, ed., Tectonics and sedimentation: SEPM Special Publication 22, p. 190-204.

Crowell, J. C., 1976, Implications for crustal stretching and shortening of coastal Ventura basin, California, in D. G. Howell, ed., Aspects of the geologic history of the California continental borderland: Pacific Section, AAPG Miscellaneous Publication 24, p. 365-382.

Crowell, J. C., 1987, Late Cenozoic basins of onshore southern California: complexity is the hallmark of their tectonic history, in R. V. Ingersoll and W. G. Ernst, eds., Cenozoic basin development of coastal California: Englewood Cliffs, New Jersey, Prentice-Hall, Inc., Rubey v. VI, p. 207-241.

Davis, T. L., J. Namson, and R. F. Yerkes, 1989, A cross section of the Los Angeles area: seismically active fold and thrust belt, the 1987 Whittier Narrows earthquake, and earthquake hazard: Journal of Geophysical Research, v. 94, n. B7, p. 9644–9664.

Department of Energy, 1985, Oil and gas field code, master list 1984: Washington D.C., Energy Information Administration, Office of Oil and Gas, Publication DOE/EIA 037C84, p. 35–41.

Driver, H. L., 1948, Genesis and evolution of the Los Angeles basin, California: AAPG Bulletin, v. 32, p. 109–125.

Engebretson, D. C., A. Cox, and R. G. Gordon, 1985, Relative motions between oceanic and continental plates in the Pacific basin: GSA Special Paper 206, 59 p.

Gardett, P. H., 1971, Petroleum potential of the Los Angeles basin, in I. H. Cram, ed., Future petroleum provinces of the United States—their geology and potential: AAPG Memoir 15, v. 1, p. 298–308.

Garrison, R. E., R. G. Douglas, K. E. Pisciotto, C. M. Isaacs, and J. C. Ingle, eds., 1981, The Monterey Formation and related siliceous rocks of California: Pacific Section, SEPM Special Publication, 219 p.

Godsen, R. M., 1981, Digital terrain map of the United States: USGS Map I-1318.

Gorsline, D. S., and R. G. Douglas, 1987, Analysis of sedimentary systems in active-margin basins: California continental borderland, in R. V. Ingersoll and W. G. Ernst, eds., Cenozoic basin development of coastal California: Englewood Cliffs, New Jersey, Prentice-Hall, Inc., Rubey v. VI, p. 64–80.

Gorsline, D. S., and K. O. Emery, 1959, Turbidity current deposits in San Pedro and Santa Monica basins off southern California: GSA Bulletin, v. 70, p. 279–290.

Hauksson, E., 1987, Seismotectonics of the Newport-Inglewood fault zone in the Los Angeles basin, southern California: Bulletin of the Seismological Society of America, v. 77, p. 539–561.

Hornafius, J. S., 1985, Neogene tectonic rotation of the Santa Ynez Range, western Transverse Ranges, California, suggested by paleomagnetic investigation of the Monterey Formation: Journal of Geophysical Research, v. 90, p. 12,503–12,522.

Hornafius, J. S., B. P. Luyendyk, R. R. Terres, and M. J. Kamerling, 1986, Timing and extent of Neogene tectonic rotation in the western Transverse Ranges, California: GSA Bulletin, v. 97, p. 1476–1487.

Howell, D. G., D. E. Champion, and J. G. Vedder, 1987, Terrane accretion, crustal kinematics, and basin evolution, southern California, in R. V. Ingersoll and W. G. Ernst, eds., Cenozoic basin development of coastal California: Englewood Cliffs, New Jersey, Prentice-Hall, Inc., Rubey v. VI, p. 242–258.

Ingle, J. C., 1981, Origin of Neogene diatomites around the North Pacific rim, in R. E. Garrison, R. G. Douglas, K. E. Pisciotto, C. M. Isaacs, and J. C. Ingle, eds., The Monterey Formation and related siliceous rocks of California: Pacific Section, SEPM Special Publication, p. 159–179.

Isaacs, C. M., 1987, Sources and deposition of organic matter in the Monterey Formation, south-central coastal basins of California, in R. F. Meyer, ed., Exploration for heavy crude oil and natural bitumen: AAPG Studies in Geology 25, p. 193–205.

Jennings, C. W., compiler, 1962, Geologic map of California, Long Beach sheet: California Division of Mines and Geology Map, 1:250,000.

Jennings, C. W., and R. G. Strand, compilers, 1969, Geologic map of California, Los Angeles sheet: California Division of Mines and Geology Map, 1:250,000.

Jeffrey, A. W. A., H. M. Alimi, and P. D. Jenden, 1991, Geochemistry of Los Angeles basin oil and gas systems, in K. T. Biddle, ed., Active margin basins: AAPG Memoir 52, p. 197–219.

Kamerling, M. J., and B. P. Luyendyk, 1979, Tectonic rotations of the Santa Monica Mountains region, western Transverse Ranges, California, suggested by paleomagnetic vectors: GSA Bulletin, v. 90, p. 331–337.

Kingston, D. R., C. P. Dishroon, and P. A. Williams, 1983, Global basin classification system: AAPG Bulletin, v. 67, p. 2175–2193.

Luyendyk, B. P., and J. S. Hornafius, 1987, Neogene crustal rotations, fault slip, and basin development in southern California, in R. V. Ingersoll, and W. G. Ernst, eds., Cenozoic basin development of coastal California: Englewood Cliffs, New Jersey, Prentice-Hall, Inc., Rubey v. VI, p. 259–283.

Luyendyk, B. P., M. J. Kamerling, and R. Terres, 1980, Geometric model for Neogene crustal rotations in southern California: GSA Bulletin, v. 91, p. 211–217.

Luyendyk, B. P., M. J. Kamerling, R. R. Terres, and J. S. Hornafius, 1985, Simple shear of southern California during Neogene time suggested by paleomagnetic declinations: Journal of Geophysical Research, v. 90, p. 12,454–12,466.

Mann, P., M. R. Hempton, D. C. Bradley, and K. Burke, 1983, Development of pull-apart basins: Journal of Geology, v. 91, p. 529–554.

Mayer, L., 1991, Central Los Angeles basin: subsidence and thermal implications for tectonic evolution, in K. T. Biddle, ed., Active margin basins: AAPG Memoir 52, p. 185–195.

McCulloh, T. H., 1960, Gravity variations and the geology of the Los Angeles basin of California: USGS Professional Paper 400-B, p. B320–B325.

McKenzie, D. P., 1978, Some remarks on the development of sedimentary basins: Earth and Planetary Science Letters, v. 40, p. 25–32.

Nicholson, C., and L. Seeber, 1989, Evidence for contemporary block rotation in strike-slip environments: examples from the San Andreas fault system, southern California, in C. Kissel and C. Laj, eds., Paleomagnetic rotations and continental rotations: Norwell, Massachusetts, Kluwer Academic Publishers, p. 247–280.

Philippe, G. T., 1965, On the depth, time and mechanism of petroleum generation: Geochimica et Cosmochimica Acta, v. 29, p. 1021–1049.

Pitman, W. C., III, and J. A. Andrews, 1985, Subsidence and thermal history of small pull-apart basins, in K. T. Biddle and N. Christie-Blick, eds., Strike-slip deformation, basin formation, and sedimentation: SEPM Special Publication 37, p. 45–49.

Rintoul, W., 1991, The Los Angeles basin: oil in an urban setting, in K. T. Biddle, ed., Active margin basins: AAPG Memoir 52, p. 25–34.

Rogers, T. H., compiler, 1965, Geologic map of California,

Santa Ana sheet: California Division of Mines and Geology Map, 1:250,000.

Rogers, T. H., compiler, 1967, Geologic map of California, San Bernadino sheet: California Division of Mines and Geology Map, 1:250,000.

Royden, L., J. G. Sclater, and R. P. von Herzen, 1980, Continental margin subsidence and heat flow: important parameters in formation of petroleum hydrocarbons: AAPG Bulletin, v. 64, p. 173–187.

Sawyer, D. S., A. T. Hsui, and M. N. Toksöz, 1987, Extension, subsidence, and thermal evolution of the Los Angeles basin—a two-dimensional model: Tectonophysics, v. 133, p. 15–32.

Sleep, N. H., 1971, Thermal effects of the formation of Atlantic margins by continental break-up: Geophysical Journal of the Royal Astronomical Society, v. 24, p. 325–350.

Stille, H., 1936, The present state of the earth: AAPG Bulletin, v. 20, p. 849–880.

Summerhayes, C. P., 1981, Oceanographic controls on organic matter in Miocene Monterey Formation, offshore California, *in* R. E. Garrison, R. G. Douglas, K. E. Pisciotto, C. M. Isaacs, and J. C. Ingle, eds., The Monterey Formation and related siliceous rocks of California: Pacific Section, SEPM Special Publication, p. 213–219.

Turcotte, D. L., and D. C. McAdoo, 1979, Thermal subsidence and petroleum generation in the southwestern block of the Los Angeles basin, California: Journal of Geophysical Research, v. 84, p. 3460–3464.

Whitney, J. D., 1865, Report of progress and synopsis of fieldwork from 1860–1864: California Geological Survey, Geology, v. 1, 498 p.

Wright, T., 1987, Geologic summary of the Los Angeles basin, *in* T. Wright and R. Heck, eds., Petroleum geology of coastal southern California: Pacific Section, AAPG Guidebook 60, p. 21–31.

Wright, T. L., 1991, Structural geology and tectonic evolution of the Los Angeles basin, California, *in* K. T. Biddle, Active margin basins: AAPG Memoir 52, p. 35–134.

Yeats, R. S., 1968, Rifting and rafting in the southern California borderland, *in* W. R. Dickinson and A. Grantz, eds., Proceedings of conference on geologic problems of San Andreas fault system: Stanford, California, Stanford University Publications, p. 307–322.

Yeats, R. S., 1987, Changing tectonic styles in Cenozoic basins of southern California, *in* R. V. Ingersoll and W. G. Ernst, eds., Cenozoic basin development of coastal California: Englewood Cliffs, New Jersey, Prentice-Hall, Inc., Rubey v. VI, p. 284–298.

Yeats, R. S., and J. M. Beall, 1991, Stratigraphic controls of oil fields in the Los Angeles basin: a guide to migration history, *in* K. T. Biddle, ed., Active margin basins: AAPG Memoir 52, p. 221–237.

Yerkes, R. F., T. H. McCulloh, J. E. Schoellhamer, and J. G. Vedder, 1965, Geology of the Los Angeles basin, California—an introduction: USGS Professional Paper 420-A, p. A1–A57.

Chapter 2

The Los Angeles Basin
Oil in an Urban Setting

William Rintoul

Bakersfield, California, U.S.A.

In the early 1890s, Edward L. Doheny, a mining prospector down on his luck, observed residents of Los Angeles gathering "brea" from the area's tarpits for use as fuel in coal-scarce California. Realizing that this crude tar was petroleum that had congealed upon contact with the open air, Doheny explored the residential neighborhood near Westlake Park, pooled resources with Charles A. Canfield, an old mining crony, and purchased a city lot for $400.

Unaware of oil-drilling methods, Doheny and Canfield began by sinking a four-by-six-foot miner's shaft, digging it out by hand with pick and shovel. They found an oil seep seven feet below the surface and kept digging, despite the presence of gas. They finally gave up at 155 feet, nearly overcome by fumes. Doheny then fashioned a crude drill from a sixty-foot eucalyptus tree trunk and continued to bore the hole. On the fortieth day of work, gas burst out of the hole and oil bubbled up into the shaft. The boom was on.

With fortunes to be made, the residential district became crowded with promoters, drillers, and derricks. Trampled gardens, chugging and wheezing pumps, flooded lawns, and other nuisances went along with the attempt to turn backyards into pay dirt. In an area bounded by Figueroa, First, Union, and Temple streets, more than 500 wells were producing oil by 1897 (Figure 1).

Drilling wells in the Los Angeles City field posed the problem of making oil production compatible with urban living. Residents had to deal with noise, dirt, traffic, odors, and waste disposal. At least one solution to the waste disposal problem proved unique. A homeowner with a rig in his backyard had no place for a sump in which to run waste water and mud. However, his house had a basement, and that's where the mud went.

As wells proliferated, the price of oil dropped to as little as ten cents a barrel. One day a producer with wells and a storage tank on the side of a hill by the intersection of Glendale and Beverly boulevards was busy near the tank when a man came up and looked at it. When the producer asked what the man wanted, the man said, "I'm from the city of Glendale, and we would like to buy a wooden storage tank, like this one." The oil operator asked how much Glendale would give for such a tank, and the man quoted a price. The oilman's ears pricked up. He quickly offered, "What about this one right here?" The man said, "Well, let me climb up and look at it." He climbed the ladder and examined the tank. He then said, "This is about what we want, and it would be just fine, except it's half full of oil." "That's all right," the oilman said. "When do you want to come after the tank?" The man said they could probably have a crew there the next day. "How soon would the tank be empty?" "I'll empty it right now," the oilman said, and that's exactly what he did. He opened the valve and let the oil run down the street.

Unfortunately, wells tended to decline rapidly. Producers did not know that they should hold the gas in the ground to keep the oil flowing. After operators had blown most of the gas out of the field, production dropped to only two or three barrels a day. Low rates provided the impetus for many one- or two-well producers to sell out, particularly because the oil had to be steamed by a set of boilers until it contained less than 2 percent sediment before it could be sold to a refinery. If a producer didn't have boilers and his neighbor did, the advantages of selling out loomed large.

The invention of the Allen patented pumping unit, which began to appear in the 1890s, spurred the concentration of ownership. The Allen pumping unit consisted of a vertical shaft driven by a bevel-gear. On the upper end of the shaft the wires, or pitmans, from various pumps were attached to an eccentric, preferably in such a manner that the pull of the pumps would balance one another. When balance could not be maintained, a counter-balance was employed. The stroke of the pump corresponded to the revolution of the eccentric—twelve to seventeen strokes a minute—according to the gravity of the oil and the amount of oil pumped at each stroke. When wells were a great distance apart, connection was made by wire cables to a reciprocating jack. In some instances, power was conveyed for as much as one-half mile. Steam and, in a few instances, gas engines furnished the motive power.

Date Submitted: 5/23/87
Date Accepted: 4/4/88

Figure 1. Within five years of the Doheny and Canfield discovery, more than 500 wells were producing oil in an area bounded by Figueroa, First, Union, and Temple streets in the Los Angeles City field.

Jack-line systems, as the Allen patented pumping units were nicknamed, could pump as many as twenty wells from one plant. If an operator had a jack-line plant and a steaming plant, he or she could make a strong pitch to acquire nearby wells. An operator with a single well on a single lot had no particular incentive to install his or her own jack-line and steaming plants. It made economic sense for a single-well producer to sell out to a larger neighbor letting a better endowed operator run a cable from his or her jack-line plant over to the small operator's well.

Mrs. Emma Summers, one of the most successful individuals to rise to a position of prominence as a producer in the Los Angeles City field, would have been regarded as a highly unlikely candidate for an oil operator by anyone's standards. She was a native of Kentucky, a piano teacher, and a graduate of the New England Conservatory of Music. She had come to Los Angeles to teach music, nine years before the discovery of oil. One of her interests was real estate, and she was an occasional investor. Given the excitement that followed the Doheny and Canfield find, it was not much of a jump from real estate to oil, and Emma Summers enthusiastically took the leap.

As a first step, she bought half interest in a well for $700. So thoroughly was she bitten by the oil bug, that before the first well could be completed, she had interests in a number of other wells and was several thousand dollars in debt. Unwilling to leave details to others, she personally oversaw the buying of tools and supplies, hired the workmen, and even superintended the actual drilling and producing operations. When the day's work was over, she hurried home to teach music, striving to earn more money so that she might increase her investment in the oil field.

In addition to becoming an operator, she attempted to corner the market on sales, succeeding to the extent of controlling more than half the production of the field before the turn of the century. At one point, she was selling more than 50,000 barrels of oil a month to downtown hotels, factories, the Pacific Light Power Company, and several southern California commuter railroads and trolley lines as well. Inevitably, she became known as the *oil queen of California*.

Other prominent figures in the Los Angeles City field were Max Whittier and Burton Green, who joined Charlie Canfield in a syndicate that decided to branch out and look for oil elsewhere. The three men purchased the Rancho Rodeo de Las Aguas to the west of the Los Angeles City field for $670,000. The ranchland was planted in beans, which seemed to be the only cash crop that would grow in the dusty soil. The bean fields, according to the consensus of Canfield, Green, and Whittier, almost certainly hid an oil field.

After they had drilled thirty dry holes, the three men concluded they had made a miscalculation. Rather than grow beans, they decided to try another approach. The Los Angeles real estate market was booming. Why not remove the beans, lay out some streets, and turn the ranch into a classy subdivision? Before they sold a single lot, they added restrictive covenants to every deed that prohibited oil exploration or the drilling of wells on the property. Because the oil boom had played havoc with one street after another in Los Angeles, they regarded the restrictions as essential protection for maintaining a respectable neighborhood. Entire city neighborhoods had vanished under forests of wooden derricks (Figure 2). These matters taken care of, there remained only one thing to be resolved: what to call their new suburban community. They named it Beverly Hills.

Although the first commercial production in the Los Angeles basin had been developed by wildcatters drilling by seeps, a new breed of oilman soon appeared on the scene, bringing the perspective that would lead to significant discoveries in the future. College-trained geologists began to enter the industry, and gradually their contributions came to be an essential part of the oil-production process. At first, however, the new professionals found themselves unappreciated. Walter English was one of the first to try to parlay a bachelor's degree in geology into a successful oil career. In Los Angeles, he spoke with Bill Orcutt, the vice-president in charge of exploration for the Union Oil Company, one of the state's biggest producers.

"Mr. Orcutt," English said, "How about giving me a job as a geologist?" Orcutt replied politely, "I'm sorry I can't give you a job. Union Oil Company already has a geologist. As a matter of fact, I am a geologist myself, so we are completely supplied with geologists." English subsequently returned to college, where he earned one of the first master's

Figure 2. Wooden derricks crowd residences for space along Court Street, Los Angeles, 1901.

degrees in paleontology awarded by the University of California.

"When I got out of school in 1913," English recounted, "Doc Merriam, who had taught me most of my stuff, sent a recommendation for me to the U.S. Geological Survey, and they hired me." English went to work in a field party. The pay was $75 a month. In the same party, there were a teamster and a cook. Their pay, English recalled, was $80 a month.

One company that was more receptive to geologists was a major oil company that initially came to California to buy production but, as time had gone on, decided to place greater reliance on exploration. When hiring geologists, Standard Oil Company looked to the University of California, and in 1912–1913, Standard hired two of Professor John C. Merriam's promising students: Percy W. Thompson and Reg Stoner. Standard Oil Company asked Thompson and Stoner to report on potential oil lands in the Los Angeles basin.

Thompson and Stoner recommended the acquisition of more than 2700 acres that Murphy Oil Company intended to sell in the Coyote Hills area. Seven wells had been drilled on the property, and while some production had been developed, water was proving a problem. Standard took a forty-year lease on all the Murphy holdings in West Coyote and on 1100 acres in East Whittier in December 1913. In exchange, the company agreed to a first payment of $1 million in cash and 2500 shares of stock, valued at $200 a share. A second payment of $1 million was made four months later, and a third payment of $4 million was spread over a five-year period. Standard also agreed to pay a royalty of one-fourth of the crude produced above 730,000 barrels annually, during the first five years, and on all crude after that. The company also agreed to keep ten strings of tools at work until production reached 10,000 barrels per day. There were those who thought Standard had lost its mind. But in five years, Standard had more than recovered its investment and operating costs, and its share of production was averaging more than 13,000 barrels a day.

In the wake of the success at Coyote Hills, Reg Stoner recommended three other likely areas in a low range of hills running northwest from Long Beach to Inglewood, including Signal Hill, Dominguez Hill, and Sunnyside Hill (Rosecrans). Management turned down the Signal Hill and Rosecrans plays, drilled two dry holes at Dominguez, then decided to look elsewhere. When Stoner recommended that the company take a land position at Santa Fe Springs, management turned him down again. Stoner's recommendation of Montebello was more favorably received. The company succeeded in leasing 712 acres from Anita Baldwin, whose father, the late E. J. "Lucky" Baldwin, had obtained the property following his successes in the Comstock lode. Drilling got under way in November 1916, with Anita Baldwin christening the bit with a bottle of California wine. The Baldwin luck held, and the initial wildcat came in three months later, allowing Montebello to join West Coyote as one of the major fields in the Los Angeles basin.

With Montebello, the stage was set for one of the greatest discovery decades in the history of American oil, featuring the discovery of a giant field on an almost one-a-year basis (Figures 3, 4). The finds included Santa Fe Springs and Richfield in 1919, Huntington Beach in 1920, Torrance in 1922, Dominguez in 1923, and Inglewood and Seal Beach in 1924. The largest of these fields would prove to be the Huntington Beach field.

Standard Oil Company proved up the Huntington Beach field in June 1920 with a discovery well whose initial production was an unspectacular seventy barrels a day. These seventy barrels were only the down payment on what would amount to one billion-plus barrels.

Among those who made fortunes from the big Huntington Beach discovery were some who realized great returns without meaning to do so, through a quirk of circumstance. In pre–World War I days, a promoter bought a few acres on the edge of what was then an unpopulated area at Huntington Beach. The promoter intended to subdivide the acres into twenty-five-foot town lots and sell them for a profit to a public eager to buy town lots almost anywhere in southern California. However, the real estate boom faded before the promoter could place his lots on the market. About the same time, a printer in New England invested in a set of plates for a then new encyclopedia. He printed and bound several thousand copies of the set, but found the public as reluctant to buy encyclopedias as town lots. A mutual friend brought the printer and the town lot promoter together. Soon afterward New England newspapers carried an advertisement informing readers that for a small sum down and an agreement to make further payments, they could obtain not only a modern encyclopedia but a city lot in the *booming* town of Huntington Beach, California, overlooking the Pacific and within a short distance of orange groves. Soon, sets of the encyclopedia appeared on New England tabletops from Bangor to Boston.

The California promoter sold out; so did the New England printer. Years passed. Grant deeds to Huntington Beach town lots were buried in the bottoms of old trunks. Encyclopedias went to secondhand booksellers.

After Standard Oil Company found oil at Huntington Beach, the boom spread into the town lot area. Lease brokers frantically searched out titles to town lots that might form the ground on which to drill a well. The search took them to New England, to the deeds that many encyclopedia buyers had forgotten they owned.

One New Englander, Ezra Hapfield, had purchased an encyclopedia for his daughter, Hattie, who was attending finishing school. When the deed arrived for a town lot in California, he put it in a desk drawer, later transferring it to an old trunk. Hattie married, bore a son, and, when her husband died, moved in with her father. Life on the farm went on. Unexpectedly, a letter from California arrived inquiring about the deed to a Huntington Beach town lot. The inquiring firm later offered Hapfield more than $300 for it, which startled him. It was more than he paid for the encyclopedia. Hapfield paid back taxes on the lot, redeemed the property, and headed west.

Hapfield's lot, like others in the "encyclopedia section," proved to be over rich oil sand. With their first oil royalties, the Hapfields bought a house on a slope overlooking the sea, with orange trees in the backyard and winter-blooming roses on the porch. It was a dream come true (Figure 5).

The Los Angeles basin discoveries, of course, were accompanied by an influx of promoters. At Signal Hill in

Figure 3. In April 1930, eleven years after its discovery, the Santa Fe Springs field was a forest of steel and wooden derricks.

Long Beach, burning flares from the natural gas released in drilling operations fostered a carnival atmosphere. The Signal Hill field became a promoter's paradise. Large red and white circus tents were set up in which bus loads of prospective investors were fed meals. Often the waiters were high-powered salesmen. Before their sales pitches ended, people were flocking to buy a share of the action.

The initial production at Signal Hill was only the first installment on what would climb in succeeding years to more than one-half million barrels of oil per acre, making it one of the richest fields in per-acre production that the world had ever seen (Figures 6, 7). In the wake of the Signal Hill find, the Long Beach Women's Club held a public discussion to debate the pros and cons of turning the city's business district into an oil field. The city manager said that if the wells were big enough, it might pay to replace business blocks and residences with derricks and refineries. However, he wisely added, the city should be assured that the wells would be big before it hurried to tear down buildings.

Some newcomers took up residence on Norwalk Boulevard between Santa Fe Springs and Norwalk. The area, known variously as the Gum Grove, Springs Slums, and Kings Camp, became famous for "girls in tents with a hot bottle of soda pop." According to one veteran of the boom, "Those girls had an ironing board and a shirt to iron. They kept ironing that same shirt while waiting for business customers."

One well-known gathering spot was May's Place, which featured gambling, drinks, and girls. May would pin "Mayday" buttons on regular customers. With the badge, customers could come in anytime. If customers wanted to give their paychecks to May, they were entitled to everything for a month, including, as one observer described it, "eats, gambling, girls, and all."

The stock market crash of 1929 and the Great Depression that followed dealt a harsh blow to California's oil industry. As bad news accumulated through the first years of the new decade, refineries went on a six-day week. To hold down production, companies laid off workers, and

Figure 4. One that got away in the booming discovery days of the 1920s in the Los Angeles basin.

exploratory drilling slowed to a standstill. Approximately 300,000 barrels a day of California production was curtailed, leaving the state with a daily output of about 500,000 barrels.

Though times were hard, a few wildcatters continued to drill. In January 1932, Ranger Petroleum drilled a well at Wilmington. The well produced 150 barrels a day of 14-gravity oil on the pump, and the company succeeded in marketing some of the crude as fuel for the Dollar Lines. It would be another five years before the discovery was designated as a new field and named Wilmington. Two months after that classification, General Petroleum Corporation discovered a new producing sand with its Terminal No. 1, which came in making 1389 barrels a day. With completion of the General Petroleum well and subsequent drilling, it became apparent that Wilmington was not only a major discovery but also the largest oil field ever found in California.

It was not until the start of World War II that drilling began to rebound in the Los Angeles basin and elsewhere, accelerating as the war continued. In Los Angeles, the resurgence led to an innovative job—a "noiseless" well.

Shell Oil Company drilled the Verne Community No. 1 from Gilmore Island, at the corner of First and Gardner streets, using fluffy rockwood and transite mineral board to wrap the derrick and machinery. The well, a deeper pool test for the old Salt Lake field, was named Verne in honor of Jules Verne, author of *20,000 Leagues under the Sea*. Verne's dreams, California Oil World reported, "once seemed fantastic but have been outdone by modern ingenuity." Shell used soundproof electric drilling equipment and isolated the well from sight by building a high fence around the location. Although the well was dry, it proved that drilling could be carried on in populated areas.

In the years that followed, the oil-drilling scene in the Los Angeles basin changed. Rigs were carefully soundproofed to minimize disturbance to urban neighbors, and drill sites were landscaped to look like everything from high-rise buildings to parks. With the new profile, wildcatters were able to continue the exploration of the Los Angeles basin. Discovery and development of new fields ranged from Sansinena near Whittier to Sawtelle in the western portion of the basin.

In Los Angeles, the city council opened the way for urban searches by adopting a comprehensive zoning plan in 1950. The plan laid down ground rules that permitted drilling from approved sites. Sites were to be enclosed, landscaped, and kept neat. Derricks and drilling equipment were to be fully soundproofed, vibrations were to be controlled, and odors were to be contained. Only electric power could be used. With such rules established, the search for oil became unlike any seen before in Los Angeles or anywhere else in the world. For one thing, it took ingenuity to run seismic lines. At one place, under a wide grid of railway tracks, a seismic crew crawled into a man-sized culvert to place recording equipment and recorded their "shots" between the trains scheduled overhead. At another site, as land was cleared for a new freeway, seismic crews enthusiastically swarmed over it ahead of construction crews and took their readings. To gather seismic data, Continental Oil Company used a technique which required a setting late at night to avoid auto traffic and other city sounds that would throw off the readings. The company's geologists, fearing that other oil companies would learn of their search and lease the land ahead of them, had seismic crews pose as street cleaners in white coveralls to throw off competitors.

With the new zoning ordinance, wildcatters got their chance. They proceeded to prove up more than 200 million barrels of new reserves beneath an area lying in a great arc from downtown Los Angeles to the waters of Santa Monica Bay and ultimately into the San Fernando Valley. They discovered eleven new oil fields and found significant amounts of oil under three existing fields. The new field discoveries included Boyle Heights, which was proved up by Richfield Oil Corporation in 1955; Cheviot Hills, Signal Oil & Gas Company, 1958; Las Cienegas, Union Oil Company of California, 1960; Los Angeles Downtown, Standard Oil Company of California, 1964; Sawtelle, Occidental Petroleum Corporation, 1965; Sherman, Standard Oil Company of California, 1965; Venice Beach, Mobil Oil Corporation, 1966; San Vicente, Standard Oil Company of California, 1968; Union Station, Standard Oil Company of California, 1967; and South Salt Lake, Standard

The Los Angeles Basin: Oil in an Urban Setting   31

Figure 5.   In the 1920s, swimmers and sunbathers shared the coastline with oil wells at Huntington Beach.

Figure 6.   A forest of derricks bristled in the wake of the boom at Signal Hill.

Figure 7. In May 1928, Richfield's well at Signal Hill (center) was the deepest producing oil well in the world, getting 4100 barrels a day from 7409 feet.

Figure 8. Occidental Petroleum Corporation's architecturally designed oil derrick at Pico Boulevard and Doheny Drive won an award from Los Angeles Beautiful, a nonprofit civic organization.

Oil Company of California, 1970. There were new pool discoveries in the Beverly Hills field by Universal Consolidated Oil Company in 1954, Occidental Petroleum Corporation in 1966 and 1967, and Standard Oil Company of California in 1967. Buttram Petroleum Company found new pool production in the Salt Lake field in 1961. Signal Oil & Gas Company proved up new pool production in the Redondo Beach area of the Torrance field in 1956.

There were ingenious approaches to the matter of camouflaging drilling and production islands. For example, the Beverly Hills field, which had been discovered in the early days of the Los Angeles basin boom, was in a state of decline in the 1950s when wildcatters decided it was time to look deeper. Occidental Petroleum Corporation concealed the rig within what appeared to be a modern ten-story office building, enclosed the drill site with a twelve-foot-high flagstone wall, and landscaped the whole area with trees, shrubs, and ground cover (Figure 8). Los Angeles Mayor Sam Yorty praised the company for "an outstanding contribution to civic beauty in a heavily populated area of the city," and Los Angeles Beautiful, a nonprofit civic organization, presented an award to Occidental for "going well beyond the call of duty in spending approximately $1,000,000 to build the novel sky-blue derrick."

A later discovery by Standard Oil Company of California led to the construction of a unique building on Pico Boulevard at Genessee Avenue in west Los Angeles. Designed to look like a modern office building, the structure, named Packard drilling structure, met design criteria so convincingly that, according to one story, two vice presidents of a rival oil company actually drove by the building three times before a city patrolman convinced them it was really Standard's oil well.

With the opening of the eastern, offshore portion of the Wilmington field in 1965, four new islands soon appeared off the Long Beach shoreline: the man-made THUMS islands served as bases from which to develop offshore production. To ensure that offshore oil operation did not offend anyone, THUMS Long Beach Company engaged the architectural firm of Linesch & Reynolds to suggest a beautification scheme that would camouflage and soundproof the operation. The resulting program included constructing drilling towers with pastel-colored balconies to look like office buildings or penthouse apartments. Mounted on rails, the towers ranged around the islands to slant drill wells six feet apart. To mask the view of drilling equipment, THUMS planted the islands with semitropical trees: Washingtonia fan palms and Canary Island date palms. For lower plantings, climate-resistant sandalwood trees, salt bushes, oleanders, acacias, and Moreton Bay fig trees were used. The ocean-dredged sand had to be washed of salt and beefed up with wood shavings and fertilizer. An elaborate water and fertilizing system was installed.

As the first island began to shape up, Long Beach citizens became enthusiastic. Planners went back to the drawing boards and came up with a design for three spectacular thirty-foot-high waterfalls. The waterfalls, which are visible from all around the bay, looked so good that city council members demanded and got floodlights for night viewing (Figure 9).

Figure 9. Night lighting of THUMS's Astronaut Islands has led some to describe the Long Beach Company's development of the seaward portion of the Wilmington field as the world's most beautiful oil field.

Chapter 3

Structural Geology and Tectonic Evolution of the Los Angeles Basin, California

Thomas L. Wright

Chevron Corporation (retired)
San Anselmo, California, U.S.A.

ABSTRACT

The Los Angeles basin formed in late Neogene time on a continental margin previously shaped by Cretaceous and early Paleogene subduction, Paleogene terrane accretion, and mid-Miocene rifting and block rotation. During Neogene time, the boundary between the Pacific and North American plates shifted progressively eastward beneath the Los Angeles region, creating the broad San Andreas transform zone. As reviewed in this paper, structures and rocks within the Los Angeles basin document each stage of that Neogene evolution.

The Los Angeles basin began to take its present shape in late Miocene time (ca. 7 Ma) by subsidence between the right-oblique Whittier and Palos Verdes fault zones and the left-oblique Santa Monica fault system. The principal phase of basin opening involved early Pliocene extension in a northwest direction, which accompanied the opening of the Gulf of California and the eastward shift of the southern San Andreas fault to its present position. Most of the structural traps that hold the basin's oil fields began to form during this latest Miocene–early Pliocene deformation.

Since mid-Pliocene time, many of these traps have been altered and enhanced—and a few have been breached—by Pasadenan deformation, involving southward shortening, the uplift of the Transverse Ranges, and the propagation of blind thrusts beneath the northern Los Angeles basin. The rapid transition from early Pliocene extension to late Pliocene contraction was associated temporally with a change in relative plate motion dated at 3.9–3.4 Ma. In analyzing Pasadenan deformation, the flake-tectonics model is more appropriate than the fold-and-thrust-belt model, although both models incorporate aseismic detachment at midcrustal depths. The flake-tectonics model is valid for all phases of Neogene deformation, both transtensional and transpressive, in the Los Angeles region.

Fields discovered to date in the Los Angeles basin will yield an ultimate 10.4 billion oil-equivalent barrels (GOEB) of petroleum. Of this, approximately 73% is trapped in faulted anticlines, 12% in simple anticlines, 10% in fault traps, and 5% in stratigraphic traps. Folding has been controlled primarily by preexisting structural hingelines and sedimentary wedge belts and secondarily by en echelon folding associated with wrench faults. Oil seeps and Quaternary topographic uplifts led to most of the discoveries prior to 1925 along the Whittier and Newport-Inglewood fault zones and in the Coyote Hills. Most later discoveries, including the 3-billion-barrel Wilmington oil field, were in structures with little or no Quaternary expression.

INTRODUCTION

The aims of this paper are to

1. illustrate and briefly describe typical and important structural traps in the Los Angeles basin

2. cite stratigraphic evidence for the time of deformation of these structures

3. use data from these local features to elucidate the nature and history of the major structural features within and bounding the basin

4. attempt a synthesis of the Neogene tectonic evolution of the Los Angeles basin

This paper comprises two principal sections, structural geology and tectonic evolution, and is organized as follows:

INTRODUCTION (including stratigraphic nomenclature) . . . 36

REGIONAL SETTING 39

    Geologic Provinces (continental borderland, Transverse Ranges, Peninsular Ranges, basement rock types) 39

    Predecessor Basins 43

    Plate Tectonic Framework (subduction zones, accretion, transform displacement, Basin-and-Range extension, rotation) 43

STRUCTURAL GEOLOGY 46

    Peripheral Basins (Chino Hingeline fault, Mahala and Prado-Corona oil fields, Capistrano syncline) 46

Date Submitted: 7/7/89
Date Accepted: 4/15/90

    Boundary Structures 47

        Whittier Fault (Whittier, Sansinena, Brea-Olinda, and Esperanza oil fields; general characteristics; vertical separation; lateral displacement; East Montebello fault; early history; Quaternary deformation)

        Palos Verdes Uplift and Fault (Beta oil field, THUMS-Huntington Beach fault)

        Santa Monica Fault System [Neogene tectonic evolution; Santa Monica fault; Sawtelle, Riviera, Cheviot Hills, East Beverly Hills, and San Vicente oil fields; Hollywood fault; North Salt Lake fault; Potrero Canyon fault; vertical displacement; left-lateral offset; active faulting; Malibu Coast and Anacapa (Dume) faults; Raymond fault]

        Northern Margin (Elysian Park anticline, Alhambra high)

        Southeastern Margin (Santa Ana Mountains, Santa Ana shelf, San Joaquin Hills, Newport Offshore ridge)

    Major Internal Structures 66

        Newport-Inglewood Fault Zone (Inglewood, Dominguez, Long Beach, Long Beach Airport, Seal Beach, Huntington Beach, Sunset Beach, and other oil fields; northern end; South Coast Offshore fault; continuing tectonic activity; Compton-Los Alamitos fault; pre-Neogene events; early to mid-Miocene uplift; mid-Miocene, late Miocene, early Pliocene, and Quaternary deformation; alternative structural concepts)

        Western Shelf (Venice Beach and Playa del Rey oil fields, Hyperion-Alondra trend, Torrance, and Wilmington-East Wilmington-Belmont Offshore oil fields)

        Central Trough

        Anaheim Nose

        Northern Shelf (Las Cienegas oil field; Las Cienegas fault; Salt Lake, South Salt Lake, Los Angeles City, Las Cienegas/Jefferson, Los Angeles Downtown, Union Station, Boyle Heights, Bandini, and East Los Angeles oil fields)

Fullerton Embayment (evolution of the northeastern Los
Angeles basin; West and East Coyote, Richfield,
Yorba Linda, Kraemer, and Olive oil fields; Peralta
Hills thrust)

North Central Area (Santa Fe Springs, Leffingwell, and
Montebello oil fields)

Questionable Structures (Norwalk, Charnock, Overland
Avenue, and Temescal faults) 89

TECTONIC EVOLUTION 90

Early Miocene Reconstruction 90

Middle Miocene Deformation 92

Santa Monica Mountains (and Point Dume)
San Joaquin Hills and Vicinity (El Modeno Volcanics)
Glendora Volcanics (and Buzzard Peak Conglomerate)
Los Angeles basin (Palos Verdes Hills, mid-Miocene
faulting)
Chronology (37–22 Ma, 22–16 Ma, 16–12 Ma,
12–9 Ma)

Late Miocene Stability 101

Tarzana and Puente Fans

Mio-Pliocene Deformation 102

Onset of Marginal Uplifts
Post–Mid-Mohnian Displacements
Basinal Deposition
Growth of Internal Structures (Newport-Inglewood fault
zone, Wilmington anticline, Western shelf)
Santa Monica, Las Cienegas, and Whittier Faults
Dynamic Setting

Pasadenan Deformation 108

Right-Lateral Displacements
Basinal Subsidence and Shelf Uplift
North Los Angeles Thrust System (blind thrusts, seg-
mentation, fold-and-thrust-belt model versus flake-
tectonics model)
Interaction between Thrusting and Right Slip

CONCLUSIONS 116
ACKNOWLEDGMENTS 118
REFERENCES CITED 118
APPENDICES 129

Abbreviations Used in Figures 129
Wells Identified in Figures 131

The main facts of Los Angeles basin stratigraphy and structure have been known for decades. Still useful summaries are provided by Woodford et al. (1954), Barbat (1958), and Gardett (1971). The lucid exposition of Yerkes et al. (1965) remains an essential reference for any serious study of the region (their annotated bibliography, arranged in chronological order, is an excellent synopsis of the first 106 years of geological studies in the Los Angeles basin). The basic data in these papers are augmented herein by details garnered during the extensive urban oil exploration of the 1960s and by data from offshore exploration. Subsequent to 1965, the concepts of plate tectonics (Atwater, 1970) have provided the theoretical framework—previously lacking—by which the complex tectonic evolution of southern California may be understood. This paper succeeds earlier ventures (Campbell and Yerkes, 1971, 1976; Wright et al., 1973) at placing the Los Angeles basin within the framework of plate tectonics and incorporates recent research on tectonic rotations (Hornafius et al., 1986; Luyendyk and Hornafius, 1987), low-stress strike-slip faulting (Zoback et al., 1987; Mount and Suppe, 1987), and regional detachments (Yeats, 1981, 1983; Davis et al., 1989).

New subsurface data undoubtedly will lead to significant future revisions of any current interpretation of Los Angeles basin geology. Despite the expense and difficulties inherent in exploration in an urban environment, future drilling will probe the depths of the central basin and the various lightly explored areas around its flanks. New seismic techniques will yield better images of subsurface structure. Studies of earthquake hazards may resolve the complex geology within the several structural "knots" on the northern edge of the basin. New or refined methods of correlation will provide a much better understanding of the temporal relationships of deformational events within the basin and will link these to specific phases in the evolution of the San Andreas transform zone. In terms of absolute time, the present "error range" of Neogene correlations in southern California is here estimated to be about 10%. The pace of current research suggests that this range may be reduced by half within the next decade.

A variety of *stratigraphic nomenclature* has been used within the Los Angeles basin (Blake, *Review of the Neogene Biostratigraphy* and *Stratigraphy of the Los Angeles Basin and Implications for Basin Evolution*, this volume, his Figure 2). In general, this paper (Figure 1) follows the provincial Miocene biostratigraphic stages of Kleinpell (1938, 1980), the middle Miocene to Pleistocene divisions of Wissler (1943), and the Pliocene–Pleistocene stages of Natland (1953; Natland and Rothwell, 1954). The paper by Blake in this volume describes the present-day application of these terms and their relationships to absolute ages and to other local and worldwide zonations and chronologies. The stratigraphic and chronologic terms used herein follow the usage of Blake as closely as possible. Three key points should be noted:

♦ The late Miocene–middle Miocene boundary is placed at 9.5 Ma, following Barron (1986, p. 106). Previously this boundary has been placed at 10.4 Ma (Berggren et al., 1985), 11 Ma (Warren, 1980), or older. The Mohnian Stage, critical to dating the evolution of the Los Angeles basin, was originally considered to be wholly within the late Miocene (Kleinpell, 1938). It is now assigned an approximate age range of 6.3–13.8 Ma (Blake, this volume, his Table 2 and Figure 2), straddling the late Miocene–middle Miocene boundary.

♦ The top of the Delmontian, traditionally taken as the top of the Miocene, is now known to extend into the lower Pliocene. Blake (this volume, his Figure 2) has indicated a significant time-transgression by the Repettian–Delmontian boundary. The Neogene biozones of California are known to transgress timelines; they are based on benthic foraminifera, which are

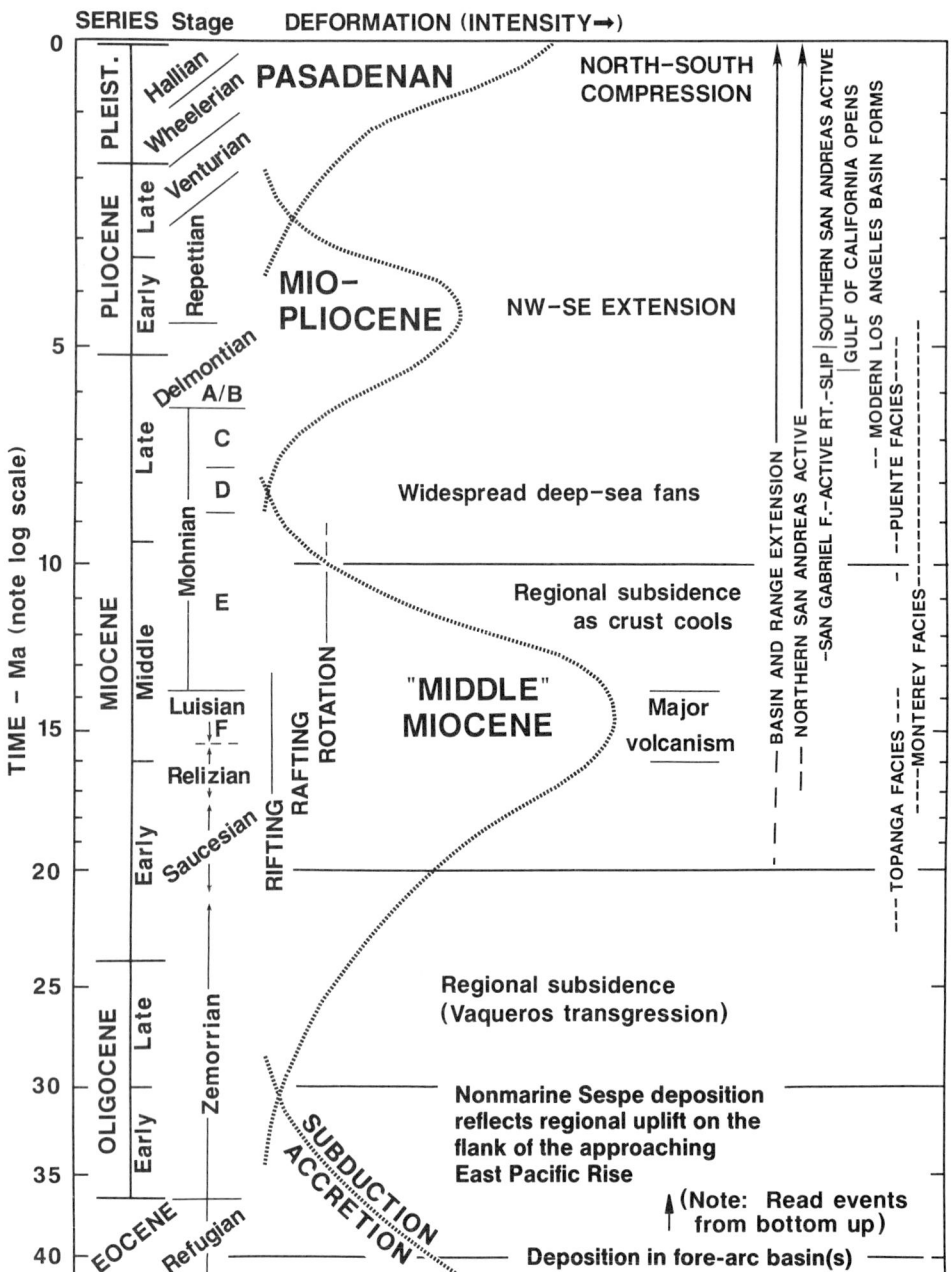

Figure 1. Chronology of major Cenozoic events in the Los Angeles region. Curve shows intensity of tectonic deformation. Local biostratigraphic zonation and major depositional facies are also shown. Boundaries between late Neogene stages are shown as slanting lines to represent the time-transgressive nature of these provincial units.

strongly influenced by environmental factors, especially water temperature and depth (Natland, 1933). However, in the Los Angeles basin, a distinctive bentonite bed occurs a few hundred feet below the Repettian-Delmontian boundary and is noted on electric logs from wells around most of the basin margins, including wells in the Whittier, Inglewood, and Huntington Beach oil fields. This marker provides a useful datum at about 4.5 Ma and suggests that this key faunal boundary is approximately synchronous around the basin.

♦ This paper follows Blake (this volume) in calling the Repetto and Pico formations. These terms have been in common use in the Los Angeles basin since the 1930s (Reed, 1936; Kew, 1937) to describe basinal rocks of Pliocene to mid-Pleistocene age. They are the standard terminology in proprietary and published oil field and well reports by oil industry geologists and by the California Division of Oil and Gas. In the Los Angeles basin, the Repetto and Pico formations were defined on the questionable basis of microfaunal zones—a problem that Natland (1953, 1954) sought to remedy by establishing the four Plio–Pleistocene stages shown in Figure 1. Blake (this volume) noted the near-synonymity of Natland's biostratigraphic stages and the older Pico–Repetto zonation. Durham and Yerkes (1964) have argued persuasively that the Repetto and (in the Los Angeles basin) Pico are not valid lithostratigraphic units (formations) and that these terms should be replaced by lower and upper Fernando Formation. This writer, however, is compelled to follow the older, dubious nomenclature because (1) the application of its biostratigraphic zonation is essential to dating the growth of structures within the Los Angeles basin; (2) conversion of the older zonation to Natland's stages without expert review of individual microfaunal lists would be misleading and may cause significant error; and (3) conversion of Pico and Repetto to upper and lower Fernando without reviewing lithologic criteria in each oil field or well would be equally erroneous, and we would lose the more precise biostratigraphic zonation. Use of quotation marks around Repetto and Pico throughout this paper to signify their informal nature has been omitted as "untidy."

Stratigraphic studies are a basic tool in deciphering the structural history of the Los Angeles basin. During nearly all of late Miocene and Pliocene time (about 9–2 Ma), submarine fans were the principal means of sedimentation in the basin (Conrey, 1967; Redin, *Oil and Gas Production from Submarine Fans of the Los Angeles Basin,* this volume). Sand deposition on the fringes of these fans maintained a uniform gradient across the basin floor. Sands and silts were swept across any slight rises on the sea bottom and accumulated in adjacent subsiding areas. Individual sandstone units, thus, thin or disappear across growing structures (Yeats and Beall, *Stratigraphic Controls of Oil Fields in the Los Angeles Basin,* this volume). On adjacent submarine slopes and sea knolls, however, the accumulation of clays and biogenic sediments was much less dependent on bathymetric relief. For that reason, interval sand-thickness maps commonly are better than isopach maps in documenting tectonic movements. Unpublished regional sand-isolith maps (T. L. Wright, 1964–1970) are cited in this paper as one kind of evidence for the time of local structural growth.

"Unconformities" indicated at various horizons within the upper Miocene and Pliocene deep-water beds of the Los Angeles basin do not result from subaerial erosion and only occasionally do they result from submarine erosion of semiconsolidated sediments. These local, angular discordances and stratigraphic lacunae result from the submarine fan sedimentation described above. Over time, these processes produce bald-headed highs with sedimentary wedges on their flanks and in the foot-wall blocks of faults. The term "unconformity" when applied in this situation may be somewhat misleading.

Facies other than these basin-plain fans provide only crude or ambiguous evidence of structural growth. In the mid to upper (proximal) parts of the fans—as in the Montebello oil field and vicinity—channel cutting and filling has obscured the record. Conglomerates, slump deposits, and sedimentary breccias commonly are localized along the lower slope of an uplifted block. The paleoboundaries of submarine slopes and sea knolls typically are marked by a lateral change from shale (perhaps organic or diatomaceous) to a deep-sea fan facies with appreciable sand content. Within the Miocene sequence of coastal California, the Monterey Formation comprises a distinctive facies assemblage that includes chert, dolomite, phosphorite, and related biogenic sediments. As shown in Figure 1, in the Los Angeles basin, the Monterey facies (which here includes the Valmonte and Altamira members in the Palos Verdes Hills and the La Vida Member of the Puente Formation) locally is laterally equivalent to parts of the Topanga Formation and to sandstone of the Soquel and younger members of the Puente Formation.

In addition to showing the chronologic placement of the local biostratigraphic units, Figure 1 summarizes the succession of major tectonic and depositional episodes described in this paper.

Within the Los Angeles basin, individual faults sometimes have been given several names. This paper generally follows the terminology of Ziony and Yerkes (1985), though other commonly used names will be indicated as appropriate. Unless otherwise cited, data on individual oil fields are from the California Division of Oil and Gas (1974).

The paper should be viewed as a progress report on the structural geology of the Los Angeles basin. It is hoped that its publication will stimulate significant additions and corrections based on proprietary data not now available to this researcher. Interpretations in this paper that differ from earlier ones may serve to stimulate new "play concepts" and perhaps lead to new discoveries. Future explorationists should be equally alert for new data that do not fit this or other current interpretations but that may lead to entirely new concepts and discoveries.

REGIONAL SETTING

Geologic Provinces

The Los Angeles basin is located at the juncture of the three primary physiographic provinces of coastal southern California (Figure 2): the Peninsular Ranges, the Transverse Ranges, and the continental borderland. The Los Angeles basin shares the geologic histories and principal geologic characteristics of all three regions.

The *continental borderland* (Vedder, 1976, 1987) is characterized by northwest-trending basins and ridges (Figure 3) that took shape during early and middle Miocene time. Neogene deformation has involved right slip on major northwest-striking faults, differential vertical displacements, folding and faulting of basinal sedimentary rocks, and apparently the rotation of local blocks. Basement rocks of the inner part of the borderland are Mesozoic blueschists, greenschists, and other rocks of the Catalina terrane (Howell and Vedder, 1981). They represent oceanic and trench

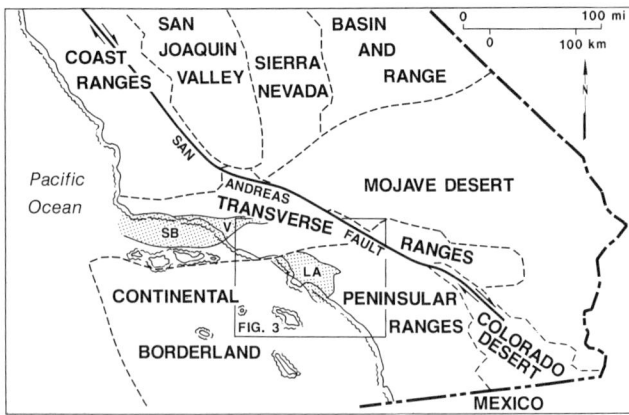

Figure 2. Geologic/geomorphic provinces of southern California. Stippled areas are the Los Angeles (LA) and Ventura–Santa Barbara (V, SB) basins. Rectangle shows area of Figure 3.

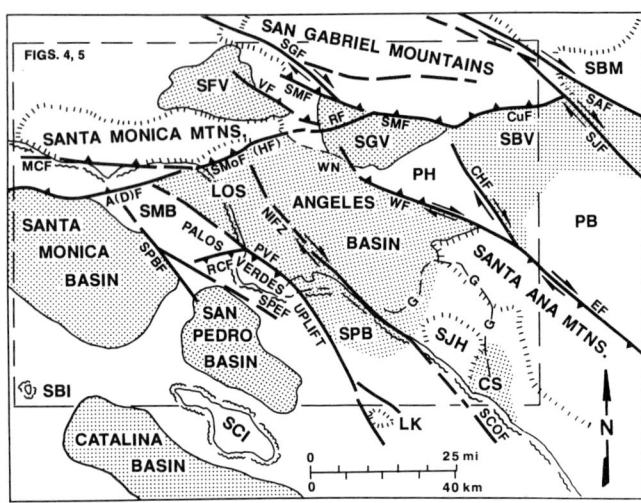

Figure 3. Los Angeles basin and vicinity, showing principal geomorphic and tectonic features. Stippled areas are Quaternary basinal areas. A(D)F = Anacapa (Dume) fault, CS = Capistrano syncline, CHF = Chino Hingeline fault, CuF = Cucamonga fault, EF = Elsinore fault, HF = Hollywood fault, LK = Lasuen Knoll, MCF = Malibu Coast fault, NIFZ = Newport-Inglewood fault zone, PB = Perris block, PH = Puente (and San Jose) Hills, PVF = Palos Verdes fault, RCF = Redondo Canyon fault, RF = Raymond fault, SAF = San Andreas fault, SBI = Santa Barbara Island, SBM = San Bernardino Mountains, SBV = San Bernardino Valley, SCI = Santa Catalina Island, SCOF = South Coast Offshore fault, SFV = San Fernando Valley, SGF = San Gabriel fault, SGV = San Gabriel Valley, SJF = San Jacinto fault, SJH = San Joaquin Hills, SMB = Santa Monica Bay, SMF = Sierra Madre fault, SMoF = Santa Monica fault, SPB = San Pedro Bay, SPBF = San Pedro Basin fault zone, SPEF = San Pedro Escarpment fault, VF = Verdugo fault, WF = Whittier fault, WN = Whittier Narrows. —G— illustrates approximate limits of San Joaquin Hills gravity high. Rectangle shows area of Figures 4 and 5.

sediments and basalts metamorphosed in mid-Cretaceous time (Platt, 1976) at relatively high pressures and low-to-moderate temperatures during initial subduction along the plate margin (Sorensen, 1984). Beneath the western Los Angeles basin (Figure 4), similar blueschists and greenschists compose the basement rock from the Newport-Inglewood fault zone westward (Woodford et al., 1954). Cretaceous and Paleogene fore-arc deposits are not found in the inner borderland but are widespread in the outer borderland, overlying basement of uncertain character. Relict northwest-trending mid-Miocene structures within the present-day Los Angeles basin resemble the basin-and-ridge features of the continental borderland. Several of these structures are evident in Figure 5, which shows the known extent and thickness of the middle Miocene sedimentary and volcanic rocks of the Topanga Group and related facies.

The west-trending *Transverse Ranges* (Bailey and Jahns, 1954; Dibblee, 1982b) are a major anomaly within the northwest to northerly trends of the western edge of North America. Recent studies of paleomagnetic declinations imply that individual crustal blocks within this province were rotated into their present positions beginning in the middle Miocene (Hornafius et al., 1986; Luyendyk and Hornafius, 1987). The rotations were almost uniformly clockwise and vary from 95° in the westernmost Transverse Ranges to 40° in the easternmost part of the province. Measured paleomagnetic declinations adjacent to the Los Angeles basin are shown in Figure 5. Other declinations measured in the Glendora volcanics and in the San Joaquin Hills (and nearby South Laguna Beach) are widely scattered, presumably owing to movements on nearby major faults (B. P. Luyendyk, oral communication, 1988).

The Transverse Ranges have been elevated during the last two to three million years by the pronounced shortening, oriented south to southeast, of the Pasadenan deformation (Crowell, 1976). The Santa Monica Mountains (Hoots, 1931; Durrell, 1954; Dibblee, 1982a) at the northern edge of the Los Angeles basin are part of the central Transverse Ranges province. The origins of the Santa Monica fault system and adjacent folds and faults within the northern Los Angeles basin are closely related to Transverse Ranges deformation.

The *Peninsular Ranges* (Jahns, 1954) extend southeastward some 900 mi (1500 km) from southern California down the length of the Baja California peninsula. They comprise mildly metamorphosed sedimentary and volcanic rocks of island-arc affinity and are mid-Jurassic to mid-Cretaceous in age (Schoellhamer et al., 1981; Gastil et al., 1981). These are intruded by quartz plutonites and gabbros (mostly mid-Cretaceous in age) of the southern California batholith.

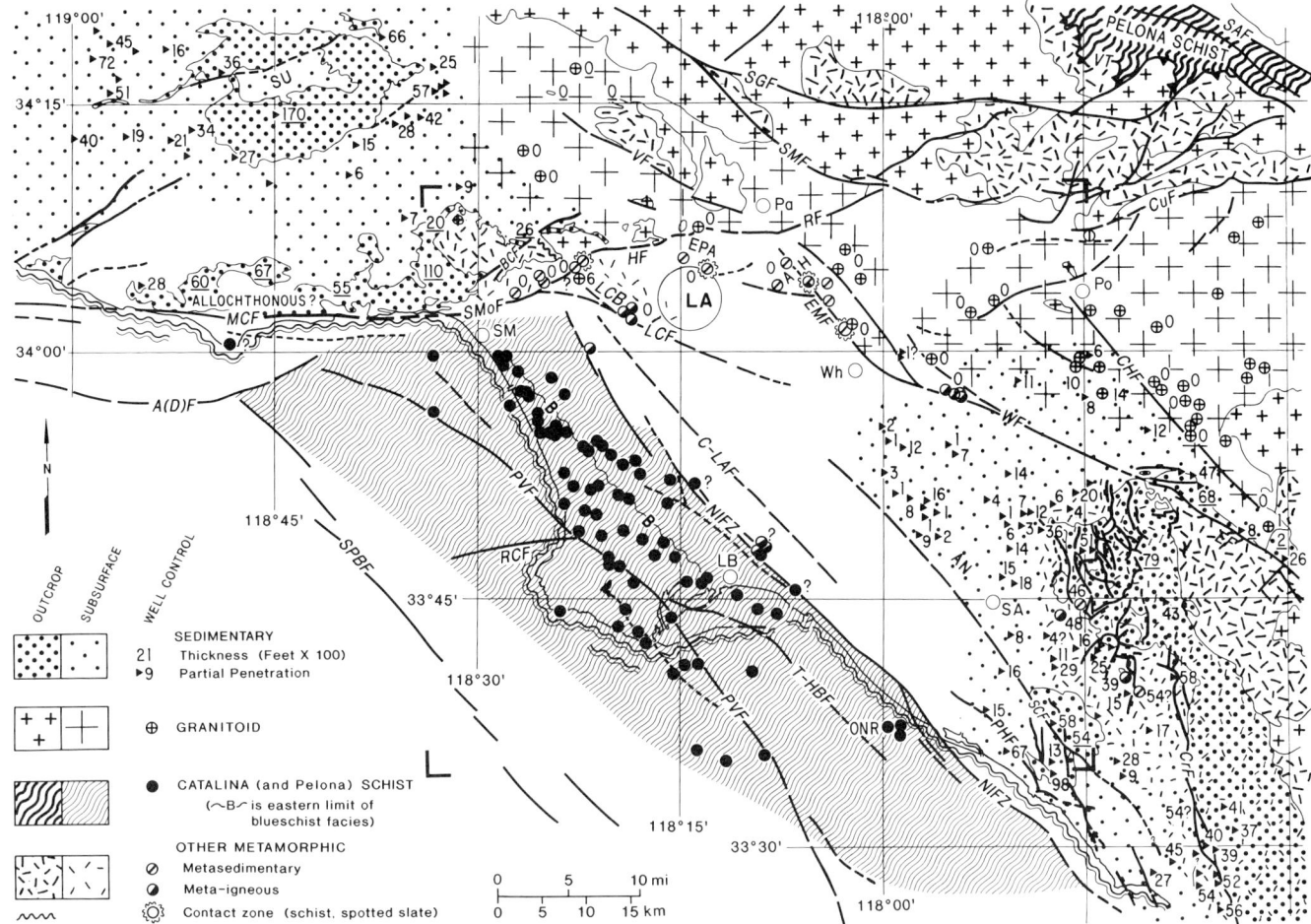

Figure 4. Distribution of basement rocks and pre-Topanga sedimentary rocks around the Los Angeles basin. Surface geology after Jennings and Strand (1969). Bold patterns and circles show outcrop and well control; subdued patterns show areas where occurrence beneath Neogene cover can reasonably be inferred (patterns overlap where basement is encountered beneath pre-Topanga sedimentary rocks). Thicknesses of Upper Cretaceous through lower Miocene section in hundreds of feet; underlined for outcrop sections. Major basement-penetrating faults are shown as footwall trace on base of Neogene. Abbreviations as for Figure 3, plus: AH = Alhambra high, AN = Anaheim nose, BCF = Benedict Canyon fault, C-LAF = Compton–Los Alamitos fault, CrF = Cristianitos fault, EMF = East Montebello fault, EPA = Elysian Park anticline, LCF = Las Cienegas fault, LCB = Las Cienegas block, ONR = Offshore Newport ridge, PHF = Pelican Hills fault, SCF = Shady Canyon fault, SU = Simi uplift, T-HBF = THUMS-Huntington Beach fault, VT = Vincent thrust. Open circles show major cities: Los Angeles (LA), Long Beach (LB), Pasadena (Pa), Pomona (Po), Santa Ana (SA), Santa Monica (SM), and Whittier (Wh). Rectangle corners show areas in Figures 7, 9, and 10.

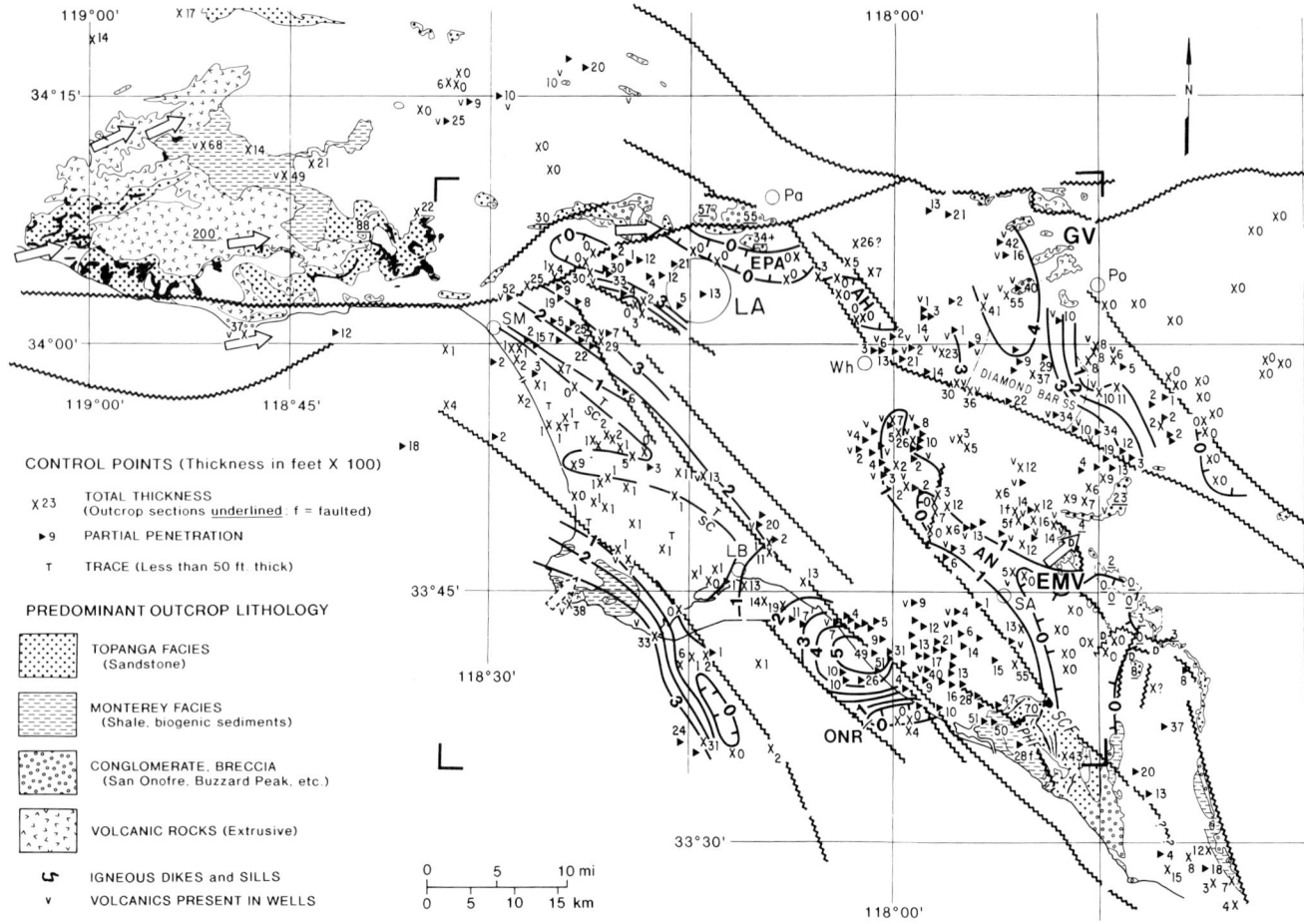

Figure 5. Distribution and thickness of the Topanga Group (and synchronous facies of Zemorrian-Luisian age) in and adjacent to the Los Angeles basin. Isopach interval = 1000 ft (isopachs limited to areas of adequate control). Hachured line bounds areas where Topanga is absent beneath younger Miocene cover. In Santa Monica–Long Beach area, dashed line is approximate facies boundary between schist conglomerate (SC) and Topanga (T). Between Whittier and Pomona, light dashes outline subsurface extent of Diamond Bar Sandstone. Large open arrows (broken where data are weak) are paleomagnetic declinations measured in volcanic and sedimentary rocks of middle Miocene age (from Kamerling and Luyendyk, 1979; B. P. Luyendyk, personal communication, 1988; Earth Sciences Board, 1988). Wavy lines are faults that affect distribution of middle Miocene rocks. Abbreviations as in Figure 4, plus: EMV = El Modeno volcanics, GV = Glendora volcanics. Rectangle corners show areas in Figures 7, 9, and 10.

Basement rocks of the Peninsular Ranges are overlain locally by Upper Cretaceous and Paleogene sedimentary rocks. In southern California and northern Baja California, these rocks have been offset during the past three million years or more by active northwest-striking right-slip faults, including the Elsinore and San Jacinto faults (Crowell, 1974). Two of these faults that extend into the area of the Los Angeles basin are the Newport-Inglewood fault zone and the Whittier fault, a northwesterly extension of the Elsinore fault. Basement rocks of the Peninsular Ranges form the core of the Santa Ana Mountains and are found in the subsurface a short distance north of the Whittier fault (Figure 4).

Basement rock types in the three provinces have many similarities. As found in outcrops or wells in the Los Angeles region, these rocks generally do not provide unambiguous evidence of fault displacements. Basement rocks of the Santa Ana Mountains and the Santa Monica Mountains and basement rock types found in wells in the northern and northeastern Los Angeles basin, all were formed from similar suites of sedimentary, volcanic, and intrusive rocks (Sorensen, 1984, 1985). All apparently were part of the same late Mesozoic calc-alkaline island-arc system and are similar to the prebatholithic terranes of the western Sierra Nevada. There and in the western Transverse Ranges they commonly are metamorphosed to slate and locally to schist rank. In the Santa Ana Mountains, these rocks exhibit a lesser degree of metamorphism: slate is noted only locally, and schist has not been reported.

A previously accepted distinction between western (Catalina) and eastern bedrock complexes (Schoellhamer and Woodford, 1951; Yerkes et al., 1965) has become blurred. Only the blueschist of the Catalina terrane is unique to a single region. Greenschists of that terrane are similar to the Pelona Schist, exposed near the northern edge of the central Transverse Ranges (Figure 4). Along the southern margin of the western Transverse Ranges, basement rocks have undergone low-pressure greenschist- to amphibolite-facies metamorphism and are further recrystallized in contact zones adjacent to Mesozoic plutons (Sorensen, 1984, 1985). Amphibolites and saussurite gabbros found in the western Transverse Ranges and beneath adjacent parts of the Los Angeles basin are also found in the Catalina terrane of the inner borderland.

Predecessor Basins

Sedimentary rocks deposited in the area before the development of the present Los Angeles basin in late Neogene time are "unrelated to the Los Angeles depositional basin. They are remnants of some basin, or basins, not geographically linked to the present basin." (Gardett, 1971). These rocks record a variety of earlier depositional and tectonic settings, which include

♦ the Late Cretaceous and Paleogene fore-arc basins that extended along the entire western edge of the North American plate. Figure 4 shows the known distribution of this sequence (designated as "Sedimentary") in the Los Angeles area. Its presence or absence beneath the deeper parts of the present Los Angeles basin has not yet been demonstrated.

♦ volcanic and sedimentary rocks (the Topanga Group of Figure 5) deposited during the early to middle Miocene rifting that opened the basins of the borderland province and rotated the blocks within the Transverse Ranges.

♦ regional subsidence during the late middle and late Miocene that facilitated the widespread deposition of biogenic sediments and clays of the Monterey and related formations.

Plate-Tectonic Framework

The Los Angeles basin has developed within the rapidly evolving San Andreas transform zone, which forms the present boundary between the Pacific and North American plates. The concepts of plate tectonics provide the basis for investigating the complex evolution of the Los Angeles region. The following brief review of the plate-tectonic model as applied in this region supplies a context for the evidence presented below that bears on the timing and style of deformation of individual structures within the basin.

Early studies had linked major right-lateral displacement on the San Andreas fault system to the opening of the Gulf of California (Hamilton, 1961), ascribed the Mesozoic batholiths of the western United States to subduction of crust and mantle (Gilluly, 1963), and related mid-Miocene rifting and volcanism in coastal southern California to collision with the East Pacific Rise (Yeats, 1968b). These ideas became part of Atwater's (1970) comprehensive plate-tectonic scenario that, with modifications (and some unresolved problems), has gained wide acceptance as the basic model for the Cenozoic tectonic evolution of western North America (Crowell, 1987). As Graham stated (1987): "... regional sedimentary-basin evolution [is] best viewed in the context of transform tectonics superposed on an older convergent-margin regime." The Atwater model relates this evolution to changes in the relative motion between the Pacific and North American plates. Figure 6 shows four "moments" in this process, as reconstructed by Engebretson et al. (1985).

For some 165 m.y., the opening of the Atlantic Ocean has driven the North American plate westward over the oceanic plates of the Pacific Ocean. California occupies a major segment of this dynamic plate boundary. During most of Mesozoic time, the boundary was formed by two successive *subduction zones* (Dickinson, 1981) that produced the granitic batholiths, capped by volcanic arcs, of the Sierra Nevada and Peninsular Ranges. In the central and northern Sierras, accreted terranes also record this convergence (Schweikert, 1981; Saleeby, 1981). A fore-arc basin (or basins) extended the length of California. Subduction of the northeasterly spreading Farallon plate beneath the North American plate (Figure 6: 56 Ma) continued throughout Paleogene time (Nilsen, 1987), reflected by Oligocene-Miocene arc volcanism in the Sierra Nevada (Graham, 1987). In the Los Angeles region, deposition in the remnant fore-

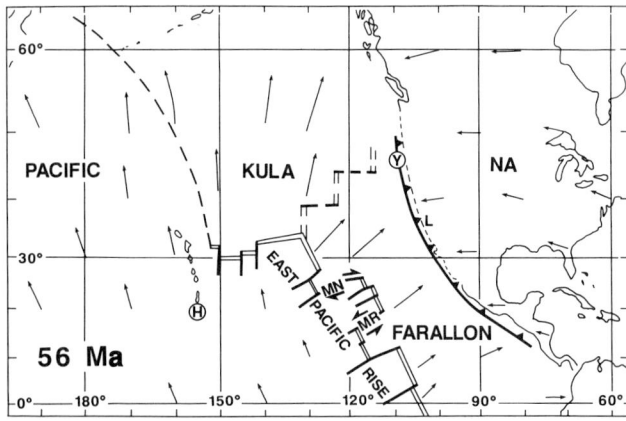
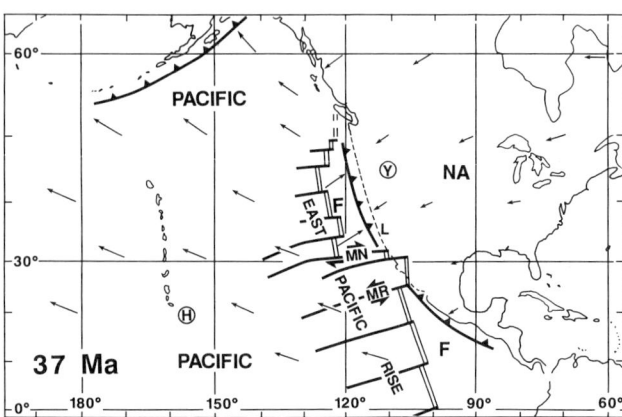
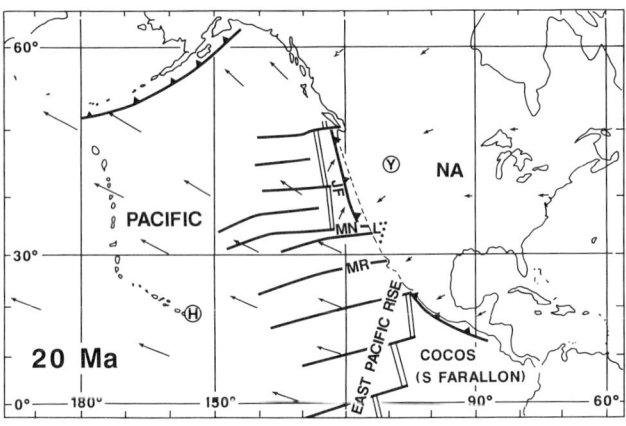
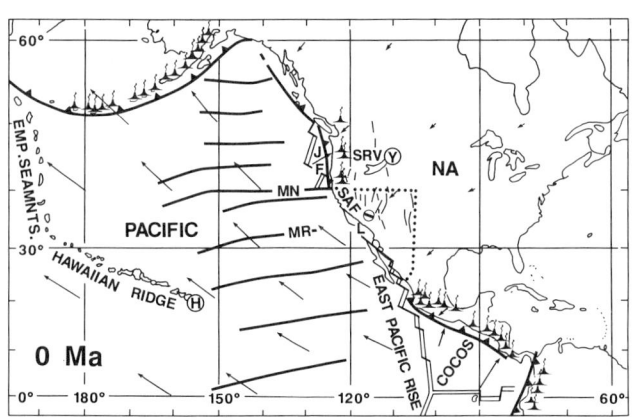

Figure 6. Cenozoic evolution of the Pacific–North American plate boundary (after Engebretson et al., 1985; local revisions and details from Dickinson, 1981). Plate positions are shown in fixed hotspot reference frame with present-day latitude-longitude grid. H and Y are the Hawaiian and Yellowstone hotspots. Vector arrows indicate 10 m.y. of motion. On the reconstructions for 56, 37, and 20 Ma, the western edge of North America has been modified to remove Basin and Range extension and the opening of the Gulf of California. L is location of Los Angeles basin. F = Farallon plate, JF = Juan de Fuca plate, MN = Mendocino fracture zone, MR = Murray fracture zone, NA = North American plate, SAF = San Andreas fault, SRV = Snake River Volcanics. (On the 37 Ma reconstruction note that motion arrows adjacent to MR apply to the easternmost segment of that fracture zone.) Dotted line on 20 Ma and 0 Ma maps shows limits of "no-slab region" (Dickinson, 1981) where subduction is no longer active. Also shown on the 0 Ma map are active arc-related volcanoes and major Holocene extensional faults. Vector of Basin and Range extension (Wernicke et al., 1988) is shown by the circled bar on the 0 Ma map.

arc basin continued through most of Eocene time. Details of basinal configurations and depositional patterns during the Late Cretaceous and Paleogene depend on specific assumptions about lateral faulting and block rotation. These assumptions are still being tested and refined through the work of numerous California geologists.

By the beginning of the Oligocene (Figure 6: 37 Ma), the North American plate had encroached upon the flank of the East Pacific Rise and had overridden the rise south of the Mendocino fracture zone. Regional emergence of coastal central and southern California, which resulted in deposition of the nonmarine Sespe Formation, has been ascribed to impingement of the continental margin onto the buoyant, young lithosphere of the rise (Dickinson et al., 1987; Nilsen, 1987). A global eustatic cycle of lower sea level (Ingle et al., 1976; Howell et al., 1987) may have accentuated this regional emergence. Motions of the two adjacent plates at this time were to the southwest (North American) and northeast (Farallon), directly opposed but not quite perpendicular to the plate boundary.

Another scenario for the Paleogene tectonic development of southern California involves the *accretion* of distinctive terranes onto the North American craton (Howell et al., 1987). Paleomagnetic data from Mesozoic sedimentary and plutonic rocks provide the basis for hypotheses that describe major northward translation of these terranes into their present positions. By this scenario, the east flank of the East Pacific Rise, moving northward with the Pacific plate, as attested by the Emperor Seamount chain (Figure 6: 56 Ma), engaged and dislocated parts of the continental margin of Central America. One cluster of terranes, the Santa Lucia-Orocopia allocthon, initially lay to the south of the Peninsular Ranges-Baja California volcanic arc; but during Late Cretaceous and early Tertiary time, the cluster moved rapidly northward and outboard from the Peninsular Ranges terrane. The Santa Lucia-Orocopia allocthon accreted to the North American craton in earliest Tertiary time (60-52 Ma) and now forms the central Coast Ranges and western Transverse Ranges.

The continuing northward motion of the oceanic plates then engaged the bypassed arc—a block some 850 mi (1400 km) in length that includes Baja California, the Peninsular Ranges, borderland, and central Transverse Ranges. Northward translation of this Baja-borderland allochthon has been dated as occurring between 45 Ma and 34 Ma (Hagstrum et al., 1987). During this process the Catalina terrane was inserted between the Peninsular Ranges and the outer borderland, which had been contiguous during the deposition of their blanket of Upper Cretaceous and lower Tertiary sedimentary rocks. Their separation is thought to have involved late Paleogene uplift of the Catalina terrane below a tensional detachment surface (Yeats, 1973; Howell et al., 1987). Although a more recent (15-5 Ma) time for the major northward movement of Baja California has been postulated (Morris et al., 1986), the preponderance of evidence indicates that since early Miocene time such movement has been limited to that associated with the late Neogene opening of the Gulf of California (Hagstrum et al., 1987). These data do not, however, preclude greater northward translations in the borderland province through Neogene time.

Though the accretion model is not incompatible with the model that emphasizes a mid-Tertiary collision with the East Pacific Rise, the two models have yet to be combined satisfactorily. Timing and position of the rise collision have been estimated through reconstruction of plate motions (Figure 6; Engebretsen et al., 1985). Major features of southern California geology, including the regional Oligocene (Sespe) emergence, block rotations within the Transverse Ranges, and mid-Miocene rifting and volcanism are best explained by overriding of the East Pacific Rise during this interval. The terrane accretion model provides a viable explanation for the pronounced stratigraphic contrasts, once ascribed to independent vertical oscillations of adjacent blocks ("elevator tectonics"), between adjacent parts of onshore and offshore southern California and within the central Coast Ranges.

By early Miocene time (Figure 6: 20 Ma), a significant segment of the Pacific plate was in direct contact with the North American plate south of the Los Angeles region. This segment was moving northwestward away from the East Pacific Rise and was nearly parallel with the plate boundary. In this area, *transform displacement* along the plate margin had already been initiated during the process of terrane accretion. To the north and south of this transform segment, remnants of the Farallon plate were preserved as the Juan de Fuca and Cocos plates, and their continued subduction formed the Cascade and Central American volcanic arcs. Adjacent to the transform segment, where subduction no longer was active, a triangular "no-slab" window began to develop beneath the continent (Dickinson, 1981).

Neogene evolution of this plate margin has been noted for the progressive transfer of parts of the North American plate, including terranes accreted during the Paleogene, to the northwest-moving Pacific plate. Initial right slip along the transform margin probably was localized along the boundary between oceanic crust and the accreted wedge of sedimentary rocks that formed the continental slope. Next involved may have been the Lucia Banks and San Gregorio-Hosgri faults in central California and faults of the borderland province off southern and Baja California (Howell et al., 1974; Vedder and Howell, 1976). Each sliver, once attached to the Pacific plate, has moved with it, and these offshore slivers should therefore be the farthest displaced. Continued eastward propagation of transform motion has created the present system of right-slip faults that is collectively termed the San Andreas transform zone (Atwater, 1970). During late Miocene and earliest Pliocene time, the eastern limit of the San Andreas transform zone in the region of the Transverse Ranges followed the San Gabriel fault (Crowell, 1979). It shifted eastward to its present location about 4 Ma, as the East Pacific Rise extended up the Gulf of California (Figure 6: 0 Ma). Faults within the San Andreas transform system may utilize relict zones of crustal weakness formed during earlier terrane accretion.

The magnitude of the total displacement across the transform zone is suggested by the northward relative movement of the Mendocino fracture zone. Since 37 Ma, it has moved from about lat. 30° to 40°N, a distance of some 700 mi (1100 km). This is about twice the apparent northward offset of the Salinian block and four times the right slip ascribed to the southern San Andreas fault (Crowell, 1979). Additional major displacement must have occurred on offshore segments of the San Andreas transform zone, including about 90 mi (150 km) on the San Gregorio-Hosgri fault (Graham and Dickinson, 1978; Clark et al., 1984).

Today in the region of San Francisco, the San Andreas transform zone extends from the Farallon escarpment eastward across the Coast Ranges some 80 mi (130 km). In southern California the zone extends from the Patton escarpment east to the southern San Andreas fault and is as much as 250 mi (400 km) in width. Deep crustal studies in the region of the Transverse Ranges show a velocity anomaly that extends east beyond the San Andreas fault (Hadley and Kanamori, 1977) and suggests that, in the mantle, the eastern limit of the transform zone lies beneath the central Mojave Desert. This and other studies lend support to the concept of crustal blocks ("flakes") carried by horizontal flow in the mantle and partially decoupled from the mantle by a ductile region in the lower crust (Gilluly, 1963; Yeats, 1968a, 1981; Oxburgh, 1972; Webb and Kanamori, 1985; Sylvester, 1988).

By this concept, the Pacific plate is being carried northwest by mantle flow driven by sinking of the cold oceanic slab into the Aleutian subduction zone and relieved by the opening of the East Pacific Rise. This mantle flow extends eastward beneath the continental crust and the zone of rupture in the brittle upper portion of the crust (Zandt and Furlong, 1982). A wide zone of viscous drag in the ductile zone adjacent to the plate boundary would account for continued slip on faults west of the most active crustal fault(s). This model of horizontal and lateral decoupling within a dynamic lithosphere is basic to the tectonic analysis contained in the second half of this paper.

Neogene *extension of the Basin and Range province* is another regional process of significance to the Los Angeles region. As shown in Figure 6 (0 Ma map), major Basin and Range faulting coincides approximately with the "no-slab" window (Dickinson, 1981) that extends east to the Rio Grande rift and north to include most of the Wasatch front. Dickinson suggests a genetic relationship between the "no-slab" region and the thermal uplift and stretching of the lithosphere in the Basin and Range province; this is but one of many alternative explanations to be advanced in recent years (Mayer, 1986). Extension of the Basin and Range province north of the Garlock fault has moved the Sierra Nevada westward (Eaton, 1932; Wright, 1976) an estimated 120-190 mi (190-300 km) (Wernicke et al., 1988), as shown by the vector bar on Figure 6 (0 Ma map). The deep crustal keel of the Sierra Nevada has impeded and deflected mantle flow along the eastern edge of the Pacific plate, resulting in the "big bend" segment of the San Andreas fault and in late Neogene uplift of the Transverse Ranges.

Clockwise *rotation* of crustal blocks within the Transverse Ranges and parts of the borderland (Luyendyk et al., 1980) occurred mostly during middle Miocene time. Such rotations, first proposed to explain the anomalous trends of Jurassic metamorphic textures in the Transverse Ranges (Jones et al., 1976), have been supported by extensive paleomagnetic measurements in middle Miocene volcanic rocks and dolomites throughout southern California (Hornafius et al., 1986; Luyendyk and Hornafius, 1987). The arrows in Figure 5 show declinations measured at localities adjacent to the Los Angeles basin. The second major section of this paper on tectonic evolution deals at length with the postulated role of block rotations in the Los Angeles region.

The abundance of three-dimensional stratigraphic and structural data from the intensely drilled Los Angeles basin affords the opportunity to measure the duration and intensity of each phase of Neogene tectonism—thus providing the fourth dimension of geologic time—and to ascertain with unusual precision the times when changing plate motions affected this part of the plate boundary. The following discussion of structural geology emphasizes the evidence so far available. The concluding section on tectonic evolution will attempt to relate this evidence to interactions between plates and rises during Neogene time.

STRUCTURAL GEOLOGY

Previous investigators have noted that "the [Los Angeles] basin began to assume its present form in the late Miocene" (Lamar, 1961) and that "the present form and structural relief of the basin were established chiefly during a phase of accelerated subsidence and deposition that was initiated in the late Miocene and continued without significant interruption into the early Pleistocene . . . The maximum rate of subsidence was attained in early Pliocene time" (Yerkes, 1972).

In this paper, the term *Los Angeles basin* is applied in a structural sense to the basin in its *present form* as described by Yerkes (1972) and refers to the area delimited by the Santa Monica and Whittier faults and the Palos Verdes uplift, in which a thick sequence of deep-water sediments accumulated during latest Miocene and Pliocene time.

Peripheral Basins

Transitional phases in the development of the present-day Los Angeles basin are recorded by Mohnian deep-sea fan deposits (Redin, this volume) in the areas immediately to the north (Figure 3): the Santa Monica Mountains, San Fernando Valley, San Jose Hills, Puente Hills, and San Gabriel Valley. The latter area has served as the primary pathway of sediment supply into the Los Angeles basin for the past seven million years. The shoreline regressed southward across the San Gabriel Valley in early Pliocene time and up to 6000 ft (1800 m) of nonmarine sediments have subsequently been deposited there.

In addition to the San Fernando and San Gabriel valleys, Figure 3 shows several other basinal areas peripheral to the Los Angeles basin. To the northeast, the San Bernardino Valley is separated from the Puente Hills by a dual zone of old and new faults. Cross sections by Durham and Yerkes (1964) show that a few miles northeast of the surficial Chino fault, marine Mohnian beds thin eastward onto basement (Figure 4) across an older hingeline (herein called the *Chino Hingeline fault*); farther to the northeast, a veneer of nonmarine Holocene overlies granitic basement. The Chino Hingeline fault is a prolongation of the N40°W strike of the Elsinore fault and may have shared an early Neogene zone of weakness along which the present Elsinore fault developed. Two small oil fields (see Biddle, *The Los Angeles Basin—An Overview*, this volume, his Figures 4 and 7) provide details on these relationships.

The *Mahala oil field*, discovered in 1921, will yield an ultimate recovery of 4 million barrels of oil (MBO) from three separate and nonoverlapping pools in Mohnian sands. One pool produces from east-dipping sands in the footwall of the Chino fault, which there dips 56° to the southwest (Michelin, 1958) and strikes N40°W. A long, narrow anticline in the hanging wall of the fault has flanks dipping about 35° to the northeast and 25° to the southwest (Durham and Yerkes, 1964). A second pool produces from a lobe of lower Mohnian sand that pinches out up the southwest flank of this fold (Castro, 1975). The field's early production came from a third pool in a small anticline on the downthrown south side of an east-trending normal fault.

The *Prado-Corona oil field*, discovered in 1966, will yield less than 2 MBO from Mohnian sands in narrow faulted anticlines along several reverse faults east of the Chino fault and parallel with it. The Chino fault and parallel faults and folds apparently formed by the compression of the wedge of Paleogene and middle and upper Miocene sedimentary rocks that pinches out against the buttress of shallow granitic basement several miles to the northeast across the earlier Chino Hingeline fault.

To the south, the *Capistrano syncline* (Figure 3) includes up to 3700 ft (1100 m) or more of middle and late Miocene deep-water beds overlying San Onofre breccia (Figure 5) and Paleogene and Cretaceous strata (Figure 4) (Bishop and Davis, 1984). Two subcommercial oil fields have produced minor amounts of oil from Cretaceous sands in traps along the Cristianitos fault, a north-trending normal fault that forms the eastern edge of the Capistrano syncline. The onshore embayment is separated from the offshore Oceanside basin by a faulted uplift believed to follow a southeastern extension of the Newport-Inglewood fault zone.

The offshore Santa Monica and San Pedro basins (Figure 3) are separated from the Los Angeles basin by the Palos Verdes schist ridge (Figure 4), thinly covered by Neogene sedimentary rocks. Neither of these basins has been tested by drilling, and conjecture about their geologic histories and stratigraphic content is beyond the scope of this paper.

Boundary Structures

Figure 7 is an index map showing the major structural features and oil fields of the Los Angeles basin and regional cross sections that illustrate these features. Seven cross sections, AA' through GG', are included in Figure 8[1]. The basin's primary southwestern boundary, the Palos Verdes fault and uplift, is shown on sections AA', DD', EE', FF', and GG'. The latter three sections also show the basin's primary northeastern boundary, the Whittier fault.

Figures 9 and 10 are structural contour maps of the two most recognizable stratigraphic horizons within the Los Angeles basin: the base of the Repetto Formation (about 4.5 Ma) and the base of the Mohnian Stage (about 14 Ma). These regional structure maps and the cross sections in Figure 8 illustrate the general pattern of the boundary and internal structures of the basin. Figures 4, 5, 8, 9, and 10 integrate the details presented in the following review of the structural geology of the Los Angeles basin and aid in defining the major basement blocks and other structural features whose evolution is described in subsequent sections. The data used to construct these maps and sections derived from numerous published sources cited in the list of references and from proprietary studies by the author from 1961 to 1972.

---

[1] The construction of balanced cross sections through regions as complex as southern California requires important simplifying assumptions. The method may not adequately accommodate elements such as syntectonic deposition, time-transgressive marker horizons, or the intersection of cross sections by major strike-slip faults with significant (or unknown) lateral displacements. When initially constructed in the 1960s, the regional cross sections of Figure 8 were "restored" according to methods then in use (Dahlstrom, 1969). The cross sections, though thoroughly revised for this paper, did not undergo further "balancing" nor were they altered by other current methods of constructing retrodeformable cross sections.

Whittier Fault

The Whittier fault (Figures 3, 7) is the best known of the three primary boundary structures. Its topographic expression is obvious, and it traverses a number of oil fields whose development began as early as 1884. The fault strikes S60°E for nearly 25 mi (40 km) and merges with the Elsinore fault in the canyon of the Santa Ana River. As documented by surface mapping and subsurface geology, the Whittier fault is a north-dipping reverse right-oblique fault (Ziony and Yerkes, 1985). In the Whittier area, the pattern of hypocenters of small earthquakes (Lamar, 1972) suggests a fault zone dipping 60° to 70°N to a depth of 4 to 6 mi (6 to 9 km).

Structural details of the Whittier fault zone are best known from the three major oil fields along its length. The *Whittier field* (Figure 8, section EE'; Figure 11) will yield an estimated 56 MBO. Most of the oil has accumulated in Repetto, Delmontian, and Mohnian sands in a southwest-dipping homocline in the footwall of the Whittier fault. Updip closure is provided by faulting, tar seals, and sand pinch-outs (Lindblom, 1975). Surface oil seeps led to the discovery in 1894 of these accumulations and others within the fault zone. The 184 anticline, concealed beneath the shallow homocline, was not discovered until 1964, 70 years after field development began. This fold has trapped oil in upper Miocene (Mohnian and Delmontian) sands. Lindblom (1975) stated that "the anticline is sharply folded and trends N70°W with the axis having a southerly migration at depth. The fold dies out quite rapidly above the upper Miocene (Delmontian) beds and reflects only a structural terrace by the Fourth zone [lower Pliocene (Repetto)]." The fold parallels the Whittier fault and is evidence of compressional deformation during latest Miocene and early Pliocene time. Thinning and pinch-out of lower Pliocene sandstones provide further evidence for uplift along this zone during that interval. Later Pasadenan uplift along the Whittier fault has brought these beds to the surface; the area of the Whittier oil field is the only locality where typical Los Angeles basin producing sands of early Pliocene age can be seen in outcrop.

The *Sansinena oil field*, discovered in 1898 and rediscovered in 1945, will yield an estimated 61 MBO chiefly from Delmontian and Mohnian sands in complex faulted folds and fault traps in the footwall of the Whittier fault (Figure 8, Section FF'). At East Sansinena, entrapment is in the updip pinch-out of Delmontian and upper and middle Mohnian sands. Woodward (1958) noted that "marked unconformities in the Miocene indicate an early dating of folding and faulting along the Whittier fault zone... Folding and faulting... very probably started early in late Miocene time, possibly in middle Miocene, and continued at intermittent intervals into Pleistocene time."

The *Brea-Olinda oil field* (Figure 8, Section GG'; Figure 12) will ultimately produce about 439 MBO. The field is divided into three structural blocks by the Whittier fault and a major southerly branch, the Tonner ("frontal") fault. Most of the production at Brea-Olinda is from sands of late Miocene and early Pliocene age in the South Flank block—the homocline south of the Tonner ("frontal") fault. Trapping mechanisms include tar seals, faults, and updip sand pinch-outs (Davis, 1975). The Tonner fault is the major

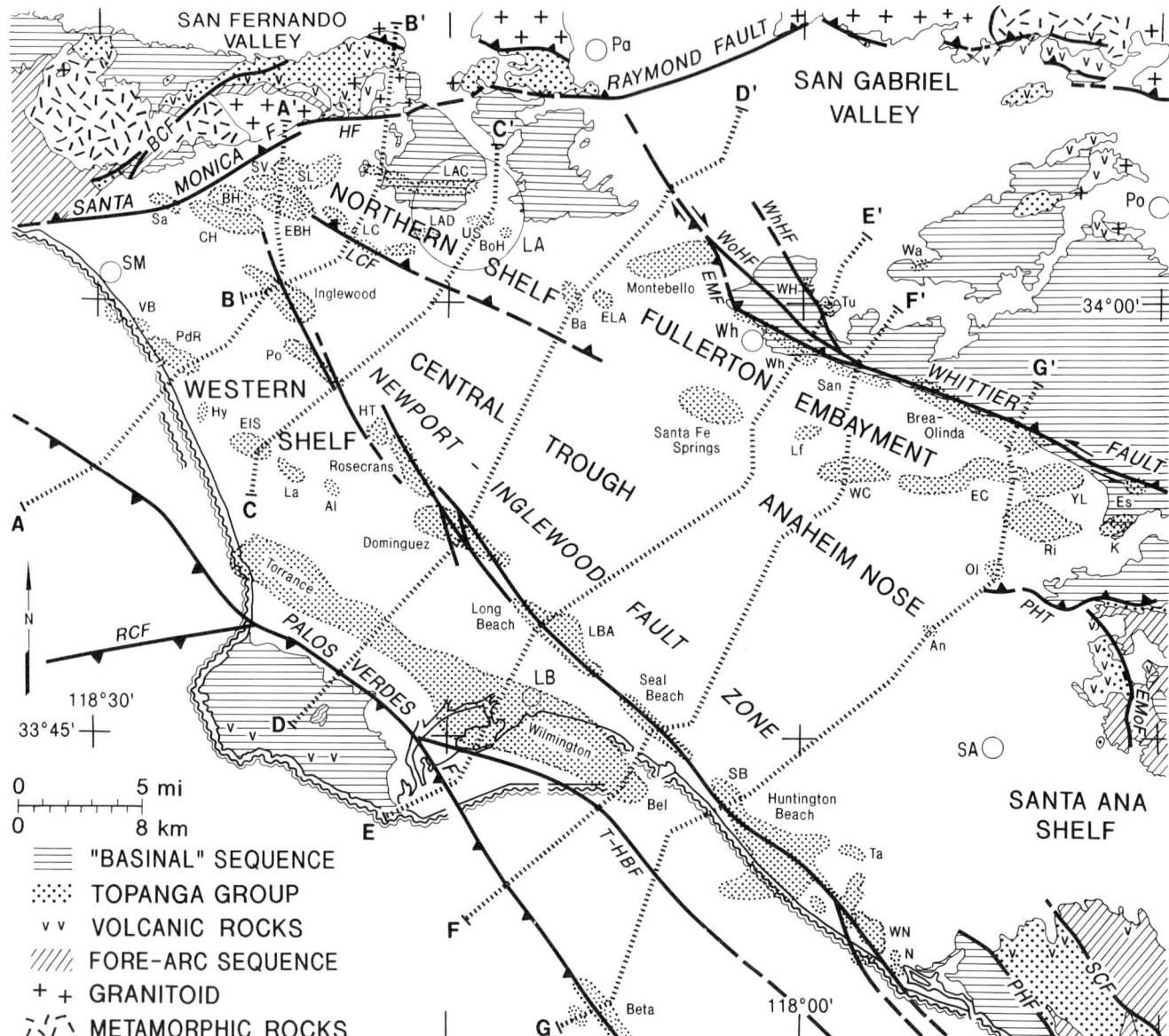

Figure 7. Index map, Los Angeles basin, showing major structural features, oil fields (stippled), and location of regional cross sections AA′ through GG′. Patterns show generalized surface geology (after Woodford et al., 1954, their plate I). "Basinal" sequence is deep-water Mohnian to Pliocene sedimentary rocks, fore-arc sequence is Upper Cretaceous to lower Miocene sedimentary rocks. Abbreviations as in Figures 3 and 4, plus EMoF = El Modeno fault, PHT = Peralta Hills thrust, WhHF = Whittier Heights fault, WoHF = Workman Hill fault. Oil-field names: Al = Alondra, An = Anaheim, Ba = Bandini, Bel = Belmont Offshore, BH = Beverly Hills, BoH = Boyle Heights, CH = Cheviot Hills, EBH = East Beverly Hills, EC = East Coyote, ELA = East Los Angeles, ElS = El Segundo, Es = Esperanza, HT = Howard Townsite, Hy = Hyperion, K = Kraemer, La = Lawndale, LAC = Los Angeles City, LAD = Los Angeles Downtown, LBA = Long Beach Airport, LC = Las Cienegas, Lf = Leffingwell, N = Newport, Ol = Olive, PdR = Playa del Rey, Po = Potrero, Ri = Richfield, Sa = Sawtelle, San = Sansinena, SB = Sunset Beach, SL = Salt Lake, SV = San Vicente, Ta = Talbert, Tu = Turnbull, US = Union Station, VB = Venice Beach, Wa = Walnut, WC = West Coyote, Wh = Whittier, WH = Whittier Heights, WN = West Newport, YL = Yorba Linda.

trapping fault, but it has been offset by a number of northeast-striking faults that divide the accumulation into separate pools.

Between the Tonner and Whittier faults, a zone of complex structure several thousand feet in width includes a wedge of Pliocene and Pleistocene beds (Durham and Yerkes, 1964) (Figure 12) infolded by lateral faulting. Within this Whittier fault block, minor production is obtained from mid-Mohnian (Soquel) sandstones (Scribner, 1958).

North of the Whittier fault, the Puente block yielded the first commercial production in the Los Angeles basin, in 1884. Drilling was stimulated by oil seeps in nearby Brea Canyon. Wells 500–1800 ft (150–550 m) in depth produced about 3 MBO from lower Mohnian (La Vida) shales and Luisian (Topanga) sandstones (Scribner, 1958). The La Vida and Topanga formations are intruded by diabase sills that are believed to be Mohnian in age (Durham and Yerkes, 1964). A number of wells in the Puente block have encountered greenschist basement at depths of about 3400–4900 ft (1040–1500 m). These are metavolcanic rocks (Sorensen, 1984, 1985) of a higher rank than those found in the Santa Ana or Santa Monica Mountains but are not comparable to the relatively high-pressure terranes of the Catalina or Pelona schists. They are, however, similar to the Santa Cruz Island Schist present on the southern edge of the western Transverse Ranges, some 100 mi (160 km) to the west.

Within the Whittier fault zone, 3.5 mi (2.2 km) southeast of Brea-Olinda, the small *Esperanza oil field* will yield 1.3 MBO. The accumulation is in mid-Mohnian sands in a narrow faulted anticline subparallel to the Whittier fault (Durham and Yerkes, 1964).

East of Esperanza, the Whittier fault forms the northern margin of the uplifted Santa Ana Mountains block and assumes a very different character. Vertical displacement is reversed, with the south side up. The fault is more nearly vertical and curves to an easterly strike for several miles as it approaches its junction with the Elsinore, Chino, and Chino Hingeline faults (Figure 4).

From the Esperanza field west, the *general characteristics* of the Whittier fault zone include

♦ a south- to southwest-dipping homocline in late Miocene to Quaternary strata. The east end of this homocline terminates in the southwest-plunging Yorba anticline.

♦ a complex footwall zone that includes significant anticlinal folds as well as large detached slivers of late Miocene and Pliocene strata.

♦ a hanging-wall block in which basement occurs at relatively shallow depths. The complex structure within and north of the Whittier fault zone has been enhanced by internal deformation within the thick La Vida and Yorba shale units. In addition, tectonic features are commonly obscured by slumping and related synsedimentary features that reflect movement down the basin slope during late Miocene and Pliocene time.

Up-to-the-north *vertical separation* of upper Miocene strata across the Whittier fault increases northwestward from about 2000 ft (600 m) near the Santa Ana River to a maximum of about 14,000 ft (4.3 km) in the Brea-Olinda oil field (Yerkes, 1972). Farther to the northwest, the separation decreases and is variable: 3500–5000 ft (1100–1500 m) at Sansinena (Woodward, 1958) where the separation is mostly flexural (Figure 8, Section FF'); 5500 ft (1700 m) at Whittier (Gourley, 1975); and perhaps only 3000 ft (900 m) at Rideout Heights, adjacent to the Whittier Narrows (Yerkes, 1972).

Lateral displacement along the Whittier fault is less easily estimated. Post-Paleocene right slip on the Elsinore-Whittier system has been estimated (Lamar, 1961) at about 19 mi (30 km), based on offset of the Paleocene upper Silverado shoreline. Using similar evidence, Sage (1973) has postulated approximately 25 mi (40 km) of right slip. These data are discussed in a later section of this paper (see Figure 36). Several (but not all) streams in the Brea Canyon area have been deflected 4000–5000 ft (1200–1500 m) in a right-lateral sense (Yerkes, 1972); but this is not definitive because differential erosion within the fault zone can mimic lateral fault movement. Near the Whittier oil field, upper Miocene lithofacies do not show any prominent right-lateral offset, though displacement of up to about 2500 ft (800 m) could be accommodated (Gourley, 1975).

Across the eastern trace of the Whittier fault, the eastern limits of thick Soquel (middle Mohnian) sandstones show an apparent right-lateral offset of about 5 mi (8 km), from the vicinity of the Esperanza field on the south side to the Scully Hill–Prado Dam area on the north side. South of the fault, the Soquel Member is more than 990 ft (300 m) thick in the Esperanza oil field (Gaede, 1959) and pinches out approximately 2 mi (3 km) to the east (Schoellhamer et al., 1981). North of the fault, the Soquel is 900 ft (275 m) thick near Scully Hill and thins eastward to a thickness of 100–200 ft (30–60 m) in the Prado-Corona and Mahala oil fields (Durham and Yerkes, 1964; California Division of Oil and Gas, 1974). This apparent offset may not represent true lateral displacement, however. The original bathymetric slope that controlled the sand edge may well have been nonlinear and related to structural features that developed semi-independently north and south of the Whittier fault.

Within the city of Whittier, the surface trace of the fault disappears beneath alluvium, but subsurface data from the Rideout Heights area of the Whittier field (Ingram, 1962) clearly show the fault to curve to an orientation of about N42°W (Figure 7). One and one-half mi (2.4 km) farther to the northwest at the eastern edge of the Montebello oil field, subsurface mapping (Stolz and Woodward, 1943; Reese, 1943b; see also Figure 35) has defined the *East Montebello fault*, striking N38°W. This fault has commonly been shown as the northwest continuation of the Whittier fault (Barbat, 1958; Yerkes et al., 1965) because it has similar right-oblique displacement with the northeast side up. Focal mechanism data from the 1987 Whittier Narrows earthquakes (Hauksson et al., 1988; Hauksson and Jones, 1989) confirmed the primary right-slip nature of the East Montebello fault. The largest aftershock ($M_L = 5.3$) was on a plane striking N30°W beneath the Alhambra high and aligned with the fault in the Montebello oil field. On the

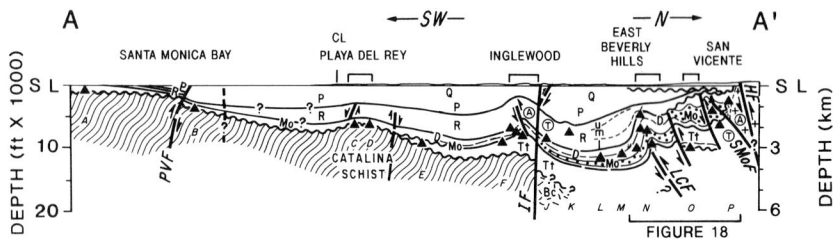

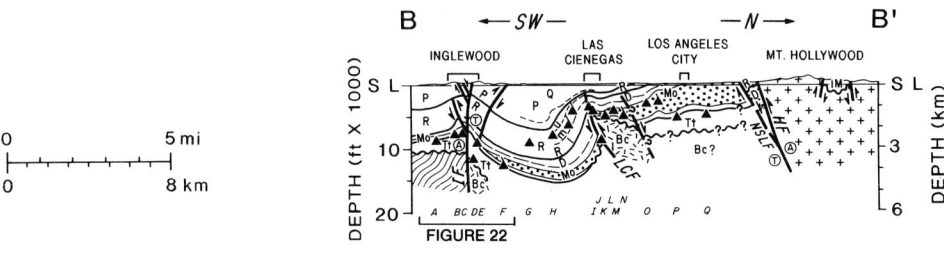

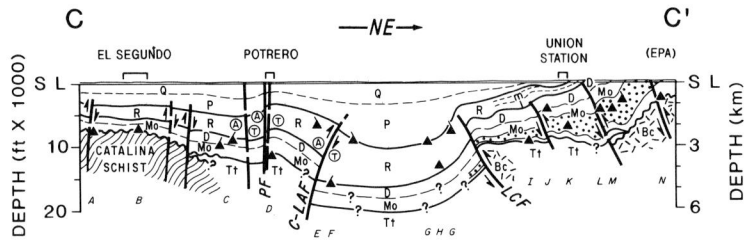

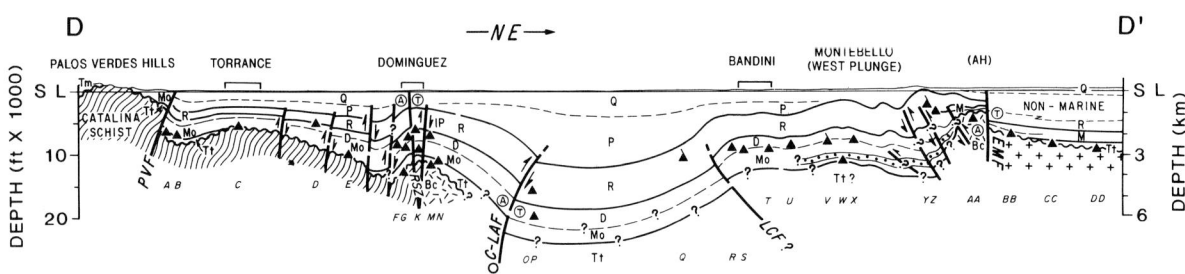

Figure 8. Regional cross sections AA' through GG', Los Angeles basin. Sections are aligned at Newport-Inglewood fault zone. Locations of sections are shown in Figure 7. Solid triangles (and letters below) indicate location and total depth of control wells listed in Appendix 2; "ears" on triangles indicate course of deviated wells. Arrows on faults show dip slip. Circled letters indicate lateral fault motion: A = away from viewer, T = toward viewer (thus, A/T is right slip, T/A is left slip). Open circles adjacent to Compton–Los Alamitos fault are projected hypocenters from Hauksson (1987).

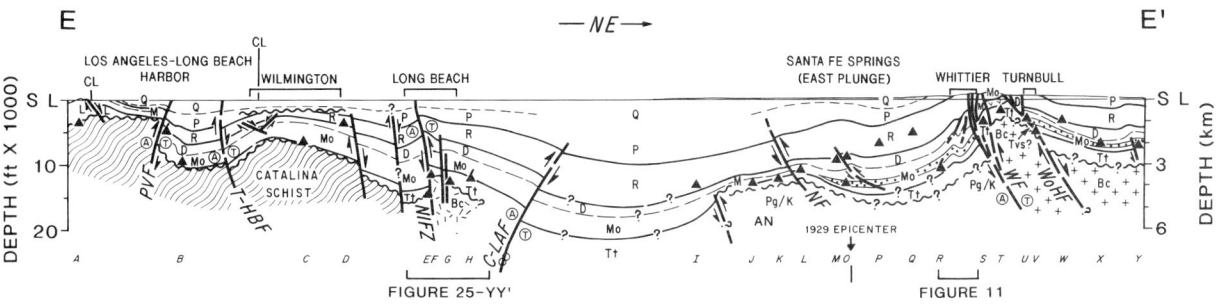

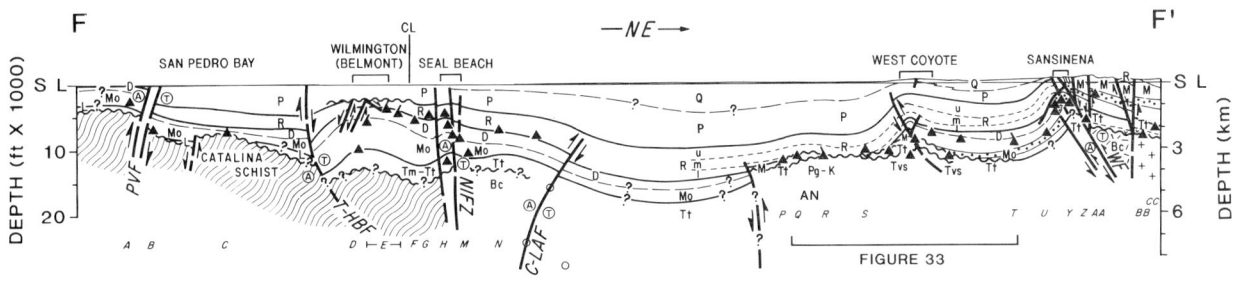

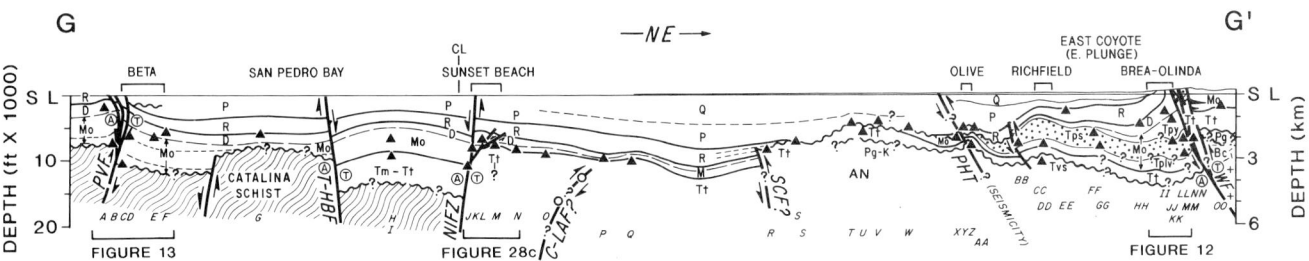

Figure 8 (continued). Abbreviations as in Figures 3, 4, and 7, plus: CHF = Cherry Hill fault, IF = Inglewood fault, NF = Norwalk fault, NSLF = North Salt Lake fault, PF = Potrero fault, RSZ = "Regional shear zone," SL = sea level, CL = coastline. Brackets above sections show limits of oil fields (labeled); brackets below sections indicate locations of detailed cross sections incorporated in the regional section. Light dashed lines are subordinate stratigraphic boundaries. Q = Quaternary, P = Pico Formation, R = Repetto Formation, D = Delmontian, Mo = Mohnian, M = undifferentiated Delmontian-Mohnian, Tm = Monterey Formation, L = Luisian, v = volcanics, Tt = Topanga Formation, Pg = Paleogene (locally includes lower Miocene), K = Cretaceous, Bc = undifferentiated metamorphic basement complex, and u, m, and l = upper, middle, and lower. Basement and volcanic rock patterns as in Figures 4 and 5. Stippled unit is middle Mohnian Soquel/"Massive" sandstones.

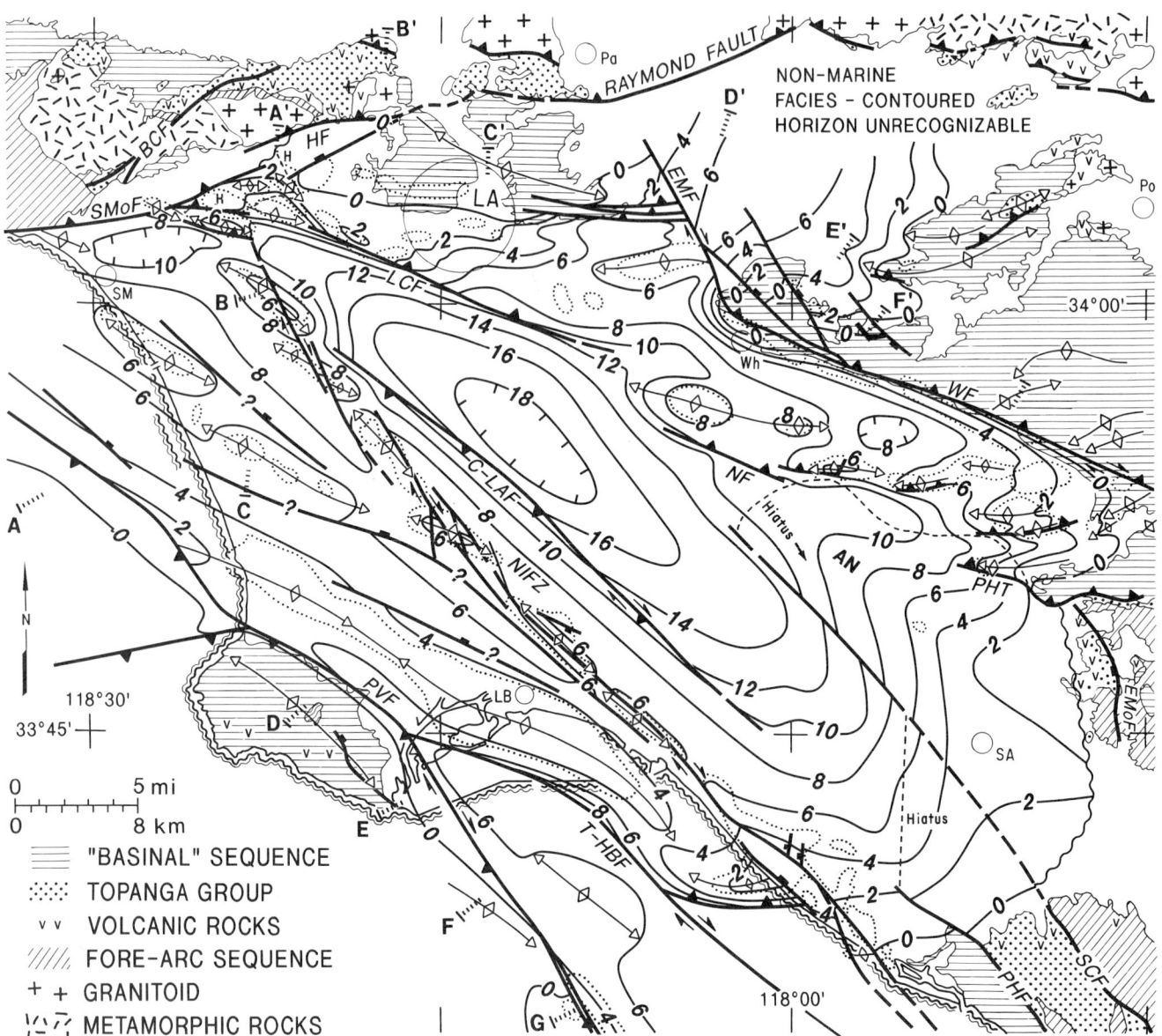

Figure 9. Structure-contour map, Los Angeles basin. Contours on "Base of Repetto," a lower Pliocene horizon (ca. 4.5 Ma). Contours in feet below sea level. Contour interval: 2000 ft (610 m). Well control (and interpolation) shown in Figure 8 (with interpolation) and in Figures 4 and 5 (wells that reached pre-Mohnian rocks). Dip direction on faults shown by barbs (thrust and reverse faults) and bars (normal faults). Faults are shown at footwall trace. Patterns and abbreviations as in Figures 3, 4, 7, and 8. Hiatus (or H) indicates area of significant stratigraphic hiatus (between upper Miocene and lower Pliocene beds) at contoured horizon. Wavy line = buried erosional edge of upper Miocene–lower Pliocene interval. Oil-field outlines are dotted.

Alhambra high, the fault has produced a vertical separation of about 2000 ft (600 m) on the top of the basement rocks. There the west side of the fault is up, opposite to the vertical separation in the Montebello oil field. This reversal of separation over a distance of only about 3 mi (5 km) seems to reflect right-lateral offset of the basement ridge (Figures 4, 5) that marks the northern boundary of the basin. These relationships suggest that the East Montebello fault might better be considered as a steeply dipping tear fault against

which the Whittier fault terminates beneath the alluvium of Whittier Narrows. Gravity and subsurface data indicate that the East Montebello fault extends northwest beneath the western edge of the San Gabriel Valley (Figure 8, Section DD'; Figures 9, 10).

This northwestern portion of the Whittier fault zone is complex and poorly understood. Two major splays diverge northwestward (Figure 7) from the Whittier fault in the vicinity of the Sansinena field. The southern splay, the

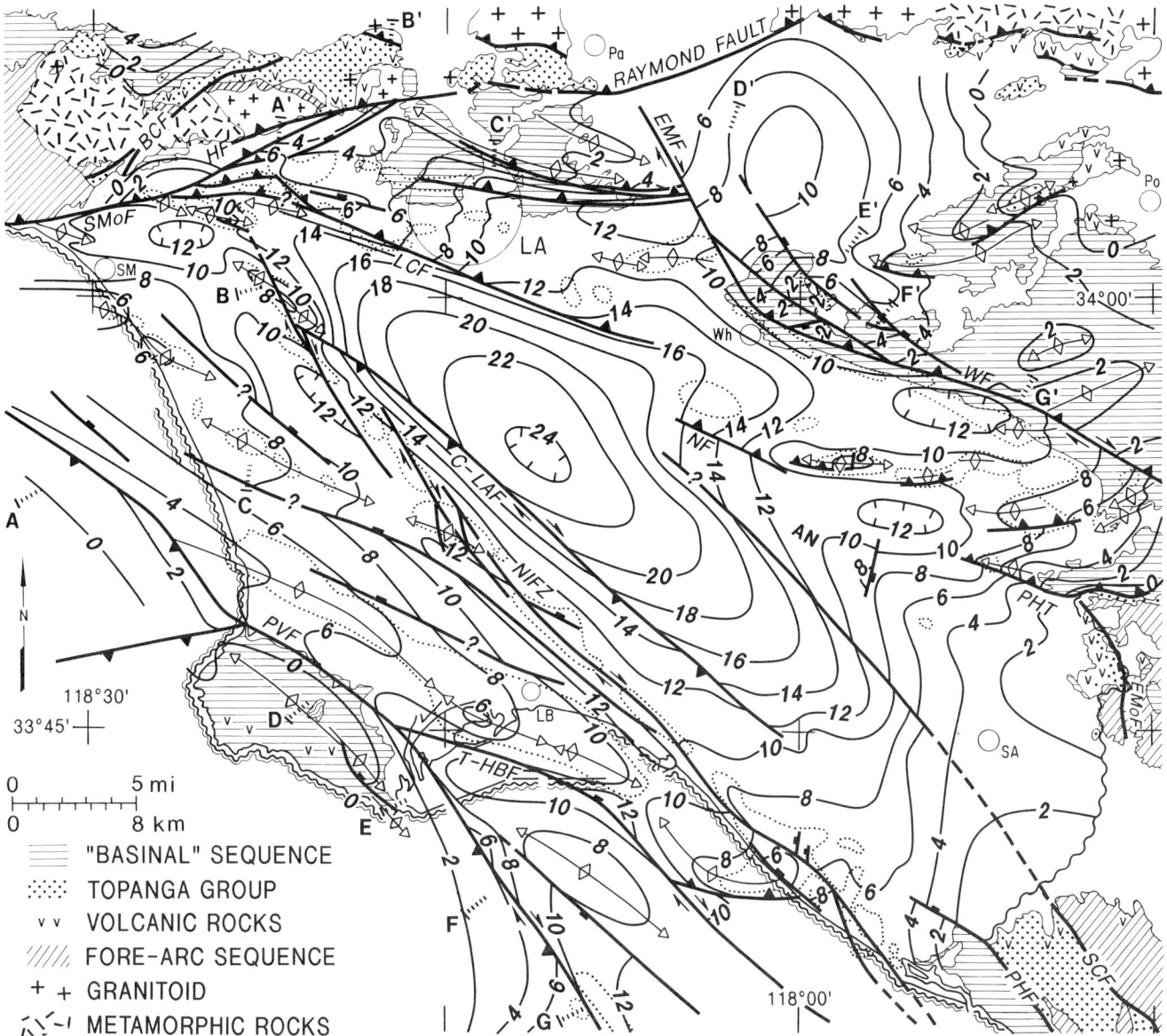

Figure 10. Structure-contour map, Los Angeles basin. Contours on "Base of Mohnian" (or younger Miocene) sediments. ("Base of Mohnian" is a late middle Miocene horizon, ca. 14 Ma). Contours in feet below sea level. Contour interval: 2000 ft (610 m). Well control, symbols, abbreviations, and patterns as in Figure 9.

Workman Hill fault, is clearly an older member of the Whittier fault zone. Near the northwestern end of its surface trace, it offsets the upper Miocene beds by 1000 ft (300 m) or more but is overlapped by unbroken mid-Pliocene strata (Daviess and Woodford, 1949). As interpreted herein (Figures 9, 10) the Whittier fault merges at depth with the Workman Hill fault. The northern splay, the Whittier Heights fault, is also an older fault that does not cut beds of middle Repetto age or younger (Gourley, 1975).

An uplifted wedge of basement rock underlies the northwesternmost Puente Hills and plunges beneath the Whittier Narrows (Figure 4; Figure 8, Section EE'). This wedge is bounded on the south by the Whittier fault and on the north by the Workman Hill–Whittier Heights fault (Yerkes, 1972). Significant thinning of mid-Mohnian Soquel sandstones across this basement uplift suggests that it is a relict of middle Miocene block faulting.

The *early history* of the Whittier-Elsinore fault system may involve a zone of weakness that existed at least since middle Miocene time, though evidence for this is not conclusive. Paleogene beds are present in the subsurface, 3–4 mi (5–6.5 km) south of the Whittier fault in the Richfield, Leffingwell, and East and West Coyote fields (Durham and Yerkes, 1964; Yerkes, 1972), but are absent north of the fault from Brea-Olinda westward, with the possible exception of the Turnbull oil field, north of the Whittier Heights fault

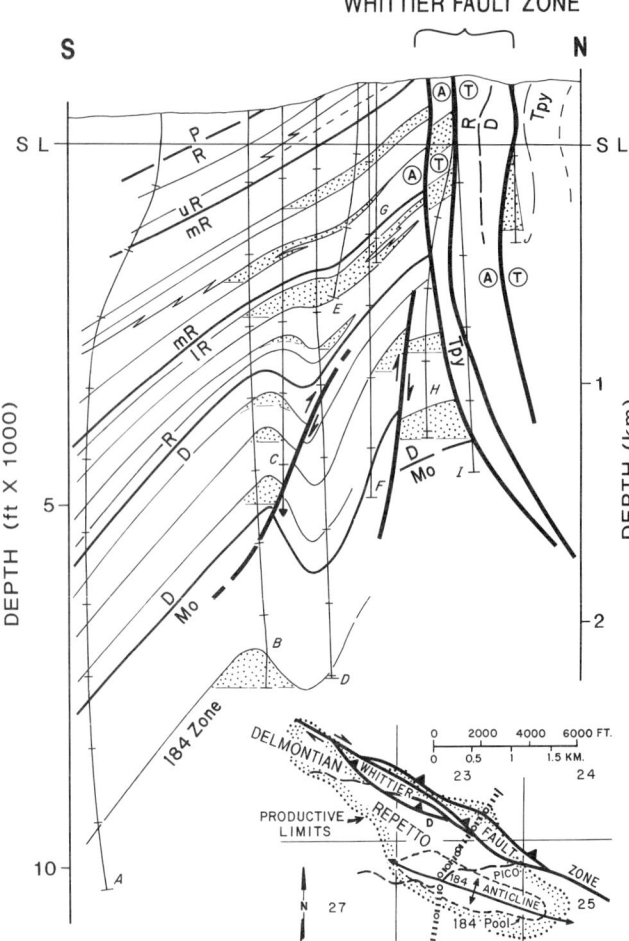

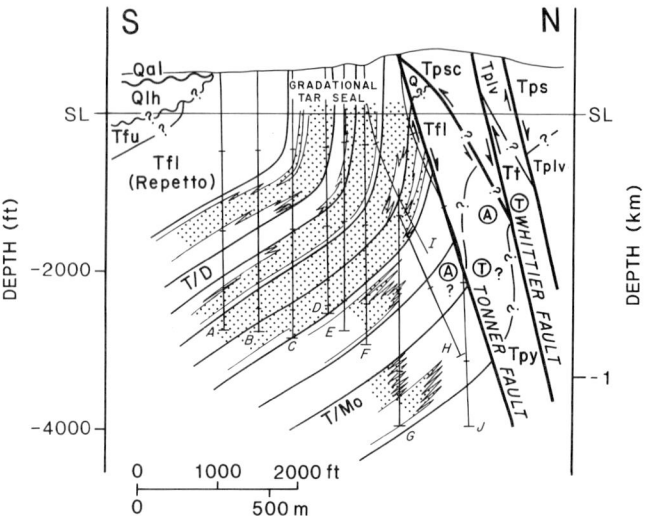

Figure 11. Cross section of the Whittier oil field. Compiled from Gaede (1964), Floyd and Maxwell (1964), California Division of Oil and Gas (1974), and Lindblom (1975). Oil-productive zones are shaded. Control wells (lettered) are listed in Appendix 2. Tpy is Yorba Shale Member of the Puente Formation. For other symbols, etc., see Figure 8. Inset map shows field limits, location of cross section, major structural features, and simplified surface geology (contacts between outcropping biostratigraphic units of Pliocene age shown by longer dashed line). Short dashes outline the 184 pool.

Figure 12. Cross section of the Brea Olinda oil field. Compiled from Davis (1975) and, north of the Whittier fault, from Durham and Yerkes (1964). Location of section shown in Figure 8, Section GG'. Oil-productive zones are shaded. Control wells (lettered) are listed in Appendix 2. T/Mo is top of Mohnian. For other symbols, etc., see Figure 8. Stratigraphic units shown are Quaternary alluvium (Qal), La Habra Formation (Qlh), upper and lower Fernando Formation (Tfu and Tfl), and members of the Puente Formation: Sycamore Canyon (Tpsc), Yorba (Tpy), Soquel (Tps), and La Vida (Tplv).

(Daviess and Woodford, 1949). Volcanic intrusives just north of the Whittier fault near Whittier and at Brea-Olinda suggest activity during latest middle Miocene time (early Mohnian, La Vida Shale). Folding of the 184 anticline at Whittier and footwall deformation at Sansinena attest to compressional deformation during latest Miocene and early Pliocene time, concurrent with uplift of the northern margin of the basin. Yeats (1987) suggested that the Whittier fault has acted as a reversing hingeline that earlier had separated the thick middle and late Miocene sequence of the Puente Hills from the Coyote Hills and the Anaheim nose, where this sequence is relatively thin. However, data from the Leffingwell oil field (Yerkes, 1972) show this thickening occurs nearly 4 mi (7 km) south of the Whittier fault (Figure 8, Sections EE' and FF'), on the northern edge of the mid-Miocene Anaheim nose.

The obvious surface features of the Whittier fault have resulted from *Quaternary deformation*. The late Pleistocene La Habra Formation has been uptilted locally all along the footwall homocline (Figure 12) to vertical dips, and at Brea-Olinda, it is strongly deformed within the fault zone. Deflected streams suggest late Quaternary right-lateral movement. It is difficult to estimate how much of the vertical separation across the fault occurred since Pliocene time, because the nearest Pliocene outcrops north of the fault are a mile or more from it across a broad faulted uplift.

Youngest displacement on the Whittier fault is dated as Holocene (past 10,000 years) southeast of the Brea-Olinda field, based on offset stratigraphy, and as late Quaternary (past 750,000 years) along its northwestern portion, based on offset stratigraphy and physiographic features (Ziony and Yerkes, 1985). The Whittier Narrows earthquakes of October, 1987 (Hauksson and Jones, 1989), were related only indirectly to the Whittier fault. The main shock occurred immediately east of the East Montebello fault and involved an east-striking thrust dipping 25° north. Projection of that thrust into the Whittier fault is difficult but not impossible (see Figure 8, Section EE'), though the 30° change in strike must be accounted for. A later section in this paper on Pasadenan deformation elaborates on these possible relationships.

Palos Verdes Uplift and Fault

The Palos Verdes uplift (Figure 3)—in this paper used in its broadest sense to include the five northwest-trending en echelon anticlinoria extending from the Santa Monica Bay southeast to Lasuen Knoll (Nardin and Henyey, 1978)—forms the southwestern edge of the Los Angeles basin. The Palos Verdes Hills constitute the central and highest of these anticlinoria. In its core, Catalina Schist is exposed at the surface (Figure 4) and is overlain by a varied sequence, up to 6800 ft (2100 m) in thickness, of middle Miocene to lower Pliocene sedimentary rocks and middle Miocene volcanic rocks (Figure 5) (Yerkes et al., 1965). Outcroppings of mid-Mohnian and younger units are notably thinner and finer grained than their equivalent intervals immediately to the northeast in the subsurface of the Los Angeles basin, thus documenting the continuing slow uplift of the Palos Verdes Hills beginning in late middle Miocene time. Clays and biogenic sediments of the Monterey facies here accumulated on a deeply submerged seaknoll at the same time as the coarser clastic Puente facies was being deposited on the basin floor in the Wilmington area, 5 mi (8 km) to the east.

The San Pedro Escarpment fault zone (Figure 3; Nardin and Henyey, 1978), along the southwest margin of the Palos Verdes uplift, was inferred initially (e.g., Emery, 1960) from the steep bathymetric scarp. Its presence is supported by gravity data (Beyer and Pisciotto, 1986). To the northwest, this fault converges with the San Pedro Basin fault zone (Figure 3) (Junger and Wagner, 1977), on which late Quaternary displacement has been mapped. The San Pedro Escarpment fault zone is not considered potentially active (Legg, 1986), although the steep slope and associated submarine slumping has precluded the mapping of shallow traces of the faults along much of this trend.

The Palos Verdes uplift is bounded on the northeast by the Palos Verdes fault. Onshore, that fault forms the northeastern boundary of the Palos Verdes Hills. Though not an obvious surface feature, it was mapped by Woodring et al. (1946) and penetrated by exploratory wells as early as 1925. Subsurface data indicate a southwest-dipping reverse fault. Vertical separation on schist basement is about 4000 ft (1200 m) (Figure 8, Section DD') in the vicinity of the Gaffey oil field, a subcommercial accumulation discovered in 1955 and localized in upper Repetto sands on the hanging wall of the fault.

The Palos Verdes fault zone, from the Anacapa fault southeast to the area of Lasuen Knoll, is about 60 mi (100 km) long. Its offshore extent has been defined by petroleum exploration, especially to the southeast in San Pedro Bay (Figure 8, Sections EE', FF', GG'). There the offshore *Beta oil field* (Section GG'; Figure 13) was discovered in 1976. It has an estimated ultimate recovery of 214 MBO. The trapping structure is a northwest-trending arch against the footwall of the Palos Verdes fault. Secondary faults with

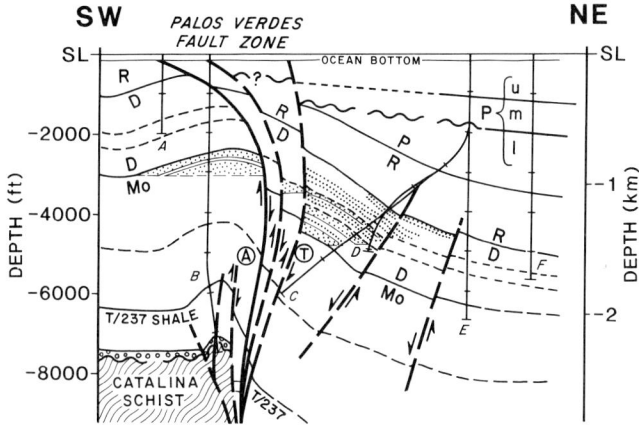

Figure 13. Cross section of the Beta oil field (from Chevron, 1980). Location of section shown in Figure 8, Section GG'. Oil-productive zones are shaded. Control wells (lettered) are listed in Appendix 2. T/237 is top of the 237 shale zone (lower? Mohnian). For other symbols, etc., see Figure 8.

small displacements divide the structure into several blocks (Chevron, 1980). Most of Beta's oil is found in a series of Delmontian sands in this footwall accumulation. Southwest of the Palos Verdes fault, a hanging-wall anticline contains significant reserves in Mohnian sands.

At Beta, the Palos Verdes fault is seen as a reverse fault with about 2000 ft (600 m) of vertical separation on basement. Reflection seismic data suggest that the shallow portion has been recurved past vertical by folding within the sedimentary section to form a typical "palm tree structure" (Sylvester, 1988). Figure 13 provides clues to the displacement history of the Palos Verdes fault at Beta. Deeper Mohnian strata are shown as thinning significantly onto the hanging-wall block, but the uppermost Mohnian and Delmontian intervals maintain a nearly constant thickness across the fault. Middle to late Pliocene deformation resulted in the progressive truncation of lower Pico beds towards the fault. Ocean-floor sampling in the Beta area has found Holocene and Pleistocene sediments northeast of the fault and Pliocene and Miocene strata southwest of it; high-resolution geophysical profiling suggests that the Palos Verdes fault reaches the sea floor.

Data on the Palos Verdes fault zone have been summarized by Fischer et al. (1987). They documented a complex zone of faulting, up to 1 mi (1.5 km) wide, with significant differences in displacement history along strike. Southeastward from San Pedro Bay, various fault segments displace late Pleistocene and Holocene deposits and locally form sea-floor scarps. Post-Repetto displacements on the central, onshore portion of the fault occurred chiefly during the past half million years. Uplift there is recorded by the thirteen major marine terraces of the Palos Verdes Hills, the highest being 1350 ft (411 m) above sea level. Measured uplift across the fault zone is 0.03–0.04 ft/yr (10–13 mm/yr), of which about 10% is believed to be tectonic (Clarke,

1987); the remainder presumably is the result of compaction of sediments. Focal mechanisms (Hauksson and Saldivar, 1986) on the Palos Verdes fault onshore and in San Pedro Bay indicate right-oblique reverse faulting.

Offshore to the north, the zone of Holocene deformation curves westward into the Redondo Canyon fault, a south-dipping reverse fault that strikes west-southwest across the southern part of the Santa Monica Bay (Yerkes et al., 1967; Nardin and Henyey, 1978). In the Santa Monica Bay, where petroleum exploration has been banned since the 1960s, the northwest extent of the Palos Verdes fault is not well defined. North of Redondo Canyon, it appears to extend northwest to the vicinity of the Anacapa (Dume) fault, but this portion of the fault is covered by up to 1000 ft (300 m) of nearly undisturbed sediments (Fischer et al., 1987) and apparently has had little or no post-Pliocene displacement. Recent small earthquakes near the Palos Verdes fault in Santa Monica Bay (Hauksson and Saldivar, 1989) with right-slip focal mechanisms imply that it is still active.

Data from focal mechanisms provide the only direct evidence for right slip on the Palos Verdes fault zone. From a more regional and long-term perspective, the en echelon anticlinoria of the Palos Verdes uplift are symptomatic of distributive dextral shear (Harding and Tuminas, 1988). It seems reasonable to conclude (Nardin and Henyey, 1978) that some of the dextral strike slip along the Palos Verdes fault is being absorbed by displacement on the Redondo Canyon fault and adjacent folding. To date, no quantitative evidence, such as offset stratigraphic features, has been presented to document the amount of right slip on the Palos Verdes fault system.

A significant fault, called the *THUMS-Huntington Beach fault* (McMurdie et al., 1973), branches southeasterly from the Palos Verdes fault zone in the San Pedro area (Figures 7 through 10). It parallels the southwest flank of the Wilmington anticline (Truex, 1974) and continues to the southeast, about 3 mi (5 km) west of the Newport-Inglewood fault zone and parallel with it (Clarke, 1987). High-resolution reflection profiling (Junger and Wagner, 1977) has traced a shallow fault, perhaps the THUMS-Huntington Beach fault, from the southeast plunge of Wilmington past the offshore Huntington Beach anticline. As mapped, it then curves easterly to converge with the Newport-Inglewood fault zone south of Newport Beach. A cluster of southeast-aligned epicenters (Legg, 1986, 1987) about 4 mi (6.5 km) off Newport Beach suggests that a western branch of the THUMS-Huntington Beach fault maintains a S45°E strike and is tectonically active. The block between the Palos Verdes fault zone and the THUMS-Huntington Beach fault has been called the Wilmington graben (Junger and Wagner, 1977).

Along the south flank of Offshore Huntington Beach, a north-dipping reverse fault extends eastward from the THUMS-Huntington Beach fault and is shown to affect Pleistocene beds (Henry, 1987) (see also Figures 27 and 28, Section dd′). This fault shows a rough symmetry with the Redondo Canyon fault about 22 mi (35 km) to the northwest. The THUMS-Huntington Beach fault and the central segment of the Palos Verdes fault appear to form a continuous, potentially active zone trending approximately N55°W.

Between the THUMS-Huntington Beach and Palos Verdes faults, the structural features shown on Figures 8, 9, and 10 have been generalized on the basis of very limited published data (Junger and Wagner, 1977; Wallis, 1981; Henry, 1987, his plate 2; Harding and Tuminas, 1988, their figure 11; Redin, this volume, his Figures 8, 11, and 14).

Santa Monica Fault System

The Santa Monica fault system, as used herein, refers to the zone of faults that extends from Santa Monica Bay eastward to the Los Angeles River and forms the northern boundary of the Los Angeles basin. It is part of an essentially continuous series of faults (Figure 3) that define the southern edge of the central Transverse Ranges and that include the Anacapa (Dume) and Malibu Coast faults to the west and the Raymond, Sierra Madre, and Cucamonga faults to the east. Major elements of the Santa Monica fault system include the Santa Monica, Potrero Canyon, and Hollywood faults (Figure 14). Figure 15 includes four cross sections across the Santa Monica fault system from Beverly Hills west to Santa Monica. Most of the main structural features in the northwestern Los Angeles basin have been described in earlier interpretations (Gardett, 1971; Erickson and Spaulding, 1975; Lang and Dreesen, 1975); it is expected that the present interpretation will be modified significantly as further data become available.

A major fault zone along the southern edge of the Santa Monica Mountains has long been inferred and was shown on the first fault map of California (Lawson et al., 1908). Direct surface evidence of faulting, however, is found only in the coastal bluffs at Pacific Palisades (McGill et al., 1987; Figure 15, Section KK′) where the Potrero Canyon fault is well exposed. Fifteen miles to the east, near the Los Angeles River, faulting has been inferred on the basis of an abrupt contact between granitic basement and upper Miocene (Mohnian) sedimentary rocks (Lamar, 1970; Dibblee, 1982a). Elsewhere the surface trace is obscured by extensive landslides along the north shore of Santa Monica Bay and by a thick alluvial apron farther east. Most of the data on the Santa Monica fault has been provided by exploratory and development drilling within the city of Los Angeles. The fault zone and its potential hazards are not well understood in the cities of Santa Monica and Beverly Hills where drilling has been prohibited for the past 50 years.

Prior to the advent of modern urban oil exploration in the 1950s, Hoots (1931) described the Hollywood fault on the basis of physiography and the truncation of the southern limb of the Santa Monica Mountains anticline. However, general acceptance of a major fault zone extending west to Santa Monica Bay did not come until the mid-1950s (Barbat, 1958) with the exploitation of deep Miocene zones in the Beverly Hills oil field in the footwall of the fault. At that time, evidence for a major north-dipping fault zone included (Lamar, 1961) the abrupt southwest-trending contact between granitic rock and Mohnian strata in the hills east of Hollywood; the separation of lower Mohnian beds across a major north-dipping fault on the north edge of the Beverly Hills oil field (Figure 15, Section HH′, wells B and C); and a steep gravity gradient parallel with the south slope of the Santa Monica Mountains that

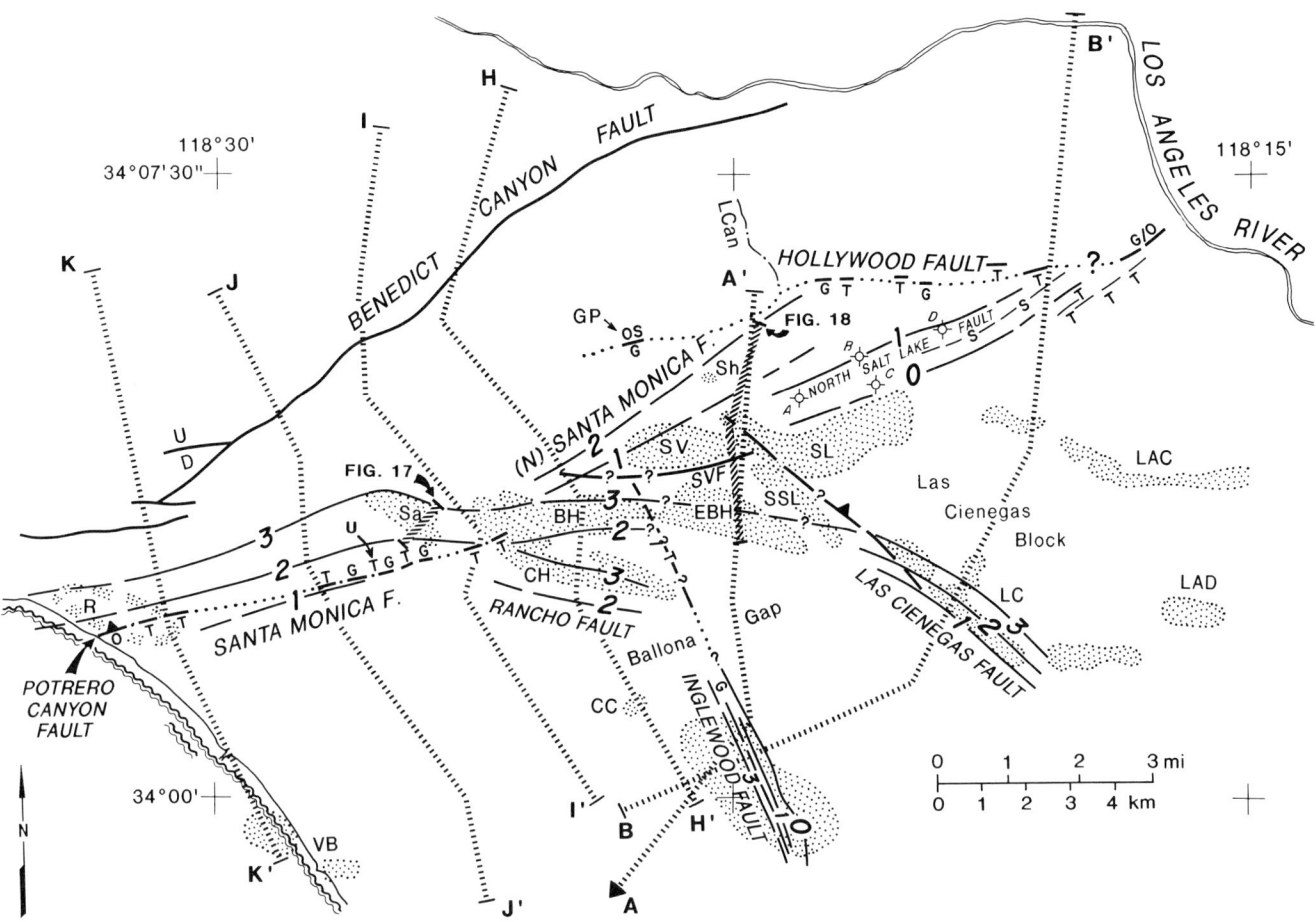

Figure 14. Map showing major faults in the northwestern Los Angeles basin. Fault-plane contours on the Inglewood, Las Cienegas, North Salt Lake, Rancho, and Santa Monica faults are based on subsurface well data. Contour interval: 1 km. Evidence of the surface traces of the Hollywood and Potrero Canyon faults and the northern extension of the Inglewood fault includes outcrop data (O), topography (T), and geotechnical data (G) from trenches and borings, plus soil contacts (S) and oil seepages (OS) (see text for sources). For other symbols, etc., see Figure 9.

Cross sections AA' and BB' are in Figure 8; cross sections HH' to KK' are in Figure 15. Also shown are locations of four core holes (listed in Appendix 2) that define the North Salt Lake fault and cross sections through the Sawtelle oil field (Figure 17), the East Beverly Hills, and San Vicente oil fields (Figure 18). Oil fields (stippled) are Beverly Hills (BH), Cheviot Hills (CH), Culver City (CC), East Beverly Hills (EBH), Inglewood (I), Las Cienegas (LC), Los Angeles City (LAC), Los Angeles Downtown (LAD), Salt Lake (SL), San Vicente (SV), Sawtelle (Sa), Sherman (Sh), South Salt Lake (SSL), Venice Beach (VB), and the undeveloped Riviera (R) discovery. GP is Greystone Park, LCan is Laurel Canyon, U is University High School.

led McCulloh (1960) to postulate a zone involving two parallel major faults. Four miles west of Beverly Hills, the existence of a major subsurface fault was confirmed in 1960 by the Mobil (now Oxy U.S.A.) Brentwood No. 1 (California Division of Oil and Gas, 1964) (well A on Section JJ', Figure 15), but this evidence was not widely known.

It is useful to review the postulated *Neogene tectonic evolution* of the northern Los Angeles basin (Figure 16) prior to examining the characteristics of the individual faults. The key element in deciphering this complex structural zone was noted early by Barbat (1958): "What had been strike-slip movement in a left-lateral sense in mid-Miocene shifted to dip-slip movement by mid-Pleistocene time." Evidence for this changing pattern of deformation is presented in this and subsequent sections of this paper.

The southern edge of the Santa Monica Mountains was defined by major faulting during the middle Miocene episode of block faulting and rotation. Movement along that zone (Figure 16A) was mostly left-lateral until the end of Miocene time (Yeats, 1968a; Campbell and Yerkes, 1976), although significant uplift of the Santa Monica Mountains began midway through the late Miocene at about 7 Ma

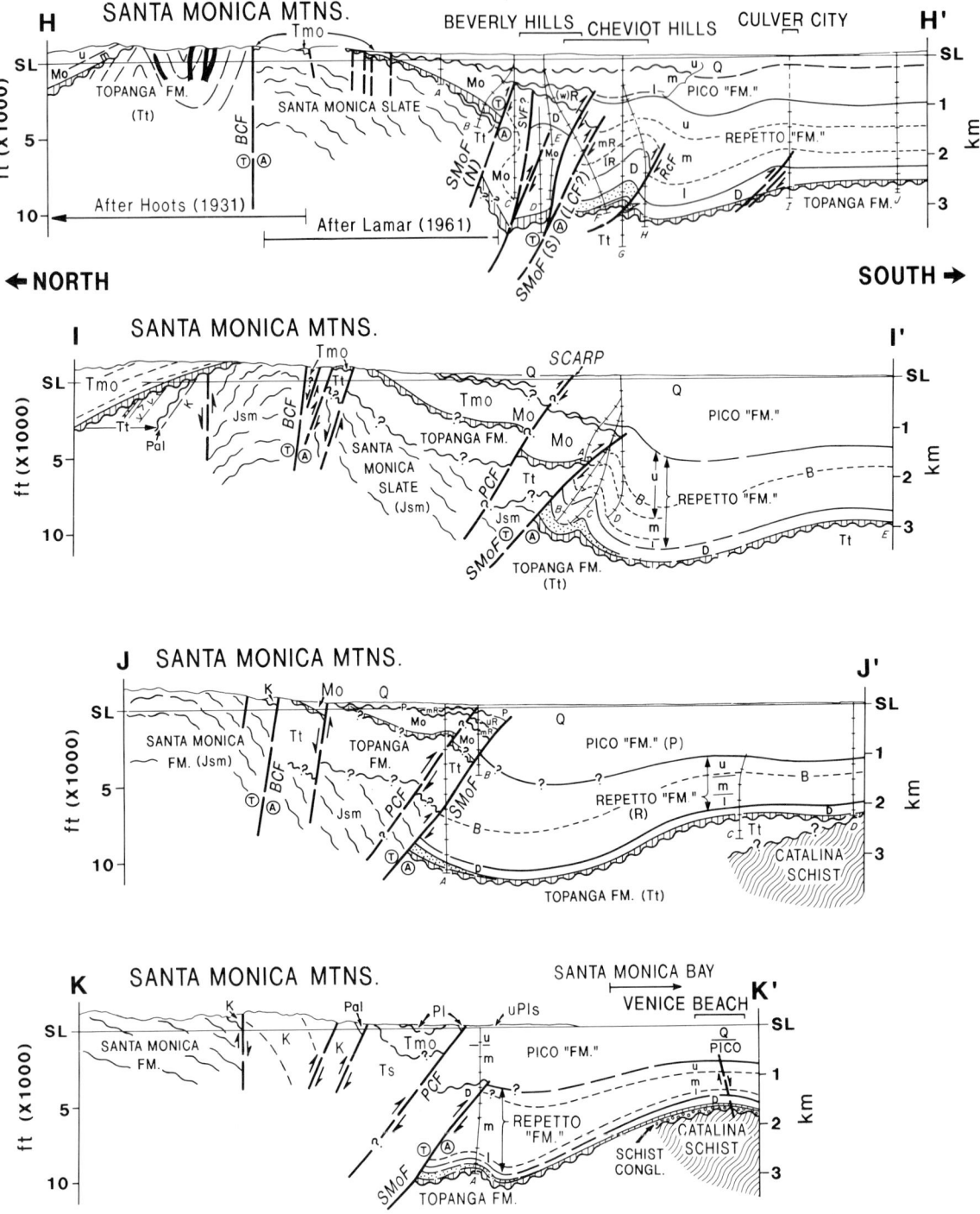

Figure 15. Cross sections HH' through KK', Santa Monica fault system. Figure 14 shows location of cross sections. Vertically lined pattern is nodular shale (lower Mohnian to Luisian), dotted pattern is Rancho sand (Delmontian), wavy lines indicate approximate bedding (or foliation) in Santa Monica Formation (Jsm). Other symbols and abbreviations as in Figure 8, plus Pl = Pliocene, Tmo = Modelo Formation, Ts = Sespe Formation, Pal = Paleocene, —B— is upper Repetto bentonite marker, W = Wolfskill zone, RcF = Rancho fault, PCF = Potrero Canyon fault, SMoF(N) and (S) are north and south branches of the Santa Monica fault, and SVF = San Vicente fault. Key wells (lettered) are listed in Appendix 2. Oil fields are named and their limits shown with horizontal brackets. Section HH' compiled from Hoots (1931), Lamar (1961), West et al. (1987), and West (personal communication, 1988); II' and JJ' after Wright (1987c); KK' compiled from California Division of Oil and Gas (1974 and public file), Durrell (1954), and Dibblee (1982a).

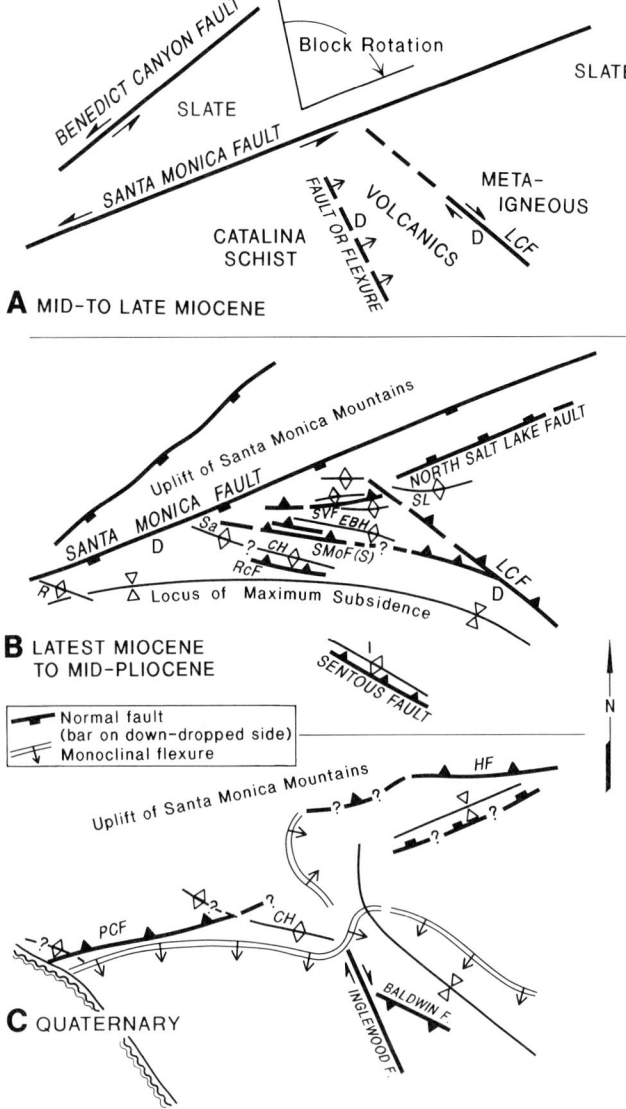

Figure 16. Sketch maps: Neogene tectonic evolution of the Santa Monica fault and related structures, northwestern Los Angeles basin. (A) middle to late Miocene (16–9 Ma); (B) latest Miocene to mid-Pliocene (7–3 Ma); and (C) Quaternary (2 Ma–Present). Abbreviations as in Figures 14 and 15. D indicates major down-faulting.

(see Figure 41). Opening of the Los Angeles basin in latest Miocene and early Pliocene time (Figure 16B) involved major vertical movement on the western part of the Santa Monica fault system and on the Las Cienegas fault. Folds in the footwall of the Santa Monica fault and at East Beverly Hills, San Vicente, and Inglewood were generated by this faulting. The North Salt Lake fault may have formed at this time by extension behind the uplifted Las Cienegas basement block that was "sagging" laterally into the deeply subsided basin to the southwest.

Since mid-Pliocene time (Figure 16C), the Potrero Canyon, Hollywood, and Inglewood faults have been superimposed on the older system. Continuing deformation on the Santa Monica and Las Cienegas faults is expressed as monoclinal flexuring with little or no offset of Pleistocene horizons. These two faults are believed to have become part of a series of north-dipping blind thrusts on which Transverse Ranges compression is being expressed beneath the northern Los Angeles basin (Davis et al., 1989). The Quaternary syncline of the basin has extended into the Beverly Hills area where it is bounded on the east by the Las Cienegas monocline and on the west by uplift localized along the extension of the Inglewood fault. Right-lateral movement on the Inglewood fault is being absorbed to the north by underthrusting and folding at Cheviot Hills and, perhaps, by rotation of the western Santa Monica Mountains to their present east-west orientation. Later this paper discusses the regional context for these distinct phases of tectonic evolution and the active blind thrusting that now involves the northern Los Angeles basin.

The major vertical stratigraphic separation across the western part of the Santa Monica fault system has occurred on the *Santa Monica fault*. That fault, initially recognized at Beverly Hills and Brentwood, was further documented by exploratory drilling (Figure 15, Section II', wells A–D) that led to the discovery of the *Sawtelle oil field* in 1965. This field, which will yield an estimated 15 MBO, produces from the Delmontian Rancho sand in the footwall of the Santa Monica fault (Eschner and Scribner, 1972). There the fault dips northward at about 40° to 50° (Figure 17), and separation of middle Miocene units is about 7000 ft (2100 m). The accumulation comprises two separate pools. In the northern pool, a fault trap, the producing zone, is overturned against slate basement in the hanging wall of the Santa Monica fault. The southern pool is contained in an asymmetrical doubly plunging anticline.

The *Riviera discovery* was made in 1966 by the Oxy U.S.A. Marquez No. 1 well (Figure 15, well A on Section KK'). Drilled 1 mi (1.6 km) east of the bluff exposure at Potrero Canyon, the well penetrated the Santa Monica fault approximately 4500 ft (1350 m) south of the projected trace of the Potrero Canyon fault (public file, California Division of Oil and Gas; McGill, 1989) and confirmed the dual nature of the Santa Monica fault system. The Oxy U.S.A. Marquez No. 1 found oil-saturated Rancho sand in a footwall anticline, and ultimate recoverable oil is estimated at 50 million barrels. Eventual development of this discovery, long delayed by local opposition, should provide significant data on the displacement history and potential seismic hazard of these faults.

East of the Sawtelle oil field (Figure 14), the Santa Monica fault appears to bifurcate (Gardett, 1971, his figure 4). A southern branch curves to the east and separates the Cheviot Hills and Beverly Hills oil fields (Figure 15, Section HH'). The *Cheviot Hills oil field*, discovered in 1958, will yield 26 MMBO from a doubly plunging, west-northwest-trending anticline. The steep south flank of the fold is broken by the Rancho thrust fault. Although some production is from Repetto sands on the east plunge of the fold, most of the oil occurs in the Delmontian Rancho sand. The oil may have accumulated originally in the sand pinch-out immediately to the south on the edge of the western shelf

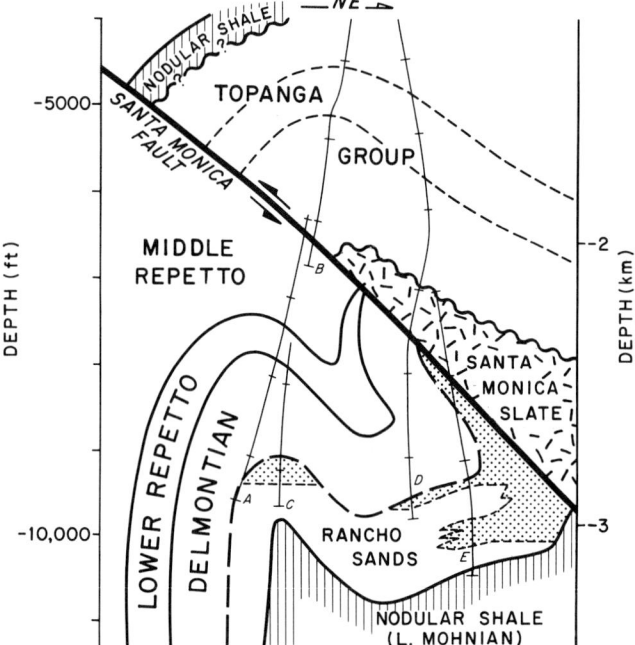

Figure 17. Cross section of the Sawtelle oil field (after Eschner and Scribner, 1972). Oil-productive zones are shaded. Control wells (lettered) are listed in Appendix 2. For other symbols, etc., see Figures 8 and 15.

(Figure 15, Section HH') and remigrated into its present location, prior to offset on the Rancho fault, as the fold developed.

The very complex geology of the *Beverly Hills oil field* is reflected in its development history. The shallow zone was discovered in 1900 and has produced 4 MBO from Repetto sands at an average depth of 2500 ft (760 m) (California Division of Oil and Gas, 1961). At this horizon the structure is relatively uncomplicated and has been described as a "small dome" on which "dips on the north flank are about 15°; those on the south flank, 45° or more" (Soper, 1943). Urban exploration in the 1950s resulted in the discovery, in 1954, of deeper Miocene zones that will ultimately produce about 30 MBO from sands of Delmontian and Mohnian age at depths of 3500–10,200 ft (1070–3100 m). Beneath the mid-Pliocene unconformity, the structure is complexly faulted (Figure 15, Section HH') (West et al., 1987), and no comprehensive structural interpretation has yet been published.

Clear evidence of mid-Pliocene uplift on the southern Santa Monica fault is provided in these two oil fields. In the Cheviot Hills field, mid-Repetto sedimentary breccias contain clasts of Mohnian shale (unpublished Chevron paleontologic studies) that were probably derived through submarine erosion of the block north of the fault beneath the Repetto unconformity. East of Cheviot Hills, the fault projects beneath the steep south flank of the East Beverly Hills field (Figure 14). It has been correlated (California Division of Oil and Gas, 1974; West et al., 1987) with the Las Cienegas fault and may well be continuous with it

at basement depths. The Las Cienegas fault (Figure 8, Section BB') (Mefferd, 1970)—discussed in more detail in the section on major internal structures—is the south-western boundary of an uplifted block north of the present basin axis. The western Santa Monica fault and the Las Cienegas fault appear to form a coherent structural trend, 14 mi (23 km) or more in length, with a similar history of major early to mid-Pliocene deformation followed by Pasadenan flexure above blind thrusts.

Between the uplifted Las Cienegas block and the Beverly Hills oil field are several oil fields discovered in the mid-1960s that provide additional insight into the displacement history of the Santa Monica fault system. The *East Beverly Hills field* (Figure 18), discovered in 1964, will yield an estimated 130 MBO from upper Miocene and lower Pliocene sands on an asymmetric, west-northwest-trending anticline (Jacobson and Lindblom, 1977). Entrapment is primarily anticlinal, augmented by stratigraphic pinch-out of some of the sands. Uppermost Mohnian sands (top Hauser Zone) pinch out around the anticlinal crest, reflecting structural growth at that time. Delmontian and lower and middle Repetto sands also thin or pinch out across the fold. Upper Pliocene beds show little or no folding, indicating that the structure developed from latest Mohnian to mid-Pliocene time. Although the steep south flank of the fold is aligned with the south branch of the Las Cienegas fault, upper Miocene beds on that steep flank are believed to be unfaulted (R. G. Lindblom, oral communication, 1987).

North of East Beverly Hills is the *San Vicente field*, discovered in 1966, with an estimated ultimate production of 21 MBO (Figure 18) (Jacobson and Lindblom, 1977). The accumulations are in (1) upper Miocene sands in several small west-trending anticlines on the north (hanging-wall) side of the San Vicente fault and (2) northward pinch-outs of unfolded Repetto reservoirs (Harding and Tuminas, 1988) that overlap the San Vicente fault onto an erosional unconformity that truncates the upper Miocene beds. The San Vicente fault may extend west into the Beverly Hills field (California Division of Oil and Gas, 1974); there a similar north-dipping reverse fault cuts the upper Miocene section and is unconformably overlapped by the lower Pliocene Wolfskill sands (Figure 15, Section HH') (West et al., 1987).

The northeast strike of the western Santa Monica fault is maintained eastward as its northern branch, which crosses the northern edge of the Beverly Hills oilfield (Figure 14; Figure 15, Section HH'). There the fault was earlier correlated with the Hollywood fault (Lamar, 1961). Farther to the east, the fault is poorly documented. Figure 14 shows contours on this portion of the fault plane based on subsurface control from wells north of the Beverly Hills field and from wells in and near the Sherman oil field. The San Vicente oil field, in the footwall of the fault, has been explored and developed by directional drilling that has yielded very little data on the fault itself.

Eastward the northern branch of the Santa Monica fault converges with the *Hollywood fault* whose surface trace is shown in Figure 14. That fault had been postulated along the steep south face of the eastern Santa Monica Mountains (Hoots, 1931; Hill et al., 1979), trending there nearly east-west. Evidence from recent geotechnical investigations (Crook et al., 1983) tends to confirm the existence of the

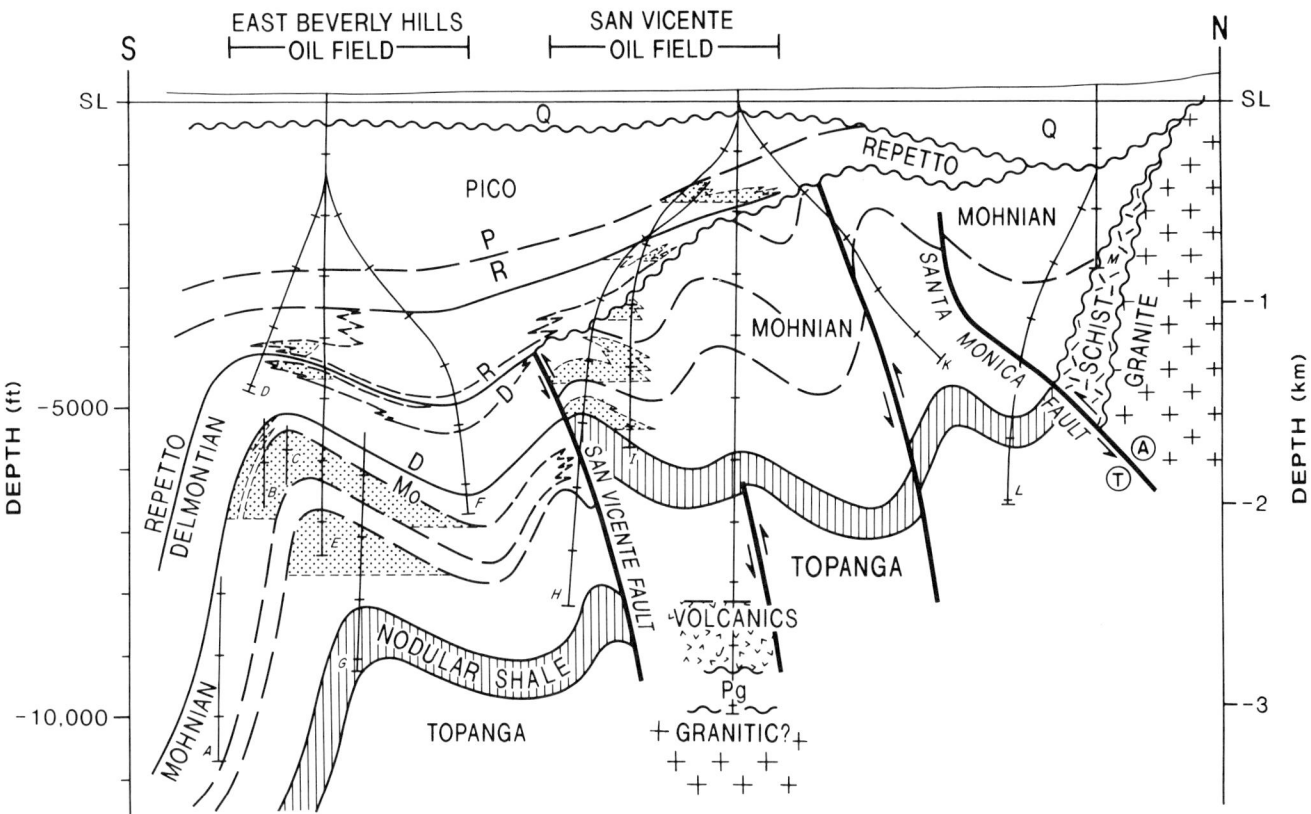

Figure 18. Cross section of the East Beverly Hills and San Vicente oil fields, after Jacobson and Lindblom (1977). Oil productive zones are shaded. Control wells (lettered) are listed in Appendix 2. For other symbols, etc., see Figures 8 and 15. Pre-Mohnian data beneath San Vicente oil field from Yerkes et al. (1965) and Yeats (1973).

Hollywood fault from Laurel Canyon eastward. Its possible extension to the west-southwest is based on weak topographic evidence and on an exposure of sheared slate and sandstone at Greystone Park where a zone of oil seeps 2000 ft (600 m) in length may reflect leakage up the fault plane (P. H. Gardett, oral communication, 1988). One mile (1.6 km) to the southeast of these seeps, the subcommercial Sherman oil field (Figure 18, well M) in the hanging wall of the Santa Monica fault produced a small amount of oil from basal Pliocene and Mohnian sands at depths of 1500-3000 ft (450-900 m). There the Santa Monica fault appears to be overlain by unfaulted Pliocene beds. Quaternary displacements in this area may be localized on the Hollywood fault, but the detailed relationship between these two faults is not known.

The *North Salt Lake fault* (Figure 14) converges eastward with the Hollywood fault. Several exploratory core holes drilled north of the Salt Lake oil field have shown the existence of this normal fault, which dips 50° to 70° north and follows the northern edge of the Salt Lake anticline. Vertical stratigraphic separation across the fault is about 1000 ft (300 m). The down-dropped block north of the fault ("Hollywood syncline" of Erickson and Spaulding, 1975) includes Pliocene strata and corresponds to a zone of active subsidence. Figure 8 (Section BB') shows the tentative eastward projection of the North Salt Lake fault based on linear topographic features (Weber, 1980) and the zone of subsidence. The fault parallels a postulated groundwater barrier in the Pleistocene beds, and active differential subsidence across its trace has been ascribed primarily to groundwater withdrawal (Hill et al., 1979). At least part of the active deformation may be tectonic.

At the western end of the Santa Monica fault system, the *Potrero Canyon fault* is exposed in the coastal bluffs at Pacific Palisades (McGill, 1980, 1989; McGill et al., 1987) where it has thrust Pliocene beds over upper Pleistocene terrace deposits. The main fault dips about 45° north, and striations on the fault are parallel to the dip. Vertical separation of about 155 ft (47 m) has been measured on the marine terrace platform, whose estimated age is about 125,000 years B.P. (McGill, 1980). The fault can be traced eastward for 6 mi (9.6 km), based on discontinuous topographic scarps (Hill, 1979), and has been confirmed by trenching at several localities (Crook et al., 1983). Figure 14 shows this surface trace, which is nearly parallel to the contours on the underlying Santa Monica fault. Figure 15 shows the relation of this surface trace to nearby subsurface control and to the Santa Monica fault.

Displacement of upper Pleistocene terrace and alluvial-fan deposits by the Potrero Canyon fault indicates that it must be considered a potentially active fault (Crook et al., 1983; McGill, 1989). Farther to the east, subsurface data at the northern edge of the Beverly Hills oil field (Figure 15, Section HH') show the upper Pliocene beds as extending unbroken across the projected alignment of the Potrero Canyon fault. There is no surface evidence that might connect the Potrero Canyon fault with the western part of the Hollywood fault or otherwise serve to indicate Holocene activity on this central part of the Santa Monica fault system.

The major *vertical displacement* across the Santa Monica fault occurred during deposition of middle and upper Repetto beds, about 4-3 Ma, and followed its southern branch into the Las Cienegas fault. Separation of the basal Mohnian nodular shale across this trend varies from about 4000-8000 ft (1200-2400 m), yet along most of its length the upper Pliocene and Pleistocene beds show little or no offset. A thick and unbroken sequence of deep-water mid-Pliocene sediments has accumulated in the footwall block. By contrast, this sequence is thin or absent north of the fault where upper Pliocene beds rest unconformably on Mohnian or Delmontian strata.

The cross sections in Figure 15 reveal significant inconsistencies in the correlation and thickness of the several Pliocene biostratigraphic zones in the footwall of the Santa Monica fault. These variations may be real, but they more probably reflect the use of differing paleontological criteria by the several sources in addition to variations in data quality. Notwithstanding this present lack of refinement, the evidence documented in these cross sections confirms the mid-Pliocene age of the major vertical displacement across the Santa Monica fault.

Estimates of *left-lateral offset* along the Santa Monica-Hollywood fault zone have varied widely. Barbat (1958) noted an 8-mi (13-km) offset of the metamorphic aureole of the Santa Monica Mountains based on subsurface data in the north-central Los Angeles basin (Figure 4). These two contact-metamorphic zones are not necessarily identical, however (Sorensen, 1984), and thus are of doubtful use as a geological line indicative of lateral slip. Dibblee (1982a) estimated cumulative left slip at about 15 mi (24 km). Between the Santa Monica Mountains and the Alhambra high, the slate-granite contact is separated by this amount on opposite sides of the fault. This might appear to be definitive but could be influenced significantly by offset on splays of the Whittier fault or by the depth of erosion into the hood of the pluton. Colburn (1973) suggested 37 mi (60 km) displacement based on comparisons of Paleocene and Upper Cretaceous facies between the Santa Monica Mountains and the Santa Ana Mountains. Subsequent research (Hornafius et al., 1986) implies that the Santa Monica Mountains have rotated up to 80° clockwise during the past 15 m.y. These new data call for a reappraisal of paleogeographic reconstructions that involve Paleogene and older rocks. Nevertheless, one such reappraisal (Link et al., 1984) adhered to the earlier estimate of Colburn.

Left-lateral offset after the middle Miocene is more easily documented by contrasts in the thickness and facies of Mohnian and Delmontian deep-water strata (Lamar, 1961). The submarine Tarzana fan, well exposed in the Santa Monica Mountains (Sullwold, 1960), has been displaced 6-9 mi (10-15 km) westward from its southern continuation that crops out north of downtown Los Angeles and extends into the subsurface of the northern Los Angeles basin. More recent detailed stratigraphic studies (Redin, this volume) support this initial interpretation. Further confirmation is provided by stratigraphic variations within the upper Miocene deep-sea fans, which are described later in this paper (see also Figures 39, 40).

As shown in Figure 15, the combination of lateral and vertical displacement on the Santa Monica fault has produced significant contrasts in the upper Miocene stratigraphy of the different blocks. On Section HH' (see also West et al., 1987), the middle and upper Mohnian section north of the fault, a part of the Tarzana fan, is 2000 ft (600 m) thick and mostly sandstone. Between the branches of the Santa Monica fault, in the Beverly Hills oil field, that same interval is about 3000 ft (900 m) thick but mostly shale. South of the Santa Monica fault in the Cheviot Hills oil field, the lower Mohnian nodular shale is overlain almost directly by the Delmontian Rancho sand, and middle and upper Mohnian beds are absent or very thin. In the Sawtelle oil field (Figure 17) (Eschner and Scribner, 1972) there is a similar abbreviated section in the footwall of the Santa Monica fault; in the hanging-wall block, most of the Mohnian interval above the nodular shale has been removed by submarine erosion during early Pliocene uplift.

Evidence of *active faulting* along the Santa Monica fault system is meager. Near the eastern part of the Santa Monica fault system, focal mechanism data on recent small earthquakes at depths of 1.25-7.5 mi (2-12 km) (Real, 1987) suggest that the zone is an active left-lateral oblique fault, with a strike of N82°E and a dip of 80° to the north. However, the epicenters tend to lie south of the mapped fault trace, and Yerkes et al. (1987) state that the Santa Monica fault zone is not clearly associated with recent seismic activity. Geotechnical investigations (Figure 14) on the Hollywood and Potrero Canyon faults are inconclusive regarding recency of movement (Crook et al., 1983), and further field exploration is severely limited by the dense urban development along the entire fault system. Significant Holocene activity on the Potrero Canyon fault is implied by the intermittent topographic scarp that displaces alluvial deposits as far east as University High School (U on Figure 14).

The Santa Monica fault system strikes westward beneath Santa Monica Bay, where two major faults have been mapped (Figure 3). The *Malibu Coast fault* has been mapped onshore from Malibu west beyond Point Dume, a distance of about 12 mi (19 km). A well drilled on Point Dume, south of the fault (Figure 4), encountered Catalina Schist (or schist breccia) at a depth of 5377 ft (1639 m) (Campbell et al., 1966). Stream deflections suggest late Quaternary left slip (Dibblee, 1982a). The *Anacapa (Dume) fault* has been mapped by marine geophysical surveys and is 2-7 mi (3-11 km) south of the Malibu Coast fault (Vedder et al., 1974). Moderate-sized earthquakes in the area of these faults occurred in 1930, 1973, and 1979. Focal mechanisms and aftershock patterns from the latter two events indicate thrust or reverse motion, with fault planes dipping 53° north in the 1979 Malibu event 11 mi (18 km) offshore (Hauksson and Saldivar, 1986, 1989) and 36° north in the 1973 Point

Mugu quake, about 17 mi (27 km) farther west (Yerkes and Lee, 1979). Most of the seismicity in eastern Santa Monica Bay appears to be located to the south of the Malibu Coast and Anacapa (Dume) faults and is tentatively attributed to blind thrust faults (Hauksson and Saldivar, 1989). It might seem logical to project this paired fault system eastward, linking the Malibu Coast fault to the Potrero Canyon fault and the Anacapa (Dume) fault to the Santa Monica fault (McGill, 1989). A definitive correlation is premature, however, due to the inadequacy of presently available data.

The Santa Monica fault system apparently extends east to join the *Raymond (Raymond Hill) fault* (Figures 4, 9, 10) (Barbat, 1958; Lamar, 1961), although the connection between the two faults is obscured by the complex structure of the Eagle Rock–Highland Park area (Lamar, 1970). The Raymond fault (Bryant, 1978; Crook et al., 1987) is a reverse left-oblique fault, uplifted on its north side. Evidence for its Holocene activity includes faulted soils observed in shallow trenches, displaced drainages, and the prominent topographic scarp that traverses alluvial-fan material of the northwestern San Gabriel Valley. The scarp has a maximum height of more than 100 ft (30 m). Subsurface data in this area are insufficient to indicate the total vertical displacement, which may range from 450–2500 ft (135–775 m) (Crook et al., 1987). Lateral displacement is unknown.

Northern Margin

The northern margin of the basin between the Santa Monica fault system and the East Montebello fault (Figure 4) is defined by the *Elysian Park anticline* (Figure 8, Section CC': EPA) and its eastern extension, the subsurface *Alhambra high* (Section DD': AH). Along this east-southeast–trending uplift, wells have encountered slate or schist basement at depths as shallow as 1370 ft (417 m) at Elysian Park and 2400 ft (730 m) at Alhambra (Figure 4). Detailed petrologic and geochemical studies of these basement rocks (Sorensen, 1984, 1985) describe them as similar to phyllites and slates of the Santa Monica Mountains and to schists and amphibolites on Santa Cruz Island.

A reverse fault is present on the south flank of this uplift at Elysian Park, and similar faulting is postulated along the Alhambra high (Figure 8, Section DD'). Thinning of Pliocene and uppermost Miocene beds northward onto this uplift indicates that it was an active positive element during the development of the present Los Angeles basin, 7–4 Ma. Levelling surveys made before and after the 1987 Whittier Narrows earthquakes show active uplift continuing in this area (Lin and Stein, 1989). East of the East Montebello fault, the area of shallow basement rock between the Whittier and Workman Hill–Whittier Heights faults (Figure 8, Section EE'; Figure 4) probably represents the offset eastern extension of the Alhambra high.

Southeastern Margin

At the southeastern end of the Los Angeles basin, a distance of more than 30 mi (50 km) separates the Whittier fault and the Palos Verdes uplift. Across this gap, the margin of the basin (Figure 3) is defined by two major uplifts— the Santa Ana Mountains and the San Joaquin Hills—and the intervening Santa Ana shelf. The northern *Santa Ana Mountains* have been uplifted on the southwestern side of the Whittier-Elsinore fault zone. Their core of slightly metamorphosed Mesozoic sedimentary and volcanic rocks is overlain by a west-dipping sequence of Upper Cretaceous and Paleogene strata whose maximum thicknesses in outcrop total nearly 13,000 ft (4000 m) (Schoellhamer et al., 1981). A north-trending syncline containing up to 2300 ft (700 m) of Miocene beds (Figure 19) separates this homocline from the Santa Ana shelf. Along the western margin of this homocline, early Miocene and older strata are broken by numerous major extensional faults that generally strike north to northeast and form a zone that trends approximately north. Most of the displacement on these faults appears to predate the late Miocene.

An apron of Quaternary alluvium as much as 1500 ft (500 m) thick covers the *Santa Ana shelf* (Figure 7). Directly beneath the Quaternary unconformity, wells have encountered sedimentary rocks ranging from Pliocene to Late Cretaceous in age (Schoellhamer et al., 1981). One well (Figure 4) reached metavolcanic basement at a depth of about 5600 ft (1700 m) subsea.

In the *San Joaquin Hills*, strata as old as Paleocene are exposed. The geology of this small uplifted area provides significant clues to the structure and stratigraphy beneath the center of the Los Angeles basin a few miles to the northwest. Figure 19 is a simplified geological map (after Vedder et al., 1957) showing several major northwest- to north-striking faults and associated diabase dikes that generally are parallel to the faults or intrude along them. Two cross sections (Figure 20) that have been restored to a top Luisian datum show the magnitude of block faulting that occurred in this area early in middle Miocene time. The Shady Canyon fault forms the northeastern boundary of a trough in which at least 7000 ft (2100 m) of lower and middle Miocene strata are preserved. Paleobathymetric studies (Ingle, 1979) indicate that initial littoral deposition gave way to bathyal conditions as the trough deepened during Relizian and Luisian time. Paleogene strata that crop out northeast of the Shady Canyon fault document a vertical separation of 5000 ft (1500 m) or more. At its southeastern end, however, the fault disappears beneath an unbroken cover of deep-water beds of Luisian age (Figure 20, Section MM'), showing that there has been no significant movement on this part of the Shady Canyon fault since middle Miocene time.

The southwestern side of this early Neogene trough steps up across the Pelican Hills fault (Figure 20, Section LL') and extends offshore to the Newport-Inglewood fault zone. By mid-Miocene time, a broad area of Catalina Schist was exposed on the uplifted block west of the Newport-Inglewood zone. From this schist highland, the San Onofre Breccia—a remarkable wedge of debris-flow and fan deposits up to 3000 ft (900 m) thick—was shed eastward onto the edge of the adjacent marine trough during middle Miocene time (Woodford, 1925; Stuart, 1979a). Later displacement that offsets the Monterey Formation has elevated the block northeast of the Pelican Hills fault, a reversal of the major mid-Miocene movement. The fault is overlapped by Pliocene beds near Upper Newport Bay (Vedder et al., 1957), implying that its last movement occurred in late Miocene time. However, data cited by Ziony

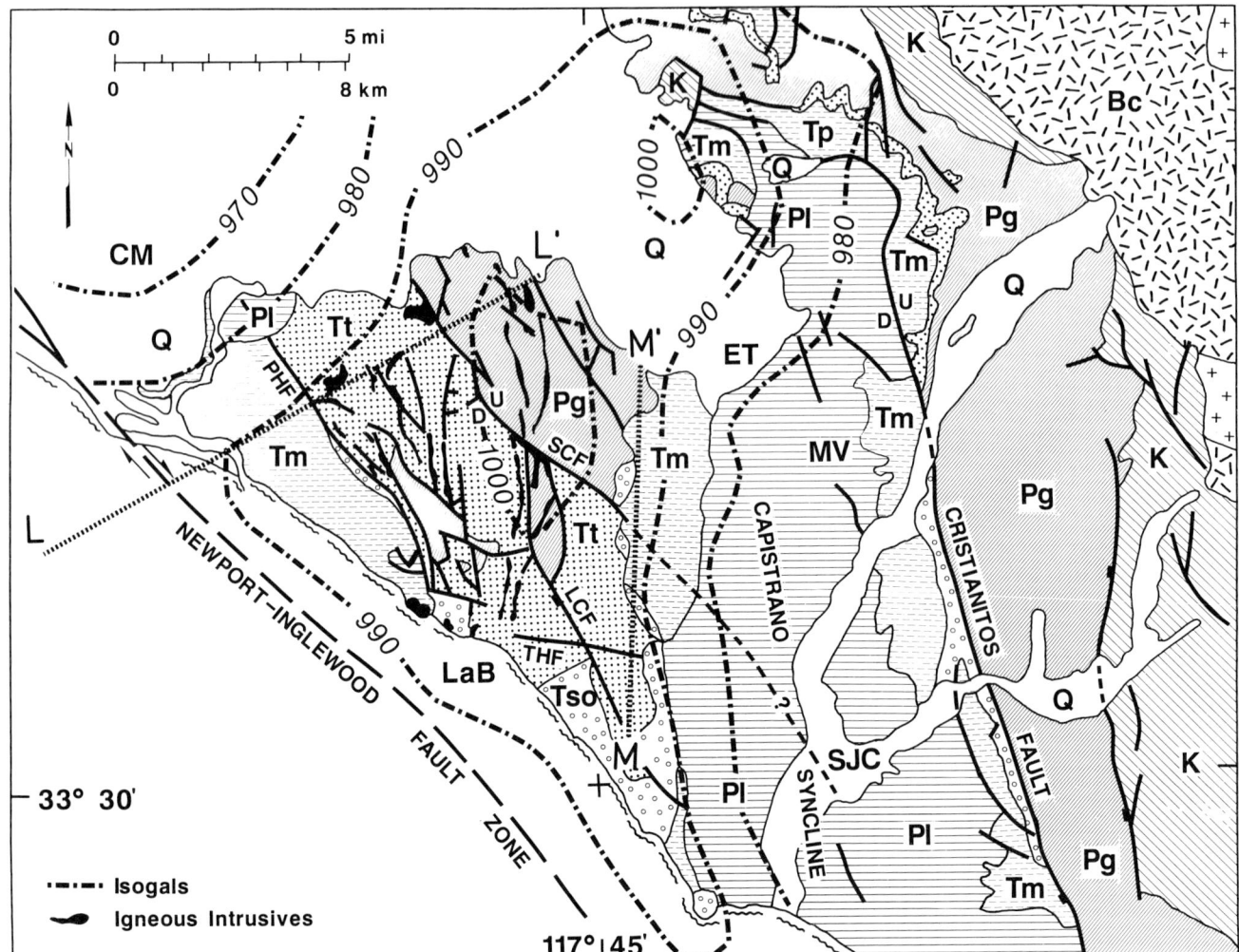

Figure 19. Geological map of the San Joaquin Hills and vicinity (compiled from Vedder et al., 1957; Rogers, 1965; Schoellhamer et al., 1981). Isogals (in 10 mgal interval) are Bouguer gravity data from Oliver et al. (1980) and proprietary sources. LL' and MM' are cross sections in Figure 20. CM is Costa Mesa, ET is El Toro, LaB is Laguna Beach, MV is Mission Viejo, SJC is San Juan Capistrano, LCF = Laguna Canyon fault, PHF = Pelican Hill fault, SCF = Shady Canyon fault, THF = Temple Hill fault, Q = Quaternary alluvium, Pl = Pliocene, Mp = Puente Formation, Mm = Monterey Formation, Tso = San Onofre Breccia, Tt = Topanga Formation, Pg = Paleogene sediments, K = Cretaceous sediments, and Bc = basement complex (patterns as in Figure 4). Other patterns identified by abbreviations in larger outcrop areas on map.

and Yerkes (1985) suggest that the Pelican Hills fault has been active during late Quaternary time.

Within the graben between the Shady Canyon and Pelican Hills faults, the Paularino Member of the Topanga Formation (Vedder et al., 1957) forms the youngest part of the trough sequence and consists of up to 1275 ft (400 m) of sandstone, siltstone, and volcanic breccia. An andesitic flow breccia at the base of the Paularino yielded a K-Ar date of 14.1–16.7 Ma (Turner, 1970). These flow breccias were probably derived from the dikes that intrude and diverge from the Shady Canyon, Pelican Hills, and Laguna Canyon faults. Field relationships suggest that faulting and intrusion were approximately coeval (Vernon et al., 1970).

Volcanic and sedimentary rocks similar in age and lithology to the Paularino crop out 10 mi (16 km) to the north at El Modeno (Figure 5) (Schoellhamer et al., 1981). Geologic features of the San Joaquin Hills thus fully document an early to middle Miocene episode of crustal extension involving block faulting and volcanism.

A positive gravity anomaly of at least 30 mgals coincides with the San Joaquin Hills and Santa Ana shelf (Figures 3, 19). Density modeling (McCulloh, 1960) shows that the anomaly involves a deep crustal mass, below 30,000 ft (9000 m), and suggests an igneous intrusive that may have been the source of the volcanics. The eastern edge of this gravity anomaly may reflect deep-seated faulting that controls the

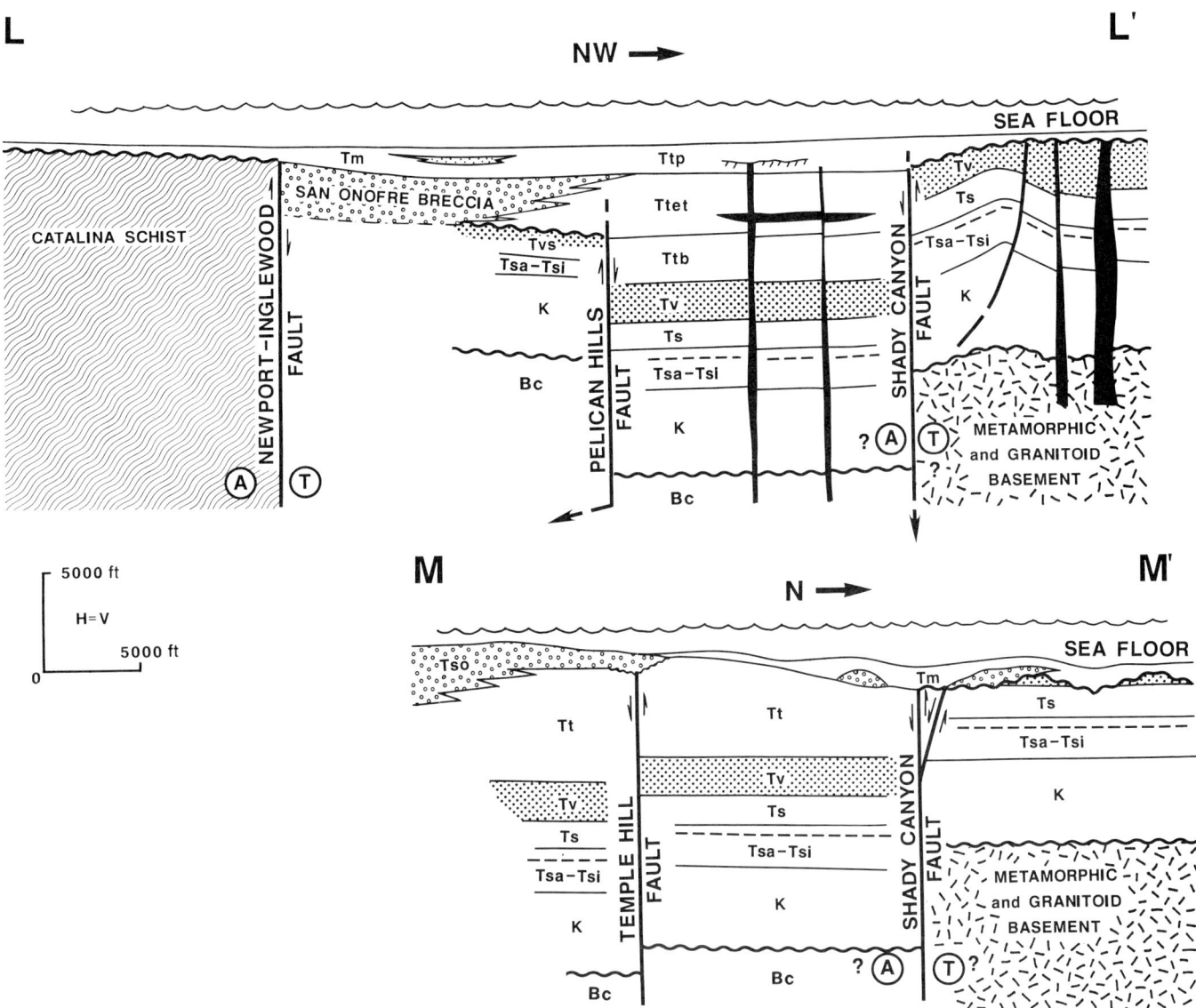

Figure 20. Restored cross sections, San Joaquin Hills. Restored to end of Luisian time (ca. 14 Ma). From Wright et al., 1973. Locations of sections shown in Figure 19. On scale bars, 5000 ft = 1524 m. Abbreviations, etc., as in Figures 8 and 19, plus members of Topanga Formation are Paularino (Ttp), Los Trancos (Ttlt), and Bommer (Ttb); Paleogene formations are Vaqueros (Tv), Sespe (Ts), undifferentiated Vaqueros/Sespe (Tvs), Santiago (Tsa), and Silverado (Tsi). Vaqueros Formation is shaded.

western edge of the Capistrano syncline. Beneath the southeastern Santa Ana shelf, several north-striking faults with combined down-to-the-east displacement of 2000 ft (600 m) or more extend southward under a cover of upper Miocene beds (Vedder et al., 1957; California Division of Oil and Gas, 1964; Schoellhamer et al., 1981). This faulting appears to die out rapidly to the south and is replaced by flexure along the west flank of the Capistrano syncline (West, 1979). A northward extension of the Capistrano syncline during Mohnian time is indicated by the presence of deep-water sandstones of the Soquel Member, more than 400 ft (120 m) in thickness, at Loma Ridge north of El Toro (Figure 19) (Schoellhamer et al., 1981).

The southernmost limits of the Los Angeles basin, located across the 10 mi (16 km) gap between the Newport-Inglewood fault zone and the Palos Verdes uplift, cannot be defined fully with data now available. Offshore core holes west of Newport Beach reached Catalina Schist basement at depths of 8750–9000 ft (2670–2745 m) (Figures 4, 5) (Henry, 1987) and define a west-trending basement ridge, herein called the *Newport Offshore ridge,* extending from the Newport-Inglewood fault zone to the THUMS-Huntington Beach fault. The upper Miocene and lower Pliocene intervals thin southward onto this ridge (Figure 28, Section dd′), which probably served as a partial sill during that time. The south flank of the Newport Offshore ridge may be reflected by the steep submarine slope that separates the shallow San Pedro Bay from a basin with water depths of 2000 ft (600 m) or more. Outcrops of Miocene beds at the sea floor on the upper parts of this slope (Clarke et al., 1987) suggest that basement may be relatively shallow in this area. Offshore gravity data (Beyer, 1987) provide supporting evidence for the Newport Offshore ridge.

Miocene sea-floor outcrops east of the Beta oil field (Vedder et al., 1986) suggest that a similar, en echelon arch may occur between the Palos Verdes fault zone and the THUMS-Huntington Beach fault. A better understanding of the structural separation between the Los Angeles basin and the offshore Oceanside basin to the southeast awaits the release of deep reflection seismic data from offshore oil exploration.

Major Internal Structures

Newport-Inglewood Fault Zone

The Newport-Inglewood fault zone (NIFZ) is the best known structural feature of the Los Angeles basin (Figure 7). Its surface expression is a range of low hills trending about N45°W and extending nearly 40 mi (64 km) from Culver City southeast to Newport Beach. Its petroleum potential was recognized at least as early as 1913 (Wright, 1987e), and a drilling boom from 1920 to 1928 found most of its seven major and eight smaller oil fields. These will yield nearly 3.2 billion barrels of oil (GBO) and 3.2 tcf of gas. Active tectonism on the NIFZ was made obvious by the 1920 Inglewood earthquake (estimated $M_L = 4.9$) and the 1933 Long Beach earthquake ($M_L = 6.3$).

The northern portion of the NIFZ is a series of en echelon faults, whereas from the Long Beach oil field south, the fault appears to be a single main strand with local splays. The cross sections in Figure 8, which are aligned on the NIFZ, provide a general view of its structure. The zone has long been considered a classical example of the development of en echelon folds and faults along a deep-seated strike-slip fault (Eaton, 1923, 1924; Ferguson and Willis, 1924; Moody and Hill, 1956; Harding, 1973). However, detailed subsurface mapping of oil fields along the NIFZ has revealed a variety of structural patterns and histories, and many of these cannot easily be reconciled with a pure strike-slip origin. Brief descriptions of the Inglewood, Dominguez, Long Beach, Seal Beach, Sunset Beach, and Huntington Beach oil fields will illustrate these variations.

The *Inglewood oil field*, the northernmost field on the NIFZ, was discovered in 1924 and will yield an estimated 400 MBO (Figures 21, 22). Productive sands are found in every stratigraphic stage from middle Miocene (Luisian) to latest Pliocene, but more than half of the total oil occurs in the mid-Pliocene Vickers zone. Folding and faulting are the principal trapping mechanisms. Oil in the Mohnian? City of Inglewood zone is trapped in a sand pinch-out up the southeast plunge of the fold. Oil in the Luisian Sentous zone is trapped by a permeability barrier caused by calcite cementation associated with volcanic intrusives that are localized in the vicinity of the Inglewood fault (W. J. Plumley, unpublished study, 1962). One well in the field east of the Inglewood fault has reached metamorphic basement that has been described as albite and chlorite-muscovite schists (Yeats, 1973). Assignment of this locality to either the Catalina terrane or the metamorphic rocks of the southern Transverse Ranges (Sorensen, 1984, 1985) must be considered ambiguous.

A structure contour map (Figure 21) on the top of the Rubel zone (top of middle Repetto) defines a simple northwest-trending anticline with a thrust fault on its southwestern flank. This fault and the anticlinal crest are offset right-laterally along the Inglewood fault. Striae in well cores within the fault zone indicate that the latest movement was mainly horizontal (Driver, 1943). Detailed isopach mapping of seven stratigraphic intervals from the lower Pliocene to lowermost Pleistocene (Wright et al., 1973; Wright, 1987d) depicts the mid-Pliocene development of a west-northwest-trending anticline and confirms that the fold axis was offset right-laterally approximately 4000 ft (1200 m) in latest Pliocene and Pleistocene time. A cross section (Figure 22) illustrates the transition from this simple fold at top Rubel depth to a complex of extensional faults that affect the shallower producing zones. These faults form a graben west of the Inglewood fault and parallel to it. Most of the faults in the uplifted eastern block are also parallel to the Inglewood fault. Their orientation suggests that these extensional faults east and west of the Inglewood fault are related not to stretching of the anticlinal crest but most probably to the right-lateral separation of the two halves of the preexisting anticline.

Varying rates of folding of the Inglewood anticline are recorded by the percentage of stratigraphic thinning from the flanks to the crest of the fold (Wright et al., 1973). During early Pliocene time, crestal thinning ranged between 10% and 25%. During the late Pliocene, it increased from 40% in the Upper Vickers zone to 75% in the uppermost Pliocene Investment zone. Isopachs of the latter interval show the first definite indications that lateral displacement on the Inglewood fault had penetrated the sediments just beneath the sea floor. In early Pleistocene time (upper Pico), a half graben formed west of the Inglewood fault, and crestal thinning had decreased to approximately 50%.

The cross section shows that the younger Pleistocene Inglewood and San Pedro formations are nearly horizontal east of the Inglewood fault. There the latest, Pasadenan phase of deformation has produced an uplifted wedge between the Inglewood fault and the young Baldwin fault (which strikes N60°W and converges with the Inglewood fault). Ironically, this late uplift of the eastern flank delayed discovery of the field by six years. The first exploratory well was drilled on the topographic high outside the eventual field limits, and the discovery well was eventually drilled in the central graben.

The *Dominguez oil field*, discovered in 1923, will produce an estimated 277 MBO from lower Pliocene and upper Miocene sands in a faulted anticlinal trap. Figure 23 is a structure contour map of the field on a horizon near the base of the Repetto. Three major faults cross the fold, dividing the field into four structural blocks (Graves, 1954). The central "Regional Shear Zone" (Avalon-Compton fault) extends northwest into the Rosecrans field and is one of the main en echelon faults of the northern NIFZ. West of the "Regional Shear Zone," the "X" fault diverges to the south, while to the east, the "B" fault zone diverges

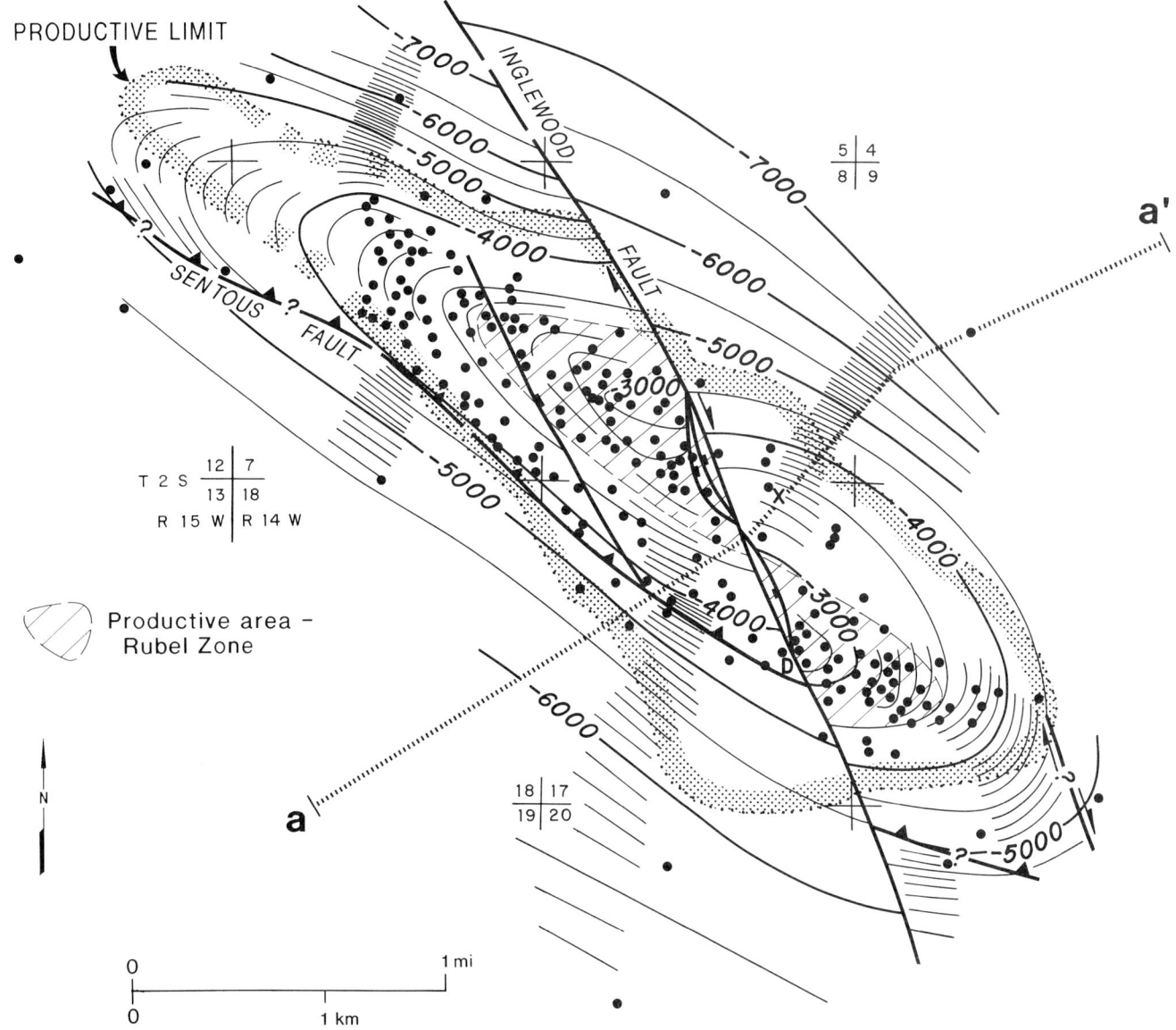

Figure 21. Subsurface structure map of the Inglewood oil field (from Wright, 1987d). Contours on a mid-Pliocene horizon (top of Rubel zone; approximate top of middle Repetto). Contour interval: 100 and 500 ft (30.5 and 152.5 m). aa' is location of cross section in Figure 22. Solid circles show control wells. D = discovery well; X = well reaching basement rock. Other symbols, etc., as in Figure 9.

in a northerly direction. Within Pliocene beds, the maximum vertical separation measured on any of these three faults is 200 ft (60 m) (Grinsfelder, 1943), but throw on the faults increases with depth (McMurdie, 1973). A fault that crosses the east plunge of Dominguez is believed to be an extension of the Cherry Hill fault, the primary strand of the NIFZ in the Long Beach field to the south. The "B" fault and the "Regional Shear Zone" define a shallow graben across the crest of the anticline. Grinsfelder (1943) noted that "each bed in the central downthrown block is relatively thicker than those in the upthrown east and west blocks," indicating structural growth during Pliocene deposition.

The fold is also cut by a series of reverse faults that strike west to west-southwest, dip 39° to 60° south, and die out as they are traced upwards (Bravinder, 1942). In a Magnitude 5.4 earthquake on October 21, 1941 (Hauksson, 1987), movement occurred on one of these faults west of the "X" fault, damaging six wells. Nine other wells in the western block were damaged, but wells penetrating the "Regional Shear Zone" were not affected. Graves (1954) described "a north-dipping zone of thrusting that strikes parallel to the axis of the fold and is encountered in the Miocene strata [below the contoured horizon of Figure 23] on the south flank of the fold." Schist basement has been reached at a depth of about 12,300 ft (3750 m) on the west plunge of the anticline and about 11,500 ft (3500 m) east of the "B" fault zone. Descriptions of these rocks (Schoellhamer and Woodford, 1951) imply that the western

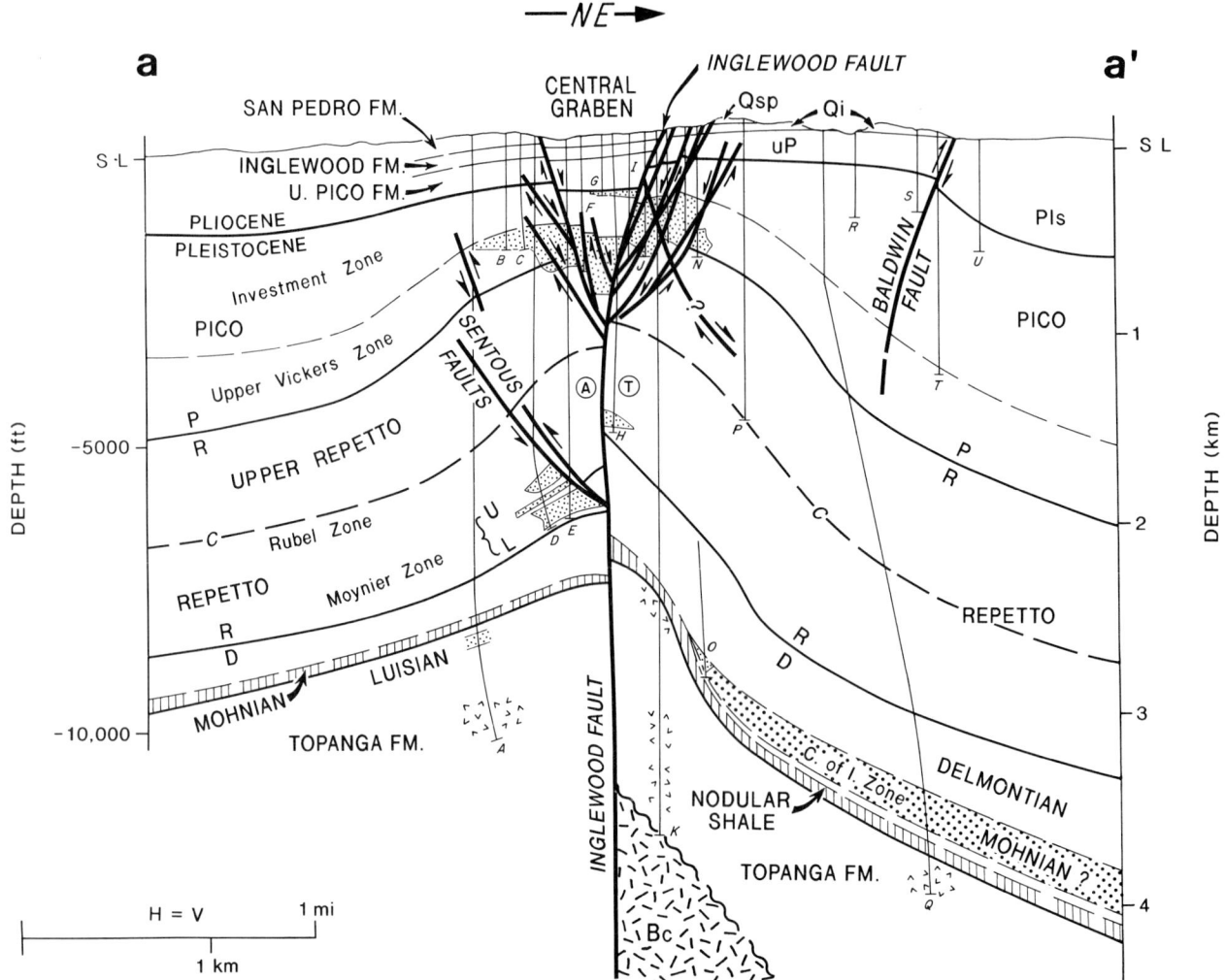

Figure 22. Cross section of the Inglewood oil field (from Wright, 1987d). Location of cross section is shown in Figure 21. Oil-productive zones are shaded. Lower Mohnian nodular shale is cross-hatched. Control wells (lettered) are listed in Appendix 2. Other symbols, etc., as in Figures 8 and 15, plus C. of I. Zone = City of Inglewood zone and Pls = Pleistocene. —C— indicates contoured horizons of Figure 21.

locality (Unocal Callendar No. 79 well) is within the greenschist facies of the Catalina terrane. To the east within the NIFZ, the Shell Reyes No. 135 well found basement described as antigorite-dolomite rock of uncertain affinity.

Based on offset of the deep north-dipping thrust, Graves (1954) suggested left-lateral movement on the "Regional Shear Zone" and this offset may continue to basement depths (McMurdie, 1973). Alternatively, similar thrusts may have formed independently on opposite sides of the "Regional Shear Zone"; details to confirm the original continuity of these faults are not available. Graves also noted that the Dominguez anticline is located on the southeast prolongation of the El Segundo-Lawndale-Alondra fold trend (see Figure 9) and suggested that structural features at Dominguez "are possibly results of alternating movements along the Newport-Inglewood shear zone and the north-dipping thrust zone on the south flank of the Dominguez structure," which he considered a possible extension of the basement ridge to the northwest. Like Inglewood, the Dominguez anticline appears to have evolved through several phases. Detailed isopach studies that might define and date these phases have not been published.

The *Long Beach oil field* (Figures 24, 25), discovered in 1921, will yield an estimated 924 MBO. Productive sands range in age from Mohnian to late Pliocene, with some crestal wells having more than 3000 net ft (900 m) of oil sand (Taylor, 1973). The estimated ultimate per acre recovery from the field, 728,000 bbl/ac, may be a world record (Troxel, 1954). The trap is a long, narrow, faulted anticline parallel to the NIFZ. On its southwest flank, where dips are about 40°, the Cherry Hill fault extends northwest towards the Dominguez oil field. Dips are 15°–20° on the northeast flank where the en echelon Northeast Flank fault extends southeast toward the Seal Beach oil field. These two faults and the north-striking Pickler fault, which curves to merge

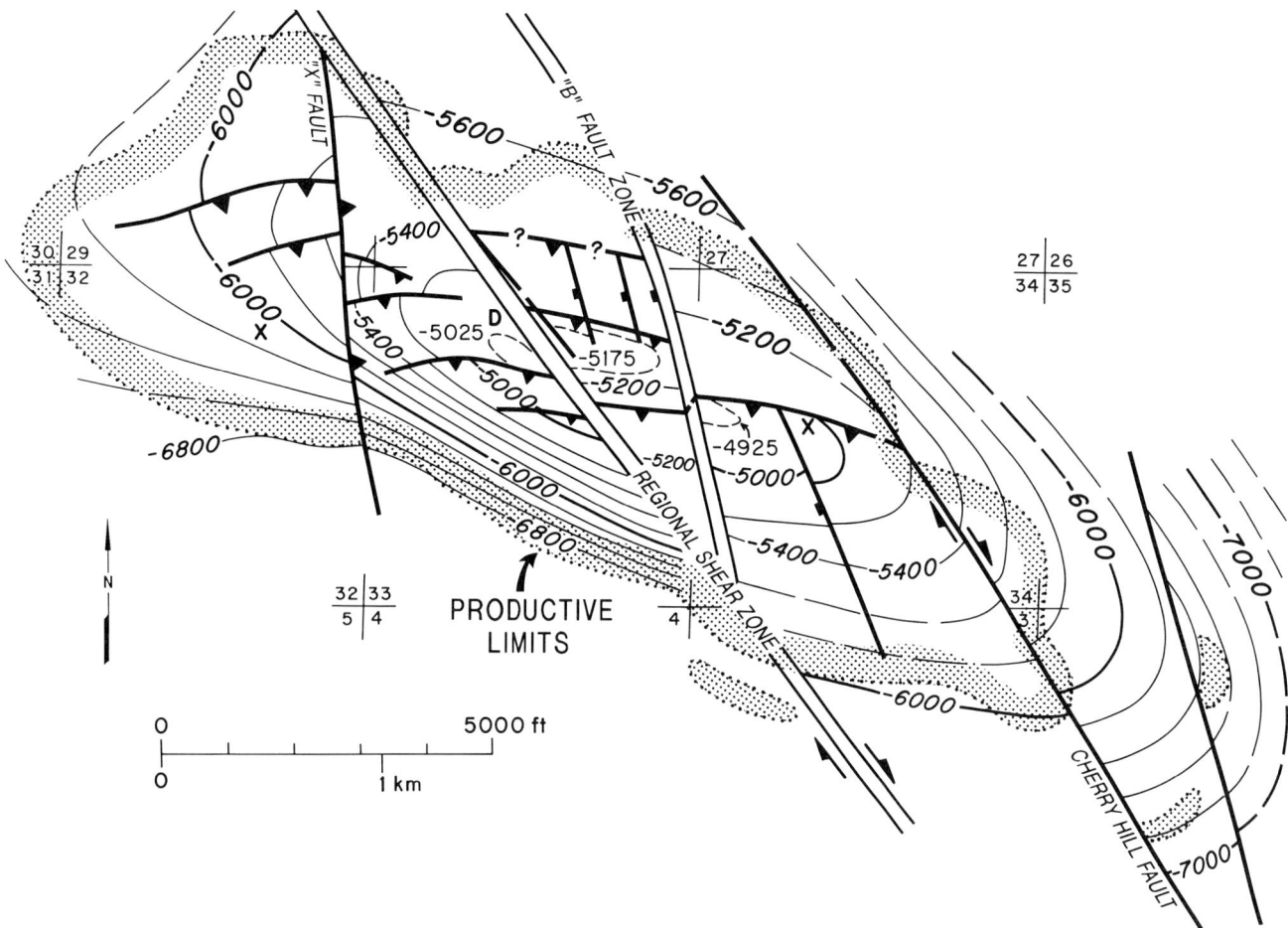

Figure 23. Subsurface structure map of the Dominguez oil field (compiled from Graves, 1954; McMurdie, 1973; California Division of Oil and Gas, 1974). Contours on top of Fifth Callendar zone (lower Pliocene). Contour interval: 200 ft (61 m); dotted supplementary contours show anticlinal crest. D = discovery well. X = wells reaching schist basement. Other symbols, etc., as in Figure 9.

with the Northeast Flank fault, bound a recently uplifted wedge in the southern part of the anticline (Figure 25, Section xx'). This wedge, carried northwest with the latest movement on the Cherry Hill fault, has overthrust the anticlinal crest along the Pickler fault (Taylor, 1973).

The Cherry Hill fault is nearly vertical down to 3500–5000 ft (1100–1500 m), but at greater depths may dip as gently as 60°. Vertical separation across the fault increases from 200 ft (60 m) in the Pleistocene section to 900 ft (275 m) in the lower Pliocene and up to 4000 ft (1220 m) in the upper Miocene (Taylor, 1973). Separation of the lower Pliocene beds decreases southward and is negligible at the southeastern limits of the field. Subsurface contours on the top of the Brown zone (mid-Pliocene) suggest a 3000-ft (900-m) right-lateral separation of the anticlinal crest across the Cherry Hill fault (Dudley, 1954), but, lacking detailed isopach control, semi-independent development of the fold on either side of the fault cannot be ruled out. Maximum vertical separation on the Northeast Flank fault is 400–500 ft (120–150 m). Deep wells show increased complexities in the upper Miocene section, but the data are too sparse to interpret. A detailed subsurface analysis that might elucidate the structural evolution of the Long Beach anticline is seriously hindered by the lack of data, and by unintentional, unmeasured well deviations on the majority of early wells.

Deep wells in the Long Beach oil field have reached basement at depths of 14,700 ft (4480 m) west of the Cherry Hill fault, 10,830 ft (3302 m) within the wedge, and 12,000 ft (3660 m) east of the Northeast Flank fault (Barrows, 1974; West et al., 1987). Basement rock west of the Cherry Hill fault (Shell Alamitos No. 48A well) is described as a quartz schist and assigned to the greenschist unit of the Catalina terrane; whereas east of that fault (Shell Alamitos No. 49A well), it is a saussurite gabbro (Sorensen, 1984, 1985). Basement rocks east of the fault, including greenschist and chlorite schist (Yeats, 1973), are of uncertain affinity.

On the north flank of the Long Beach anticline, the *Long Beach Airport oil field* was discovered in 1954 and will yield an estimated 12 MBO. The accumulation is in Delmontian sands in an east-plunging anticline closed against the northwest-trending Wardlow-Airport fault zone (Figure 24)

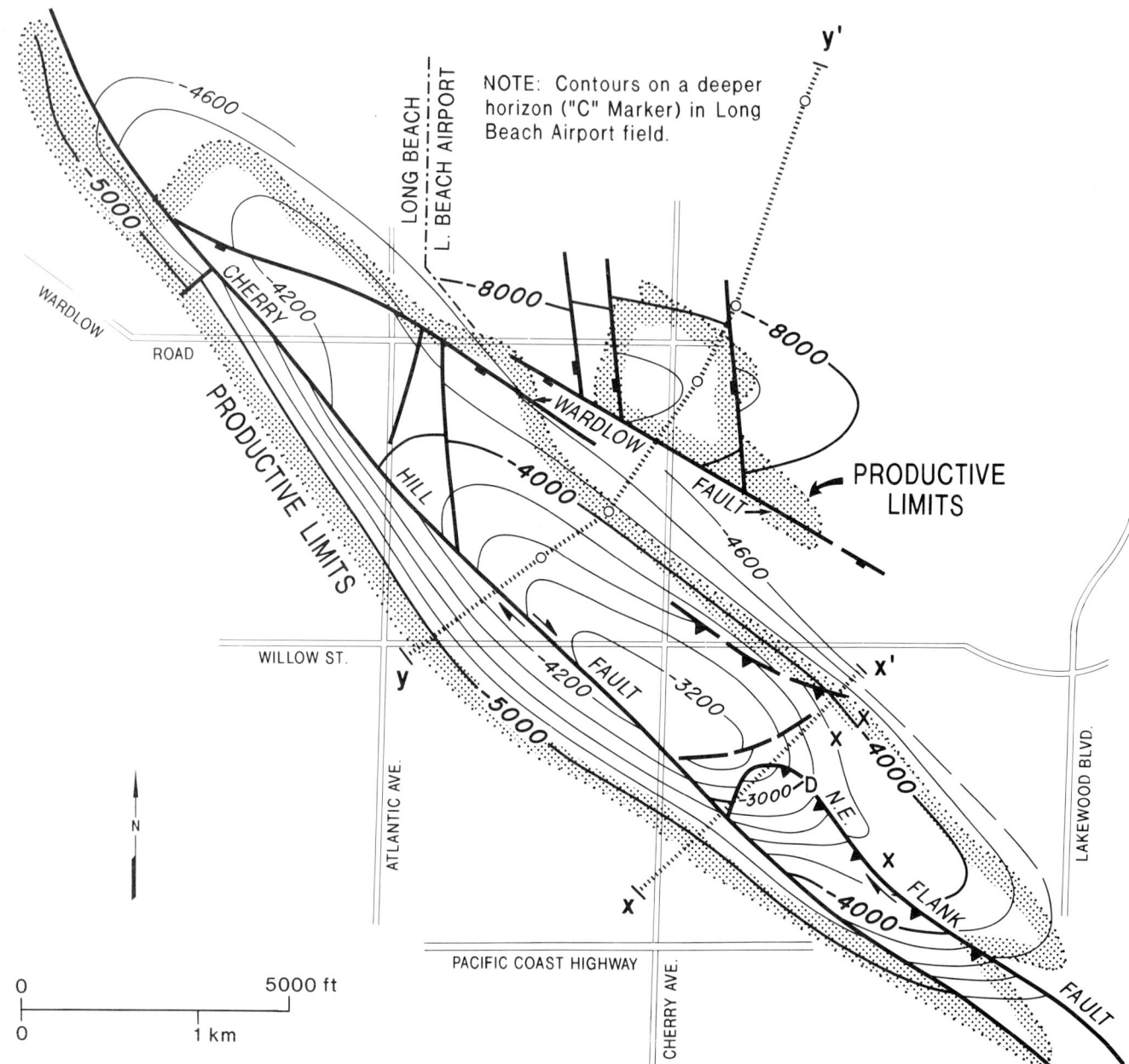

Figure 24. Subsurface structure map of the Long Beach and Long Beach Airport oil fields (compiled from Bauer et al., 1966; Taylor, 1973; M. E. Wright, 1987). Contours on top of Brown zone (lower Pliocene) at Long Beach and on "C" marker (upper Miocene) at Long Beach Airport. Contour interval: 200 ft (61 m). D = discovery well. X = wells reaching schist basement. Other symbols, etc., as on Figure 9. xx' and yy' are locations of cross sections in Figure 25.

(Harris, 1958). Section yy' (Figure 25) shows that the fold dies out within the Delmontian interval and that the fault zone dies out within the lowermost Repetto beds.

The *Seal Beach oil field* was discovered in 1924 and will ultimately yield 225 MBO from upper Miocene and lower Pliocene sands. The trap is an anticline that is traversed by and parallel to the NIFZ (Figure 26). In contrast to the Long Beach anticline, the fold axis at Seal Beach (and about 70% of the oil accumulation) is southwest of the major faulting. Subsurface mapping indicates that vertical separation of Delmontian beds across the zone varies between 100 and 700 ft (30-200 m). Bowes (1943), noting that folding increases downward as indicated by steeper dips at depth and by pronounced thickening of the Repetto section down the northeast flank of the anticline, inferred that "diastrophic activity began before the deposition of lower Pliocene sediments." A detailed subsurface interpretation is available only for the east extension of the north

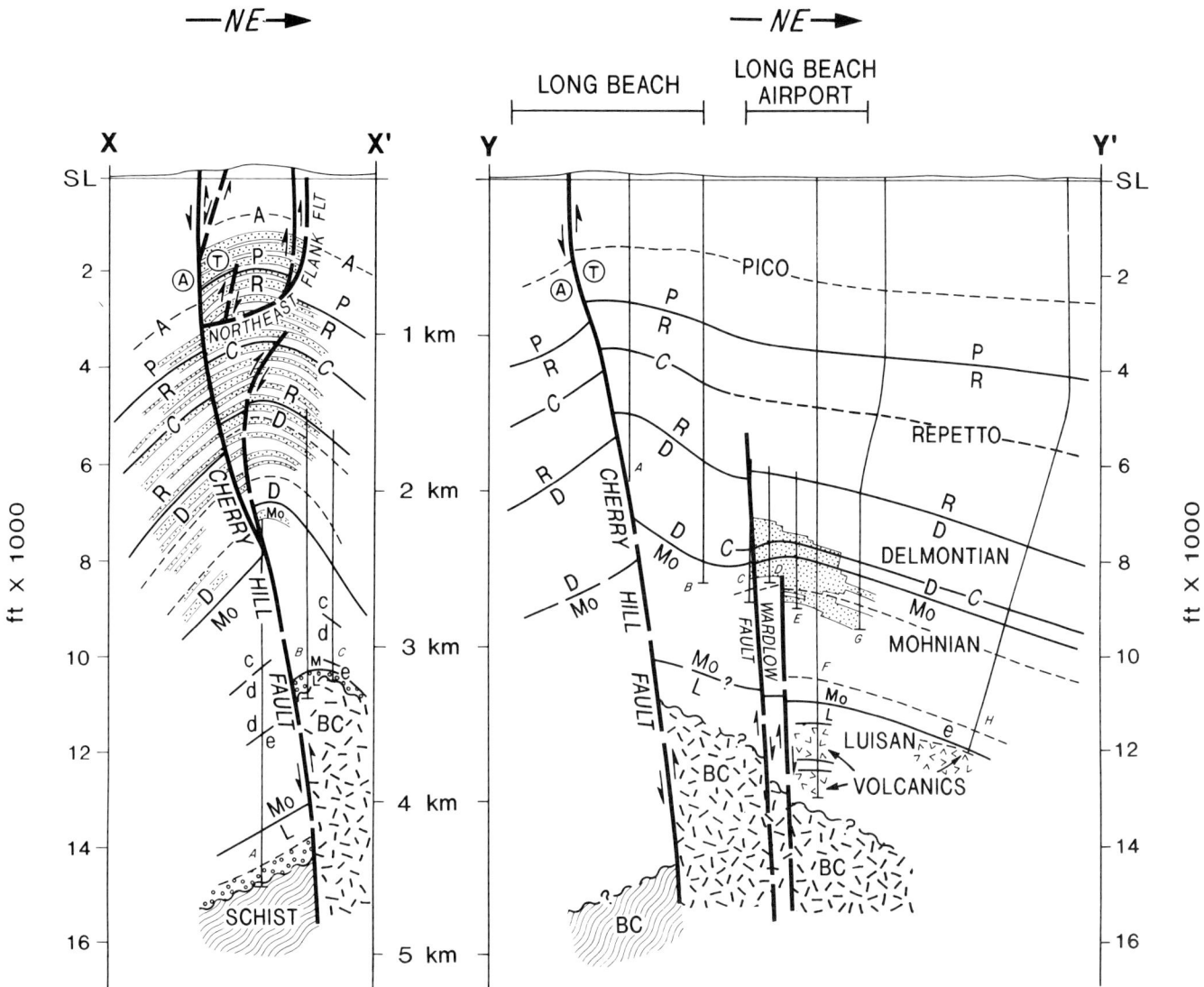

Figure 25. Cross sections of the Long Beach and Long Beach Airport oil fields. Locations of cross sections are shown in Figure 24. Solid line labelled "C" indicates contoured horizons of Figure 24. Control wells (lettered) are listed in Appendix 2. Other symbols, etc., as in Figure 8, plus c, d, and e are Wissler microfaunal zones. Section xx', across highest part of Long Beach oil field, compiled from Taylor (1973), M. E. Wright (1987), and West et al. (1987). Deep well control projected from 600–2800 ft (180–850 m) southeast of section. Productive intervals are shaded. Section yy', compiled from Harris (1958), shows relation of Long Beach Airport oil field to Long Beach anticline. Productive intervals (shaded) shown for Long Beach Airport field only. Approximate top of basement based on regional control.

block (Ingram, 1966); this shows some faults terminating at the base of the Repetto and others that extend higher in the section.

The Continental 62 Bixby well, west of the NIFZ in the Seal Beach field, reportedly was in Catalina Schist at a total depth of 12,162 ft (3708 m) (California Division of Oil and Gas, 1974; M. E. Wright, 1987), although West et al. (1987) show this well as bottoming in beds of Luisian–Relizian age.

Late Pleistocene structure at Seal Beach is clearly expressed by topography (Figure 26) (Reed and Hollister, 1936; Bowes, 1943; M. E. Wright, 1987). Landing Hill shows the southeast plunge of the fold, and Alamitos Heights shows its northwest plunge. Erosion by the antecedent San Gabriel River has erased the central portion of the uplifted Pleistocene surface. West of Alamitos Heights, a northeast-facing topographic scarp suggests the surface trace of a member of the NIFZ. Its principal trace, the Seal Beach

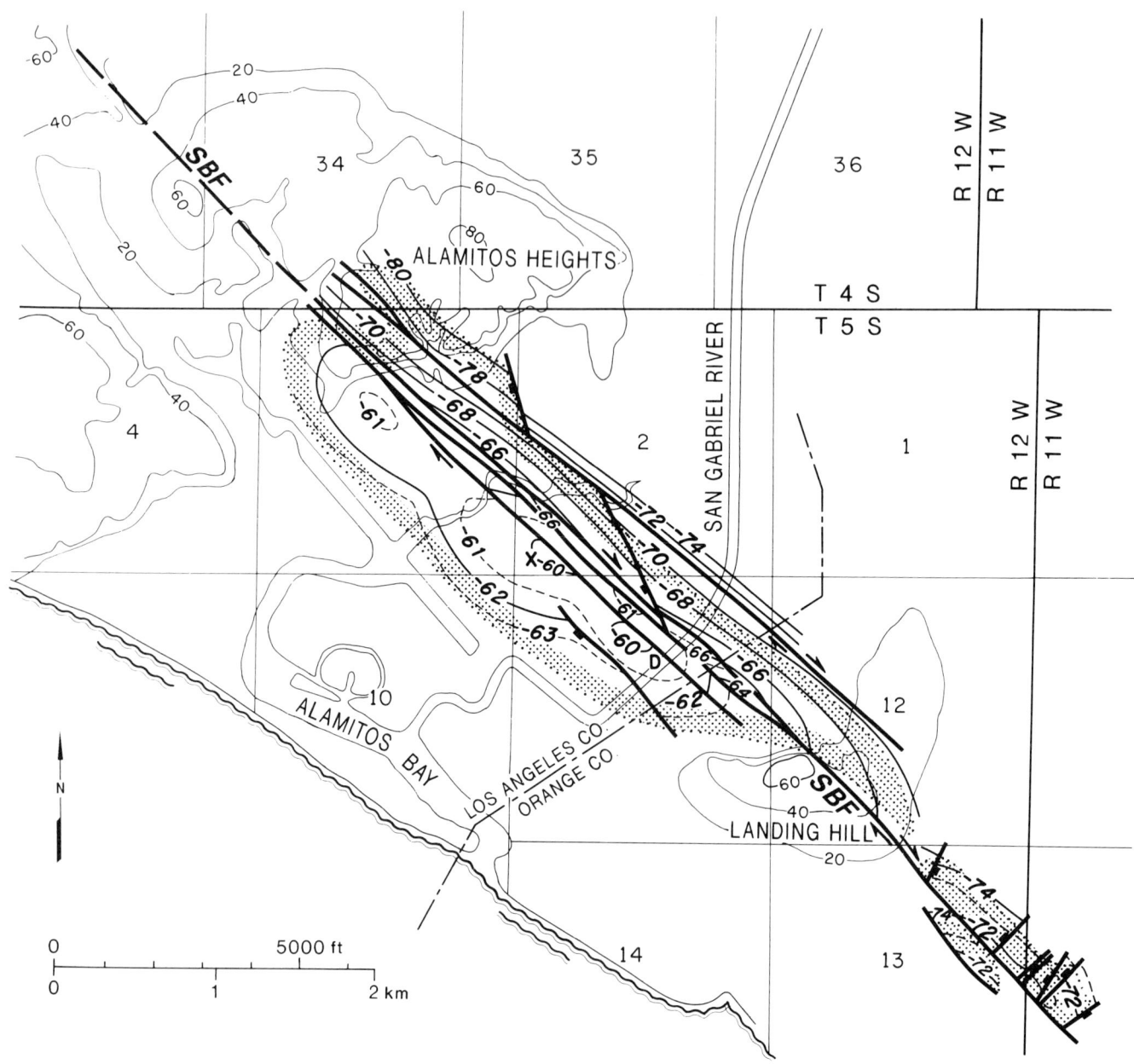

Figure 26. Map of topography and subsurface structure of the Seal Beach oil field and vicinity (compiled from Reed and Hollister, 1936; Bowes, 1943; California Division of Oil and Gas, 1974). Subsurface contours (heavy lines) on a Delmontian horizon (top of Lower McGrath zone). Contour interval: 200 ft (61 m) (solid) and 100 ft (30.5 m) (dashed); zeros are omitted. Topographic contour interval: 20 ft (6 m). SBF is Seal Beach fault; D = discovery well; X = well reaching? schist basement (Conoco Bixby No. 62 well). Limits of oil production are shaded. Other symbols, etc., as in Figure 9.

fault, has been mapped at the surface (Poland et al., 1956) across the western slope of Alamitos Heights and directly above the main subsurface trace (Figure 26), intersecting the crest of Landing Hill. Barrows (1974) stated that "on opposite sides of the Seal Beach fault, the northeastern part of Landing Hill has been dropped down 15 feet [4.6 m] relative to the southwestern part."

The *Huntington Beach oil field* was the first field discovered on the Newport-Inglewood trend, in 1920. It is the largest field on the trend, with an ultimate recovery estimated at 1119 MBO. Production is from sands ranging in age from middle Mohnian to middle Pliocene (lower Pico). Modes of structural entrapment differ between the three main areas of the field (Figure 27) (Hazenbush and Allen, 1958), and are

♦ a broad northeast-dipping arched monocline on the northeast side of the Newport-Inglewood fault

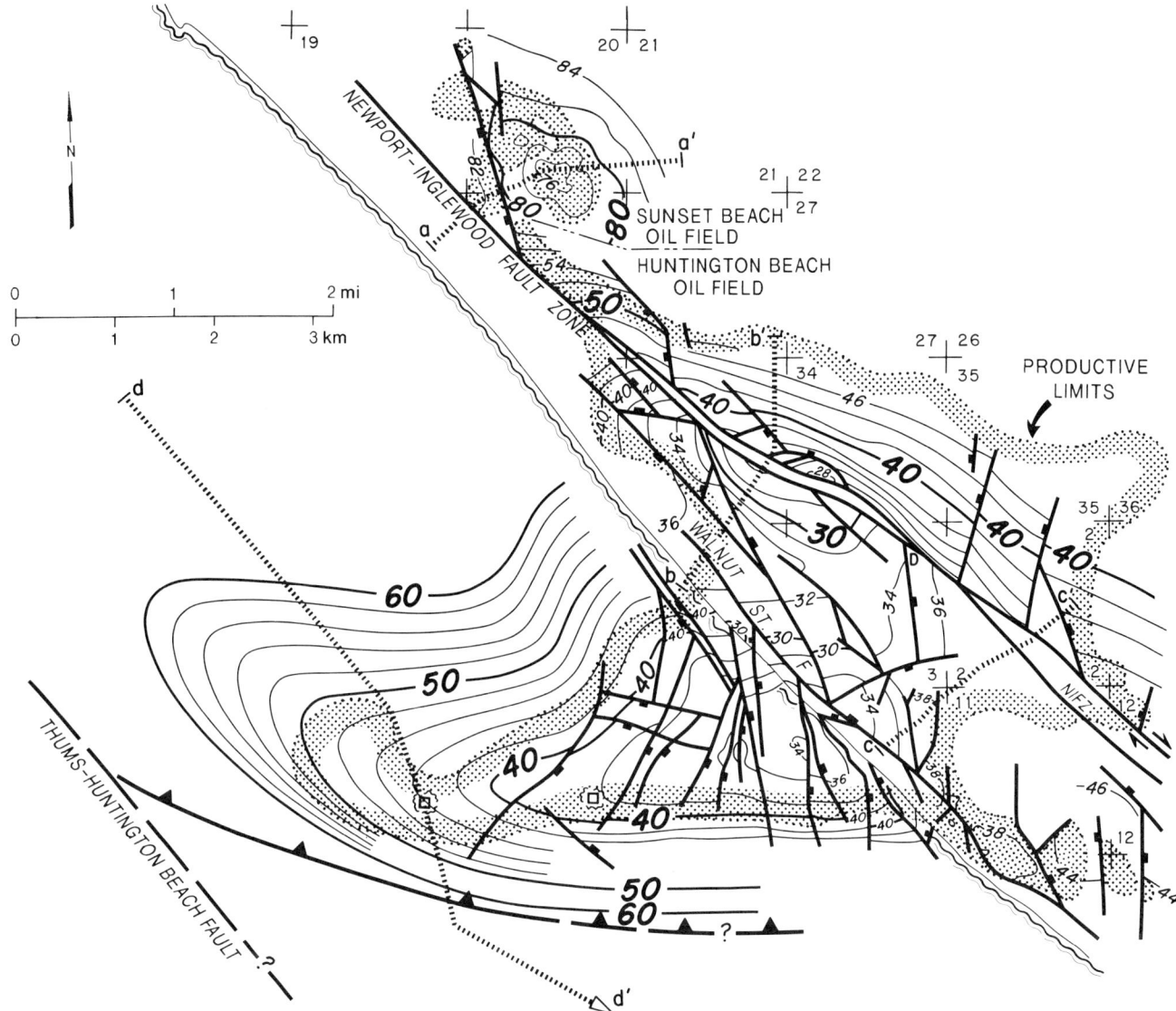

Figure 27. Subsurface structure map of the Huntington Beach and Sunset Beach oil fields (compiled from Allen and Hazenbush, 1957; Hazenbush and Allen, 1958; Frame, 1960; Bauer et al., 1966; Clarke and Henderson, 1987). This map is best viewed as a *form-line map*, because contoured horizon varies: Sunset Beach is upper Luisian [top of Lomita Lands (Main) zone; Huntington Beach Onshore is Delmontian (top of Jones sand); Huntington Beach Offshore is top of middle Mohnian (top of Lower Main zone). Contour interval: 200 ft (61 m). (—30— is contours in hundreds of feet.) Limits of oil production are shaded. aa', bb', cc', and dd' are cross sections in Figure 28. D is discovery well (Huntington Beach). Hollow squares are offshore oil platforms. Other symbols, etc., as in Figure 9.

- a complexly faulted central portion lying between the Newport-Inglewood fault and the zone of the Walnut Street fault, beneath the shoreline

- a relatively unfaulted offshore anticline that trends east-west and plunges west from the Walnut Street fault

Subsurface mapping (Hazenbush and Allen, 1958) indicates that in the southeastern part of the field, post-Repetto movement on the Newport-Inglewood fault has shifted to a more gently dipping eastern strand (Figure 28, Section cc'). This Holocene fault merges with the pre-Pico fault at a depth of about 5000 ft (1500 m) to form a shallow wedge that opens to the southeast. North of the area of convergence, the Newport-Inglewood fault in the subsurface is a shatter zone several hundred feet wide. Vertical separation across the fault is about 1200 ft (365 m) in the southeastern part of the field, decreasing northward to a negligible offset in the central field. Farther

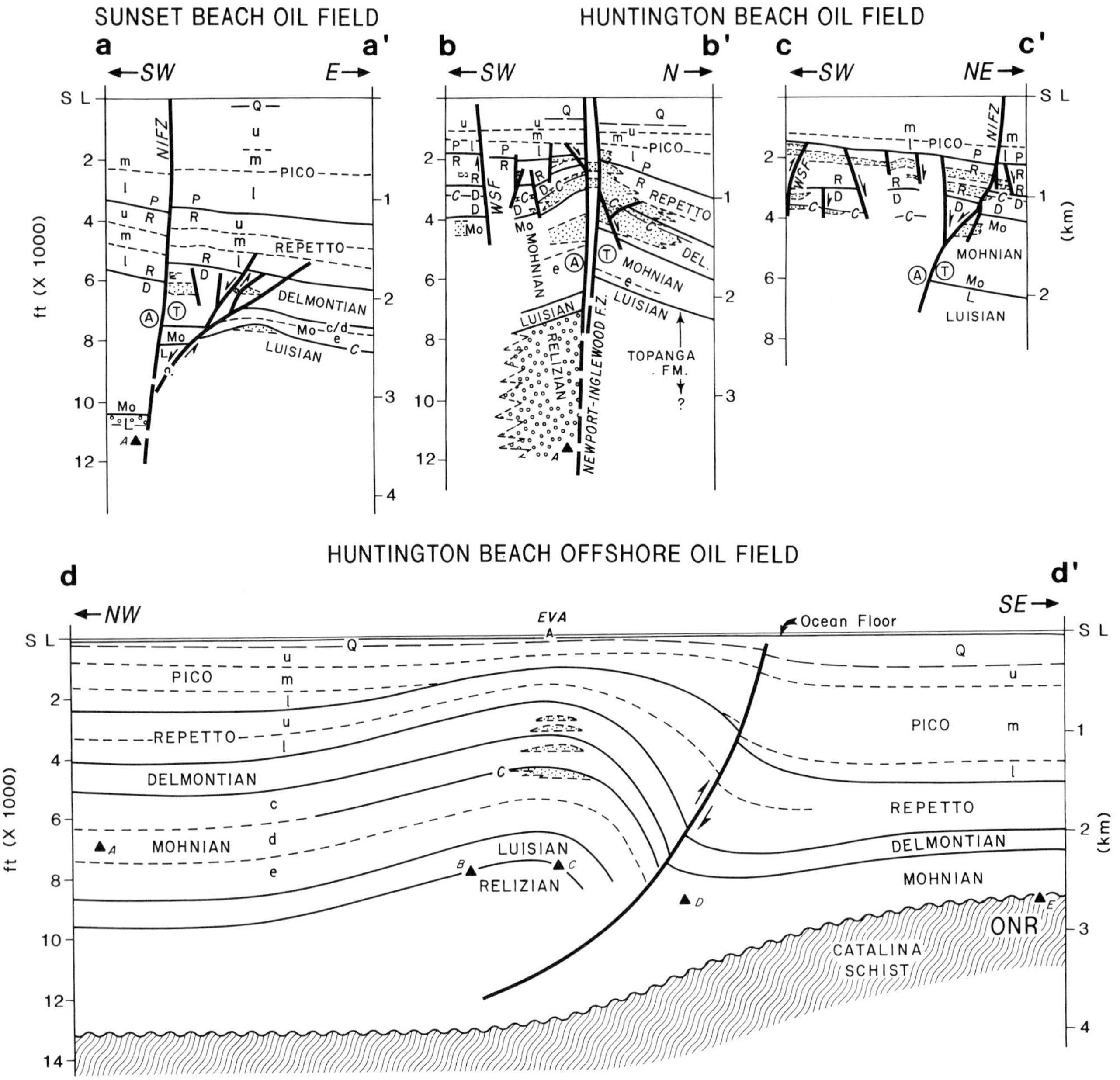

Figure 28. Cross sections of the Huntington Beach and Sunset Beach oil fields. Locations of sections are shown in Figure 27. Oil-producing zones are shaded. —C— shows contoured horizons of Figure 27. Other symbols, etc., as in Figures 8 and 25. Control wells (lettered) for Section dd' are listed in Appendix 2. For maximum clarity, well control has been omitted from Sections aa', bb', and cc', but is shown on the original California Division of Oil and Gas figures cited. aa': Sunset Beach oil field (compiled from Allen and Hazenbush, 1957; Yeats, 1974; West et al., 1987). aa' and cc': Huntington Beach oil field (compiled from Hazenbush and Allen, 1958; Henry, 1987; West et al., 1987). dd': Offshore Huntington Beach oil field (based on well data from Henry, 1987; mapping by Bauer et al., 1966).

to the north, the top of the Delmontian is separated at least 1000 ft (300 m) while upper Pliocene beds show very little offset (Figure 28, Section bb'). This cross section shows that most of the deformation onshore, other than the Quaternary Newport-Inglewood faulting, occurred before late Pliocene time. The major folding and faulting apparently took place during deposition of the lower Repetto beds, although the topography of the Huntington Beach area reflects mild Quaternary folding.

In the southern part of the field (Figure 27; Figure 28, Section cc'), complex faulting of the axial portion during Delmontian (and Repetto?) time strongly affected depositional patterns; correlations and structural interpretations in this area are highly uncertain. To the northwest within the central block of complex structure, the thickness of the Bolsa sands (Repetto) shows wide and commonly abrupt lateral variations, especially close to the Newport-Inglewood fault. Mid-Mohnian sands also are much thicker in this block than north of the fault. Harding (1973) cited these variations in stratigraphic thickness across the Newport-Inglewood fault as a criterion for wrench faulting, but they also provide evidence of a long history of deformation. Using such evidence, Hazenbush and Allen (1958) suggested that right-lateral movement on the Newport-Inglewood fault at Huntington Beach may total 6 mi (10 km) since middle Miocene time. Hill (1971) pointed out the questionable nature of this estimate, noting that "the lithofacies control for slip is considered to be equivocal."

The anomalous westerly trend of the Offshore Huntington Beach anticline evidently has resulted from compression of the thick wedge of middle and upper Miocene strata against the Offshore Newport ridge (Figure 28, Section dd'). The faulted south flank of the fold dips up to 65°, while dips on the north flank average 10°. Production is from upper Miocene sands. On the western third of the offshore structure, upper Mohnian (Division C) sands thin and shale out in an eastward direction up the anticlinal plunge, forming a stratigraphic trap that will yield an estimated 40 MBO (Noble, 1967). Stratigraphic thinning across the crest of the fold indicates that most of its growth occurred in early to middle Pliocene time, although the south flank of the fold has continued to develop during the Quaternary. Position of the THUMS-Huntington Beach fault west of Offshore Huntington Beach is not closely constrained by available data, but may coincide with Holocene faults mapped on the sea floor (Junger and Wagner, 1977).

Immediately northwest of the Huntington Beach field, the small *Sunset Beach oil field* provides additional evidence of early Pliocene deformation (Figure 28, Section aa'). Willis (1958) noted three different periods of structural development:

♦ late Mohnian folding

♦ earliest Pliocene faulting on north-striking normal faults that dip west and have maximum displacements exceeding 600 ft (180 m)

♦ Quaternary displacement on the Newport-Inglewood fault

The north-striking normal faults have been cited as an example of wrench-related extension (Harding, 1973), but they could also be slump structures in the zone where late Miocene strata thicken rapidly westward across the boundary of an older uplifted block.

Most of the *other oil fields* along the Newport-Inglewood trend (Figure 7) are similar in geologic structure to the fields described above. South of Huntington Beach, the Newport and West Newport oil fields include a complexly faulted zone southwest of the Newport-Inglewood fault and a north-dipping monocline northeast of it; West Newport's offshore production is from a west-trending anticline on the Offshore Newport ridge. The Potrero oil field is south of the Inglewood field and similar to its deeper structure: a northwest-trending anticline is offset in a right-lateral sense by a pair of en echelon faults (Johnson, 1961). The Rosecrans oil field, northwest of Dominguez, is unlike other fields along the trend. It is a broad anticline west of the Newport-Inglewood fault and grossly parallel to it. Harding (1973) has noted that two individual folds within the broad uplift diverge northwestward from the fault in typical wrench-fold style. The upper Miocene and lower Pliocene section at Rosecrans is cut by west-striking, south-dipping thrust faults (Foster, 1954) that are similar to those at Dominguez.

At its *northern end,* the NIFZ commonly is extended to include the Cheviot Hills and even the Beverly Hills structures (Soper, 1943; Hill, 1971; Yeats, 1973; Barrows, 1974). These structures are more closely related to the Santa Monica fault system, and, in this paper, the NIFZ is restricted to the zone south of Ballona Gap (Figure 14). Notwithstanding, a northwest-trending topographic lineation near the eastern edge of the Cheviot Hills is aligned with the Inglewood fault and could indicate its northward continuation (Barrows, 1974). Development drilling in the East Beverly Hills oil field has not shown its steep south flank to be cut by any northwest-striking fault that might correspond to the Inglewood fault (California Division of Oil and Gas, 1974). Beneath the field, however, a recent small earthquake (Hauksson, 1987) at a depth of 25,000 ft (7.5 km) yielded a focal mechanism showing right slip on a nodal plane striking N35°W, which is consistent with motion on the NIFZ. North of Ballona Gap, the influence of late Quaternary right-lateral movement on the NIFZ is expressed by the topographic boundary between the downwarped alluvial slope of Beverly Hills on the east and the uplifted, dissected terrain of Cheviot Hills and Westwood on the west (Figure 16C).

The interpretation in Figure 14 suggests that, within the sedimentary section, the Inglewood fault terminates against the south branch of the Santa Monica fault in the vicinity of the saddle that separates the East and West Beverly Hills oil fields. Northwestward movement of the block west of the Inglewood fault is being absorbed by the folding and thrusting seen in the Cheviot Hills oil field. An alternative interpretation (Lang and Dreessen, 1975) would have the Inglewood fault curve westward and become the Rancho fault on the south flank of the Cheviot Hills anticline. Continuing to the west, that fault would displace the Santa Monica fault near the Sawtelle oil field. The postulated trace of that fault would have to intersect one or more deep exploratory wells (Chevron Desilu No. 1 and Pauley Alladin Community No. 1) that were directionally drilled in the area between the Inglewood and Rancho faults; data from those wells show no evidence for the proposed fault connection. Interaction between the NIFZ and the Santa Monica fault system is discussed further in the section on Pasadenan deformation.

The extension of the NIFZ offshore to the southeast has long been debated. Three miles (5 km) beyond good subsurface control in the Newport field, the alignment of the NIFZ coincides with a steep submarine scarp having 1200–1500 ft (360–460 m) of vertical relief. Deep-penetration seismic reflection profiles and other geophysical data (cited in Barrows, 1974) have revealed a feature called the *South*

Coast Offshore fault (Figure 3), which follows that submarine scarp. Barrows compared the known features of this offshore fault with the NIFZ and concluded that it represents the continuation of the NIFZ some 45 mi (70 km) southeastward toward the vicinity of Carlsbad. A recent compilation of offshore geological data (Clarke et al., 1987) shows at least four fault segments along the length of the submarine scarp. The central segments are indicated as cutting strata of Quaternary age. A postulated connection with the Rose Canyon fault in San Diego (Emery, 1960) is not supported by more recent work (Junger, 1976; Vedder, 1987) that shows a right-stepping en echelon relationship between the north end of the Rose Canyon fault and the south end of the South Coast Offshore fault.

Continuing tectonic activity along the NIFZ is demonstrated by numerous historic earthquakes, of which the largest (Hauksson, 1987) were the 1920 Inglewood earthquake ($M_L = 4.9$), the 1933 Long Beach earthquake ($M_L = 6.3$) and its $M_L = 5.4$ aftershock, and the 1941 earthquakes at Dominguez (5.4 and 5.0). An excellent study of seismicity on the NIFZ from 1973 to 1985 (Hauksson, 1987) provided hypocenters for 64 events of $M_L = 2.5$ or greater, and focal mechanisms for 39 of these. Two to nine of these small, felt earthquakes occur each year along the NIFZ. Fewer than 20% of these focal mechanisms in the general area of the NIFZ are consistent with its commonly accepted style of right-lateral motion on northwest-striking faults. Events along and west of the surface NIFZ include nine north-trending right-lateral faults and five reverse faults dipping mostly south at 35° to 70°. Hauksson interprets the 39 focal mechanisms as documenting compression on the NIFZ from Dominguez north and extension on the segment from Long Beach south.

One-third of the events studied by Hauksson (1987) are located 2-3 mi (3-5 km) east of the line of surface deformation (see Figure 43) in a near-vertical zone at depths of 13,000-50,000 ft (4-15 km). This microseismicity appears to confirm the existence of a subsurface fault—called the Compton fault by McMurdie et al. (1973) and the Los Alamitos fault by Ziony et al. (1974)—parallel to the NIFZ at least from Seal Beach to Rosecrans and perhaps the entire length of the zone. It is herein suggested that the *Compton-Los Alamitos fault* (Figures 9, 10) is a major deep-seated element of the NIFZ and represents the eastern edge of the uplifted basement block of the western Los Angeles basin (Figure 4). This boundary previously had been inferred, on the basis of gravity data, to occur about 1.5 mi (2.4 km) northeast of the NIFZ (Yerkes et al., 1965). Some of the microseismicity in this zone yields focal mechanisms showing normal faults dipping to the east or southeast (Hauksson, 1987). More than half of the focal mechanisms along this line indicate right-lateral movement on near-vertical planes that strike NS to N15°W—suggestive of a zone of en echelon tension joints or gash fractures along the edge of the basement block. One such tensional event occurred beneath the northern end of the Pelican Hills fault, which might be the southern extension of the Compton-Los Alamitos fault.

Two deep exploratory wells drilled along the trend of the Compton-Los Alamitos fault encountered steep dips suggestive of faulting. Northeast of the Dominguez oil field (Figure 8, Section DD'), the Chevron Carlin Community No. 1 well logged northeast dips averaging 47° in Pliocene beds from 6700 ft (2050 m) to total depth of 15,978 ft (4871 m) (California Division of Oil and Gas, public file). Maximum dip of 58° was encountered at 11,000 ft (3350 m). The American Petrofina Central Corehole No. 1 well—the basin's deepest well—was drilled 2 mi (3.2 km) northwest of the Carlin well to a total depth of 21,215 ft (6466 m). Dipmeter data (California Division of Oil and Gas, public file) indicate northeast dips of 45° to 65° between 6000-11,000 ft (1800-3350 m), with faulting striking north-northwest at depths of 4500 ft (1370 m) and 11,150-11,700 ft (3400-3570 m). Below 12,000 ft (3660 m), dips are 10°-20° to the northeast. On cross section DD' (Figure 8), this fault or kink band (Suppe, 1985) has been projected downward into a near-vertical zone of microseismicity recorded from depths of 16,000-40,000 ft (5-12 km) (Hauksson, 1987).

A similar relationship is shown east of the Potrero oil field (Figure 8, Section CC'), in the deep American Petrofina Manchester C. H. No. 1 and 2 wells (Davis et al., 1989, their plate 1). The inferred position of the Compton-Los Alamitos fault west of that well would indicate that it converges into the main NIFZ just south of the Inglewood oil field (Figures 9, 10). As indicated on these maps and on the cross sections of Figure 8, the Compton-Los Alamitos fault is interpreted to be a right-oblique fault along which the steep northeast flank of the Newport-Inglewood uplift has overthrust the axial portion of the central trough.

Pre-Neogene events are believed to have localized the NIFZ as a significant crustal feature. It has been suggested to be the boundary between the western (Catalina Schist) and eastern (Mesozoic plutonic and metamorphic) bedrock complexes (Woodford, 1925; Woodford et al., 1954), a granite-Franciscan contact (Reed and Hollister, 1936), and the Mesozoic subduction zone along which oceanic basement underthrust the North American plate (Hill, 1971). The original nature of this contact almost certainly has been telescoped and obscured during Paleogene terrane accretion (Howell et al., 1987), when the Catalina terrane was emplaced. Evidence now available cannot resolve whether the eastern limit of the Catalina terrane follows the main NIFZ or extends to the buried Compton-Los Alamitos fault. Well data cited above confirm that Catalina greenschists extend to the shallow NIFZ at Long Beach (Sorensen, 1984) and perhaps at Dominguez and Newport Offshore. However, basement rocks found within or east of the NIFZ in the Inglewood, Dominguez, and Long Beach oil fields consist of metamorphic rock types found both in the Catalina terrane and in the southern Transverse Ranges.

Along the southern NIFZ and the South Coast Offshore fault, major *early to mid-Miocene uplift* of the block west of these faults is recorded by the San Onofre Breccia. This distinctive detrital formation, Relizian in age (Stuart, 1979b), crops out along the coast from Laguna Beach south to Oceanside and inland some 10-15 mi (16-24 km) along both flanks of the Capistrano syncline and in the San Joaquin Hills (Figure 5). East of the Cristianitos fault, the San Onofre Breccia is entirely nonmarine, but to the west it includes shallow-marine interbeds and also forms occasional lenses within the Monterey Formation (Woodford et al., 1954; Stuart, 1979b). Later in the Neogene, the emergent block southwest of the South Coast Offshore fault foundered to form the Oceanside basin.

A great thickness of the original overburden on the Catalina terrane must have been removed prior to its accretion either tectonically or by erosion at its present location. Continuing erosion of the offshore Catalina terrane is documented by variations in clast lithology of the San Onofre Breccia in the Laguna Beach area (Stuart, 1979b). Basal, graded sandstones contain a high proportion of granitic components. If derived from the west, this suggests that a veneer of Paleogene and/or Cretaceous rocks or perhaps the western lip of the Vincent thrust (J. C. Crowell, written communication, 1989) once overlay the Catalina terrane. Upwards in the section, pebbles are almost entirely greenschist for an interval of several hundred meters before the first significant blueschist detritus occurs. [Blueschist clasts also are common in deep-water beds of the upper Los Trancos Member, of Relizian age, in the trough east of the Pelican Hills fault (Vedder, 1970).] This sequence suggests that a sheath of greenschist, perhaps tectonically emplaced (Platt, 1975), covered the core of blueschist.

The occurrence of schist basement along the NIFZ and west of it strongly suggests that the uplifted basement block that shed the San Onofre Breccia extended northwest to the Inglewood-Playa del Rey area and that significant middle Miocene vertical separation extended the entire length of the NIFZ. The apron of San Onofre Breccia has been found east of the NIFZ in the subsurface of the Huntington Beach field (Yeats, 1973) and may well extend to the northern end of the NIFZ, deeply buried beneath younger beds.

Evidence of localized *mid-Miocene deformation* on the NIFZ includes the significant variations in thickness of the Luisian interval across the uplifted basement wedge in the Long Beach oil field (although these variations may be due to subsequent lateral displacement on the bounding faults) and Luisian intrusive volcanic rocks in the Seal Beach (Truex, 1972) and Inglewood oil fields. Volcanic rocks have also been described along the NIFZ in the Dominguez oil field (Graves, 1954), but, like the Relizian volcanic rocks at Inglewood, these are extrusive rocks and are thought to be part of the extensive middle Miocene volcanic sequence and not localized on the NIFZ.

Relizian sedimentary and volcanic rocks in the Inglewood field (Castle and Yerkes, 1976) include conglomerates but are a Topanga facies; schist detritus has not been described. Elsewhere along the NIFZ, drilling has not found strata older than Luisian. Schist breccias interbedded with Luisian beds are reported in wells west of the NIFZ in the Seal Beach (Truex, 1972) and Sunset Beach (Yeats, 1974) fields. Schist-bearing basal conglomerates underlying deep-water Mohnian and Luisian strata are widespread on the western shelf and at Wilmington and are considered to be younger than the San Onofre Breccia (R. S. Yeats, written communication, 1989). On the Palos Verdes Peninsula, early Mohnian sandstones and breccias within the Monterey Formation are rich in Catalina Schist detritus and contain sedimentary structures that indicate deposition in a south-directed submarine fan (Russell, 1987). This evidence implies that the western shelf persisted as a submarine high during middle Miocene regional submergence.

Flanking this middle Miocene submarine high were two local depocenters (Figure 5). On its southwest margin, the Monterey Formation accumulated to a thickness of about 4000 ft (1200 m) in the Palos Verdes Hills (Yerkes et al., 1965). On its southeast plunge, a deep graben formed beneath offshore Wilmington and Huntington Beach. Bounded on the southwest by the THUMS-Huntington Beach fault, this local depocenter contains more than 5000 ft (1500 m) of Relizian and Luisian black shales of the Monterey facies (Truex, 1972; Henry, 1987). Very sparse data (Figure 5) (see also Redin, this volume, his Figures 7, 8) suggest that the block to the west, between the Palos Verdes fault and the THUMS-Huntington Beach fault, collected only a limited thickness of sediments during middle and early late Miocene time. This interval also thins eastward from offshore Wilmington, where it is 1300-2000 ft (400-600 m) thick, to about 300 ft (90 m) in the Long Beach and Seal Beach oil fields, west of the NIFZ. North of the Long Beach field, however, there is no evidence for such up-to-the-east displacement on the NIFZ during middle Miocene time.

The southern NIFZ continued to act as a depositional hingeline into *late Miocene* time. Mohnian sands thicken westward across the zone from 200-400 ft (60-120 m) thick in the Sunset Beach and Huntington Beach oil fields to 1200-1600 ft (370-490 m) thick in the deep, offshore trough. Localized deformation during this time is recorded by thinning of sandstones over the crests of the Seal Beach, Long Beach, and Dominguez oil fields and over the southeast plunge of the Inglewood anticline. Except at Inglewood (and perhaps Rosecrans), sand-isolith maps do not show local structural growth during Delmontian time. Indeed, from Long Beach to Sunset Beach a belt of thicker Delmontian rocks follows the NIFZ.

Evidence for *early Pliocene deformation* along the NIFZ has been cited for the six representative oil fields discussed above. Detailed subsurface studies document thinning across the Inglewood anticline during early, middle, and late Repetto time. Regional sand-isolith maps show structural growth at Dominguez during early Repetto, at Long Beach during middle Repetto, and at Huntington Beach during early and middle Repetto (parallel to the Huntington Beach Offshore anticline) and late Repetto (parallel to the NIFZ). These conclusions, based on the existing regional isolith mapping, would undoubtedly be refined and modified by detailed stratigraphic studies in the latter four oil fields.

Quaternary deformation at Landing Hill and Alamitos Heights, Signal Hill, Dominguez, and Inglewood is documented by prominent hills (see Figure 43) formed in the soft, easily eroded Pleistocene beds. Uplift at Inglewood (Baldwin Hills) during the past 30,000 years has averaged about 1.65 ft (0.5 m) per hundred years (Bandy and Marincovich, 1973; Wright, 1987d). Topographic scarps reflect faulting at Inglewood, Potrero, Signal Hill (Long Beach), Seal Beach, and Sunset Beach (Barrows, 1974). Faults cutting late Pleistocene beds are visible in outcrop at Inglewood, Long Beach, Huntington Beach, Dominguez (Barrows, 1974), Seal Beach, and West Newport (Guptill and

Heath, 1981). In the latter area, the strikes of individual fault breaks are about 10°–15° west of north, parallel to many of the planes defined by focal mechanisms.

Several *alternative structural concepts* have been applied to the NIFZ in recent years. Barrows (1974) provided a thorough review of prior investigations and hypotheses and of the earthquake history of the zone. Yeats (1973; see also Platt and Stuart, 1974; Yeats, 1974) carefully reviewed the pattern of subsurface faulting within and adjacent to the NIFZ and noted a large number of faults that cut no formations younger than early Pliocene. Harding (1973) examined the subsurface structure of the NIFZ "in the context of basic wrench tectonics." He adhered to the traditional view that the individual anticlines and individual faults within the NIFZ are en echelon and that they essentially formed by the action of a deep-seated right-lateral fault on the overlying sedimentary blanket. While noting that variations in the classic wrench pattern may "result from the presence of buried slices, wedges, or blocks whose shapes cause them to be forced upward during the lateral movements," Harding concluded that the structures within the zone "may be taken as a unit and related dynamically to one type of deformation—wrenching."

There are dissenting views to this interpretation. Yeats (1973) found it "satisfactory for the late Pliocene and Quaternary history of the Newport-Inglewood zone, but too simple for the late Miocene and early Pliocene. It does not account for the fact that most of the more diversely oriented normal and reverse faults (except for Inglewood oil field) became inactive during the Pliocene whereas the en echelon right-lateral slip faults of the Newport-Inglewood zone continued to be active through the Pleistocene." According to Yeats (1973), "the Miocene and early Pliocene fault pattern is explained by right-lateral simple shear of the entire western Los Angeles Basin rather than shear limited to a single fault zone."

Nardin and Henyey (1978) provided a thoughtful analysis of en echelon folding as applied to the borderland and western Los Angeles basin. Citing Garfunkel (1966), they noted that "simple shear along wrench faults cannot produce en echelon drag folds because the amount of extension required parallel with the fold axes is unrealistically large" and that "a compressional component normal to wrench fault, or in other words, convergence, is necessary to produce en echelon folding." They concluded that "en echelon folds are not a direct consequence of wrench faulting but are related genetically to the same stresses which led to the formation of the fault."

A summary of the NIFZ might begin with the simple conclusion of Reed and Hollister (1936): "in the blanket of Upper Miocene and later strata the Newport-Inglewood structure is a domed, slightly faulted anticline; at a depth of several thousand feet it may be a granite-Franciscan contact that has been a locus of movement since the Middle Miocene and perhaps much longer. The more recent disturbances may have involved a slight shearing tendency which puckered the strata into small domes and fractured them with minor en echelon faults."

Indeed, the NIFZ reflects a major basement boundary that has served as the locus of recurrent deformation in several varied tectonic regimes. Each of the Neogene episodes has probably involved regional right-lateral simple shear, but along the NIFZ itself, total right-lateral slip since the middle Miocene has not exceeded 3 km (Yeats, 1973), or about 1–2 mi. Evidence of this from the subsurface is compatible with the estimates of 0.5 mm/year during the past 5 m.y. (Guptill and Heath, 1981) and 0.4–0.8 mm/year (Bird and Rosenstock, 1984).

Classic wrench-fault deformation, however, is not the primary cause of most of the anticlinal features along the NIFZ. In the preceding discussion we have seen that many of these structures do not conform to a pattern of en echelon folding, but are related to local basement geometry and perhaps to a wide zone of pervasive shear within the basement. Along the southern NIFZ, the Long Beach, Seal Beach, and Huntington Beach (onshore) structures are block-edged force folds (Harding and Tuminas, 1988) forced along the middle to late Miocene block boundary. Offshore Huntington Beach has been constricted against the Offshore Newport ridge. Dominguez is a part of the El Segundo–Lawndale–Alondra fold trend, complicated by offset on the NIFZ. Inglewood (and perhaps Potrero) formed in concert with uplift of the Las Cienegas block that buckled the sedimentary wedge against shallow basement of the western shelf (Wright, 1987d).

Western Shelf

The western shelf (Figure 7) extends from the NIFZ to the Palos Verdes uplift and south to San Pedro Bay. It is floored by Catalina Schist (Sorensen, 1984), which has been gently folded into three en echelon anticlines (Figures 9, 10) that trend about N60°W and plunge southeast into the basin. Deep-water clastic sediments gradually transgressed westward onto the shelf from Luisian to Pleistocene time, and unpublished isolith mapping suggests that these folds grew almost continuously from late middle Miocene through middle Repetto time.

The *Venice Beach and Playa Del Rey oil fields* (Figure 29) are on the northernmost of these three folds. Playa del Rey, discovered in 1929, will yield an ultimate 63 MBO. One-third of the oil is found in thin middle Repetto sands on the west flank of the field. The remainder is from lenses of basal schist conglomerate and sandstone that pinch out toward the crest of the fold. The irregular basement surface suggests that it was modified by subaerial erosion, accounting for the variable distribution and thickness of the productive basal unit (Metzner, 1943). The Venice Beach field, discovered in 1964, will yield an estimated 4 MBO from basal schist sandstone and conglomerate. Oil is anticlinally trapped in the offshore pool, and onshore it is trapped by a combination of pinch-out and faulting down the southeast plunge of the fold.

Four oil fields have been found on the *Hyperion-Alondra trend*, the central of the western shelf's three folds. Hyperion, on the northwest, will yield less than 1 MBO from schist conglomerate, fractured schist, and basal Mohnian nodular shale. At El Segundo, these same three zones will yield about 15 MBO. The conglomerate pinches out around the top of the irregular basement high; schist production is widespread. Pliocene sands in a shallow dome on the northeast flank of the basement fold have yielded 23 bcf of gas. The Lawndale oil field will yield 4 MBO, mostly from schist conglomerate but with minor production from

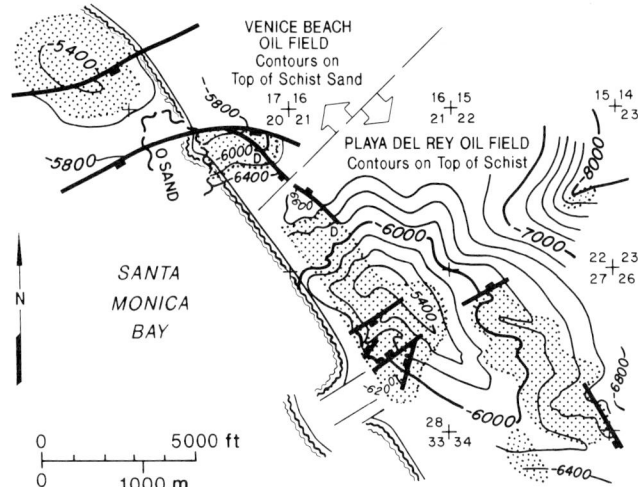

Figure 29. Subsurface structure map of the Playa del Rey and Venice Beach oil fields (compiled from California Division of Oil and Gas, 1974). Contour interval: 200 ft (61 m). Limits of oil productive areas are shaded. Other symbols, etc., as in Figure 9.

thin Delmontian and lower Repetto sands. The Alondra field has produced 2 MBO from schist conglomerate. These four fields were discovered between 1928 and 1946.

On the western shelf's southern fold, the *Torrance oil field* was discovered in 1922 and was extended offshore in 1956. An estimated 226 MBO will ultimately be recovered from Mohnian, Delmontian, and lower to middle Repetto sands that thin gradually westward up the 8-mi (13-km) plunge of this fold. Total net sand thickness averages 210 ft (64 m). Though minor cross-faults and structural saddles are present, much of the entrapment apparently is due to updip shale-out of the sands. Schist conglomerate is present at Torrance but is unproductive.

Development drilling down the southeast plunge of Torrance is credited with the discovery of the *Wilmington oil field* in 1932, though the existence of a new major field was not immediately recognized. The plunge reverses to the east, rising about 900 ft (275 m) to a structural high beneath Long Beach Harbor (Figure 30). A seismic survey in 1936 (Mayuga, 1970) defined this high and led to the drilling of the Mobil Terminal No. 1 well, which triggered a major drilling boom. Offshore development in the 1950s and 1960s extended the field to its full length (including *Belmont Offshore*) of nearly 13 mi (21 km). With an estimated ultimate recovery (including Belmont) of 2.9 billion barrels of oil (GBO) and 1.2 tcf of gas, Wilmington will yield one-third of the total petroleum in the Los Angeles basin. It is the third largest oil field in the United States.

Sands in each interval from middle Mohnian to upper Repetto thicken eastward from the Torrance field (Figure 31, Section ll'; Henry, 1987). All of these intervals, in addition to fractured schist basement, schist conglomerate, and black shales of late Luisian and early Mohnian age (Truex, 1972), are productive at Wilmington, where per-acre productivity is more than six times that at Torrance. Total net sand thickness averages 1570 ft (479 m), and the aggregate of maximum sand thicknesses may be as much as 2800 ft (850 m) (Mayuga, 1970). Average porosities of 30-40% in the Delmontian and Repetto zones are an important factor in the field's high productivity and in the subsidence problems that plagued its early history, until mitigated by a massive waterflood program.

The Wilmington anticline is a broad, doubly plunging asymmetric fold (Figures 30, 31). Dips on its north flank rarely exceed 20°, while south-flank dips commonly reach 60° to 70° (Clarke, 1987). The axial plane migrates slightly to the south in the upper Pliocene and upper Miocene intervals (Truex and Hunter, 1973; Truex, 1974), apparently owing to southward thinning of these intervals. In map view the anticline is slightly sigmoidal: at its northwestern and southeastern ends the axis trends N50°W to N60°W, while in the central portion the axial trend is N75°W. The high-angle THUMS-Huntington Beach fault, which parallels the south flank of the anticline, was revealed by a seismic survey in 1969 (Truex and Hunter, 1973).

The Wilmington anticline is broken by a series of transverse normal faults (Mayuga, 1970) and by oblique normal faults on its eastern and western plunges. Dips on the fault planes range from 45° to 70°. (Section aa' of Figure 31 intersects these faults at acute angles and thus shows deceptively low apparent fault dips). Many of the faults terminate towards the flanks of the structure, though a few seem to extend across its entire width. The five largest faults, with vertical separations greater than 200 ft (60 m), offset the basement surface (Figure 31, Section ll') (Clarke, 1987). The Long Beach Unit fault, with vertical separation of 400–600 ft (120–180 m), is the largest of these faults and the only one to displace upper Pliocene (Pico) horizons. A component of right-lateral movement has been suggested for this and perhaps other faults (Truex, 1974; Clarke, 1987), but the evidence is not compelling; offset isopachs on the steep south flank may be the result of unequal flexure on opposite sides of the fault, as isopachs on the north flank and in deeper intervals are not offset.

Several alternative explanations have been advanced for the genesis of these transverse and oblique faults. They have been described as "formed primarily by vertical components of tensional forces acting along the axis of the anticline" (Mayuga, 1970) and similarly as "related to extension of sedimentary cover rocks during later doming of the Catalina Schist basement at the core of the structure" (Platt and Stuart, 1974). Figure 30 shows that these transverse faults are clustered in three zones along the anticlinal axis. Two of these zones coincide with reversals in plunge at basement depth (Figure 31, Section ll'); basement control for the easternmost cluster, in the Belmont Offshore field, is lacking. Yeats (1973, 1974), however, noted that "the doming trends northwest and the faults trend north, oblique to the axis of the Wilmington structure. Minimum principal stresses related to arching should be normal to the axis of doming, not oblique to it, and faults resulting from such stresses should be parallel with the axis of doming." He suggested that "anisotropy in the Catalina schist controls fault trends in the overlying sedimentary rocks" and "fault displacement [is] controlled by the right-lateral simple-shear model."

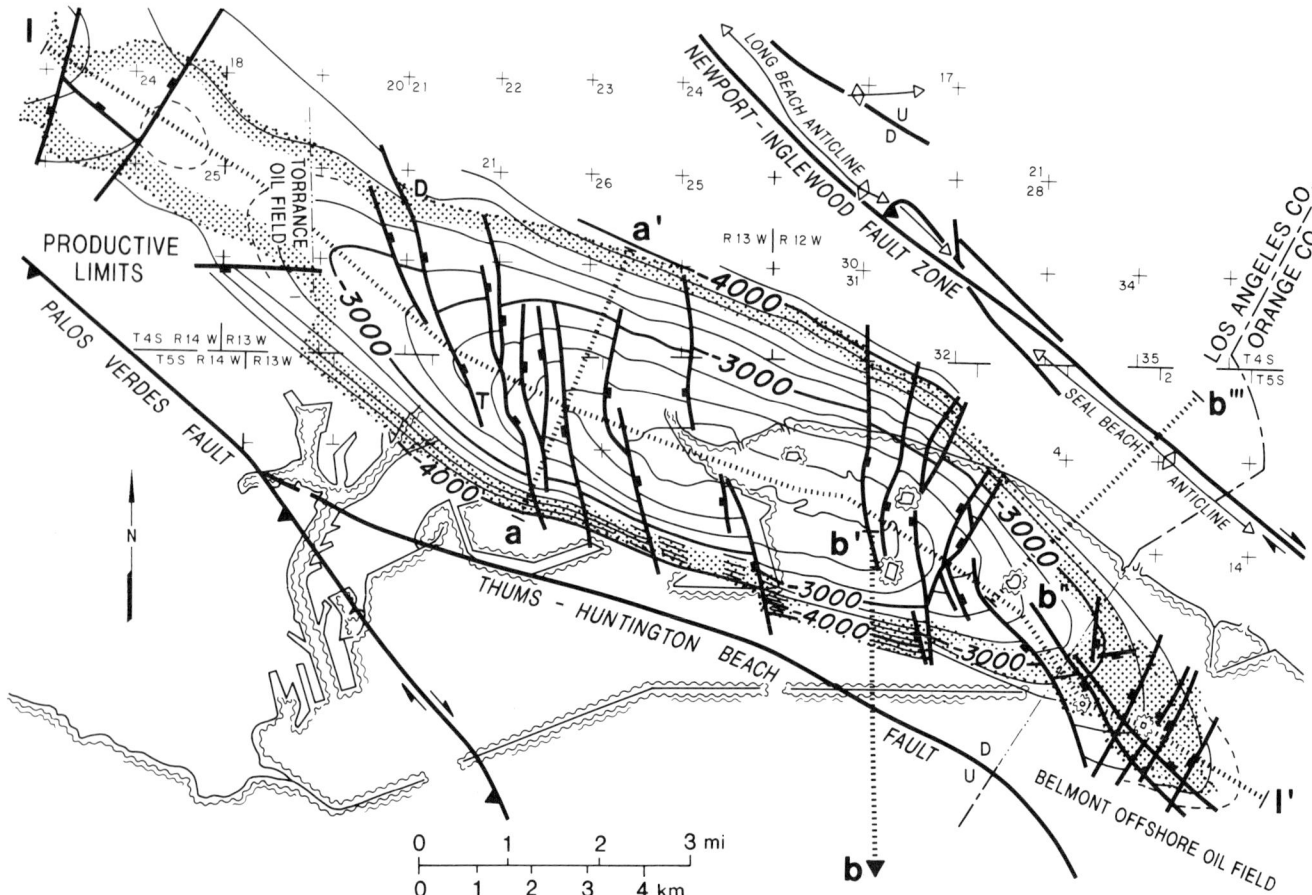

Figure 30. Subsurface structure map of the Wilmington anticline (compiled from Bauer et al., 1966; California Division of Oil and Gas, 1974; Olsen, 1975; Clarke and Henderson, 1987). Contours are on a lower Pliocene horizon (top of Ranger zone). Contour interval: 200 ft (61 m) (supplementary 100-ft contours are dashed). aa', bb'b"b''' and ll' are cross sections of Figure 31. Outline of oil-productive area is shaded. D is discovery well; T is Mobil 1 Terminal well. Other symbols, etc., as in Figure 9. For clarity, details of Inner Harbor are omitted from base map. The southeast and northwest ends of the Wilmington anticline include the Belmont Offshore oil field and part of the Torrance oil field, respectively.

A third hypothesis (Clarke, 1987) relates the Torrance-Wilmington anticline and associated faulting to "simple right-lateral shear that was developed between the Newport-Inglewood Structural Trend and the Palos Verdes Fault and/or the THUMS-Huntington Beach Fault." Convergence of the latter fault with the NIFZ has "heightened and exaggerated" the Wilmington anticline. Certainly the THUMS-Huntington Beach fault has played an even more significant role in the increased folding at Wilmington as compared to Torrance. The downdropped block northeast of that fault accumulated a thick sequence of sedimentary rocks during middle and early late Miocene time, and these subsequently were compressed to form the steep south flank of the Wilmington anticline (Figure 31, Section b"b''').

The structural evolution of the Wilmington anticline has been documented more completely than for any other structure in the Los Angeles basin. Early reports on the field (Winterburn, 1943) noted that "folding and faulting occurred progressively during the deposition of the Repetto beds [but] are not reflected in the younger sediments above the (basal Pico) unconformity." Truex (1972, 1974) provided isopach maps for the Long Beach unit (East Wilmington) that show ten intervals from the upper Luisian-lower Mohnian "black shale member" to the Upper Ranger zone at the top of the lower Repetto. These maps and accompanying discussion permit the brief analysis that follows.

Luisian and lower Mohnian black shales onlap basement toward the northwest up the east plunge of Wilmington; schist and schist conglomerate on the crest of the fold were exposed during part of early Mohnian time. Depositional relationships suggest the burial of a preexisting basement high, perhaps with minor structural growth in latest middle Miocene time. From mid-Mohnian through early Delmontian time (237, Ford, Union Pacific, and Lower and Upper

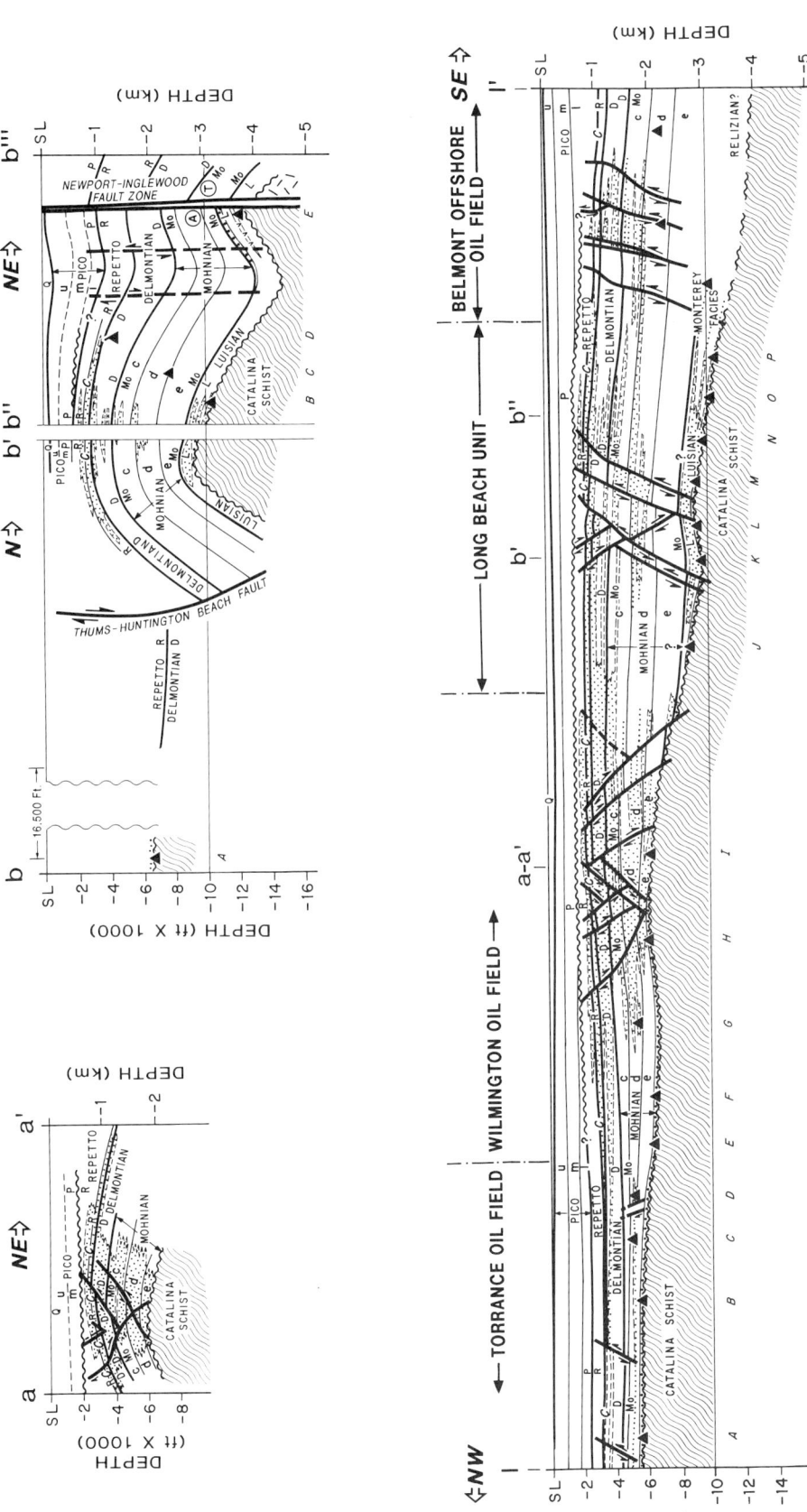

Figure 31. Cross sections of the Wilmington anticline. Solid triangles show control wells (lettered), which are listed in Appendix 2. Other symbols, etc., as in Figures 8 and 25, plus dotted unit is schist conglomerate. —C— indicates contoured horizon of Figure 30. Oil-productive zones are shaded. aa': Crest of Wilmington anticline (after Mayuga, 1970). bb'b"b"': East plunge of Wilmington anticline [compiled from Truex, 1972 (south flank), 1974 (north flank); with minor revisions from Mayuga, 1970; California Division of Oil and Gas, 1974; Clarke and Henderson, 1987; Henry, 1987; West et al., 1987]. Transverse faults omitted. Note that section is offset 8200 ft (2.6 km) at crest of anticline. ll': Longitudinal section along anticlinal axis (compiled from Murray-Aaron and Pfeil, 1984; Schoellhamer and Woodford, 1951; Allen, 1966; Mayuga, 1970; Truex, 1972; California Division of Oil and Gas, 1974; Olsen, 1975; Clark and Henderson, 1987; Henry, 1987).

Terminal zones), sedimentary rocks thinned to the southwest and south across the East Wilmington area. Isopachs are subparallel to the THUMS–Huntington Beach fault, showing only minor irregularities and no evidence for anticlinal growth. Where the Belmont and Long Beach Unit faults cross the present axis, isopach "thins" in the black shale member and the Ford zone (middle Mohnian) suggest folding parallel (and perhaps precursory) to the later faulting; the northerly trend of the Ford zone "thin" is notable. The longitudinal section (Figure 31, Section ll') shows the present-day local basement culmination in this area.

Isopachs of the lower Pliocene Ranger zone strongly reflect structural form, showing 300–400 ft (90–120 m) of folding, more than 200 ft (60 m) of southeast plunge, and about 100 ft (30 m) of vertical separation on the Long Beach Unit fault during this interval. Folding began in late Delmontian time and continued through early and middle Repetto deposition. Effects of the Long Beach Unit fault are seen on the south flank in the upper Delmontian (or perhaps even earlier in the Terminal zone), do not disturb the axial portion until late early Repetto (upper Ranger F-X interval), and are negligible on the north flank. These relationships suggest that this fault originated as a tear fault on the south flank, separating the steeper flank to the west from the more gently dipping segment to the east.

Isopach maps of Repetto intervals above the Ranger zone are not provided, but Truex (1974) stated that these indicate that "continuous uplift probably occurred throughout the Repettian Stage" and show post-Ranger displacement of about 150–250 ft (45–75 m) on the Long Beach Unit fault. All studies of the Wilmington field have noted that folding and faulting terminate at the base of the upper Pliocene Pico Formation (see Figure 31, Section aa'); the uppermost 1800–2000 ft (550–600 m) of strata are nearly planar, though tilted about 6° to the south (Clarke, 1987). Only the Long Beach Unit fault offsets Pico beds. (It should be noted that the absence of post-Repetto folding on the eastern part of the Wilmington anticline is documented primarily by seismic data, because this part of the oil field was developed by directional drilling and, in contrast to onshore Wilmington, shallow well control is not widely distributed.)

The Pico-Repetto contact at Wilmington is commonly called an unconformity and is ascribed to "a period of emergence" (Mayuga, 1970) in mid- to late Pliocene time. But evidence from foraminiferal studies clearly shows that beds above and below this contact were deposited in waters several thousand feet deep (Natland and Rothwell, 1954). The "strong angular discordance" described at the base of the Pico at Wilmington is typical of basin-plain sedimentation across growing structures, as described in the introduction to this paper. At Wilmington, the absence of the lower Pico and perhaps the uppermost part of the Repetto in addition to part of the middle Pico (Mayuga, 1970) indicates that structural growth continued into late Pliocene time.

At Wilmington, Truex (1974) described a local exception to this basin-plain model. Part of the uppermost Ranger zone (mid-Repetto F_0–F interval) consists of graded sandstones, up to 200 ft (60 m) thick, that have cut a southwest-trending channel into the underlying beds. Such a channel could result from headward erosion once the basin plain filled to sill depth. But its position at Wilmington suggests that it may be due to a focusing of tractionite/ turbidite flow through an ancestral Alamitos Gap between the growing Seal Beach and Long Beach (Signal Hill) anticlines. If so, it would record an overlap between the waning deformation at Wilmington and the onset of Pasadenan deformation along the NIFZ. The extent and relationships of this F_0 sand body merit further study.

Growth of the Wilmington anticline is apparently the result of "a couple produced by simultaneous movement of (the Newport-Inglewood) fault and on (THUMS– Huntington Beach) fault on the south flank" (Truex, 1974). Shallow basement south of the latter fault served as a buttress for folding of the eastern part of the anticline. Summarizing the geologic record at Wilmington, Yeats (1974) suggested that "the Wilmington field may contain indications of the Miocene extensional event accompanying volcanism (quartz and calcite veins in basement cores), the late Miocene and early Pliocene simple-shear episode throughout the western Los Angeles Basin (early Pliocene age of faulting [and folding]), and the localization of right-lateral shear at the Newport-Inglewood zone in late Pliocene time onward (termination of faulting at the unconformity at the base of the upper Pico)." During the Pasadenan, deformation of the block containing the Wilmington anticline has involved gentle southward tilting, but no folding.

Wilmington and adjacent oil fields—Long Beach, Seal Beach, and Huntington Beach—will yield 5.2 GBO, about 58% of the basin's total resource; yet the area encompassed by these fields occupies no more than 10% of the basin. Here all factors are optimal. A thick localized depocenter of Monterey-type organic sediments (Figure 5) provides source rock. A high geothermal gradient of 3.29°F/100 ft (60°C/km) of depth at Wilmington (Beyer, 1988) and similar high temperatures as recorded in the Seguro Petroleum Seguro No. 1 well at Huntington Beach have contributed to maturation of hydrocarbons at a relatively shallow depth and to their subsequent mobilization. Upper Miocene and lower Pliocene sandstones, deposited in an extensive suprafan system (Redin, this volume), are thick and porous. Deformation during latest Miocene and early to middle Pliocene time created anticlinal traps and faults along which the oil migrated from the middle Miocene source beds (Figure 31). Finally, late Pliocene and Quaternary deposition provided a seal of finer grained sediments that has been broken only along the primary faults of the NIFZ.

Central Trough

The central deep (Woodford et al., 1954) or trough (Figure 7) is the axial portion of the Los Angeles basin, north of the NIFZ, and trends N50°W. It is 30–35 mi (48–56 km) in length, extending from the Santa Ana shelf northwest to the convergence of the NIFZ with the Santa Monica fault system. Maximum width between the Dominguez and Santa Fe Springs oil fields is about 8 mi (13 km). There, Pliocene rocks are downfolded below 15,000 ft (4600 m) (Figure 8), and basal Mohnian beds are believed to be at least 24,000 ft (7300 m) deep (Figure 10). The deepest well in the basin to date is the American Petrofina Central Core Hole No. 1. Located 2 mi (3.2 km) northeast of the Rosecrans oil field, it was drilled in 1975 to a total depth of 21,215 ft (6466 m) [subsea depth is approximately 20,700 ft (6311 m)] in sedimentary rocks of Delmontian (Bostick et al., 1978; Henry, 1987) or lowermost Repetto (West et al., 1988) age.

Top of basement has been inferred at a depth of 30,000 ft (9100 m) beneath the central trough (Yerkes et al., 1965) and has generally been assumed to be Peninsular Ranges basement overlain by Paleogene and Upper Cretaceous sedimentary rocks. An alternative view (Crowell, 1974) suggests that neither Peninsular Ranges nor Catalina basement occurs beneath the central trough, but that it is floored by a complex of middle Miocene volcanic and sedimentary rocks formed during the "pull-apart" origin of the basin. Wells on the margins of the central trough have penetrated 3000–4000 ft (900–1200 m) of this sedimentary-volcanic complex (Figure 5) at Inglewood (above schist basement), West Coyote (above Paleogene beds), and on the north flank of Huntington Beach.

Within the central trough, estimates of the total thickness of basin fill that are calculated on the basis of gravity data (McCulloh, 1960; Yeats and Beall, this volume) assume that the basin is floored with Paleogene–Cretaceous sedimentary rocks (density 2.6+ g/cm^3) and basement rocks (density 2.75 g/cm^3). However, if the trough is floored with a mid-Miocene sedimentary-volcanic mix (with an average density somewhere between 2.6 and 2.75 g/cm^3), estimates of upper Miocene sedimentary thicknesses based on these gravity data would require recalculation. The residual gravity anomaly noted by McCulloh (1960) beneath the central trough may lend support to this alternative interpretation.

Anaheim Nose

The Anaheim nose (Figure 7) (Driver, 1948; Woodford et al., 1954) is a major structural feature in the subsurface of the eastern Los Angeles basin (Figure 8, Sections EE', FF', GG'; Figures 9, 10, 33). Entirely concealed beneath the thick upper Pliocene and Pleistocene section, it was discovered by geophysical surveys and exploratory drilling initiated in the late 1930s. These revealed a northwest-plunging buried ridge of Paleogene rocks (Figure 4) that are overlain unconformably by basinal strata of Luisian to mid-Pliocene age. From 1942 to 1946, three subcommercial oil fields were discovered on the northwest plunge of the Anaheim nose. Traps at depths of 8900–11,900 ft (2700–3600 m) are stratigraphic pinch-outs in Repetto and Delmontian sands and in Topanga sands beneath the unconformity. Up the plunge of the nose, 5 mi (8 km) to the southeast, another small field at Anaheim produced from middle Pico sands trapped against a north-striking fault.

The Anaheim nose apparently is a mid-Miocene fault block approximately 18 mi (29 km) long and 3–4 mi (5–6.5 km) wide. Its southwestern edge coincides fairly closely with a northwestern projection of the Shady Canyon fault, and the timing of the two structures is quite similar. No comparable fault has been associated with the northeast flank of the nose, which may well be a tilted fault block ("half-horst"). On the eroded subcrop surface, strata of early Miocene (Vaqueros Formation), Oligocene (Sespe Formation), and Eocene age have been encountered in drilling. Unpublished isopach/isolith maps document the relative position of this ridge as the basin subsided, but they are imprecise: well control is sparse on the flanks of the Anaheim nose, and biostratigraphic correlations in the Pliocene may be skewed by shallower water depths in this portion of the basin.

Sedimentary and volcanic rocks of the Topanga Formation (Figure 5) cover all of the Anaheim nose except for the southeasternmost, updip portion and a small block midway down the nose, where middle and upper Repetto beds rest on Oligocene and Eocene rocks west of a north-striking fault. During late Miocene time, the nose remained positive in relation to adjacent areas: Mohnian strata are missing from a zone 1–2 mi (1.6–3.2 km) wide along the crest of the feature, while continued offlap resulted in an area 2.5–5 mi (4–8 km) wide that is bare of Delmontian beds. Progressive onlap of lower, middle, and upper Repetto beds record the burial of the Anaheim nose during middle Pliocene time. The locus of upper Repetto thinning is more than 1 mi (1.6 km) north of the comparable Mohnian axis, reflecting southwesterly tilting of the nose during Pliocene subsidence of the broad basin.

Northern Shelf

The northern shelf (Figure 7) occupies the northern part of the Los Angeles basin, west of the Whittier Narrows. Its western segment, the uplifted block northeast of the Las Cienegas fault, was not defined until the discovery of the *Las Cienegas oil field* in 1960. That field will produce an estimated 65 MBO from upper Miocene sands in several small block-edge forced anticlines (Harding and Tuminas, 1988) on the hanging wall of the Las Cienegas fault. Basement rocks are found at depths of 4000–5000 ft (1200–1500 m) within the field and are described as amphibolite (Yeats, 1973; Sorensen, 1984, 1985) with similarities to the Willows Plutonic Complex on Santa Cruz Island and to the saussurite gabbros of coastal and offshore southern California.

The *Las Cienegas fault* (Figure 14; Figure 8, Section BB') (Mefferd, 1970) is a northwest-striking reverse fault with at least 4000 ft (1200 m) of vertical separation on Mohnian beds, yet the upper Pliocene interval has not been offset. The fault bifurcates towards the northwest (Gardett, 1971, his figure 4; California Division of Oil and Gas, 1974). The northern branch continues in a northwesterly direction and, at basement depths, may underlie the steep southwest flank of the Salt Lake oil field (see Figure 14). The southern branch curves westward towards the Beverly Hills oil field and, at basement depths, may connect with the southern branch of the Santa Monica fault. As described previously (Figure 16), the major vertical displacement on the Las Cienegas fault occurred during early to middle Pliocene time; Pasadenan deformation has been expressed as monoclinal flexure at the tip of a postulated blind thrust.

The northern shelf includes two of the oldest oil fields in the basin. The *Salt Lake oil field*, discovered in 1902 by wells drilled north of the Rancho La Brea tar pits, has an estimated ultimate recovery of 54 MBO from Mohnian sands. Early production was from west-plunging flexures that are closed to the east by a small north-striking fault and are limited on the north by the North Salt Lake fault. Additional production was developed in the early 1960s from the same and deeper sands in a narrow, northwest-trending anticline along the southwest edge of the old field. This fold, localized along the northern branch of the Las Cienegas fault (Figure 14), has dips up to vertical on its southwest flank (Wright, 1987b) and 30°–60° on its northeast flank. On the south flank of this fold, the *South Salt Lake oil field*, discovered in 1964, will yield 10 MBO from upper Delmontian and lower Repetto sand pinch-outs (Jacobson and Lindblom, 1977; Samuelian, 1984). These pinch-outs indicate that uplift of the Las Cienegas block began in late Miocene time, after deposition of the mid-Mohnian sands at Salt Lake.

The *Los Angeles City oil field*, discovered in 1892 by a hand-dug well located near oil seeps, will yield 24 MBO from Mohnian sands on the south flank of the Elysian Park anticline. Entrapment is by faulting and stratigraphic pinch-out.

Along the southwestern edge of the Las Cienegas block (Figure 14), the basement surface is as shallow as 4000 ft (1200 m), but northward it dips, or is downfaulted, to perhaps 8000 ft (2400 m) (Figure 8, Section BB′) before rising beneath the Elysian Park anticline. Along the southeastern edge of the block, Delmontian and Repetto beds thicken rapidly eastward and the basal Mohnian horizon steps downward some 3000 ft (900 m) or more (Figure 10). This step may reflect a northeast-striking fault at basement depths, aligned with a zone of surface faults that transect the axis and north flank of the Elysian Park anticline between Alhambra and Highland Park (Lamar, 1970).

Four small oil fields have been developed in east-west-trending anticlines localized along this southeastern edge of the Las Cienegas block. From southwest to northeast, these oil fields are

- the *Jefferson area of the Las Cienegas field*, discovered in 1965, which will yield about 22 MBO from Delmontian and mid-Mohnian sands
- the *Los Angeles Downtown field*, discovered in 1963, which will yield 15 MBO from Delmontian and upper and middle Mohnian sands, plus 23 bcf gas from lower Repetto sands
- the *Union Station field*, discovered in 1967, which will yield 2 MBO from Delmontian and upper and middle Mohnian sands
- the *Boyle Heights field*, discovered in 1955, which has produced 0.3 MBO from Delmontian sands (Jamison and Malloy, 1958b).

These folds are believed to have been initiated by buckling in the wedge of Mohnian and Topanga strata that thins southward against the east-plunging crest of the Las Cienegas basement high (Figure 8, Section CC′). North-dipping reverse faults cut the Jefferson and Los Angeles Downtown folds, and the Jefferson area is also traversed by several northeast-striking faults that are perhaps related to the postulated basement fault.

East of the Las Cienegas block, the northern shelf plunges gradually southeastward some 12 mi (19 km) to the Whittier Narrows area. The shelf's southern edge, a sharp flexure in upper Pliocene and Quaternary beds, is aligned along a possible easterly extension of the Las Cienegas fault (Figure 8, Section DD′; Figures 9, 10). Exploration to date along this portion of the northern shelf has been relatively unsuccessful, yielding only two small oil fields. The *Bandini field*, discovered in 1953, will ultimately produce 6 MBO from Delmontian and upper Mohnian sands at depths of 4200–8400 ft (1300–2600 m) (Jamison and Malloy, 1958a). The trap is a broad north-northwest–trending anticline whose eastern flank is downdropped 350–600 ft (100–180 m) on a fault that parallels the fold axis. The *East Los Angeles field* was discovered in 1946 and will yield 7 MBO from Delmontian sands in a gently folded dome. In both fields, an element of stratigraphic entrapment may be provided by channel walls and levee/overbank deposits that are prevalent in the inner fan facies of the Delmontian and Repetto sections in this part of the basin. This discontinuous, cut-and-fill facies extends along the northern shelf from Boyle Heights to Santa Fe Springs and is a significant impediment to correlation and exploration in this area.

Fullerton Embayment

In the northeastern Los Angeles basin, the Fullerton embayment (Figure 7), between the Anaheim nose and the Whittier fault, contains four major oil fields and several smaller ones. Most of the oil is trapped in west-trending anticlines. Largest of these is the Coyote Hills uplift (Figure 32A), more than 7 mi (11 km) long, where topographic relief of 250–300 ft (75–90 m) reflects folding that has continued into the late Quaternary.

Evolution of the northeastern Los Angeles basin was summarized by Schoellhamer et al. (1981) in the following manner:

> Uplift of the Anaheim nose began during middle Miocene time, and the distribution of the Puente Formation sandstone bodies indicates that the growing fold acted as a barrier to the sand that was carried in from the north and east. These sandstone bodies were deposited in a northwest-trending trough or basin ... Northeastward migration of the trough during late Miocene time is indicated by the progressive shifting of the maximum thickness lines of younger and younger sand units in that direction. In this area, accentuation of northwest-trending folds continued at least to the end of the Miocene. The more westerly folds are much younger.

Within the broad, central part of this northward-thickening trough, folding is distributed across the Olive-Burruel Ridge, the north and south Richfield anticlines, and the eastern East Coyote (Anaheim) anticline. Westward, the upper Miocene thins beneath western East Coyote (Hualde dome) and the West Coyote anticline, accentuating the sedimentary wedge south of the Whittier fault and focusing the folding on the single major uplift of the Coyote Hills. The influence of basin geometry on the spacing and intensity of folding is clearly demonstrated in this part of the Los Angeles basin.

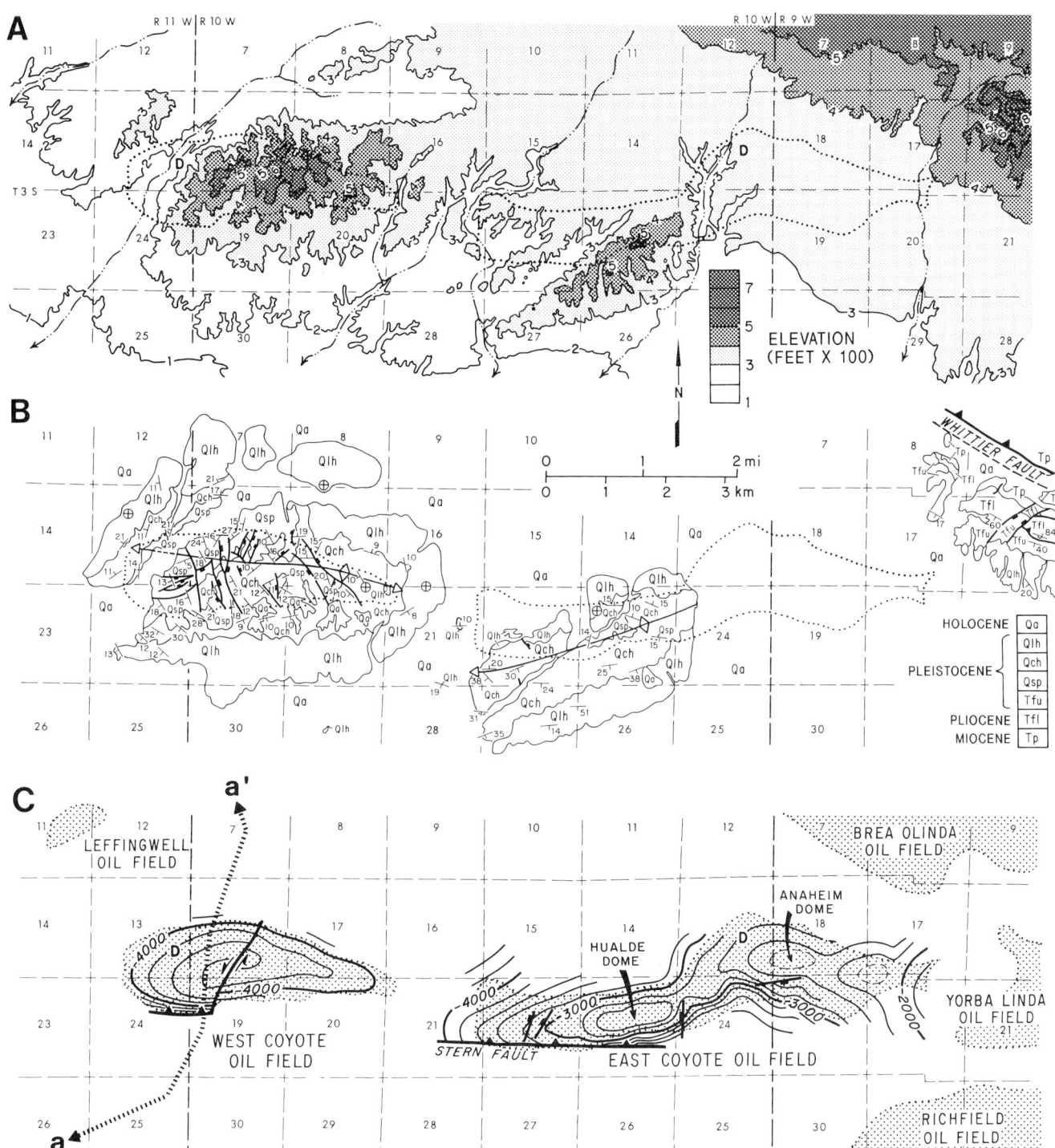

Figure 32. Maps showing topography, surface geology, and subsurface structure of the Coyote Hills. D shows discovery wells for the East and West Coyote oil fields. Oil fields are stippled on Figure 31C and shown with dotted outlines on Figures 31A and B. (A) *Topography*. Contour interval: 100 ft (30 m). Higher elevations shaded for emphasis. (B) *Surface geology*, including selected bedding attitudes (from Yerkes, 1972; Durham and Yerkes, 1964). Formations shown are Alluvium (Qa), La Habra Formation (Qlh), Coyote Hills Formation (Qch), San Pedro Formation (Qsp), upper and lower Fernando Formation (Tfu and Tfl), and Puente Formation (Tp). (C) *Subsurface structure* [compiled from Ybarra et al., 1960 (East Coyote); Mefferd and Cordova, 1962 (West Coyote); California Division of Oil and Gas, 1974]. Contours on a lower Pliocene (upper/middle Repetto) horizon ("J" marker in East Coyote field, top of Upper 99 zone in West Coyote field) (California Division of Oil and Gas, 1974). Contour interval: 200 ft (61 m) (dashed supplementary 100-ft contours). Limits of oil production are shaded. Other symbols, etc., as in Figure 9. aa' is cross section of Figure 33.

Discovery of the *West Coyote oil field* in 1909 was based on oil and gas shows in water wells, but its early development led to the realization that the low hills that rose above the plain of the Los Angeles basin may well reflect young anticlinal folds (Yerkes, 1972). This concept eventually led to the discovery of the East Coyote field and the major fields of the Newport-Inglewood trend. The West Coyote field will ultimately produce an estimated 258 MBO from lower, middle, and upper Repetto sands. It is the westernmost of three distinct anticlinal culminations that differ in their surface and subsurface characteristics (Figure 32). The West Coyote anticline, nearly 3 mi (5 km) long, is an asymmetric fold with a steep, reverse-faulted south flank (Figure 33). That fault has produced 600–700 ft (180–215 m) of vertical separation in middle Repetto beds but dies out in the upper Pliocene interval. Lower Repetto beds on the north flank are cut by a south-dipping reverse fault with 400 ft (120 m) of vertical separation (Mefferd and Cordova, 1962); that fault dies out within the middle Repetto.

At upper Repetto depths, the fold is simple and slightly arcuate in map view, transected by one significant fault that strikes N30°E; west of this fault the anticlinal crest is 250 ft (75 m) higher and 600 ft (180 m) south of the crest in the east block. At the surface (Figure 32B; Yerkes, 1972), Pleistocene beds show a more complex structure: the axial region is cut by normal and reverse faults that strike mostly north-northeast to northwest. Beds dip 9°–12° on both flanks, though southwest dips up to 36° are associated with faulting on the steep west plunge of the fold. The axial plane is nearly vertical despite the asymmetry of the deeper structure.

The anticlinal high at West Coyote plunges eastward across a nonproductive structural saddle nearly 1 mi (1.6 km) wide into the north flank of the Hualde dome (Figure 32C), which is the western culmination of the *East Coyote oil field*. Discovered in 1911, this field will yield an estimated 122 MBO from Mohnian, Delmontian, and lower, middle, and upper Repetto sands (Ybarra et al., 1960). At the surface (Figure 32B) (Yerkes, 1972), the Hualde dome is clearly defined in Quaternary outcrops. South-flank dips average 24° and are as high as 51°, while the north flank averages 15° with a maximum of 24°. Surface faulting is insignificant. At its west plunge, the crest is offset about 3500 ft (1070 m) south from the surface axis of West Coyote, although at Repetto depths, this offset is no more than about 1500 ft (450 m). Unlike West Coyote, where the surface axis and the topographic high coincide closely, the highest topography at East Coyote is formed by resistant marine Pleistocene beds on the south flank of the fold (Figures 32A, 32B).

Subsurface mapping (Ybarra et al., 1960) has defined a north-dipping reverse fault, the Stern fault, along the south flank of the Hualde dome. Its vertical displacement is as much as 930 ft (285 m) at the west end of the dome, where it is believed to cut beds as young as earliest Pleistocene, but it dies out to the east. Subsurface geometry, though poorly controlled, implies that the Stern fault is not continuous with the comparable south-flank fault at West Coyote; these faults are believed to have developed independently on the two anticlinal culminations. In the western part of the Hualde dome, transverse and longitudinal faults have displaced upper Miocene beds but are overlapped by undisturbed lower Repetto sands. Uplift west of a pair of northerly striking faults has elevated mid-Mohnian beds some 1200 ft (365 m) and has created a late Miocene sea-floor high, at least 1000 ac (4.16 km²) in area, that is bare of Delmontian strata.

Folding becomes less intense eastward along the Hualde dome as it plunges northeast into the structural saddle that separates it from the Anaheim dome (Figure 32C). This culmination, where the initial East Coyote discovery was made, is covered by old and young Quaternary alluvium and has only slight topographic relief (Figures 32A, 32B). Both flanks dip about 15° at the base of the Pleistocene (Durham and Yerkes, 1964). The crest of the Anaheim dome is about 2000 ft (600 m) north of the eastern end of the Hualde dome (Ybarra et al., 1960). The western part of its south flank is broken by a small normal fault with less than 200 ft (60 m) of vertical separation, but the Anaheim dome otherwise is not faulted.

Deep wells have reached Paleogene strata at depths of about 10,000 ft (3050 m) beneath West Coyote and 8500 ft (2600 m) beneath East Coyote (Figure 4) (Yerkes, 1972). These are overlain by 500–1500 ft (150–450 m) of Topanga beds (Figure 5), with the addition at West Coyote of up to 1100 ft (335 m) of volcanic rocks. Unpublished isolith maps document structural growth at West and East Coyote during Mohnian and Delmontian time and early Repetto growth on the Anaheim dome. Structural details reviewed above indicate latest Miocene and early Pliocene faulting at West Coyote and the Hualde dome. Southward thickening of upper Pliocene beds down the south flank of the Hualde dome (Ybarra et al., 1960) indicates some structural growth during that time. Perhaps this early deformation was sufficient for initial entrapment of hydrocarbons, but detailed isopach/isolith studies are necessary to confirm this.

According to Yerkes (1972), "The Coyote Hills structure dates from at least late Miocene time, when folding and faulting, concentrated in the West Coyote area, caused prominent local unconformities and southeastward thinning in the upper Miocene section." He also noted that structural relief at West Coyote is about the same—1000–1200 ft (300–365 m)—on basal Repetto, mid-Pliocene, and basal Pleistocene horizons and that the "late Pliocene . . . does not thin over the structural highs at the East Coyote field, indicating that the uplift is probably post-Pliocene in age." On a more regional scale, however, the great increase in thickness of the Pico interval from the north flank of West Coyote to its south flank (Figure 33) indicates that the fold was a major hinge during late Pliocene time.

This cross section implies that the fold has formed during Pasadenan deformation by buckling within the 5000-ft (1500-m) thickness of middle and upper Miocene strata that wedge out southward against the Anaheim nose. Offsets of the anticlinal axes along the Coyote Hills structure and lateral variations in fold geometry apparently are related to major relief on the mid-Miocene surface and to consequent abrupt changes in the thickness of middle and upper Miocene sedimentary rocks.

Two miles (3 km) southeast of the Coyote Hills is the *Richfield oil field*. Discovered in 1919, it will yield 217 MBO from Mohnian and basal Repetto sands. Entrapment is in

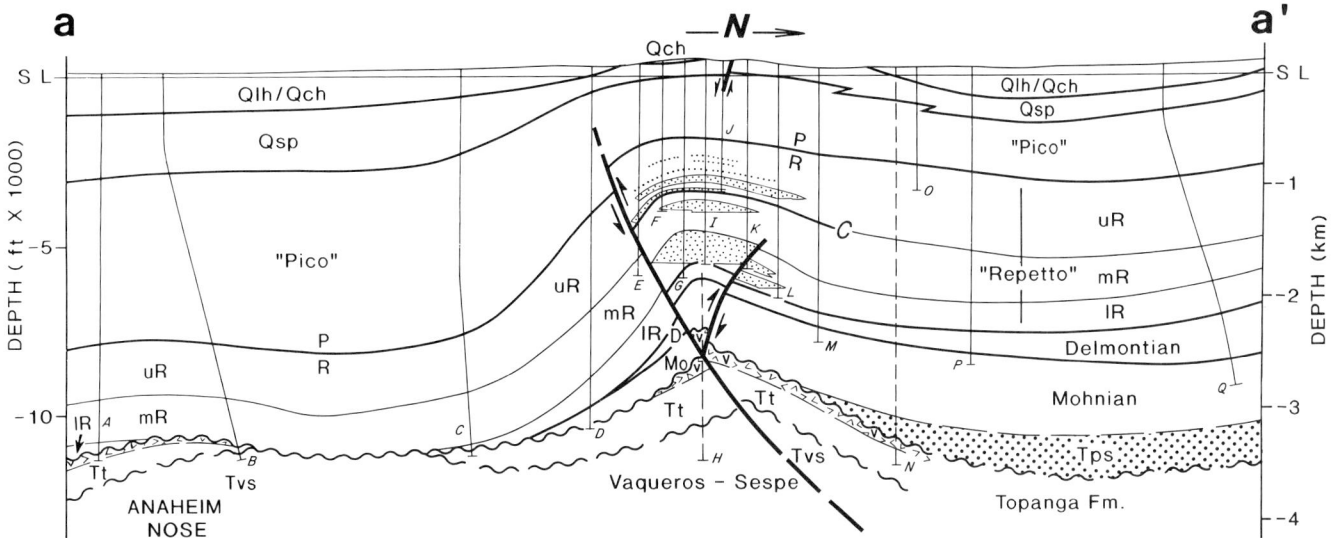

Figure 33. Cross section of the West Coyote oil field and Anaheim nose (compiled from Mefferd and Cordova, 1962; Yerkes, 1972; California Division of Oil and Gas, 1974; and proprietary data). Location of section is shown in Figure 32C. Control wells (lettered) are listed in Appendix 2. Oil-productive zones are shaded. —C— shows contoured horizons in Figure 32C. Formation abbreviations as in Figure 32B, plus Tps = Soquel Member of Puente Formation, Tvs = undifferentiated Vaqueros-Sespe formations. Other symbols, etc., as in Figure 8.

two parallel anticlines that trend east-northeast (Ingram, 1961) (see also Figure 8, Section GG'). The main northern fold is a broad dome that narrows eastward. Dips are 15°-20° on the north flank and 20°-30° on the south flank. Faulting is minor. The narrower southern fold is broken by a pair of north-dipping reverse faults, with 100-500 ft (30-150 m) of vertical displacement on Mohnian sands, that are shown as terminating against the basal Repetto sands. Isolith mapping suggests anticlinal growth during middle Repetto deposition. Significant thinning of the total lower Pliocene interval across the Richfield area has been attributed (Durham and Yerkes, 1964) to uplift of the anticlines in Pliocene time; the upper Pliocene also thins across the field. Noting that late Pleistocene strata (La Habra Formation) are folded on the southern and western flanks of the field, Durham and Yerkes (1964) concluded that "the fold is at least in part of late Pleistocene or younger age." It has no obvious surface expression, however.

The *Yorba Linda oil field*, north of Richfield and east of East Coyote, is the only major stratigraphic trap in the Los Angeles basin. Discovered in 1937, it will ultimately yield 95 MBO from lenticular sands of late Pliocene, middle and early Repetto, and Delmontian age on a homocline dipping 10°-15° to the southwest (Heath, 1958; Cook, 1975). Most of the accumulation is in a southwest-directed submarine fan-channel system of Pliocene age, where entrapment is controlled by silt barriers and channel walls.

The *Kraemer oil field*, east of Richfield, was discovered in 1918 and will yield 4 MBO from Delmontian and Mohnian sands on a small southwest-trending anticline. At the surface, the fold axis is expressed in Delmontian strata and trends S55°W (Schoellhamer et al., 1981). In the subsurface (Reese, 1943) the axis trends S70°W. The fold is asymmetric,

with dips up to 80° on the south flank and averaging 25° on the north flank. Kraemer is one of four small anticlines, 3000-4500 ft (900-1400 m) apart, that plunge southwest from the Whittier fault zone in the area where it converges on the uplifted Santa Ana Mountains block. Lamar (1961) noted that "North of the Kraemer oil field the folds in the uppermost portion of the Puente formation appear to die out up section. This could be explained by assuming that the folds were gradually growing as a result of movement on the Whittier fault during the deposition of the Puente formation." In the footwall of the Whittier fault, middle and lower Miocene beds are exposed up-plunge from the two eastern folds. Lower Pliocene beds are involved in this folding, and at Kraemer, the older Quaternary alluvium has been uplifted, suggesting that some of the deformation there is Pasadenan.

The small *Olive oil field*, 2 mi (3 km) southwest of Richfield, was discovered in 1953 and will yield 2 MBO from a conglomeratic Repetto sand that shales out across a west-plunging faulted nose (Saunders, 1958; Gaede, 1958). The fold is thought to be the westward extension of an anticline that is seen in outcrop at Burruel Ridge on the westernmost end of the Santa Ana Mountains (Schoellhamer et al., 1981). It is interpreted (Figure 8, Section FF') as having formed in the hanging wall of the *Peralta Hills thrust* (Figure 7) (Fife et al., 1980; Bryant and Fife, 1982). This fault is exposed along the south flank of Burruel Ridge, where it dips north at 30°-70°. There the fault has juxtaposed lower Mohnian beds over Pleistocene terrace deposits and earlier Miocene rocks and apparently follows a detachment zone in shales and silts of the lower Mohnian La Vida Member. Slip on this detachment zone may have facilitated folding over much of the northern shelf. Alternatively, the Peralta Hills thrust projects downdip toward a zone of microseismicity that dips 60° north at depths of 14,000–20,000 ft (4–6 km) (Bryant and Fife, 1982). The possible relationship of the Peralta Hills thrust with deep-seated blind thrusts along the northern margin of the Los Angeles basin is discussed later in this paper.

North Central Area

The *Santa Fe Springs oil field,* fourth largest in the basin, is located where the Fullerton embayment opens westward toward the central trough. This field will ultimately produce 622 MBO from Delmontian and upper, middle, and lower Repetto sands. The 1919 discovery well (point D on Figure 34) was a marginal producer, and intensive development did not begin until the Unocal Bell No. 1 well (point B on Figure 34) was completed as a flush producer late in 1921. The field's deepest well reached a total depth of 13,541 ft (4128 m) in strata of Mohnian age. The trap, perhaps the least complex structure in the Los Angeles basin, was described as "a relatively flat elongated dome with the axis trending approximately N70°W" by Winter (1943). On a broader scale, the fold has been shown as occurring within an uplifted wedge between east-striking reverse faults on its northern and southern flanks (Yerkes, 1972; McCulloh et al., 1978).

Plunge of the fold is about 6°–7° to the west. Westward thickening of uppermost Miocene beds has shifted the structural high to the southeast in the deeper zones (Elmore, 1958). At basal and mid-Pliocene horizons, McCulloh (1967, his figure 1) showed the structural highs to coincide. Two minor cross-faults die out beneath the middle Repetto sands. The flanks of the fold dip 10°–12° except in Delmontian beds along the southwest flank, where dips of 20°–25° are evident (Yerkes, 1972). These two lines of evidence suggest that folding had begun by earliest Pliocene time. Unpublished isolith mapping also suggests some Delmontian and early Repetto growth.

Citing evidence for significant growth on the Santa Fe Springs fold during both late Pliocene and latest Pleistocene times, Yerkes (1972) noted that "local structural relief at the base of the Pleistocene is less than 500 feet, whereas that at the base of the upper and lower Pliocene is more than 1200 feet." Yerkes also noted that "at least 200 feet of folded upper Pleistocene, pre-Gaspur [earliest Holocene] strata are present on the flanks of the dome but are missing at the crest." At the surface, the fold is reflected by a very slight topographic dome. This topographic anomaly, in addition to the presence of gas in water wells, was the basis for discovery of the field.

The regional setting of the Santa Fe Springs anticline suggests that it was formed by shortening between the northwest-plunging Anaheim nose and the uplifted hanging-wall block of the Whittier fault. The axis plunges gently eastward some 3 mi (5 km) into the *Leffingwell oil field,* which was discovered in 1946 and yielded less than 1 MBO before its abandonment. Production was from Delmontian sands closed to the west by a north-striking fault of pre-Pliocene age (Gaede, 1957). Deep wells indicate that the Mohnian interval thins abruptly southwest of the field onto a middle Miocene high (Yerkes, 1972). Isolith

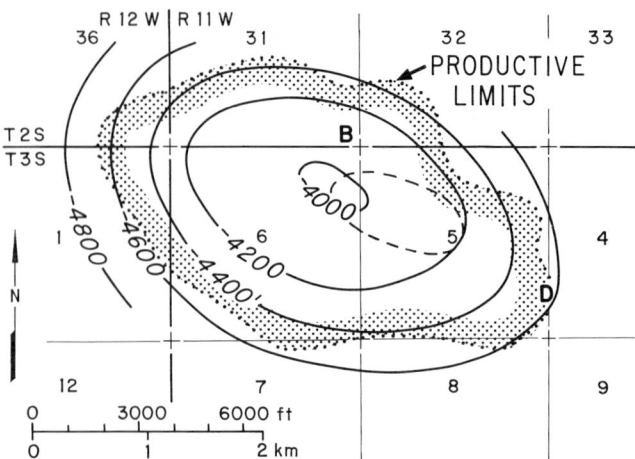

Figure 34. Subsurface structure map of the Santa Fe Springs oil field (compiled from Elmore, 1958; California Division of Oil and Gas, 1974). Contours on a mid-Pliocene horizon (top of Meyer zone, basal upper Repetto). Contour interval: 200 ft (61 m). D is discovery well; B is Unocal Bell No. 1 well. Dashed line is outline of uppermost Miocene (Delmontian) production (Santa Fe and Bell 100 zones).

mapping implies Mohnian and Delmontian growth on this fold.

The *Montebello oil field* was discovered in 1917 and will yield 194 MBO. Production is from Delmontian and lower and middle Repetto sands and conglomerates. The Repetto strata were deposited at the apex of the primary submarine fan system that supplied most of the sediments to the Los Angeles basin during latest Miocene and early Pliocene time (Conrey, 1967). Consequently, they are "so lenticular that reliable stratigraphic marker beds are lacking" (Reese, 1943b), and "except where wells are closely spaced and detail logs are available, correlations within [the Repetto] are very hazardous." The original central portion of the Montebello field was developed largely before the advent of electric logging, and no modern studies of the field have been published. However, unpublished studies reviewed by this writer confirm the general conclusions of early published reports (Reese, 1943b; Stolz and Woodward, 1943).

Three separate structural traps form the Montebello field (Figure 35). The original discovery in 1917 was based on topography and surface geologic mapping of an east-trending anticline that curves to the southwest at its west end. As mapped in lower Pliocene beds, the fold is asymmetric, with north-flank dips up to 50° and dips on the south flank varying from 15° near the west end to 35° at the east end. East plunge of 25° is reported on this Pliocene fold, and some evidence of cross-faulting in its central part is noted.

The East Montebello accumulation, discovered in 1933, is within a west-plunging nose mapped in upper Miocene beds and closed to the east against the East Montebello fault. Within the field, that fault is an east-dipping, north-

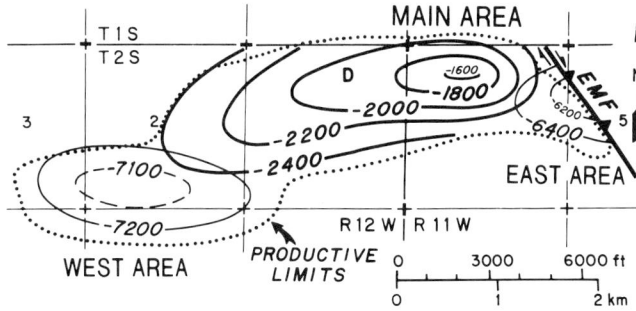

Figure 35. Subsurface structure map of the Montebello oil field (compiled from California Division of Oil and Gas, 1974, after Stolz and Woodward, 1943). Contours in West area on an upper Miocene horizon (top of 8th zone; Delmontian); in Main area, on a lower Pliocene horizon (top of 1st zone; middle Repetto); and in East area, on a basal Pliocene horizon (top of Farmer sand; upper Delmontian). Contour interval: 200 ft (61 m) (with supplementary 100-ft contour dashed). D is discovery well. EMF is East Montebello fault.

northwest-striking reverse fault. Regionally (as noted in the discussion of the Whittier fault zone) it is seen to be a right-oblique fault against which the Whittier fault appears to terminate. Production at East Montebello is from the Delmontian and lowermost Repetto. As described by Reese (1943b), the Repetto thickens rapidly to the northwest so that the west plunge of the East Montebello nose, in basal Repetto beds, is overlain in the upper Repetto by east plunge of the main anticline.

The West Montebello accumulation, discovered in 1938, produces from Delmontian and lower Repetto sands on a separate anticline southwest of the main fold. This deeper structure trends east and plunges 5° east and 4° west. The south flank dips 6°-8°, while dips on the north flank are 2°-3° at the top of the Delmontian and increase with depth to 7°-8°. As described by Stolz and Woodward (1943), "sediments of lowermost Pliocene [lower Repetto] and upper Miocene [Delmontian] ages have been folded into an anticline that does not conform with that developed in the Pliocene formations of the old Montebello field. It would appear that this condition is due to a phase of crustal folding that started during the late Miocene and continued for a short time into the early Pliocene . . . A post-Pliocene period of folding in which the stresses may have acted from a somewhat different direction resulted in the development of the main anticlinal structure of the old Montebello field."

The deeper folds at East and West Montebello are believed to be similar in origin to the 184 anticline at Whittier: short-wavelength folds that developed in the wedge of sedimentary rocks south of the Alhambra high and the Whittier fault, controlled more by vertical uplift on those features than by right-lateral shear. The section now being compressed against the Alhambra high by Pasadenan right-lateral movement on the Whittier fault and by southward thrusting on deep Transverse Range faults (Davis et al., 1989) is much thicker than in early Pliocene time. As a consequence, the shallow main Montebello anticline is much broader than the earlier folds (Ferguson and Willis, 1924). In the Montebello-Whittier Narrows area, complex depositional patterns associated with the late Miocene and Pliocene submarine canyon and fan system (Gourley, 1975) have obscured the details that might otherwise document the late Neogene evolution of the northern margin of the Los Angeles basin.

Questionable Structures

Geological and structural maps of the Los Angeles basin commonly include several structural features for which the evidence is problematic. Chief among these is the *Norwalk fault* (Figures 9, 10). This feature was first mentioned by Richter (1958) to explain the epicentral location of the 1929 Whittier earthquake, which Richter placed in the Norwalk area 7.5 mi (12 km) southwest of the Whittier fault and 3 mi (5 km) south of the Santa Fe Springs oil field. Richter noted that many years after the earthquake "evidence establishing the Norwalk fault was . . . brought forward, chiefly from commercial oil exploration work." On maps, Richter and others extended the Norwalk fault eastward across the northern flank of the Anaheim nose, then curved it southeast to follow the south flank of the Olive-Burruel Ridge anticline, where the Peralta Hills thrust was later described. More recently, Harding and Tuminas (1988, their figure 15) projected the Norwalk fault along a somewhat more northerly course to connect with reverse faults on the south flank of the Richfield oil field.

Yerkes (1972) evaluated deep wells between the northwest plunge of the Anaheim nose and the Santa Fe Springs and West Coyote oil fields in an attempt to locate the Norwalk fault in the subsurface. Some 3 to 5 mi (5 to 8 km) east of Richter's revised epicenter, Yerkes reported that "subsurface structure indicates the presence of a buried, north-dipping reverse fault that trends generally east-southeastward . . . and perhaps is coincident with the Norwalk fault of Richter" (see Yerkes, 1972, his plate 2, sections BB', CC', and DD'). In these deep wells west of the West Coyote field, four of six fault intercepts are consistent with the fault plane interpreted by Yerkes, as is one of the two such intercepts south of the Santa Fe Springs field (where he related steeper south-flank dips of 20°-25° to the Norwalk fault). This fault dips approximately 60° north (Figure 8, Section EE').

Yerkes (1972) pointed out that "the linearity of the south margin of the Coyote Hills . . . suggests that it may be an eroded fault scarp" and noted that it "coincides with the Norwalk fault of Richter." However, he could not trace the subsurface fault (south of Santa Fe Springs) eastward through a cluster of deep wells south of West Coyote to tie in with this topographic scarp. Detailed subsurface correlations by this investigator (Figure 33) similarly failed to find evidence for the Norwalk fault south of the West Coyote anticline and indicate that the thrust fault on the south flank of that fold is approximately 4500 ft (1400 m) north of an eastward projection of the Norwalk fault. Yerkes (1972) concluded that "although the epicenters of the 'Whittier' earthquake can be associated spatially with a buried fault, the only possible surface expression of faulting in this area (the West Coyote 'scarp') that can be associated with the Norwalk fault of Richter cannot be correlated with any recognized subsurface feature."

Hutton (1987) used improved velocity models to relocate the epicenter of the 1929 event, placing it at the east end of the Santa Fe Springs field, some 4 mi (6.5 km) northeast of Richter's corrected location and 3.5 mi (5.6 km) southwest of the Whittier fault. No hypocentral depth or focal mechanism was cited. The subsurface fault plane defined by Yerkes (1972), projected to depth (Figure 8, Section EE'), would pass beneath Hutton's epicenter at a depth of 2.5 to 5 mi (4 to 8 km) in the upper part of the seismogenic zone defined by Hauksson (1987).

The 1929 event may have occurred on a north-dipping blind thrust—perhaps the same one responsible for the 1987 Whittier Narrows earthquake—that appears to be aligned with the Peralta Hills thrust (Figure 8, Section GG') farther to the east. As noted above, a connection between these two faults cannot be carried through the shallow sedimentary section. Perhaps the reverse faults on the south flanks of the West Coyote anticline and the Hualde dome are en echelon segments of the Norwalk fault—tips of the blind thrust refracted upward by the buried northward salient of the Anaheim nose.

In the northwestern Los Angeles basin, the *Charnock and Overland Avenue faults*, which strike N35°W, were mapped on the basis of groundwater barriers (Poland et al., 1959). During the 1960s, intensive reflection seismic surveying and exploratory drilling in the area of these postulated faults found no evidence for any significant displacements that might coincide with the Charnock or Overland Avenue faults. Perhaps the groundwater barriers might be related to upper Pleistocene channels formed by streams flowing south-southeast from the uplifted Santa Monica Mountains.

North of Santa Monica in the Santa Monica Mountains, the *Temescal fault* was mapped (Hoots, 1931; Durrell, 1954) as the northwest-trending boundary between Santa Monica slate and adjacent Cretaceous strata (Figures 4, 9). More recent studies (Fritsche, 1973) have shown that the contact is not a fault but is part of the unconformity at the base of the Upper Cretaceous strata (Dibblee, 1982a).

THE TECTONIC EVOLUTION OF THE LOS ANGELES BASIN

The Los Angeles basin has formed within the rapidly evolving San Andreas transform zone of southern California. The genesis of the basin and of most of its surrounding and internal structures are closely linked to the development of that transform zone over the past 30 m.y. Major aspects of the tectonic evolution of southern California are still the subject of active research and debate. New data and fresh concepts build on each other to challenge existing interpretations and stimulate alternative hypotheses. Not for some years may we expect a clear consensus on the regional events that controlled the formation of the Los Angeles basin. For that reason alone, significant aspects of this paper are both speculative and preliminary.

Reconstruction of the Neogene evolution of the Los Angeles basin provides an opportunity to test the hypothesis that the basin opened as a consequence of clockwise rotation of crustal blocks that form the present Transverse Ranges (Luyendyk et al., 1980; see also Biddle, this volume, his Figures 10, 11, and 12). Data reviewed in the present study are compatible with the rotational hypothesis but do not provide conclusive proof, nor do they bear upon any potential problems outside the area studied. To stimulate further critical research, the reconstruction that follows will attempt to explain Neogene tectonics according to this hypothesis of block rotation and (localized) rafting. In most other respects, it is in general accord with several recent chronologies of regional or local geologic events (Crowell, 1987; Schwartz and Colburn, 1987; Vedder, 1987). This paper does not attempt to develop comprehensive regional restorations within the interval between 20 Ma and the present, as the rotational hypothesis is still open to significant modification based on new data and on testing against paleogeologic syntheses.

Early Miocene Reconstruction

Significant stages in the Neogene evolution of the Los Angeles basin are shown on a series of palinspastic maps (Figures 36, 37, 39, and 42). Most of the evidence on which these reconstructions are based has been presented in the preceding sections of this paper. The initial map (Figure 36) is an early Miocene reconstruction taken from Luyendyk and Hornafius (1987, their figure 11-11). Pre-Topanga lithologies and some additional faults have been added from Figure 4 and serve as the basis for minor adjustments to that reconstruction.

Luyendyk and Hornafius (1987) used apparent offset of the early Paleocene shoreline (from Sage, 1975) to estimate backslip of 36 mi (60 km) on the Malibu Coast–Santa Monica–Raymond fault system and 24 mi (40 km) on the Whittier-Elsinore fault. On the latter feature, Lamar (1961) estimated 19 mi (30 km) of slip across the Elsinore fault; Figure 36 shows about 18 mi (29 km). Slip of the granite-slate contact across the Santa Monica–Raymond fault system (Figure 4) is approximately 15 mi (24 km). Neither of the features cited provides a unique linear feature or geological "line" (Crowell, 1962) capable of yielding an accurate estimate of total slip. A linear Paleocene shoreline cannot be taken as proven, and the granite-slate contact is a function of the depth of erosion of individual basement blocks. The reconstruction in Figure 36 shows about 30 mi (48 km) of backslip on the Santa Monica–Raymond fault system. The amount of backslip on this fault system is unconstrained by the rotational model (Luyendyk and Hornafius, 1987) and could be more or less than is shown here. Also, these reconstructions do not allow for crust that may have been carried beneath the Transverse Ranges by the downwelling flow postulated by Bird and Rosenstock (1984).

The palinspastic reconstruction of Figure 36 reflects the conventional picture of early Miocene paleogeology along

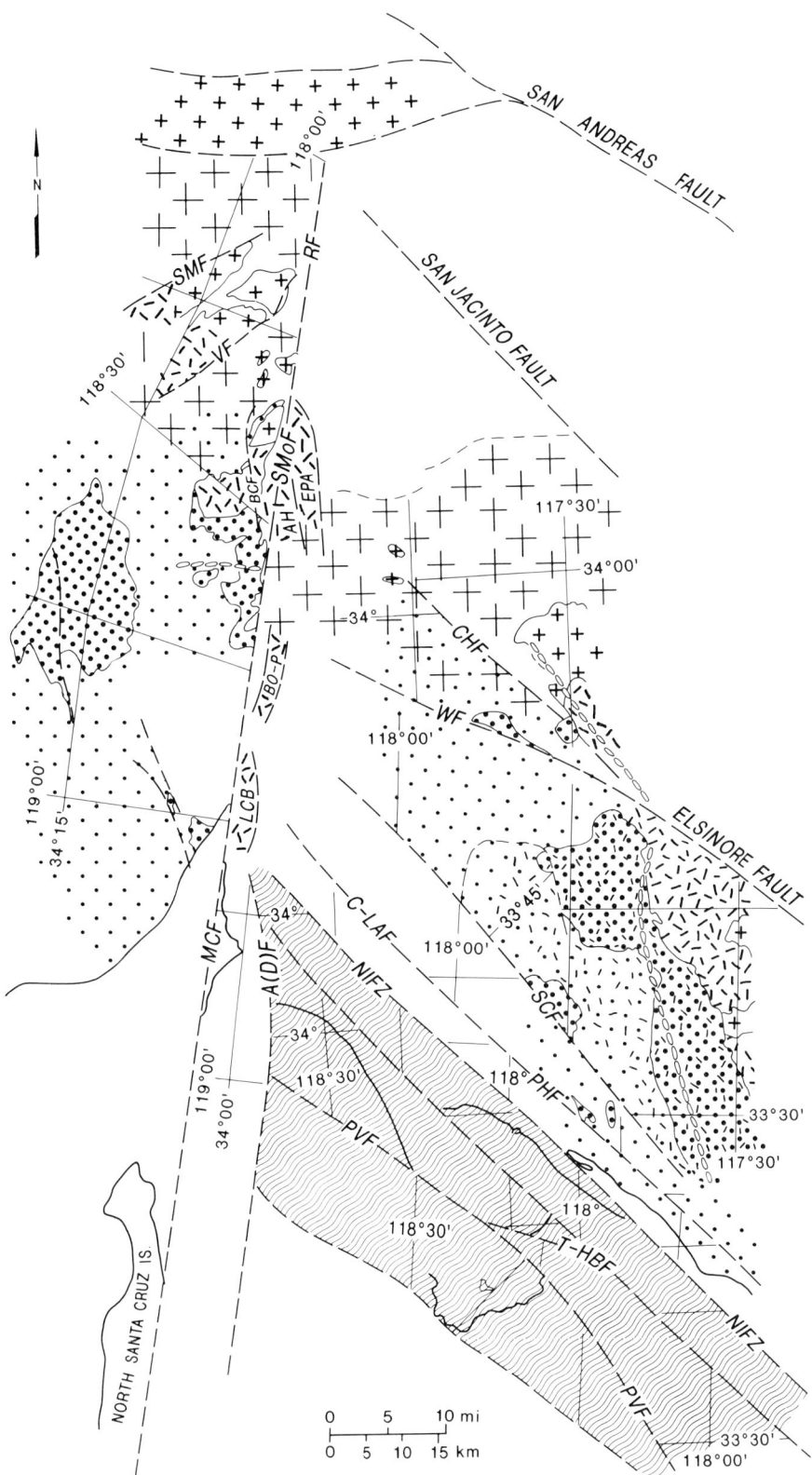

Figure 36. Early Miocene (ca. 20 Ma) reconstruction for the Los Angeles region (after Luyendyk and Hornafius, 1987). Faults and outcrop and subcrop areas of basement rocks and pre-Topanga strata have been added from Figure 4, resulting in local modifications to the reconstruction. Line of ovals indicates Paleocene shoreline. Symbols and abbreviations as in Figures 3 through 9, plus BO-P = Puente block of Brea-Olinda oil field. Present-day coastline and latitude-longitude grid shown for reference.

the western margin of the North American plate. The innermost of the northwest-trending belts consists of Cretaceous granitoid rocks, flanked by mildly metamorphosed sedimentary and volcanic rocks of Jurassic and Early Cretaceous age. These are overlain by Late Cretaceous and Paleogene strata of the fore-arc basin. In the northeast part of the area studied, these beds progressively onlap the granitic basement: Cretaceous and Paleocene strata have not been found north of the Whittier fault, and Eocene beds are overlapped by the Oligocene Sespe Formation that directly overlies basement in the area south of Pomona (Durham and Yerkes, 1964). In the eastern Santa Monica Mountains adjacent to the Benedict Canyon fault, the Sespe is absent and Topanga sedimentary and volcanic rocks overlap a thin section of Paleocene and Cretaceous strata (Durrell, 1954, 1956; Dibblee, 1982a).

The Newport-Inglewood fault zone (including the Compton-Los Alamitos fault) reflects a Mesozoic subduction zone (Hill, 1971) that probably was offset and reduced in width during Paleogene terrane accretion. North of the Transverse Ranges, the same terrane boundary may be represented by the Nacimiento (southern Rinconada) fault (Page, 1970). Within the Transverse Ranges, this boundary is buried beneath a thick sequence of Upper Cretaceous and Paleogene beds that overlie Franciscan rocks in the western Transverse Ranges and are present in the western part of the borderland province (Howell and Vedder, 1981).

Middle Miocene Deformation

Uplift and stripping of the inner borderland and western shelf is the earliest obvious manifestation of the major middle Miocene deformation in the Los Angeles region. In the offshore, "Neogene tectonism reached its peak during middle Miocene time." (Vedder and Howell, 1976). The salient features of this event—major block faulting, volcanism, and localized thick conglomerates—have been recognized for more than 50 years (Woodford, 1925; Reed and Hollister, 1936). The resulting sedimentary record of this "geologic turmoil" (Gardett, 1971) makes up the Topanga Formation (or Group, in the central Santa Monica Mountains). That unit includes (Figure 5) the volcanic rocks plus terrigenous clastics that were derived from granitoid terranes to the north and northeast, sedimentary terranes to the east and northwest, and Catalina Schist to the southwest.

A palinspastic map (Figure 37) provides a middle Miocene (ca. 14 Ma) reconstruction showing fault blocks and dominant local depositional facies. Backslip on the major faults is estimated by interpolation between the early Miocene reconstruction (Figure 36) after Luyendyk and Hornafius (1987) and a late Miocene reconstruction (Figure 39) adapted from Redin (this volume, his Figure 9). The various local features included in Figure 37 represent a composite of significant middle Miocene events and may cover a time span as broad as from 18 to 12 Ma. This has unavoidably obscured local details of that deformation. For example, the Palos Verdes Hills and offshore Wilmington were first uplifted and stripped of their cover before deposition of the Monterey facies shown on this map. Two diagrammatic cross sections (Figure 38) from Campbell and Yerkes (1976) show the style of middle Miocene tectonism and sedimentation in the Los Angeles region and borderland province.

Middle Miocene volcanic rocks in the Los Angeles region are described as "calc-alkaline, low-to-medium-K basalt to andesite" (Weigand, 1982). Most of the thickest volcanic sequences, those over 2000 ft (600 m) thick, are localized along the southern margin of the western Transverse Ranges, at Santa Cruz Island, the Santa Monica Mountains, and Glendora (Wright, 1987a). Radiometric dating of these volcanic rocks was first used to determine absolute ages for the provincial Miocene foraminiferal stages (Turner, 1970). Dates of the volcanic rocks range from 17 to 12 Ma (Crowell, 1987), but they are concentrated between 14 and 16 Ma (Weigand, 1982) in the region from the Santa Cruz and San Clemente islands to the Santa Monica Mountains, El Modeno, and the San Joaquin Hills. Beyond this region an older, early Miocene episode of volcanism, which peaked at about 24 Ma, is recorded at San Miguel Island and elsewhere along the western edge of the borderland (Howell et al., 1987), north of the Los Angeles basin at Neenach and Vasquez Rocks (Weigand, 1982) and northeast of the basin in the Mojave Desert (Nason et al., 1979).

The mid-Miocene deformation is now seen to reflect regional transtension associated with collision of the continental margin against the East Pacific Rise. High heat flux from the rise (Henyey, 1976) softened the continental crust, facilitating rifting, rotation, right-lateral shear, and perhaps crustal folding (Junger, 1976). Several processes have been suggested that might account for these effects. Persistence of divergent mantle flow associated with the overridden rise (Yeats, 1968b) is one such process. Another is initiation of right slip along the transform boundary south of the Mendocino fracture zone (Dickinson, 1981). Northwestward migration of the Pacific plate carried with it—beneath the edge of the North American plate—the subducted slab north of that fracture zone. In its wake, the northward opening of the "no-slab window" produced a diapiric upwelling of hot asthenosphere (Campbell and Yerkes, 1976; Dickinson and Snyder, 1979) whose present position is beneath the Coast Ranges of northern California (Zandt and Furlong, 1982). Plate reconstructions (Figure 6: 20 Ma) suggest that this "slabless window" had opened beneath the Los Angeles region sometime during the early Miocene.

It is tempting to ascribe the initial clockwise rotation of the Transverse Ranges to opposed mantle flow north and south of the Mendocino fracture zone, along that segment that separates the offset portions of the East Pacific Rise (Figure 6; 37 and 20 Ma). The resulting, locally intense, right-lateral mobility might readily produce the rapid regional rotation proposed by Hornafius et al. (1986). Current plate reconstructions (Engebretson et al., 1985) (see also Figure 6), however, show the Mendocino fracture zone as having moved past the region of the Transverse Ranges at least 5 m.y. prior to the period of major rotation.

The Murray fracture zone, on the other hand, is conveniently positioned to have affected this rotation. Geologists (Menard, 1955; Barbat, 1958; Emery, 1960) have long sought to find a genetic link between the offshore Murray fracture zone and the onshore Transverse Ranges, both major west-trending co-linear features. The history

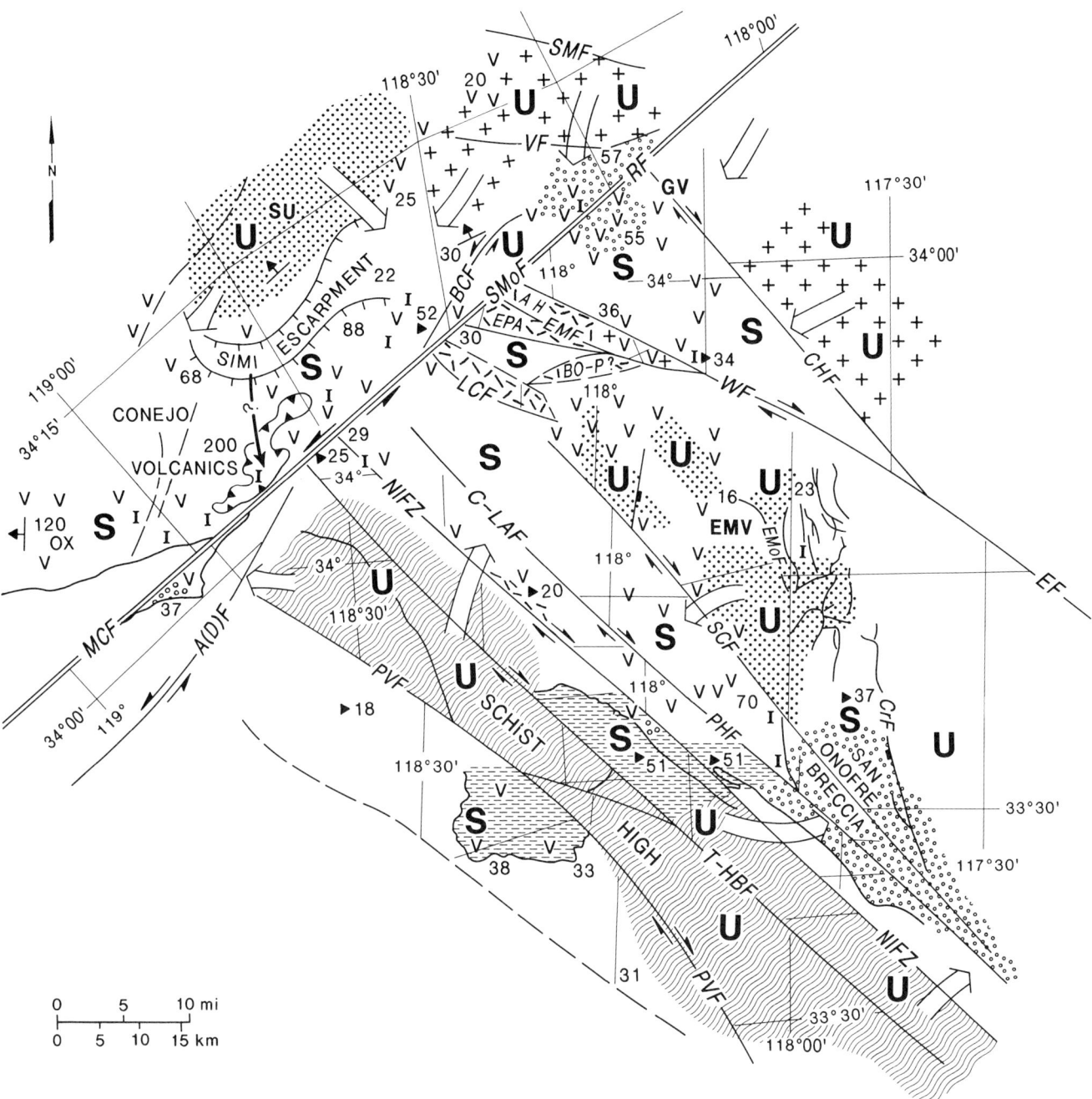

Figure 37. Middle Miocene (ca. 14 Ma) reconstruction for the Los Angeles region (see text for basis of reconstruction). (Simi escarpment from Luyendyk and Hornafius, 1987.) Patterns show uplifted blocks (U) and adjacent source terranes (after Figures 4 and 36) and selected depositional facies (after Figure 5). S = postulated areas of subsidence. Open arrows show inferred directions of sediment transport. Local maximum thicknesses (bold numbers; in hundreds of feet) of sedimentary and volcanic rocks (V) excerpted from Figure 5. I = areas of mid-Miocene igneous intrusive rocks. Arrowed dip symbols (northwest of Santa Monica fault) show dip of strata underlying mid-Miocene angular unconformities. Solid arrow shows postulated movement of allochthonous blocks from Simi escarpment. Unnamed faults in east-central part of map are known to have been active in mid-Miocene time. Symbols and abbreviations as in Figures 3 through 9 and 36, plus OX = Oxnard oil field. Present-day coastline and latitude-longitude grid shown for reference.

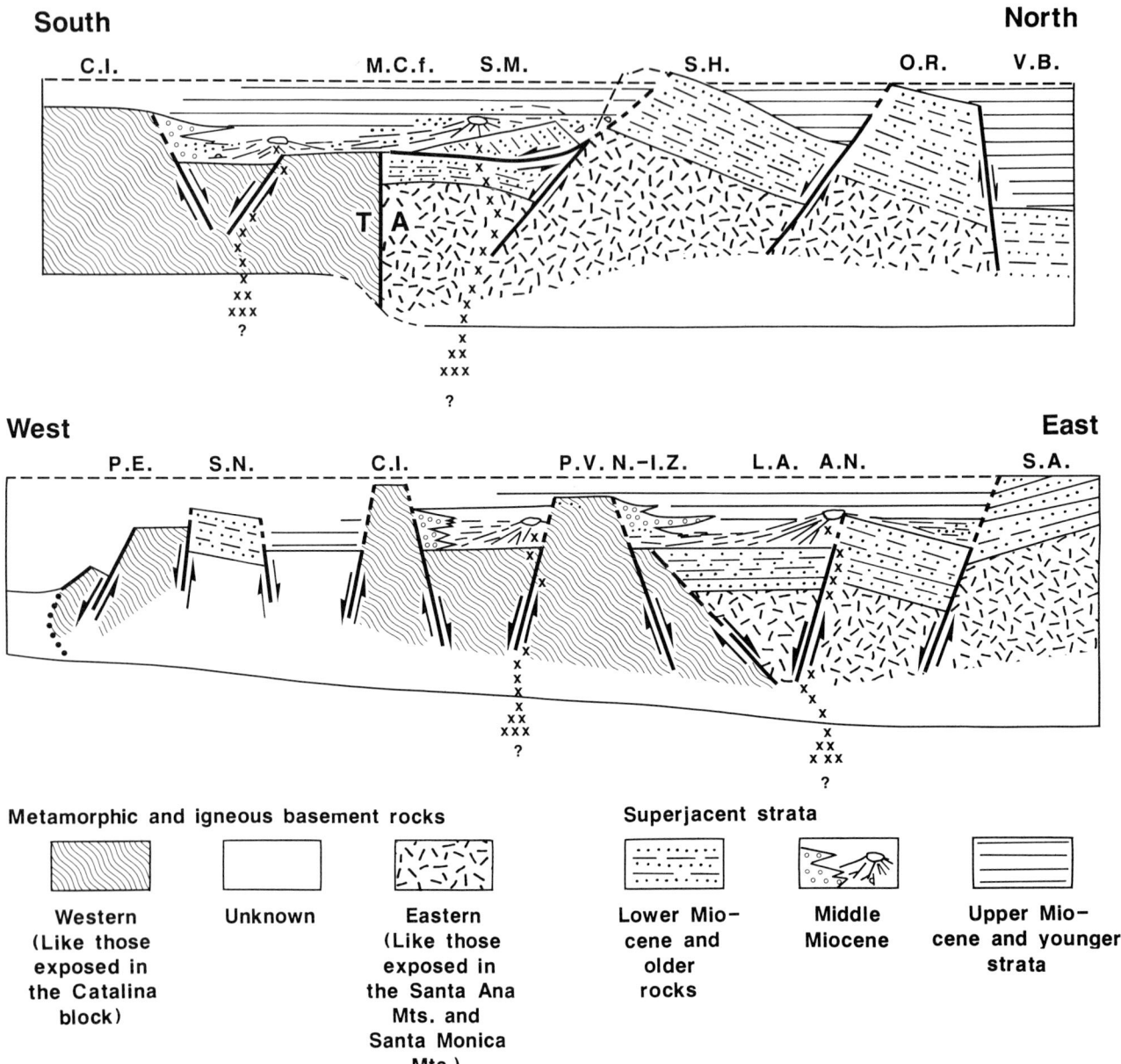

Figure 38. Diagrammatic sections illustrating suggested style of crustal extension north-south through the Santa Monica Mountains (upper section) and east-west through the Santa Ana Mountains and Los Angeles basin (from Campbell and Yerkes, 1976). C.I. = Santa Catalina Island; M.C.f = Malibu Coast fault; S.M. = Santa Monica Mountains; S.H. = Simi Hills; O.R. = Oak Ridge; V.B. = Ventura basin; P.E. = Patton Escarpment; S.N. = San Nicholas Island; P.V. = Palos Verdes Hills; N.-I.Z. = Newport-Inglewood zone; L.A. = Los Angeles basin (central trough), A.N. = Anaheim nose; S.A. = Santa Ana Mountains. X's are igneous intrusions.

of the San Gabriel fault (Crowell, 1975, 1979) suggests that by the end of the middle Miocene the central and western Transverse Ranges had been transferred to the Pacific plate and were moving with it, approximately in line with the Murray fracture zone. The Murray fracture zone is thus suitably located to have caused mid-Miocene rotations within the Transverse Ranges but—unfortunately for such a theory—offset of the rise across the Murray prior to its collision with the continental margin (Figure 6: 37 Ma) was such that it produced left slip and counterclockwise rotation.

Another long-standing and unresolved question concerns the relationship between middle Miocene transtensional events in the region of Los Angeles and crustal extension in the Basin and Range province. Extension in the latter region occurred principally during the past 15 m.y. (Wernicke et al., 1988), although an earlier phase of extension was active locally by 30 Ma (Zoback et al., 1981). Large-scale extension in parts of the Basin and Range province was associated with the development of metamorphic core complexes (Crittenden, 1980; Rehrig, 1986). It is not known whether this process played a role in the tectonic evolution of coastal southern California, although J. C. Crowell (written communication, 1989) noted that some of the faults in the basement rock areas of the San Gabriel Mountains and northern Peninsular Ranges may be low-angle detachment structures.

Extant theories for the origin of Basin and Range extension have been reviewed by Stewart (1978) and Mayer (1986) and include the following: right-slip faulting across a broad belt between the North American and Pacific plates; back-arc spreading that was accentuated above the "no-slab" window; subduction of the East Pacific Rise; and mantle plumes. Formulation of a definitive hypothesis for Neogene plate interactions in the region of southern California awaits further insights into the nature of mantle flow and improved reconstructions of the relative positions (at specific times) both of the plates and of discrete crustal blocks along the plate margin.

Santa Monica Mountains

Effects of the middle Miocene deformation are well exposed in outcrop areas adjacent to the Los Angeles basin. Within the Santa Monica Mountains, Dibblee (1982a) cited evidence for "very rapid accumulation of enormous amounts of sedimentary and volcanic material in a local basin that subsided deeply under conditions of extreme tectonic turbulence" and "local but severe angular unconformities within the Miocene sequence [that] indicate the severity of tectonic interruptions." In the central and western parts of the range, the Topanga Group is up to 20,000 ft (6100 m) in thickness (Yerkes and Campbell, 1979). There it is divided into three formations. Its lower unit, the Topanga Canyon Formation, is Relizian and perhaps partly Saucesian in age and includes a lower marine member (chiefly siltstone or mudstone), a medial nonmarine sandstone member, and an upper marine member of sandstone and siltstone. This formation is overlain by the Conejo Volcanics, which range in age from 15.5 ± 0.8 m.y. to 13.9 ± 0.4 m.y. (Turner and Campbell, 1979). These volcanic rocks are 6000 ft (1800 m) thick in the Oxnard oil field, up to 9800 ft (3000 m) thick in the western Santa Monica Mountains, and thin eastward to near zero in the Old Topanga Canyon area (Yerkes and Campbell, 1979). The upper unit of the Topanga Group, the Calabasas Formation (Yerkes and Campbell, 1979), is "a thick, widespread heterogeneous sequence of sandstone, siltstone and sedimentary breccia that intertongues with, and locally unconformably overlies, the Conejo Volcanics." It is up to 4000 ft (1200 m) thick and mostly Luisian in age; locally, as much as 800 ft (245 m) of the upper Calabasas extends into the lower Mohnian Stage.

The Topanga Canyon Formation rests conformably on lower Miocene marine sedimentary rocks of the Vaqueros Formation in the western Santa Monica Mountains and on the equivalent nonmarine beds of the Sespe Formation to the east (Yerkes and Campbell, 1979). In the eastern Santa Monica Mountains, lower Topanga beds overlie Eocene strata without noticeable angular discordance (Durrell, 1956). Sharp angular discordance between units of the Topanga Group or at the base of its upper formations record episodes of synsedimentary tectonism (Figure 37). In the Oxnard oil field (Dosch and Mitchell, 1964), the basal Topanga dips east at a 35° discordance with underlying northwest-dipping Oligocene and Eocene strata. On the southern edge of the Simi uplift, Cretaceous beds dipped up to 80° northward at the time they were overlapped by Luisian sedimentary rocks of the Calabasas (or Modelo) Formation, but they have since been rotated to lower dips by southward tilting of the Calabasas (Campbell, 1973).

Locally, sharp discordance at the base of the Conejo Volcanics indicates "that significant deformation preceded the outbreak of volcanism" (Yerkes and Campbell, 1979). Younger unconformities confirm the continuing nature of middle Miocene tectonism. South of the Simi uplift, the Luisian Calabasas Formation unconformably overlies Conejo Volcanics with great angular discordance (Dibblee, 1982a), and east of Topanga Canyon it unconformably overlies the Topanga Canyon and Sespe formations. Late middle Miocene deformation is recorded by the common angular discordance between Topanga beds and the overlying Modelo Formation. This discordance increases eastward; adjacent to the Benedict Canyon fault (Figure 15, Sections HH' and II'), lower Mohnian Modelo beds overlap a tightly folded Topanga sequence, Paleocene and Cretaceous strata, and the underlying Santa Monica Slate.

In outcrop the Conejo Volcanics and adjacent sedimentary rocks provide insights into the nature of deposition in an opening rift basin. The volcanic rocks are predominately of submarine origin and include "alternating interbedded andesitic and basaltic breccias, mudflow breccias, flows, pillow breccias, and aquagene tuffs with minor interbedded locally fossiliferous volcanic sandstone and siltstone, limestone, and tuff" (Yerkes and Campbell, 1979). In Malibu Canyon (Crowell, 1976), "the cliffs display irregular masses of andesite or basalt and sedimentary rocks suggesting marked extension of strata at the time the volcanic rocks were emplaced." At Point Mugu (Link and Dibblee, 1988), deep-water shale and sandstone of Saucesian age (lower Topanga) is intruded by diabase dikes and sills and also by sandstone dikes and sills. These clastic injections and associated slump structures are indicative of downslope movement on an unstable basin floor. In the Oxnard area, subsurface relationships (Yeats, 1987) imply that "the basin subsided by caldera collapse as the subjacent magma chamber was emptied." Isotope data (Weigand, 1982) suggest that "magma supplying the Conejo volcanics was little contaminated by continental crust, if at all, during ascent." This lends support to a hypothesis that the western Santa Monica Mountains may be underlain by oceanic crust.

In the central Santa Monica Mountains, another aspect of middle Miocene tectonism involves postulated detachment faults (low-angle normal faults) of Paleogene strata that broke away from the southern flank of the Simi uplift and slid south into the deep rifted Topanga basin (Campbell et al., 1966; Campbell, 1973; Campbell and Yerkes, 1976; Yeats, 1987). These allochthonous blocks are shown in Figures 4 and 37 and in the diagrammatic cross section (Figure 38, upper section) across the Santa Monica Mountains. They have been mapped nearly as far east as

Santa Monica (McGill, 1989). The allochthonous nature of these detachment fault sheets is not universally accepted, however (Truex, 1976, 1977; Dibblee, 1982a).

South of the Malibu Coast fault at *Point Dume*, outcrop and subsurface data (Yerkes and Campbell, 1979) describe a stratigraphic sequence similar to that in the western Santa Monica Mountains but with noteworthy differences. A volcanic sequence (Zuma Volcanics), similar to the Conejo Volcanics, interfingers with marine sandstones and mudstones of Saucesian and Relizian (or Luisian) age that also include sedimentary breccias composed of Catalina Schist detritus, similar to the San Onofre Breccia. A well drilled on Point Dume penetrated 950 ft (290 m) of this sedimentary sequence and 2790 ft (850 m) of volcanic rocks. It bottomed in Catalina Schist (or schist breccia) directly underlying the volcanic interval. Deep-water Monterey Shale of Luisian and Mohnian age overlies the volcanic/sedimentary sequence. Outcropping Zuma Volcanics at Point Dume were dated radiometrically at 14.1 ± 1 m.y. (Berry et al., 1976) and yield magnetic declinations averaging 77.5° (Earth Science Board, 1988), which is consistent with the age and declination of the Conejo Volcanics. The Dume block evidently rotated along with the Santa Monica Mountains (Figure 5) after the volcanic episode.

The apparent absence of pre-Topanga strata and the comparatively thin interval of middle Miocene sedimentary and volcanic rocks (20–30% as thick as sequences to the north and northwest, as shown in Figure 37) suggest that the Dume block was a positive area marginal to the deep Topanga trough of the western Santa Monica Mountains. North of the Malibu Coast fault, minor schist detritus has been noted in Topanga sandstones in the southwesternmost Santa Monica Mountains (Stuart, 1979a), implying derivation from the schist high off the western shelf or beneath Santa Monica Bay. Chemical similarities between the Zuma and Conejo volcanics may constrain major lateral separation on the Malibu Coast fault (Wiegand, 1982).

San Joaquin Hills and Vicinity

Early phases of the middle Miocene deformation in the Santa Monica Mountains are repeated in the San Joaquin Hills and adjacent area, though on a smaller scale. In the San Joaquin Hills, however, intensive tectonism ceased before the end of middle Miocene time except along the Newport-Inglewood fault zone. The general geology of the San Joaquin Hills has been described above (see also Figures 19 and 20). A Topanga sequence 7000 ft (2100 m) in thickness fills a northwest-trending trough between the Shady Canyon fault and the NIFZ. There its lower contact with shallow-marine sandstones of the Vaqueros Formation (here early Miocene and Oligocene in age) is described as gradational and without apparent discordance (Vedder, 1979). Within the trough, the Topanga comprises the lower Bommer Member, sandstone-conglomerate of Saucesian? and Relizian age; Relizian siltstone (and sandstone) of the Los Trancos Member; and the upper Paularino Member, which includes sandstone, siltstone, and volcanic breccia of Relizian and early Luisian age (Vedder et al., 1957; Vedder, 1979). The lower part of this flow breccia yielded a radiometric date of 15.4 ± 1.3 m.y. (Turner, 1970).

Overlying the Topanga Formation is the San Onofre Breccia, which is up to 3000 ft (900 m) thick along the coast but thins northeastward across the San Joaquin Hills (Stuart, 1979b). The Monterey Formation, in this area up to 1500 ft (450 m) thick and late Luisian to early Delmontian in age, overlaps the San Onofre, Topanga, and Paleogene strata. In basal Monterey beds that overlap the Shady Canyon fault and adjacent masses of San Onofre Breccia, the writer has mapped a nodular phosphorite horizon of Luisian age that marks a "hardground" developed on the top of a deeply submerged seaknoll or bank. The San Joaquin Hills block remained a mildly positive area through the remainder of Neogene time.

The Monterey Formation extends undisplaced across the Shady Canyon and other faults, facilitating the construction of cross sections restored to a Top Luisian datum (Figure 20). Early and middle Miocene activity on the Shady Canyon fault produced at least 5000 ft (1500 m) of vertical separation of Paleogene formations. The Topanga is cut by other faults that strike in a northward direction (Figure 19). Diabase dikes that intrude the Topanga and older beds (and perhaps the lower Monterey Formation) form a radiating pattern north and northwest from Laguna Beach (Vedder, 1979) where their radiometric dates average 15.8 m.y. (Weigand, 1982). Magnetic declinations from these and other volcanic rocks in the San Joaquin Hills are scattered (B. P. Luyendyk, oral communication, 1988), presumably owing to subsequent major faulting along the NIFZ and related faults.

To the north in the foothills of the Santa Ana Mountains, a zone of major extensional faults (Figure 4) is overlapped by lower Mohnian beds that show little or no displacement. The predominant northerly strike of these middle Miocene faults implies that the absence of Topanga strata on the eastern Santa Ana shelf (Figure 5) is the result of middle Miocene uplift of a north-trending block in the hanging wall of the Shady Canyon fault (Figure 37). Shallow-marine beds of Relizian age, about 400 ft (120 m) in thickness, form the basal Monterey Formation (Fife, 1979) on the east flank of this block and suggest attenuated deposition on a submerged uplift concurrent with filling of the deep Topanga trough immediately to the south. East of the San Joaquin Hills, the north-striking Cristianitos fault was apparently active during Relizian and Luisian time, forming a depositional hingeline between a wholly nonmarine San Onofre Breccia to the southeast and a partially marine San Onofre facies within the Capistrano syncline (Stuart, 1979b).

Volcanic rocks on the southeastern rim of the Los Angeles basin are similar to those in the Santa Monica Mountains. Within the Topanga trough of the San Joaquin Hills, the lower part of the Paularino Member (Topanga Formation) includes andesitic flows and flow breccias up to 80 ft (25 m) thick. Vedder (1979) described these as "large pods and lenses of breccia that contain a large amount of glassy andesite [that] are intertongued with and possibly are injected into the sandstone beds . . . Bedding in the containing sandstone is highly deformed near some of the breccia masses suggesting that the breccia may have formed along the distal edges of nearby submarine lava flows that were partially injected into the unconsolidated sediments."

A few miles to the north, the *El Modeno Volcanics* (EMV on Figures 5 and 37) occur along the northern and western flanks of the postulated north-trending mid-Miocene high. In outcrop, the El Modeno Volcanics (Schoellhamer et al., 1981) are up to 750 ft (230 m) thick and include basalt flows, palagonite tuff and tuff-breccia, andesite flows and flow breccia, and minor interbedded sedimentary rocks. Evidence for submarine accumulation includes basalt pillows in a siltstone matrix; other intervals have textures suggestive of subaerial accumulation. Minor andesite and basalt dikes intrude the El Modeno Volcanics and adjacent Topanga and Eocene beds. The El Modeno Volcanics rest on a surface eroded into Topanga strata of questionable Relizian age and are overlain by upper Luisian and lower Mohnian beds. An andesite flow provided a radiometric date of 14.1 ± 1.6 m.y. (Weigand, 1982, recalculated from Turner, 1970). Magnetic declinations (B. P. Luyendyk, oral communication, 1988) show a fairly consistent 45° clockwise rotation (Figure 5) that is believed to be the result of rotation of a small local fault block (Figures 36, 37).

Viewed as a whole, faulting and volcanism in the San Joaquin Hills (and the adjacent area to the north) provide convincing evidence of middle Miocene transtensional deformation. North and northwest-striking faults and dikes reflect east-west extension and right-lateral transform motion (Wright et al., 1973). The El Modeno Volcanics and those of the San Joaquin Hills and Laguna Beach mark the eastern edge of middle Miocene volcanism adjacent to the Los Angeles basin. North-striking mid-Miocene extensional faulting (Cristianitos fault and faults in the foothills of the Santa Ana Mountains) occur east of these volcanic rocks, parallel to the gravity high described previously. These features suggest that right slip on the Shady Canyon fault, and perhaps other faults involved with mid-Miocene rotation of the Transverse Ranges, may have been at least partially absorbed across a broad zone of extensional faulting above an area of crustal attenuation and mantle upwelling (McCulloh, 1960). In marked contrast to the northern edge of the Los Angeles basin, where later Miocene rotation and left slip has been followed by major thrusting, the post-Topanga history of the San Joaquin Hills and Santa Ana shelf reflects only mild vertical movements.

Glendora Volcanics

Northeast of the Los Angeles basin in the Puente and San Jose Hills and in the San Gabriel Valley, surface and subsurface data provide limited evidence relating to the middle Miocene deformation. The Glendora Volcanics (GV on Figures 5 and 37) crop out immediately south of the postulated zone of rifting and left slip that marks the exposed southern edge of the Transverse Ranges. In outcrop these rocks include flows, flow breccias, tuffs, and tuff breccias that are predominately andesitic but range in composition from basaltic to rhyolitic (Shelton, 1954). Internal textures and structures indicate that extrusion was both submarine (lava pillows) and subaerial (tops of flows are vesicular and oxidized). No radiometric dates have been reported from the Glendora Volcanics. They interfinger with and are overlain by Topanga beds dated as Luisian (and Relizian?) (Shelton, 1955). Magnetic declinations measured in the Glendora Volcanics show a wide scatter (B. P. Luyendyk, oral communication, 1988) that is ascribed to complex local faulting associated with rifting and left slip and that may also involve late Neogene thrusting on the Sierra Madre and related faults.

In their area of outcrop, the Glendora Volcanics rest directly on crystalline basement (Shelton, 1954). Exploratory wells a few miles to the southwest (Figures 4, 5) penetrated up to 3700 ft (1130 m) of this volcanic sequence, its maximum known thickness. There, Topanga (and locally Paleogene) beds are present between the volcanic and basement rocks. Beneath the eastern margin of the San Gabriel Valley, these (or similar) volcanic rocks extend southwest to the Whittier fault. No such rocks have been reported within the thin Topanga sequence beneath the southeastern Puente Hills (Figure 5). Their absence may be due to late middle Miocene erosion but may also mark the eastern edge of rift volcanism as noted farther south, east of the San Joaquin Hills and El Modeno areas. South of the Glendora Volcanics outcrop, proprietary gravity mapping (see also Oliver et al., 1980) shows a north-northeast-trending gravity high of about 15 mgals relief that is flanked on the east by a south-southeast-trending low. Relief on the basement surface seems insufficient to explain these features and they may be due to crustal extension and attenuation like that postulated beneath the San Joaquin Hills.

The *Buzzard Peak Conglomerate*, which is up to 3000 ft (915 m) thick, overlies the Glendora Volcanics in the San Jose Hills and is considered part of the Topanga Formation (Woodford et al., 1946). It is poorly sorted and its clasts range up to boulders in size and are mostly granitoid and related plutonic rocks. To the south and southwest, it thins and grades into sandstones and shales. Its southern equivalent, the Diamond Bar Sandstone (Figure 5) (Durham and Yerkes, 1964), is up to 1650 ft (500 m) thick and resembles later Miocene submarine fans in its depositional setting. The Buzzard Peak Conglomerate was probably derived from nearby uplifted granitoid blocks within the central Transverse Ranges, just across the rifted margin. As shown on the middle Miocene restoration (Figure 37), backslip on this fault system places the Buzzard Peak Conglomerate adjacent to the similar, thick Topanga conglomerates of the easternmost Santa Monica Mountains. Also juxtaposed in this restoration, thick conglomerates north of the Raymond fault in the Eagle Rock area (Lamar, 1970) are composed almost entirely of a biotite-hornblende-rich quartz diorite detritus not seen in the Buzzard Peak Conglomerate. These conglomerates in the Eagle Rock area may be younger than Topanga (Dibblee, 1989). Further detailed comparisons of clast lithology and provenance areas may well provide useful constraints on fault movements.

Los Angeles Basin

The Topanga Formation and its equivalents are deeply buried throughout most of the Los Angeles basin, and well data provide only limited insights into patterns of mid-Miocene faulting and deposition. Topanga sedimentary and volcanic rocks thin across the Anaheim nose (Figure 5), northeast of the projected extension of the Shady Canyon fault. On the crest of the nose, they are locally missing west of a north-striking fault whose presence is inferred from an abrupt contrast in Paleogene formations encountered beneath the unconformity. On the northeast flank of the nose, maximum thicknesses do not exceed 1600 ft (500 m), except at West Coyote where an anomalous thickness (Yerkes, 1972) is believed to be the result of steep dip and thrust repetition. Local variations in thickness suggest that the Anaheim nose may be a complex of small fault blocks similar to those mapped east of El Modeno. The Topanga may thicken northward significantly into the footwall of the Whittier fault, below the depth of well control. That formation is up to 3600 ft (1100 m) thick on the north side of that fault, and it has been suggested (Yeats, 1987) that the Whittier fault served as a depositional hingeline, subsiding to the north in the Puente Hills block, during most of middle Miocene time.

Down the northwest plunge of the Anaheim nose, at the limits of exploratory drilling to date, Topanga volcanic and sedimentary rocks are seen to increase in thickness. Along the flank of the Newport-Inglewood trend, northeastward thickening and an increase in the proportion of volcanic rocks is observed at Inglewood, Dominguez, and Long Beach, where Topanga overlies metamorphic basement east of the NIFZ (Figures 4, 5). East of Huntington Beach, wells have penetrated up to 2100 ft (640 m) of Topanga without reaching its base. There, extrusive volcanic rocks are encountered only down the basin flank, on trend with the Paularino volcanic sequence of the San Joaquin Hills trough. At the northern end of the central basin, a deep well in the San Vicente oil field (Figure 18, well J) penetrated 1055 ft (322 m) of volcanics overlying questionable Sespe red beds (Yerkes et al., 1965; Yeats, 1973) or nonmarine Topanga beds of the Ferndale Member (T. R. Redin, written communication, 1989). At all of these localities, the volcanic-sedimentary sequence is thickening toward the axis of the later Neogene basin.

Beneath the central deep of the Los Angeles basin, the Topanga Formation is overlain by 16,000-24,000 ft (4.9-7.3 km) of younger beds (Figure 10) and has not yet been reached by exploratory drilling. Its lithology, thickness, and distribution and the nature of the underlying sedimentary section and basement are matters for conjecture. One interpretation suggests that Sespe red beds may extend beneath most of the basin northeast of the NIFZ, overlying a sequence of older Eocene, Paleocene, and Cretaceous strata that extends from the Anaheim nose beneath the entire northeast flank of the basin but not beneath its south flank (Yerkes et al., 1965, their figures 6 through 8). Later hypotheses postulate rifting above a schist floor (Yeats, 1968) or a volcanic-floored pull-apart basin (Crowell, 1974, 1976) overlying attenuated and broken crust. More recently, the rotational hypothesis (Luyendyk et al., 1980; Luyendyk and Hornafius, 1987) interprets the central Los Angeles basin as a deep, triangular "sphenochasm" opened along the join between rotated and unrotated blocks.

The present reconstruction, which applies the rotational hypothesis, does not require the creation of a single large area of "new" basin floor during the middle Miocene deformation. Instead, most of the necessary extension is accommodated in a series of smaller rifted openings beneath the present northern margin of the Los Angeles basin and closely adjacent to the "join" (Malibu Coast-Santa Monica-Raymond fault system). By this model, the basin floor resembles that suggested by Crowell (1974, 1976):

fragmented blocks of the Paleogene-Cretaceous-basement sequence, engulfed in a mixture of sedimentary and volcanic rocks similar to exposures in the central and western Santa Monica Mountains and San Joaquin Hills.

Volcanic rocks have not been found anywhere on the schist high of the western shelf. On its southwest flank in the *Palos Verdes Hills*, thick sills of andesite and basalt intrude Relizian and lower Luisian strata of the Monterey Formation (Woodring et al., 1946), and extrusive rocks (pillow basalts) have been reported (Conrad and Ehlig, 1987). An exploratory well drilled on the southwestern tip of the Palos Verdes Peninsula penetrated a significant interval of volcanic rocks in the lower half of a 3800-ft (1160-m) section of Monterey overlying schist basement (California Division of Oil and Gas, 1964). Those volcanic rocks may be associated with intrusion along the San Pedro Escarpment fault zone or may represent the edge of the volcanic sequence thought to underlie the offshore Santa Monica basin (Junger and Wagner, 1977). A tuff associated with the volcanic rocks in the Palos Verdes Hills (but not necessarily of local origin) has been dated radiometrically at 14.5±1.1 m.y. (Turner, 1970). Magnetic declinations obtained from these volcanic rocks provide fair evidence for 45° clockwise rotation that is ascribed to local rotation of the Palos Verdes block, perhaps between the Palos Verdes and San Pedro Escarpment faults (B. P. Luyendyk, oral communication, 1988).

Volcanic rocks have not been reported in several other parts of the Los Angeles basin where drilling has provided an adequate sample of the Topanga Formation or other facies of the same age. On the northern shelf, in the narrow trough (or graben) between the Las Cienegas basement high and the Elysian Park anticline (Figure 5), six wells have penetrated Topanga intervals of 1200-3300 ft (365-915 m) without encountering volcanic rocks. In the thick sequence of "Black Shale" (Monterey facies) on the southeast plunge of the Wilmington anticline (Figures 5, 31), no volcanic rocks have been encountered except for a thin interval within the Luisian section, immediately west of the NIFZ in the Seal Beach oil field (Truex, 1972). This occurrence may be associated with local igneous intrusion along the fault zone.

Intrusive igneous rocks are reported in outcrop and in a number of wells within and north of the Whittier fault zone (Durham and Yerkes, 1964; Yerkes, 1972), west from the vicinity of the Kraemer oil field. These rocks are generally diabasic and occur as sill-like intrusions as much as 750 ft (200 m) thick and extending as far as 2 mi (3.2 km) north of the Whittier fault. They intrude the Topanga Formation and, more commonly, the early Mohnian La Vida Shale; in several wells they are overlain by late Mohnian siltstone. Intrusion evidently took place in the interval 13-9 Ma, near the end of middle Miocene time. Similar intrusive rocks localized near the Inglewood fault cut the Luisian Sentous zone in the Inglewood oil field. These fault-related intrusions are somewhat younger than the peak period of volcanism in the Los Angeles region. They appear to represent the final phase of emptying of the magma chambers along still active faults.

Other effects of *mid-Miocene faulting* have been noted along these major northwest-striking faults. In the Long Beach oil field (Figure 25, Section xx'), the Topanga interval is

110 ft (335 m) thick south of the Cherry Hill fault (NIFZ) and as thin as 100 ft (30 m) on the uplifted schist block north of the fault. At the northwest end of the Whittier fault, the hanging-wall block is underlain by a wedge of basement rock that continues west of the East Montebello fault into the Alhambra high (Figure 4). Topanga sedimentary and volcanic rocks are missing from the central part of this basement wedge (Figure 5). On the Alhambra high, an extension of the Workman Hill fault separates slate basement from granitoid rocks that extend northeast beneath the San Gabriel Valley. Three wells that reached granitoid basement in this area show the Topanga Formation to thicken off the north flank of the Alhambra high (Figure 8, Section DD'), but they did not encounter volcanic rocks.

Sedimentary and volcanic rocks of the Topanga Formation also are absent from the slate core of the Elysian Park anticline. In the present reconstruction (Figures 36, 37), this and other narrow basement highs—the Las Cienegas block, the Alhambra high, and perhaps even the schist block in the hanging wall of the Whittier fault at Brea-Olinda—are visualized as crustal slivers that fragmented from the edges of major blocks adjacent to the join in an early stage of rifting. During the peak interval of volcanism and rotation, these slivers "floated" into new positions along the edges of these blocks or within the small, newly opened "sphenochasms." Figure 36 shows possible initial locations of the four "slivers" cited. Figure 37 shows possible intermediate positions during the period of maximum middle Miocene mobility.

Less speculative features shown on the middle Miocene reconstruction (Figure 37) include the two Monterey depocenters—Palos Verdes Hills and Wilmington East Plunge—superimposed on the schist high of the western shelf. These two basinal areas formed on the schist surface as it subsided following early Miocene uplift and erosion. During the late Relizian and Luisian, as terrigenous clastics and submarine volcanic rocks of the Topanga Formation onlapped the lower northeastern margin of the western shelf, these two "perched" basins accumulated the biogenous and hemipelagic strata of the Monterey facies.

Chronology

The chronology of tectonism outlined in Figure 1 follows other recent chronologies (Crowell, 1987; Luyendyk and Hornafius, 1987; Schwartz and Colburn, 1987; Vedder, 1987) in most respects. For the Los Angeles region, the data and hypotheses outlined above suggest the following scenario for early and middle Miocene deformation.

37 to 22 Ma. Oligocene emergence of coastal California is recorded by the widespread distribution of the nonmarine Sespe Formation and is believed to be due principally to tectonic uplift related to the approach of the Pacific-Farallon spreading ridge (Nilsen, 1987). Initial contact of the North American plate with the Pacific plate has been estimated at about 37 Ma (Engebretson et al., 1985) (see also Figure 6) to 29 Ma (Crowell, 1987). Sespe deposition was followed by a marine transgression that shows marked local variations in its timing and in the thickness of the resulting Vaqueros Formation and its correlatives. These variations provide evidence for the earliest stages of the crustal extension and rifting of the middle Miocene deformation. Maximum thicknesses have been measured in the western Santa Monica Mountains (Yerkes and Campbell, 1979), where a 2600-ft (800-m) Vaqueros sequence includes a lower siltstone-mudstone member of Zemorrian age and an upper sandstone member of Saucesian age. In the San Joaquin Hills, the deep Topanga trough is underlain by a 3800-ft (1160-m) Vaqueros interval of Zemorrian and Saucesian age (Vedder, 1979). Both of these local rift basins had begun to form by earliest Miocene time.

The sedimentary fill in these early basins does not contain schist debris, implying that the inner borderland and western shelf either were not yet uplifted and shedding sediments or had a cover of Paleogene and Cretaceous beds from which the Oligocene-Miocene clastics were derived. In the outer borderland, the reported occurrence of early Miocene volcanics (Howell et al., 1987) suggests that rifting in coastal southern California first affected that area and progressed eastward. By early Miocene time, these borderland terranes were attached to the Pacific plate and were involved in right-slip faulting along the widening transform segment of the plate boundary.

22 to 16 Ma. Sometime during the early Miocene, high heat flow and transtension began to affect most of the Los Angeles region, promoting the development of local basins. As described by Luyendyk and Hornafius (1987), "Crustal attenuation resulting from the extension presumably caused subsidence within these basins due to isostatic compensation within a thinned continental crust (McKenzie, 1978). The creation of new crust due to magmatic injection at depth within the basins in the areas of greatest crustal extension would also result in subsidence due to the ensuing cooling and thermal contraction (Turcotte and McAdoo, 1979)."

The early Miocene restoration (Figure 36) suggests that a singular narrow trough may have extended from the San Joaquin Hills to the western Santa Monica Mountains. Within this trough, the locus of the Vaqueros deposition cited above Paleogene and Cretaceous strata may be preserved, at least locally. Along its southwestern edge, continued uplift exposed the Catalina Schist on the western shelf and inner borderland. The oldest schist debris derived from this uplifted area is found (Stuart, 1979a) in upper Saucesian strata (ca. 18 Ma) on Santa Cruz and Anacapa islands along the western (now northern) edge of the source terrane and in Relizian deep-water sedimentary rocks (ca. 16 Ma) along its northeastern edge in the San Joaquin Hills. These relationships provide further evidence for the eastward progression of the middle Miocene deformation. Within the inner borderland, other local rift basins, including the Santa Monica, San Pedro, Catalina, and Oceanside basins, may have been initiated during early Miocene time, leaving intervening high blocks as local sources of the San Onofre Breccia.

As yet, no evidence has been found to indicate significant early Miocene tectonism north or northeast of the Shady Canyon fault. In the Santa Ana Mountains and eastern Puente Hills (Durham and Yerkes, 1964; Schoellhamer et al., 1981), the nonmarine Sespe and marine Vaqueros are mapped as alternating facies within a single undifferentiated formation, 700-3000 ft (215-900 m) in thickness, reflecting an oscillating shoreline on a relatively stable basin margin. A similar relationship has been noted in the central Santa Monica Mountains (Schoellhamer et al., 1981), thus extending this stable marginal block northwest across the later Santa Monica-Raymond fault system (Figure 36). Within this block, indications of the onset of major deformation, including volcanic rocks, coarse granitic detritus, and other elements of the Topanga facies, do not appear to be older than middle Miocene.

16 to 12 Ma. Rotation and left slip opened the "join" along the boundary between the Transverse Ranges and the Los Angeles basin and borderland province. The prerift reconstruction (Figure 36) (Luyendyk and Hornafius, 1987) shows that this "join"—the Malibu Coast-Santa Monica-Raymond fault trend—was initially oriented nearly north-south, subparallel to the trend of the East Pacific Rise, which had been overridden in this region a few million years earlier (Figure 6: 20 Ma). Along this "join," volcanic rocks welled up within a series of small "sphenochasms," extending from Glendora to Santa Cruz Island, and mingled with floods of clastic sediments derived from nearby uplifts. The Conejo Volcanic trough extended eastward into the northern Los Angeles basin, perhaps as far as the Anaheim nose (Figure 37). Crustal slivers may have fragmented from the margins of adjacent rotating blocks to be rafted into new locations within the trough or against its other bounding blocks. Fragmentation was greatest near the "join," but other more distant pieces, such as the El Modeno block and the Palos Verdes Hills, appear to have rotated in place as transform shear pervaded the entire region.

The Glendora Volcanics and the easternmost Santa Monica Mountains may form a smaller sphenochasm that could have opened the San Gabriel embayment on the west flank of the San Jose Hills. Away from the "join," igneous rocks rose along the Whittier, Shady Canyon, Pelican Hills, and Newport-Inglewood faults, though in smaller amounts. Igneous intrusions into upper Luisian beds at Inglewood and into lower Mohnian beds along the Whittier fault signify the final emptying of the magma chamber.

Right slip between the nonrotating blocks occurred mostly on these major fault zones but also was accommodated throughout triangular regions of attenuated crust. These appear to fit the model proposed by Luyendyk and Hornafius (1987) in which "sphenochasms . . . south of the rotating blocks [Transverse Ranges] are bounded by northwest-southeast-trending right-slip faults, north-south-trending normal faults, and east-west-trending oblique-slip faults." The equivalent faults for the central part of the Los Angeles region (Figure 37) are, on the southwest, the NIFZ and parallel faults; on the east, the Cristianitos fault, the El Modeno and related faults, and the north-trending faults in the San Joaquin Hills; on the north, the Malibu Coast and Santa Monica faults. A smaller triangular region to the north is bounded by the Whittier fault, the Chino Hingeline fault, and the Raymond fault.

By about 13 Ma, the peak period of volcanism and tectonic activity had slackened. Granitic detritus continued to flood into basinal areas in the northern and northeastern parts of the region, as the Topanga Formation infilled topography on older rocks (Schwartz and Colburn, 1987). Few or none of these terrigenous clastics reached the southwestern parts of the region, however, where biogenous and hemipelagic sediments of the Monterey facies had begun to accumulate in the areas of the Oxnard shelf, Palos Verdes Hills (Altimira Shale Member), East Wilmington ("Black Shale"), and San Joaquin Hills. Intervals of schist sand or breccia within Monterey beds of Luisian age at Seal Beach (Truex, 1972) and the Palos Verdes Hills (Russell, 1987) may have been redistributed from submarine highs adjacent to these Monterey depocenters.

As left slip progressed along the southern edge of the Transverse Ranges, the "join" lost its initial irregular form and was sheared into a linear feature. Extreme structural linearity is the most distinctive characteristic of major strike-slip faults (Sylvester, 1988), and in the reconstructions of Figures 36, 37, 39, and 42, this attribute has been imposed on the Malibu Coast-Santa Monica-Raymond fault zone. As previously noted, the middle Miocene restoration (Figure 37) is a composite of local events that occurred during the peak of middle Miocene tectonic activity from about 16-13 Ma. Intermediate restorations would clarify the relationships between these events, but available data are inadequate for this task.

12 to 9 Ma. After the cessation of volcanism, cooling of the lithosphere was aided by horizontal heat transfer into crustal blocks adjacent to the rifted areas (Turcotte and McAdoo, 1979). The subsidence that accompanied this cooling was thus fairly uniformly distributed throughout the entire coastal region during the latter part of middle Miocene time. The shoreline regressed to the vicinity of the present-day San Andreas fault. In the Transverse Ranges, the northeastern limit of the San Andreas transform zone was localized along the San Gabriel fault beginning about 12 Ma (Crowell, 1982). West of it, the entire deeply subsided Los Angeles region was now moving with the Pacific plate.

East of the San Gabriel-San Andreas zone, source terranes in the Colorado and Sonoran deserts (Crowell, 1987) had been disrupted by rifting. Along this zone, the Cuyama, Soledad, Ridge, Punchbowl, and Diligencia basins formed a series of basins "inboard and adjacent to the primary strandline [that formed] sediment trap[s] throughout the Miocene, allowing accumulation of distinctively organic-rich, terrigenous-poor sediments in the more outboard basins" (Lagoe, 1987).

In the Los Angeles area, fine-grained deep-water sedimentary rocks of late middle Miocene age extend from the Palos Verdes Hills (Valmonte Diatomite Member) and Laguna Beach northeast and north to the Glendora area where early Mohnian shales of the La Vida Member crop out within the Sierra Madre fault zone (Proctor et al., 1970). A comparison of the La Vida facies of the Puente Hills and Santa Ana Mountains, the nodular shale of the northern Los Angeles basin and western shelf, and the biogenous-hemipelagic Monterey facies (see Redin, this volume, his Figure 7) shows the chief distinction to be the southwestward decrease in the proportion of silt and clay. Over most of the sea floor, anoxic conditions prevailed to varying degrees (Schwartz and Colburn, 1987) and fostered the preservation of abundant organic material in what would become the most important petroleum source rocks in the region.

These strata provide an unsatisfactory record of concurrent deformation, as they blanket banktops, slopes, and basin floors with little discernable variation in composition or texture. Differences in thickness may be controlled less by sea floor topography than by patterns of bottom currents. In a few areas, such as East Wilmington, anomalous thicknesses attest to areas of local tectonic subsidence. The earlier horizontal movements undoubtedly continued, including rotations within the Transverse Ranges, left slip along that area's southern boundary, and right slip on the NIFZ, Whittier-Elsinore fault, and other major northwest-striking faults of the Los Angeles region. A declination-versus-age curve for the Santa Monica Mountains and northern Santa Cruz Island (Hornafius et al., 1986) postulates that rotation maintained a uniform high rate from 15–10 Ma and slowed markedly over the remainder of the Neogene.

Late Miocene Stability

A period of late Miocene stability followed the middle Miocene deformation. As drainage patterns were reestablished on the landmass east and northeast of the Los Angeles region, sediments filled the nearshore basins. Submarine fans could then extend across a gently sloping bathyal plain into the area of the present Los Angeles basin (Redin, this volume) and be thickest and most continuous in middle ("early late") Mohnian time (ca. 8 Ma). Mid-fan and lower-fan facies provide a useful record of basin shape and of local vertical movements, during most of late Miocene and Pliocene time. Figure 39 is a restoration for the end of middle Mohnian time. It has been slightly revised from the late Mohnian palinspastic map of Redin (this volume, his Figure 9), with details added from Wright (1987c, his figure 4) and backslip on faults adjusted to allow for a linear Santa Monica–Raymond fault zone and minor post-Mohnian rotation of the Santa Monica Mountains.

Tarzana and Puente Fans

Two major submarine fans fed the Los Angeles region during most of middle and late Mohnian time (Wissler's Divisions D and C; ca. 9–6.5 Ma). The Tarzana fan (Sullwold, 1960) is a sequence of deep-water sandstones within the Modelo Formation. This fan was built southward across the San Fernando Valley, perhaps as overflow from the thick sand accumulations of the Ridge basin and eastern Ventura basin. The fan crossed the site of the Santa Monica Mountains, whose late Neogene uplift had not yet begun, and flowed into the northwestern part of the present Los Angeles basin. It lapped around the flanks of the Las Cienegas high and the Inglewood portion of the western shelf. On the southwest flank of the Elysian Park anticline, the Mohnian sand sequence is nearly 5000 ft (1525 m) in thickness (Lamar, 1970).

The Puente fan (Gourley, 1975) is essentially equivalent to the Soquel Member of the Puente Formation. During middle Mohnian time, it extended southwestward from the region of the eastern San Gabriel Mountains across the Puente Hills area and the Whittier fault and onlapped the northeast flank of the submerged Anaheim nose. A possible southward connection into the Capistrano syncline is suggested by a 400-ft (120-m) interval of Soquel sandstones east of El Modeno (Schoellhamer et al., 1981).

Beneath the eastern Puente Hills (Durham and Yerkes, 1964), the Soquel Member is 2000–3000 ft (600–900 m) in thickness; it wedges out against the Chino Hingeline fault, suggesting that activity on this fault continued into the late Miocene. South of the Whittier fault, maximum thicknesses are about 2000 ft (600 m) in the East Coyote oil field and about 2800 ft (850 m) thick in the Richfield oil field (Yerkes, 1972). Thicknesses, thus, are roughly comparable on both sides of the Whittier fault and indicate that it may not have been a significant depositional hingeline at this time. Local variations in thickness and sand content may be due either to fan morphology or to structural growth; data in most areas are insufficient to evaluate these alternatives. Near the western end of the Whittier fault, a very thin Soquel interval north of the fault (Yerkes, 1972) extends northwest onto the Alhambra high and suggests that this part of the fault zone was an uplifted sill (or perhaps a submarine slope) at this time. Farther to the southeast along the Whittier fault (Redin, this volume, his Figures 8, 9), similar features have been implied.

The Puente and Tarzana fans merge in the area between the Anaheim nose and the Las Cienegas high (Figure 39). Mohnian sandstones cross the NIFZ on the southeast plunge of the Inglewood high, where the City of Inglewood pool is trapped in a sand pinch-out. Farther south, these sandstones onlap the western shelf, cross the plunge of the Torrance anticline, and extend beneath San Pedro Bay. There they branch around the uplifted block southwest of the THUMS-Huntington Beach fault, which remained positive into the middle Pliocene (Redin, this volume). Similar thicknesses of early and middle Mohnian beds are found on both sides of the Palos Verdes fault in the Beta oil field (Figure 13), implying that any movement on the fault during this time had little or no vertical component. In the Long Beach and Seal Beach oil fields, significant thinning of Mohnian sandstones reflects either a residual high or continued deformation on this part of the NIFZ.

Reconstruction of the Tarzana fan provides strong evidence for approximately 9 mi (15 km) of post-Mohnian left slip on the Santa Monica–Raymond fault system (Lamar, 1961; Redin, this volume). Figure 39 shows the general continuity of the fan. A more detailed comparison (Figure 40) of the distribution of late Miocene sandstone units north and south of the Santa Monica fault lends additional support. In both panels, thick sandstones in the eastern channel (points C and C') are of middle Mohnian age and did not persist into late Mohnian. These sandstones thin or pinch out westward onto basement highs (points B and B'). The western channel (points A and A') contains sandstones of both middle and late Mohnian age. Farther west, Mohnian sandstones are absent and a thick Delmontian sandstone represents the final vestige of the Tarzana fan.

Figures 39 and 40 show an alignment, within the Tarzana fan, between the Las Cienegas basement high and the slate

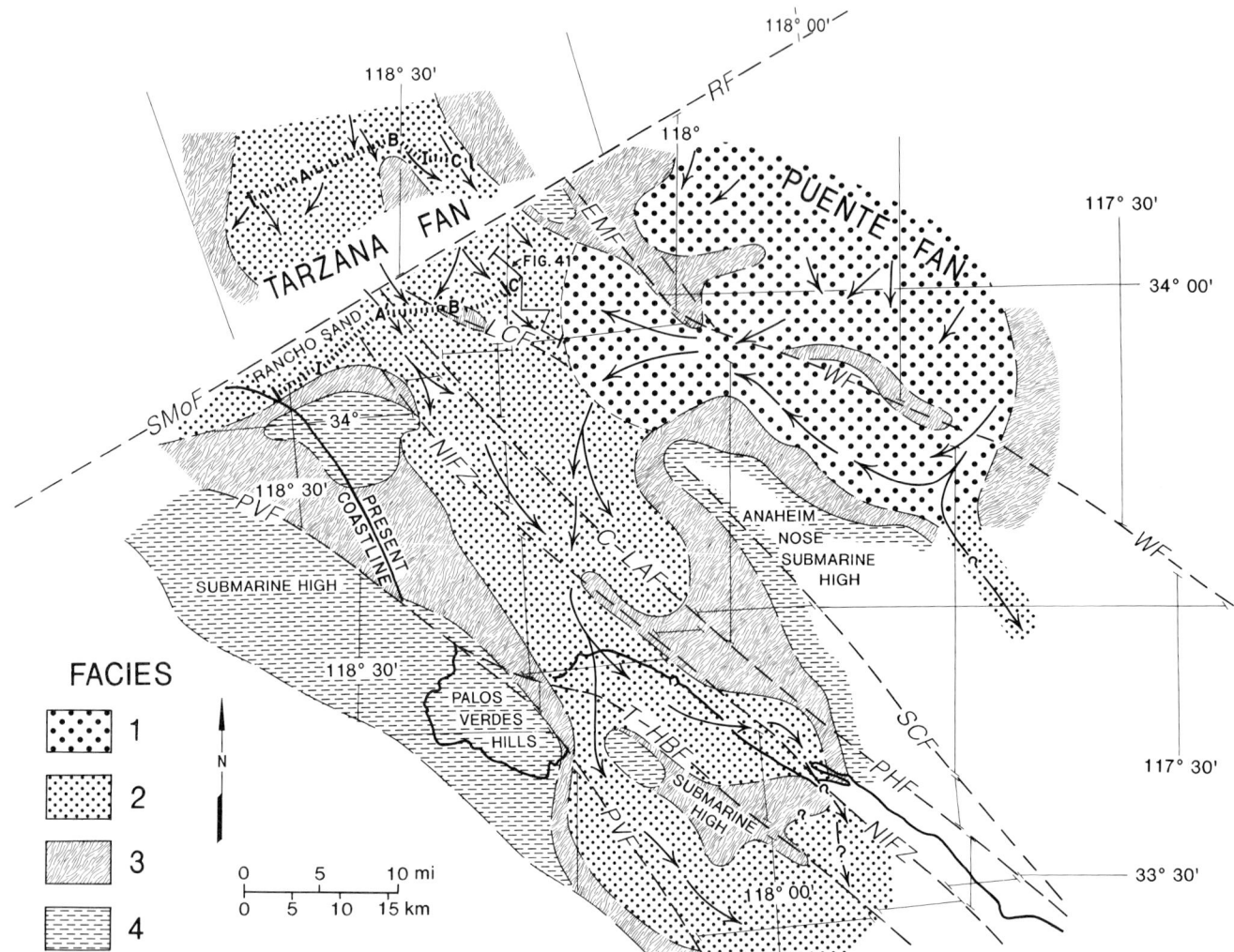

Figure 39. Late Miocene (ca. 8 Ma) reconstruction for the Los Angeles basin showing submarine fans and related facies (from Redin, this volume, his Figure 9, with minor revisions as described in text). Facies 1 = mid to upper channelized fan (conglomeratic); facies 2 = "supra fan" (high sand %); facies 3 = overbank, shelf slope, and distal turbidites (low sand %); facies 4 = submarine high (hemipelagic and biogenic sedimentation). Arrows show inferred directions of sediment transport. ABC and A'B'C' are stratigraphic cross sections of Figure 40 (I = intersection with Interstate 405 in each section). Location of Figure 41 also shown. Present-day coastline and latitude-longitude grid shown for reference. Abbreviations, etc., as in Figures 3 through 5.

high near Encino. Metamorphic rock types are dissimilar, and these two localities do not necessarily reflect a single block that moved within the "join." It seems likely, however, that the two temporarily adjacent blocks may have been involved in a single uplift as transverse folding within the zone of left slip divided the fan into two channels.

Mio-Pliocene Deformation

Onset of Marginal Uplifts

A distinctive phase of deformation during latest Miocene and early Pliocene time shaped the present Los Angeles basin by major vertical separation on the Santa Monica, Las Cienegas, and Whittier faults and renewed elevation of the Palos Verdes uplift. The onset of these vertical movements is clearly reflected in the stratigraphic record. Along the northwestern edge of the Los Angeles basin, the Tarzana fan was cut off by uplift of the Santa Monica Mountains during late Mohnian time (Wright et al., 1973). A detailed subsurface stratigraphic section (Figure 41) in the northwestern Los Angeles basin documents this major change in basinal configuration at approximately 7 Ma. The early Mohnian nodular shale (La Vida Member) and the middle Mohnian (Division D) "Massive sands" of the Soquel Member thin southward onto the plunge of the Las Cienegas high. Total thickness of the late Mohnian (Division C) interval does not vary significantly, but late Mohnian and younger sandstones thicken southward into the

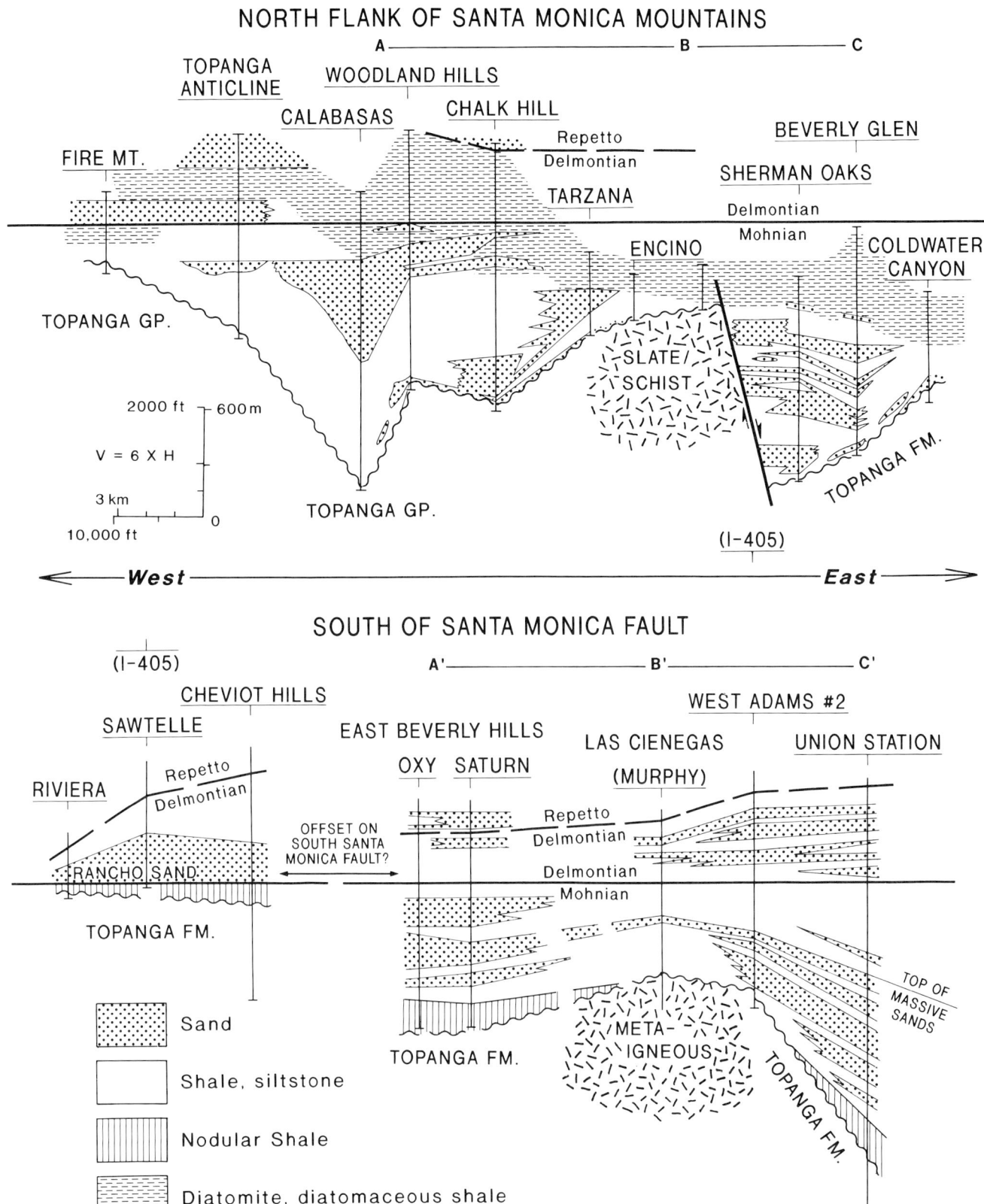

Figure 40. Comparative stratigraphic cross sections of the Tarzana fan across the Santa Monica–Raymond fault. Locations are shown in Figure 39. Vertical exaggeration is six times. Upper cross section from Sullwold (1960), revised to a top of Mohnian datum. Lower section based on oil-field data from Eschner and Scribner (1972); California Division of Oil and Gas (1974; public file); Jacobson and Lindblom (1977). Position of Interstate 405 (I-405) shows present-day alignment across Santa Monica fault.

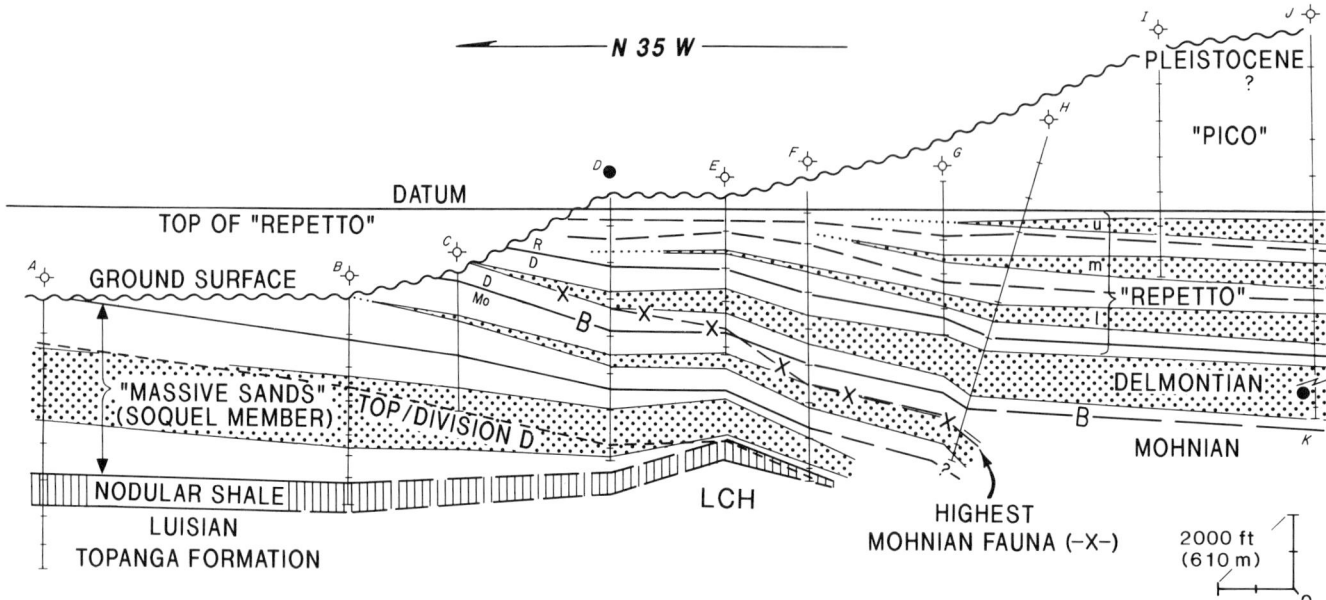

Figure 41. Subsurface stratigraphic diagram across the northern shelf of the Los Angeles basin. Location of diagram, shown in Figure 39, is immediately north and east of downtown Los Angeles. Datum is top of Repetto (lower Fernando). Vertical and horizontal scales are the same. Stratigraphic intervals in wells have been corrected for effects of steep dips and show true stratigraphic thickness. Dotted pattern shows composite sand thickness within each interval. The first (highest) occurrence of Mohnian foraminifera in wells (—X— on diagram) transgresses a bentonite horizon (—B—) that has been used in this local area as a synchronous top of Mohnian marker. LCH = Las Cienegas high. Total length of section is about 6.5 mi (10.4 km).

present Los Angeles basin and pinch out onto the northern shelf.

If foraminiferal evidence for the top of Division D provides an approximate timeline, the reversal in sea floor gradient occurred at the very beginning of late Mohnian time. Other evidence in this area, however, attests to the nonsynchronous nature of these faunal zones. A distinctive multiple bentonite marker (—B— on Figure 41) is found near the top of the Mohnian interval in this part of the basin. Detailed paleontological studies show that the latest (highest) occurrence of the Mohnian marker fauna (—X— on Figure 41) crosses the bentonite timeline. These species evidently survived on the shallower, less turbid basin slope after they had become extinct on the basin floor. Lamar (1970) described a similar possible "time-transgressive fauna" marking the top of the Luisian in this same area.

Post-Mid-Mohnian Displacements

The stratigraphic sections of Figure 40 imply that uplift of the Santa Monica Mountains began at their eastern end and progressed westward, shifting the Tarzana fan into the Calabasas-East Beverly Hills areas during the late Mohnian and into the Fire Mountain-Sawtelle area during the Delmontian. The block north of the Santa Monica fault was raised above the influence of submarine fan systems, as indicated by the change in facies to diatomite and diatomaceous shale. This facies was deposited in the eastern part of the area during much of late Mohnian time but not until Delmontian to the west. Evidence already cited documents major vertical separation across the Santa Monica fault system during latest Miocene and early Pliocene time (about 7-3 Ma). Most compelling is the juxtaposition (Figure 15) between thick intervals of early and middle Pliocene strata in the footwall of these faults and the erosional hiatus on the hanging wall that places Mohnian or older rocks unconformably beneath middle Pliocene or younger beds.

Major vertical movement on the Whittier fault during this time was described long ago by Woodward (1958): "The alignment of the distribution of coarser sediments strongly suggests fault movement or folding concurrent with deposition during late Miocene time. Foraminiferal studies of Pliocene sediments reveal considerable Miocene landslide material in the down-thrown south block." Yerkes (1972) stated that "activity along the zone clearly dates back to early Pliocene time, when it formed a scarp from which Puente detritus was shed into the lower Fernando sea."

The early Pliocene reconstruction (Figure 42) shows the limits of basinal sand deposition during Delmontian and Repetto "time." It also shows folds and faults for which there is evidence of structural growth during that period (ca. 7-3 Ma). Backslip on faults is based on various estimates of Quaternary rates of displacement, as reviewed earlier. The Santa Monica-Raymond fault system is shown not in

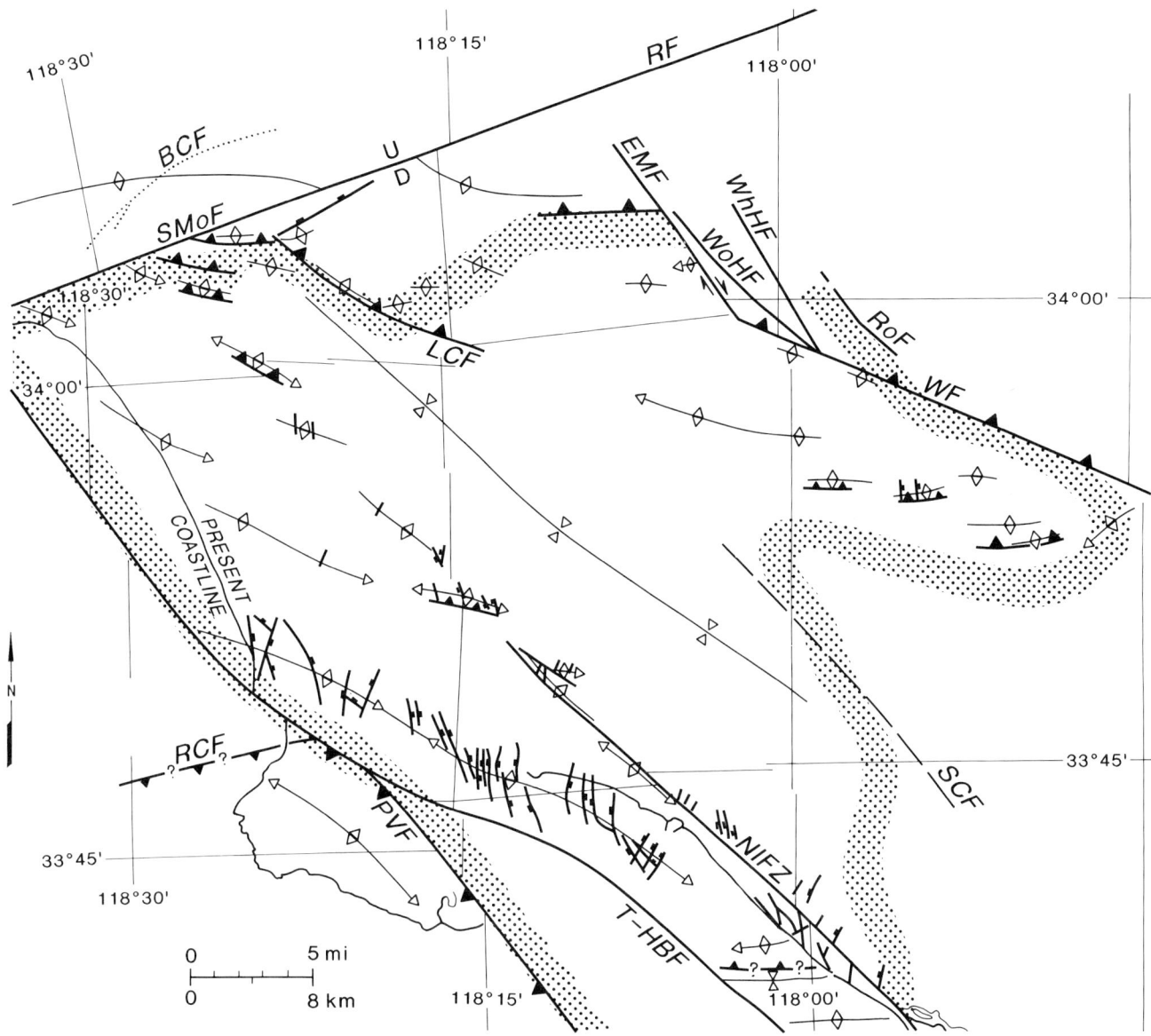

Figure 42. Early Pliocene reconstruction for the Los Angeles basin showing folding and faulting during the interval between 7 and 3 Ma. Palinspastic restoration to circa 4 Ma. Stippled pattern shows limits of significant Delmontian–Repetto sands within the basin. Abbreviations, etc., as in Figures 3, 7, and 9, plus RoF = Rowland fault. Present-day coastline and latitude-longitude grid shown for reference.

its present orientation, but as a linear feature, with Pasadenan deflections removed. Also, about 1.3 mi (2 km) of northeast-southwest shortening across the basin—the estimated effect of late Pliocene and Quaternary folding—has been restored. In this reconstruction, events during about 4 m.y. of geologic time have been composited, with the consequent loss of significant temporal and spatial variations. Although subsurface data in oil company files are probably sufficient to permit the more precise subdivision and mapping of this important tectonic episode, the task would require consistent basinwide paleontological correlations, and detailed isopach/isolith studies of many individual structures.

North of the submarine slopes that delimited the Delmontian–Repetto basin, scattered remnants of shallow-water early Pliocene sedimentary rocks are preserved. Fossiliferous pebbly sandstones near the western end of the Raymond fault (Lamar, 1970) are dated as Pliocene (or possibly late Miocene); just north of the fault these sandstones are only 10 ft (3 m) thick, while south of the fault and 3 mi (2 km) west they are more than 1600 ft (500 m) thick and are interbedded within conglomerates and coarse sandstones that are nonmarine in part.

The areas of the San Fernando Valley and the Puente Hills, though already being uplifted on the Santa Monica and Whittier faults, were still deeply submerged during early Pliocene time. On the north flank of the Santa Monica Mountains at Chalk Hill (Figure 40), lower Pliocene shale, diatomaceous shale, and siltstone (Hacker, 1969) contain upper Delmontian and Repetto foraminifera deposited in relatively deep water. Deep-water Repetto beds are also found north of the western part of the Whittier fault (Yerkes, 1972) and include fine- to medium-grained sandstones indicative of deposition on the basin floor. This, and evidence cited earlier, suggests that the Rowland, Whittier Heights, and Workman Hill faults were active during early Pliocene time and localized this portion of the bathymetric scarp. To the west, exposures in the Repetto Hills and on the northern edge of downtown Los Angeles and portions of the Repetto interval in the Las Cienegas oil field (Mefferd, 1970) are massive siltstones (Lamar, 1970) with occasional lenses of conglomerate or mollusks ascribed to submarine slumping onto the lower slope.

Basinal Deposition

By the end of Miocene time, the area of the San Gabriel Valley and Puente Hills was the sole pathway for terrigenous clastics into the Los Angeles basin. Mobilized by their rapid plunge from the shelf into waters of lower bathyal depth (Blake, this volume), sands and entrained fine sediments moved as turbidites and tractionites into the farthest reaches of the basin floor. Clays suspended in the overlying waters were carried across the basin by currents and deposited on the western shelf and the Palos Verdes uplift. In the Palos Verdes Hills, the increased turbidity is recorded by the change from the Valmonte Diatomite, which was deposited in clear waters of the late Mohnian sea, to the Delmontian Malaga Mudstone. On the proximal edge of the basin, Delmontian sandstones and conglomerates of the Sycamore Canyon Member (Puente Formation) were localized in two major feeder systems (Redin, this volume, his Figure 11): a primary channel-canyon complex in the area of the Whittier Narrows (Gourley, 1975), including the Montebello and Whittier oil fields, and a secondary channel in the southeastern Puente Hills and Santa Ana Canyon.

Within the basin, Delmontian and Repetto sandstones progressively onlapped the relict highs of the Anaheim nose and the block southwest of the THUMS-Huntington Beach fault and offlapped the block northeast of the Las Cienegas fault. On the western shelf, cycles of persistent distal turbidite sandstones thin gradually westward. The San Joaquin Hills remained a positive area whose east flank was onlapped by clastic rocks of the Capistrano syncline and whose west flank controlled the edge of thick sedimentation in the adjacent portion of the Los Angeles basin during late Miocene and early Pliocene time.

The uppermost Delmontian interval around most of the Los Angeles basin is almost entirely shale or siltstone. This suggests a period when the central basin subsided faster than it could be filled, leaving much of the basin floor above the level of sand deposition. Local unconformities or hiatuses in various parts of the basin during early Pliocene time record submarine erosion, either from the tops of uplifted blocks as in the Palos Verdes Hills where the lower Repetto section is missing (Blake, this volume) or in the axes of proximal channels as in the Santa Ana Canyon where conglomerates of the basal Fernando Formation are unconformable on Miocene strata (Schoellhamer et al., 1981; Lamar, 1961).

Growth of Internal Structures

Submarine fans swept across the *Newport-Inglewood fault zone* during latest Miocene and early Pliocene time (Conrey, 1967; Redin, this volume, his Figures 11 and 14) and recorded the varying styles of deformation shown in Figure 42. Yeats (1973) noted that "Sunset Beach, Huntington Beach, and West Newport oil fields . . . are characterized by north-trending normal faults with no movement younger than early Pliocene. Dominguez, Rosecrans, and Howard Townsite oil fields are characterized by west-trending reverse faults of the same age as the normal faults." Isopach studies of the Inglewood anticline (Wright et al., 1973; Wright, 1987d) show that folding began in early Pliocene time (perhaps accompanied by thrusting on the Sentous fault system) and peaked in the late Pliocene. Evidence for folding and faulting of the Potrero and Rosecrans anticlines during latest Miocene and early Pliocene time is considered problematical. At Dominguez (Graves, 1954), the down-plunge extension of the El Segundo-Lawndale anticline is cut by a zone of north-dipping thrusts that involve Miocene beds on the south flank of the fold but do not affect the younger horizons. At Long Beach (Figure 25, Section YY'), displacement on the Wardlow-Airport fault occurred during early Pliocene time, coeval with growth on the Long Beach and Long Beach Airport anticlines and probably on the Cherry Hill fault as well. On the north flank of the Seal Beach anticline (Bowes, 1943), significant downdip thickening of the Repetto interval records early Pliocene structural growth. At Sunset Beach and onshore and offshore Huntington Beach, except on the active strand of the NIFZ, evidence cited above indicates that the folding and faulting that controls hydrocarbon accumulation in these oil fields took place mostly in early Pliocene time.

The *Wilmington anticline* is undoubtedly the best documented of the Mio-Pliocene structures in the Los Angeles basin. Detailed isopach studies (Truex, 1974) indicate that "an ancestral Wilmington anticline began evolving near the initiation of G sand deposition, in the upper part of Wissler's Division A, Delmontian Stage, of the Miocene. In the lower Repetto, the deposition of the F sand was followed by a period of tectonic activity that included the initiation of the major faults of the Wilmington field, partial erosion of the uplifted blocks, and the development, by bottom currents, of a submarine channel . . . Continuous uplift probably occurred throughout the Repettian Stage . . . The axis of folding migrated northward to its present . . . location. Earlier faults . . . were reactivated, and some new faulting probably occurred. The Wilmington anticline was eroded actively between the close of the Repettian and the beginning of the Wheelerian(?) Stage. The Pico and San Pedro Formations were deposited unconformably on the truncated Repetto."

Figure 42 shows the principal faults on the Wilmington anticline. Up-plunge on the Torrance anticline and on the parallel El Segundo-Lawndale and Venice-Playa del Rey trend of the *western shelf* folding and faulting are believed to have occurred at the same time as at Wilmington, although this has not been documented except by unpublished regional stratigraphic studies (Wright, 1964–1970). In the Palos Verdes Hills, folding of Miocene beds has been variously dated as early Pliocene (Yerkes et al., 1965, 1967) and as late Pliocene–early Pleistocene: "between about 3.0 m.y. and 1.0 m.y.B.P." (Nardin and Henyey, 1978). All of this folding west of the NIFZ involved schist basement that was still at least partly ductile and connotes "right-lateral simple shear of the entire western Los Angeles Basin rather than shear limited to a single fault zone" (Yeats, 1973).

Santa Monica, Las Cienegas, and Whittier Faults

In the northwestern Los Angeles basin, compelling evidence for early to mid-Pliocene faulting on the Santa Monica and Las Cienegas faults and folding on the East Beverly Hills and related anticlines has already been reviewed (Figure 16B). Reverse faulting on the Las Cienegas and Whittier faults may have been augmented by basinward rotation of these faults owing to lateral sag of uplifted basement blocks. The same mechanism probably produced north dip and apparent reverse motion on the Santa Monica fault, despite its extensional nature during this period.

Folding in the footwall of the Whittier and associated faults has been documented at Montebello (Figure 35) and Whittier (Figure 11). To the east in the Kraemer area, Lamar (1961) noted that "the angular unconformity at the base of the Fernando group in this area indicates major pre-Pliocene movement on the Whittier fault," and he described folds in the uppermost Puente Formation that "appear to die out up section." [The Delmontian Stage and, therefore, the upper Sycamore Canyon Member of the Puente Formation have since been recognized as extending into the early Pliocene (Blake, this volume)]. South of the Whittier fault, data previously cited indicate latest Miocene and/or early Pliocene faulting and folding in the Richfield and in the East and West Coyote oil fields and suggest earliest Pliocene growth on the Santa Fe Springs dome. All of these structures warrant further detailed isopach/isolith studies.

Dynamic Setting

The shaping of the modern Los Angeles basin during latest Miocene and early Pliocene time is thought to be the product of three distinct mechanisms:

- continued moderate right or right-oblique slip on the Palos Verdes, NIFZ, and Whittier-Elsinore fault systems. Left slip on the Santa Monica-Raymond fault system and, perhaps, the final phase of regional rotation of the Santa Monica Mountains block contributed to folding of the East Beverly Hills and San Vicente anticlines during deposition of the lower Repetto.

- subsidence along loci extending northwest within the central trough and Fullerton embayment and northeast through the Whittier Narrows into the San Gabriel Valley. Both were areas of significant mid-Miocene volcanism. Cooling of the underlying intruded crust produced isostatic subsidence due to increased densities, augmented by thermal contraction (Turcotte and McAdoo, 1979). Subsidence extended uniformly across the Anaheim nose and into other areas outside the zones of igneous intrusion, because of the finite flexural strength within the lithosphere (Karner and Dewey, 1986).

- stretching of the crust in a north-northwest direction has been proposed (Crowell, 1976) to explain the opening of the Santa Clara trough (western Ventura basin; see Figure 2) during early Pliocene time. The Santa Clara trough is north of the western Santa Monica Mountains; its southern edge is about 15 mi (24 km) north of the Malibu Coast-Santa Monica fault system and approximately parallel to it. Yeats (1978) dated the acceleration of subsidence rates in the Ventura basin at about 4 Ma, plus or minus 1 m.y., and suggested a similar history for the Los Angeles basin. In the northwestern Los Angeles basin, the profound Pliocene subsidence south of the Santa Monica fault (as documented in an earlier section of this paper) occurred between 4 and 3 Ma.

Yeats (1978) called attention to the temporal coincidence between the early Pliocene acceleration in subsidence rates of southern California basins and the opening of the Gulf of California. Right slip on the San Gabriel fault ceased about 5 Ma (Crowell, 1982) and the eastern edge of the San Andreas transform zone moved across the San Gabriel Mountains to the present San Andreas fault on the edge of the Mojave Desert. All of Baja California and the Peninsular Ranges were being pulled in a northwest direction by the Pacific plate. Late Miocene extension and subsidence in the region of the Gulf of California (Karig and Jensky, 1972) culminated in the separation of Baja California from the Mexican mainland during early Pliocene time. The East Pacific Rise opened the mouth of the gulf at about 4.9 Ma (Curray and Moore, 1984), and by about 3.2 Ma, it had extended to its northern end (Figure 6: 0 Ma). This period coincides closely with the opening of the Santa Clara trough and the northwestern Los Angeles basin, suggesting a brief episode of regional transtension that ended with the final separation of Baja California. Within the rotated block of the central Transverse Ranges, the Malibu Coast-Santa Monica fault system and the faults that bounded the Santa Clara trough were ideally oriented to relieve that tension.

The pattern of Mio-Pliocene deformation shown in Figure 42 was associated with that phase of regional transtension. The abrupt end of that phase coincided with a major change in relative plate motion along the boundary between the Pacific and North American plates. This was described by Engebretson et al. (1985) as "a change in the motion of the Pacific plate relative to hotspots at 5 Ma from a west-northwesterly direction to a more northwesterly direction." The timing of this change has since been more precisely placed at 3.9–3.4 Ma (Harbert and Cox, 1989).

The Pacific plate, whose northwesterly motion now extended to the western edge of the Mojave Desert, had begun to encounter the deep crustal roots of the Sierra Nevada and southeastern San Joaquin Valley. By late

Pliocene time, this apposition was to create the uplift of the Transverse Ranges, the great bend of the San Andreas fault in that region, and the present tectonic setting of the Los Angeles basin.

Pasadenan Deformation

As summarized by Woodford et al. (1954), the Pasadenan deformation was named by Stille (1936) on the basis of the middle Pleistocene unconformity that was emphasized by Reed (1933). It is one of the best-known features of the Pleistocene rocks in southern California. The term has gained general acceptance (Van Eysinga, 1975), though it is little used by the current generation of geologists in southern California. In this study the term *Pasadenan* is applied to the phase of deformation that has produced major uplift and shortening of the Transverse Ranges province. This episode began in the mid-Pliocene (ca. 3.5 Ma), accelerated in mid-Pleistocene time, and still is continuing.

No lengthy period of tectonic quiescence separated the Pasadenan deformation from the Mio-Pliocene deformation that preceded it. The two may have overlapped in time, as the onset of Pasadenan deformation was not necessarily synchronous throughout the region. Several hypotheses have been advanced recently that might account for Pasadenan deformation. Evaluating the San Andreas fault system in central California, Zoback et al. (1987) suggested that "the ~20° clockwise change in relative plate motion 4 to 5 million years ago . . . caused a ~90° change in stress direction and tectonic style within about 100 km of the San Andreas—from fault-parallel basin development and fault-normal extension, to the uplift and fault-normal compression observed today." As noted above, this change in relative plate motion has since been dated at 3.9-3.4 Ma by Harbert and Cox (1989). They described it as "sudden" and stated that it "corresponds to a change from strike-slip to transpressive motion along the coastal California Pacific–North American plate boundary."

Regional geophysical and geological studies have illuminated a long-suspected cause for Pasadenan north-south compression in southern California (Ferguson and Willis, 1924). Seismic travel-time data (Hadley and Kanamori, 1977) indicate that the Transverse Ranges are underlain by a high-velocity keel that extends east beyond the present San Andreas fault into the Mojave Desert. This has been interpreted (Bird and Rosenstock, 1984; Humphreys et al., 1984) as a slab of colder, denser "mantle lithosphere" that is descending into the asthenosphere. Descent of the slab accompanies crustal spreading beneath the Salton trough and northern Gulf of California and northwesterly motion of the mantle in the intervening region of the Peninsular Ranges. The folded and uplifted blocks of the Transverse Ranges, decoupled from this "conveyor belt" on a zone of midcrustal detachment (Yeats, 1981), are being compressed against the granitic root of the Sierra Nevada and southern Great Valley. The central Transverse Ranges are being thrust northward over the southern San Joaquin Valley and southward over the Los Angeles basin and adjacent areas.[2] The resulting pattern of active tectonics and seismicity has been summarized by Yerkes (1985) for the Los Angeles region.

The ongoing Pasadenan deformation is reflected by the topography within the Los Angeles basin (Figure 43), faults with Holocene activity, and the epicenters of modern earthquakes. Figure 43 shows potentially active faults as listed in Ziony and Yerkes (1985) and/or discussed in this report. Epicenters shown (with focal mechanisms where available) are those discussed in the text. Only epicenters located on the basis of the most recent velocity models (e.g., Hauksson, 1987) are considered of sufficient accuracy for use in defining active faults within the seismogenic zone.

Right-Lateral Displacements

Holocene deformation of the Los Angeles region is dominated by two discrete systems: northwest-striking right-slip or right-oblique faults extending from the Peninsular Ranges and south-vergent thrust or left-oblique faults along the southern margin of the Transverse Ranges. Geologic data on fault slip rates compiled by Bird and Rosenstock (1984) record a progressive westward decrease in rates of right slip: San Andreas fault (south of Cajon Pass), 10–40 mm/yr; San Jacinto fault, 8–18 mm/yr; Whittier-Elsinore fault, 1–1.5 mm/yr; Newport-Inglewood, 0.4–0.8 mm/yr. This relationship implies that shear within the Pacific plate increases with proximity to the North American plate and suggests a broad zone of viscous drag within the mantle. West of the NIFZ, the documented absence of folding on the Wilmington anticline after middle Pliocene time (Truex, 1974) and the apparent cessation of folding on the other basement-cored anticlines of the western shelf suggest decreased ductility of the schist basement perhaps owing to further cooling.

The Palos Verdes Hills are a classic locale of Pleistocene deformation in southern California. On the east plunge of the uplift, lower? Pleistocene marl and siltstone dip 17° to 23° easterly and are unconformably overlain by nearly horizontal upper Pleistocene marine terrace deposits (Woodring et al., 1946; Yule and Zenger, 1987). Late Pliocene uplift (and possible emergence) is indicated by the absence of the Pico Formation. Renewed uplift during the past 1 m.y. (Schwartz and Colburn, 1987) is recorded by 13 emergent marine terraces. The highest terrace, 1280 ft (390 m) above sea level, is perhaps as old as middle or early Pleistocene (Bryant, 1987). Along the northeastern edge of the hills, folding parallel to the adjacent Palos Verdes fault involves all of these Pleistocene formations; the lowest (youngest) terrace deposits locally dip 26° (Yerkes et al., 1965). Deformation of these terrace levels (Bryant, 1987) suggests that during Quaternary time the Palos Verdes Hills were uplifted as a block, tilted slightly to the south. Folding

---

[2] Possible relationships between the mid-Pliocene change in plate motion, the postulated lithospheric slab beneath the Transverse Ranges, and the onset of collision tectonics in that region were not addressed by Harbert and Cox (1989), but they promised an intriguing avenue for future research. Harbert and Cox described the variation in plate motion as "unrelated to terrane collision."

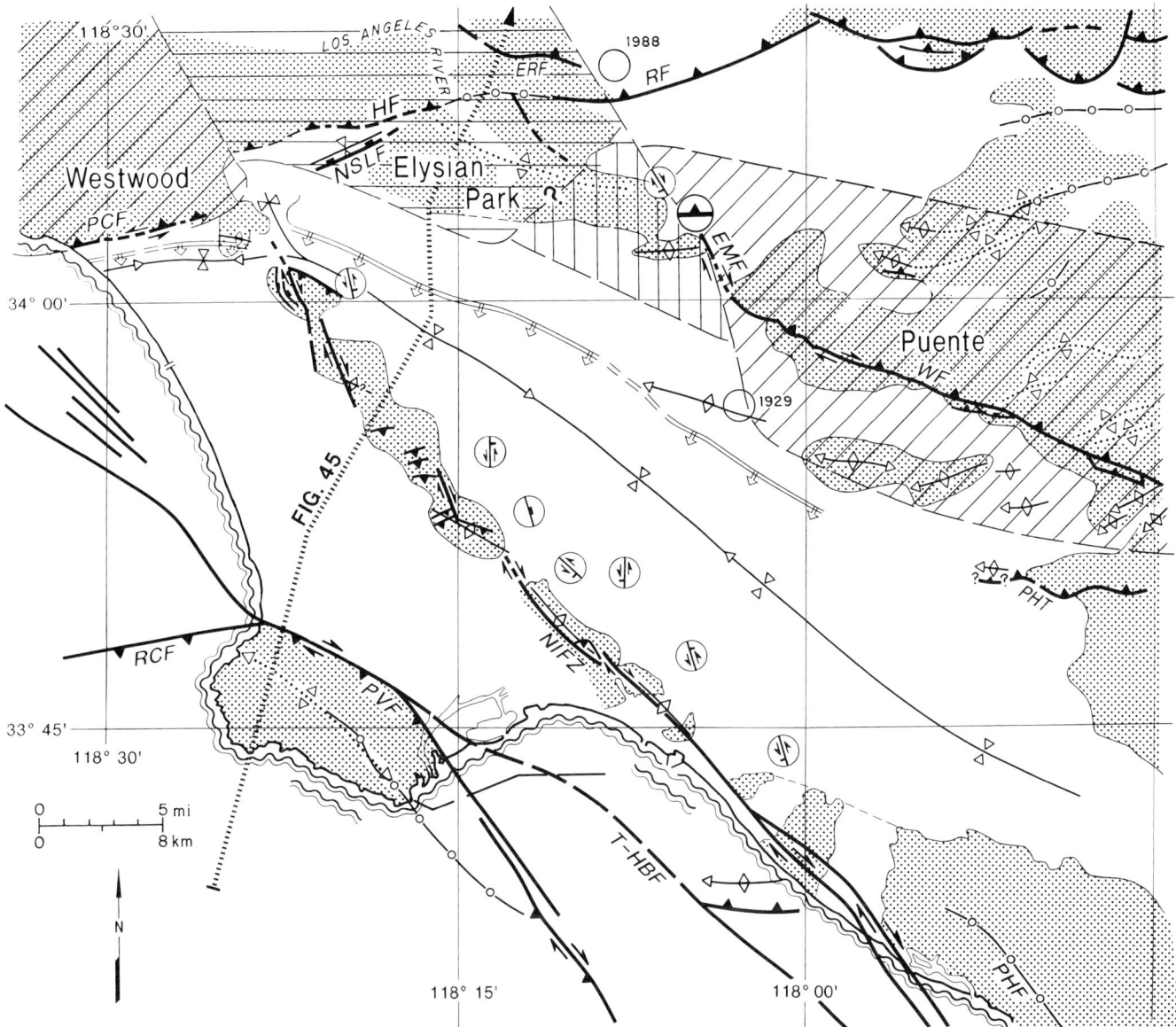

Figure 43. Holocene deformation of the Los Angeles basin and vicinity. Areas of late Quaternary uplift are stippled. Anticlines, synclines, and faults are those affecting late Quaternary sediments. Westwood, Elysian Park, and Puente are postulated segments (shown within the seismogenic zone by lined patterns) of the North Los Angeles thrust system. Dotted lines show anticlines where Quaternary folding is suspected but cannot be proven owing to the absence of latest Neogene sediments on one or both flanks. Double lines with arrows indicate the late Quaternary flexure along the northeastern and northern edges of the basin. —o—o— indicates faults on which Holocene activity is reported by Ziony and Yerkes (1985) but is not otherwise discussed in this paper. Circles show selected earthquake epicenters (see text). Abbreviations, etc., as in Figures 3, 7, and 9, plus ERF = Eagle Rock fault. Location of Figure 45 is indexed.

on its northwestern edge is associated with right-oblique slip on the Palos Verdes fault. Most of the lateral component of this motion is taken up by north-vergent thrusting on the offshore Redondo Canyon fault (Nardin and Henyey, 1978).

In an inverse symmetry with the Redondo Canyon fault, the Offshore Huntington Beach anticline (Figure 28, Section dd′) and its south flank fault show continued Quaternary growth and are evidently absorbing some right slip west of the NIFZ. This would appear to require active right slip on the THUMS-Huntington Beach fault; around its southeastern end, a cluster of epicenters (Legg, 1986, 1987) indicates continued activity. High-resolution geophysical surveys across the THUMS-Huntington Beach fault (Junger and Wagner, 1977) are not of sufficient detail to indicate the nature of Quaternary deformation. To the north, slip on this fault is transferred onto the Palos Verdes fault, whose maximum Holocene expression is along its onshore portion, northwest of the area of convergence.

Pasadenan right slip on the NIFZ is on the order of 4000 ft (1200 m), according to data from the Inglewood oil field. Quaternary faulting is en echelon in the northern part of the NIFZ and continuous in the southern part (Figure 43), and individual structures display a wide variety of responses. At Inglewood, a fault-parallel graben has developed west of the Inglewood fault, while the east block is an uplifted wedge. The Dominguez anticline is being folded en echelon to the NIFZ and is cut by active south-dipping thrust faults. Subsurface mapping (Foster, 1954; Matthews, 1954) and topographic form suggest that similar north-vergent thrusts play a significant role in the Rosecrans, Howard Townsite, and, perhaps, Potrero oil fields. The Long Beach and Seal Beach anticlines have retained their fault-parallel orientation during Quaternary deformation. In the Huntington Beach area, a broad uplift parallel to the NIFZ plunges northwest from the preexisting San Joaquin Hills high.

Focal mechanisms of recent small earthquakes along the NIFZ (not shown in Figure 43 to avoid clutter) indicate significant differences between the northern and southern parts of the zone (Hauksson, 1987). South of Dominguez, eight of nine events show right slip on near-vertical planes, but only two of these are subparallel to the NIFZ, while six are oriented approximately north-south. From Dominguez north, six of nine events show right slip on planes oriented in a northwestern direction, though dip and rake are variable. Two events indicate reverse faulting. These data were interpreted by Hauksson (1987) as indicating that "the north segment is characterized by compression while the south segment is characterized by tension." Thus, the geological differences between northern (en echelon) and southern (continuous) segments are paralleled by seismotectonic differences. Note also that the cross sections in Figure 8, viewed as a group, clearly show the progressive increase in compressive shortening toward the northwestern end of the basin.

Seismicity on the Compton-Los Alamitos fault is similar to that on the southern segment of the NIFZ. Figure 43 shows representative epicenters and focal mechanisms along this zone (Hauksson, 1987), which were located on the basis of an improved velocity model for the western Los Angeles basin. Eight of 14 recent small events involved right slip on planes striking north-south to north-northwest, one involved right slip parallel to the NIFZ, and five were tensional on northeasterly planes. The Compton-Los Alamitos fault plays a key role in shaping the steep northeast flank of the NIFZ and in the Quaternary subsidence of the central trough (Figure 8).

Basinal Subsidence and Shelf Uplift

Concurrent with Pasadenan deformation, the deep-marine Los Angeles basin of early Pliocene time has been filled with a flood of clastic sediments. These clastics are being shed from the uplifted San Gabriel Range and channeled through the Whittier Narrows and, in a much lesser volume, the Santa Ana Canyon. Rates of deposition have exceeded the rate of deformation, except in a few local areas, and the gentle, planar, southwestern slope of the basin floor was generally maintained until mid-Pleistocene time. The Palos Verdes uplift served as a sill that empounded all but the clay-sized sediments. The progressive shallowing of the basin as it filled is recorded by the benthic foraminifera preserved in the Pliocene-Pleistocene sequence (Blake, this volume). Most of the species used to zone this sequence are presently living along the adjacent continental margin, and the succession of faunas that initially were thought to have chronologic significance are now accepted as indications of water depth or general environmental conditions (energy, O_2, etc.). Thus, the Repettian (Repetto) assemblage is lower bathyal (>2000 m), the Venturian (lower Pico) is lower middle bathyal (1500-2000 m), the Wheelerian (middle Pico) is upper to upper middle bathyal (150-1500 m), and the Hallian (upper Pico) is neritic (<150 m). These faunal horizons provide good stratigraphic markers within individual oil fields and similar local areas. Because of their time-transgressive nature, however, they are of little use in establishing a basinwide tectonic chronology.

An approximation of Pasadenan uplift and subsidence within the Los Angeles basin can be obtained from the present depths of key faunal horizons. The Pico-Repetto (Venturian-Repettian) horizon has been mapped throughout the basin on evidence from benthic foraminifera and is generally accepted as occurring within the upper Pliocene sequence. This horizon represents the boundary between lower middle bathyal and lower bathyal depths; that is, at a subsea depth of about 2000 m (6500 ft) at the site of deposition (Ingle, 1979). Figure 44 shows the approximate uplift or subsidence of this horizon since its deposition at 2-3 Ma, based on its present depth and the assumed 2000 m (6500 ft) original depth of the biofacies boundary. This is not a structural contour map; the horizon transgresses both time and bedding planes. Also, no adjustment has been made for compaction of sediments. Nevertheless, the map provides useful insights into Pasadenan deformation.

In the central trough, net Pasadenan subsidence has exceeded 4000 ft (1200 m). Late Pliocene downwarp extended that syncline southeast across the Santa Ana shelf, where exploratory wells have penetrated upper Fernando beds unconformably overlying Paleogene red beds (Schoellhamer et al., 1981). The Anaheim nose has subsided nearly 2000 ft (600 m). In the northwestern part of the basin, the downwarp extended the central syncline northward across the East Beverly Hills area. Relative to sea level, the western shelf and San Pedro Bay areas have been uplifted regionally a minimum of 2000 ft (600 m) and about 4000 ft (1200 m) in the area from the Torrance-Wilmington anticline southwest to the Palos Verdes fault.

Along the northern margin of the basin, local outcrops of middle and lower bathyal strata of mid-Pliocene age imply Pasadenan uplift on the order of 6000 ft (1800 m). In well-documented sections in downtown Los Angeles and the Repetto Hills (Natland and Rothwell, 1954), siltstones bearing a Wheelerian fauna directly overlie Repettian

Structural Geology and Tectonic Evolution of the Los Angeles Basin   111

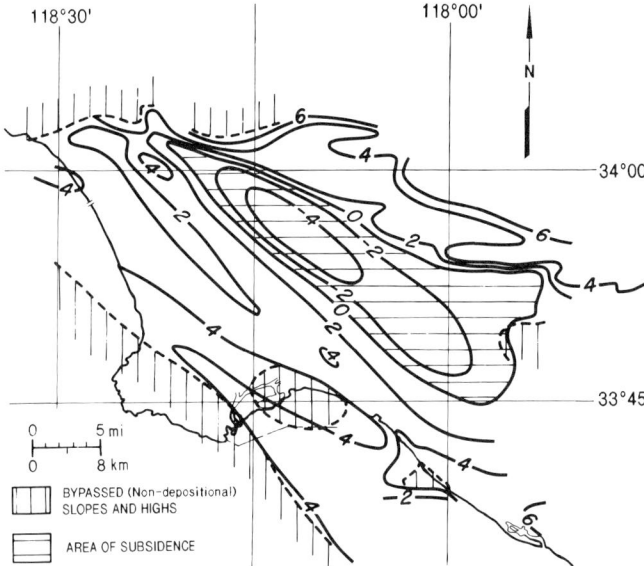

Figure 44. Approximate relative uplift and subsidence of the Los Angeles basin during Pasadenan deformation. Form lines show approximate postdepositional change in elevation of the boundary between the lower–middle-bathyal and lower-bathyal faunal zones (Pico/Repetto or Venturian/Repettian), deposited during late Pliocene time. Form line interval: 2000 ft (610 m).

siltstones (Lamar, 1970) and the Venturian stage is absent. Sediments moving down the proximal slope evidently bypassed these areas and the block north of the Las Cienegas fault (Figure 44), during the time the northern edge of the basin was being uplifted through the lower middle bathyal zone.

The Venturian (lower Pico) biozone is also missing from the crests of the Wilmington (Figure 31) and Offshore Huntington Beach (Figure 28, Section dd') anticlines. Bathymetric relief on these mid-Pliocene structures apparently was sufficient for them to have been scoured by bottom currents sweeping across the basin floor, precluding sedimentation until sometime after regional uplift had raised the area above the lower middle bathyal zone. On the Palos Verdes Hills, both Venturian and Wheelerian beds are absent; the area was a seaknoll during late Pliocene and early Pleistocene time. It is capped by the Lomita Marl, of early or perhaps middle Pleistocene age (Yule and Zenger, 1987), which was deposited in water depths of 150–600 ft (46–183 m) and lies unconformably on strata of late Miocene and mid-Pliocene (Repetto) age (Yerkes et al., 1965).

Major localized uplift along the NIFZ and in the Coyote Hills occurred after those areas had been blanketed by deep-water upper Pliocene sedimentary rocks. Continued influx of sediments during late Pleistocene time has filled the Los Angeles basin behind the uplifted Newport-Inglewood zone. Several antecedent streams have incised gaps across portions of this uplift (Figure 43): the San Gabriel River

across the Seal Beach anticline, the Santa Ana River across the plunge of the San Joaquin Hills, and a third channel (now abandoned) between those two rivers. The Los Angeles River has alternated between its present course through the structural saddle between Dominguez Hill and Signal Hill and a westerly course through the Ballona Gap (Figure 14) north of the Inglewood uplift. It last followed this westerly course in 1884. Continuing subsidence of the central trough may well require future major expenditures to avert urban flooding, inasmuch as upstream dams and debris basins are now entrapping the sediments that would otherwise maintain this area at elevations above sea level.

North Los Angeles Thrust System

Active tectonism in the northern and northeastern parts of the Los Angeles basin reflects the interplay between northwestward motion of the Pacific plate and compressive shortening of the Transverse Ranges. As described by Campbell and Yerkes (1976), "the present style of deformation is accommodated by right-lateral tears in a southern structural block that is being thrust under a northern structural block." Along the southern edge of the western San Gabriel Range, the dominant reverse or thrust nature of the faulting has long been recognized (Hill, 1930). In outcrops along the Sierra Madre fault zone (Crook et al., 1987), basement rocks have overthrust Quaternary beds on planes that dip northerly at angles of 10° to 85°. The Eagle Rock fault (Weber, 1979) "dips 15° to 30° northward and places basement rocks over conglomerate of the Topanga Formation and possibly over older alluvium." On the eastern part of this fault, Lamar (1970) has mapped 85° north dip. On the Raymond fault, evidence of reverse or thrust faulting includes an outcrop near its western end where alluvium is cut by faults dipping 14° to 58° north and exploratory trenches farther east in which north dips of 57° to 83° have been mapped on the fault (Bryant, 1978; Crook et al., 1978). Focal mechanisms of recent small earthquakes on the Sierra Madre–Cucumonga fault system imply that left slip is still active on this part of the southern Transverse Ranges boundary.

Reverse faulting is now dominant on the Santa Monica, Las Cienegas, and Whittier faults, as shown by subsurface data (Barbat, 1958; Gardett, 1971) (Figures 8, 11, 12, 15, 17). Until recently, however, the nature of Holocene activity on these faults has been poorly understood. The relative importance of active right slip on the Whittier fault has perhaps been over-estimated (Bird and Rosenstock, 1984). The Santa Monica fault appears to be overlapped by undisplaced middle Pico (early Pleistocene?) beds at its eastern and western ends. The Potrero Canyon fault, in outcrop a reverse fault that dips 43° to 55° north and offsets upper Pleistocene terrace deposits (McGill et al., 1987), evidently dies out eastward as it approaches the NIFZ. This fault may be the Holocene expression of renewed thrusting on the Santa Monica fault zone. Alternatively, it might represent a rotated shallow normal fault along which the lip of the hanging-wall block has sagged into the adjacent deep sedimentary basin; in this case the fault would have no seismogenic potential.

Recent studies have suggested that the Santa Monica and Las Cienegas faults and other structures within or adjacent to the northern Los Angeles basin reflect active north-dipping *blind thrusts* associated with shortening and uplift of the Transverse Ranges (Davis et al., 1989). Blind thrusts are part of the fold-and-thrust-belt model first applied to the Transverse Ranges by Davis (1986). Preliminary studies using this model in the Los Angeles basin (Davis, 1987) gained support from the 1 October 1987 Whittier Narrows earthquake ($M_L = 5.9$), whose focal mechanism and aftershock pattern (Figure 43) (Hauksson et al., 1988) indicated a blind thrust fault located at depths of 36,000–52,500 ft (11–16 km), dipping north at 25°, and roughly coincident with the Elysian Park thrust fault earlier postulated by Davis (1987).

A cross section through the northwestern Los Angeles basin (Figure 45) (after Davis et al., 1989) illustrates the structure as interpreted by the method of balanced, retrodeformable cross sections and a fold-and-thrust-belt model. In this interpretation: "The Los Angeles area is undergoing seismically active convergence causing the development of a basement-involved fold and thrust belt represented by the Transverse Ranges and the northern Peninsular Ranges. The major anticlinoria and uplifts are associated with thrust ramps off a regional detachment(s) at 10–15 km depth. The detachment(s) coincides with the base of seismicity and probably the change from brittle to ductile deformational processes in the crust" (Davis et al., 1989). The three major compressional uplift trends are the Palos Verdes anticlinorium and western shelf, the Santa Monica Mountains anticlinorium, viewed as part of an active fold that extends eastward to include the Elysian Park anticline and the hanging-wall block of the Whittier fault, and the Verdugo Mountains–San Rafael Hills and San Gabriel Mountains. Thrust ramps associated with the latter uplift reach the surface as the Verdugo–Eagle Rock fault and the Sierra Madre fault. The more southerly uplifts are ascribed to blind thrusts.

The depth distribution of hypocenters beneath the western Los Angeles basin (Figure 45) (Hauksson, 1987) "suggests that the lower seismogenic depth limit or the brittle ductile transition zone [is] at 11 to 13 km depth" (7–8 mi). Three of the total of 64 events during the 1973–1985 period were deeper, about 9 mi (15 km). Other recent studies in southern California (Webb and Kanamori, 1985) show the base of seismicity as a fairly even boundary extending as deep as about 12 mi (20 km).

The blind-thrust model provides a valuable tool for interpreting Pasadenan structure in the northern portion of the Los Angeles basin. The northeastern and northern margins of the Holocene central syncline are distinctive flexures (Figure 43) that involve Pleistocene strata and resemble fault-propagation folds. These folds are driven by a series of blind thrusts, herein termed the North Los Angeles thrust system (NLATS), which includes the Elysian Park thrust fault. Thrust faults in this system ramp up from a regional midcrustal detachment zone beneath the brittle-ductile transition.

Segmentation of the North Los Angeles thrust system may occur along the NIFZ, the East Montebello fault, and other preexisting major northwest-striking right-slip faults, including the Palos Verdes and Chino Hingeline faults, that penetrate to the ductile zone. Between these right-slip faults, discrete segments of the NLATS may be tentatively defined on the basis of surficial Holocene structures, topography, and well-located recent epicenters (Figure 43).

The Westwood segment incorporates reactivation of the Santa Monica fault as a blind thrust (further rotated from its mid-Pliocene dip) that terminates upward into a west-trending flexure (Figure 15). The Potrero Canyon fault, if an active thrust, is a parallel splay in the Westwood segment. This segment accommodates underthrusting of the western shelf beneath the Santa Monica Mountains. Its eastern edge is formed by a deep-seated northerly extension of the NIFZ inferred from active microseismicity (Hauksson, 1987). At the surface, this boundary is marked by (1) the abrupt transition between the uplifted terrain of Cheviot Hills and the Beverly Hills oil fields (Century City) on the west, the smooth subsiding alluvial plain to the east, and (2) by the offset between Holocene elements of the Santa Monica fault system: the Potrero Canyon and Hollywood faults. The 1930 Santa Monica Bay earthquake ($M_L = 5.2$; Hauksson and Saldivar, 1986) probably occurred on this segment, about 5 mi (8 km) west of the area of Figure 43; data are not sufficient to determine its depth or focal mechanism.

The Elysian Park segment has reactivated the Las Cienegas fault, now a blind thrust also, to drive the steep southeast-trending Holocene flexure that forms the northeastern margin of the central syncline in that area (Figure 8, Section BB'; Figure 45). A subparallel flexure involves upper Pliocene (and perhaps Pleistocene) strata along the southern flank of the Elysian Park anticline and Repetto Hills (Lamar, 1970). Holocene north dip on this antiform cannot be confirmed owing to the absence of beds younger than early Pliocene. The Hollywood fault and very likely the Verdugo–Eagle Rock fault ramp up from the detachment zone within this segment. The Elysian Park segment may be further subdivided by a northeast-striking basement fault postulated to underlie the down-to-the-east step of the northern shelf.

The Santa Monica Mountains block has remained unbroken west of Cahuenga Pass (or perhaps the Los Angeles River) while being uplifted by underthrusting of the northern and western shelves. The postulated segmentation of the North Los Angeles thrust system by the NIFZ evidently is confined to the footwall block of the uppermost thrust in that system. Inasmuch as thrusting here is driven by the footwall blocks, this segmentation should serve to limit the magnitude of earthquakes generated by the NLATS in this area.

A third major element of the NLATS, herein termed the Puente segment, underlies the Puente Hills and extends south beneath the Coyote Hills. The hypocenter of the 1 October 1987 Whittier Narrows earthquake was located on the westernmost part of this segment. Its western limit is the East Montebello fault, locus of the largest (4 October 1987) aftershock, which was 2 mi (3 km) to the northwest of the primary event (Hauksson et al., 1988). Southeasterly extension of this tear fault at depth is suggested by the recalculated location of the 1929 "Norwalk" event (Hutton, 1987) and by the abrupt plunge of the Anaheim nose (Figure 10). East of this tear fault, the Coyote Hills anticlinorium has been compressed and upfolded by south-vergent Holocene thrusting that may be represented at the surface as the Peralta Hills thrust (Bryant and Fife, 1982). To the north, reverse faulting and uplift on the Whittier fault is believed to ramp up from the Puente segment of the NLATS. The eastern limit of the Puente segment appears to coincide

Structural Geology and Tectonic Evolution of the Los Angeles Basin 113

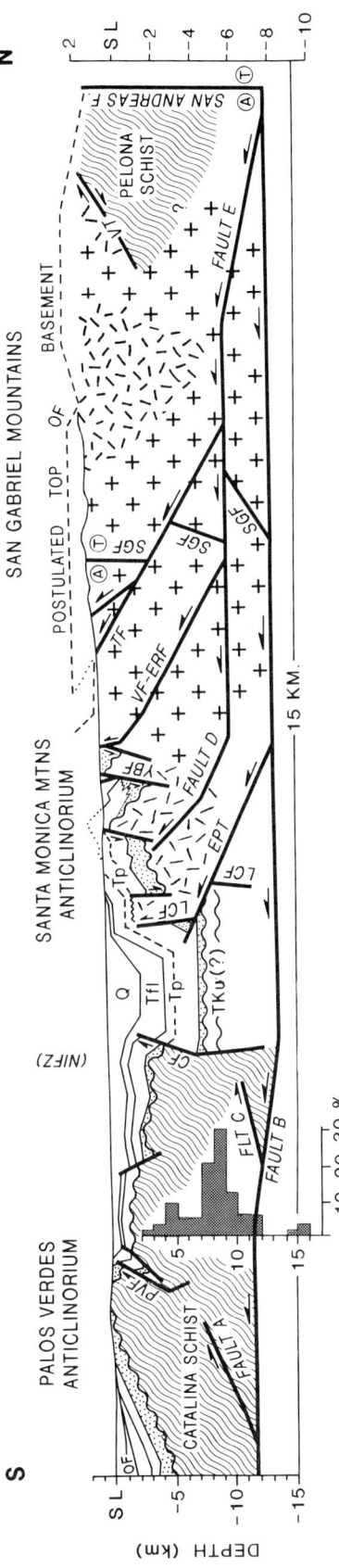

Figure 45. Retrodeformable cross section across the Los Angeles basin (from Davis et al., 1989; well control and construction lines omitted). Horizontal and vertical scales are equal. Histogram (from Hauksson, 1987) shows depth distribution of 64 small earthquakes beneath the western Los Angeles basin during 1973–1985.
Abbreviations as in Figure 3, plus CF = Compton fault, EPT = Elysian Park thrust, ERF = Eagle Rock fault, TF = Tujunga fault, VT = Vincent thrust, YBF = York Boulevard fault. Basement rock types as in Figure 4, except "other metamorphics" include Paleozoic and Precambrian in San Gabriel Mountains (from Jennings and Strand, 1969; Ehlig, 1981). Dotted pattern is Topanga Group. TKu = undifferentiated pre-Topanga sedimentary rocks, Tp = Puente Formation, Tfl = lower Fernando Formation. [Note: Monterey Formation shown as Puente and Topanga formations in Palos Verdes Hills.] OF = ocean floor, SL = sea level. Location of cross section shown in Figure 43.
[Note: This cross section is located along approximately the same alignment as CC' in Figure 8. Stratigraphic terminology used on the two sections is not identical.]

approximately with the Chino Hingeline fault. Farther to the east (Figure 3), south-vergent thrusting of the San Gabriel Mountains along the Cucamonga fault has depressed the northern edge of the Perris block. That block's southwestern edge may be underthrusting the Santa Ana Mountains along the Elsinore fault, which there dips steeply south (Schoellhamer et al., 1981, their section LQ).

On the west side of the Los Angeles basin, a similar apposition of Transverse Ranges and Peninsular Ranges tectonics is evident west of the Palos Verdes fault. The westward extension of the North Los Angeles thrust system is here represented by the south-vergent Malibu Coast and Anacapa (Dume) faults. Immediately south of the latter fault, the focal mechanism of the 1979 Malibu earthquake ($M_L = 5.0$) is interpreted (Hauksson and Saldivar, 1989) as indicating a thrust dipping 53° north at a depth of 7.5 mi (12 km). Right slip on the Palos Verdes fault has resulted in north-vergent thrusting of the Palos Verdes Hills on the Redondo Canyon fault (Yerkes et al., 1967; Nardin and Henyey, 1978), with little apparent Holocene deformation of the footwall block. It should be noted that, in analyzing the 1979 Malibu earthquake and other recent seismicity in Santa Monica Bay (Hauksson and Saldivar, 1989), north-dipping nodal planes were chosen a priori from thrust mechanisms. The alternative nodal planes for the 1979 Malibu earthquake and other events are thrusts dipping south, parallel to the Redondo Canyon fault. Aftershock patterns would seem to permit this alternative interpretation. In either case, the opposing thrust systems are closely juxtaposed.

The structural solution of Davis et al. (1989) (Figure 45) incorporates northeasterly directed thrusting along the Palos Verdes fault, associated with Holocene folding of the Palos Verdes Hills, Torrance and Alondra (Lawndale) anticlines, while neglecting the active role of the Redondo Canyon fault. Data previously cited indicate that Holocene deformation of the Palos Verdes Hills and Wilmington anticlines has been characterized by minor tilting but no perceptible folding (except locally along the Palos Verdes fault). Farther north on the western shelf, Torrance and the El Segundo–Alondra trend show little or no evidence of Pasadenan folding (Figure 8, Sections DD' and CC'), whereas such folding at Playa del Rey (Figure 8, Section AA') may reflect proximity to the NLATS. Active or Holocene thrusting has been noted in southern Santa Monica Bay (Hauksson and Saldivar, 1989), on the Redondo Canyon fault (Yerkes et al., 1967), in the Dominguez oil field (Bravinder, 1942), and on the south flank of the Offshore Huntington Beach anticline (Figure 28, Section dd'). These and other local thrust faults along the southwestern edge of the Los Angeles basin are subsidiary structures of the adjacent strike-slip fault zones and their associated folds. They do not indicate a southern extension of Transverse Ranges structure nor the presence of a separate fold-and-thrust belt (Hauksson and Saldivar, 1989).

The *fold-and-thrust-belt model and flake tectonics* provide alternative explanations for low-angle thrusting in southern California. Defining the latter model, Oxburgh (1972) described the shearing of "large sheet-like masses of material ('flakes')" from the top of one continent and driven over another during collision. In this process, "patterns of contemporary collision zone seismicity ... may be expected to become diffuse through the dominance of movements on low angle crustal faults." Oxburgh (1972) applied this concept to contemporary deformation in southern California, suggesting that "the seismicity of the area is associated with movements on crustal faults which must completely bound small detached flakes which slip with respect to each other in response to motion of the more coherent (but probably less rigid) lithosphere beneath."

Yeats (1981) further developed the flake-tectonics model for the California Transverse Ranges. In this region, the collision is between the relatively stable Great Valley–Sierra Nevada salient of the North American plate and the Peninsular Ranges blocks that are being carried by the Pacific plate to underthrust the flakes that comprise the Transverse Ranges. Each phase of Cenozoic deformation, beginning with Paleogene terrane accretion, has contributed to the creation of crustal flakes in southwestern California. Yeats (1981) postulated that the midcrustal zone of detachment beneath the Transverse Ranges consists of ductile Pelona Schist or Catalina Schist.

The fold-and-thrust-belt model has been explicated for map-scale folds in sedimentary sequences (Suppe, 1983). As defined by Suppe (1985): "The structure of [continental] fold-and-thrust belts is fundamentally dominated by thrust faults at depth that dip toward the interior of the mountain belt and run along bedding planes over much of their length. Thrust sheets in the interiors of mountain belts commonly involve ... basement rocks ..." in their root zones. The model includes several mechanisms of folding (Suppe, 1983): "If the rocks are layered they may fold in response to riding over a bend in a fault ... this mechanism [is called] fault-bend folding." In overthrust belts, this occurs when the decollement steps up, beneath the fold, to a higher stratigraphic level. Fault-propagation folding is "caused by compression in front of a fault tip during fault propagation" (Suppe, 1983). Bedding-plane shear on multiple surfaces within the hinge zone is an essential element of both types of folding. In many fold-and-thrust belts, the geometry inherent in parallel stratification has permitted the use of quantitative theoretical methods to estimate the location and displacement of blind thrusts—faults that do not penetrate to the surface and that commonly have no unequivocal subsurface evidence to confirm their existence.

In the Los Angeles basin, the fold-and-thrust-belt model has been used to predict the existence and location of a number of active blind thrusts (Davis et al., 1989). These thrusts (Faults A, B, C, D, and E in Figure 45) are predicated solely on the basis of the method's geometric analysis, without other supporting evidence. The postulated regional detachment (Fault B) lies close to the brittle-ductile transition suggested in other studies (e.g., Hauksson, 1987). The 10 October 1987 Whittier Narrows event was ascribed by Davis et al. (1989) to the Elysian Park thrust (EPT), although the hypocenter of that event (Hauksson and Jones, 1989) was east of the East Montebello fault and slip appears to have been restricted to the Puente segment. Above Fault B, these thrusts, including an extensive horizontal detachment thrust (Figure 45), are within the seismogenic zone, in brittle crystalline basement riven with preexisting high-angle faults. The resulting cross sections, restored by quantitative geometric analysis, have been used to predict slip rates and, thus, seismic recurrence intervals on the Elysian Park thrust fault.

As applied in the Los Angeles region, the fold-and-thrust-belt and flake-tectonics models have several major elements in common. Both incorporate a basal decollement or detachment zone associated with the brittle-ductile transition that forms the lower boundary of the seismogenic zone. Both involve the folding and thrusting of crystalline basement rocks at shallow depths, as seen in outcrop along the southern margin of the Transverse Ranges. The latter element provides one important means of choosing between the two models.

Along the southern edge of the Transverse Ranges, application of the fold-and-thrust-belt model requires that postulated thrust faults and flexural-slip folding extend through large zones of brittle, nonlayered basement rock within the seismogenic zone (Figure 45). These faults do not "run along bedding planes for most of their length" (Suppe, 1985), and the rocks lack the stratified geometry implicit in the mechanisms of fault-bend and fault-propagation folding, as cited above. Fault-bend folds appear to require layered rock beneath the fold. Suppe and Medwedeff (1984), in describing fault-propagation folding, noted that "if and when inappropriate stratigraphy is encountered this mode of propagation ceases." Thus, the fact that the postulated fold-and-thrust belts in the Los Angeles region are "basement-involved" even in their upper frontal portions limits the utility of the model in that area.

In the structural analysis of fold-and-thrust belts, the use of retrodeformable cross sections in a quantitative sense requires certain simplifying assumptions that, as noted earlier, may not be valid in the Los Angeles region. For example, such cross sections should not be intersected by faults with significant strike-slip displacement, and indeed, the East Montebello (Whittier Extension) fault has been omitted from one such cross section through the Whittier Narrows area (Davis et al., 1989, their figure 9), though the fault is shown on the accompanying map (their figure 5). Preparation of retrodeformable cross sections assumes that bed lengths remain constant during deformation, yet the mechanism of "layer-parallel shortening" during deformation (Medwedeff, 1989) appears to be acceptable in Tertiary basins of California and in the Appalachian Plateau. Also, calculations that require measurement of bed lengths are difficult to achieve where extensive areas of crystalline basement are involved in the folding.

Construction of retrodeformable cross sections in the Los Angeles region must also deal with multiple and dissimilar episodes of deformation. The Elysian Park anticline, for example, is not simply the result of late Pliocene and Quaternary south-directed shortening. Its significant middle Miocene growth is documented by Topanga beds absent from the anticlinal crest but 2000 to 5000 ft (600 to 1500 m) thick on both flanks beneath a cover of early Mohnian deep-water shale (Figure 5) (Lamar, 1970). Late Miocene accentuation of its south flank is documented by stratigraphic thickening of upper Mohnian and Delmontian strata (Figure 41). Analyses of structural growth must either incorporate these earlier episodes or exclude them by relying solely on the geometry of younger strata. Even then, the active axial surface of a fold may be influenced more strongly by that fold's relict geometry than by the active local state of stress.

For these two basic reasons—the compound history of Neogene deformation and the absence of layered geometry in major portions of the folded rocks—the fold-and-thrust-belt model cannot be applied rigorously in the Los Angeles region, although some of its elements are present. The flake-tectonic model is entirely applicable in rocks of any geometry, layered or nonlayered. Within its context, folding of nonlaminar basement rock occurs by pervasive intercrystalline shear. As noted earlier, the development of crustal flakes is enhanced by multiple episodes of deformation. The flake-tectonic model is a less demanding one that does not claim to locate unseen faults or to predict the magnitude or recurrence intervals of earthquakes. In the future, semiquantitative interpretations of that type might be provided by new techniques, perhaps including growth fault-bend folding (Medwedeff and Suppe, 1986; Medwedeff, 1989), which utilizes the thinning of syntectonic sediments to determine the growth rate of folds.

Interaction between Thrusting and Right Slip

Interaction between Transverse Ranges thrusting and Peninsular Ranges right slip has deformed the previously linear left-oblique fault system that forms the southern margin of the Transverse Ranges (Figure 43). A 25° bend at the west end of the Raymond fault is west of the projected intersection of the East Montebello fault and is tentatively ascribed to right slip on that fault. Moderate right slip along the Elsinore-Whitter fault has moved the northern shelf and rotated the Hollywood fault about 15° clockwise from the assumed pre-Pasadenan trend of the Raymond–Santa Monica fault system. The analysis of similar rotations west of the NIFZ (Figure 3) is obscured by uncertain relationships between onshore faults and the Malibu Coast and Anacapa (Dume) faults.

Three active modes of faulting may well occur along the northern edge of the Los Angeles basin. In addition to south-vergent thrusting of the NLATS and right-oblique motion on the East Montebello and Inglewood faults, some left slip may be continuing on the Santa Monica–Raymond fault system (Ziony and Yerkes, 1985; Real, 1987). In attempting to resolve these latter two modes, earthquake focal mechanisms provide an ambiguous tool. For example, the 3 December 1988 event ($M_L = 5.0$) beneath Pasadena (Figure 43) yielded nodal planes showing left slip parallel to the Raymond fault or right slip parallel to the East Montebello fault, which projects into the Raymond fault in the area of the hypocenter. Similar choices are involved to the west, where Real (1987) has interpreted left slip along the Hollywood fault and, in the immediately adjacent Elysian Park-Repetto Hills area, right slip on northwesterly trending faults. The same ambiguity is to be expected where the NIFZ projects northward into the terminations of the Santa Monica and Hollywood faults.

Resolution of this problem could be advanced significantly by additional seismograph stations and by the development of crustal velocity models that more accurately represent the three-dimensional variations in velocity structure beneath the Los Angeles region. These improvements could facilitate the assignment of hypocenters to specific faults and permit a better understanding of aftershock patterns. The accuracy of most existing data is estimated at 30–60 mi (50–100 km) for pre-1900 epicenters, 3–9 mi (5–15 km) for events from 1932 to 1970, and 3 mi (5 km) for events after 1970 (Simila and Fischer, 1982). The western Los Angeles basin model developed by Hauksson (1987) approaches the necessary degree of accuracy but is applicable only for that local area.

Focal mechanisms for recent small earthquakes along the NIFZ (Hauksson, 1987), combined and inverted, yield a maximum principal stress that is horizontal and has an azimuth of 25° east of north. The flexure that separates the central trough from the northern shelf (Figure 43) is orthogonal to that axis of stress. A crude vector analysis suggests that this local stress field might be the sum of south-directed Transverse Ranges shortening and northwest-directed right slip at a ratio of about 2:1. In the southwestern Los Angeles basin, by contrast, the Transverse Ranges influence is subordinate to right slip, as shown by the extensional nature of the southern NIFZ and adjacent Compton–Los Alamitos fault (Hauksson, 1987).

CONCLUSIONS

Radical processes, including the accretion of multiple terranes during Paleogene time and major rotations of blocks within the Transverse Ranges during the early Neogene, have been postulated to explain aspects of the geology of southern California. The preceding analysis of the Los Angeles region has shown that those processes are compatible with one detailed scenario for the tectonic evolution of that region. Although this lends support to the validity of the hypotheses of terrane accretion and block rotation in southern California, it does not constitute proof, as alternative scenarios may be put forward that also fit the evidence.

Many earlier analyses of the structural geology of southern California have been strongly influenced by the striking regional patterns provided by the San Andreas, Garlock, and Transverse Ranges faults and, more locally, the Newport-Inglewood fault zone. These patterns have suggested that the tectonics of southern California can be interpreted in the context of a broad and uniform regional stress field. Any such two-dimensional approach to the complex geology of this region is quite incomplete. Tectonic analyses must also include the third dimension of depth—the configuration of underlying basement blocks and of deeper lithospheric features—and the fourth dimension of geologic time.

Recent data on the low-stress nature of the San Andreas fault also call for a reconsideration of traditional concepts. Zoback et al. (1987) concluded that "the San Andreas and its related faults seem to represent pronounced zones of weakness, possibly through the entire lithosphere, that localize deformation and reorient tectonic stresses." Examining the state of stress near the San Andreas fault, Mount and Suppe (1987) concluded that "transpressive tectonics in central California can be better described as decoupled transcurrent and compressive deformation" and that "the thrust structures are largely decoupled from the strike-slip fault."

These new concepts indicate the importance of looking beneath the regional or local crustal stresses, present or past, in an attempt to divine the mantle processes that produced them. In many regions, those may be predominantly thermal processes of a slow and static nature. In southern California, however, they are dynamic processes that involve the interaction of nonuniform, opposing mantle motion at the plate margin. A consideration of the evolution of structures in the Los Angeles and Ventura basins (Crowell, 1976, 1987; Yeats, 1983) suggests that our ability to understand past and present tectonism in this region might be improved by an alternative conceptual approach. This concept sees deformation as driven by persistent mantle processes that progressively engage and carry overlying tectonic plates or flakes.

During Cenozoic time (Figure 6), those mantle processes have involved the northwestward motion of the Pacific plate, spreading beneath the East Pacific Rise, and the westward expansion of the Basin and Range province (Wright, 1976). A significant early result was the segmentation of the crust into tectonic flakes (Oxburgh, 1972; Yeats, 1981). This was initiated during Late Cretaceous–Paleogene terrane accretion and was augmented by mid-Miocene rifting. The boundaries between those flakes constitute zones of weakness that have localized major faulting during all subsequent episodes of deformation.

In southern California, the fundamental element of Neogene tectonic evolution has been the progressive eastward propagation of Pacific plate motion—within the mantle—beneath the crust of the North American plate. Part of the western margin of that plate was first rifted and rotated in middle Miocene time by the waning phases of mantle upwelling and spreading associated with the overridden East Pacific Rise. The rifted corner of the North American plate was subsequently engaged by the northwest-moving mantle flow that carries the Pacific plate. One effect was a brief episode of regional Mio-Pliocene transtension that ended when Baja California separated from the Mexican mainland. Each eastward step of the San Andreas transform zone has been accompanied by a change in the pattern of deformation in the Los Angeles region. Northwestward flow of Pacific plate mantle is now driving the collision between the Peninsular Ranges block and the Sierra–Great Valley salient of the North American plate.

Pasadenan deformation of the Los Angeles region involves northwestward underthrusting of Peninsular Ranges blocks beneath the south-vergent (and perhaps westward moving) blocks of the Transverse Ranges. The primary fault blocks are tectonic flakes that, as the Tranverse Ranges are approached, decouple from the downwelling mantle (Bird and Rosenstock, 1984) at a detachment zone beneath the brittle-ductile transition. From that decollement, major thrusts (both blind and outcropping) and right-slip faults extend upward into the seismogenic zone. Within that collage of tectonic flakes, the axial part of the Los Angeles basin has continued to subside as its northern, northeastern, and southwestern flanks are uplifted.

Within the mantle, Pacific plate (northwestward) motion extends east into the Mojave Desert area, and viscous drag against the edge of the North American plate produces right slip on northwest-striking faults of the Peninsular Ranges and continental borderland. Slip rates on these faults increase systematically eastward, reflecting the underlying concentration of mantle shear as the deep plate margin is approached. Of these faults, only the San Andreas has broken through the collision zone of the Transverse Ranges. Regionally, the eastward increase in right slip and northward increase in compressive shortening portends a future intensification of tectonic activity in the western Mojave Desert. As a corollary, both modes of tectonism are least active in the southwestern part of the Los Angeles region, where seismicity is relatively low.

The transition between the strong compressive shortening of the Transverse Ranges and the moderate right slip of Peninsular Ranges blocks occurs systematically across the Los Angeles basin. Those relationships are clearly indicated in the cross sections of Figure 8, which show contrasting structural styles on the two sides of the basin. The northeast flank is dominated by blind thrusts of the Transverse Ranges system that flatten with depth. The southwest flank features right-oblique faults of the Peninsular Ranges system that steepen into near-vertical zones of active seismicity. Viewed from south to north (GG' to AA'), these cross sections confirm the gradual change from extension at the southern end of the basin to compressive shortening at the northern end, as also revealed by focal mechanisms of recent small earthquakes (Hauksson, 1987) along the Newport-Inglewood fault zone (NIFZ). Faulting in oil fields along the NIFZ mimic those focal mechanisms: north-striking normal faults are found in the Huntington Beach (Figure 27) and Long Beach Airport (Figure 24) fields, and west-striking thrusts occur in the Howard Park and Rosecrans fields and at Dominguez (Figure 23), where they are known to be active.

Studies of earthquake focal mechanisms have shown conclusively that the North Los Angeles thrust system and the Compton-Los Alamitos fault are important structural features of the Los Angeles region. As data from future earthquakes are assimilated, such studies will continue to serve as a primary tool in studies of faults and tectonic patterns in this region. The precision of these studies is being enhanced by the use of improved three-dimensional velocity models. In this field, the technique of tomographic inversion offers great promise, and might ultimately lead to the definition of the discrete crustal flakes involved in Pasadenan (and perhaps older) deformation. Mapping of major faults within the seismogenic zone might also be aided significantly by a grid of deep seismic reflection profiles of the Cal/Crust and COCORP type.

Relative motion between crustal flakes may involve rifting and separation, transform movement, or collision and shortening, combinations of these, and superposition of several modes over time. Local structures are shaped not by regional stress fields embracing areas hundreds of miles across but by the interaction of adjacent tectonic flakes, creating local or subregional stresses. These stresses do not act upon a uniform crust—the "clay blanket" of laboratory experiments—but upon a three-dimensional mosaic produced by previous tectonic episodes and containing basement blocks and sedimentary wedges that may differ significantly in their densities, ductilities, and thermal characteristics. In shaping local structure, the influence of these internal features of the shallow crust may be as important as the orientation of the stresses being applied.

As one example, consider the early Pliocene deformation west of the southern Newport-Inglewood fault zone (NIFZ) (Figure 9). There the Wilmington anticline was folded along a trend of N50°W to N75°W, while the nearby Offshore Huntington Beach anticline was folded on an east-west trend. Each fold is parallel to an adjacent basement high (Figure 5). In forming a structure, the shape of the mold counts for as much as how the hammer is swung.

The trap-forming structures of the Los Angeles basin have thus been formed by a variety of processes. On the western shelf, the en echelon Torrance, Lawndale-El Segundo, and Venice-Playa del Rey anticlines (Figure 8, Sections AA', CC', and DD') are the result of distributive simple shear of ductile schist basement. The Dominguez anticline (Figure 8, Section DD', and Figure 23) was initiated by this mechanism and was later modified by right slip on the NIFZ. The Wilmington anticline (Figure 30; Figure 31, Section bb'), a buttress fold formed in the graben between the NIFZ and the THUMS-Huntington Beach fault, also involves folding of ductile basement. Buckling within a compressed sedimentary wedge was the primary cause of folding at Inglewood (Figure 8, Section AA'; Figures 21, 22), where it later was modified by right-slip faulting and wedge uplift; at Offshore Huntington Beach (Figure 27; Figure 28, Section dd'), at Santa Fe Springs (Figure 8, Section EE'); and in the small anticlines of the Jefferson pool in the Las Cienegas, the Los Angeles Downtown, the Union Station (Figure 8, Section CC'), and the Boyle Heights oil fields. In the Coyote Hills (Figure 8, Section FF'; Figure 33) and at Richfield (Figure 8, Section GG'), this style of folding may have evolved into fault-propagation folds above the Peralta Hills thrust and into an equivalent blind thrust farther to the west.

Fault-parallel anticlines, resulting from block-edge forced folding (Harding and Tuminas, 1988), occur along the southern NIFZ at Long Beach (Figure 8, Section EE'; Figures 24, 25), Seal Beach (Figures 8, Section FF'; Figure 25), and (at least in part) Huntington Beach (Figure 27; Figure 28, Section bb'). A similar type of folding is seen in the central and western parts of the Las Cienegas anticline (Figure 8, Section BB'; Figure 9) and perhaps also at East Beverly Hills (Figure 8, Section AA') and the Salt Lake anticline.

Fault traps and commonly anticlinal traps were formed in the footwall of major faults (Harding and Tuminas, 1988). Along the Whittier fault these include Brea-Olinda (Figure 8, Section GG'; Figure 12), Sansinena (Figure 8, Section FF'), and Whittier (Figure 8, Section EE'; Figure 11). Adjacent to the Santa Monica fault, traps of this type include Sawtelle (Figure 15, Section II'; Figure 17), West Beverly Hills, and Cheviot Hills (Figure 15, Section HH'). Along the Palos Verdes fault, only the Beta oil field (Figure 8, Section GG'; Figure 13) has been discovered to date. The Los Angeles City oil field appears to be a similar footwall trap, though on a much smaller scale.

All of those structures developed within a wide region of pervasive right slip associated with the evolving San Andreas transform zone. Nevertheless, strike-slip folding caused by displacement along an individual fault is not a dominant factor in the genesis of structures in the Los

Angeles basin, though it may well have contributed to deformation along the northern NIFZ (Figure 9) and perhaps to deformation along the Santa Monica fault (by left slip). That mechanism and other classic patterns of fold and fault development have been modified substantially and, in some instances, have been nullified by the effects of preexisting basement blocks and sedimentary wedges.

From late middle Miocene to late Pliocene time, continual syntectonic deposition on the deeply submerged floor of the Los Angeles basin provided a sensitive record of structural development. At Wilmington and Inglewood, detailed multi-interval isopach studies have greatly advanced our knowledge of the timing and mechanism of local deformation. The results also provide significant insights into the changing patterns of regional deformation.

Similar studies of other oil fields in the basin, especially the Dominguez, Playa del Rey, El Segundo–Lawndale, Cheviot Hills, Santa Fe Springs, West and East Coyote, Richfield, Huntington Beach (onshore), and Seal Beach fields, should significantly advance our understanding not only of local trap-forming mechanisms but of the changing patterns of regional deformation. In some of the basin's oil fields, however, detailed isopach studies may be difficult or impossible to achieve. In the oldest fields, for example, Salt Lake, few wells have been drilled since the advent of electric logging. In some older fields, such as Long Beach and Rosecrans, well courses have not generally been surveyed and significant accidental deviations are thought to be common. Where wells have been drilled directionally from platforms or urban drill sites, as at Offshore Huntington Beach, Cheviot Hills, and West and East Beverly Hills, well control in the upper part of the sedimentary section may not be distributed widely enough for meaningful isopach studies. In the proximal end of the basin, as at Montebello, the extreme variability of fan/channel facies quite likely would preclude the precise local correlations necessary for detailed isopaching.

Detailed studies, including isopach mapping, of local structural growth have recently been initiated or proposed as a method for estimating rates of fault movement and earthquake recurrence intervals. Such methods must be used with caution. The quantitative application of the fold-and-thrust-belt model is believed to be inappropriate in the Los Angeles region, although variants, such as growth fault-bend folding, may be valid for semi-quantitative studies. The utility of all such methods for studies of earthquake hazards in the Los Angeles area is conditional on significantly improved time-stratigraphic correlations in the late Pliocene and Quaternary sediments of that area.

ACKNOWLEDGMENTS

The basic data and knowledge embodied in this paper were amassed during a decade of active urban oil exploration between 1961 and 1971 and owe much to my colleagues in Chevron's Los Angeles Basin District in those years, to the late Thomas J. McCroden, District Geologist, and to Chevron's Southern Division staff and paleontologists. Geologists, friendly competitors over the decades, at Unocal, Arco, Mobil, Signal, Oxy, and other oil companies and Jack C. West and other consultants have generously shared ideas and data. The present paper has grown out of an earlier summary of the Neogene tectonic evolution of the Los Angeles basin, which was presented at the 1973 annual meeting of AAPG. The undertaking to expand and update that paper for this publication resulted in no small part from the steady encouragement of John C. Crowell, who also supplied very helpful comments for the revision. Suggestions from other reviewers, including G. C. (Butch) Brown, David C. Bushnell, Tod P. Harding, Egil Hauksson, Tom Redin, and especially Robert S. Yeats and Robert F. Yerkes, greatly improved the structure and content of the final version, though responsibility for any remaining errors, omissions, and misinterpretations is mine alone. I am indebted to Kevin T. Biddle for his skilled and patient editorial guidance, to Chevron U.S.A. and Exxon International for substantial assistance in drafting the figures, and to Nancy Angle of Chevron for her drafting expertise. Chevron (San Ramon) and the United States Geological Survey (Menlo Park) provided essential library services. A most profound debt is due to Louise, my wife, for all manner of sustenance.

This paper is dedicated to the memory of Edwin S. Parker and Robert C. Erickson, co-authors of the 1973 paper, whose friendship and geologic insights are sorely missed.

REFERENCES CITED

Allen, D. R., 1966, Long Beach offshore area, longitudinal-diagrammatic section showing estimated oil and water zones, in W. E. Bauer, A. A. Cohn, Jr., J. LeConte, E. S. Pickett, and J. N. Truex, A tour of the coastal oil fields of the Los Angeles basin in and adjacent to San Pedro Bay, California: Guidebook, Pacific Sections, AAPG–SEPM–Society of Exploration Geophysicists, Joint Annual Field Trip, inset to map.

Allen, D. R., and G. C. Hazenbush, 1957, Sunset Beach oil field: California Division of Oil and Gas, California Oil Fields—Summary of Operations, v. 43, n. 2, p. 47–50.

Atwater, T., 1970, Implications of plate tectonics for the Cenozoic tectonic evolution of western North America: GSA Bulletin, v. 81, p. 3513–3536.

Bailey, T. L., and R. H. Jahns, 1954, Geology of the Transverse Range province, southern California, in R. H. Jahns, ed., Geology of southern California: California Division of Mines Bulletin 170, p. 83–106.

Bandy, O. L., and L. Marincovich, Jr., 1973, Rates of late Cenozoic uplift, Baldwin Hills, Los Angeles, California: Science, v. 181, p. 653–655.

Barbat, W. F., 1958, The Los Angeles basin area, California, in L. G. Weeks, ed., Habitat of oil: AAPG, p. 62–77.

Barron, J. A., 1986, Updated diatom biostratigraphy for the Monterey Formation of California, in R. E. Casey and J. A. Barron, eds., Siliceous microfossils and microplankton of the Monterey Formation and modern analogs: Pacific Section, SEPM Special Publication 45, p. 105–119.

Barrows, A. G., 1974, A review of the geology and earthquake history of the Newport-Inglewood structural zone, southern California: California Division of Mines and Geology Special Report 114, 115 p.

Bauer, W. E., A. A. Cohn, Jr., J. LeConte, E. S. Pickett, and J. N. Truex, 1966, Coastal oil fields of the Los Angeles basin in and adjacent to San Pedro Bay, California, in Pacific Sections, AAPG-SEPM-Society of Exploration Geophysicists, 1966 Annual Field Trip, map, scale 1 in:3000 ft.

Berggren, W. A., D. V. Kent, J. J. Flynn, and J. A. Van Couvering, 1985, Cenozoic geochronology: GSA Bulletin, v. 96, p. 1407-1418.

Berry, A. L., G. B. Dalrymple, M. A. Lanphere, and J. C. VonEssen, eds., 1976, Summary of miscellaneous potassium-argon age measurements, U.S. Geological Survey, Menlo Park, California, for the years 1972-74: USGS Circular 727, 13 p.

Beyer, L. A., 1987, Bouger gravity and magnetic anomaly map of the inner-southern California continental margin, in H. G. Greene and M. P. Kennedy, eds., Geology of the inner-southern California continental margin: California Division of Mines and Geology, map, scale 1:250,000.

Beyer, L. A., 1988, Summary of geology and petroleum plays used to assess undiscovered recoverable petroleum resources of Los Angeles basin province, California: USGS Open-File Report 88-450L, 62 p.

Beyer, L. A., and K. A. Pisciotto, 1986, Bouger gravity and magnetic anomaly map of the mid-southern California continental margin, in H. G. Greene and M. P. Kennedy, eds., Geology of the mid-southern California continental margin: California Division of Mines and Geology, map, scale 1:250,000.

Biddle, K. T., 1991, The Los Angeles basin—an overview, in K. T. Biddle, ed., Active margin basins: AAPG Memoir 52, p. 5-24.

Bird, P., and R. W. Rosenstock, 1984, Kinematics of present crust and mantle flow in southern California: GSA Bulletin, v. 95, p. 946-957.

Bishop, C. C., and J. F. Davis, 1984, Southern California region—correlation of stratigraphic units of North America (COSUNA) project: AAPG, chart.

Blake, G. H., 1991, Review of the Neogene biostratigraphy and stratigraphy of the Los Angeles basin and implications for basin evolution, in K. T. Biddle, ed., Active margin basins: AAPG Memoir 52, p. 135-184.

Bostick, N. H., S. M. Cashman, T. H. McCulloh, and C. T. Waddell, 1978, Gradients of vitrinite reflectance and present temperature in the Los Angeles and Ventura basins, California, in D. F. Oltz, ed., Low temperature metamorphism of kerogen and clay minerals: Pacific Section, SEPM, p. 65-96.

Bowes, G. H., 1943, Seal Beach oil field: California Division of Mines and Geology Bulletin 118, p. 325-328.

Bravinder, K. M., 1942, Los Angeles basin earthquake of October 21, 1941, and its effect on certain producing wells in Dominguez field, Los Angeles County, California: AAPG Bulletin, v. 26, p. 388-399.

Bryant, M. E., 1987, Emergent marine terraces and Quaternary tectonics, Palos Verdes peninsula, California, in P. J. Fischer, ed., Geology of the Palos Verdes peninsula and San Pedro Bay: Pacific Section, SEPM Guidebook 55, p. 63-78.

Bryant, M. E., and D. L. Fife, 1982, The Peralta Hills fault, a Transverse Ranges structure in the northern Peninsular Ranges, southern California, in D. L. Fife and J. A. Minch, eds., Geology and mineral wealth of the California Transverse Ranges: Santa Ana, California, South Coast Geological Society, p. 403-409.

Bryant, W. A., 1978, The Raymond Hill fault—an urban geological investigation: California Geology, v. 31, n. 6, p. 127-142.

California Division of Oil and Gas, 1961, California oil and gas fields, Part 2, Los Angeles—Ventura basins and central coastal regions: San Francisco, California, California Division of Oil and Gas, p. 495-913.

California Division of Oil and Gas, 1964, Exploratory wells drilled outside of oil and gas fields in California to December 31, 1963: San Francisco, California, California Division of Oil and Gas, 320 p.

California Division of Oil and Gas, 1974, California oil and gas fields, Volume 2, south, central coastal and offshore California: Sacramento, California, California Division of Oil and Gas, unpaginated.

Campbell, R. H., 1973, Generalized geologic map and structure sections, Santa Clara River to Santa Monica Bay, southern California, in I. P. Colburn and A. E. Fritsche, eds., Cretaceous stratigraphy of the Santa Monica Mountains and Simi Hills, southern California: Pacific Section, SEPM, Fall Field Trip Guidebook, map with text.

Campbell, R. H., and R. F. Yerkes, 1971, Cenozoic evolution of the Santa Monica Mountains-Los Angeles basin area: II, relation to plate tectonics of the northeast Pacific Ocean: GSA Abstracts with Programs, v. 3, p. 92.

Campbell, R. H., and R. F. Yerkes, 1976, Cenozoic evolution of the Los Angeles basin area—relation to plate tectonics, in D. G. Howell, ed., Aspects of the geologic history of the California continental borderland: Pacific Section, AAPG Miscellaneous Publication 24, p. 541-560.

Campbell, R. H., R. F. Yerkes, and C. M. Wentworth, 1966, Detachment faults in the central Santa Monica Mountains, California, in Geological Survey research, 1966: USGS Professional Paper 550-C, p. C1-C11.

Castle, R. O., and R. F. Yerkes, 1976, Recent surface movements in the Baldwin Hills, Los Angeles County, California: USGS Professional Paper 882, 125 p.

Castro, M. J., 1975, Mahala field, in J. N. Truex, ed., A tour of the oil fields of the Whittier fault zone, Los Angeles basin California: Pacific Sections, AAPG-Society of Exploration Geophysicists-SEPM, Joint Annual Field Trip, April 26, 1975, p. 73-76.

Chevron U.S.A., Inc., 1980, Development and production plan for proposed Platform Edith, San Pedro Bay, offshore southern California, Federal OCS Lease P 0296: Chevron U.S.A., Inc. (November 24, 1980).

Clark, J. C., E. E. Brabb, H. G. Greene, and D. C. Ross, 1984, Geology of Point Reyes Peninsula and implications for San Gregorio fault history, in J. K. Crouch and S. B. Bachman, eds., Tectonics and sedimentation along the California margin: Pacific Section, SEPM, v. 38, p. 67-85.

Clarke, D. D., 1987, Structure of the Wilmington oil field, in D. D. Clarke and C. P. Henderson, eds., Geologic field guide to the Long Beach area: Pacific Section, AAPG Guidebook n. 58, p. 43-55.

Clarke, D. D., and C. P. Henderson, 1987, Coastal oil fields of the Los Angeles basin in and adjacent to the city of Long Beach, California, *in* D. D. Clarke and C. P. Henderson, eds., Geologic field guide to the Long Beach area: Pacific Section, AAPG Guidebook 58, map, scale 1 in:5000 ft.

Clarke, S. H., H. G. Greene, M. P. Kennedy, and J. G. Vedder, 1987, Geologic map of the inner-southern California continental margin, *in* H. G. Greene and M. P. Kennedy, eds., Geology of the inner-southern California continental margin: California Division of Mines and Geology, map, scale 1:250,000.

Colburn, I. P., 1973, Stratigraphic relations of the southern California Cretaceous strata, *in* I. P. Colburn and A. E. Fritsche, eds., Cretaceous stratigraphy of the Santa Monica Mountains and Simi Hills: Pacific Section, SEPM, Fall Field Trip Guidebook, p. 45–73.

Conrad, C. L., and P. L. Ehlig, 1987, The Monterey Formation of the Palos Verdes Peninsula, California—an example of sedimentation in a tectonically active basin within the California continental borderland, *in* P. J. Fischer, ed., Geology of the Palos Verdes Peninsula and San Pedro Bay: Pacific Section, SEPM Guidebook 55, p. 17–30.

Conrey, B. L., 1967, Early Pliocene sedimentary history of the Los Angeles basin, California: California Division of Mines and Geology Special Report 93, 63 p.

Cook, D. L., 1975, Synopsis of geology and production history of Shell Oil's Yorba Linda field, *in* J. N. Truex, ed., A tour of the oil fields of the Whittier fault zone, Los Angeles basin California: Pacific Sections, AAPG-Society of Exploration Geophysicists-SEPM, Joint Annual Field Trip, April 26, 1975, p. 66–72.

Crittenden, M. D., Jr., 1980, Metamorphic core complexes of the North American Cordillera: summary, *in* M. D. Crittenden, Jr., P. J. Coney, and G. H. Davis, eds., Cordilleran metamorphic core complexes: GSA Memoir 153, p. 485–490.

Crook, R., Jr., R. J. Proctor, and C. E. Lindvall, 1983, Seismicity of the Santa Monica and Hollywood faults determined by trenching: Menlo Park, California, Technical report to the USGS Under Contract 14-08-0001-20523, 26 p.

Crook, R., C. R. Allen, B. Kamb, C. M. Payne, and R. J. Proctor, 1987, Quaternary geology and seismic hazard of the Sierra Madre and associated faults, western San Gabriel Mountains: USGS Professional Paper 1339, p. 27–63.

Crowell, J. C., 1962, Displacement along the San Andreas fault, California: GSA Special Paper 71, 61 p.

Crowell, J. C., 1974, Origin of late Cenozoic basins in southern California, *in* W. R. Dickinson, ed., Tectonics and sedimentation: SEPM Special Publication 22, p. 190–204.

Crowell, J. C., 1975, The San Andreas fault in southern California: California Division of Mines and Geology Special Report 118, 272 p.

Crowell, J. C., 1976, Implications of crustal stretching and shortening of coastal Ventura basin, California, *in* D. G. Howell, ed., Aspects of the geologic history of the California continental borderland: Pacific Section, AAPG Miscellaneous Publication 24, p. 365–382.

Crowell, J. C., 1979, The San Andreas fault system through time: Journal of the Geological Society of London, v. 136, p. 293–302.

Crowell, J. C., 1982, The tectonics of Ridge basin, southern California, *in* J. C. Crowell and M. H. Link, eds., Geologic history of Ridge basin, southern California: Pacific Section, SEPM, p. 25–42.

Crowell, J. C., 1987, Late Cenozoic basins of onshore southern California: complexity is the hallmark of their tectonic history, *in* R. V. Ingersoll and W. G. Ernst, eds., Cenozoic basin development of coastal California: Englewood Cliffs, New Jersey, Prentice-Hall, Inc., Rubey, v. VI, p. 207–241.

Curray, J. R., and D. G. Moore, 1984, Geologic history of the Gulf of California, *in* J. K. Crouch and S. B. Bachman, eds., Tectonics and sedimentation along the California margin: Pacific Section, SEPM, v. 38, p. 17–35.

Dahlstrom, C. D. A., 1969, Balanced cross-sections: Canadian Journal of Earth Sciences, v. 6, p. 743–754.

Daviess, S. N., and A. O. Woodford, 1949, Geology of the northwestern Puente Hills, Los Angeles County, California: USGS Oil and Gas Investigations Preliminary Map 83, scale 1 in:1000 ft.

Davis, R. A., Jr., 1975, Brea-Olinda oil field, *in* J. N. Truex, ed., A tour of the oil fields of the Whittier fault zone, Los Angeles basin, California: Pacific Sections, AAPG-Society of Exploration Geophysicists-SEPM, Joint Annual Field Trip, April 26, 1975, p. 47–56.

Davis, T. L., 1986, A structural outline of the San Emigdio Mountains, *in* T. L. Davis and J. A. Namson, eds., Geologic transect across the western Transverse ranges: Pacific Section, SEPM Field Trip Guidebook, p. 23–32.

Davis, T. L., 1987, Subsurface study of the late Cenozoic structural geology of the Los Angeles basin: USGS Open-File Report 87-374, p. 143–147.

Davis, T. L., J. Namson, and R. F. Yerkes, 1989, A cross section of the Los Angeles area: seismically active fold and thrust belt, the 1987 Whittier Narrows earthquake, and earthquake hazard: Journal of Geophysical Research, v. 94, p. 9644–9664.

Dibblee, T. W., Jr., 1982a, Geology of the Santa Monica Mountains and Simi Hills, southern California, *in* D. L. Fife and J. A. Minch, eds., Geology and mineral wealth of the California Transverse Ranges: Santa Ana, California, South Coast Geological Society, p. 94–130.

Dibblee, T. W., Jr., 1982b, Regional geology of the Transverse Ranges province of southern California, *in* D. L. Fife and J. A. Minch, eds., Geology and mineral wealth of the California Transverse Ranges: Santa Ana, California, South Coast Geological Society, p. 7–26.

Dibblee, T. W., Jr., 1989, Mid-Tertiary conglomerates and sandstones on the margins of the Ventura and Los Angeles basins and their tectonic significance, *in* I. P. Colburn, P. L. Abbott, and J. Minch, eds., Conglomerates in basin analysis: a symposium dedicated to A. O. Woodford: Pacific Section, SEPM, v. 62, p. 207–226.

Dickinson, W. R., 1981, Plate tectonics and the continental margin of California, *in* W. G. Ernst, ed., The geotectonic development of California: Englewood Cliffs, New Jersey, Prentice-Hall, Inc., Rubey v. I p. 1–28.

Dickinson, W. R., and W. S. Snyder, 1979, Geometry of subducted slabs related to San Andreas transform: Journal of Geology, v. 87, p. 609–627.

Dickinson, W. R., et al., 1987, Geohistory analysis of rates of sediment accumulation and subsidence for selected California basins, *in* R. V. Ingersoll and W. G. Ernst, eds.,

Cenozoic basin development of coastal California: Englewood Cliffs, New Jersey, Prentice-Hall, Inc., Rubey v. VI, p. 1–23.

Dosch, M. W., and W. S. Mitchell, 1964, Oxnard oil field: California Division of Oil and Gas, California Oil Fields—Summary of Operations, v. 50, n. 1, p. 21–33.

Driver, H. L., 1943, Inglewood oil field: California Division of Mines and Geology Bulletin 118, p. 306–309.

Driver, H. L., 1948, Genesis and evolution of Los Angeles basin, California: AAPG Bulletin, v. 32, p. 109–125.

Dudley, P. H., 1954, Geology of the Long Beach oil field, Los Angeles County, in R. H. Jahns, ed., Geology of southern California: California Division of Mines Bulletin 170, map sheet 34.

Durham, D. L., and R. F. Yerkes, 1964, Geology and oil resources of the eastern Puente Hills area, southern California: USGS Professional Paper 420-B, 62 p.

Durrell, C., 1954, Geology of the Santa Monica Mountains, Los Angeles and Ventura counties, in R. H. Jahns, ed., Geology of southern California: California Division of Mines Bulletin 170, map sheet 8.

Durrell, C., 1956, Preliminary report on the geology of the Santa Monica Mountains, in Los Angeles forum, Pacific petroleum geologist [newsletter]: Pacific Section, AAPG, v. 10, n. 4, p. 1–3.

Earth Sciences Board, 1988, Paleomagnetism of the Zuma Volcanics, Point Dume, Los Angeles County, California: California Geology, v. 41, n. 11, p. 243–247.

Eaton, J. E., 1923, Structure of the Los Angeles basin and environs: Oil Age, v. 20, n. 6, p. 8–9, 52.

Eaton, J. E., 1924, Structure of the Los Angeles basin and environs: Oil Age, v. 21, n. 1, p. 16–18, 52, 54.

Eaton, J. E., 1932, Decline of Great Basin, southwestern United States: AAPG Bulletin, v. 16, p. 1–49.

Ehlig, P. L., 1981, Origin and tectonic history of the basement terrane of the San Gabriel Mountains, central Transverse Ranges, in W. G. Ernst, ed., The geotectonic development of California: Englewood Cliffs, New Jersey, Prentice-Hall, Inc., Rubey v. I p. 253–283.

Elmore, W. Z., 1958, Santa Fe Springs oil field, in J. W. Higgins, ed., A guide to the geology and oil fields of the Los Angeles and Ventura regions: Pacific Section, AAPG, p. 100–104.

Emery, K. O., 1960, The sea off southern California, a modern habitat of petroleum: New York, New York, John Wiley & Sons, 366 p.

Engebretson, D. C., A. Cox, and R. G. Gordon, 1985, Relative motions between oceanic and continental plates in the Pacific basin: GSA Special Paper 206, 59 p.

Erickson, R. C., and A. O. Spaulding, 1975, Urban oil production and subsidence control—a case history, Beverly Hills (East) oil field, California: Society of Petroleum Engineers of the American Institute of Mining, Metallurgical, and Petroleum Engineers Paper SPE 5603, 13 p.

Eschner, S., and M. K. Scribner, 1972, Discovery and development of the Sawtelle field, in R. R. Morrison, preprint chm., Technical program preprints: Pacific Sections, AAPG-Society of Exploration Geophysicists, 47th Annual Meeting, Bakersfield, California, March 9–10, 1972, unpaginated.

Ferguson, R. N., and C. G. Willis, 1924, Dynamics of oil-field structure in southern California: AAPG Bulletin, v. 8, p. 576–583.

Fife, D. L., 1979, The "basal member of the Monterey Formation," lower Aliso Creek area, Orange County, California, in D. L. Fife, ed., Geologic guide of San Onofre nuclear generating station and adjacent regions of southern California: Pacific Sections, AAPG-SEPM-Society of Exploration Geophysicists, Guide Book 46, p. A17–A24.

Fife, D. L., R. A. Hoffman, M. E. Bryant, R. J. Rushing, R. W. Ruff, S. A. Santarcangelo, and M. E. Unruh, 1980, The Peralta Hills thrust, southern California: GSA Abstracts with Programs, v. 12, p. 106.

Fischer, P. J., R. H. Patterson, and A. C. Darrow, 1987, Palos Verdes fault zone—onshore to offshore, in P. J. Fischer, ed., Geology of the Palos Verdes Peninsula and San Pedro Bay: Pacific Section, SEPM Guidebook 55, p. 91–133.

Floyd, J. W., and E. A. Maxwell, 1964, Whittier oil field, Los Angeles County [internal report]: Standard Oil Company of California, Unpublished Review Team Study 1-18-18.

Foster, J. F., 1954, Rosecrans and South Rosecrans oil fields: California Division of Oil and Gas, California Oil Fields—Summary of Operations, v. 40, n. 2, p. 5–15.

Frame, R. G., 1960, California offshore petroleum development: California Division of Oil and Gas, California Oil Fields—Summary of Operations, v. 46, n. 2, p. 5–46.

Fritsche, A. E., 1973, Bibliography of previous work and geologic map compilation for the Cretaceous strata of the Santa Monica Mountains and Simi Hills, in Cretaceous stratigraphy of the Santa Monica Mountains and Simi Hills: Pacific Section, SEPM, Fall Field Trip Guidebook, p. 3–13.

Gaede, V. F., 1957, Leffingwell oil field: California Division of Oil and Gas, California Oil Fields—Summary of Operations, v. 43, n. 2, p. 35–38.

Gaede, V. F., 1958, Olive oil field: California Division of Oil and Gas, California Oil Fields—Summary of Operations, v. 44, n. 2, p. 65–68.

Gaede, V. F., 1959, Esperanza oil field: California Division of Oil and Gas, California Oil Fields—Summary of Operations, v. 45, n. 2, p. 13–16.

Gaede, V. F., 1964, Central area of Whittier oil field: California Division of Oil and Gas, California Oil Fields—Summary of Operations, v. 50, n. 1, p. 59–67.

Gardett, P. H., 1971, Petroleum potential of Los Angeles basin, California: AAPG Memoir 15, v. 1, p. 298–308.

Garfunkel, Z., 1966, Problems of wrench faults: Tectonophysics, v. 3, p. 457–473.

Gastil, G., G. Morgan, and D. Krummenacher, 1981, The tectonic history of peninsular California and adjacent Mexico, in W. G. Ernst, ed., The geotectonic development of California: Englewood Cliffs, New Jersey, Prentice-Hall, Inc., Rubey v. I, p. 285–306.

Gilluly, J., 1963, The tectonic evolution of the western United States: Quarterly Journal of the Geological Society of London, v. 119, p. 133–174.

Gourley, J. W., 1975, Upper Miocene Puente Formation in the Whittier Hills and surrounding region, in J. N. Truex, ed., A tour of the oil fields of the Whittier fault zone, Los Angeles basin, California: Pacific Sections, AAPG-Society of Exploration Geophysicists-SEPM, Joint Annual Field Trip, April 26, 1975, p. 13–24.

Graham, S. A., 1987, Tectonic controls on petroleum occurrence in central California, in R. V. Ingersoll and W. G. Ernst, eds., Cenozoic basin development of coastal California: Englewood Cliffs, New Jersey, Prentice-Hall, Inc., Rubey v. VI, p. 47-63.

Graham, S. A., and W. R. Dickinson, 1978, Apparent offsets of on-land geologic features across the San Gregorio-Hosgri fault trend, in E. A. Silver and W. R. Normark, eds., San Gregorio-Hosgri fault zone, California: California Division of Mines and Geology Special Report 137, p. 13-23.

Graves, D. T., 1954, Geology of the Dominguez oil field, in R. H. Jahns, ed., Geology of southern California: California Division of Mines Bulletin 170, map sheet 32.

Grinsfelder, S., 1943, Dominguez oil field: California Division of Mines and Geology Bulletin 118, p. 318-319.

Guptill, P. D., and E. G. Heath, 1981, Surface faulting along the Newport-Inglewood zone of deformation: California Geology, v. 34, n. 7, p. 142-148.

Hacker, R. N., chm., 1969, Geology and oilfields of coastal areas, Ventura and Los Angeles basins, California: Pacific Sections, AAPG-Society of Exploration Geophysicists-SEPM, 44th Annual Meeting, Los Angeles, California, Guidebook, 60 p.

Hadley, D., and H. Kanamori, 1977, Seismic structure of the Transverse Ranges, California: GSA Bulletin, v. 88, p. 1469-1478.

Hagstrum, J. T., M. G. Sawlan, B. P. Hausback, J. G. Smith, and C. S. Gromme, 1987, Miocene paleomagnetism and tectonic setting of the Baja California Peninsula, Mexico: Journal of Geophysical Research, v. 92, p. 2627-2639.

Hamilton, W., 1961, Origin of the Gulf of California: GSA Bulletin, v. 72, p. 1307-1318.

Harbert, W., and A. Cox, 1989, Late Neogene motion of the Pacific plate: Journal of Geophysical Research, v. 94, p. 3052-3064.

Harding, T. P., 1973, Newport-Inglewood trend, California—an example of wrenching style of deformation: AAPG Bulletin, v. 57, p. 97-116.

Harding, T. P., and A. C. Tuminas, 1988, Interpretation of footwall (lowside) fault traps sealed by reverse faults and convergent wrench faults: AAPG Bulletin, v. 72, p. 738-757.

Harris, P. B., 1958, Long Beach Airport pool, in J. W. Higgins, ed., A guide to the geology and oil fields of the Los Angeles and Ventura regions: Pacific Section, AAPG, p. 75-77.

Hauksson, E., 1987, Seismotectonics of the Newport-Inglewood fault zone in the Los Angeles basin, California: Seismological Society of America Bulletin, v. 77, p. 539-561.

Hauksson, E., and L. M. Jones, 1989, The 1987 Whittier Narrows earthquake sequence in Los Angeles, southern California: seismological and tectonic analysis: Journal of Geophysical Research, v. 94, p. 9569-9589.

Hauksson, E., and G. V. Saldivar, 1986, The 1930 Santa Monica and the 1979 Malibu, California, earthquakes: Seismological Society of America Bulletin, v. 76, p. 1542-1559.

Hauksson, E., and G. V. Saldivar, 1989, Seismicity and active compressional tectonics in Santa Monica Bay, southern California: Journal of Geophysical Research, v. 94, p 9591-9606.

Hauksson, E., et al., 1988, The 1987 Whittier Narrows earthquake in the Los Angeles metropolitan area, California: Science, v. 239, p. 1409-1412.

Hazenbush, G. C., and D. R. Allen, 1958, Huntington Beach oil field: California Division of Oil and Gas, California Oil Fields-Summary of Operations, v. 44, n. 1, p. 13-25.

Heath, E. G., 1958, Yorba Linda oil field, in J. W. Higgins, ed., A guide to the geology and oil fields of the Los Angeles and Ventura regions: Pacific Section, AAPG, p. 105.

Henry, M. J., 1987, Los Angeles basin—an overview, in D. D. Clarke and C. P. Henderson, eds., Geologic field guide to the Long Beach area: Pacific Section, AAPG Guidebook 58, p. 1-29.

Henyey, T. L., 1976, Heat flow and tectonic patterns on the southern California borderland, in D. G. Howell, ed., Aspects of the geologic history of the California continental borderland: Pacific Section, AAPG Miscellaneous Publication 24, p. 427-448.

Hill, M. L., 1930, Structure of the San Gabriel Mountains, north of Los Angeles, California, with a Foreword by F. S. Hudson: University of California Publications, Department of Geological Sciences Bulletin, v. 19, n. 6, p. 137-170.

Hill, M. L., 1971, Newport-Inglewood zone and Mesozoic subduction, California: GSA Bulletin, v. 82, p. 2957-2962.

Hill, R. L., 1979, Potrero Canyon fault and University High School escarpment, in J. R. Keaton, chm., Field guide to selected engineering geologic features—Santa Monica Mountains: Southern California Section, Association of Engineering Geologists, Annual Field Trip, May 19, 1979, p. 83-103.

Hill, R. L., E. C. Sprotte, J. H. Bennett, C. R. Real, and R. C. Slade, 1979, Location and activity of the Santa Monica fault, Beverly Hills—Hollywood area, California, in Earthquake hazards associated with faults in the greater Los Angeles metropolitan area, Los Angeles County, California, including faults in the Santa Monica-Raymond, Verdugo-Eagle Rock, and Benedict Canyon fault zones: California Division of Mines and Geology Open-File Report 79-16 LA, p. B1-B43.

Hoots, H. W., 1931, Geology of the eastern part of the Santa Monica Mountains, Los Angeles County, California: USGS Professional Paper 165-C, p. 83-134.

Hornafius, J. S., B. P. Luyendyk, R. R. Terres, and M. J. Kamerling, 1986, Timing and extent of Neogene tectonic rotation in the western Transverse Ranges, California: GSA Bulletin, v. 97, p. 1476-1487.

Howell, D. G., and J. G. Vedder, 1981, Structural implications of stratigraphic discontinuities across the southern California borderland, in W. G. Ernst, ed., The geotectonic development of California: Englewood Cliffs, New Jersey, Prentice-Hall, Inc., Rubey v. I, p. 535-558.

Howell, D. G., D. E. Champion, and J. G. Vedder, 1987, Terrane accretion, crustal kinematics, and basin evolution, southern California, in R. V. Ingersoll and W. G. Ernst, eds., Cenozoic basin development of coastal California: Englewood Cliffs, New Jersey, Prentice-Hall, Inc., Rubey v. VI, p. 242-258.

Howell, D. G., C. J. Stuart, J. P. Platt, and D. J. Hill, 1974, Possible strike-slip faulting in the southern California borderland: Geology, v. 2, p. 93-98.

Humphreys, E., R. W. Clayton, and B. H. Hager, 1984, A tomographic image of mantle structure beneath southern California: Geophysical Research Letters, v. 11, p. 625-627.

Hutton, L. K., 1987, 1987 Whittier Narrows earthquake—prior seismicity and comparison with the 1929 "Whittier" sequence: Pasadena, California, Seismological Laboratory, California Institute of Technology, 3 p.

Ingle, J. C., Jr., 1979, Biostratigraphy and paleoecology of early Miocene through early Pleistocene benthonic and planktonic foraminifera, San Joaquin Hills-Newport Bay-Dana Point area, Orange County, California, in C. J. Stuart, ed., A guidebook to Miocene lithofacies and depositional environments, coastal southern California and northwestern Baja California: Pacific Section, SEPM, p. 53-77.

Ingle, J. C., Jr., S. A. Graham, and W. R. Dickinson, 1976, Evidence and implications of worldwide late Paleogene climatic and eustatic events [abstract]: GSA Abstracts with Programs, v. 8, p. 934-935.

Ingram, W. L., 1961, Richfield oil field: California Division of Oil and Gas, California Oil Fields—Summary of Operations, v. 47, n. 2, p. 55-63.

Ingram, W. L., 1962, Rideout Heights area of Whittier oil field: California Division of Oil and Gas, California Oil Fields—Summary of Operations, v. 48, n. 2, p. 93-96.

Ingram, W. L., 1966, Seal Beach oil field, north block—east extension: California Division of Oil and Gas, California Oil Fields—Summary of Operations, v. 52, n. 1, p. 63-68.

Jacobson, J. B., and R. G. Lindblom, 1977, Timing—geological and governmental—in urban oil development, Los Angeles, California: Oil and Gas Journal, v. 75, n. 6, p. 160-175.

Jahns, R. H., 1954, Geology of the Peninsular Range province, southern California and Baja California, in R. H. Jahns, ed., Geology of southern California: California Division of Mines Bulletin 170, p. 29-52.

Jamison, H. C., and R. J. Malloy, 1958a, Bandini oil field, in J. W. Higgins, ed., A guide to the geology and oil fields of the Los Angeles and Ventura regions: Pacific Section, AAPG, p. 96-99.

Jamison, H. C., and R. J. Malloy, 1958b, Boyle Heights oil field, in J. W. Higgins, ed., A guide to the geology and oil fields of the Los Angeles and Ventura regions: Pacific Section, AAPG, p. 98-99.

Jennings, C. W., and R. G. Strand, 1969, Geologic map of California, Los Angeles sheet: California Division of Mines and Geology, scale 1:250,000.

Johnson, R. A., 1961, East area of Potrero oil field: California Division of Oil and Gas, California Oil Fields—Summary of Operations, v. 47, n. 2, p. 65-74.

Jones, D. L., M. C. Blake, and C. Rangin, 1976, The four Jurassic belts of northern California and their significance to the geology of the southern California borderland, in D. G. Howell, ed., Aspects of the geologic history of the California continental borderland: Pacific Section, AAPG Miscellaneous Publication 24, p. 343-362.

Junger, A., 1976, Tectonics of southern California borderland, in D. G. Howell, ed., Aspects of the geologic history of the California continental borderland: Pacific Section, AAPG Miscellaneous Publication 24, p. 418-426.

Junger, A., and H. C. Wagner, 1977, Geology of the Santa Monica and San Pedro basins, California continental borderland: USGS Miscellaneous Field Studies Map MF-820, 10 p., 5 sheets, scale 1:250,000.

Kamerling, M. J., and B. P. Luyendyk, 1979, Tectonic rotations of the Santa Monica Mountains region, western Transverse Ranges, California, suggested by paleomagnetic vectors: GSA Bulletin, v. 90, p. 331-337.

Karig, D. E., and W. Jensky, 1972, The proto-Gulf of California: Earth and Planetary Science Letters, v. 17, p. 169-174.

Karner, G. D., and J. F. Dewey, 1986, Rifting: lithospheric versus crustal extension as applied to the Ridge basin of southern California, in M. T. Halbouty, ed., Future petroleum provinces of the world: AAPG Memoir 40, p. 317-337.

Kew, W. S. W., 1937, Los Angeles basin excursion—Santa Monica Mountains-Inglewood field: Guidebook—Field Excursions—Southern California, AAPG, 22nd Annual Meeting, p. 7-8.

Kleinpell, R. M., 1938, Miocene stratigraphy of California: AAPG, 450 p.

Kleinpell, R. M., 1980, The Miocene stratigraphy of California revisited: AAPG Studies in Geology 11, p. 1-182.

Lagoe, M. B., 1987, Middle Cenozoic basin development, Cuyama basin, California, in R. V. Ingersoll and W. G. Ernst, eds., Cenozoic basin development of coastal California: Englewood Cliffs, New Jersey, Prentice-Hall, Inc., Rubey v. VI, p. 172-206.

Lamar, D. L., 1961, Structural evolution of the northern margin of the Los Angeles basin [Ph.D. dissertation]: Los Angeles, California, University of California, Los Angeles, 106 p.

Lamar, D. L., 1970, Geology of the Elysian Park-Repetto Hills area, Los Angeles County, California: California Division of Mines and Geology Special Report 101, 45 p.

Lamar, D. L., 1972, Microseismicity and recent tectonic activity in the Whittier fault area: Menlo Park, California, Technical Report to USGS under Contract 14-08-00001-12288, 40 p.

Lang, H. R., and R. S. Dreessen, 1975, Subsurface structure of the northwestern Los Angeles basin: California Division of Oil and Gas Technical Report TP01, p. 15-22.

Lawson, A. C., et al., 1908, The California earthquake of April 18, 1906, Report of the State Earthquake Investigation Committee: Carnegie Institution of Washington Publication 87, v. 1, parts 1-2, 451 p.

Legg, M. R., 1986, Earthquake epicenters and selected fault plane solutions of the mid-southern California continental margin, in H. G. Greene and M. P. Kennedy, eds., Geology of the mid-southern California continental margin: California Division of Mines and Geology, map, scale 1:250,000.

Legg, M. R., 1987, Earthquake epicenters and selected fault plane solutions of the inner-southern California continental margin, in H. G. Greene and M. P. Kennedy, eds., Geology of the inner-southern California continental margin: California Division of Mines and Geology, map, scale 1:250,000.

Lin, J., and R. S. Stein, 1989, Coseismic folding, earthquake recurrence, and the 1987 source mechanism at Whittier Narrows, Los Angeles basin, California: Journal of

Geophysical Research, v. 94, p. 9614-9632.

Lindblom, R. G., 1975, The Whittier oil field (central area), in J. N. Truex, ed., A tour of the oil fields of the Whittier fault zone, Los Angeles basin, California: Pacific Sections, AAPG-Society of Exploration Geophysicists-SEPM, Joint Annual Field Trip, April 26, 1975, p. 25-41.

Link, M. H., and T. W. Dibblee, Jr., 1988, Field trip guide to the Ventura basin area, in M. H. Link, ed., Ventura basin: geologic introduction and field trip guidebook: Pacific Section, AAPG and Los Angeles Basin Geological Society Guidebook 63, p. 23-36.

Link, M. H., R. L. Squires, and I. P. Colburn, 1984, Slope and deep-sea fan facies and paleogeography of Upper Cretaceous Chatsworth Formation, Simi Hills, California: AAPG Bulletin, v. 68, p. 850-873.

Luyendyk, B. P., and J. S. Hornafius, 1987, Neogene crustal rotations, fault slip, and basin development in southern California, in R. V. Ingersoll and W. G. Ernst, eds., Cenozoic basin development of coastal California: Englewood Cliffs, New Jersey, Prentice-Hall, Inc., Rubey v. VI, p. 259-283.

Luyendyk, B. P., M. J. Kamerling, and R. Terres, 1980, Geometric model for Neogene crustal rotations in southern California: GSA Bulletin, v. 91, p. 211-217.

Matthews, J. F., Jr., 1954, Howard Townsite oil field: California Division of Oil and Gas, California Oil Fields—Summary of Operations, v. 40, n. 2, p. 17-22.

Mayer, L., ed., 1986, Extensional tectonics of the southwestern United States: a perspective on processes and kinematics: GSA Special Paper 208, 122 p.

Mayuga, M. N., 1970, Geology and development of California's giant—Wilmington oil field, in Geology of giant petroleum fields—symposium: AAPG Memoir 14, p. 158-184.

McCulloh, T. H., 1960, Gravity variations and the geology of the Los Angeles basin of California: USGS Professional Paper 400-B, p. B320-B325.

McCulloh, T. H., 1967, Mass properties of sedimentary rocks and gravimetric effects of petroleum and natural gas reservoirs: USGS Professional Paper 528-A, p. A1-A50.

McCulloh, T. H., S. M. Cashman, and R. J. Stewart, 1978, Diagenetic baselines for interpretive reconstructions of maximum burial depths and paleotemperatures in clastic sedimentary rocks, in D. F. Oltz, ed., Low temperature metamorphism of kerogen and clay minerals: Pacific Section, SEPM, p. 18-46.

McGill, J. T., 1980, Recent movement on the Potrero Canyon fault, Pacific Palisades area, Los Angeles, in Geological Survey research, 1980: USGS Professional Paper 1175, p. 258-259.

McGill, J. T., 1989, Geologic maps of the Pacific Palisades area, Los Angeles, California: USGS Miscellaneous Investigations, Map I-1828, 2 sheets, scale 1:4,800.

McGill, J. T., D. L. Lamar, R. L. Hill, and E. D. Michael, 1987, Potrero Canyon fault, landslides and oil drilling site, Pacific Palisades, Los Angeles, California, in M. L. Hill, ed., Centennial field guide: Cordilleran Section, GSA, p. 213-216.

McKenzie, D. P., 1978, Some remarks on the development of sedimentary basins: Earth and Planetary Science Letters, v. 40, p. 25-32.

McMurdie, D. S., 1973, Geology and general history of the Dominguez oil field in an urban environment—Los Angeles County, California: Pacific Sections, AAPG-SEPM-Society of Exploration Geophysicists, Guidebook, Trip 1, p. 25-31.

McMurdie, D. S., J. C. Taylor, and J. N. Truex, compilers, 1973, Contours on top Miocene, southern Los Angeles basin, in Metropolitan oil fields and their environmental impact: Pacific Sections, AAPG-SEPM-Society of Exploration Geophysicists, Guidebook, Trip 1, map, approximate scale 1 in:9300 ft.

Medwedeff, D. A., 1989, Growth fault-bend folding at southeast Lost Hills, San Joaquin Valley, California: AAPG Bulletin, v. 73, p. 54-67.

Medwedeff, D. A., and J. Suppe, 1986, Growth fault-bend folding—precise determination of kinematics, timing, and rates of folding and faulting from syntectonic sediments [abstract]: GSA Abstracts with Programs, p. 692.

Mefferd, M. G., 1970, Las Cienegas oil field: California Division of Oil and Gas, California Oil Fields—Summary of Operations, v. 56, n. 1, p. 5-13.

Mefferd, M. G., and S. Cordova, 1962, West Coyote oil field: California Division of Oil and Gas, California Oil Fields—Summary of Operations, v. 48, n. 1, p. 37-46.

Menard, H. W., 1955, Deformation of the northeastern Pacific basin and the west coast of North America: GSA Bulletin, v. 66, n. 9, p. 1149-1198.

Metzner, L. H., 1943, Playa del Rey oil field: California Division of Mines and Geology Bulletin 118, p. 292-294.

Michelin, J., 1958, Mahala oil field, in J. W. Higgins, ed., A guide to the geology and oil fields of the Los Angeles and Ventura regions: Pacific Section, AAPG, p. 152-153.

Moody, J. D., and M. J. Hill, 1956, Wrench-fault tectonics: GSA Bulletin, v. 68, p. 1207-1246.

Morris, L. K., S. P. Lund, and D. J. Bottjer, 1986, Palaeolatitude drift history of displaced terranes in southern and Baja California: Nature, v. 321, p. 844-847.

Mount, V. S., and J. Suppe, 1987, State of stress near the San Andreas fault: implications for wrench tectonics: Geology, v. 15, p. 1143-1146.

Murray-Aaron, E. R., and A. W. Pfeil, 1948, Recent developments in the Wilmington oil fields: California Division of Oil and Gas, California Oil Fields—Summary of Operations, v. 34, n. 2, p. 5-13.

Nardin, T. R., and T. L. Henyey, 1978, Pliocene-Pleistocene diastrophism of Santa Monica and San Pedro shelves, California continental borderland: AAPG Bulletin, v. 62, p. 247-272.

Nason, G. W., T. E. Davis, and R. J. Stull, 1979, Cenozoic volcanism in the Newberry Mountains, San Bernardino County, California, in J. M. Armentrout, M. R. Cole, and H. TerBest, Jr., eds., Cenozoic paleogeography of the western United States—Pacific coast paleogeography symposium 3: Pacific Section, SEPM, p. 89-95.

Natland, M. L., 1933, Temperature and depth ranges of some recent and fossil Foraminifera in the southern California region: Scripps Institute of Oceanography Technical Series Bulletin, v. 3, p. 225-230.

Natland, M. L., 1953, Pleistocene and Pliocene stratigraphy of southern California [Ph.D. dissertation]: Los Angeles,

California, University of California, Los Angeles, 165 p.

Natland, M. L., and W. T. Rothwell, Jr., 1954, Fossil Foraminifera of the Los Angeles and Ventura regions, California, in R. H. Jahns, ed., Geology of southern California: California Division of Mines Bulletin 170, p. 33-42.

Nilsen, T. H., 1987, Paleogene tectonics and sedimentation of coastal California, in R. V. Ingersoll and W. G. Ernst, eds., Cenozoic basin development of coastal California: Englewood Cliffs, New Jersey, Prentice-Hall, Inc., Rubey v. VI, p. 81-123.

Noble, F. J., 1967, Exploitation of California offshore field, Parcels 14 and 20A, Huntington Beach, California [abstract]: AAPG Bulletin, v. 51, p. 476.

Oliver, H. W., R. H. Chapman, S. Biehler, S. L. Robbins, W. F. Hanna, A. Griscom, L. A. Beyer, and E. A. Silver, 1980, Gravity map of California and its continental margin: California Division of Mines and Geology, map, scale 1:750,000.

Olsen, L. J., 1975, Belmont offshore oil field: California Division of Oil and Gas Technical Report TP01, p. 1-14.

Oxburgh, E. R., 1972, Flake tectonics and continental collision: Nature, v. 239, p. 202-204.

Page, B. M., 1970, Sur-Nacimiento fault zone of California: continental margin tectonics: GSA Bulletin, v. 81, p. 667-689.

Platt, J. P., 1975, Metamorphic and deformational processes in the Franciscan Complex, California: some insights from the Catalina Schist terrane: GSA Bulletin, v. 86, p. 1337-1347.

Platt, J. P., 1976, The significance of the Catalina Schist in the history of the southern California borderland, in D. G. Howell, ed., Aspects of the geologic history of the California continental borderland: Pacific Section, AAPG Miscellaneous Publication 24, p. 47-52.

Platt, J. P., and C. J. Stuart, 1974, Newport-Inglewood fault zone, Los Angeles basin, California: discussion: AAPG Bulletin, v. 58, p. 877-883.

Poland, J. F., A. A. Garrett, and A. Sinnott, 1959, Geology, hydrology, and chemical character of ground waters in the Torrance-Santa Monica area, California: USGS Water-Supply Paper 1461, 425 p.

Poland, J. F., et al., 1956, Ground water geology of the coastal zone, Long Beach-Santa Ana area, California: USGS Water-Supply Paper 1109, 162 p.

Proctor, R. J., C. M. Payne, and D. C. Kalin, 1970, Crossing the Sierra Madre fault zone in the Glendora tunnel, San Gabriel Mountains, California: Engineering Geology, v. 4, n. 1, p. 5-63.

Real, C. R., 1987, Seismicity and tectonics of the Santa Monica-Hollywood-Raymond Hill fault zone and northern Los Angeles Basin, in Recent reverse faulting in the Transverse Ranges, California: USGS Professional Paper 1339, p. 113-124.

Redin, T., 1991, Oil and gas production from submarine fans of the Los Angeles basin, in K. T. Biddle, ed., Active margin basins: AAPG Memoir 52, p. 239-259.

Reed, R. D., 1933, Geology of California: AAPG, 355 p.

Reed, R. D., and J. S. Hollister, 1936, Structural evolution of southern California: AAPG, 157 p.

Reese, R. G., 1943a, Kraemer area of the Richfield oil field: California Division of Mines and Geology Bulletin 118, p. 361.

Reese, R. G., 1943b, Montebello area of the Montebello oil field: California Division of Mines and Geology Bulletin 118, p. 340-342.

Rehrig, W. A., 1986, Processes of regional Tertiary extension in the western Cordillera: insights from the metamorphic core complexes, in L. Mayer, ed., Extensional tectonics of the southwestern United States: a perspective on processes and kinematics: GSA Special Paper 208, p. 97-122.

Richter, C. F., 1958, Elementary seismology: San Francisco, California, W. H. Freeman, 768 p.

Rogers, T. H., 1965, Geologic map of California—Santa Ana sheet: California Division of Mines and Geology, scale 1:250,000.

Russell, P., 1987, The Point Fermin fan: a small, late middle Miocene age fan within the Monterey Formation, in P. J. Fischer, ed., Geology of the Palos Verdes Peninsula and San Pedro Bay: Pacific Section, SEPM, p. 31-46.

Sage, O. G., Jr., 1973, Paleocene geography of the Los Angeles region, in R. L. Kovach and A. Nur, eds., Proceedings of the conference on tectonic problems of the San Andreas fault system: Stanford University Publications, Geological Sciences, v. XIII, p. 348-357.

Sage, O. G., Jr., 1975, Upper Cretaceous and Paleocene sedimentation and tectonic implications, Simi Hills, Ventura County, California, in D. W. Weaver, G. R. Hornaday, and A. Tipton, eds., Conference on future energy horizons of the Pacific Coast, Paleogene symposium and selected technical papers: Pacific Sections, AAPG-SEPM-Society of Exploration Geophysicists, Annual Meetings, p. 417-438.

Saleeby, J., 1981, Ocean floor accretion and volcanoplutonic arc evolution of the Mesozoic Sierra Nevada, in W. G. Ernst, ed., The geotectonic development of California: Englewood Cliffs, New Jersey, Prentice-Hall, Inc., Rubey v. I, p. 132-181.

Samuelian, R. H., 1984, South Salt Lake oil field: California Division of Oil and Gas Technical Report TR32, p. 17-25.

Saunders, N. E., 1958, Olive oil field, in J. W. Higgins, ed., A guide to the geology and oil fields of the Los Angeles and Ventura regions: Pacific Section, AAPG, p. 149-151.

Schoellhamer, J. E., and A. O. Woodford, 1951, The floor of the Los Angeles basin, Los Angeles, Orange, and San Bernardino counties, California: USGS Oil and Gas Investigations Map OM 117, 2 sheets, scale approximately 1 in:2 mi.

Schoellhamer, J. E., and A. O. Woodford, J. G. Vedder, R. F. Yerkes, and D. M. Kinney, 1981, Geology of the northern Santa Ana Mountains, California: USGS Professional Paper 420-D, 109 p.

Schwartz, D. E., and I. P. Colburn, 1987, Late Tertiary to Recent chronology of the Los Angeles basin, southern California, in P. J. Fischer, ed., Geology of the Palos Verdes Peninsula and San Pedro Bay: Pacific Section, SEPM, p. 5-16.

Schweickert, R. A., 1981, Tectonic evolution of the Sierra Nevada range, in W. G. Ernst, ed., The geotectonic development of California: Englewood Cliffs, New Jersey,

Prentice-Hall, Inc., Rubey v. I, p. 87–131.

Scribner, M. K., 1958, Brea Canyon area, in J. W. Higgins, ed., A guide to the geology and oil fields of the Los Angeles and Ventura regions: Pacific Section, AAPG, p. 106–108.

Shelton, J. S., 1954, Miocene volcanism in coastal southern California, in R. H. Jahns, ed., Geology of southern California: California Division of Mines Bulletin 170, p. 31–36.

Shelton, J. S., 1955, Glendora volcanic rocks, Los Angeles basin, California: GSA Bulletin, v. 66, p. 45–90.

Simila, G. W., and P. J. Fischer, 1982, Seismicity and earthquake focal mechanisms of the Transverse Ranges, in D. L. Fife and J. A. Minch, eds., Geology and mineral wealth of the California Transverse Ranges: Santa Ana, California, South Coast Geological Society, p. 410–414.

Soper, E. K., 1943, Beverly Hills oil field: California Division of Mines and Geology Bulletin 118, p. 287.

Sorensen, S. S., 1984, Petrology of basement rocks of the California continental borderland and the Los Angeles basin [Ph.D. dissertation]: Los Angeles, California, University of California, Los Angeles, 423 p.

Sorensen, S. S., 1985, Petrologic evidence for Jurassic, island-arc-like basement rocks in the southwestern Transverse Ranges and California continental borderland: GSA Bulletin, v. 96, p. 997–1006.

Stewart, J. H., 1978, Basin and Range structure in western North America: a review: GSA Memoir 152, p. 1–13.

Stille, H., 1936, Present tectonic state of the earth: AAPG Bulletin, v. 20, p. 849–880.

Stolz, H. P., and A. F. Woodward, 1943, West Montebello area of the Montebello oil field: California Division of Mines and Geology Bulletin 118, p. 335–339.

Stuart, C. J., 1979a, Middle Miocene paleogeography of coastal southern California and the California borderland—evidence from schist-bearing sedimentary rocks, in J. M. Armentrout, M. R. Cole, and H. TerBest, Jr., eds., Cenozoic paleogeography of the western United States: Pacific Section, SEPM, Pacific Coast Paleogeography Symposium 3, p. 29–44.

Stuart, C. J., 1979b, Lithofacies and origin of the San Onofre Breccia, coastal southern California, in C. J. Stuart, ed., A guidebook to Miocene lithofacies and depositional environments, coastal southern California and northwestern Baja California: Pacific Section, SEPM, p. 25–42.

Sullwold, H. H., Jr., 1960, Tarzana fan, deep submarine fan of late Miocene age, Los Angeles County, California: AAPG Bulletin, v. 44, p. 502–512.

Suppe, J., 1983, Geometry and kinematics of fault-bend folding: American Journal of Science, v. 283, p. 684–721.

Suppe, J., 1985, Principles of structural geology: Englewood Cliffs, New Jersey, Prentice-Hall, Inc., 537 p.

Suppe, J., and D. A. Medwedeff, 1984, Fault-propagation folding [abstract]: GSA Abstracts with Programs, p. 670.

Sylvester, A. G., 1988, Strike-slip faults: GSA Bulletin, v. 100, p. 1666–1703.

Taylor, J. C., 1973, Recent developments at Signal Hill Long Beach oil field: Pacific Sections AAPG, SEPM, and Society of Exploration Geophysicists, Guidebook, Trip 1, p. 16–25.

Troxel, B. W., 1954, Geologic guide no. 3, Los Angeles basin, in R. H. Jahns, ed., Geology of southern California: California Division of Mines Bulletin 170, 46 p.

Truex, J. N., 1972, Fractured shale and basement reservoir, Long Beach unit, California: AAPG Bulletin, v. 56, p. 1931–1938.

Truex, J. N., 1974, Structural evolution of Wilmington, California, anticline: AAPG Bulletin, v. 58, p. 2398–2410.

Truex, J. N., 1976, Santa Monica and Santa Ana Mountains—relation to Oligocene Santa Barbara basin: AAPG Bulletin, v. 60, p. 65–86.

Truex, J. N., 1977, Santa Monica and Santa Ana Mountains—relation to Oligocene Santa Barbara basin: reply: AAPG Bulletin, v. 61, p. 264–269.

Truex, J. N., and W. J. Hunter, 1973, Development of the offshore East Wilmington oil field: Pacific Sections, AAPG-SEPM-Society of Exploration Geophysicists, Guidebook, Trip 1, p. 3–9.

Turcotte, D. L., and D. McAdoo, 1979, Thermal subsidence and petroleum generation in the southwestern block of the Los Angeles basin, California: Journal of Geophysical Research, v. 84, p. 3460–3464.

Turner, D. L., 1970, Potassium-argon dating of Pacific Coast Miocene foraminiferal stages, in Radiometric dating and paleontologic zonation: GSA Special Paper 124, p. 91–129.

Turner, D. L., and R. H. Campbell, 1979, Age of the Conejo Volcanics, in R. F. Yerkes and R. H. Campbell, Stratigraphic nomenclature of the central Santa Monica Mountains, Los Angeles County, California: contributions to stratigraphy: USGS Bulletin 1457-E, p. E18–E22.

Van Eysinga, F. W. B., 1975, Geological time table, 3rd ed.: Amsterdam, Elsevier.

Vedder, J. G., 1970, Summary of geology of San Joaquin Hills, in Geologic guidebook, southeastern rim of the Los Angeles basin, Orange County, California: Newport Lagoon-San Joaquin Hills-Santa Ana Mountains: Pacific Sections, AAPG-SEPM-Society of Exploration Geophysicists, p. 15–19 and road-log narrative.

Vedder, J. G., 1976, Precursors and evolution of the name California continental borderland, in D. G. Howell, ed., Aspects of the geologic history of the California continental borderland: Pacific Section, AAPG, Miscellaneous Publication 24, p. 6–11.

Vedder, J. G., 1979, The Topanga Formation in the San Joaquin Hills, Orange County, California, in C. J. Stuart, ed., A guidebook to Miocene lithofacies and depositional environments, coastal southern California and northwestern Baja California: Pacific Section, SEPM, p. 19–24.

Vedder, J. G., 1987, Regional geology and petroleum potential of the southern California borderland, in D. W. Scholl, A. Grantz, and J. G. Howell, eds., Geology and resource potential of the continental margin of western North America and adjacent ocean basins—Beaufort Sea to Baja California: Houston, Texas, Circum-Pacific Council for Energy and Mineral Resources, Earth Science Series, v. 6, p. 403–447.

Vedder, J. G., and D. G. Howell, 1976, Review of the distribution and tectonic implications of Miocene debris from the Catalina Schist, California continental borderland and adjacent coastal areas, in D. G. Howell, ed., Aspects of the geologic history of the California continental borderland: Pacific Section, AAPG, Miscellaneous Publication 24, p. 326–340.

Vedder, J. G., R. F. Yerkes, and J. E. Schoellhamer, 1957, Geologic map of the San Joaquin Hills–San Juan Capistrano area, Orange County, California: USGS Oil and Gas Investigations Map OM-193, scale 1:24,000.

Vedder, J. G., L. A. Beyer, A. Junger, G. W. Moore, A. E. Roberts, J. C. Taylor, and H. C. Wagner, 1974, Preliminary report on the geology of the continental borderland of southern California: USGS Map MF-624, 34 p., scale 1:500,000.

Vedder, J. G., H. G. Greene, S. H. Clarke, and M. P. Kennedy, 1986, Geologic map of the mid-southern California continental margin, in H. G. Greene and M. P. Kennedy, eds., Geology of the mid-southern California continental margin: California Division of Mines and Geology, map, scale 1:250,000.

Vernon, J. W., A. D. Warren, and J. L. Wildharber, 1970, Road log—southeastern rim, Los Angeles basin, in Geologic guidebook, southeastern rim of the Los Angeles Basin, Orange County, California—Newport Lagoon–San Joaquin Hills–Santa Ana Mountains: Pacific Sections, AAPG-SEPM-Society of Exploration Geophysicists, p. 1–14.

Wallis, W. S., 1981, Oil and gas exploration, offshore southern California, in M. T. Halbouty, ed., Energy resources of the Pacific region: AAPG Studies in Geology 12, p. 383–389.

Warren, A. D., 1980, Calcareous nannoplankton biostratigraphy of Cenozoic marine stages in California, in R. M. Kleinpell, ed., The Miocene stratigraphy of California revisited: AAPG Studies in Geology 11, p. 60–69.

Webb, T. H., and H. Kanamori, 1985, Earthquake focal mechanisms in the eastern Transverse Ranges and San Emigdio Mountains, southern California, and evidence for a regional detachment: Seismological Society of America Bulletin, v. 75, p. 737–757.

Weber, F. H., Jr., 1979, Geology and surface features related to character and recency of movement along faults, north-central Los Angeles area, Los Angeles County, California—a preliminary report, in R. L. Hill, principal investigator, Earthquake hazards associated with faults in the greater Los Angeles metropolitan area, Los Angeles County, California, including faults in the Santa Monica–Raymond, Verdugo–Eagle Rock, and Benedict Canyon fault zones: California Division of Mines and Geology Open-File Report 79-16 LA, p. D1–D56.

Weber, F. H., Jr., 1980, Geological features related to character and recency of movement along faults, north-central [Los Angeles area] Los Angeles County, California, in F. H. Weber, Jr., J. H. Bennett, R. H. Chapman, G. W. Chase, and R. B. Saul, eds., Earthquake hazards associated with the Verdugo–Eagle Rock and Benedict Canyon fault zones, Los Angeles County, California: California Division of Mines and Geology Open-File Report 80-10 LA, p. B1–B116.

Weigand, P. W., 1982, Middle Cenozoic volcanism of the western Transverse Ranges, in D. L. Fife and J. A. Minch, eds., geology and mineral wealth of the California Transverse Ranges: Santa Ana, California, South Coast Geological Society, p. 170–188.

Wernicke, B., G. J. Axen, and J. K. Snow, 1988, Basin and Range extensional tectonics at the latitude of Las Vegas, Nevada: GSA Bulletin, v. 100, p. 1738–1757.

West, J. C., 1979, Structural map and sections on top of Cretaceous in Capistrano syncline area, in D. L. Fife, ed., Geologic guide of San Onofre nuclear generating station and adjacent regions of southern California: Pacific Sections, AAPG-SEPM-Society of Exploration Geophysicists, Guide Book 46, map, scale approximately 1:96,600.

West, J. C., T. W. Redin, J. H. Luksch, R. C. Blaisdell, and K. E. Green, 1987, Correlation section across Santa Monica Mountains—Beverly Hills to Newport—San Joaquin Hills, Los Angeles and Orange Counties, California: Pacific Section AAPG, scale 1 in:4000 ft horizontal, 1 in:1000 ft vertical.

West, J. C., T. W. Redin, G. C. Manings, W. A. Bartling, and K. E. Green, 1988, Correlation section across Los Angeles basin from Palos Verdes Hills to San Gabriel Mountains: Pacific Section, AAPG, Correlation Section CS 3R, scale 1 in:4000 ft horizontal, 1 in:1000 ft vertical.

Willis, R., 1958, Sunset Beach oil field, in J. W. Higgins, ed., A guide to the geology and oil fields of the Los Angeles and Ventura regions: Pacific Section, AAPG, p. 135–141.

Winter, H. E., 1943, Santa Fe Springs oil field: California Division of Mines and Geology Bulletin 118, p. 343–346.

Winterburn, R., 1943, Wilmington oil field: California Division of Mines and Geology Bulletin 118, p. 301–305.

Wissler, S. G., 1943, Stratigraphic formations of the producing zones of the Los Angeles basin oil fields: California Division of Mines and Geology Bulletin 118, p. 209–234.

Woodford, A. O., 1925, The San Onofre Breccia—its nature and origin: University of California Publications, Bulletin of the Department of Geological Sciences, v. 15, n. 7, p. 159–280.

Woodford, A. O., T. G. Moran, and J. S. Shelton, 1946, Miocene conglomerates of Puente and San Jose Hills, California: AAPG Bulletin, v. 30, p. 514–560.

Woodford, A. O., J. E. Schoellhamer, J. G. Vedder, and R. F. Yerkes, 1954, Geology of the Los Angeles basin, in R. H. Jahns, ed., Geology of southern California: California Division of Mines Bulletin 170, p. 65–81.

Woodring, W. P., M. N. Bramlette, and W. S. W. Kew, 1946, Geology and paleontology of Palos Verdes Hills, California: USGS Professional Paper 207, 145 p.

Woodward, A. F., 1958, Sansinena oil field, in J. W. Higgins, ed., A guide to the geology and oil fields of the Los Angeles and Ventura regions: Pacific Section, AAPG, p. 109–118.

Wright, L., 1976, Late Cenozoic fault patterns and stress fields in the Great Basin and westward displacement of the Sierra Nevada block: Geology, v. 4, p. 489–494.

Wright, M. E., 1987, Long Beach, Recreation Park, and Seal Beach oil fields—Los Angeles County, California, in D. D. Clarke and C. P. Henderson, eds., Geologic field guide to the Long Beach area: Pacific Section, AAPG Guidebook 58, p. 37–42.

Wright, T. L., 1987a, Geologic evolution of the petroleum basins of southern California, in T. Wright and R. Heck, eds., Petroleum geology of coastal southern California: Pacific Section, AAPG Guidebook 60, p. 1–19.

Wright, T. L., 1987b, Geological setting of the Rancho La

Brea tar pits, *in* T. Wright and R. Heck, eds., Petroleum geology of coastal southern California: Pacific Section, AAPG Guidebook 60, p. 87–91.

Wright, T. L., 1987c, Geologic summary of the Los Angeles basin, *in* T. Wright and R. Heck, eds., Petroleum geology of coastal southern California: Pacific Section, AAPG Guidebook 60, p. 21–31.

Wright, T. L., 1987d, The Inglewood oil field, *in* T. Wright and R. Heck, eds., Petroleum geology of coastal southern California: Pacific Section, AAPG Guidebook 60, p. 41–49.

Wright, T. L., 1987e, Petroleum development in the Los Angeles basin—a historical perspective, *in* T. Wright and R. Heck, eds., Petroleum geology of coastal southern California: Pacific Section, AAPG Guidebook 60, p. 57–65.

Wright, T. L., E. S. Parker, and R. C. Erickson, 1973, Stratigraphic evidence for the timing and nature of late Cenozoic deformation in the Los Angeles region [abstract]: AAPG Bulletin, v. 57, p. 813.

Ybarra, R. A., M. W. Dosch, and A. D. Stockton, 1960, East Coyote oil field: California Division of Oil and Gas, California Oil Fields—Summary of Operations, v. 46, n. 1, p. 71–76.

Yeats, R. S., 1968a, Rifting and rafting in the southern California borderland, *in* W. R. Dickinson and A. Grantz, eds., Proceedings of the conference on geologic problems of the San Andreas fault system: Stanford University Publications, Geological Sciences, v. XIII, p. 307–322.

Yeats, R. S., 1968b, Southern California structure, sea-floor spreading, and history of the Pacific basin: GSA Bulletin, v. 79, p. 1693–1702.

Yeats, R. S., 1973, Newport-Inglewood fault zone, Los Angeles basin, California: AAPG Bulletin, v. 57, p. 117–135.

Yeats, R. S., 1974, Newport-Inglewood fault zone, Los Angeles basin, California: reply: AAPG Bulletin, v. 58, p. 884–888.

Yeats, R. S., 1978, Neogene acceleration of subsidence rates in southern California: Geology, v. 6, p. 456–460.

Yeats, R. S., 1981, Quaternary flake tectonics of the California Transverse Ranges: Geology, v. 9, p. 16–20.

Yeats, R. S., 1983, Large-scale Quaternary detachments in Ventura basin, southern California: Journal of Geophysical Research, v. 88, p. 569–583.

Yeats, R. S., 1987, Changing tectonic styles in Cenozoic basins of southern California, *in* R. V. Ingersoll and W. G. Ernst, eds., Cenozoic basin development of coastal California: Englewood Cliffs, New Jersey, Prentice-Hall, Inc., Rubey v. VI, p. 284–298.

Yeats, R. S., and J. M. Beall, 1991, Stratigraphic controls of oil fields in the Los Angeles basin: guide to maturation history, *in* K. T. Biddle, ed., Active margin basins: AAPG Memoir 52, p. 221–237.

Yerkes, R. F., 1972, Geology and oil resources of the western Puente Hills area, southern California: USGS Professional Paper 420-C, 63 p.

Yerkes, R. F., 1985, Geologic and seismologic setting, *in* J. I. Ziony, ed., Evaluating earthquake hazards in the Los Angeles region—an earth-science perspective: USGS Professional Paper 1360, p. 25–41.

Yerkes, R. F., and R. H. Campbell, 1979, Stratigraphic nomenclature of the central Santa Monica Mountains, Los Angeles County, California: contributions to stratigraphy, with a section on age of the Conejo Volcanics by D. L. Turner and R. H. Campbell: USGS Bulletin 1457-E, 31 p.

Yerkes, R. F., and W. H. K. Lee, 1979, Faults, fault activity, epicenters, focal depths, and focal mechanisms, 1970–75 earthquakes, western Transverse Ranges, California: USGS Miscellaneous Field Studies Map MF-1032, 2 sheets, scale 1:250,000.

Yerkes, R. F., T. H. McCulloh, J. E. Schoellhamer, and J. G. Vedder, 1965, Geology of the Los Angeles basin, California—an introduction: USGS Professional Paper 420-A, 57 p.

Yerkes, R. F., D. F. Gorsline, and G. A. Rusnak, 1967, Origin of Redondo Submarine Canyon, southern California: USGS Professional Paper 575-C, p. 95–105.

Yerkes, R. F., P. Levine, and W. H. K. Lee, 1987, Contemporary tectonics along south boundary of central and western Transverse Ranges, southern California [abstract]: EOS, v. 68, p. 1510.

Yule, J. D., and D. H. Zenger, 1987, Late Pleistocene angular unconformity at San Pedro, California, *in* M. L. Hill, ed., GSA Centennial Field Guide—Cordilleran Section: GSA p. 175–178.

Zandt, G., and K. P. Furlong, 1982, Evolution and thickness of the lithosphere beneath coastal California: Geology, v. 10, p. 376–381.

Ziony, J. K., and R. F. Yerkes, 1985, Evaluating earthquake and surface faulting potential, *in* J. I. Ziony, ed., Evaluating earthquake hazards in the Los Angeles region—an earth-science perspective: USGS Professional Paper 1360, p. 43–91.

Ziony, J. K., C. M. Wentworth, J. M. Buchanan-Banks, and H. C. Wagner, 1974, Preliminary map showing recency of faulting in coastal southern California: USGS Miscellaneous Field Studies Map MF 585, 7 p.

Zoback, M. D., et al., 1987, New evidence on the state of stress of the San Andreas fault system: Science, v. 238, p. 1105–1111.

Zoback, M. L., R. E. Anderson, and G. E. Thompson, 1981, Cainozoic evolution of the state of stress and style of tectonism of the Basin and Range province of the western United States: Royal Society of London Philosophical Transactions, v. 300A, p. 407–434.

Appendix 1

Abbreviations Used in Figures

STRATIGRAPHIC TERMS

(u, m, and l = upper, middle, and lower)
(T/ = top of)
(H = significant stratigraphic hiatus)

- Q = Quaternary
  - Qa = Alluvium
  - Qsp = San Pedro Formation
  - Qi = Inglewood Formation
  - Qlh = La Habra Formation
  - Qch = Coyote Hills Formation
- Pls = Pleistocene
- Pl = Pliocene
- P = "Pico Formation"
  - uP = upper "Pico Formation"
  - mP = middle "Pico Formation"
  - lP = lower "Pico Formation"
- Tfu = upper Fernando Formation
- Tfl = lower Fernando Formation
- R = "Repetto Formation"
  - uR = upper "Repetto Formation"
  - mR = middle "Repetto Formation"
  - lR = lower "Repetto Formation"
- D = Delmontian
- M = Miocene
- Tp = Puente Formation
  - Tpsc = Sycamore Canyon Member
  - Tpy = Yorba Member
  - Tps = Soquel Member
  - Tplv = La Vida Member
- Mo = Mohnian
  - (c, d, and e are Wissler divisions)
  - uMo = upper Mohnian
  - mMo = middle Mohnian
  - lMo = lower Mohnian
- NS = nodular shale
- Tmo = Modelo Formation
- Tm = Monterey Formation
- L = Luisian
- Tt = Topanga Formation
  - Ttp = Paularino Member
  - Ttlt = Los Trancos Member
  - Ttb = Bommer Member
- Tso = San Onofre Breccia
- Re = Relizian
- V = volcanic rocks
- Gv = Glendora Volcanics
- EMV = El Modeno Volcanics
- Pg = Paleogene
- Tv = Vaqueros Formation
  - (Tvs = undifferentiated Vaqueros/Sespe)
- Ts = Sespe Formation
- Tsa = Santiago Formation
- Pal = Paleocene
- Tsi = Silverado Formation
- K = Cretaceous
- Bc = Basement complex
  - Gr = granitoid rocks
  - Sch = Schist
  - CAT = Catalina terrane

STRUCTURAL AND GEOGRAPHIC NAMES

(—C— or —CH— is contoured horizon)
- A(D)F = Anacapa (Dume) fault
- AH = Alhambra high
- AN = Anaheim Nose
- BCF = Benedict Canyon fault
- BO-P = Brea Olinda-Puente block
- CF = Compton fault
- CHF = Chino Hingeline fault
- ChHF = Cherry Hill fault
- C-LAF = Compton-Los Alamitos fault
- CM = Costa Mesa
- CrF = Cristianitos fault
- CuF = Cucamonga fault
- EF = Elsinore fault
- EMF = East Montebello fault
- EMoF = El Modeno fault
- EPA = Elysian Park anticline
- EPT = Elysian Park thrust
- ERF = Eagle Rock fault
- ET = El Toro
- GP = Greystone Park
- HF = Hollywood fault
- IF = Inglewood fault
- LA = Los Angeles
- LaB = Laguna Beach
- LB = Long Beach

LCan	=	Laurel Canyon
LCF	=	Las Cienegas fault
LCH	=	Las Cienegas high
LK	=	Lasuen Knoll
MCF	=	Malibu Coast fault
MV	=	Mission Viejo
NF	=	"Norwalk fault"
NIFZ	=	Newport-Inglewood fault zone
NSLF	=	North Salt Lake fault
ONR	=	Offshore Newport ridge
OX	=	Oxnard
Pa	=	Pasadena
PB	=	Perris block
PCF	=	Potrero Canyon fault
PH	=	Puente (and San Jose) Hills
PHF	=	Pelican Hills fault
PHT	=	Peralta Hills thrust
Po	=	Pomona
PoF	=	Potrero fault
PVF	=	Palos Verdes fault
RCF	=	Redondo Canyon fault
RcF	=	Rancho fault
RF	=	Raymond fault
RoF	=	Rowland fault
RSZ	=	Regional shear zone
SA	=	Santa Ana
SAF	=	San Andreas fault
SBF	=	Seal Beach fault
SBI	=	Santa Barbara Island
SBV	=	San Bernardino Valley
SCF	=	Shady Canyon fault
SCI	=	Santa Catalina Island
SCOF	=	South Coast Offshore fault
SFV	=	San Fernando Valley
SGF	=	San Gabriel fault
SGV	=	San Gabriel Valley
SJC	=	San Juan Capistrano
SJF	=	San Jacinto fault
SJH	=	San Joaquin Hills
SM	=	Santa Monica
SMB	=	Santa Monica Bay
SMF	=	Sierra Madre fault
SMoF	=	Santa Monica fault
SPB	=	San Pedro Bay
SPBF	=	San Pedro Basin fault
SPEF	=	San Pedro Escarpment fault
SU	=	Simi uplift
SVF	=	San Vicente fault
T-HBF	=	THUMS-Huntington Beach fault
TF	=	Tujunga fault
THF	=	Temple Hill fault
VF	=	Verdugo fault
WF	=	Whittier fault
Wh	=	Whittier
WhHF	=	Whittier Heights fault
WN	=	Whittier Narrows (Figure 3) or West Newport (Figure 7)
WoHF	=	Workman Hill fault
YBF	=	York Boulevard fault

Appendix 2

Wells Identified in Figures

Figure 8, Section AA'

A   Humble H9R-2
B   Humble H9R-4
C   So. Calif. Gas Treasure No. 8
D   So. Calif. Gas Royalty Service Del Rey 12 No. 1
E   Superior Inglewood Extension No. 1
F   Standard (Chevron) Vickers 2 No. 18
G   Standard (Chevron) Vickers 2 No. 22
H   Standard (Chevron) Vickers 2 No. 21
I   Getty Vickers No. 73
J   Royalty Service Marlow Burns No. 1
K   Standard (Chevron) Dept. of Rec. and Parks C.H. No. 1
L   Union (Unocal) Occidental-Genessee C.H. No. 1
M   Standard (Chevron) Adamson C.H. No. 1
N   Standard (Chevron) Saturn C.H. No. 1 (refer to Figure 18 for additional well control)
O   Morgan-Brown U-6 No. 1
P   Standard (Chevron) Laurel No. 2 and 2A

Figure 8, Section BB'

A   Standard (Chevron) Vickers 2 No. 18
B   Standard (Chevron) Vickers 1 No. 36
C   Standard (Chevron) Rubel 2 No. 28
D   Standard (Chevron) Baldwin-Cienega No. 105
E   Standard (Chevron) Baldwin-Cienega No. 318A
F   Shell Baldwin Hills Community No. 1
G   Signal Standard (Chevron)-La Tijera E.H. No. 1
H   Standard (Chevron) Dublin C.H. No. 1
I   Union (Unocal) C.H. No. 24
J   Union (Unocal) 4th Ave. No. 1
K   Union (Unocal) 4th Ave. No. 16
L   Union (Unocal) Union-Signal-Texam U-19 No. 1
M   Union (Unocal) Las Cienegas C.H. No. 25
N   Union (Unocal) 4th Ave. No. 5A
O   Standard (Chevron) Wilton C.H. No. 1
P   Standard (Chevron) Hobart C.H. No. 1
Q   Union (Unocal) Paramount U-14 No. 1

Figure 8, Section CC'

A   Dominguez Estates Sunset No. 1
B   Block (Chevron) El Segundo No. 17
C   Union (Unocal) Lennox E.H. No. 1
D   Getty Cypress No. 7 and Standard (Chevron) Hardy Comm. No. 3 (10,300–12,800 ft)
E   Standard (Chevron) Bohlinger C.H. No. 1
F   American Petrofina Manchester C.H. No. 1 (projected 3800 ft from SE)
G   Union (Unocal) Standard (Chevron)-La Tijera E.H. No. 1
H   American Petrofina Vernon No. 1 (projected 3900 ft from SE)
I   Standard (Chevron) Southern Pacific 57 No. 1
J   Richfield (Arco) Los Angeles River Fee No. 1
K   Standard (Chevron) Challenge Creamery C.H. No. 1
L   Atlantic Los Angeles River Shops No. 1
M   Standard (Chevron) Edison C.H. No. 1
N   Seaboard C.H. No. 5

Figure 8, Section DD'

A   Tidelands Chandler-McBurney No. 1
B   Texaco Seaboard-Bandini-Sidebotham No. 1
C   Superior Torrance No. 66
D   Goldberg Marigold No. 1
E   Standard (Chevron) Del Amo No. 1
F   Shell Reyes No. 130
G   Union (Unocal) Callendar No. 79
H   Shell Reyes No. 95
I   Shell Reyes No. 94
J   Shell Reyes No. 150
K   Union (Unocal) Hellman No. 49
L   Union (Unocal) Hellman No. 68
M   Shell Reyes No. 135
N   Union (Unocal) Hellman No. 58
O   Standard (Chevron) Carlin Community No. 1
P   American Petrofina Central C.H. No. 1 and redrill (projected 8500 ft from NW)
Q   Standard (Chevron) Occidental-Wickes No. 1
R   Standard (Chevron) Weiser No. 1
S   General Exploration CWOD No. 16 (ex-Standard WOD No. 1)
T   Atlantic Jarvis No. 2
U   Richfield (Arco) Union Pacific Unit No. 2
V   Standard (Chevron) Sadler Community No. 1
W   Union (Unocal) Howard and Smith No. 3
X   Ridge Golf Club No. 1
Y   Miller Garvey Hills No. 1

Z	RFD Haig No. 1
AA	Humble (Exxon) South San Gabriel Unit No. 1
BB	Humble (Exxon) Rosemead Unit No. 1 (projected 4200 ft from SE)
CC	American Petrofina El Monte No. 1
DD	Standard (Chevron) Live Oak C.H. No. 1

Figure 8, Section EE'

A	Point Fermin Oil and Gas Co. No. 1
B	Standard (Chevron) Pauley-Los Angeles Harbor No. 2
C	Mobil M-415-D
D	Mobil M-904-E (refer to Figure 25, Section YY' for additional well control)
E	Shell Alamitos No. 48A
F	Shell Alamitos No. 49A
G	Texaco Long Beach Airport NCT-1 No. 1 (redrill)
H	Standard (Chevron) Weingart No. 1
I	Abraham Bloomfield Community No. 1
J	Standard (Chevron) Ravera Community No. 1
K	General Petroleum (Mobil) La Mirada Community 46 No. 1
L	Westates S.E. Santa Fe Springs No. 1
M	Gulf (Chevron) Newgate Unit B No. 1
N	Gulf (Chevron) Newgate Unit A No. 1
O	Union (Unocal) Bell No. 107
P	Ryan and Morrow No. 1
Q	CCMO East Whittier Community No. 1
R	Standard (Chevron) Murphy-Whitter No. 101 (refer to Figure 11 for additional well control)
S	Standard (Chevron) Central Fee No. 135
T	Los Angeles Brewing Jones Community No. 1
U	Pacific Lighting Gas Supply Turnbull Community No. 1
V	Pacific Lighting Gas Supply Turnbull Community No. 2
W	Flintkote Turnbull No. 1
X	Jergins (Exxon) Handorf No. 1
Y	Texaco McGinnis No. 1

Figure 8, Section FF'

A	Standard (Chevron) 10R-30
B	Standard (Chevron) 10R-18
C	Humble (Exxon) H10R-7 (projected 4500 ft from SE)
D	Humble (Exxon) State PRC 186 No. D-3
E	Humble (Exxon) State PRC 186 Nos. 1, 2
F	Getty No. 1
G	Seal Beach No. 1
H	Conoco Bixby No. 18
I	Conoco Bixby No. 62
J	Getty Bryant No. 14
K	Getty Bryant No. 25
L	Shell Bryant No. 7
M	Shell Bryant 4 No. 1
N	Union (Unocal) Nieto No. 1
O	South Basin SBJA No. 1
P	General Petroleum (Mobil) Heath No. 2
Q	General Petroleum (Mobil) Heath No. 1
R	Standard (Chevron) Pacific Community No. 1
S	Barnsdall Emery Trust No. 1 (redrill) (refer to Figure 33 for additional well control)
T	Union (Unocal) Milhous No. 1
U	Union (Unocal) Sansinena 10A No. 3 (projected 8000 ft from E)
V	Union (Unocal) Sansinena No. 10
W	Union (Unocal) Sansinena No. 11
X	Bering York No. 1
Y	Union (Unocal) Sansinena 1B No. 15
Z	Bender Mayo No. 1
AA	Willard Lawrence Estate No. 1
BB	Morton and Sons Rowland Estate 3 No. 1
CC	Morton and Sons Rowland Estate 3 No. 2

Figure 8, Section GG'

A	Standard (Chevron) 10R-26
B	Chevron P-0296 No. 7
C	Chevron P-0296 No. 1 (refer to Figure 13 for additional well control)
D	Chevron P-0296 No. 9
E	Chevron P-0296 No. 11
F	Mobil SP-20
G	Humble (Exxon) H10R-1 (projected 16,000 ft from SE)
H	Mobil State PRC 3119 No. 2
I	Shell State PRC 426 C.H. No. 1 (5500–7930 ft) (projected 12,000 ft from SE)
J	Standard (Chevron) Lomita Land and Water 3 No. 1
K	Atlantic Aston No. 2 (refer to Figure 28, Section aa' for additional well control)
L	Atlantic Ramser No. 10
M	Atlantic McIvor No. 6
N	Hedges Airport No. 2
O	Shell Kettler No. 1
P	Shell Lewis No. 1
Q	Standard (Chevron) Hazard No. 1
R	Standard (Chevron) Stanley Community No. 1
S	Superior Garden Grove Unit No. 1
T	Sunray Foiles No. 1
U	Shell Mathis No. 1
V	British-American Bank of America No. 1
W	Standard (Chevron) Wagner Community No. 1
X	Texaco Hodges No. 1
Y	Texaco Huarte No. 1
Z	Texaco Ruff No. 1
AA	Atlantic Thompson-Steward Unit No. 1
BB	Texaco Dowling No. 1
CC	Union (Unocal) Chapman No. 29 (6000–10,400 ft) (projected 4200 ft from E)
DD	Union (Unocal) Chapman No. 61
EE	Shell Hemphill No. 1
FF	Standard (Chevron) Lemke Trustee No. 1
GG	Union (Unocal) Graham-Loftus No. 1 (projected 4300 ft from W)
HH	Union (Unocal) Naranjal No. 42A (refer to Figure 12 for additional well control)
II	Union (Unocal) Stearns No. 92
JJ	Union (Unocal) Stearns No. 101
KK	Union (Unocal) Stearns No. 109

LL General Petrol. (Mobil) Olinda No. 101 (2550–7761 ft) (projected 4800 ft from SE)
MM Shell Menchengo No. 12 (6600–10,013 ft) (projected 7500 ft from WNW)
NN Shell Keeler Community No. 1 (projected 4500 ft from NW)
OO Shell Brea Canyon C.H. No. 3

Figure 11

A Standard (Chevron) Murphy-Whittier No. 101
B Standard (Chevron) Murphy-Whittier No. 184
C Standard (Chevron) Murphy-Whittier No. 237
D Standard (Chevron) Murphy-Whittier No. 162
E Standard (Chevron) Murphy-Whittier No. 96
F Standard (Chevron) Murphy-Whittier No. 73
G Standard (Chevron) Murphy-Whittier No. 229
H Standard (Chevron) Central Fee No. 103
I Standard (Chevron) Central Fee No. 76
J Standard (Chevron) Central Fee No. 117

Figure 12

A Union (Unocal) Stearns No. 32
B Union (Unocal) Stearns No. 147
C Union (Unocal) Stearns No. 92
D Union (Unocal) Stearns No. 159
E Union (Unocal) Stearns No. 91
F Union (Unocal) Stearns No. 144
G Union (Unocal) Stearns No. 101
H Union (Unocal) Stearns No. 137
I Union (Unocal) Stearns No. 152
J Union (Unocal) Stearns No. 149

Figure 13

A Standard (Chevron) 10R-26
B Standard (Chevron) P-0296 No. 7
C Standard (Chevron) P-0296 No. 9
D Standard (Chevron) P-0296 No. 8 sidetrack
E Standard (Chevron) P-0296 No. 11
F Mobil SP-20

Figure 14

A Standard (Chevron) Fairfax High School C.H. No. 1
B Standard (Chevron) Vista C.H. No. 1
C Standard (Chevron) Los Angeles City Fee C.H. No. 1
D Texaco Hollywood C.H. No. 1

Figure 15, Section HH'

A Hudson Los Angeles Country Club C.H. No. 1
B Hudson Los Angeles Country Club No. 3
C Hudson Los Angeles Country Club No. 1
D Gulf (Chevron)-Universal Fox No. 9 (original hole)
E Gulf (Chevron)-Universal Fox No. 12a
F Gulf (Chevron)-Universal Fox No. 9a
G Signal-Richfield (Hillcrest-Beverly) Rancho No. 1
H Signal-Richfield (Hillcrest-Beverly) Rancho No. 2
I Conoco MGM No. 1
J Standard (Chevron) Vickers 2 No. 18

Figure 15, Section II'

A Signal-Richfield-Conoco U-67 No. 1, redrill 4
B Signal-Richfield-Conoco U-67 No. 1, redrill 2
C Signal-Richfield-Conoco U-67 No. 1, redrill 1
D Signal-Richfield-Conoco U-67 No. 1, original hole
E Standard (Chevron) Basin Builders C.H. No. 1 (projected 4100 ft from SW, and updip)

Figure 15, Section JJ'

A Mobil Brentwood No. 1, redrill 1
B Santa Monica-Sawtelle Birch No. 1
C Texaco Texam No. 1
D Pauley and Frankel Kidson No. 1

Figure 15, Section KK'

A Occidental Marquez E.H. No. 1

Figure 17

A Occidental (Cities Serv.) Dowlen-Federal No. 1, R/D 2
B Occidental (Cities Serv.) Dowlen-Federal No. 1, R/D 1
C Occidental (Cities Serv.) Dowlen-Federal No. 1, R/D 3
D Occidental (Cities Serv.) Dowlen-Federal No. 3, OH
E Occidental (Cities Serv.) Dowlen-Federal No. 3, R/D 1

Figure 18

A Standard (Chevron) P-(Packard) No. 12
B Standard (Chevron) P-(Packard) No. 10
C Standard (Chevron) P-(Packard) No. 3
D Standard (Chevron) Saturn C.H. No. 1 (redrill 1)
E Standard (Chevron) Saturn C.H. No. 1 (original hole)
F Standard (Chevron) Saturn C.H. No. 1 (redrill 2)
G Standard (Chevron) P-(Packard) No. 6
H Standard (Chevron) S-(San Vicente) No. 7
I Standard (Chevron) S-(San Vicente) No. 11
J Morgan-Brown U-6 No. 1
K Standard (Chevron) Dorothy Hay No. 3
L Standard (Chevron) Laurel C.H. No. 2A
M Standard (Chevron) Laurel C.H. No. 2

Figure 22

A Standard (Chevron) Los Angeles Investment 3 No. 1
B Standard (Chevron) Los Angeles Investment 1 No. 226
C Standard (Chevron) Los Angeles Investment 1 No. 220
D Standard (Chevron) Los Angeles Investment 1 No. 195
E Standard (Chevron) Los Angeles Investment 1 No. 191
F Standard (Chevron) Los Angeles Investment 1 No. 140
G Standard (Chevron) LAI-BC Line Well No. 15

H Standard (Chevron) LAI-BC Line Well No. 313
I Standard (Chevron) Baldwin-Cienega No. 14 (and 314)
J Standard (Chevron) Baldwin-Cienega No. 114
K Standard (Chevron) Baldwin-Cienega No. 105 (and 325)
L Standard (Chevron) Baldwin-Cienega No. 125
M Standard (Chevron) Baldwin-Cienega Nos. 236 and 237
N Standard (Chevron) Baldwin-Cienega No. 37
O Standard (Chevron) Baldwin-Cienega No. 318A
P Standard (Chevron) Cienega No. 2
Q Shell Baldwin Hills Community No. 1
R Standard (Chevron) Prospect No. 3
S Standard (Chevron) Prospect No. 2
T Standard (Chevron) Baldwin No. 2
U Standard (Chevron) Prospect No. 1

Figure 25, Section XX'

A Shell Alamitos No. 48A
B Shell Alamitos No. 49A
C Shell Dolley No. 2

Figure 25, Section YY'

A Texaco Harlow-Kent No. 19
B Davis Investment Smith No. 2
C Texaco Long Beach Airport NCT-1 No. 12
D Texaco Long Beach Airport NCT-1 No. 5
E Texaco Long Beach Airport NCT-1 No. 2
F Texaco Long Beach Airport NCT-1 No. 1 (redrill)
G Richfield (Arco) Richfield Camp No. 1
H Standard (Chevron) Weingart No. 1

Figure 28, Section aa'

A Standard (Chevron) Lomita Land and Water 3 No. 1

Figure 28, Section bb'

A Shell Bolsa S-1-D

Figure 28, Section dd'

A Mobil State PRC 3119 No. 2
B Shell State PRC 426 C.H. No. 1
C Union (Unocal) C.H. 1-14
D Union (Unocal) C.H. 2-14
E Humble (Exxon) State PRC 1550 No. 3

Figure 31, Section bb'

A Humble (Exxon) H10R-7
B THUMS C No. 416
C THUMS C No. 822-I
D THUMS C No. 129-I
E Conoco Bixby No. 62

Figure 31, Section ll'

A Jerman Torrance No. 1
B Superior Torrance No. 66
C Superior Torrance No. 69
D Sievers Stevens Woodman No. 1 (projected 2500 ft from NE)
E Atlantic Johnson No. 41 (projected 3500 ft from NE)
F Zephyr (Exxon) Banning 8 No. 1 (projected 3300 ft from NE)
G Selegna Selegna No. 10
H Mobil Terminal No. 1
I Long Beach Oil Development W-133
J THUMS A No. 403
K THUMS D No. 600
L THUMS D No. 605
M THUMS D No. 109
N THUMS C No. 609
O THUMS C No. 416
P THUMS C No. 403
Q Humble (Exxon) State PRC 186 No. D-3
R Standard (Chevron) Surfside No. 2
S Standard (Chevron) Surfside No. 1

Figure 33

A General Petroleum (Mobil) Heath No. 1
B Standard (Chevron) Pacific Community No. 1
C Barnsdall Emery Trust No. 1 (redrill)
D General Exploration Emery No. 2
E Standard (Chevron) Murphy-Coyote No. 243
F Standard (Chevron) Murphy-Coyote No. 208
G Standard (Chevron) Murphy-Coyote No. 300
H Standard (Chevron) Emery No. 92 (projected 1100 ft from west)
I Standard (Chevron) Murphy-Coyote No. 120
J Standard (Chevron) Murphy-Coyote No. 231
K Standard (Chevron) Murphy-Coyote No. 122
L Standard (Chevron) Murphy-Coyote No. 258
M Standard (Chevron) Murphy-Coyote No. 319
N Standard (Chevron) Stern Community 2 No. 1 (projected 1900 ft from west)
O Whittier Des Moines No. 1
P Dunlap Apex No. 1
Q Union (Unocal) Milhous No. 1

Figure 41

A Richfield (Arco) Silver Lake Community No. 1
B Seaboard Los Angeles Brick No. 1 (0-1780 ft is Ventura Figuroa No. 1)
C Standard (Chevron) Miller C.H. No. 1
D Standard (Chevron) Challenge Creamery No. 1
E Signal Standard-Exley C.H. No. 1
F Standard (Chevron) Southern Pacific 57 No. 1
G Union (Unocal) San Antonio E.H. No. 1
H Signal Union-Budd E.H. No. 1
I Neaves Vernon No. 1
J Occidental Vernon No. 1
K Atlantic Jarvis No. 2

Chapter 4

Review of the Neogene Biostratigraphy and Stratigraphy of the Los Angeles Basin and Implications for Basin Evolution

Gregg H. Blake

Unocal
Ventura, California, U.S.A.

ABSTRACT

The Los Angeles basin is one of the Neogene basins along the California continental margin that was formed by extension related to complex wrench-fault mechanisms. These extensional mechanisms caused the pull-apart tectonics that resulted in rapid deepening of many of the continental margin basins in the late Oligocene to early Miocene.

For the last 60 years, the biostratigraphic framework for correlation between the oil fields of the Los Angeles basin has been based on the benthic foraminiferal zones and divisions first published by Wissler (1943, 1958). These biostratigraphic units are used to correlate the upper middle Miocene to upper Pliocene clastic reservoirs across the basin. The benthic foraminiferal zones of Wissler (1943, 1958) were correlated to the Miocene benthic foraminiferal zones of Kleinpell (1938, 1980) and the Pliocene–Pleistocene stages of Natland (1952).

With the utilization of other microfossil disciplines (e.g., siliceous and calcareous planktonic microfossils), a more refined biostratigraphic scheme has been developed. Correlation of plankton biostratigraphies with the radiometric time scale has in turn allowed correlation and calibration of benthic foraminiferal zonations. Application of this new biostratigraphic-chronostratigraphic scheme can now be used to constrain the timing and magnitude of tectonic and depositional events marking basin development.

Benthic foraminiferal assemblages from three stratigraphic sections, located around the margins of the Los Angeles basin, were correlated to siliceous microfossils, calcareous nannofossils, planktonic foraminifers, and radiometric dates to determine the age relationships of the different benthic foraminiferal zonations.

Three regional cross sections were constructed based on the thickness of the Neogene sedimentary package and based on the chronostratigraphic relationships between the benthic foraminiferal zones and the other fossil groups. These regional cross sections illustrate the shifting of the central basin depocenter through the late Neogene. The pre-late Miocene deposits are thickest along the basin margin, especially in the northwestern part of the central block, the Puente Hills, and the Capistrano embayment. The sediments that were deposited during the late Miocene through the early Pleistocene were deposited in the central trough, bounded by the Newport-Inglewood, Santa Monica-Raymond Hill, and the Whittier fault systems.

Unconformities and hiatuses are associated with each tectonic block. Some of these unconformities are local events related to structural growth or fault movement. The hiatuses are regional events caused by changes in ocean chemistry or velocity of bottom currents.

The unconformities that occur in the late Miocene to early Pliocene are related to local structural growth in the Palos Verdes Hills area and the Anaheim nose area of the southeastern portion of the basin. Other unconformities that occur in the late Miocene are related to movement along the Newport-Inglewood fault and the Whittier fault.

In the late middle Miocene a hiatus occurs within the Monterey Formation in the Palos Verdes Hills and the Newport Bay area. This hiatus is thought to be related to climatic and oceanic events associated with continental and oceanic glaciation.

INTRODUCTION

The Los Angeles basin is one of the most prolific oil-producing basins in California. It is representative of the continental margin basins formed in the late Oligocene to early Miocene as a result of the collision between the Pacific spreading ridge and the North American plate and the initiation of wrench tectonics along the continental margin (Atwater, 1970).

Across the basin, there is a large variability in the depositional setting, facies relationships, and areal distribution of middle Miocene through Pleistocene strata that constitute potential source and reservoir units. This variability was caused by the dynamic interaction between oceanography, sedimentation, and tectonics that controlled the individual basinal history.

For the last 60 years, benthic foraminiferal assemblages have been used as the primary tool for correlating these middle Miocene to Pleistocene source beds and clastic reservoirs within the Los Angeles basin. Increased understanding of the oceanographic, sedimentologic, and tectonic aspects of the basin history allow benthic foraminiferal correlations to be established that provide both stratigraphic relationships and environmental interpretations. Advances in other microfossil disciplines (e.g., siliceous microfossils and calcareous nannoplankton) provide the potential to create a more refined and better calibrated biostratigraphic framework for correlating lithologic units across the basin.

The purpose of this paper is to review and describe the Neogene microfossil biostratigraphy of the Los Angeles basin and to propose a correlation between the benthic foraminiferal biostratigraphy and other microfossil disciplines. This discussion consists of five parts:

1. a review of the formational units and their stratigraphic setting

Date Submitted: 8/16/87
Date Accepted: 1/10/89

2. a review of Neogene biostratigraphic schemes used in the Los Angeles basin
3. the correlation of the Neogene benthic foraminiferal stages and divisions with the planktonic microfossil zonations providing a chronostratigraphic framework
4. an examination of regional biostratigraphic correlations across the basin
5. a discussion of the regional Neogene biostratigraphic correlations and their relationship to basin history and evolution

REGIONAL GEOLOGIC SETTING

The Los Angeles basin constitutes one of several basins that formed the Neogene continental borderland of southern California during the middle Miocene to Holocene. The basin extended from the San Joaquin Hills and Santa Ana Mountains to the south and merged with the Ventura basin on the north (Yerkes et al., 1965). This basin includes areas now incorporating the Santa Monica Mountains, the southern foothills of the San Gabriel Mountains, the Palos Verdes Hills, the Puente Hills, much of the northern Santa Ana Mountains, and the San Joaquin Hills (Figure 1).

The Los Angeles basin can be conveniently subdivided into four distinct fault-bounded blocks that were originally defined by Yerkes et al. (1965) (Figure 1). The major faults bounding each block are responsible for different basement configurations, local stratigraphic sequences, and depositional histories.

Northwestern Block

The northwestern block includes parts of the east-trending Santa Monica Mountains and the San Fernando Valley. It is separated from the other blocks by the Santa Monica–Raymond Hill–Cucamonga fault system. Basement rocks of this block (e.g., Santa Monica Slate) are exposed in the eastern Santa Monica Mountains and locally in intervening areas. The stratigraphic sequence in this block consists of Late Cretaceous to Pleistocene marine clastic rocks together with middle Miocene volcanic rocks.

The dominant topographic feature of this block is the Santa Monica Mountains, which form a broad west-plunging anticline. The south flank of the anticline is truncated by, and upthrown along, the Santa Monica fault zone; the north flank dips into the San Fernando Valley.

Southwestern Block

As defined, the southwestern block of the Los Angeles basin is the exposed onshore part of a much larger fault block that forms a portion of the submerged margin. This block is separated from the central block by the Newport-Inglewood fault (Yerkes et al., 1965). The Palos Verdes Hills are the most prominent topographic feature of the block.

The basement in this block is the Mesozoic Catalina Schist. The superjacent sequence contains mainly middle Miocene to Recent marine clastic and hemipelagic rocks; locally there are middle Miocene igneous rocks (Yerkes et al., 1965).

Northeastern Block

The northeastern block of the basin is a fault-bounded triangular wedge (Figure 1). It is separated from the central block by the Whittier fault. The Raymond Hill–Cucamonga fault separates it from the northwestern block.

Basement in this block is granitic rock that represents the northern extension of the Peninsular Ranges. The superjacent rocks consist of 24,000 ft (7300 m) of Late Cretaceous to Pleistocene fine- to coarse-grained marine clastic rocks. In the eastern portion of the block, there are up to 4000 ft (1200 m) of middle Miocene volcanic rocks as well as Paleocene to lower Miocene nonmarine sedimentary rocks.

Central Block

The central block forms the central depositional and structural trough of the basin. It is bounded on the north by the left-oblique Santa Monica–Raymond Hill fault; on the east by the Whittier fault, together with a series of minor north-trending normal faults (e.g., Cristianitos fault); and on the west by the dextral Newport-Inglewood fault (Figure 1). Prominent topographic features include the Santa Ana Mountains at the southeastern margin and the San Joaquin Hills at the southern margin.

The basement for this block is assumed to be granitic rocks of the Peninsular Ranges. The Newport-Inglewood fault is believed to coincide with a concealed fault boundary between the Catalina Schist and the granitic basement (Conrad and Ehlig, 1983). The superjacent rocks for the entire block attain a maximum thickness of at least 32,000 ft (9750 m) and probably consist of Late Cretaceous through Pleistocene marine and nonmarine clastic rocks and interbedded middle Miocene volcanic rocks (Yerkes et al., 1965).

REGIONAL STRATIGRAPHY

The stratigraphy of the Los Angeles basin and environs has been described in detail by Hoots (1931), Woodring et al. (1946), Durham and Yerkes (1964), Yerkes et al. (1965), Yerkes (1972), Yerkes and Campbell (1979), and Schoellhamer et al. (1981). Figure 2 is a generalized stratigraphic column for the Los Angeles basin and surrounding areas. Tables 1 through 5 describe the formational nomenclature, lithology, and paleobathymetry for the marginal areas and central portion of the Los Angeles basin.

Uplifted and exposed sequences around the basin margin (Santa Monica Mountains, Puente Hills, Santa Ana Mountains, and San Joaquin Hills) include Upper Cretaceous to Pleistocene sediments. However, some of the older units have not been encountered in the central portion of the basin (Figure 2). Therefore, the majority of discussion in this paper will be limited to those units known to occur within the subsurface portion of the basin.

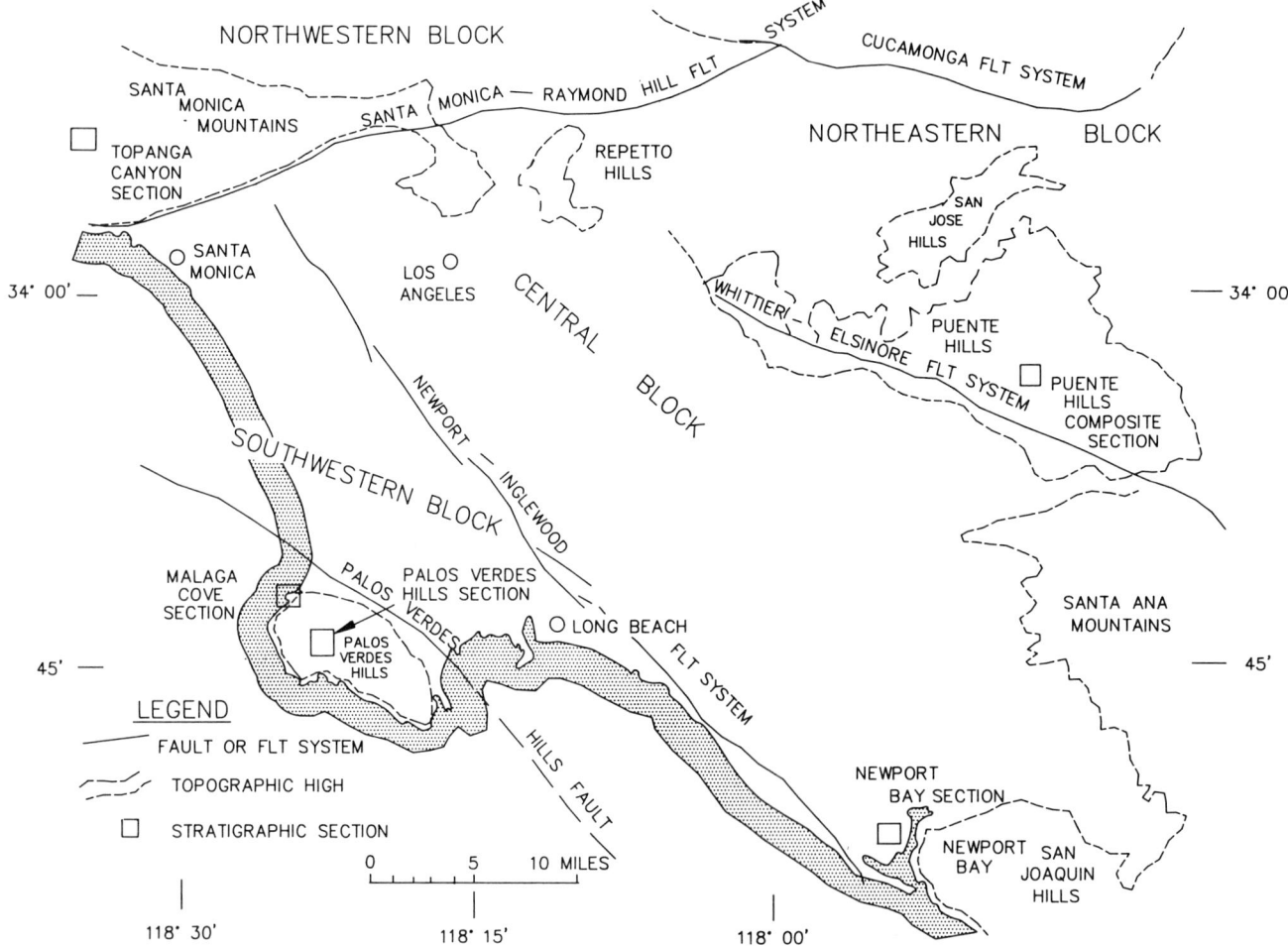

Figure 1. Location map of the Los Angeles basin. Shown are important topographic features and stratigraphic sections. Faults and structural blocks are based on Yerkes et al. (1965).

Basement

The dominant basement in the northwestern block, exposed in the Santa Monica Mountains, consists of the Santa Monica Slate and intruded Cretaceous rocks (Table 1). The Catalina Schist, basement of the southwestern block, was derived from graywacke, chert, and basaltic volcanics and gabbro similar to the Franciscan melange of the California Coast Ranges (Table 2) (Woodring et al., 1946). The only onshore surface exposure is in the Palos Verdes Hills, with the best exposures on Catalina Island to the west. The central block is floored by a basement complex consisting of granodiorite and biotite-quartz diorite rocks, representing the northern extension of the southern California batholith (Yerkes et al., 1965). This granitic basement is exposed in the Santa Ana Mountains, where it unconformably underlies the Santigo Peak Volcanics (Table 3). The granitoid plutonic rocks of the central block are also exposed in the northeastern block at the Puente and San Jose Hills and have yielded a radiometric age of 153 ± 3 m.y. (Yerkes, 1972).

Late Cretaceous Series

A 2600 ft (800 m) thick Upper Cretaceous marine sequence is exposed in the Santa Monica Mountains of the northwestern block. Yerkes and Campbell (1979) named this unit the Tuna Canyon Formation (Table 1). Another thick section of Upper Cretaceous rocks is exposed in the Santa Ana Mountains of the central block (Tables 3, 4). This 2500-ft- (760-m-) thick sequence is composed of (in ascending order) the Turonian Trabuco Formation, the Turonian to Campanian Ladd Formation, and the Campanian Williams Formation (Figure 2).

Paleocene to Middle Eocene Series

There are two Paleocene formations encountered around the Los Angeles basin. The Coal Canyon Formation crops out in the central Santa Monica Mountains (northwestern block) (Figure 2). The Silverado Formation crops out in the Santa Ana Mountains (southern central block) (Table 3).

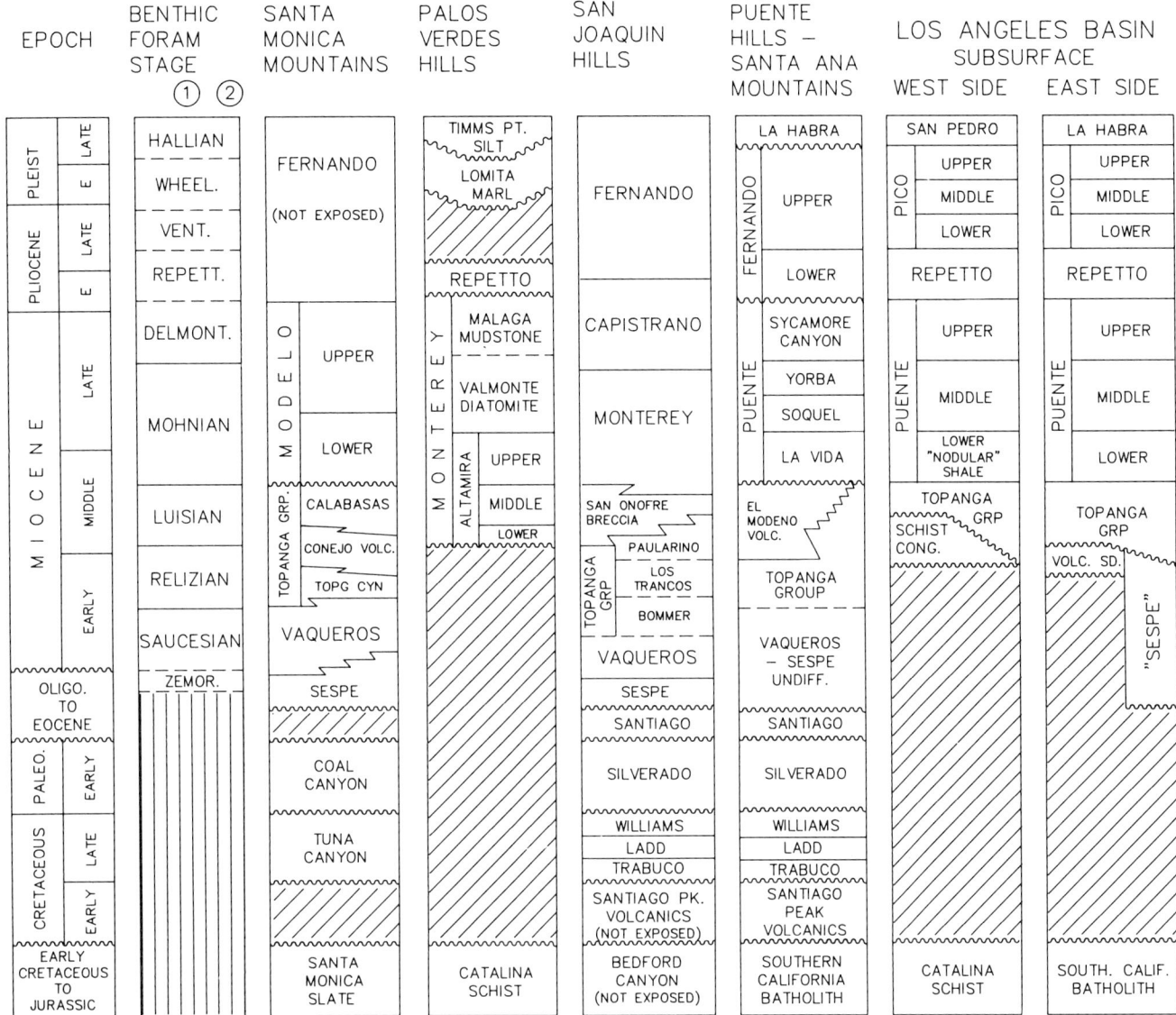

Figure 2. Generalized stratigraphic chart for the Los Angeles basin and surrounding margin areas. For detailed lithologic and stratigraphic information for individual columns, see Tables 1 through 5. The formation terminology is from published reports and unpublished data. Benthic foraminiferal stages are from (1) Kleinpell (1938) and (2) Natland (1952).

The middle Eocene Santiago Formation overlies the Silverado Formation in the Santa Ana Mountains (Figure 2) (Durham and Yerkes, 1964).

Upper Eocene to Lower Miocene Series

Sespe Formation

In the Santa Monica Mountains, the Sespe Formation is a widespread, 3300-ft- (1000-m-) thick, nonmarine sequence of sandstone, pebbly sandstone, and mudstone that interfingers with the overlying Vaqueros Formation (Figure 2) (Yerkes and Campbell, 1979). Its late Eocene, Oligocene, and early Miocene age assignment is based on stratigraphic position. However, vertebrate faunas collected from the Sespe Formation outside of Los Angeles basin indicate a late Eocene to early Miocene age range (Howard, 1987).

Found throughout the southern part of the Los Angeles basin is a thick upper Eocene? to lower Miocene nonmarine red-bed clastic section assumed to be equivalent to the Sespe Formation of the Ventura basin and Santa Monica Mountains (Tables 3, 4) (Yerkes et al., 1965).

Vaqueros Formation

In the Santa Monica Mountains and San Joaquin Hills, the Vaqueros Formation is bounded below by the Sespe Formation and above by the Topanga Group (Figure 2); it characteristically contains the *Turritella inezana* fauna of

Text continues on p. 143.

Table 1. Detailed compilation of stratigraphic information for the Santa Monica Mountains (northwestern block).

Formation		Age	Thickness	Lithology	Environment
Topanga Group	Upper	Late Miocene ("Delmontian" to Mohnian stages)	5000 ft (1525 m)	Soft, diatomaceous shales and claystones	Upper bathyal
	Lower	Late to middle Miocene (Mohnian Stage)		Alternating units of siliceous platy shale and silty claystone interbedded with medium- to coarse-grained sandstone. Lower part—phosphatic shale and conglomerate	Upper to middle bathyal
Modelo	Calabasas	Middle Miocene (Luisian stage)	3950 ft (1200 m)	Medium- to thick-bedded, medium- to coarse-grained sandstone and interbedded silty shale	Upper to middle bathyal
	Conejo Volcanics	Middle Miocene (13.9 ± 0.4 to 15.5 ± 0.8 m.y.)	650 ft (200 m)	Interbedded andesitic and basaltic breccia, mudflow breccia, flow, pillow breccia, and tuff	Subaerial
	Topanga Canyon	Early Miocene (Relizian Stage)	2200 ft (675 m)	Medium- to coarse-grained silty sandstone, sandy siltstone, and pebbly sandstone	Ranges from non-marine to neritic
Vaqueros		Early Miocene to late Oligocene [Saucesian to Zemorrian(?) stages]	2600 ft (790 m)	Medium- to coarse-grained, thin- to thick-bedded biotitic sandstone. Lower part is dark gray-to-black sandy siltstone and mudstone	Nonmarine to nearshore marine
Sespe		Early Miocene to late Eocene	3300 ft (1000 m)	Sandstone, pebbly sandstone, and mudstone	Fluvial
Coal Canyon		Paleocene	1500 ft (450 m)	Sandstone, pebble conglomerate siltstone, and rare algal limestone	Neritic
Tuna Canyon		Campanian Stage	2600 ft (800 m)	Coarse-grained, massive, lithic sandstone, siltstone, and mudstone	Bathyal
Santa Monica Slate		Late Jurassic	Unknown	Intensely jointed dark gray-to-black slate with minor amounts of conglomerate, metasiltstone, and metasandstone	Unknown

Benthic foraminiferal stages are from Kleinpell (1938, 1980) for the Miocene and from Natland (1952) for the Pliocene-Pleistocene. Environment divisions are based on Ingle (1980): inner neritic (0 to 50 m); outer neritic (50 to 150 m); upper bathyal (150 to 500 m); middle bathyal (500 to 2000 m); and lower bathyal (2000 to 2800 m). Stratigraphic data compiled from Kew (1924), Hoots (1931), Imlay (1963), and Yerkes and Campbell (1979).

Table 2. Detailed compilation of stratigraphic information for the Palos Verdes Hills (southwestern block).

Formation			Age	Thickness	Lithology	Environment
Timms Pt Silt			Late Pleistocene	120 ft (37 m)	Massive sandy silt and silty sand	Inner neritic
Lomita Marl			Late Pleistocene	100 ft (31 m)	Unconsolidated calcareous sand and marl	Outer neritic
Repetto Siltstone			Late to early Plioc. (Repettian Stage)	150 ft (46 m)	Massive, glauconitic foraminiferal siltstone with minor amounts of schist breccia and sandstone lenses	Lower bathyal
Monterey	Malaga Mudstone Member		Early Pliocene to late Miocene ("Delmontian" Stage)	330 ft (100 m)	Dark brown radiolarian mudstone and fine-grained siltstone with minor amounts of dolomite, diatomaceous shale, and laminated diatomite	Middle bathyal
Monterey	Valmonte Diatomite Member		Late Miocene (Mohnian Stage)	500 ft (152 m)	Laminated diatomite and phosphatic diatomaceous shale with minor amounts of porcelanite, chert, and dolomite	Upper to middle bathyal
Monterey	Altamira Shale Member	Upper	Late to middle Miocene (Mohnian Stage)	1400 ft (415 m)	Dominated by phosphatic shale and organic-rich shale; minor intervals of porcelaneous shale, limestone, and calcareous sandstone; interval of thick, blue glaucophane schist conglomerate sandstone and brecciated shale	Upper to middle bathyal
Monterey	Altamira Shale Member	Middle	Middle Miocene (Luisian Stage)		Porcelaneous shale and minor amounts of diatomaceous rocks; lesser amounts of tuffs, calcareous phosphatic siltstone, and calcareous mudstones	Upper to middle bathyal
Monterey	Altamira Shale Member	Lower	Middle Miocene (Luisian Stage)		Silty and sandy shales with lesser amounts of silty porcelaneous shale, porcelaneous mudstone, and tuffs	Upper to middle bathyal
Catalina Schist			Early Cretaceous to Late Jurassic	Unknown	Extensively metamorphosed schist derived from greywacke, chert, volcanics, and gabbro	Unknown

Benthic foraminiferal stages are from Kleinpell (1938, 1980) for the Miocene and from Natland (1952) for the Pliocene-Pleistocene. Environment divisions are based on Ingle (1980): inner neritic (0 to 50 m); outer neritic (50 to 150 m); upper bathyal (150 to 500 m); middle bathyal (500 to 2000 m); and lower bathyal (2000 to 2800 m). Stratigraphic data compiled from Wissler (1943), Woodring et al. (1946), and Conrad and Ehlig (1983, 1987).

Table 3. Detailed compilation of stratigraphic information for the San Joaquin Hills and Newport Bay area (central block).

Formation		Age	Thickness	Lithology	Environment
Fernando		Early Pleistocene to early Pliocene (Hallian to Repettian stages)	4600 ft (1400 m)	Massive, medium-grained sandstone, pebbly sandstone, sandy conglomerate, siltstone, and claystone	Neritic to lower bathyal
Capistrano		Early Pliocene to late Miocene (Repettian to U. Mohnian stages)	1900 ft (580 m)	Highly fractured siliceous siltstone and shale that grade upward into micaceous siltstone and sandstone	Middle to lower bathyal
Monterey		Late to middle Miocene ("Delmontian" to Luisian stages)	984 ft (300 m)	Contorted siliceous shale that grades upward into punky, diatomaceous silt, laminated diatomite, and mudstone	Upper to middle bathyal
San Onofre Breccia		Middle Miocene (Luisian Stage)	7052 ft (2150 m)	Massive, poorly sorted Catalina schist-derived breccia and sandstone with siltstone and tuff	Upper to middle bathyal
Topanga Group	Paularino	Middle to early Miocene (lower Luisian to upper Relizian stages)	1500 ft (460 m)	Interbedded sandstone and siltstone with lenses of Catalina Schist-bearing siltstone and sandstone and volcanic flow breccia	Upper to middle bathyal
	Los Trancos	Early Miocene (Relizian Stage)	3100 ft (945 m)	Thin-bedded to massive-bedded clayey siltstone and micaceous fine- to medium-grained sandstone	Middle bathyal
	Bommer	Early Miocene (Relizian Stage)	2400 ft (730 m)	Lenticular, thick-bedded pebbly sandstone interbedded with finer grained sandstone and siltstone	Upper to middle bathyal
Vaqueros		Early Miocene to late Oligo. (Zemor. to Saucesian Stages)	3800 ft (1158 m)	Fine- to coarse-grained sandstone and sandy siltstone	Inner neritic
Sespe		Early Miocene to late Eocene(?)	2450 ft (750 m)	Variegated sandstone and conglomeratic sandstone with sandy claystone	Fluvial
Santiago		Middle Eocene	350 ft (105 m)	Pebble and cobble conglomerate grading into sandstone and sandy siltstone	Nonmarine to inner neritic

Kew (1924). The Vaqueros Formation consists of shallow-water marine sandstones and sandy siltstones (Yerkes, et al., 1965).

Sespe/Vaqueros Undifferentiated

Although Kew (1924) and other workers have differentiated the Vaqueros and Sespe formations in areas north of the Los Angeles basin, the two formations have generally been mapped as one unit in the Puente Hills and Santa Ana Mountains (Table 4) (English, 1926; Schoellhamer et al., 1954; Durham and Yerkes, 1964). In the Puente Hills, strata containing marine fossils are interbedded with the Sespe Formation but cannot be mapped as a separate unit, excluding typical Sespe red beds (Durham and Yerkes, 1964).

Lower to Middle Miocene Series

Topanga Group

As originally defined, Kew (1924) applied the name "Topanga Formation" to the rocks above the Vaqueros Formation and below the Modelo Formation that contained the *Turritella ocoyana* fauna. English (1926) applied the name first to similar rocks exposed in the Santa Ana Mountains and the Puente Hills.

Yerkes and Campbell (1979) formally applied the name "Topanga Group" to the entire 20,000 ft (6100 m) sequence that was previously defined by Kew (1924) as the "Topanga Formation." They divided the Topanga Group into three formations (in ascending order): Topanga Canyon, Conejo Volcanics, and Calabasas (Figure 2) (Table 1).

In the central San Joaquin Hills, the Topanga Group, as originally defined, consists of a thick sequence of beds dominated by sandstone. The formation is differentiated into three members in this area; in ascending order they are the Bommer Member, the Los Trancos Member, and the Paularino Member (Figure 2).

The Topanga Group in the Puente Hills conformably overlies the undifferentiated Vaqueros/Sespe formations. The Topanga is overlain by either volcanics or by the Puente Formation (Figure 2).

In the subsurface, the Topanga Group is a thick, heterogeneous sequence of well-indurated sandstone interbedded with shale and dark gray-to-black siltstone (Table 5) (Wissler, 1943; Durham and Yerkes, 1964).

During the Relizian Stage, the Topanga Group was represented by a littoral facies in the northern and eastern parts of the Los Angeles basin. These rocks contain *Pecten nevadanus* and *Turritella temblorensis* as well as shallow neritic benthic foraminiferal species and ostracodes (Natland and Rothwell, 1954). In the Puente Hills, the Topanga Group contains shallow-water marine fossils and thus is thought to represent inner neritic-to-outer neritic environments. During this time in the western portion of the basin, the Topanga Group was deposited in middle bathyal water depths; that process continued during the Luisian Stage.

In most areas, there is a sharp lithologic break between coarse clastics of the Topanga Group and the overlying finer grained lower Mohnian rocks. Because of this break,

Text continues on p. 146.

Table 3 (continued).

Formation	Age	Thickness	Lithology	Environment
Silverado	Paleocene	1170 ft (350 m)	Fine- to coarse-grained sandstone, silty sandstone, and pebbly sandstone	Nonmarine
Williams	Late Cretaceous (Campanian Stage)	2500 ft (760 m)	Medium- to coarse-grained sandstone, siltstone, mudstone, and lenses of cobble conglomerate	Inner neritic
Ladd	Late Cretaceous (Turon. to Camp. stages)		Pebble to boulder conglomerate, pebbly sandstone, coarse-grained sandstone, silty mudstone	Neritic to upper bathyal
Trabuco	Late Cretaceous (Turonian Stage)		Red, pebble to boulder conglomerate and coarse-grained, pebbly sandstone	Fluvial

Benthic foraminiferal stages are from Kleinpell (1938, 1980) for the Miocene and from Natland (1952) for the Pliocene-Pleistocene. Environment divisions are based on Ingle (1980): inner neritic (0 to 50 m); outer neritic (50 to 150 m); upper bathyal (150 to 500 m); middle bathyal (500 to 2000 m); and lower bathyal (2000 to 2800 m). Stratigraphic data compiled from White (1956), Yerkes et al. (1965), Ingle (1973), Stuart (1979), and Yerkes (1979).

Table 4. Detailed compilation of stratigraphic information for the Puente Hills and Santa Ana Mountains (northeastern and central blocks).

Formation			Age	Thickness	Lithology	Environment
La Habra			Late Pleist. (Hallian Stage)	1350 ft (412 m)	Reddish brown sandstone, silty conglomerate, and silt	Nonmarine to inner neritic
Fernando	Upper		Late Plio. to early Pleist. (Wheel. to Vent. Stages)	3400 ft (1037 m)	Sandstone, pebbly sandstone, sandy conglomerate	Upper to Middle bathyal
Fernando	Lower		Late to early Pliocene (Repettian Stage)	2600 ft (793 m)	Silty sandstone, siltstone, and claystone	Lower bathyal
Puente		Sycamore Canyon Member	Early Pliocene to late Miocene ("Delmontian")	3500 ft (1067 m)	Fine- to coarse-grained sandstone, pebble conglomerate lenses, micaceous sandy siltstone, and platy, siliceous siltstone	Middle bathyal
Puente		Yorba Member	Late Miocene (Mohnian Stage)	3000 ft (914 m)	Diatomaceous siltstone, sandy siltstone, and mudstone interbedded with sandstone and dolomite	Upper to middle bathyal
Puente		Soquel Member	Late Miocene (Mohnian Stage)	3000 ft (914 m)	Medium- to coarse-grained, feldspathic sandstone, pebbly sandstone, pebble conglomerate, and interbedded siltstone	Upper to middle bathyal
Puente		La Vida Member	Late to middle Miocene (Mohnian Stage)	3300 ft (1000 m)	Interbedded micaceous siltstone, platy laminated calcareous siltstone, and silty feldspathic sandstone	Middle bathyal
El Modeno Volcanics			Middle Miocene	1200 ft (365 m)	Andesitic tuff, tuff-breccia, and flow breccia	Subaerial to shallow marine
Topanga Group			Middle to early Miocene (Relizian to Luisian stages)	2100 ft (640 m)	Fine-grained feldspathic sandstone with lenses of conglomerate, silty sandstone, pebbly sandstone, and indurated black siltstone	Inner to outer neritic
Vaqueros-Sespe Undifferentiated			Early Miocene to late Eocene	3000 ft (914 m)	Reddish brown cobble and boulder conglomerate, feldspathic sandstone, coarse-grained to conglomeratic sandstone, and dark gray carbonaceous siltstone	Nonmarine to inner neritic

Table 4 (continued).

Formation	Age	Thickness	Lithology	Environment
Santiago	Middle Eocene	350 ft (105 m)	Pebble and cobble conglomerate grading into sandstone and sandy siltstone	Nonmarine to inner neritic
Silverado	Paleocene	1170 ft (350 m)	Fine- to coarse-grained sandstone, silty sandstone, and pebbly sandstone	Nonmarine
Williams	Late Cretaceous (Campanian Stage)	545 ft (166 m)	Medium- to coarse-grained sandstone, pebbly sandstone, siltstone, and silty mudstone	Inner neritic
Ladd	Late Cretaceous (Turon. to Camp. stages)	1250 ft (357 m)	Pebble, cobble, and boulder conglomerate coarse-grained sandstone, siltstone, and claystone	Neritic to upper bathyal
Trabuco	Late Cretaceous (Turonian Stage)	385 ft (117 m)	Red, pebble to boulder conglomerate and coarse-grained pebbly sandstone	Fluvial
Santiago Peak Volcanics	Early Cretaceous	Unknown	Andesite auto-breccia and dacite lapilli tuff, locally metamorphosed	Unknown
Southern California Batholith	Early Cretaceous to Jurassic	Unknown	Granodiorite and biotite-quartz diorite	

Benthic foraminiferal stages are from Kleinpell (1938, 1980) for the Miocene and from Natland (1952) for the Pliocene–Pleistocene. Environment divisions are based on Ingle (1980): inner neritic (0 to 50 m); outer neritic (50 to 150 m); upper bathyal (150 to 500 m); middle bathyal (500 to 2000 m); and lower bathyal (2000 to 2800 m). Stratigraphic data compiled from Durham and Yerkes (1964), Yerkes et al. (1965), Bottjer et al. (1982), and Cooper et al. (1982).

it is assumed that Topanga deposition was terminated at an unconformity of the Luisian-lower Mohnian Stage boundary (Yerkes et al., 1965). In the Santa Monica Mountains and Puente Hills, this unconformable contact can be seen in outcrop.

San Onofre Breccia

The San Onofre Breccia is a distinctive unit that unconformably overlies the Topanga Group and unconformably underlies the Monterey Formation in the San Joaquin Hills and adjacent offshore areas (Vedder and Howell, 1976; Stuart, 1979). The exotic San Onofre Breccia suggests chaotic deposition of coarse debris down a slope into a deep middle Miocene (Luisian) basin and resultant formation of a depositional wedge between the Topanga Group and the Monterey Formation (Ingle, 1979).

Middle to Upper Miocene Series

The middle to late Miocene rocks are represented by several stratigraphic units whose names vary around the basin. These stratigraphic units represent upper to lower middle bathyal deposition of hemipelagic and clastic sediments.

Modelo Formation

The Modelo Formation was originally defined by Eldridge and Arnold (1907) as a sequence of shale and sandstone above the fossiliferous Vaqueros Formation and below the Pico Formation in the northeastern Ventura basin (Yerkes and Campbell, 1979). Kew (1924) redefined the Modelo to include rocks mapped as Vaqueros by Eldridge; thus, the formation in the type area totals about 9000 ft (2745 m) in thickness.

In the Santa Monica Mountains, the middle to late Miocene Modelo Formation is 5000 ft (1525 m) thick (Table 1) (Hoots, 1931). The Modelo unconformably overlies the Topanga Group or older rocks (Figure 2). The Modelo is unconformably overlain by Pico Formation, the basal division of the Fernando Group.

Monterey Formation

The Monterey Formation represents hemipelagic deposition in middle to upper bathyal environments. It crops out in the Palos Verdes Hills, Newport Bay area, and San Joaquin Hills. In the Palos Verdes Hills, the Monterey Formation ranges in age from late middle Miocene to early Pliocene (Luisian to "Delmontian") (Woodring et al., 1946). It rests unconformably on Catalina Schist and is overlain by the early Pliocene Repetto Formation (Figure 2). At Newport Bay and in the San Joaquin Hills, the Monterey Formation rests on the San Onofre Breccia and ranges in age from Luisian to Delmontian (Ingle, 1973). It is overlain by the late Miocene to early Pliocene Capistrano Formation.

Puente Formation

The Puente Formation is equivalent in age to the upper portion of the Monterey Formation; it is, however, distinctly different in lithology and depositional environment. The formation was named by Eldridge and Arnold (1907) for its exposures in the Puente Hills. They recognized a lower shale member, an intermediate sandstone member, and an upper shale member. English (1926) mapped the formation in the Puente Hills and also divided it into a lower shale, a middle sandstone, and an upper member that included a varied sequence of siltstone, sandstone, and conglomerate beds.

Daviess and Woodford (1949) informally divided the formation into four members that were subsequently formally named by Schoellhamer et al. (1954): the La Vida Member, the Soquel Member, the Yorba Member, and the Sycamore Canyon Member (Figure 2). They gave the names to the units corresponding to the lower shale, middle sandstone, and upper member, respectively, and adopted the term Sycamore Canyon for the fourth, uppermost member. There is considerable variation in thickness of these units due to numerous internal unconformities, intertonguing, and gradational contacts (Yerkes, 1972). The Puente Formation ranges from the lower Mohnian Stage to the "Delmontian" Stage (middle to late Miocene). The youngest beds of the Sycamore Canyon Member may be of early Pliocene age (Table 3) (Durham and Yerkes, 1964).

In the subsurface of the Los Angeles basin, the stratigraphic interval between the Topanga and Repetto formations has been designated the Puente Formation (Figure 2). In the western part of the basin, the bathyal phosphatic nodular shale of the lower Puente overlies the Topanga Group; in the eastern part of the basin, this interval is represented by bathyal clastic deposits and overlies the Topanga Group (Table 5).

Capistrano Formation

The Capistrano Formation occurs in the San Joaquin Hills area of southwestern Los Angeles basin. It ranges in age from late Miocene to early Pliocene (upper Mohnian to lower Repettian) and is stratigraphically equivalent to the Malaga Mudstone Member of the Monterey Formation (Table 4).

Pliocene to Pleistocene Series

The Pliocene rocks of the Los Angeles basin have a complicated nomenclature history. The name Fernando Formation was introduced by Eldridge and Arnold (1907) for the Pliocene strata of the Los Angeles and Ventura basins. Kew (1924) raised the Fernando to group rank and included his Pico and Saugus formations. After 1924, the entire Pliocene sedimentary sequence of the Los Angeles basin was assigned to the Pico Formation by geologists working in the basin. English (1926) used the term Fernando Group in the Puente Hills for the Pliocene rocks and part of the underlying Sycamore Canyon Member.

Actually, the Pliocene to Pleistocene sequence could not be divided into formational units based on lithologic criteria. This sequence could, however, be divided into subdivisions based on two major benthic foraminiferal faunal assemblages. These subdivisions were referred to as upper and lower Pliocene, upper and lower Fernando, or upper and lower Pico (Wissler, 1943).

Table 5. Detailed compilation of stratigraphic information for the subsurface of the Los Angeles basin (central and southwestern blocks).

Formation		Age	Thickness	Lithology	Environment
La Habra		Late Pleist. (Hallian Stage)	1350 ft (412 m)	Reddish brown sandstone, silty conglomerate, and silt	Nonmarine
San Pedro		Late Pleist. (Hallian Stage)	400 ft (122 m)	Yellow brown silt, sand, clay, and gravel	Nonmarine to inner neritic
Pico	Upper Member	Late to early Pleist. (Hallian to Wheelerian stages)	1000 to 3000 ft (305 to 915 m)	Massive, micaceous siltstone and claystone interbedded with silty sand	Neritic to upper bathyal
Pico	Middle Member	Early Pleistocene		Shale, sandy micaceous claystone, and siltstone interbedded with sand	Middle bathyal
Pico	Lower Member	Late Plioc. (Venturian Stage)		Fine- to coarse-grained sand interbedded with sandy micaceous siltstone and shale	Middle bathyal
Repetto		Late to early Pliocene (Repettian Stage)	18000 ft (5490 m)	Interbedded fine- to coarse-grained sandstone, pebbly sandstone, sandy micaceous shale, siltstone, and claystone	Lower bathyal
Puente	Upper Member	Early Pliocene to late Miocene (Delmontian Stage)	2500 ft (762 m)	Fine- to coarse-grained sandstone interbedded with micaceous sandy siltstone and platy siliceous siltstone with pebble conglomerate	Upper to middle bathyal
Puente	Middle Member	Late Miocene (Mohnian Stage)	3800 ft (1160 m)	Medium- to coarse-grained, feldspathic sandstone interbedded with sandy siltstone and diatomaceous siltstone with lenses of pebble conglomerate	Upper to middle bathyal
Puente	Lower Member	Late to middle Miocene (Mohnian Stage)	1500 ft (458 m)	Micaceous siltstone, platy laminated calcareous siltstone, silty medium-grained feldspathic sandstone. In western part of basin, phosphatic nodular shale	Upper to middle bathyal
Topanga Group		Middle to early Miocene (Relizian to Luisian stages)	2100 ft (640 m)	Well-indurated massive coarse-grained sandstone and pebbly sandstone interbedded with sandy shale and dark siltstone	Inner neritic to middle bathyal

Table 5 (continued).

Formation	Age	Thickness	Lithology	Environment
Volcanic Sandstone	Middle to early Miocene (Relizian to Luisian stages)	1100 ft (335 m)	Andesitic tuffaceous sandstone	Subaerial to inner neritic
Schist Conglomerate	Middle to early Miocene (Relizian to Luisian stages)	120 ft (37 m)	Schist-bearing conglomerate, schist conglomerate sandstone, and brecciated shale	Neritic to upper bathyal
Sespe	Early Miocene to late Eocene	1200 ft (365 m)	Variegated sandstone and conglomerate sandstone interbedded with sandy claystone	Alluvial fans to braided stream systems
South California Batholith	Early Cretaceous	Unknown	Granodiorite and biotite-quartz diorite	Unknown
Catalina Schist	Early Cretaceous to Late Jurassic	Unknown	Metamorphosed schist derived from chert greywacke, basaltic volcanics, and gabbro	Unknown

Benthic foraminiferal stages are from Kleinpell (1938, 1980) for the Miocene and from Natland (1952) for the Pliocene-Pleistocene. Environment divisions are based on Ingle (1980): inner neritic (0 to 50 m); outer neritic (50 to 150 m); upper bathyal (150 to 500 m); middle bathyal (500 to 2000 m); and lower bathyal (2000 to 2800 m). Stratigraphic data compiled from Wissler (1943), Durham and Yerkes (1964), Yerkes et al. (1965), and Yerkes (1972).

The formational terminology became so involved that the Pacific Section of the Society of Economic Paleontologists and Mineralogists (SEPM) appointed a committee to establish a uniform classification. In 1930 this committee proposed the name Repetto Formation for the lower Pliocene subdivision and a restricted use of Pico for the upper Pliocene (Wissler, 1943).

A type section for the Repetto Formation was designated in the Repetto Hills. The 2000 ft (609 m) of exposed Pliocene outcrop was originally described by Reed (1932). There was never agreement on an upper and a lower boundary for the Repetto Formation at the type section because the section cannot be divided on lithologic character. The basal siltstone of the Repetto rests on diatomaceous shales of the Puente Formation. The upper contact of the Repetto was originally placed at the top of three coarse feldspathic sandstone beds, ranging in thickness from a few inches to several feet. The overlying 600 ft (183 m) of siltstone, although lithologically similar to the Repetto Formation, carries foraminiferal species thought to represent the lower Pico Formation. However, further sample examination showed that this interval also contains several species characteristic of the Repetto. Thus, this 600 ft (183 m) interval was later correctly assigned to the Repetto Formation (Wissler, 1943).

In the subsurface of the Los Angeles basin, the Pliocene strata cannot be divided into formations based on lithologic criteria. The formations can only be separated into biostratigraphic units based chiefly on benthic foraminiferal assemblages. The Repetto and Pico formations were recognized by six major faunal divisions, four in the Pico and two in the Repetto (Figure 2). Wissler (1943) presented a classification, originally advocated by the committee, that was based on three faunal divisions instead of the two previously proposed for the Repetto.

It has been suggested that the name Repetto is unsatisfactory as a formational unit because it is based on biostratigraphic rather than a lithologic criteria (Durham and Yerkes, 1964). For this paper the terms Pico and Repetto will be retained for the subsurface.

Fernando Formation

The Fernando Formation includes over 6000 ft (1829 m) of Pliocene siltstone, sandstone, pebbly sandstone, and conglomerate. In the Puente Hills and the eastern Los Angeles basin, the Fernando has been divided into lower and upper members on the basis of an extensive erosional unconformity and lithologic variations (Yerkes et al., 1965). The lower member contains abundant Pliocene benthic foraminifers that are indicative of the lower bathyal Repettian Stage; it is correlative to the upper Capistrano Formation and the Repetto Formation (Figure 2).

The benthic foraminiferal faunas in the upper member suggest a depositional environment that ranges from middle bathyal at the base to inner neritic at the top. This member is equivalent to the Venturian and Wheelerian stages and the Pico Formation.

Repetto Formation

The lower Pliocene Repetto Formation is the most extensive Pliocene unit in the Los Angeles basin (Wissler, 1943). The formation ranges in thickness within the basin from 1000 to more than 18,000 ft (305 to 5488 m) and contains benthic foraminiferal faunas indicative of lower bathyal environments.

This formation represents a major clastic influx over late Miocene sandstones and diatomaceous shales of the Puente and Monterey formations. There is a gradual increase in grain size and in percentage of sand from the base to the top of the lower Repetto sequence in the central part of the basin; this has been interpreted as a gradual increase in topographic relief of the source areas during early Pliocene (Yerkes, 1972).

The formation is representative of the Pliocene depositional history of the Los Angeles basin. Miocene subsidence and deposition continued without interruption into and through the Pliocene in the central Los Angeles basin area. The maximum rate of subsidence was attained during deposition of the lower Repetto. These clastic units represent submarine fan deposition in lower bathyal environments (see Redin, *Oil and Gas Production from Submarine Fans of the Los Angeles Basin*, this volume).

In the central plain of the basin, subsidence and deposition continued without interruption into late Pliocene time; however, the rate of deposition gradually overtook the rate of subsidence, and, subsequently, water depth began to decrease. In marginal areas (e.g., Palos Verdes Hills and Puente Hills), unconformities within or at the base of the upper member of the Repetto Formation indicate tectonic activity along boundary faults (e.g., Newport-Inglewood and Whittier) (Yerkes, 1972).

In the Palos Verdes Hills, the Repetto Formation is thought to disconformably overlie the Malaga Mudstone (Wissler, 1943). Wissler (1943) proposed that the entire lower half of the Repetto Formation is missing at this location. However, this interpretation is not accepted by the author based on foraminifers found in the Repetto at Malaga Cove. (See the description of the Palos Verdes Hills below for a more detailed discussion).

At Newport Bay, the upper Capistrano and lower Fernando formations compose the lower Pliocene stratigraphic interval. In the Puente Hills, the Repetto Formation was deposited on a prograding shelf and slope environment. The shelf-slope break was situated along the present southwestern edge of the Puente Hills (Conrey, 1967).

In the subsurface, sandstones and siltstones of the Repetto Formation extend throughout the basin. This stratigraphic interval is more than 18,000 ft (5500 m) thick in the center of the basin.

Pico Formation

Upper Pliocene to upper Pleistocene sediments included in the Pico Formation contain lower middle bathyal to neritic deposits (Figure 2). In the Palos Verdes Hills, the Pico Formation is replaced by the upper Pleistocene outer neritic Lomita Marl and Timms Point Silt. In the Newport Bay area, the Pico is correlated with the upper Fernando Formation and consists of upper bathyal to inner neritic environments (Ingle, 1973).

In subsurface, the upper Pliocene to upper Pleistocene sediments are included in the Pico Formation and represent a middle bathyal to inner neritic basin-filling sequence.

San Pedro/La Habra Formations

Upper Pleistocene to Holocene units in the Los Angeles basin represent the transition from inner neritic to nonmarine deposition. The different lithologic units have been grouped into the San Pedro Formation for the western Los Angeles basin and the La Habra Formation for the eastern part (Figure 2).

The La Habra Formation was originally included in the Fernando Formation by English (1926). However, it was redefined by Durham and Yerkes (1964) who assigned a late Pleistocene age to the La Habra Formation based on its stratigraphic position.

EXISTING BASIN BIOSTRATIGRAPHIC SCHEMES

Kleinpell (1938, 1980)

Kleinpell (1938) established six provincial stages and associated benthic foraminiferal zones for the upper Oligocene to upper Miocene of California (Figure 3). These stages and zones are Oppelian time-stratigraphic units based on designated type sections located in several basins situated along the Neogene continental margin. The benthic foraminiferal zones that are associated with the Miocene stages are based on Oppelian concurrent range zones. [For a more detailed description and discussion of definitions of the Miocene stages and zones, see Kleinpell (1938, 1980)].

The upper Miocene Delmontian Stage is the youngest stage established by Kleinpell (1938). The stratigraphic interval represented by the Delmontian occurs throughout the Los Angeles basin and includes very productive reservoirs in many oil fields of the basin.

The type locality of the Delmontian is within the Monterey Formation and is located in the northern Salinas basin, east of Monterey, California. However, the type locality of the *Bolivina obliqua* zone (originally assumed to be representative of the lower Delmontian Stage) is in the Modelo Formation (Hoots's units 17 and 18) of the Topanga Canyon section in the Santa Monica Mountains (Figure 4).

Kleinpell (1980) later defined the *Bolivina foraminata* zone as the lower Delmontian and the *Bolivina obliqua* as the upper Delmontian (Figure 3). The Topanga Canyon section is the type locality for both zones.

Kleinpell (1938) chose Hoots's units 1 through 16 of the Modelo Formation at the Topanga Canyon section as the type section for the middle to upper Miocene Mohnian Stage (Figure 4). This section is also the type locality for the zones within the Mohnian Stage. Strata representative of this provincial stage occur throughout the Los Angeles basin and include many of the important producing intervals (Wissler, 1943).

As originally defined, the Mohnian Stage was composed of three benthic foraminiferal zones (Figure 3). The *Bolivina modeloensis* zone was defined as Hoots's units 1 to 5; unit 6 was designated the *Bulimina uvigerinaformis* zone; and *Bolivina hughesi* zone was defined as units 7 through 16 (Figure 4). Since the original zones were established, there has been controversy concerning the validity of the type zones within this stage (Pierce, 1956, 1972; Ford, 1972).

Based on examination of surface samples from stratigraphic sections east of the Topanga Canyon locality, Pierce (1956) proposed that the Mohnian Stage was representative of the entire upper Miocene, implying that the "Delmontian," as defined by Kleinpell (1938), is equivalent to the upper Mohnian Stage (Figure 4). Pierce established four Mohnian benthic foraminiferal zones.

Ford (1972) retained the original Kleinpell (1938) definition of the Mohnian Stage; however, he redefined the upper two benthic foraminiferal zones. Based on Ford (1972), the *Bulimina uvigerinaformis* zone ranges from Hoots's unit 6 to the base of unit 9. His definition of the *Bolivina hughesi* zone is from the base of unit 9 to the base of unit 10. The poorly preserved foraminiferal faunas in units 10 through 16 were reported by Ford (1972) to contain nondiagnostic species. Kleinpell (1980) redefined the Mohnian Stage with a new set of zones that include, from younger to older, *Bolivina goudkoffi, Bolivina wissleri, Bolivina barbarana,* and *Bolivina modeloensis* (Figure 4).

There has been further controversy concerning the stratigraphic relationship between the Delmontian and Mohnian stages. Other microfossil disciplines have demonstrated that the "Delmontian" Stage at its type locality in central California is equivalent to strata of the lower Mohnian Stage at the Topanga Canyon section (Pierce, 1972; Barron, 1976).

For this paper, a modified version of Ford's (1972) zonal definitions are used. This version is described in more detail in the following discussion of the Topanga Canyon section.

Because of the controversy associated with the biostratigraphic definition of the Delmontian Stage, Delmontian is used in this paper to refer to the stratigraphic interval between the top of the Mohnian Stage and the base of the lower Pliocene Repettian Stage within the Los Angeles basin; it includes the strata encompassing the *Bolivina obliqua* zone, as defined by Kleinpell (1938) (Figure 4).

The provincial middle Miocene Luisian Stage is the oldest widely distributed stage in the Los Angeles basin and surrounding margin areas. The type section for this stage is in the Monterey Formation and is located in southern Salinas basin, central California. However, the type locality for the zones within this stage (*Siphogenerina collomi, Siphogenerina nuciformis,* and *Siphogenerina reedi*) is at the type locality for the underlying lower to middle Miocene Relizian Stage.

The type section for the Relizian Stage is in the Monterey Formation, which is located in the northern Salinas basin, central California. This section is also the location of the type zones *Siphogenerina branneri* and *Siphogenerina hughesi.*

Billman and Hopkins (1980) proposed a redefinition of the boundary between the Luisian and Relizian stages based on the apparent miscorrelation of the zones between the type locality of the Luisian Stage in the southern Salinas basin and the locality of the type zones in the northern portion of the basin. They postulated that the *Siphogenerina collomi* zone is representative of the upper Luisian and that the *Siphogenerina reedi* zone is representative of the lower Luisian Stage. The *Siphogenerina nuciformis* zone is unrecognizable and is probably equivalent to the upper

Text continues on p. 153.

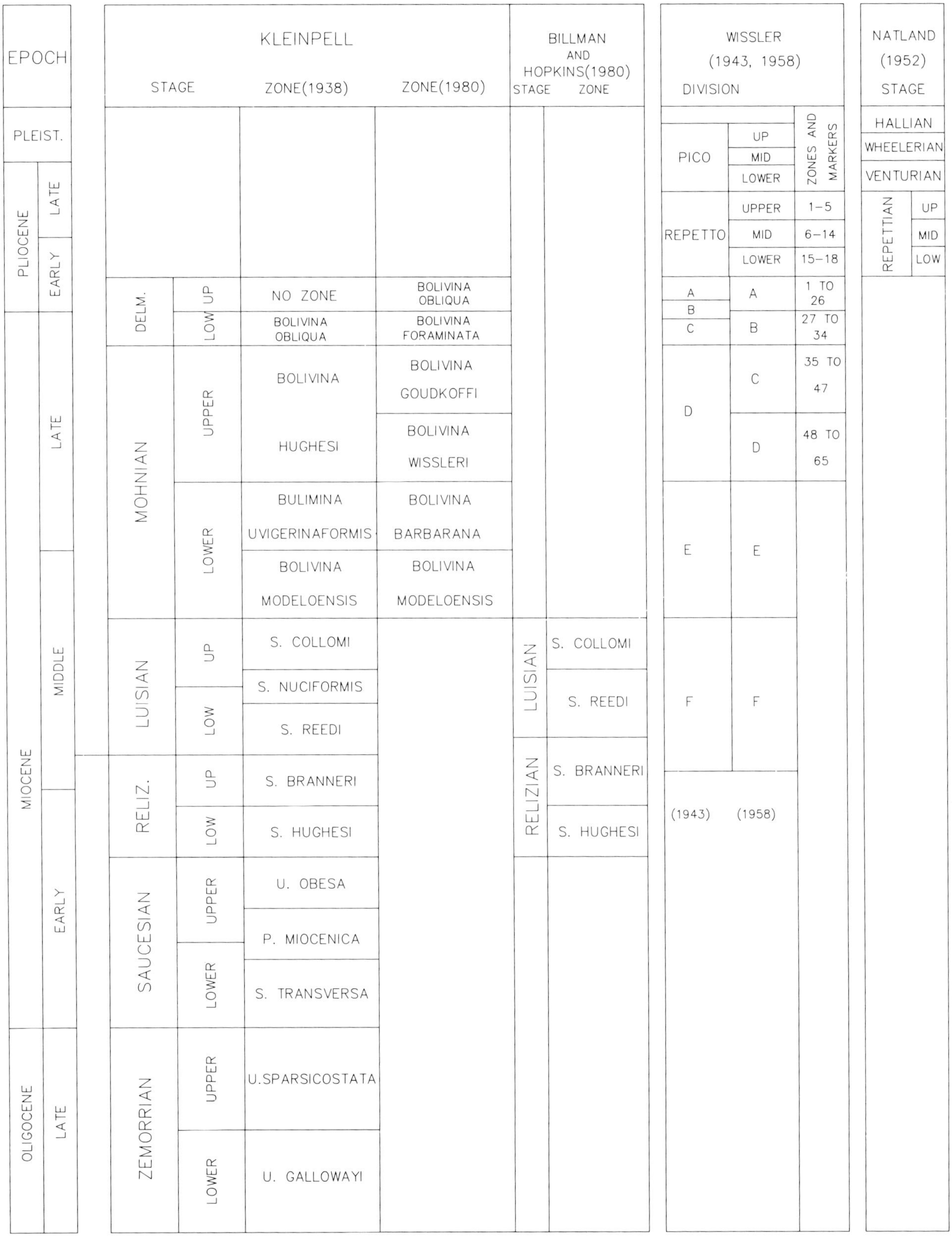

EPOCH		KLEINPELL			BILLMAN AND HOPKINS (1980)		WISSLER (1943, 1958)			NATLAND (1952)	
		STAGE	ZONE (1938)	ZONE (1980)	STAGE	ZONE	DIVISION		ZONES AND MARKERS	STAGE	
PLEIST.							PICO	UP		HALLIAN	
								MID		WHEELERIAN	
								LOWER		VENTURIAN	
PLIOCENE	LATE						REPETTO	UPPER	1–5	REPETTIAN	UP
								MID	6–14		MID
	EARLY							LOWER	15–18		LOW
MIOCENE	LATE	DELM.	LOW UP	NO ZONE	BOLIVINA OBLIQUA			A	A	1 TO 26	
				BOLIVINA OBLIQUA	BOLIVINA FORAMINATA			B		27 TO 34	
								C	B		
		MOHNIAN	UPPER	BOLIVINA HUGHESI	BOLIVINA GOUDKOFFI			D	C	35 TO 47	
					BOLIVINA WISSLERI				D	48 TO 65	
			LOWER	BULIMINA UVIGERINAFORMIS	BOLIVINA BARBARANA			E	E		
				BOLIVINA MODELOENSIS	BOLIVINA MODELOENSIS						
	MIDDLE	LUISIAN	UP	S. COLLOMI		LUISIAN	S. COLLOMI	F	F		
			LOW	S. NUCIFORMIS			S. REEDI				
				S. REEDI							
	EARLY	RELIZ.	UP	S. BRANNERI		RELIZIAN	S. BRANNERI	(1943)	(1958)		
			LOW	S. HUGHESI			S. HUGHESI				
		SAUCESIAN	UPPER	U. OBESA							
				P. MIOCENICA							
			LOWER	S. TRANSVERSA							
OLIGOCENE	LATE	ZEMORRIAN	UPPER	U. SPARSICOSTATA							
			LOWER	U. GALLOWAYI							

Figure 3. Correlation of benthic foraminiferal biostratigraphic zonations for the late Oligocene through Pleistocene of California. Zonations are calibrated to epochs as currently interpreted.

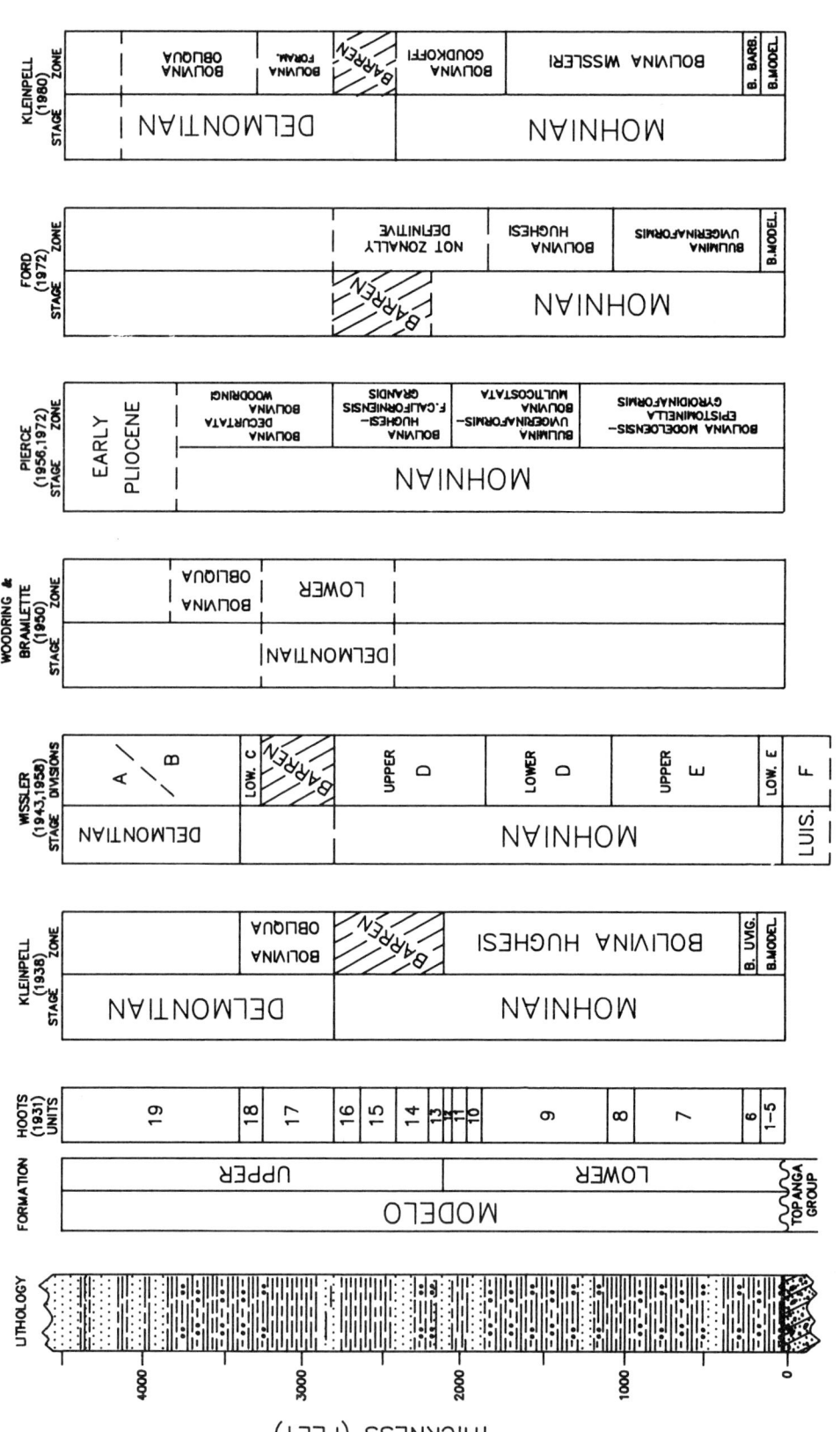

Figure 4. Stratigraphic column of the Modelo Formation for the Topanga Canyon section, Santa Monica Mountains, showing application and comparison of published benthic foraminiferal zonations. See Figure 1 for location. Barren intervals are shown by lined pattern. Zone and division definitions are based on the presence of characteristic species and benthic foraminiferal assemblages. Zonal boundaries associated with the barren intervals were usually assigned based on stratigraphic position.

part of the *S. reedi* zone (Billman and Hopkins, 1980). In addition, these authors demonstrated that the lower 225 ft (68 m) of the type Luisian section is actually equivalent to the Relizian Stage.

For this paper, the redefinition of the Luisian and Relizian stages by Billman and Hopkins (1980) will be used for biostratigraphic correlations within the Los Angeles basin (Figure 3). Species representative of this stage have rare occurrences in the central portion of the Los Angeles basin and in surrounding margins.

Faunas characteristic of the lower Miocene Saucesian and upper Oligocene to lower Miocene Zemorrian stages are not present in the Los Angeles basin. However, relatively unfossiliferous strata equivalent to these stages are present in the outer edges of the basin.

Wissler (1943, 1958)

Biostratigraphic Units

Wissler (1943) published a Neogene benthic foraminiferal biostratigraphic framework that had been used by the Union Oil Company of California in the Los Angeles basin since the late 1920s. This biostratigraphic scheme was applied to the entire middle Miocene to upper Pleistocene section throughout the basin; it was specifically designed for subsurface downsection sample examination. This scheme is based on abundance variations of important foraminiferal species, assemblage zones, and the concurrent range zones of key benthic foraminiferal species.

Wissler (1943) divided the upper Pliocene to Pleistocene Pico Formation into three time-stratigraphic units (Figure 3), each defined by abundance variations in benthic foraminiferal assemblages. The upper Pico is characterized by two major assemblages: the upper Pico *Uvigerina peregrina* fauna and the lowermost upper Pico *Cibicides mckannai-Gyroidina altiformis* fauna. The middle Pico consists of a *Uvigerina peregrina* fauna. The lower Pico is defined by the highest persistent stratigraphic occurrence of *Bulimina subacuminata*.

The lower Pliocene Repetto Formation is also divided by Wissler (1943) into three time-stratigraphic units, including eighteen benthic foraminiferal zones (Figure 3). The upper Repetto (zones 1-5) is characterized by the first occurrence downsection of *Plectofrondicularia californica* and/or *Melonis pompilioides*. However, the precise top of the upper Repetto used by the oil industry is placed at a slightly higher stratigraphic level based on a subtle morphologic variation in *Cibicides mckannai*. The top of the middle Repetto (zones 6-14) is defined by the first occurrence downsection of *Karreriella milleri*. The first occurrence downsection of *Liebusella pliocenica* marks the top of the lower Repetto (zones 15-18).

Wissler (1943) separated the Miocene of the Los Angeles basin into six major foraminiferal faunal divisions (A through F), which are defined by several benthic foraminiferal horizons (Figure 3). In addition to these divisions, Wissler subdivided the upper five divisions into sixty-five subzones. These subzones are defined primarily on benthic foraminiferal assemblage trends (e.g., abundance variations in species). It is important to note that several species used to define the Miocene divisions of Wissler (1943) are also designated as key representative species of the Delmontian, Mohnian, and Luisian stages of Kleinpell (1938).

The top of Division A is defined by the first downsection occurrence of *Rotalia garveyensis* and the reappearance of *Uvigerina hootsi*. This division is defined by markers 1-26 and consists of relatively well-preserved calcareous species in the upper portion of the division. Several of these species are characteristic of the zones in the overlying Repetto units. Crushed calcareous benthic foraminiferal species occur on bedding planes in the middle portion of Division A. In the lower portion of this division the microfossil assemblages are dominated by radiolarians and diatoms to the exclusion of the calcareous benthic foraminiferal species.

Division B (markers 27-34) is characterized by a foraminiferal assemblage similar to Division A and the addition of the first downsection occurrence of large, crushed *Bulimina* sp. This *Bulimina* sp. is now considered to be either a large *Globobulimina* sp. or the equivalent of *"Ellipsoglandulina" fragilis* (Woodring and Bramlette, 1950). *E. fragilis* occurs in the "Delmontian" of the Santa Maria basin and Topanga Canyon section. In addition, a small, crushed *Globobulimina* sp. and several arenaceous species are common to abundant within this division.

It is difficult to establish the top of this division because the fauna is relatively rare, consists almost entirely of crushed specimens, and is otherwise poorly preserved. Therefore, in many wells and sections, this stratigraphic interval is designated Division A/B.

Divisions A and B of Wissler (1943) and the "Delmontian" stage of Kleinpell (1938, 1980) contain similar bathyal faunas but are defined by different index benthic foraminiferal species. The divisions are based on species characteristic of upper Miocene strata of the Los Angeles basin, whereas the assemblage used to define the "Delmontian" consist of several bolivinid species common throughout the upper Miocene to lower Pliocene strata in many Neogene basins of California. However, these two biostratigraphic units appear to be correlative based on the occurrence of similar species (*"Ellipsoglandulina" fragilis*) and the stratigraphic position.

The top of Wissler's (1943) Division C (markers 35-47) is based on the first common to abundant downsection occurrence of the large, crushed *Bulimina* sp. (*Globobulimina* sp. or *"Ellipsoglandulina"* fragilis) and a downsection increase in the occurrence of *Gyroidina soldanii rotundimargo*. The top of this division is not always a sharp boundary because evaluating the abundance of large, crushed *Bulimina* sp. along bedding surfaces is dependent upon sample recovery, specimen preservation, and lithologic content (e.g., shale versus sandstone). The top of this division also contains the first downsection occurrence of *Bolivina hughesi, Bolivina decurtata, Bolivina barbarana,* and *Bolivina girardensis*, which are associated with the upper part of the Mohnian Stage of Kleinpell (1938). In many sections and wells, calcareous foraminiferal faunas are abundant in the shales of Division C, particularly in the lower part.

Wissler (1958) redefined the criteria used to identify the top of this division as the first downsection occurrence of the *Bolivina hughesi* and *Bolivina decurtata* assemblage, and he correlated the division to the upper unit of the upper Mohnian. In addition, the occurrence of *Gyroidina soldanii rotundimargo* and *Bulimina* sp. (large, crushed) are still used to recognize this division.

Division C, as originally defined by Wissler (1943), was correlated to the lower "Delmontian" Stage. The lower Delmontian assemblage in Hoots's unit 18 was considered to be equivalent to that found in the lower part of Division C (Figure 4) (Wissler, 1943). After the division was redefined using the *Bolivina hughesi* assemblage, Wissler (1958) stated that it was equivalent to the upper portion of the *Bolivina hughesi* zone of the upper Mohnian Stage. This later definition of Division C resulted in a stronger correlation between this division and the upper Mohnian Stage. Several upper and middle bathyal species common to Division C are characteristic of the upper Mohnian benthic foraminiferal assemblages.

Division D (markers 48–65) was originally defined on the *Bolivina hughesi* assemblage. However, Wissler (1958) redefined the top of this division on faunal marker 48, which is the first downsection occurrence of a crushed foraminiferal species *"Renulina."* This benthic foraminiferal specimen is thought to be the crushed remains of *Cassidulinella renulinaformis* (Natland and Rothwell, 1954). It is the only species restricted to this division. Therefore, Division D, like Division B, can be difficult to identify. Without the occurrence of *Cassidulinella renulinaformis*, this division contains several calcareous foraminiferal species also common to Division C and indicative of the upper Mohnian of Kleinpell (1938).

Division D was originally considered equivalent to the upper Mohnian based on its stratigraphic position and on its upper Mohnian assemblage. Wissler (1958) correlated the upper part of this division to Hoots's units 10 through 16 based on the occurrence of *Cassidulinella renulinaformis* (Figure 4). This species occurs within the *Bolivina hughesi* zone of the upper Mohnian of the Topanga Canyon section (Natland and Rothwell, 1954) and the Newport Bay section (Warren, 1972).

Division E is defined by the first downsection occurrence of *Bulimina uvigerinaformis*, *Baggina californica*, and other species representative of the late middle Miocene of California. The lower portion of this division is represented by the first downsection occurrence of *Epistominella gyroidinaformis*, a species indicative of the lower Mohnian Stage (Kleinpell, 1938). Recognizable specimens are rare within the Division E interval in the majority of core samples from the Los Angeles basin. In the northern part of the basin and around the basin margins relatively well-preserved foraminifers allow for easy recognition of Division E.

Strata designated as Division E also contain a well-developed phosphatic nodular shale (Figure 2). Within the Los Angeles basin, this phosphatic nodular shale has been used as a lithochronozone. The occurrence of this nodular shale in wells or sections has been used as an indicator of Division E.

Division E and the lower Mohnian Stage appear to be synonymous (Figure 3). Both biostratigraphic units are defined by the same key index species and middle bathyal benthic foraminiferal assemblage. The upper portion of the division represented by *Bulimina uvigerinaformis* and *Baggina californica* is equivalent to the upper part of the *Bulimina uvigerinaformis* zone of the lower Mohnian.

Division F is the oldest benthic foraminiferal zone, or division, recognized by Wissler (1943); it is based on the occurrence of *Valvulineria californica* and other upper to middle bathyal species representative of the early middle Miocene. This division contains several species representative of the Luisian Stage. Rare species that are characteristic of the Relizian Stage are also present. In many parts of the basin, this division is difficult to recognize because benthic foraminifers are rare.

The top of Division F and the Luisian Stage are based on the same characteristic middle Miocene species and contain similar upper to middle bathyal assemblages. These two biostratigraphic units are synonymous and represent the same stratigraphic intervals.

Well-developed Relizian faunas have been found in surrounding basin margins (Woodring et al., 1936; Yerkes, 1972). Wissler (1943) considered any species occurring below Division F to be associated with the Topanga Formation. Benthic assemblages of both the Luisian and Relizian are very similar and probably represent similar paleobathymetric environments (Blake, 1985). For this paper, the lower portion of Division F will be considered equivalent to the upper Relizian Stage (Figure 3).

Sample Examination Procedures

Because of the paucity of well-preserved calcareous foraminiferal specimens in the subsurface of the Los Angeles basin, the Miocene biostratigraphic scheme of Wissler (1943) was based on identification of "crushed" foraminiferal faunas in unprocessed sediment samples. To obtain the best results, cored shale samples were split along bedding planes, and those surfaces were then examined under the microscope. The crushed foraminiferal species are flattened and poorly preserved. Because of poor specimen preservation, some of Wissler's Miocene divisions were defined by foraminiferal faunas identified only to the generic level.

Until the late 1950s and early 1960s, most Los Angeles basin wells were cored continuously. Bedding surfaces of shales were examined at one foot intervals for biostratigraphic analyses. Post–early 1960 drilling has seen a drastic reduction in coring, and biostratigraphic analysis has been confined to ditch or well cuttings. Because of downhole contamination, these samples have not been reliable for application of the crushed foraminiferal zonation. Since identification of divisions and markers was commonly based on abundance variations in a particular species, this approach only worked well through foot-by-foot core sample analysis.

If shales containing crushed foraminifers are washed using standard paleontological procedures, the majority of specimens are destroyed. The only specimens recovered are those small enough or strong enough to withstand the sample processing. This, of course, creates an artificial bias in faunal compositions.

As a result of post-1960 drilling practices, Divisions A, C, E, and F, delineated on the presence or absence of uncrushed benthic foraminiferal specimens, are still recognizable. In contrast, Divisions B and D, based on crushed specimens, are difficult to distinguish with the well cuttings.

Natland (1952)

Natland (1952) established four stages for the marine Pliocene-Pleistocene of southern California based on benthic foraminifers (Figure 3). The stages are defined by a series of abundance zones and assemblage zones based on faunal variations in benthic foraminiferal species. Although these stages have type sections located in the Ventura and Los Angeles basins, they are "environmental" stages because a large majority of species defining the stages are still living along off the modern California continental margin (Blake, 1987). Therefore, these stages are not based on evolutionary events, but rather on faunal variations reflecting basic changes in bathymetric and depositional environments.

The type section for the Pleistocene Hallian Stage is in the upper member of the Pico Formation in the western Ventura basin. The characteristic fauna consist of *Ammonia beccarii, Cassidulina limbata, Cassidulina tortuosa,* and several *Elphidium* species. These species are representative of inner to outer neritic environments [0 to 300 ft (0 to 92 m) water depth].

The type section for the lower Pleistocene Wheelerian Stage is in the upper and middle members of the Pico Formation in the western Ventura basin. *Bolivina interjuncta, Bolivina spissa, Epistominella pacifica,* and *Uvigerina peregrina* are the species defining this stage. These species are indicative of upper to middle bathyal environments [water depths of 300 to 3280 ft (92 to 1000 m)] and especially of low-oxygen slopes and basins, similar to present-day borderland basins.

The Venturian Stage has its type section in the lower member of the Pico Formation in the western Ventura basin. Characteristic species defining this stage include *Bolivina sinuata, Bulimina hebespinata, Bulimina subacuminata,* and *Uvigerina hispidocostata*. The assemblage in this stage is indicative of lower middle to lower bathyal environments [3280 to 6560 ft (1000 to 2000 m) water depth].

The type section for the Repettian Stage is located in the Repetto Hills in the Repetto Formation. The Repettian Stage consists of concurrent range zones that define the major subdivisions in the Los Angeles basin.

The upper Repettian begins at the first abundant occurrence of *Cibicides mckannai* and ends at the first downsection occurrence of *Karreriella milleri* (Natland, 1952). In addition to *C. mckannai*, other species characteristic of the upper Repettian include *Bulimina rostrata, Melonis pompilioides,* and *Plectofrondicularia californica*. The middle part of the stage begins with *Karreriella milleri* and ends at the first downsection occurrence of *Liebusella pliocenica*. The lower Repettian includes beds with *Liebusella pliocenica* and large *Hopkinsina nodosa*. The base of the Repettian Stage is placed at the highest occurrence of *Rotalia garveyensis*. Recovery of foraminifers in the lower part of the Repettian section is poor; consequently, the extent of the lower Repettian section is often indiscernible (Natland, 1952).

The stratigraphic succession represented by the Repettian, Venturian, Wheelerian, and Hallian stages is indicative of a basin-filling cycle (Ingle, 1980; Blake, 1985). The Repettian assemblage characterizes lower bathyal environments, the Venturian assemblage indicates lower to lower middle bathyal, the Wheelerian assemblage represents upper middle to upper bathyal, and the Hallian assemblage indicates neritic environments.

Wissler (1943) used similar assemblages to define biostratigraphic subdivisions within the Pico and Repetto formations. The upper Pico faunal zone of Wissler (1943) is equivalent to the Hallian Stage based on similarities in benthic foraminiferal assemblages. The middle and lower Pico faunal zones of Wissler and the Wheelerian and Venturian stages of Natland are synonymous because both are defined by the same key index species and assemblages.

The species characterizing the upper, middle, and lower parts of the Repettian are the same as those designated by Wissler to represent the Repetto faunal zones. The upper Repetto and upper portion of the Repettian are defined by the same species. The middle Repetto faunal zone and middle portion of the Repettian Stage are based on the same species (*Karreriella milleri*) and similar benthic foraminiferal assemblages. The lower Repetto of Wissler and the lower Repettian of Natland are both also defined by the same species, the first occurrence downsection of *Liebusella pliocenica*.

CORRELATION OF BENTHIC STAGES/DIVISIONS TO PLANKTONIC MICROFOSSIL ZONATIONS AND RADIOMETRIC DATING

For the last ten years, planktonic microfossils have provided an improved chronostratigraphy for West Coast Neogene stratigraphy. Recent studies of Kleinpell's (1938) type Miocene Stage sections have successfully correlated these provincial stages to deep-sea planktonic zonations (Barron, 1976; Poore et al., 1981, Baldauf and Barron, 1982; Arends and Blake, 1986).

In California, diatoms are the most reliable means of dating and correlating middle Miocene to Pliocene strata (Barron, 1986a). The northeastern Pacific diatom zonation of Barron (1981, 1986a) is currently the most widely used Neogene planktonic zonation in California.

In addition, the low-latitude coccolith zonation of Okada and Bukry (1980) is useful in lower to lower middle Miocene stratigraphic intervals throughout California. However, in the late middle Miocene, climatic reorganization of surface-water masses established a low-diversity temperate assemblage that cannot be correlated to low-latitude coccolith zonation (Bukry, 1981). Consequently, late middle Miocene and Pliocene temperate coccolith zones that have been defined span broader time intervals than equivalent low-latitude zones (Bukry, 1973). The temperate coccolith zones are used in this paper because the low-latitude coccolith zonation is generally not recognizable in southern California, except for the late Pliocene and Pleistocene.

Planktonic foraminiferal zonations based on low-latitude assemblages have not been widely utilized for correlation within the Los Angeles basin. However, zonal calls from distribution of index species have been made on outcrop sections such as the Newport Bay section (Ingle and Barron, 1978; Keller and Barron, 1982). In addition, first appearance datums (FAD) and last appearance datums (LAD) of selected species and coiling direction variations in

Neogloboquadarina pachyderma have been used for intrabasinal correlation (Ingle, 1967, 1972; Lagoe and Thompson, 1988)

As discussed in the previous section, it is possible to correlate the zones and divisions of Wissler (1943) to the stages of Kleinpell (1938, 1980) and Natland (1952) because of the similarities between their characteristic benthic foraminiferal assemblages. The correlation between benthic foraminiferal biostratigraphic zonations of the Los Angeles basin and early middle Miocene to early Pliocene coccolith and diatom zones can be established by comparing their stratigraphic relationships within basin-margin outcrop sections. The Topanga Canyon section, Palos Verdes Hills sequence, Newport Bay section, and a Puente Hills composite section have been utilized for this purpose (Figure 1).

Calibration of the microfossil zones can also be estimated by correlating the zones to available radiometric ages. Because of the widespread occurrence of volcanic units within California Neogene strata, several published radiometric dates can be used to determine an absolute age framework of the provincial benthic stages and divisions.

With the use of calcareous and siliceous planktonic microfossils and radiometric ages, a more refined chronostratigraphic correlation can be established for the benthic foraminiferal stages and divisions.

Topanga Canyon Section

Stratigraphy

Hoots (1931) used the 4500 ft (1372 m) section exposed along Topanga Canyon in the Santa Monica Mountains as representative of the Modelo Formation. As stated in previous sections, Hoots divided the Modelo into nineteen lithologic units (Figure 5). In this stratigraphic sequence, the Modelo Formation unconformably overlies the Topanga Group.

The lower member of the Modelo, units 1 through 12, is composed of alternating platy, siliceous shales and soft shales interbedded with coarse-grained sandstone (Hoots, 1931). The upper member, units 13 through 19, consists of soft, diatomaceous shales interbedded with fine-grained sandstone.

Radiometric Dates

There is a fission-track (F-T) age of 11.5 ± 1.2 m.y. from a bentonite bed within the upper part of Hoots's unit 5 (uppermost *Bolivina modeloensis* Zone) (Obradovich et al., 1978). This age agrees with the age range established by subzone CN5b. A zircon from a bentonite bed within Hoots's unit 9 (approximately midway in the *Bolivina hughesi* Zone and within the upper part of Division D) was dated by the fission-track method at 7.8 ± 0.9 m.y. (Obradovich and Naeser, 1981). This ash occurs within an interval that is barren of siliceous microfossils and contains nondiagnostic calcareous nannofossils. However, the age is in agreement with the age range based on diatom and calcareous nannofossil control above and below the zircon dates.

Benthic Foraminifers

The benthic foraminiferal assemblage in units 1 through 5 is of the *Bolivina modeloensis* Zone of the lower Mohnian and is representative of the lower half of Division E. The *Bulimina uvigerinaformis* Zone and the upper half of Division E were originally thought to be restricted to unit 6 (Kleinpell, 1938). However, as Ford (1972) reported, the first downsection occurrence of *Bulimina uvigerinaformis* is at the top of unit 8. This extends the *B. uvigerinaformis* Zone, top of lower Mohnian and Division E, to the top of unit 8 (Figure 5).

Natland and Rothwell (1954) reported *Cassidulinella renulinaformis* on bedding surfaces of mudstones and shales in the upper part of unit 9 within the *Bolivina hughesi* Zone (upper Mohnian). Wissler (1943) originally proposed that units 10 through 16 were equivalent to the upper part of Division D. However, the first downsection occurrence of *Cassidulinella renulinaformis* in the upper part of unit 9 is currently considered the top of Division D (Wissler, 1943).

Benthic foraminiferal species characteristic of the *Bolivina hughesi* Zone and Division C occur within the stratigraphic interval from the upper part of unit 9 to the base of unit 13 (Figure 5). Although the first downsection occurrence of *B. hughesi* is in the upper part of unit 9, the top of the zone and Division C are placed at the base of unit 13 to coincide with the first rare downsection occurrence of species (e.g., *Bolivina woodringi*) thought to be restricted to the Mohnian Stage (Ford, 1972). Units 14 through 17 are either barren or contain sparse benthic foraminiferal faunas that are not zonal or division definitive.

As previously stated, the Mohnian-Delmontian contact is arbitrarily placed at the top of unit 16, because of the relatively barren interval between unit 15 and 18. The interval from unit 13 to the upper part of unit 15 is equivalent to Divisions B and C based on *"Ellipsoglandulina" fragilis*, a form equivalent to *Bulimina* sp. (large, crushed). Unit 18 and the lower part of unit 19 contain species indicative of the *Bolivina obliqua* Zone of the "Delmontian." This interval is equivalent to Divisions A and B based on faunal composition and stratigraphic position.

Diatoms

At this time, there are five definable diatom zones and subzones in the Topanga Canyon section. Zonal calls for unit 17 through lower unit 19 are from Barron (1976) and R. G. Arends of Unocal (unpublished data). Zones and subzones in units 7 through 10 were interpreted by R. G. Arends of Unocal. Only approximate chronostratigraphic correlations and ages for the provincial stages and divisions are possible because of limited diatom zone control.

The interval from the base of unit 7 to the lower part of unit 9 is assigned to subzones b–d undifferentiated of the *Denticulopsis hustedtii–Denticulopsis lauta* Zone. The interval from the upper part of unit 9 to the top of unit 10 is assigned to *Denticulopsis hustedtii* Zone. The flora in the lower part of unit 17 is characteristic of subzone b of the *Thalassiosira antiqua* Zone. The upper part of unit 17 is assigned to subzone a of the *Nitzschia reinholdii* Zone (Figure 5). Unit 18 and the lower part of unit 19 are assigned to subzone b of the *Nitzschia reinholdii* Zone.

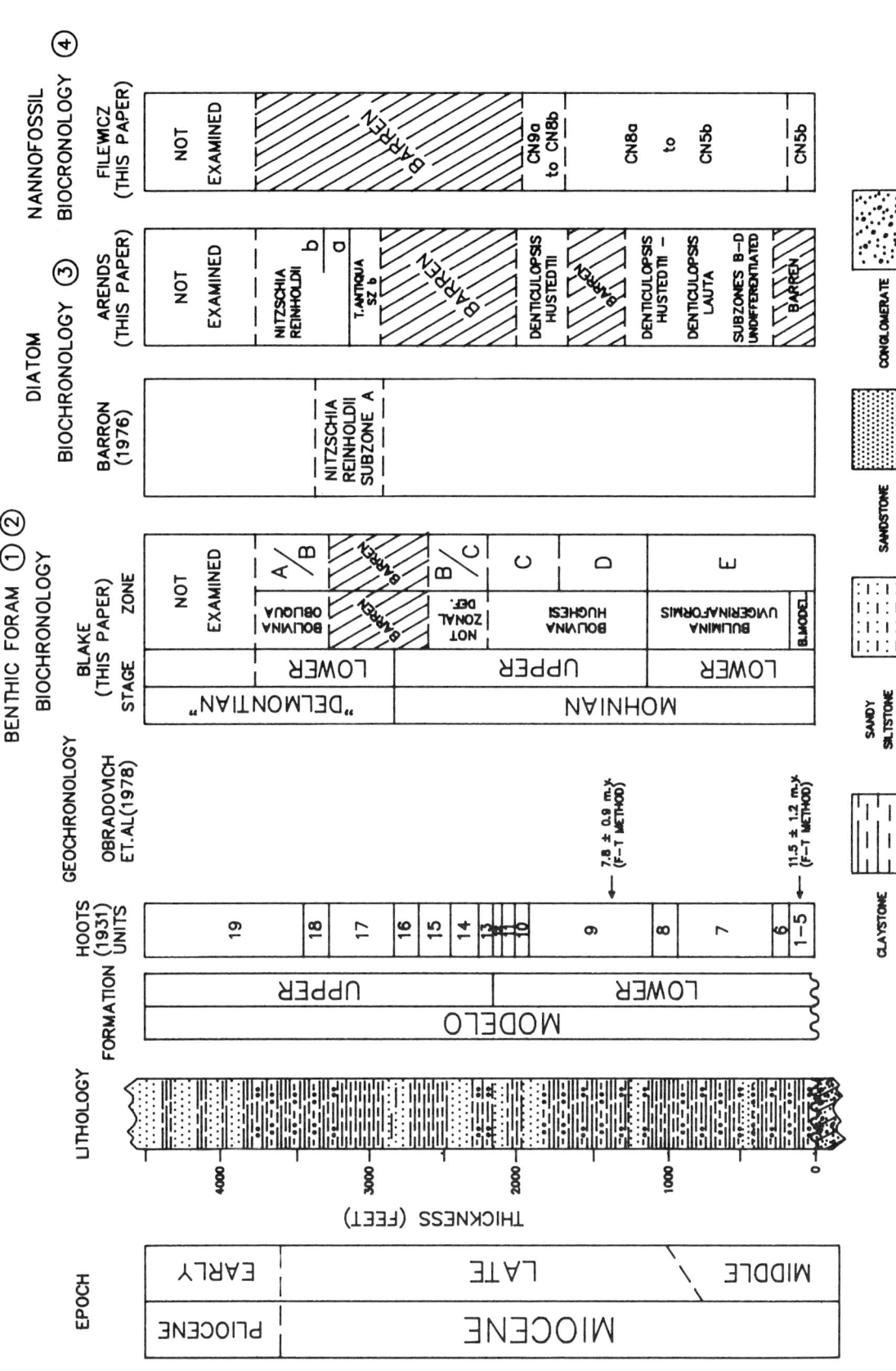

Figure 5. Composite stratigraphic column of the Modelo Formation for the Topanga Canyon section as measured and divided by Hoots (1931). Barren intervals are shown by lined pattern. Zonal boundaries associated with the barren intervals were usually assigned based on stratigraphic position. (1) Benthic foraminiferal stages based on the Miocene provincial stages of Kleinpell (1938, 1980). (2) Foraminiferal divisions are from Wissler (1943, 1958). (3) Diatom zones based on Barron (1981, 1986b). (4) Calcareous nannofossil zones based on Okada and Bukry (1980). Biostratigraphic correlations from various microfossil disciplines are discussed in the text.

The top of the *Bulimina uvigerinaformis* Zone and the boundary between Divisions E and D occur in subzones b–d undifferentiated of the *Denticulopsis hustedtii–Denticulopsis lauta* Zone (12.7 to 8.4 Ma) (Figure 6). The boundary between Divisions D and C occurs within the interval assigned to the *Denticulopsis hustedtii* Zone (8.4 to 7.6 Ma).

The contact between the Mohnian and "Delmontian" was originally placed by Barron (1976) in subzone a of the *Nitzschia reinholdii* Zone, but has since been placed in subzone b of the *Thalassiosira antiqua* Zone (7.0 to 6.7 Ma). The upper part of the *Bolivina obliqua* Zone and Division A/B undifferentiated is in subzone b of the *Nitzschia reinholdii* Zone (6.1 to 5.1 Ma).

Calcareous Nannofossils

The calcareous nannofossils from the Topanga Canyon section were examined and interpreted by M. V. Filewicz of Unocal (unpublished data). Only three zones could tentatively be identified; therefore, this fossil group can only provide approximate chronostratigraphic correlations and ages for the benthic foraminiferal stages and divisions. The CN5b subzone spans units 1 to 5 in the lower member of the Modelo Formation; thus, the boundary between the Division F (Luisian) and Division E (early Mohnian) is the CN5b subzone or older (older than 10.8 Ma) (Figure 6). In addition, the *Bolivina modeloensis* Zone and the lower part of Division E are equivalent in age to at least part of the CN5b subzone (approximately 11.8 to 10.8 Ma).

The interval between unit 5 and the upper part of unit 9 contains nondefinitive calcareous nannofossil species. This interval ranges from subzones CN5b through CN8b based on stratigraphic position. The upper part of Division E (lower Mohnian) and Division D (upper Mohnian) occurs within the interval assigned to CN5b through CN8b (approximately 11.8 to 7.3 Ma) (Figure 6).

The upper part of unit 9 contains species of calcareous nannofossils characteristic of the CN8b through CN9a subzones (Figure 5). The boundary between Divisions D and C occurs within the CN8b to CN9a zonal interval (8.0 to 6.4 Ma) (Figure 6).

Summary

To summarize, the boundary between Division F (Luisian) and Division E (upper Mohnian) occurs below the nannoplankton zone CN5b. The top of Division E and the *Bulimina uvigerinaformis* Zone occurs in the interval assigned to the undifferentiated subzones b–d of the *Denticulopsis hustedtii–Denticulopsis lauta* Zone (Figure 6). The boundary between Divisions C and D is defined by the first downsection occurrence of *Cassidulinella renulinaformis* and occurs within the *Denticulopsis hustedtii* Zone and nannoplankton subzones CN8b to CN9a. The top of Division C occurs between the *Denticulopsis hustedtii* Zone and subzone b of the *Thalassiosira antiqua* Zone. This top represents the first occurrence downsection of the benthic foraminiferal assemblage characteristic of the *Bolivina hughesi* Zone.

The top of Division A/B undifferentiated and the *Bolivina obliqua* Zone occurs in the upper part of subzone b of the *Nitzschia reinholdii* Zone.

Palos Verdes Hills Section

Stratigraphy

The Palos Verdes Hills are underlain mainly by Miocene through Pleistocene marine clastic rocks. These units rest upon Mesozoic Catalina Schist that crops out in a small area near the center of the hills.

As previously discussed, the Monterey Formation of the Palos Verdes Hills is divided into three members: the Altamira Shale, the Valmonte Diatomite, and the Malaga Mudstone (Woodring et al., 1946).

Radiometric Dates

There are five reliable radiometric dates within the Palos Verdes Hills section. Turner (1970) dated the Portuguese Tuff Bed in the Altamira Shale at 14.9 ± 1.1 m.y., using the fission-track method. The age of the Miraleste Tuff within the lower part of the middle Altamira Shale Member is dated by fission-track between 14.5 and 14.2 m.y. (Conrad and Ehlig, 1983).

A thin ash in the upper part of the Valmonte Diatomite Member at Malaga Cove has a fission-track date of 6.9 ± 0.6 m.y. (Obradovich and Naeser, 1981).

A 4.42 ± 0.57 Ma date was obtained from an ash in the upper portion of the Malaga Mudstone (Obradovich and Naeser, 1981). This ash coincides with the occurrence of a diatom assemblage representative of the *Thalassiosira oestrupii* Zone, which ranges from 5.1 to 3.6 m.y. (Figure 8).

Ashes near the base of the Repetto Siltstone at Malaga Cove have a zircon fission-track date of 3.4 ± 0.3 m.y. (Obradovich and Naeser, 1981). This date provides an age for the middle Repetto Formation because at this locality the lower Repetto is missing

Benthic Foraminifera

Benthic foraminiferal assemblages are very abundant and diverse throughout the Altamira Shale Member (Woodring et al., 1946). The lower and middle Altamira Shale contains several species indicative of the Luisian Stage and Division F (Figure 7). Woodring et al. (1946) reported the occurrence of benthic foraminiferal species representative of the Relizian Stage in the lower portion of the Altamira Shale. However, the species listed by Woodring et al. (1946) are indicative of the lower Luisian Stage and do not include any species restricted to the Relizian Stage.

The upper Altamira Shale is lower Mohnian (Division E) based on a well-developed benthic foraminiferal assemblage. This interval is equivalent to units 1 through 8 of the Modelo Formation in the Topanga Canyon section. The schist-bearing sand in the southeastern Palos Verdes Hills also contains lower Mohnian faunas. This sandstone is in the same general stratigraphic position as the basal schist-bearing conglomerates underlying Division E of the western part of the basin (Wissler, 1943).

The Valmonte Diatomite occurs locally along the crest and along parts of the northeastern and eastern flanks of the Palos Verdes Hills. The benthic foraminiferal fauna in the Valmonte Diatomite is characteristic of the upper

Text continues on p. 161.

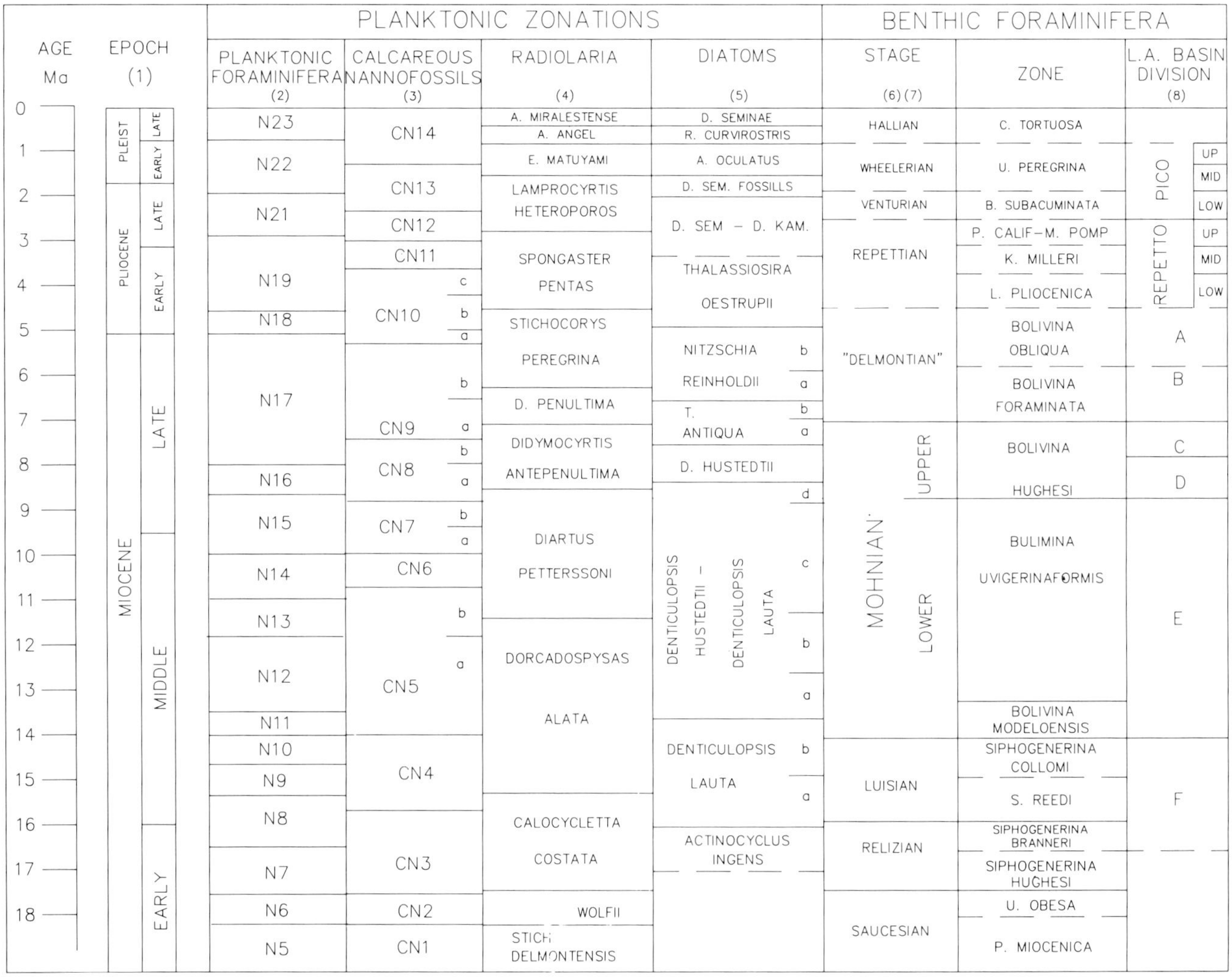

AGE Ma	EPOCH (1)	PLANKTONIC ZONATIONS				BENTHIC FORAMINIFERA		L.A. BASIN DIVISION (8)
		PLANKTONIC FORAMINIFERA (2)	CALCAREOUS NANNOFOSSILS (3)	RADIOLARIA (4)	DIATOMS (5)	STAGE (6)(7)	ZONE	
0	PLEIST LATE	N23	CN14	A. MIRALESTENSE / A. ANGEL	D. SEMINAE / R. CURVIROSTRIS	HALLIAN	C. TORTUOSA	PICO UP
1	PLEIST EARLY	N22		E. MATUYAMI	A. OCULATUS	WHEELERIAN	U. PEREGRINA	PICO MID
2	PLIOCENE LATE		CN13	LAMPROCYRTIS HETEROPOROS	D. SEM. FOSSILS	VENTURIAN	B. SUBACUMINATA	PICO LOW
3	PLIOCENE LATE	N21	CN12		D. SEM – D. KAM.	REPETTIAN	P. CALIF–M. POMP	REPETTO UP
			CN11	SPONGASTER PENTAS	THALASSIOSIRA OESTRUPII		K. MILLERI	REPETTO MID
4	PLIOCENE EARLY	N19	CN10 c				L. PLIOCENICA	REPETTO LOW
5		N18	CN10 b / a	STICHOCORYS PEREGRINA		"DELMONTIAN"	BOLIVINA OBLIQUA	A
6	MIOCENE LATE				NITZSCHIA REINHOLDII b / a		BOLIVINA FORAMINATA	B
7	MIOCENE LATE	N17	CN9 a b	D. PENULTIMA	T. ANTIQUA b / a	UPPER MOHNIAN	BOLIVINA HUGHESI	C
8	MIOCENE LATE	N16	CN8 b / a	DIDYMOCYRTIS ANTEPENULTIMA	D. HUSTEDTII			D
9		N15	CN7 b / a	DIARTUS PETTERSSONI	d	LOWER MOHNIAN	BULIMINA UVIGERINAFORMIS	E
10	MIOCENE MIDDLE	N14	CN6		DENTICULOPSIS HUSTEDTII – DENTICULOPSIS LAUTA c / b / a			
11	MIOCENE MIDDLE	N13	b					
12	MIOCENE MIDDLE	N12	CN5 a	DORCADOSPYSAS ALATA				
13	MIOCENE MIDDLE						BOLIVINA MODELOENSIS	
14		N11						
15		N10 / N9	CN4		DENTICULOPSIS LAUTA b / a	LUISIAN	SIPHOGENERINA COLLOMI / S. REEDI	F
16		N8		CALOCYCLETTA COSTATA	ACTINOCYCLUS INGENS	RELIZIAN	SIPHOGENERINA BRANNERI	
17	EARLY	N7	CN3				SIPHOGENERINA HUGHESI	
18	EARLY	N6	CN2	WOLFII		SAUCESIAN	U. OBESA	
		N5	CN1	STICH. DELMONTENSIS			P. MIOCENICA	

Figure 6. Chronostratigraphic framework for the Miocene through Pleistocene of California showing correlation of radiolarian, diatom, and benthic foraminiferal stages and divisions in the Los Angeles basin to the planktonic foraminiferal, calcareous nannofossil, radiolarian, and diatom zones. (1) Berggren et al., 1985, and Barron et al., 1985. (2) Blow, 1969. (3) Okada and Bukry, 1980. (4) Riedel and Sanfilippo, 1978, and Barron et al., 1985. (5) Barron, 1981, 1986b. (6) Kleinpell, 1938, 1980. (7) Natland, 1952. (8) Wissler, 1943, 1958.

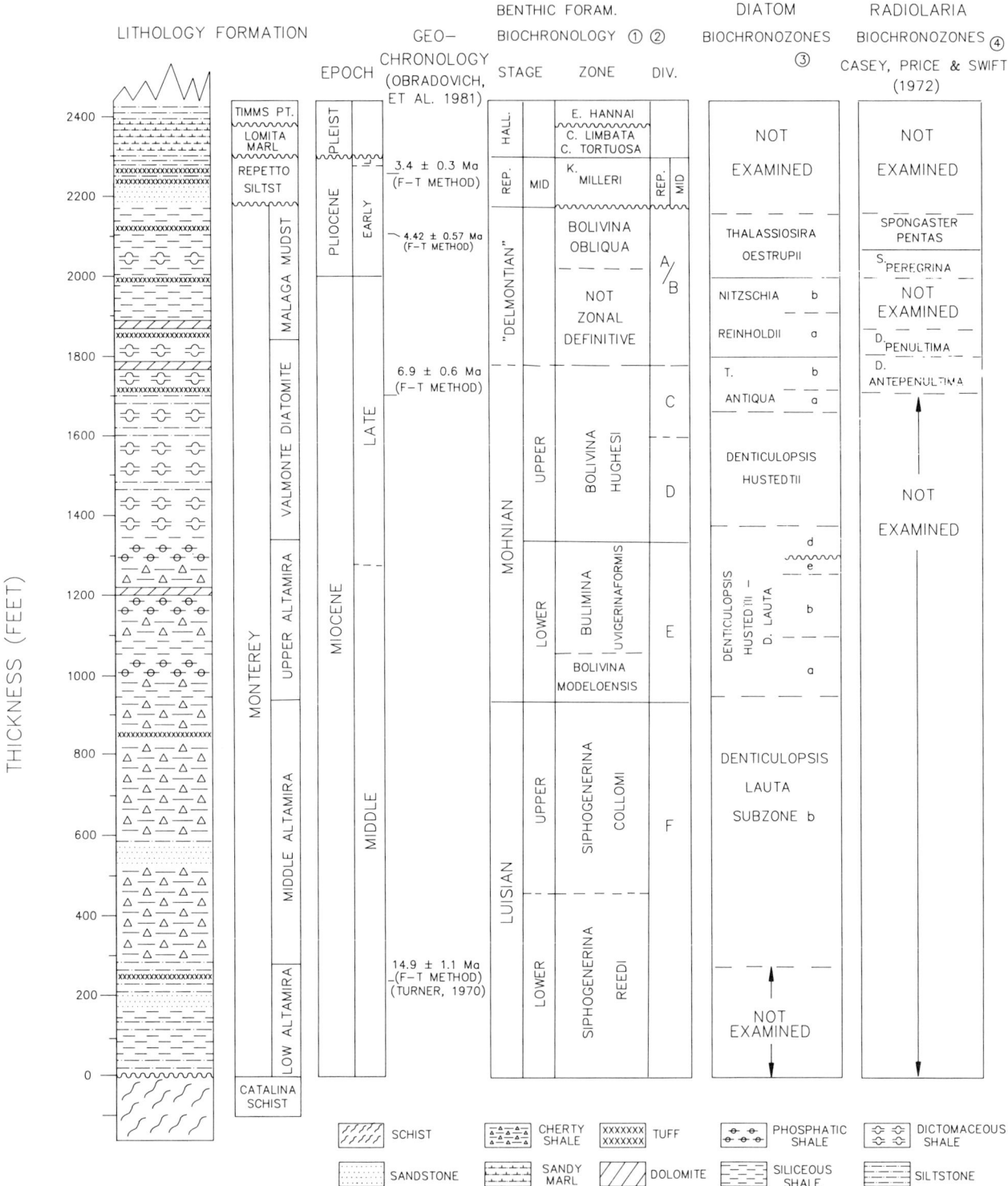

Figure 7. Composite stratigraphic column for the Palos Verdes Hills section. Lithologic units are from Woodring et al., 1946. See Figure 1 for location. Note the hiatus subzone c of the *Denticulopsis hustedtii–D. lauta* diatom zone. (1) Benthic foraminiferal stages based on the Miocene provincial stages of Kleinpell (1938, 1980) and Pliocene-Pleistocene provincial stages of Natland (1952). (2) Foraminiferal divisions from Wissler (1943, 1958). (3) Diatom zones based on Barron (1981, 1986b). (4) Radiolarian zones modified from Riedel and Sanfilippo (1978).

Mohnian Stage. It is equivalent to units 9 through 13 of the Modelo Formation in Topanga Canyon. Wissler (1958) correlated a sample from the middle part of the Valmonte Diatomite to a subsurface marker horizon 200 to 400 ft (60 to 122 m) below the top of Division D in the southwestern block. *Cassidulinella renulinaformis* has been reported from this section in the eastern part of the Palos Verdes Hills (A. D. Warren, personal communication). The remaining part of the Valmonte Diatomite is within Division C and the *Bolivina hughesi* Zone based on the occurrence of *Bolivina hughesi* and associated species. In the Malaga Cove section, the uppermost part of the Valmonte Diatomite contains a rare benthic foraminiferal assemblage that is equivalent to the "Delmontian" and Division A/B undifferentiated (Figure 8).

Outcrops of the Malaga Mudstone are confined to small patches in and near San Pedro, along the north side of the hills, and in the type locality at Malaga Cove (Woodring et al., 1946). The Malaga Mudstone contains a "Delmontian" benthic foraminiferal assemblage.

The foraminiferal assemblages in the Malaga Mudstone are rare, poorly preserved, and usually have relatively low diversity. The species are characteristic of Divisions A and B. A sample taken from the lower Malaga Mudstone contains several large *Globobulimina* species and *Gyroidina soldanii rotundimargo*; it may be representative of Division B or C, but most likely it is equivalent to Division B. As previously stated, the top of Division C is located within the upper part of the Valmonte Diatomite (Figure 7).

The Malaga Mudstone is equivalent to units 14 through 19 of the Modelo at Topanga Canyon. However, because of poor foraminiferal recovery, it is not possible to make any direct zonal correlations.

The Repetto Siltstone disconformably overlies the Malaga Mudstone Member of the Monterey Formation and is unconformably overlain by Pleistocene strata (Woodring et al., 1946). The Repetto Siltstone is exposed at Malaga Cove and at other localities near the northern boundary of the hills. A diverse Repettian foraminiferal assemblage indicative of lower bathyal environments occurs within the Repetto Siltstone at Palos Verdes Hills (Ingle, 1967). This Repettian assemblage is composed of several species characteristic of the *Karreriella milleri* fauna of the middle Repettian. *Karreriella milleri*, restricted by Natland (1952) to the middle and lower portions of the Repettian Stage, occurs throughout the Repetto Siltstone of the Malaga Cove section (Figure 8). However, *Liebusella pliocenica* and *Hopkinsina nodosa*, which are restricted to the lower Repettian, do not occur within this section. Therefore, the Repetto Siltstone at the Malaga Cove section and at other areas on the north slopes of the Palos Verdes Hills is representative only of the middle part of the Repetto Formation. This is in contrast to Wissler (1943) who reported that both the upper and middle parts of the Repetto Formation were present in the Palos Verdes Hills.

The Pleistocene units are principally exposed on the eastern and northern flanks of the Palos Verdes Hills. The Repettian beds are unconformably overlain by the Hallian Lomita Marl. The Timms Point Silt and the San Pedro Sandstone also contain species representative of the Hallian Stage. This implies that the Venturian and Wheelerian stages [lower, middle, and upper Pico foraminiferal units of Wissler (1943)] are missing at this locality. The unconformity between the Repetto Siltstone and Pleistocene units and the associated absence of Venturian and Wheelerian foraminiferal assemblages indicate structural growth and uplift of the Palos Verdes Hills in early Pleistocene.

There are no published studies of calcareous nannofossils on the Palos Verdes Hills section. However, both diatoms and radiolarians have been studied from the Monterey Formation in this area (Casey et al., 1972; Rowell, 1981; Sloan, 1987; Arends, unpublished data, reported in this paper).

Diatoms and Radiolarians

Casey et al. (1972) studied the radiolarian assemblages in the Valmonte Diatomite and Malaga Mudstone members at the Malaga Cove section (Figure 8). Rowell (1981) examined and interpreted the siliceous microfossils from several different locations in the Palos Verdes area, including Malaga Cove. The siliceous microfossils from the upper Altamira Shale were described by Sloan (1987). Arends of Unocal has reinterpreted the data presented by Casey et al. (1972) and Rowell (1981) using more current zonal schemes of Barron (1986a).

The middle part of the Altamira Shale is assigned to subzone b of the *Denticulopsis lauta* Zone (Figure 7). The upper part of the Altamira Shale is representative of subzones a through d of the *Denticulopsis hustedtii-Denticulopsis lauta* Zone. Rowell's (1981) study was based on a composite of numerous structurally complex sections in the Palos Verdes area. Based on a reinterpretation, Arends (unpublished data, reported in this paper) documented a minor hiatus within the upper part of subzone c and the lower part of subzone d of the *Denticulopsis hustedtii-Denticulopsis lauta* diatom zone. This hiatus occurs in the upper part of the Altamira Shale, which is predominantly organic-rich phosphatic shale similar to the nodular shale unit in the western Los Angeles basin subsurface. It is possible that this hiatus may represent a sediment-starved condensed section. However, in other southern California sequences, this hiatus represents nondeposition and/or deep-water erosional events (Barron, 1986b; Arends and Blake, 1986).

The overlying Valmonte Diatomite Member ranges from the *Denticulopsis hustedtii* Zone to the lower part of subzone a of the *Nitzschia reinholdii* Zone (Figure 7). The Malaga Mudstone Member ranges from subzone a of the *Nitzschia reinholdii* Zone to at least the *Thalassiosira oestrupii* Zone (Figure 8). However, the uppermost part of the Malaga Mudstone was not examined for diatoms.

The radiolarian species in the Valmonte Diatomite are assigned to the *Didymocyrtis antepenultima* Zone and the lower part of the *Didymocyrtis penultima* Zone (Figures 7, 8). The lower Malaga Mudstone is in the *D. penultima* Zone, and the upper part is assigned to the *Stichocorys peregrina* Zone and *Spongaster pentas* Zone (Figure 8). The uppermost Malaga Mudstone was not studied for radiolarians.

Planktonic Foraminifera

Ingle (1967) originally reported on planktonic foraminiferal faunas from the upper Malaga Mudstone and Repetto

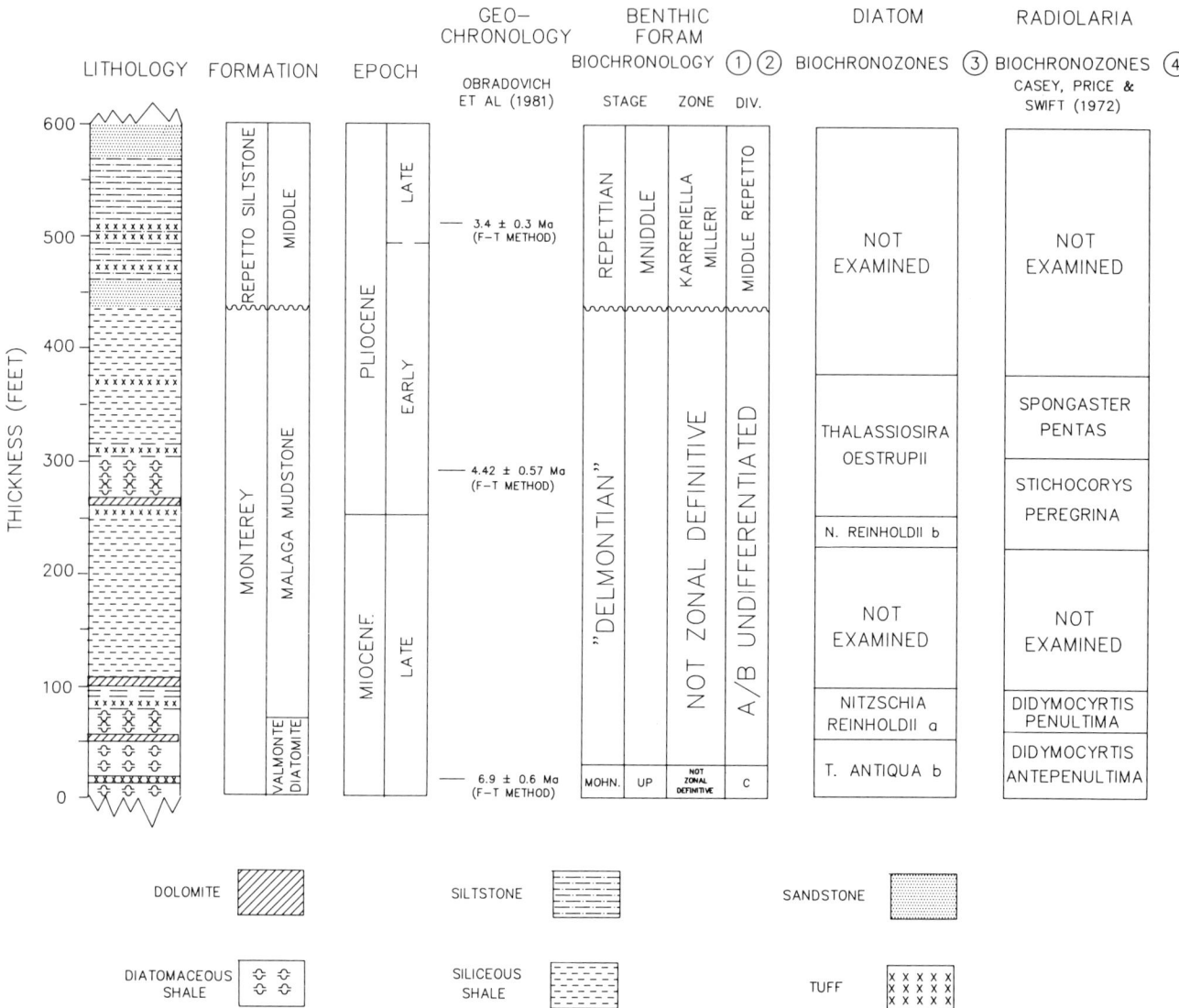

Figure 8. Composite stratigraphic column for the Malaga Cove Section, Palos Verdes Hills. See Figure 1 for location. Lithologic units are from Woodring et al. (1946), Ingle (1967), and Casey et al. (1972). Stage boundaries associated with the intervals of poor foraminiferal preservation were usually assigned based on changes in the benthic foraminiferal assemblage. (1) Benthic foraminiferal stages of Kleinpell (1938, 1980) for Miocene and Natland (1952) for Pliocene-Pleistocene. (2) Foraminiferal divisions from Wissler (1943, 1958). (3) Diatom zones based on Barron (1981, 1986b). (4) Radiolarian zones modified from Riedel and Sanfilippo (1978).

Siltstone. Using first appearance datums (FAD), last appearance datums (LAD), and coiling direction shifts in *Neogloboquadrina pachyderma*, it is possible to correlate these assemblages to the well-documented tropical zonations of Blow (1969) (Figure 9) (Ingle, 1967; Lagoe and Thompson, 1988).

In the Malaga Cove section, the radiolarian *Prunopyle titan* LAD (5.2 Ma), the *Sphaeroidinella dehiscens* FAD (4.8 Ma), and the sinistral coiling direction of *Neogloboquadrina pachyderma* (4.6 to 4.2 Ma) occur in the upper Malaga Mudstone. The *Globorotalia inflata* FAD (3.2 Ma) and another sinistral coiling shift in *Neogloboquadrina pachyderma* (2.5 to 2.4 Ma) occur within the Repetto Siltstone.

Summary

The benthic stages and divisions can be dated through chronostratigraphic correlation of the diatom zonation, planktonic foraminiferal datums, and radiometric dates of the ashes. The Luisian Stage and Division F are equivalent to subzone b of the *Denticulopsis lauta* diatom zone (15.0 to 13.8 Ma) (Figure 7). The boundary between the Division F (Luisian) and Division E (lower Mohnian) is in the uppermost subzone b of the *Denticulopsis lauta* Zone with an approximate age range of 15.0 to 13.8 Ma. The boundary between the Division E (lower Mohnian) and Division D (lower upper Mohnian) occurs in subzone d of the

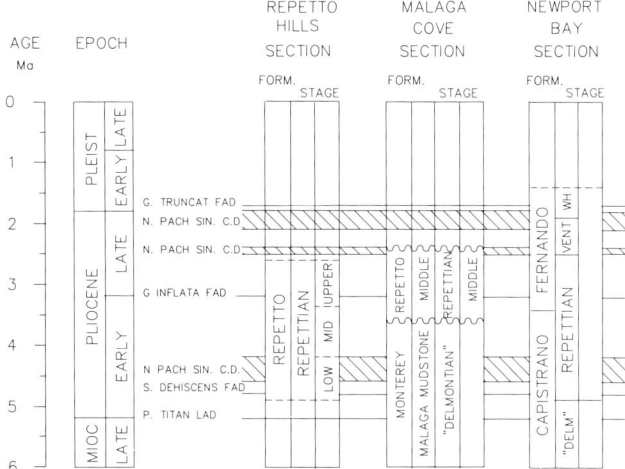

Figure 9. Comparison of stratigraphic columns from the Repetto Hills, Malaga Cove, and Newport Bay showing the correlation of various planktonic foraminiferal datums and variations in coiling direction of *Neogloboquadrina pachyderma*. See Figure 1 for location of sections. FAD = first appearance datum; LAD = last appearance datum; SIN C.D. = sinistral coiling direction of *N. pachyderma*; *G. truncat* = *Globorotalia truncatulinoides*; *N. pach* = *Neogloboquadrina pachyderma*; *S. dehiscens* = *Sphaeroidinella dehiscens*; *P. titan* = *Prunopyle titan*. Estimated ages for biostratigraphic datums are from Ingle (1967) and Lagoe and Thompson (1988).

shifts (Figure 9). The middle Repettian section at Malaga Cove is age-equivalent to the uppermost middle and upper Repettian stage at Repetto Hills. This would imply that either the middle part of the Repettian Stage is time-transgressive or that the *Karreriella milleri* Zone is not restricted to the middle and lower parts of the Repettian Stage. Because the characteristic foraminifers of the Repettian stage are strongly controlled by environmental facies (Blake, 1985), the discrepancy in correlation is probably related to environmental differences.

Newport Bay Section

Stratigraphy

Biostratigraphers have studied the Newport Bay section extensively for more than forty years, and, thus, this section constitutes a major Neogene reference section for the Los Angeles basin. The Neogene units include the middle to upper Miocene Monterey Formation, the upper Miocene to lower Pliocene Capistrano Formation, and the Pliocene through Pleistocene Fernando Formation.

Benthic foraminiferal assemblages have been described in detail by Kleinpell (1938), Ingle (1967, 1971), and Warren (1972). Planktonic foraminifers from the Miocene to Pleistocene have been described by Lipps (1964) and Ingle (1967, 1971). Radiolarians have been discussed by Casey (1972) and Casey et al. (1972). The diatom biostratigraphy has been reported by Barron (1975, 1976, 1981, 1986b). Calcareous nannoplankton from the Monterey Formation at Newport Bay was analyzed by Lipps (1968), Wilcoxon (1969), Warren (1972, 1980), and Carlos (1985).

Benthic Foraminifera

The lower part of the Monterey Formation contains abundant benthic foraminifers indicative of the *Siphogenerina collomi* Zone of the upper Luisian Stage and Division F (Figure 10). This interval is equivalent to the middle part of the Altamira Shale in the Palos Verdes Hills and the Topanga Group at the Topanga Canyon section.

The middle part of the Monterey Formation is assigned to the *Bolivina modeloensis* and *Bulimina uvigerinaformis* zones of the lower Mohnian Stage (equivalent to Division E), as defined by Warren (1972). This same interval was defined by Kleinpell (1980) as the *Bolivina modeloensis* Zone, *Bolivina barbarana* Zone, and the *Bolivina wissleri* Zone of the lower Mohnian (Figures 10, 11).

An abundant benthic foraminiferal assemblage indicative of the *Bolivina hughesi* Zone of the upper Mohnian occurs in the upper part of the Monterey Formation. *Cassidulinella renulinaformis* defines Division D within the lower part of this sequence. Division C foraminifers are present in the overlying upper Mohnian interval. Kleinpell (1980) has interpreted this upper Mohnian interval to be equivalent to the *Bolivina goudkoffi* Zone (Figure 11). The upper part of the Monterey Formation at Newport Bay is equivalent to the Valmonte Diatomite Member of the Monterey in the Palos Verdes Hills and units 8 through 13 in the Modelo Formation of Topanga Canyon.

Denticulopsis hustedtii–Denticulopsis lauta Zone (8.9 to 8.4 Ma). The top of Division D is in the upper part of the *Denticulopsis hustedtii* Zone (8.4 to 7.6 Ma) (Figure 6).

The boundary between Division C (upper Mohnian) and Division A/B undifferentiated ("Delmontian") is in subzone b of the *Thalassiosira antiqua* Zone (7.0 to 6.7 Ma) and the *Didymocyrtis antepenultima* radiolarian zone (8.6 to 7.0 Ma). Both Division A/B and the "Delmontian" Stage range up to at least the *Thalassiosira oestrupii* Zone (5.1 to 3.6 Ma) and the *Spongaster pentas* radiolarian zone (4.5 to 2.9 Ma).

Based on the radiometric dating of ashes in the uppermost Malaga Mudstone and the lower Repetto Siltstone, the age of their unconformable contact is between 4.42 ± 0.57 m.y. and 3.4 ± 0.3 m.y. The correlation of the planktonic foraminiferal datums and coiling direction shifts places the age of the unconformity between 4.2 and approximately 3.2 Ma (Figure 9).

The *Karreriella miller* Zone of the middle Repettian ranges in age at least from 3.5 to 2.4 Ma based on the occurrence of the *Globorotalia inflata* FAD (3.2 Ma) and the change in coiling direction of *Neogloboquadrina pachyderma*. This age range coincides with the 3.4 Ma ash date in the lower part of the Repetto Siltstone.

The Malaga Cove section can be correlated to the type Repetto in the Repetto Hills and the Newport Bay section using planktonic foraminiferal datums and coiling direction

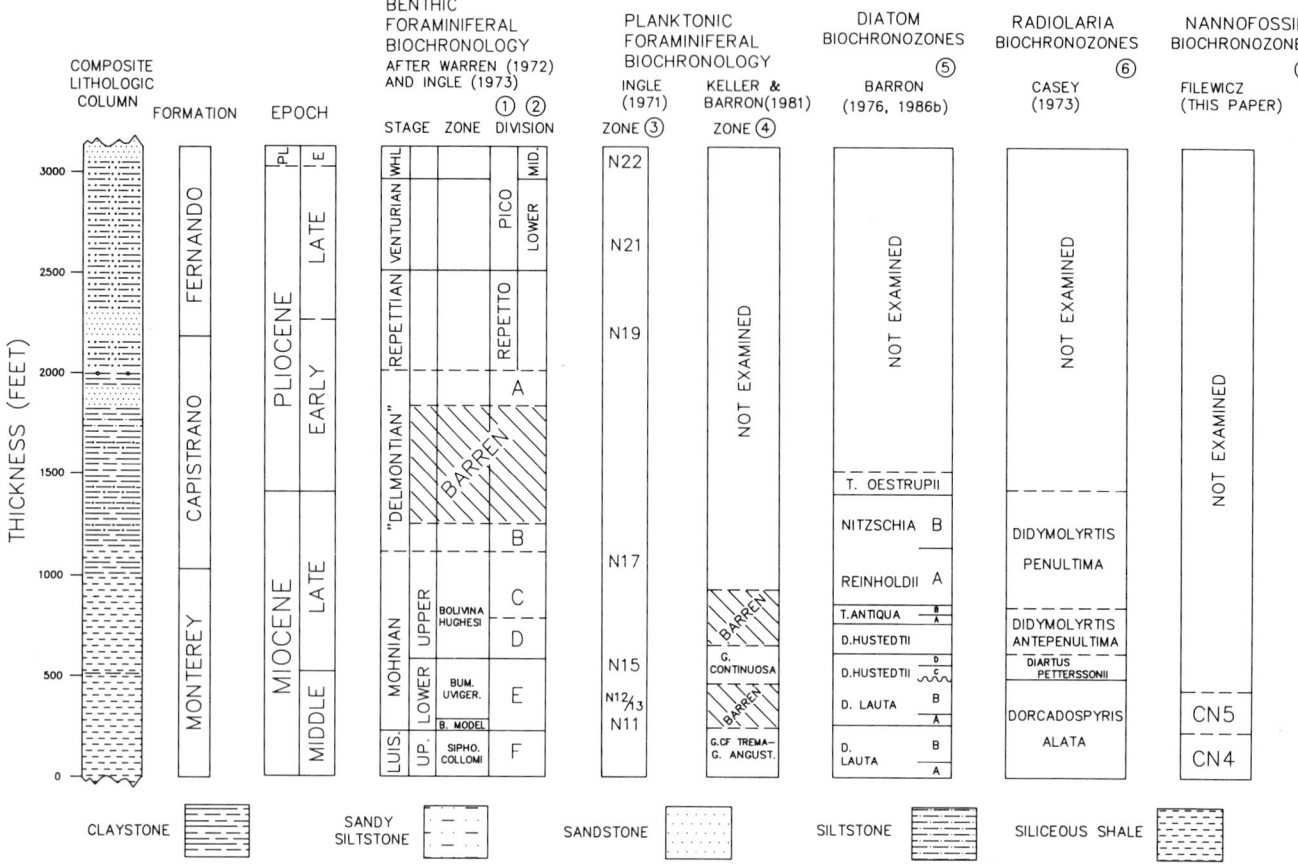

Figure 10. Stratigraphic column for the Newport Bay section. Lithologic descriptions are from Ingle (1973) and Ingle and Barron (1978). Barren intervals are shown by lined pattern. Stage and/or divisional boundaries associated with the barren intervals were usually assigned based on stratigraphic position. Note the hiatus subzone c of the *Denticulopsis hustedtii–D. lauta* diatom zone. (1) Benthic foraminiferal stages are based on the Miocene provincial stages of Kleinpell (1938, 1980) and the Pliocene provincial stages of Natland (1952). (2) Foraminiferal divisions are from Wissler (1943, 1958). (3) Planktonic foraminiferal zonation is based on Blow (1969). (4) Planktonic foraminiferal zonation is based on Keller and Barron (1982). (5) Diatom zones are based on Barron (1981, 1986b). (6) Radiolarian zonation is based on Riedel and Sanfilippo (1978). (7) Calcareous nannofossil zones are based on Okada and Bukry (1980).

The uppermost Monterey and the lower part of the overlying Capistrano Formation contain a benthic foraminiferal assemblage assigned to the "Delmontian" (Ingle, 1971; Kleinpell, 1980). However, Warren (1972) interpreted the fauna within this interval to be representative of the *Bolivina hughesi* Zone of the upper Mohnian (Figure 11). For this paper, the interpretation of Warren (1972) will be used.

Radiolarians are common to abundant throughout the lower mudstones of the Capistrano Formation. Foraminifers are absent in this interval except for those found within the diatomites at the contact between the Monterey and Capistrano formations (Ingle, 1973). The foraminiferal fauna from the lower Capistrano Formation is assigned to the "Delmontian" Stage and is representative of Division B (Figure 10). This sequence is equivalent to the lower part of the Malaga Mudstone and units 13 through 18 of the Modelo Formation at Topanga Canyon.

The sandy siltstones of the upper Capistrano Formation contain abundant benthic foraminifers that are indicative of the upper "Delmontian" and Repettian stages. The upper "Delmontian" faunas are representative of Division A. The uppermost Capistrano interval is equivalent to the Repetto Siltstone of the Palos Verdes Hills and the Repetto Formation of the central portion of the Los Angeles basin. Lower bathyal Repettian foraminiferal faunas are present in both the upper Capistrano and lower Fernando formations.

The Fernando Formation contains benthic foraminiferal faunas representative of the Repettian, Venturian, and Wheelerian stages. The progressive upsection trend from lower bathyal (Repettian) to middle bathyal (Venturian)

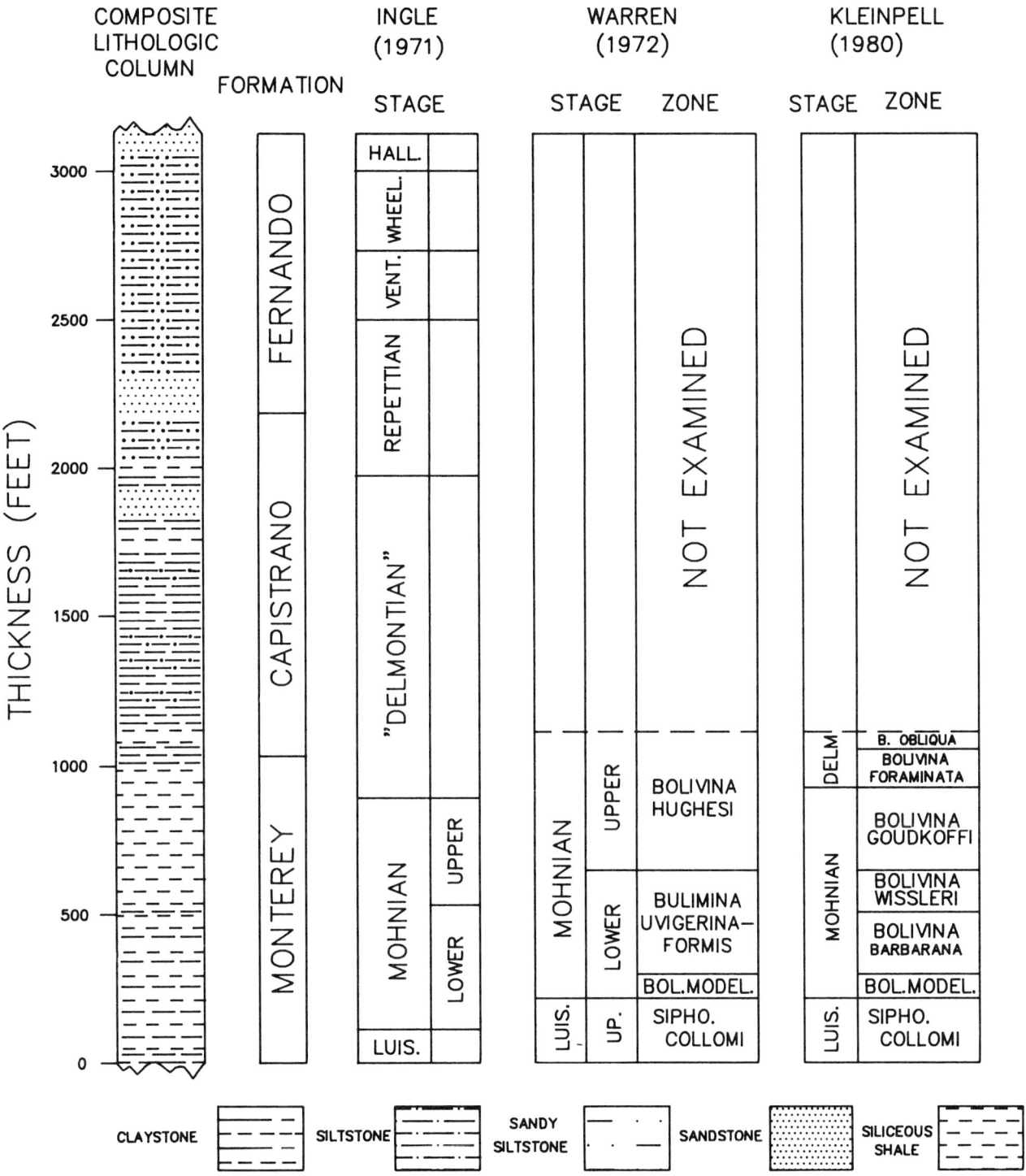

Figure 11. Stratigraphic column for the Newport Bay section showing the correlation between the benthic foraminiferal stage and zone assignments of Ingle (1971), Warren (1972), and Kleinpell (1980). Zone and division definitions are based on the presence of characteristic species and benthic foraminiferal assemblages. The age of the boundary between the Capistrano and Monterey formations varies between the "Delmontian" and Mohnian stages, depending on the benthic foraminiferal stage definition used. For this paper, the stage designation of Warren (1972) is used.

to upper bathyal (Wheelerian) to neritic (Hallian) foraminiferal assemblages in the Fernando Formation indicates rapid basin infilling; this trend is seen elsewhere along the margin of the Los Angeles basin in the Pliocene-Pleistocene interval (Ingle, 1973). This Fernando sequence is also equivalent to the upper Repetto and Pico formations of the central portion of the Los Angeles basin.

Planktonic Foraminifers

Keller and Barron (1982) defined two planktonic foraminiferal zones within the Newport Bay section (Figure 10). Both of these zones are in the Monterey Formation. Also in the Monterey Formation are rare occurrences of the important middle to late Miocene index species *Globorotalia mayeri*, *Globorotalia menardii*, *Neogloboquadrina pachyderma*, "*Sphaeroidinella subdehiscens*," and *Orbulina universa*; they allow for the recognition of the Neogene planktonic zones of Blow (1969) from N11 through N15 (Figure 10) (Ingle, 1971).

Planktonic foraminifers found within the lower Fernando Formation are correlated with Pliocene planktonic zones N19 and N20 (Ingle, 1971). The first appearance datum (FAD) of *Globorotalia inflata* with the FAD of *Globorotalia truncatulinoides* is in the upper Fernando Formation, thus approximating the Pliocene-Pleistocene boundary (N21/N22) (Ingle and Barron, 1978).

The Pliocene-Pleistocene boundary, as defined by the initial appearance of the planktonic foraminifer *Globorotalia truncatulinoides*, occurs in sediments containing a Wheelerian benthic foraminiferal stage (Ingle, 1972). The Pliocene-Pleistocene boundary in California is usually placed between the provincial Wheelerian and Hallian stages of Natland (1952), but, as previously stated, these are environmental stages, not chronostratigraphic units.

Based on the occurrence of the planktonic foraminiferal FADs and shifts in *Neogloboquadrina pachyderma* coiling direction, this section can be compared to the Malaga Cove section (Figure 9). The Capistrano Formation is almost directly age-equivalent to the Malaga Mudstone. However, the Malaga Mudstone does not contain species indicative of the Repettian Stage. The lower Fernando Formation is equivalent to the Repetto Siltstone of the Palos Verdes Hills and the upper part of the type Repetto Formation.

Diatoms

Barron (1976, 1986a) reported on a normal sequence of diatoms from the upper portion of subzone a of the *Denticulopsis lauta* Zone to the lower portion of the *Thalassiosira oestrupii* Zone (Figure 10). The zones in the Monterey Formation range from subzone a of the *Denticulopsis lauta* Zone to subzone a of the *Nitzschia reinholdii* Zone (Figure 10). Based on recent data, however, most of subzone c of *Denticulopsis hustedtii–Denticulopsis lauta* Zone appears to be missing due to a hiatus; and only the upper portion of this subzone is present (Barron, 1986b). This hiatus is equivalent to the hiatus noted in the Palos Verdes section.

The Monterey Formation of the Newport Bay section is equivalent to the Altamira Shale and Valmonte Diatomite members of the Monterey Formation at the Palos Verdes Hills. The Capistrano Formation contains subzones a and b of the *Nitzschia reinholdii* Zone and the *Thalassiosira oestrupii* Zone. This sequence is similar to the Malaga Mudstone Member.

Radiolaria

Casey (1972) assigned the radiolarian assemblages from the Monterey and Capistrano formations to four zones (Figure 10). The Monterey Formation includes the *Dorcadospyris alata* Zone, the *Diartus petterssonii* Zone, the *Didymolyrtis antepenultima* Zone, and the lower part of the *Didymolyrtis penultima* Zone. The upper part of the *D. penultima* Zone is within the lower Capistrano Formation.

Calcareous Nannofossils

Several calcareous nannofossil zones have been reported from the Newport Bay section by Lipps (1968) and Wilcoxon (1969). However, studies by Carlos (1985) and Filewicz (unpublished data) can only corroborate the CN4 and CN5 zones within the lower part of the Monterey Formation (Figure 10). The younger zonal assignments cannot be substantiated with the available data.

Summary

To summarize, the boundary between Division F (the *Siphogenerina collomi* Zone of the Luisian Stage) and Division E (the *Bolivina modeloensis* Zone of the lower Mohnian Stage) is in the subzone b of the *Denticulopsis lauta* Zone (15.0 to 13.8 Ma) and in the lowermost part of calcareous nannofossil zone CN5 (14.0 to 10.8 Ma) (Figure 10). The top of the Division E and the lower Mohnian are in the subzone d of the *Denticulopsis hustedtii–Denticulopsis lauta* Zone (8.9 to 8.4 Ma) and in the *Diartus petterssonii* Zone (11.5 to 8.6 Ma). The boundary between Division D and Division C is in the subzone a of the *Thalassiosira antiqua* Zone (7.6 to 7.0 Ma) and in the upper part of *Didymolyrtis antepenultima* Zone (8.6 to 7.0 Ma).

The top of Division C and the *Bolivina hughesi* Zone of the upper Mohnian Stage is in subzone a of the *Nitzschia reinholdii* Zone (6.7 to 6.1 Ma) and in the *Didymolyrtis penultima* Zone (7.0 to 6.2 Ma). Division B ranges from subzone a to subzone b of the *Nitzschia reinholdii* Zone (6.7 to 5.1 Ma) and the *Didymolyrtis penultima* Zone (7.0 to 6.2 Ma). Division A is either the same age or younger than the *Thalassiosira oestrupii* Zone (5.1 to 3.6 Ma) and is either the same age or older than planktonic foraminiferal zone N19 (5.0 to 3.0 Ma). This age fits the correlation of this division with the *Thalassiosira oestrupii* Zone in the Palos Verdes Hills section.

The top of the "Delmontian" and Division A occurs between the LAD of the radiolarian *Prunopyle titan* (5.2 Ma) and the FAD of Sphaeroidinella dehiscens (4.8 Ma) (Figure 9). This would imply that the age of the boundary between the Division A ("Delmontian") and the Repettian Stage at the Newport Bay section is approximately 5.0 ± 0.2 m.y.

The top of the Repettian Stage in this section occurs above the FAD of *Globorotalia inflata* (3.2 Ma) and at the base of a *Neogloboquadrina pachyderma* sinistral coiling direction shift (2.5 to 2.4 m.y.). In the Newport Bay section, the top of the Repettian is approximately 2.5 Ma (Figure 9).

The boundary between the Venturian and Wheelerian stages occurs within another *Neogloboquadrina pachyderma* sinistral coiling direction event (2.1 to 1.8 Ma) and directly below the FAD of *Globorotalia truncatulinoides* (1.7 Ma). The age of the top of the Venturian stage in this section is approximately 1.95 ± 0.15 m.y. The top of the Wheelerian stage is younger than 1.7 m.y.

Puente Hills Composite Section

Stratigraphy

The Puente Hills are located in the northeastern part of the Los Angeles basin. An aggregate of approximately 18,000 ft (5500 m) of Tertiary rocks overlie a Mesozoic plutonic basement (Yerkes, 1972). Because of the economic interest associated with the Neogene sequence in the Puente Hills and surrounding areas, this part of the basin has been studied extensively (Durham and Yerkes, 1964; Yerkes et al., 1965; Yerkes, 1972; and Schoellhamer et al., 1981).

Benthic Foraminifers

The middle Miocene Topanga Formation disconformably overlies the undifferentiated early Eocene to early Miocene Vaqueros/Sespe formations. This sequence contains poorly preserved middle Miocene molluscs and a Relizian benthic foraminiferal assemblage (Figure 12).

The Topanga Formation is interbedded with and overlain by middle Miocene volcanic rocks. These rocks are correlated with the El Modeno volcanics of the northern Santa Ana Mountains (Durham and Yerkes, 1964). In the southern part of the Puente Hills, an andesite flow, correlated with the El Modeno volcanic sequence, yielded a 13.7 ± 1.6 m.y. date using the K-Ar method (Turner, 1970). The volcanic unit also overlies Luisian sediments of the Topanga Formation. In locations where the volcanic rocks are interbedded with sandstones of the Topanga, Luisian benthic species are present. This stratigraphic unit correlates with the Topanga Group at the Topanga Canyon section.

The upper middle to upper Miocene Puente Formation unconformably overlies the Topanga Formation and volcanic rocks. The Puente Formation is divided into four members: in ascending order they are the La Vida, Soquel, Yorba, and Sycamore Canyon. The La Vida Member contains benthic foraminiferal faunas indicative of Division E and the lower Mohnian Stage. The lower part of the member contains a fauna equivalent to the *Bolivina modeloensis* Zone. The upper part of the member is assigned to the *Bulimina uvigerinaformis* Zone. This member is equivalent to units 1 through 8 of the Modelo Formation in Topanga Canyon, the upper Altamira Shale of Palos Verdes Hills, and the middle part of the Monterey Formation of the Newport Bay section.

Conformably overlying the La Vida Member is the Soquel Member containing a Division D (upper Mohnian) fauna (Figure 12). *Cassidulinella renulinaformis*, an index species for Division D, has not been reported from this section. Therefore, the boundary between Divisions C and D is interpolated, based on minor faunal variations. A K-Ar plagioclase date from a bentonite bed within this member gave a minimum age of 9 Ma (Turner, 1970); this is considered a minimum age owing to possible alteration of the dated plagioclase.

There is a gradational contact between the Soquel and the overlying Yorba Member. The Yorba Member contains abundant benthic foraminiferal assemblages characteristic of the *Bolivina hughesi* Zone of the upper Mohnian and Division C. The Soquel and Yorba members combined are stratigraphically equivalent to units 9 through 13 of the Modelo Formation in Topanga Canyon, the Valmonte Diatomite in the Palos Verdes Hills, and the upper part of the Monterey Formation at Newport Bay.

Conformably overlying the Yorba is the Sycamore Canyon Member. Benthic foraminiferal faunas in this unit are representative of the "Delmontian" Stage and Divisions A and B. Also within this member are a few species characteristic of the Repettian stage. The Repettian species are indicative of a lower bathyal environment and probably represent a deepening of the basin during Sycamore Canyon deposition. Units 13 through 19 of the Modelo Formation at Topanga Canyon, the Malaga Mudstone of the Palos Verdes Hills, and the lower part of the Capistrano Formation are equivalent to this member.

The Puente Formation is unconformably overlain by the Pliocene–Pleistocene Fernando Formation. This formation has been divided into two members that are equivalent to the Repettian and Pico formations, based on benthic foraminiferal assemblages. The lower member is equivalent to the Repetto Formation and contains abundant lower bathyal Repettian fauna.

The lower member of the Fernando Formation is overlain both conformably and unconformably by the upper member of the Fernando (equivalent to the Pico Formation). The upper member contains benthic foraminiferal assemblages indicative of the lower middle bathyal Venturian, the upper bathyal Wheelerian, and the neritic Hallian stages. The Fernando Formation is overlain by the late Pleistocene La Habra Formation. The Fernando Formation is equivalent to the Repetto Siltstone, Lomita Marl, and Timms Point Silt of the Palos Verdes Hills and the upper Capistrano and Fernando formations of the Newport Bay section. As in the Newport Bay section, the Fernando Formation in the Puente Hills is indicative of a basin-filling sequence.

Regional Correlation of Benthic Stages and Divisions to Planktonic Zonations and Radiometric Dating

The correlation between the benthic foraminiferal stages of Kleinpell (1938, 1980) and Natland (1952) and the divisions of Wissler (1943, 1958) with the various planktonic zonations are shown in Figure 6. Boundary correlations between benthic foraminiferal stages and divisions and calcareous and siliceous microfossils appear to be relatively consistent between the Topanga Canyon, Palos Verdes Hills, and Newport Bay stratigraphic sections. There is also close agreement between the age of the provincial stage boundaries in these three sections and how they relate to other Neogene basins of California (Blake, 1985; Arends and Blake, 1986; Barron, 1986a).

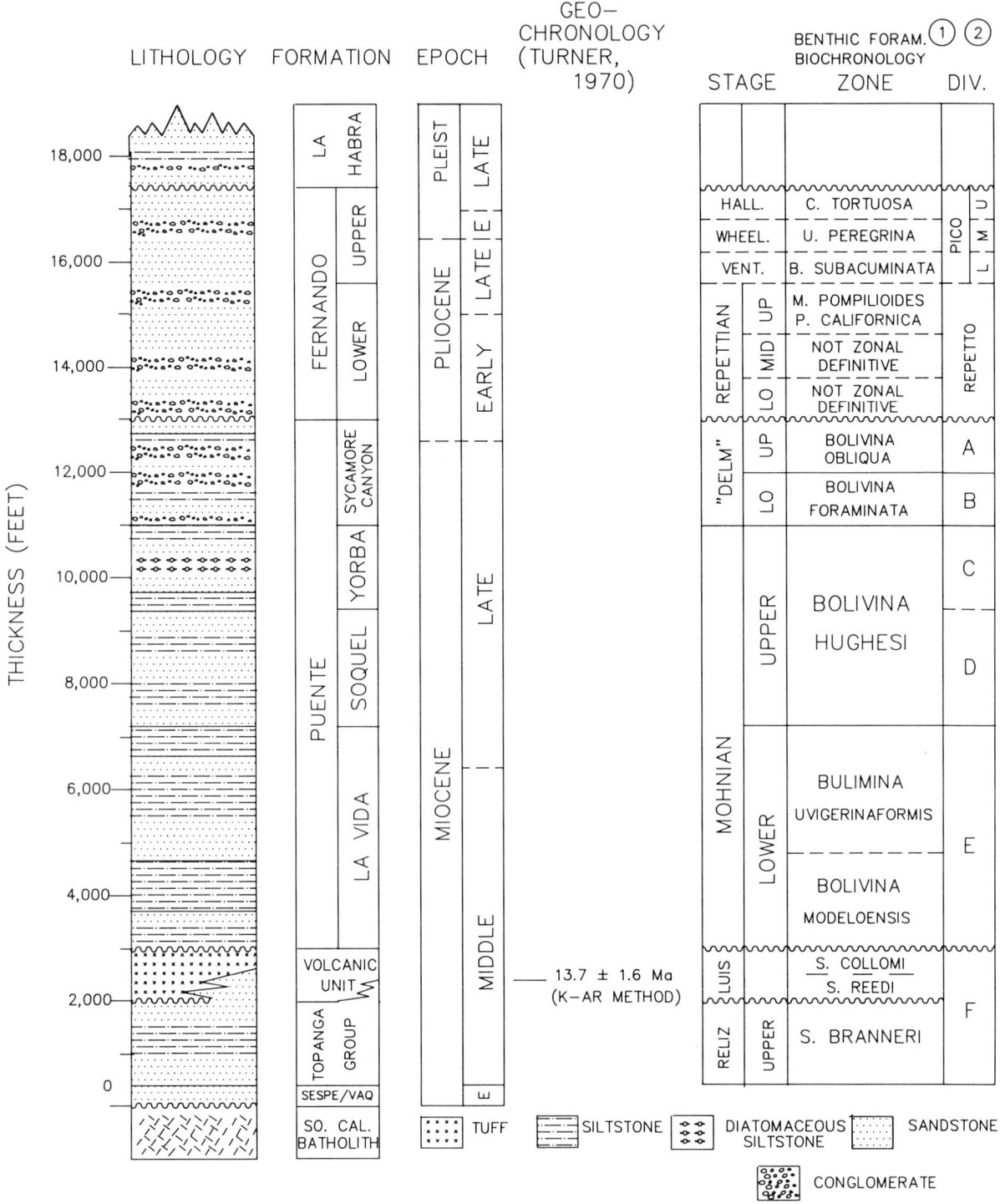

Figure 12. Composite stratigraphic column for the Puente Hills section. Lithologic descriptions are from Durham and Yerkes (1964) and Yerkes (1972). (1) Benthic foraminiferal stages are based on the Miocene provincial stages of Kleinpell (1938, 1980) and the Pliocene provincial stages of Natland (1952). (2) Foraminiferal divisions are from Wissler (1943, 1958).

The boundary between Division F and the *Siphogenerina collomi* Zone of the Luisian Stage and Division E and the *Bolivina modeloensis* Zone of the lower Mohnian Stage occurs in two sections. In the Topanga Canyon section, this boundary is an unconformable contact; it occurs below or in the lower part of nannofossil zone CN5b (11.8 Ma to 10.8 Ma). In the Newport Bay section, the boundary occurs in nannofossil zone CN5 (14.0 to 10.8 Ma) and in the subzone b of the *Denticulopsis lauta* Zone (15.0 to 13.8 Ma). The interpolated age for this boundary is 13.9 Ma, based on the overlap of these two zones.

The top of the *Bolivina modeloensis* Zone (lower Division E) occurs in the nannofossil subzone CN5b (11.8 to 10.8 Ma) and below subzones b–d undifferentiated of the *Denticulopsis hustedtii–Denticulopsis lauta* Zone (12.7 to 8.4 Ma) in the Topanga Canyon section and in subzone a of the *Denticulopsis hustedtii–Denticulopsis lauta* Zone (13.8 to 12.7 Ma) in the Palos Verdes Hills and Newport Bay sections. In the Newport Bay section, the top of this benthic foraminiferal zone occurs in the N11 planktonic foraminiferal zone (14.0 to 13.6 Ma). The interpolated age for the top of the *Bolivina modeloensis* Zone is 13.7 ± 0.1 m.y., based on the overlap of the diatom, planktonic foraminiferal, and nannofossil zones in the Palos Verdes Hills and Newport Bay section. In the Topanga Canyon section, the age is 10.8 Ma, approximately 3 m.y. younger when correlated to the calcareous nannofossil zone CN5b. However, if the top is older than the diatom zone (12.7 Ma), then the age would be closer to the age determined at the other two sections. Because of the relatively poor recovery and preservation of the siliceous and calcareous microfossils in the part of the Topanga Canyon section, the interpolated age used in this paper will be 13.7 Ma.

The top of Division E and the *Bulimina uvigerinaformis* Zone of the lower Mohnian occurs within subzone d of the *Denticulopsis hustedtii–Denticulopsis lauta* Zone (8.9 to 8.4 Ma) in two of the stratigraphic sections and in undifferentiated subzones b–d of the *Denticulopsis hustedtii–Denticulopsis lauta* Zone (12.7 to 8.4 Ma) in the Topanga Canyon section. In the Newport Bay section, the top also occurs in the *Diartus petterssoni* Zone (11.5 to 8.6 Ma). This would imply that the age of this top is approximately 8.75 ± 0.15 m.y. This date is similar to ages assigned to top of the lower Mohnian elsewhere in central and southern California (Blake, 1985; Barron, 1986a).

The top of Division D occurs within the *Denticulopsis hustedtii* Zone (8.4 to 7.6 Ma) in both the Topanga Canyon and Palos Verdes Hills sections. In the Newport Bay section, it occurs in subzone a of the *Thalassiosira antiqua* Zone (7.6 to 7.0 Ma). In the Topanga Canyon section, the first downsection occurrence of *C. renulinaformis* (Division D index species) occurs within the calcareous nannofossil subzone CN9a to CN8b (8.0 to 6.4 Ma) and is above an ash bed dated at 7.8 ± 0.9 m.y. In the Newport Bay section, the top of Division D occurs in the *Didymolyrtis antepenultima* Zone (8.6 to 7.0 Ma). Based on the overlap of these siliceous and calcareous microfossil zones and their location above the ash bed, the age for the top of this division is interpolated to be 7.4 ± 0.4 m.y. As previously discussed, the boundary between Divisions C and D is based on a crushed foraminiferal species, *C. renulinaformis*, the only species restricted to this division. Miscorrelation of the top of this division between sections could result from variable fossil preservation.

The top of Division C and the *Bolivina hughesi* Zone of the upper Mohnian occurs in subzone b of the *Thalassiosira antiqua* Zone (7.0 to 6.7 Ma) and the *Didymocyrtis antepenultima* Zone (8.6 to 7.0 Ma) in the Palos Verdes Hills section. In the Newport Bay section, the top is found stratigraphically higher in subzone a of the *Nitzschia reinholdii* Zone (6.7 to 6.1 Ma) and in the *Didymolyrtis penultima* Zone (7.0 to 6.2 Ma). In the Topanga Canyon section, the top of the type Mohnian Stage occurs in the subzone b of the *Thalassiosira antiqua* diatom Zone (7.0 to 6.7 Ma) or older diatom zones (Figure 5). Because of poor preservation of the siliceous microfossils in the Topanga Canyon section, it is not possible to correlate the top of the type Mohnian to a particular diatom subzone or zone.

For the Topanga Canyon and the Palos Verdes sections, the age for the top of Division C and the upper Mohnian is interpolated to be 6.8 ± 0.1 m.y. For the Newport Bay section, the top of this division and stage is dated at approximately 6.5 Ma. This discrepancy in age between the sections is probably the result of poor foraminiferal preservation in the Topanga Canyon and Palos Verdes Hills sections. In both of these sections, the benthic foraminiferal faunas are not zonally definitive at the top of the upper Mohnian.

The Newport Bay section presents a third possible correlation; as previously discussed, there are two interpreted tops of the Mohnian (Figure 11). If the top designated by Ingle (1971) and Kleinpell (1980) is used, it occurs at the boundary between subzone b of *Thalassiosira antiqua* Zone and subzone a of *Nitzschia reinholdii* Zone, giving an approximate age of 6.7 Ma.

All three of these dates for the top of the Mohnian are within 0.3 m.y., well within the dating error associated with this technique of second- and third-order age interpolations. This paper accepts the interpretation of Warren (1972) and will use an age of 6.5 Ma for the top of Division C and the Mohnian Stage.

As previously stated, it is difficult to distinguish between Divisions A and B; thus, these two divisions will be combined for this discussion on chronostratigraphic correlations. The top of the "Delmontian" Stage and Divisions A and B is not exposed at the Topanga Canyon section. It is an unconformity at the Malaga Cove section of the Palos Verdes Hills. In the Newport Bay section, it is a conformable contact with the overlying Repettian stage.

In the Topanga Canyon section, the top of the "Delmontian" and Division A–B undifferentiated is either in subzone b of *Nitzschia reinholdii* Zone (6.1 to 5.1 Ma) or in a younger diatom zone (Figure 5). At the Malaga Cove section, the unconformable top of the "Delmontian" occurs either in the *Thalassiosira oestrupii* Zone (5.1 to 3.6 Ma) or in a younger diatom zone (Figure 8). This top is above the *S. dehiscens* FAD (4.8 Ma) and the sinistral coiling direction shift in *N. pachyderma* (4.6 to 4.2 Ma) (Figure 9). However, it must be older than the 3.4 Ma date for the tuffaceous layer of the Repetto Siltstone.

The top of the "Delmontian" and Division A–B undifferentiated in the Newport Bay section is either in the *Thalassiosira oestrupii* Zone (5.1 to 3.6 Ma) or in a younger diatom zone (Figure 10). The top occurs between the LAD of the radiolarian species *Prunopyle titan* (5.2 Ma) and the FAD of *Sphaeroidinella dehiscens* (4.8 Ma) (Figure 9). This would imply that the age of the top of the "Delmontian" and Division A–B undifferentiated in the Newport Bay section is 4.95 ± 0.15 m.y. This is similar to the age of the boundary between the "Delmontian" and the Repettian that is documented in the Repetto Hills (Figure 9).

The Repettian, Venturian, Wheelerian, and Hallian stages together only occur in the Newport Bay section. Therefore, the dates associated with the planktonic foraminiferal datums, siliceous microfossil datums, and coiling direction shifts will be used for age determinations. The top of the Repettian Stage is approximately 2.5 Ma. The top of the Venturian stage ranges in age from 2.1 to 1.8 Ma and is interpolated at 2.0 Ma. The top of the Wheelerian stage is not exposed in the stratigraphic sections, but must be younger than 1.6 Ma (Figure 9).

REGIONAL BIOSTRATIGRAPHIC CORRELATIONS

Correlation of provincial benthic foraminiferal stages and divisions with planktonic zonations at basin margins provides a chronostratigraphic framework for evaluating the consistency of the foraminiferal correlations across the basin. Comparison of any stratigraphic differences between the four structural blocks of the Los Angeles basin will show any variations in the depositional histories. To complete a regional evaluation, these variations should then be compared to regional tectonic and eustatic events associated with the evolution of the Los Angeles basin.

Three regional cross sections anchored by the previously described surface sections serve to illustrate lithologic and biostratigraphic relationships across the basin (Table 6) (Figure 13). These three cross sections are tied by common wells and stratigraphic sections; wells are plotted by depth to demonstrate regional depositional thickness trends within lithologic units (Figures 14, 16, and 18).

Several major hiatuses have been documented for the middle and early late Miocene (ca. 14 to 9 Ma) and the late Miocene to early Pliocene (ca. 7 to 5 Ma) within portions of the California continental basins. Hoskins and Griffiths (1971), Crouch (1979), Arends and Blake (1986), and Barron (1986a) have reported the occurrence of widespread late middle Miocene unconformities along the continental margins off southern and central California.

To examine the age and duration of the hiatuses in the Los Angeles basin, the three regional cross sections have also been plotted with time as the vertical axis (Figures 15, 17, and 19). The age of the upper and lower boundaries of the hiatuses are approximated to show the relative position of the boundaries to depositional units. The biochronology used for correlation is based on the relationship between the provincial stages and divisions and the planktonic zonations established in the selected stratigraphic outcrop sections. From the examination of the three regional sections, timing and extent of the hiatuses will be determined and, if possible, related to significant tectonic and depositional events.

Section A-A'

This is an east-west cross section from the Palos Verdes Hills in the southwestern block to the Puente Hills section on the northeastern block. It should be noted that for two wells in this section, Division E (lower Mohnian) is not determined by benthic foraminiferal faunas but is instead based on the presence of organic-rich phosphatic, nodular shale (Figure 14). This shale is thought to be equivalent to the upper Altamira Shale Member of the Monterey and to the lower Puente shale in the subsurface (Figure 2).

On the southwestern block, the Palos Verdes section, the G. P. Terminal #1, and the Shell Dolly #2 illustrate the Neogene unconformably overlying the Catalina Schist basement. There is a west to east thinning of Divisions E and F in the southwestern block. In contrast, Divisions A through D thin from the east to west (Figure 14). A Pliocene-Pleistocene unconformity occurs in both the Palos Verdes Hills and in Terminal #1 and very likely records the tectonic uplift of the Palos Verdes Hills in the late Pliocene-early Pleistocene.

In the central and northeastern blocks, Neogene sediments rest unconformably on volcanic breccia or undifferentiated Sespe/Vaqueros. In this section, A-A', the schist basement does not occur east of the Newport-Inglewood fault.

In the G. P. La Mirada #46-1, the interval from the lower part of Division A to Division E and between Divisions A through D in the Union Tousseau #2 is either very thin or missing owing to a hiatus or structural growth. For these two wells, the entire middle and upper Miocene is approximately 1500 to 3100 ft (457 to 945 m) thick. However, in the Union Chapman #29 and in the Puente Hills, Divisions A through F greatly thicken across the central and northeastern blocks, indicating that the middle and late Miocene depositional trough was located in the eastern part of the Los Angeles basin, east of the Whittier fault.

In contrast, there is a dramatic increase in the thickness of the Pliocene-Pleistocene Repetto and Pico formations in the central block and a dramatic thinning of the Repetto-Pico sequence from west to east in the central block. This interval represents the basin-filling sequence in the late Neogene. The Pliocene-Pleistocene depositional trough is assumed to be located in this portion of the central block.

The thickness of the Pliocene-Pleistocene sequence varies from over 16,000 ft (4880 m) in the Norwalk-Bellflower #2 to a possible minimum of 2900 ft (885 m) in the Chapman #29. This thinning is also present in the Puente Hills section on the northeastern block. This rapid thinning is possibly related to vertical movement on the Whittier fault that was initiated in the late Miocene (ca. 6.6 Ma) (Schwartz and Colburn, 1987).

A phosphatic nodular shale occurs from the Palos Verdes Hills east to possibly the Union Chapman #29. East of this well, Division E is dominated by clastic influx that resulted in dilution of hemipelagic deposition of the phosphatic nodular shales.

There are several hiatuses and unconformities within this cross section (Figure 15). The oldest unconformity is between the undifferentiated Sespe/Vaqueros and the underlying southern California batholith. The oldest Neogene unconformity is between the Topanga Group and the underlying Catalina Schist on the southwestern block and an unnamed volcanic sandstone and undifferentiated Sespe/Vaqueros on the central and northeastern blocks (Figure 15). This contact represents the initial Neogene marine deposition on either basement or nonmarine sediments. The upper boundary of this time gap, Division F, occurs in the Relizian to Luisian stages (ca. 16.4 to 14.0 Ma) (Figure 15). In the Puente Hills section, there is an

Table 6. Regional cross sections.

Block	Well/Section	Location	Well no.
	Cross Section A–A′		
SW	Palos Verdes Hills Section		
SW	General Petro. (Mobil) Terminal #1	Sec. 4-5S/13W	1
SW	Shell Dolly #2	Sec. 29-4S/12W	2
C	Union Norwalk–Bellflower E. H. #2	Sec. 23-3S/12W	3
C	General Petro. (Mobil) La Mirada #46-1	Sec. 15-3S/11W	4
C	Union Tousseau #2	Sec. 22-3S/10W	5
C	Union Chapman #29	Sec. 29-3S/9W	6
NE	Puente Hills Section		
	Cross Section B–B′		
NE	Topanga Canyon Section		
C	Signal Rancho #1	Sec. 25-1S/15W	7
SW	Standard (Chevron) Baldwin Cienega #105	Sec. 17-2S/14W	8
SW	Union Westchester E. H. #1	Sec. 31-2S/14W	9
C	Occidental West Athens E. H. #1	Sec. 1-3S/14W	10
SW	Union Hellman #58	Sec. 27-3S/13W	11
SW	General Petro. (Mobil) Terminal #1	Sec. 4-5S/13W	1
SW	Palos Verdes Hills Section		
	Cross Section C–C′		
NE	Topanga Canyon Section		
C	Shell Verne Community #1	Sec. 21-1S/14W	12
C	Signal Homestead E. H. #1	Sec. 25-1S/14W	13
C	Western Gulf Pacific Electric #1	Sec. 33-2S/13W	14
C	Standard Houghton Community #1	Sec. 36-2S/12W	15
C	Gen. Petro. (Mobil) La Mirada #46-1	Sec. 15-3S/11W	4
C	Standard Pacific Community #1	Sec. 26-3S/11W	16
C	Shell Harbeson #1	Sec. 9-4S/10W	17
C	Standard Hazard #1	Sec. 13-5S/11W	18
C	Jergins Ellis #1	Sec. 32-5S/10W	19
C	Newport Bay Section		

Identification and location of the wells and surface sections used in the regional cross sections. SW = southwestern block; C = central block; NE = northeastern block; and NW = northwestern block. All locations are San Bernardino Base Meridian (SBBM).

additional unconformity between the Topanga Group and the overlying Puente Formation.

The next hiatus occurs in Division E (lower Mohnian Stage). This hiatus is detectable only by diatom biostratigraphy in the Palos Verdes Hills section. In the Dolly #2 and the La Mirada #46-1, the missing section is from a thin interval of nodular shale to a thin Division C/D interval and a thin interval of Division A, respectively. In the Tousseau #2, this hiatus extends from Division D up to the upper part of Division A. In several other wells within the basin, Division E consists of nondiagnostic species and nodular shales. This hiatus may be present within these

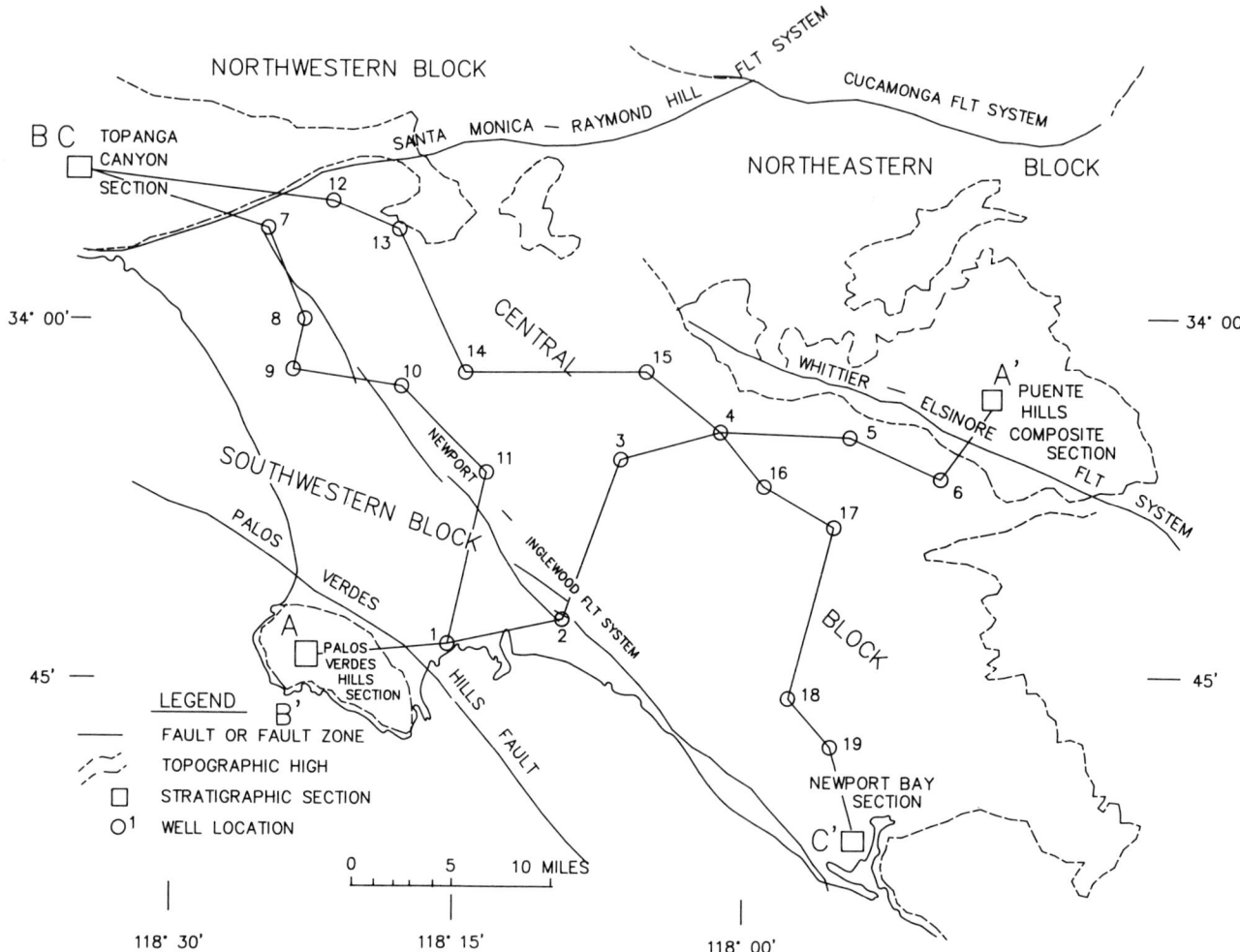

Figure 13. Map of the Los Angeles basin with location of the three regional cross sections (A-A′, B-B′, and C-C′). The boundaries between the tectonic blocks are based on Yerkes et al. (1965). Each regional cross section is anchored by previously described composite stratigraphic section. See Table 6 for the identification of wells used in the cross sections.

intervals, but cannot be detected by biostratigraphic analyses. This hiatus in the Dolly #2 and the Palos Verdes Hills section is related to oceanic or other regional factors. However, the hiatus in the La Mirada #46-1 is probably related to the structural uplift of the Anaheim nose (Wright, 1987).

In the early Pliocene to early Pleistocene, there are two unconformities within the Palos Verdes Hills section and the Terminal #1. These two unconformities are related to structural uplift of the Palos Verdes Hills that began in the late Miocene to early Pliocene.

Section B-B′

This cross section starts on the northwestern block with the Topanga Canyon section, where the Modelo Formation (Divisions A though E) unconformably overlies Topanga Group (Division F). The Pliocene-Pleistocene Repetto and Pico formations are not exposed in outcrop at this section.

South of the Santa Monica-Raymond Hill fault, the Signal Rancho #1 (northern extension of the central block) contains a 1500 ft (460 m) sequence of Topanga Group (Division F) (Figure 16). This well contains a thick interval of Puente Formation; however, Divisions C and D are missing. The Repetto and Pico formations are over 7000 ft (2135 m) thick in this well and probably represent the northern margin of the Pliocene-Pleistocene depositional trough.

The northern extension of the Newport-Inglewood fault passes between the Rancho #1 (central block) and the Standard Baldwin-Cienega #105 (southwestern block). The Baldwin-Cienega #105 contains over 2600 ft (792 m) of the Topanga Group and Division F. The thickest accumulation of the Topanga Group occurs in the northern portion of the basin. The Topanga Group rests on a thick volcaniclastic unit that, in turn, unconformably overlies Catalina Schist. The Puente Formation is relatively thin [approximately 1900 ft (580 m)]. There is a barren interval and/or an unconformity between Divisions A and E. As in the Rancho #1, Divisions C and D appear to be missing in the Baldwin-Cienega #105.

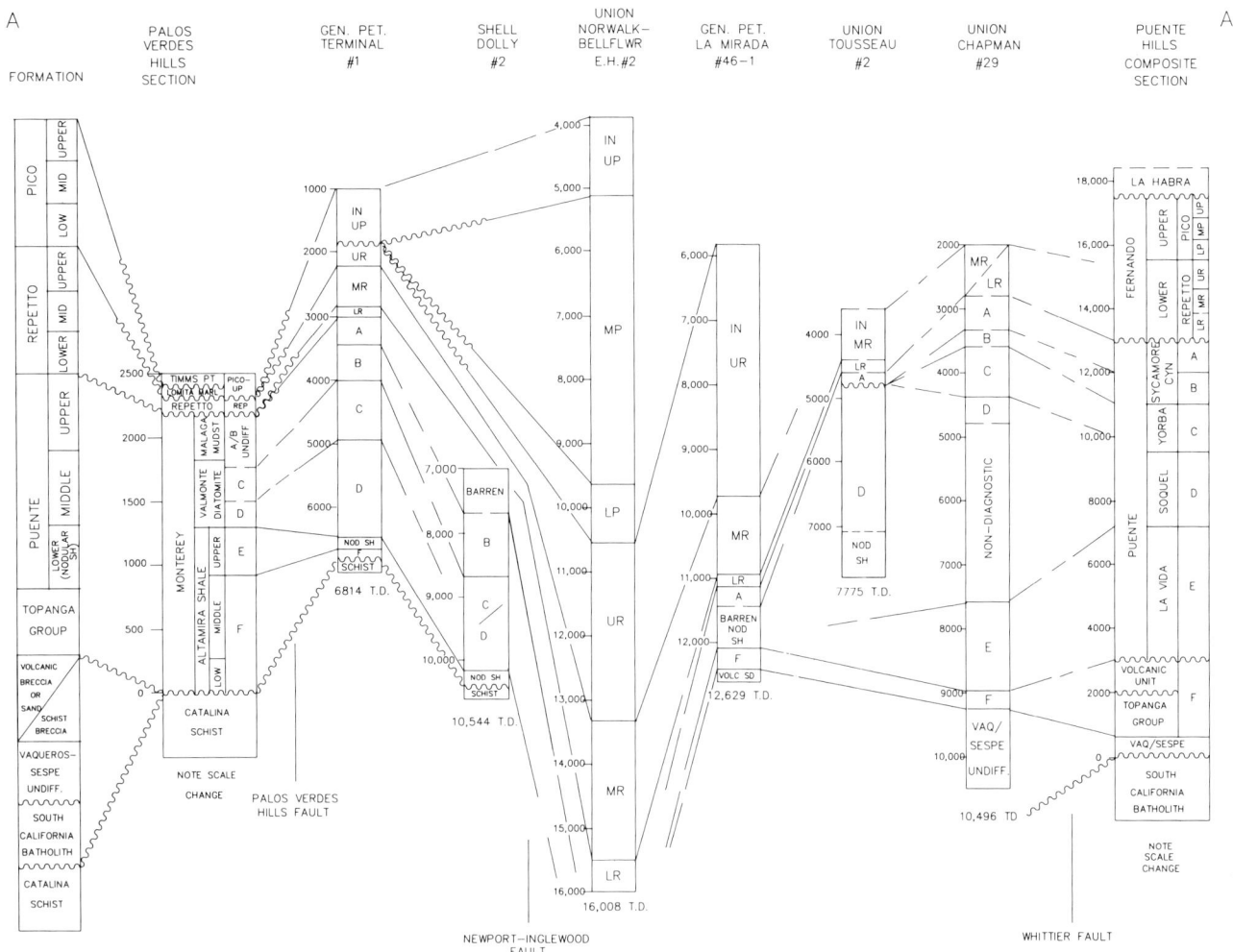

Figure 14. Regional cross section A–A'. Foraminiferal divisions are based on Wissler (1943, 1958). Depths are in feet. Abbreviations for lithologic units and formational subdivisions are as follows: UP—upper Pico; MP—middle Pico; LP—lower Pico; UR—upper Repetto; MR—middle Repetto; LR—lower Repetto; NOD SH—nodular shale unit of the lower Puente Formation; TOPG GRP—Topanga Group; VOLC SD—volcanic sand interval; and VAQ/SESPE—Vaqueros/Sespe undifferentiated. Dashed lines represent inferred boundaries. In intervals that are barren of microfossils, lithologic terms are used (e.g., Nodular shale and Volcanic sand).

There is over 6000 ft (2440 m) of Repetto and Pico section in the Baldwin-Cienega #105. It appears that this well was situated along the western part of the Pliocene–Pleistocene depositional trough, but east of the basin margin (Figure 16).

The Union Westchester E. H. #1 has a sequence similar to the Baldwin-Cienega #105. The Topanga Group (Division F), however, is not present in the Westchester E. H. #1. The Puente Formation is thin and unconformably overlies a schist breccia composed of Catalina Schist clasts. There is a major hiatus between Division A and the nodular shale of Division E, with Divisions B, C, and D missing. The schist breccia is barren of microfossils and is the equivalent to the Pt. Fermin Sandstone unit of the Palos Verdes Hills and possibly equivalent to the San Onofre Member of the upper Topanga Formation of the San Joaquin Hills. The breccia occurs in several oil fields of the southwestern block (e.g., Playa del Rey, El Segundo, and Dominguez).

Conformably overlying the Puente Formation are the Pliocene-Pleistocene Repetto and Pico formations, which are over 6500 ft (1980 m) thick. This thickness implies that this well is still within the western part of the Pliocene–Pleistocene depositional trough.

The Occidental West Athens E. H. #1 is on the central block, east of the Newport-Inglewood fault. This well is in the Pliocene-Pleistocene depositional trough of the central block. Similar to other wells on the central block, over 9000 ft (2743 m) of Repetto and Pico sediments are present. This late Neogene depositional trough formed in the late Miocene, based on a thick interval [over 2000 ft (610 m)] of Divisions A and B in the West Athens E. H. #1; a similar thickness is present in the Rancho #1.

The sequence in the Union Hellman #58, west of the Newport-Inglewood fault, is characteristic of the southwestern block. The Puente Formation (Divisions A through

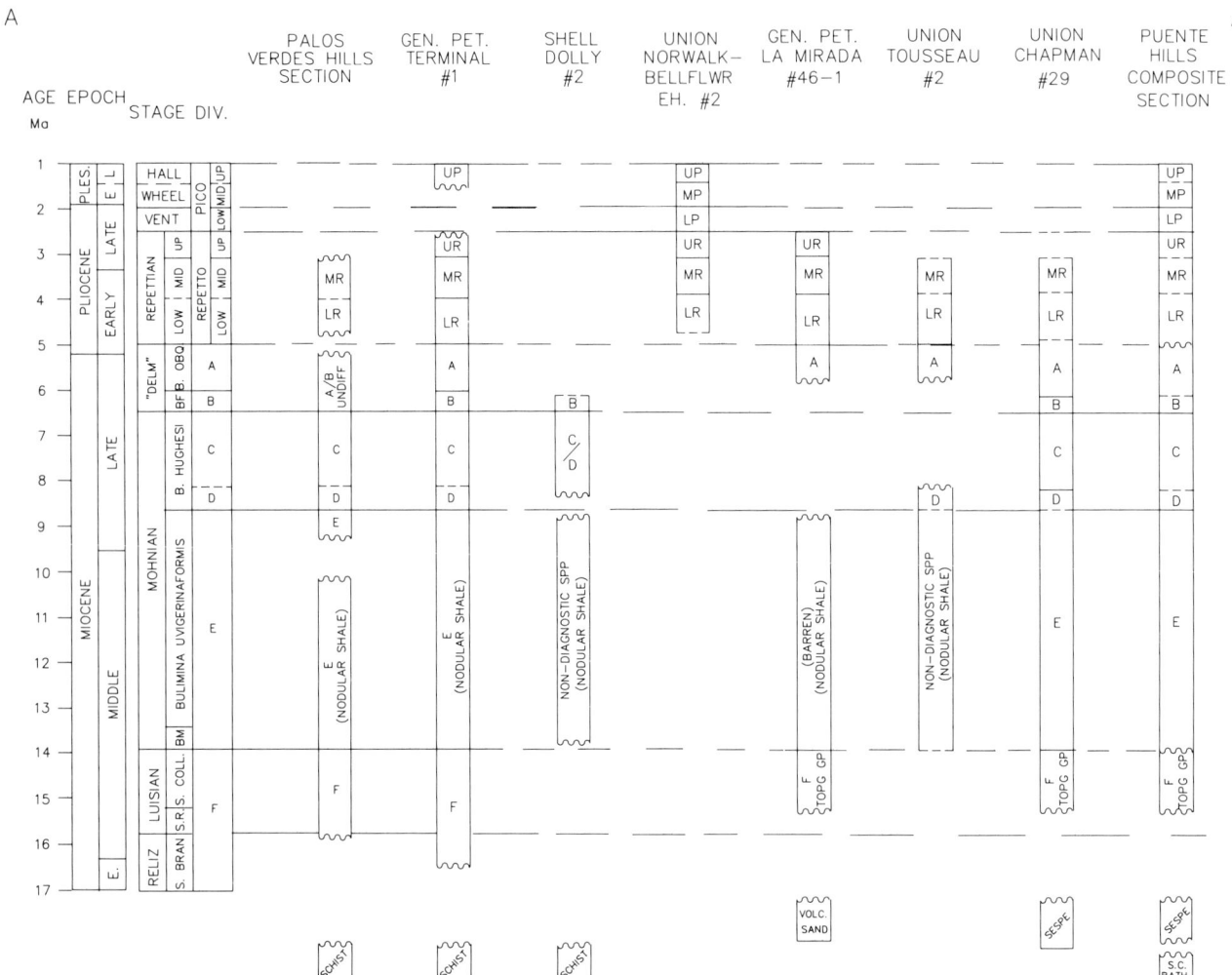

Figure 15. Regional cross section A-A' plotted as a function of time. Benthic foraminiferal stages are based on the Miocene provincial stages of Kleinpell (1938, 1980) and the Pliocene provincial stages of Natland (1952). Foraminiferal divisions are from Wissler (1943, 1958). Abbreviations for lithologic units and formational subdivisions are as follows: UP—upper Pico; MP—middle Pico; LP—lower Pico; UR—upper Repetto; MR—middle Repetto; LR—lower Repetto; TOPG GRP—Topanga Group; VOLC SD—volcanic sand; SESPE—Vaqueros/Sespe undifferentiated; and S.C. BATH.—Southern California Batholith. Dashed lines represent inferred boundaries. The duration of the hiatuses and unconformities are approximated. In intervals that are barren of microfossils, only the lithologic terms are used (e.g., Nodular shale and Volcanic sand).

E) unconformably overlies schist breccia (Figure 16). As in the Westchester E. H. #1, the Topanga Group (Division F) is missing. The Puente Formation was deposited without any apparent hiatuses; however, Division E is barren of microfossils and is represented by the nodular shale. There are over 6500 ft (1980 m) of Repetto and Pico sediments, implying a location on the western part of the Pliocene-Pleistocene depositional trough similar position to the Baldwin-Cienega #105 and the Westchester #1.

The Palos Verdes Hills stratigraphic section and the G. P. Terminal #1 have both been discussed in the previous section. Division F unconformably overlies Catalina Schist,

and approximately 2000 ft (610 m) of the Puente Formation or the Monterey Formation is present. The hiatuses associated with the deposition of the Repetto and Pico have resulted in significant missing section.

In the southwestern block, pre-Topanga Group units are characterized by the occurrence of the Catalina Schist and schist breccia (Figure 16). Basement rocks were not encountered in the subsurface of the central block of this area.

The Topanga Group occurs only in the northern parts of the southwestern and central blocks. The Puente Formation is relatively thick [3000 to 4000 ft (915 to 1220

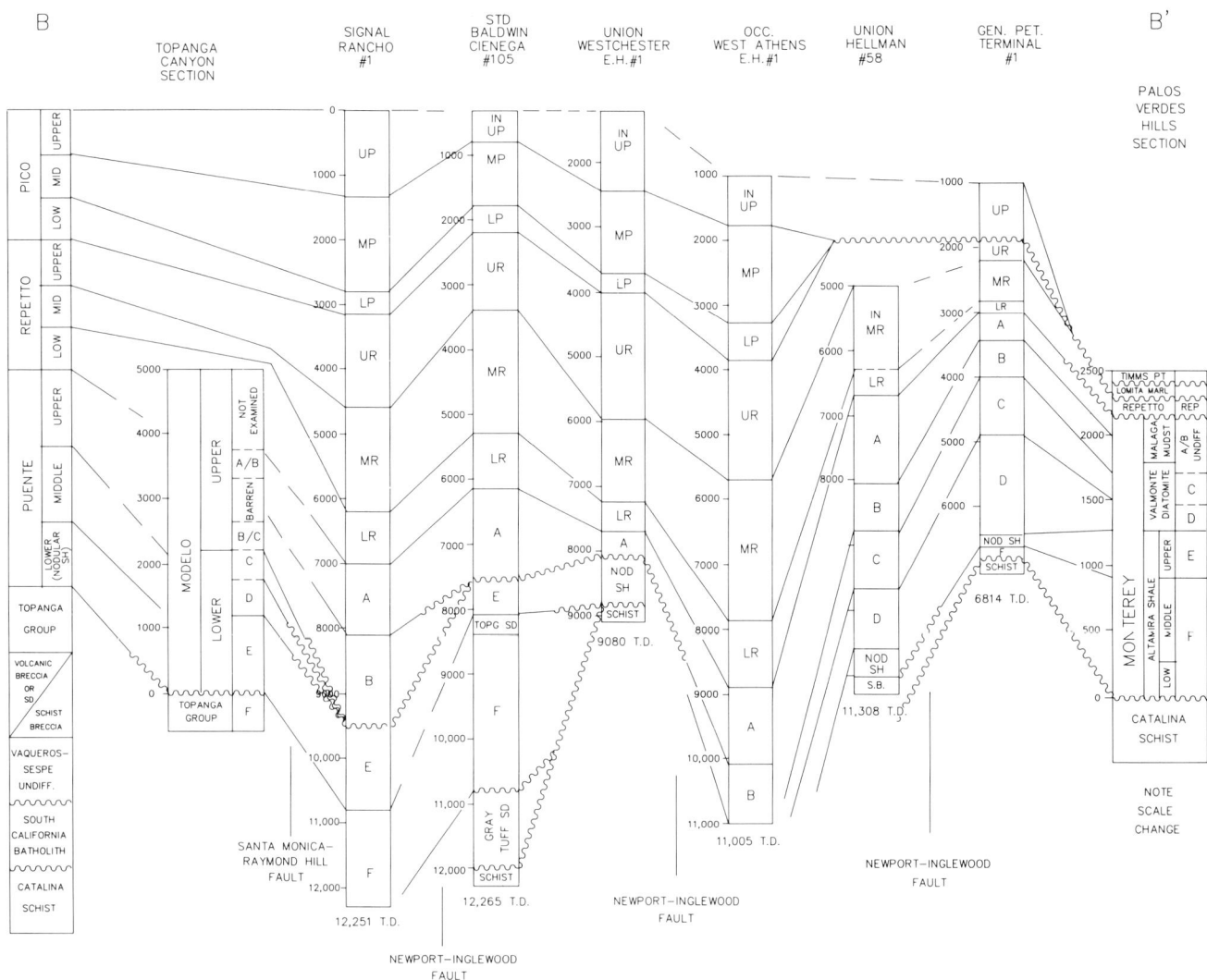

Figure 16. Regional cross section B–B'. Foraminiferal divisions are based on Wissler (1943, 1958). Depths are in feet. Abbreviations for lithologic units and formational subdivisions are as follows: UP—upper Pico; MP—middle Pico; LP—lower Pico; UR—upper Repetto; MR—middle Repetto; LR—lower Repetto; NOD SH—nodular shale unit of the lower Puente Formation; TOPG GRP—Topanga Group; GRAY TUFF SD—volcanic sand interval; and S.B.—schist breccia. Dashed lines represent inferred boundaries. In intervals that are barren of microfossils, lithologic terms are used (e.g., Nodular shale and Topanga sand).

m)] in most of the wells, although portions or all of Divisions B through D are missing in the northern part of the two blocks.

The Repetto and Pico formations have a combined thickness of 6500 to 9000 ft (1980 to 2743 m) for the northern parts of both blocks. There is a considerable increase in thickness east of the Newport-Inglewood Fault (e.g., West Athens #1), indicating that the Pliocene and Pleistocene basin was subsiding faster east of the Newport-Inglewood fault. However, sediments were being deposited in lower bathyal environments on both sides of the fault during the late Neogene.

The thickness of the Repetto and Pico sediments on both the central and southwestern blocks imply that this part of the basin was situated along the western edge of the Pliocene-Pleistocene depositional trough, but east of the basinal margin.

There are five hiatuses or unconformities within the stratigraphic sections and wells along this transect (Figure 17). As in section A–A', the oldest unconformity is associated with the initiation of Neogene deposition over the Catalina Schist. In the southern part of the southwestern block, sediments deposited over the schist basement are Division F (Relizian to Luisian stages). In the northern part of the block, this unconformity is between schist basement and the overlying schist breccia and tuffaceous units. Above these units is an unconformity between the schist breccia and volcanic tuffaceous units and the overlying Topanga

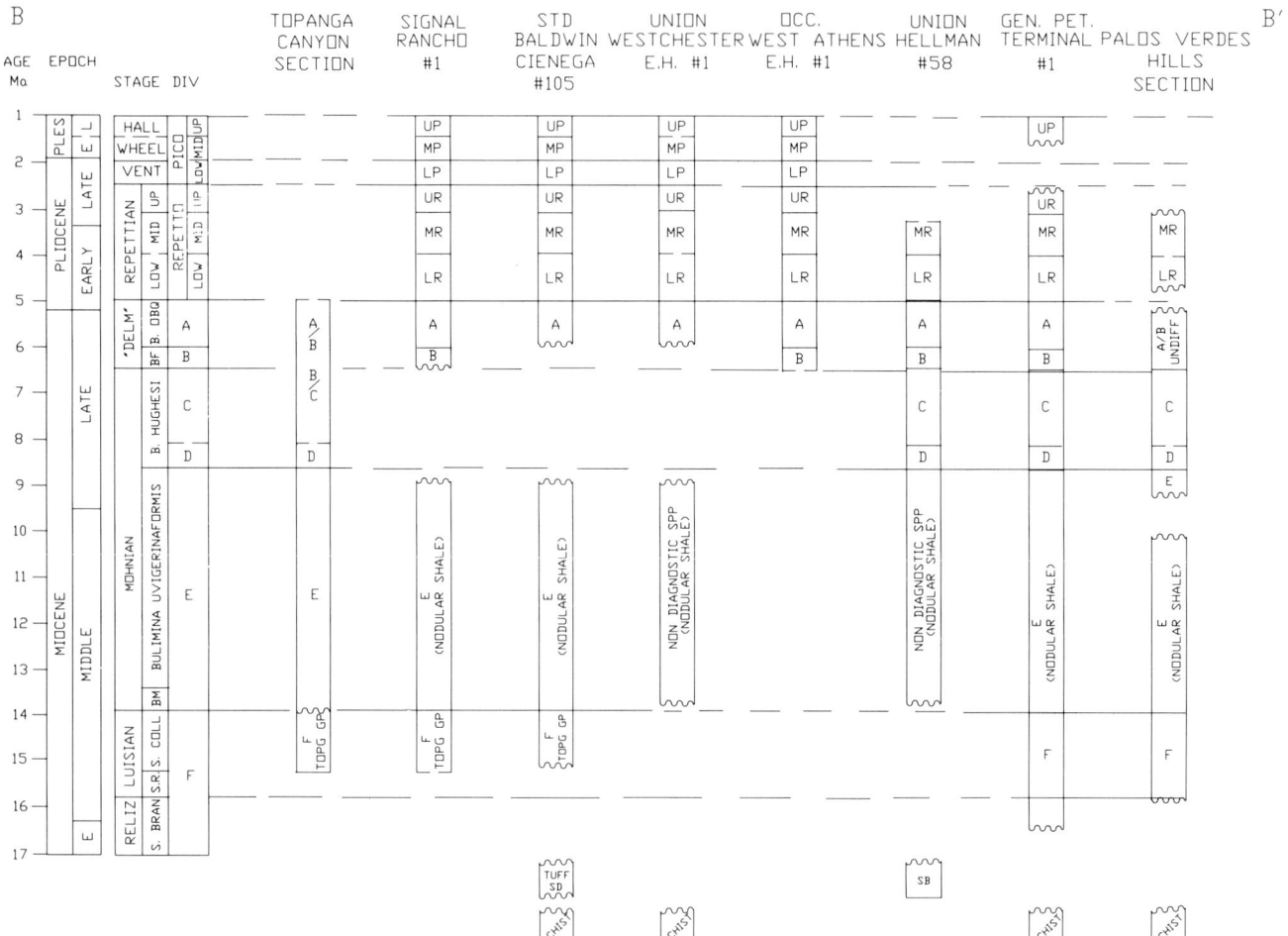

Figure 17. Regional cross section B-B' plotted as a function of time. Benthic foraminiferal stages are based on the Miocene provincial stages of Kleinpell (1938, 1980) and the Pliocene provincial stages of Natland (1952). Foraminiferal divisions are from Wissler (1943, 1958). Abbreviations for lithologic units and formational subdivisions are as follows: UP—upper Pico; MP—middle Pico; LP—lower Pico; UR—upper Repetto; MR—middle Repetto; LR—lower Repetto; TOPG GRP—Topanga Group; and TUFF SD—volcanic sand interval. Dashed lines represent inferred boundaries. The duration of the hiatuses and unconformities are approximated. In intervals that are barren of microfossils, only the lithologic terms are used (e.g., Nodular shale and Topanga Group).

Group and Puente Formation. The age of the unconformity between the volcanic tuff and the Topanga Group also is within the Division F (Luisian Stage). The contact between the schist breccia and the Puente Formation is within the nodular shale (equivalent to Division E). In the Topanga Canyon section, this hiatus is represented by an angular unconformity between the Topanga Group and the Modelo Formation. This unconformity is probably related to local structural growth; however, it is the same age as the unconformity between the schist breccia and the Puente Formation and could be related to a regional tectonic event.

The oldest hiatus in the Palos Verdes Hills section occurs within Division E (lower Mohnian Stage). As previously stated, this hiatus is detectable only by diatom biostratigraphy.

The next youngest hiatus within the cross section occurs in the northern part of both the central and southwestern blocks. The missing section associated with this hiatus ranges from the upper part of Division E through Division B in the central block and to Division A in the southwestern block. In the central block, this missing interval ranges in age from approximately 9 to 6.5 Ma. In the southwestern block, the age of the missing interval is from approximately 9 to 6 Ma (Figure 17). This hiatus may be related to structural growth of the southwestern and northern part of the central block associated with movement of the Santa Monica-Raymond fault and the Newport-Inglewood fault in the late Miocene. The early Pliocene and Pleistocene unconformities in the Terminal #1 and the Palos Verdes Hills section have already been discussed in previous sections.

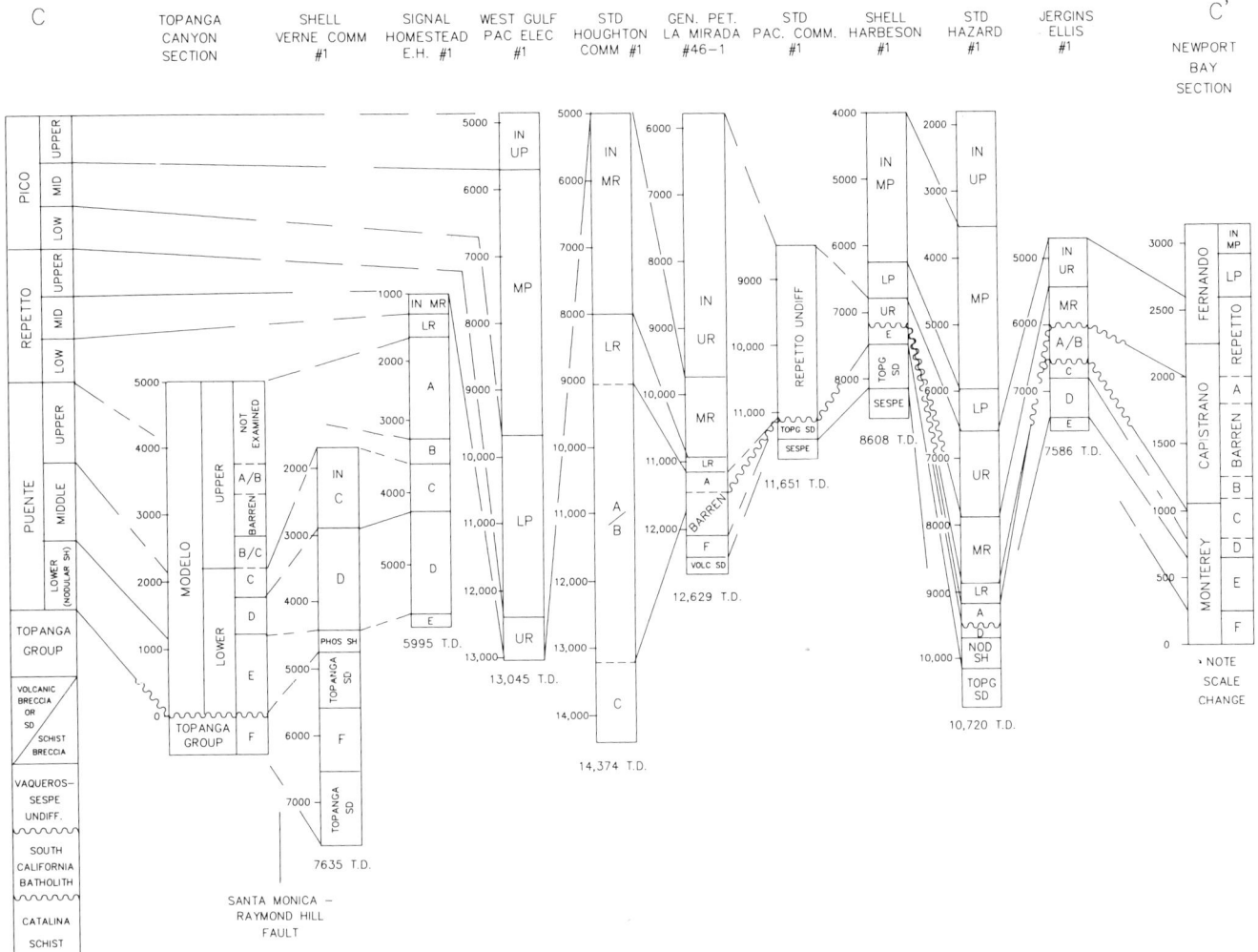

Figure 18. Regional cross section C–C'. Foraminiferal divisions based on Wissler (1943, 1958). Depths are in feet. Abbreviations for lithologic units and formational subdivisions are as follows: UP—upper Pico; MP—middle Pico; LP—lower Pico; UR—upper Repetto; MR—middle Repetto; LR—lower Repetto; PHOS SH and NOD SH—nodular shale unit of the lower Puente Formation; TOPG GRP—Topanga Group; VOLC SD—volcanic sand interval; and SESPE—Vaqueros/Sespe undifferentiated. Dashed lines represent inferred boundaries. In intervals that are barren of microfossils, lithologic terms are used (e.g., Nodular shale and Volcanic sand).

Section C–C'

This northwest-south section traverses the axis of the central block and forms a transect through the main Miocene and Pliocene-Pleistocene depositional troughs of the basin. This section begins at the Topanga Canyon section with the Modelo Formation (Division A to E) unconformably overlying the Topanga Group (Division F) (Figure 18). The transect crosses the Santa Monica-Raymond Hill fault south to the Shell Verne Comm. #1, which contains a thick sequence of Topanga Formation and Division F similar to the Signal Rancho #1 and is evidence for an early to middle Miocene depositional trough for this part of the basin.

In the Verne Comm. #1 and the Signal Homestead E. H. #1, Division E is thin whereas overlying divisions are relatively thick [approximately 3500 ft (1067 m) thick]. Unlike wells to the west, there are no apparent intervals of missing sections or hiatuses within the Puente Formation in these two wells.

Four of the wells have Topanga Group overlying either volcanic breccia or undifferentiated Sespe/Vaqueros (Figure 18). The Topanga Group is barren of age-diagnostic species in all wells except the La Mirada #46-1.

From the La Mirada #46-1 to the Standard Hazard #1, the thin Puente Formation is also essentially barren of microfossils, except for rare occurrences of benthic foraminiferal faunas characteristic of Divisions B, C, and

E (Mohnian Stage). In addition, there are several hiatuses within this interval that account for the missing Puente Formation and the lower and middle Repetto.

From the Western Gulf Pacific Electric #1 to the Standard Hazard #1, the Repetto and Pico formations are extremely thick across the central block. For this part of the central block, the interval for the Repetto and Pico formations ranges in thickness from 10,000 to 13,000 ft (3000 to 4000 m)].

The Jergins Ellis #1 and the Newport Bay surface section contain similar stratigraphic sequences representing the southern margin of the Neogene depositional trough. Divisions A through E are present in the Puente Formation of the Ellis #1. The Monterey Formation of the Newport Bay section contains Divisions A through F. In both sequences, the Division B interval is either barren or contains poorly preserved specimens and has been grouped with Division A in Ellis #1. The Repetto and Pico formations are approximately 6000 ft (1830 m) thick in the Ellis #1. The Fernando Formation (Repetto and Pico equivalent) is only about 1100 ft (335 m) thick in the Newport Bay section. This thinning of the Pliocene–Pleistocene indicates that the Newport Bay section is located along the southern margin of the Pliocene–Pleistocene depositional trough.

Section C–C' shows the depositional trough throughout the central block from its apparent northern margin at the Santa Monica–Raymond Hill fault to its southern margin at the Newport Bay section. The Modelo, Puente, and Monterey formation thicknesses vary over the central block from the thick Puente [over 6000 ft (1830 m)] in the northern part of the block to the thick Monterey in the Newport Bay section [approximately 1100 ft (457 m)]. However, in the southeastern part of the central block, major stratigraphic intervals are missing between Division E and the middle Repetto owing to structural growth.

The Pliocene–Pleistocene depositional trough was actively subsiding throughout the central block. Extremely high rates of sedimentation during the last 4.5 m.y. have resulted in combined Repetto, Pico, and La Habra formational thicknesses that range from over 9000 to 14,000 ft (2743 to 4268 m).

Cross section C–C' contains four major hiatuses or unconformities, some which have been previously discussed for the other sections (Figure 19). The oldest unconformity for this section involves the initiation of Neogene deposition on the Catalina Schist. The initial Neogene deposition is over the Sespe/Vaqueros undifferentiated and a volcanic breccia unit. However, this unconformity occurs in Division F and appears to be consistent with the age from the other sections.

The next major unconformity occurs at the boundary between the Topanga Group and the overlying Modelo and Repetto formations. In the Topanga Canyon section, this hiatus occurs between 14 to 13.6 Ma. However, in the Standard Pacific Community #1, the time gap ranges from the Topanga Group to the lower Repetto with an age of 14 to approximately 4.6 Ma (Figure 19).

There are two hiatuses associated with Division E in the central block. The oldest occurs in the Newport Bay section and represents an interval of missing section detectable only by diatom biostratigraphy. It is equivalent to the hiatus in the Palos Verdes Hills section. The second hiatus occurs in the La Mirada #46-1 and has been previously discussed in section A–A'. This hiatus ranges in age from approximately 9 to 5 Ma.

There is an unconformity in the Shell Harbeson #1 and Standard Hazard #1. The time gap ranges from the upper part of Division E or lowest Division D to either Division A or the upper Repetto (approximately 13 to 3 Ma). This unconformity is probably associated with the other intervals of missing sections in this part of the central block. They are the result of structural growth of the Anaheim nose.

Discussion

As demonstrated by these three cross sections, Neogene deposition varied between and within the four discrete tectonic blocks of the Los Angeles basin. In addition, the depositional history of individual stratigraphic units varied across the basin, likely representing a response to variations in the tectonic evolution of the individual blocks.

Each tectonic block contains a different pre-Neogene "basement" and an overlying sedimentary section. In the southwestern block, west of the Newport–Inglewood fault, the Catalina Schist is the basement. In the central and the northeast blocks, the Sespe/Vaqueros formations are the deepest units penetrated in the subsurface, so the basement rocks are unknown. It is assumed that the basement rocks are probably granitic, related to the southern California batholith.

The Topanga Group is thickest in the Santa Monica Mountains of the northwestern block and in the northern part of the central and southwestern blocks. It thins to the south in all blocks. In the Palos Verdes Hills and Newport Bay, the Topanga Group is represented by sediments characteristic of the biogenic-rich Monterey Formation. In the central part of the southwestern block, the Topanga Formation is missing and has been replaced with a schist breccia, possibly representative of the Pt. Fermin Sandstone unit and the San Onofre Breccia.

The thickest sections of the Puente/Monterey/Modelo formations are along the basin margins and the northern part of the central block. These formations display a depositional pattern similar to the Topanga Group, except in the northern part of the southwestern block where Divisions B, C, and D of the Puente Formation are either very thin or missing because of major hiatuses. This may indicate that the Newport–Inglewood fault had started to affect deposition in this area during post–Topanga Group deposition.

The Puente Formation is not encountered in the wells of the central block owing to the extremely thick interval of Pliocene–Pleistocene sediments. In the southeast part of the central block, major intervals of the stratigraphic section are missing between the Topanga Group and the Repetto Formation along different unconformable surfaces. These missing intervals are possibly related to the structural growth of the Anaheim nose feature (Wright, 1987).

The regional depositional pattern for the Puente and Monterey formations includes thickening toward the basin margins and away from the Pliocene–Pleistocene depositional trough. This implies that the Miocene formations and

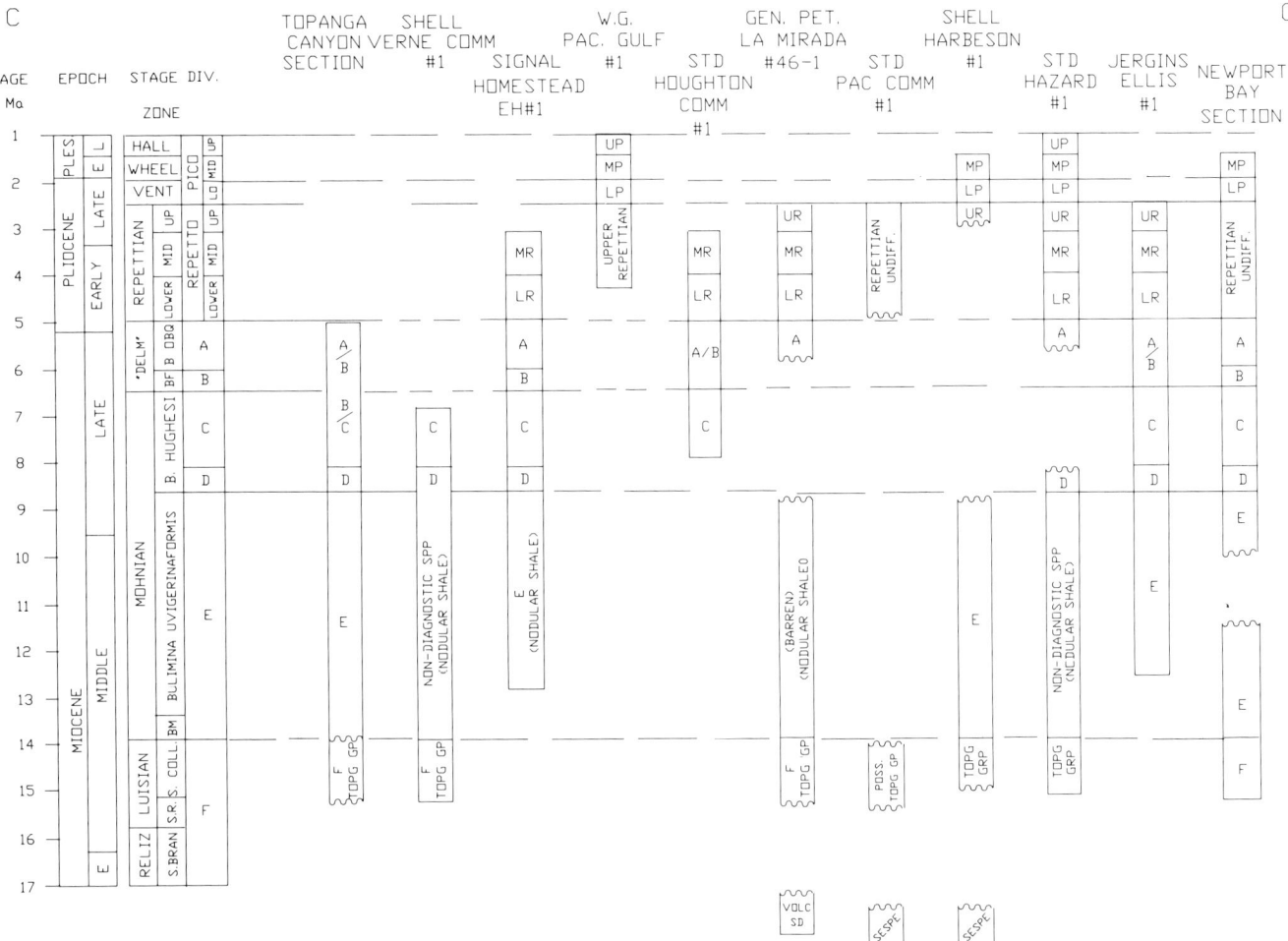

Figure 19. Regional cross section C–C' plotted as a function of time. Benthic foraminiferal stages are based on the Miocene provincial stages of Kleinpell (1938, 1980) and the Pliocene provincial stages of Natland (1952). Foraminiferal divisions are from Wissler (1943, 1958). Abbreviations for lithologic units and formational subdivisions are as follows: UP—upper Pico; MP—middle Pico; LP—lower Pico; UR—upper Repetto; MR—middle Repetto; LR—lower Repetto; TOPG GRP—Topanga Group; VOLC SD—volcanic sand interval; and SESPE—Vaqueros/Sespe undifferentiated. Dashed lines represent inferred boundaries. The duration of the hiatuses and unconformities are approximated. In intervals that are barren of microfossils, only the lithologic terms are used (e.g., Nodular shale and Topanga Group).

Pliocene-Pleistocene formations have different depositional histories. The change in depositional pattern occurred during the late Miocene to early Pliocene (ca. 6 to 5 Ma) (see Yeats and Beall, this volume).

The dominant depositional trend displayed by the Pliocene-Pleistocene Repetto and Pico formations is opposite of the Miocene trends. In contrast to the middle Miocene, the Repetto and Pico thicken dramatically into the present center of the Los Angeles basin, especially east of the Newport-Inglewood fault. The northern margin of the depositional trough is the Santa Monica-Raymond Hill fault and the southern margin is near the area around the San Joaquin Hills. The eastern margin is at the Whittier fault zone. The Repetto and Pico units thin from the center of the trough to the east in the Tousseau #2 and Chapman #29, both located just west of the Whittier fault. The depositional trough appears to be structurally controlled by the Newport-Inglewood, Santa Monica-Raymond Hill, and Whittier faults. The depositional history of the Pliocene-Pleistocene units is related to the structural evolution of these faults.

There are several widespread and important unconformities throughout the Los Angeles basin and several of only local extent. The regional unconformities are the likely expressions of regional tectonic and depositional events affecting the continental margin and the basin.

The initial marine Neogene deposition within the central portion of the basin occurred in Division F (upper Relizian to Luisian stages) (approximately 16 to 14 Ma). This age is younger than the surrounding basinal margins and implies that subsidence for the central part of the basin did not occur until early middle Miocene. The strata deposited at the basin margins before this subsidence event represent predominantly shallow-marine to nonmarine environments (e.g., undifferentiated Sespe/Vaqueros formations).

A schist breccia occurs between the Catalina Schist and the Puente Formation in the southwestern block. Where this breccia occurs, the Topanga Group is usually absent. The unconformity between the schist breccia and the Puente Formation represents an interval of time from approximately 16 to 10 Ma. Therefore, the breccia is the likely equivalent of the San Onofre Breccia of the Topanga Group. The San Onofre Breccia ranges in age from 16.8 to 13.8 Ma. The Pt. Fermin Sandstone unit within the Altamira Shale Member of the Monterey at Palos Verdes Hills is composed of schist debris and ranges in age from 14.0 to 10.8 Ma (Conrad and Ehlig, 1983). This breccia represents the uplift and erosion of the Catalina Schist on the western and southern margins of the basin.

As previously discussed, in surrounding basin-margin areas, an unconformity occurs between the Topanga Formation and the overlying Modelo and Puente formations at the boundary of Divisions F and E. This unconformity is probably related to either a tectonic event and/or a eustatic event. There is a major, eustatic sea-level drop between 14 and 13 Ma (Haq et al., 1987).

There is a hiatus within Division E of the Puente and Monterey formations around the basin margin. This hiatus is evidenced by diatom biostratigraphy and is equivalent to a regional late Miocene hiatus reported by Barron (1986b) and Arends and Blake (1986). It occurs at approximately 10 Ma, during a major drop in eustatic sea-level in response to a period of global glaciation (Woodruff, 1985). This hiatus could also be the result of deep-sea erosion through an increase of deep bottom-water current velocity (Barron and Keller, 1983). In the southeastern part of the central block, around the Coyote Hills, several wells have cut an unconformity between Division E and either Division A or the Repetto Formation. This unconformity is the probable result of structural growth of the Anaheim nose feature. This structural nose is a northern subsurface extension of the Santa Ana Mountains. In addition, this unconformity may be related to the onset of Whittier fault movement dated at approximately 6.6 Ma (Schwartz and Colburn, 1987).

In the northern part of the central and southwestern blocks, there is a regional hiatus between the upper part of Division E and either Division B or Division A. This hiatus ranges in age from approximately 10 to 6 or 5 Ma, depending on the location. This widespread latest Miocene to early Pliocene event coincides with a major drop in eustatic sea level (Haq et al., 1987). However, this event may also be related to a tectonic event. Doyle and Gorsline (1977) state that much of the relief and bathymetry of the present southern California continental margin dates from the latest Miocene. A change in the Pacific plate motion at about 6 Ma is reported by Jackson et al. (1975). During the latest Miocene to early Pliocene, the central depositional trough underwent major subsidence as evidenced by the thick sequence of the Repetto Formation and the occurrence of the lower bathyal Repettian assemblage. The subsidence event and the unconformity or hiatus could be the product of the change in the Pacific plate motion and the opening of the Gulf of California (Ingle, 1980).

In addition, there are several unconformities in the latest Miocene to early Pleistocene of the Palos Verdes and Puente Hills that are the result of structural growth of these two areas. These unconformities usually affect the Repetto and Pico formations.

CONCLUSIONS

The Los Angeles basin is composed of four structural blocks that have individual depositional histories in response to regional tectonic and depositional events. By comparing the biostratigraphic relationships across the basin, it is possible to establish and document many of the regional and local events that have controlled deposition within the basin.

Benthic foraminiferal assemblages from three stratigraphic sections located around the basin margins were correlated to siliceous and calcareous planktonic microfossil zonations and radiometric dates to determine the age relationships of the benthic foraminiferal zonations and divisions. The benthic foraminiferal biostratigraphies could be calibrated using the various planktonic microfossil groups. Application of this biostratigraphic-chronostratigraphic scheme was used to constrain the timing and magnitude of tectonic and depositional events marking basin development.

The deposition of the Cretaceous and Paleogene units occurred around the margins of the basin (i.e., in the Santa Monica Mountains, the Puente Hills, the Santa Ana Mountains, and San Joaquin Hills). These units have not been penetrated in the subsurface of the central portion of the basin. The oldest sediments found there have been assigned to the Sespe Formation or the undifferentiated Sespe/Vaqueros formations. This would imply that the majority of the present basin, excluding the central trough where the basement has not been penetrated, was a positive feature until the late Paleogene to early Neogene.

Marine Neogene deposition into the Los Angeles basin began in Division F (upper Relizian to Luisian stages). This deposition is represented by the shallow-water Topanga Group. Based on the occurrence of upper Relizian Stage (ca. 16 to 15 Ma) benthic foraminiferal assemblages, the oldest Neogene marine deposition south of the Santa Monica–Raymond fault was in the Palos Verdes Hills area, the San Joaquin Hills area, and the Puente Hills. Within these areas there was rapid basinal subsidence to approximately 3000 ft (1000 m) water depth. In the central portion of the basin, east of the Palos Verdes fault and west of the Puente Hills, marine deposition and rapid basinal subsidence did not occur until the upper Division F or the Luisian Stage (15 to 13 Ma).

In the southwestern block, initial deposition is represented by a schist breccia composed of eroded Catalina Schist debris. It is not known whether or not this unit is marine because it is barren of marine fossils.

The major depocenters for Division F sediments were in the Santa Monica Mountains, the northern portions of the southwestern and central blocks, the Palos Verdes Hills, and the southwestern part of the central block (San Joaquin Hills). In other parts of the basin, Division F is relatively thin or missing owing to erosion.

The middle and upper Miocene sediments of Division E through A are represented by the Monterey, Modelo, and Puente formations. In the western part of the basin, the southwestern block, the southwest part of the central block (i.e., Newport Bay area), and parts of the northwestern block, the depositional units are characterized by biogenic-rich, siliceous, and phosphatic sediments of the Monterey and Modelo formations. In contrast, other portions of the Los Angeles basin are dominated by the clastic units of the Puente Formation.

In the biogenic units of the western basinal margin, there is a regional hiatus in Division E (lower Mohnian Stage) (approximately 10 Ma). This hiatus is probably related to climatic and oceanic interactions and the associated drop in eustatic sea level caused by the early late Miocene glaciation event.

The major depocenters for Division E to A sediments were located in the Santa Monica Mountains, the northern part of central block, the Palos Verdes Hills area, and, most importantly, the Puente Hills area where there is over 10,000 ft (3050 m) of Puente Formation. Within the central trough of the basin, the area east of the Newport-Inglewood fault and west of the Whittier fault, these sediments are usually relatively thin or have not been totally penetrated.

In the late Miocene to early Pliocene (ca. 6 to 5 Ma), there is major subsidence associated with the initiation of structural growth along the Los Angeles basin margins. The subsidence is represented by the thick sequences of the Repetto Formation in the central block and along the eastern margin of the southwestern block. At approximately the same time, there is uplift and erosion in the Santa Monica Mountains, the Palos Verdes Hills area, and the Anaheim nose area. In addition, there is evidence of onset of movement along the Newport-Inglewood fault and the Whittier fault.

This movement along the Newport-Inglewood fault occurs in the northern part of the southwestern and central blocks as evidenced by erosion of Divisions A to E in this area. The movement along the Newport-Inglewood fault forms the relatively sharp western boundary of the late Miocene to Pleistocene depositional trough.

Movement along the Whittier fault controlled deposition along the eastern margin of the central block. The Repetto and Pico units thicken dramatically across the fault from the northeastern block to the central block.

The movement of the boundary faults and subsequent uplift of the basin margins resulted in a shift of the major depocenters from the basin margins to the central depositional trough. The location of the latter depocenter existed until the basin was filled in the late Pleistocene.

ACKNOWLEDGMENTS

I would like to thank Unocal for permission to publish this review. I would like to express my thanks to R. G. Arends for providing the siliceous microfossil age determinations and to M. V. Filewicz for providing the calcareous nannofossil age determinations. I would also like to thank C. K. Davis and W. L. Schneider for their excellent drafting assistance. I am grateful to E. C. Thomas, R. G. Arends, and J. P. Chauvel for their suggestions on improving the manuscript. The manuscript benefitted from the reviews of K. T. Biddle, J. C. Ingle, Jr., A. D. Warren, and R. Wright.

REFERENCES CITED

Almgren, A. A., 1982, Foraminiferal paleoenvironmental interpretation of the Late Cretaceous Holz Shale, Santa Ana Mountains, California, in D. J. Bottjer, I. P. Colburn, and J. D. Cooper, eds., Late Cretaceous depositional environments and paleogeography, Santa Ana Mountains, southern California: Pacific Section, SEPM, p. 45-58.

Arends, R. G., and G. H. Blake, 1986, Biostratigraphy and paleoecology of the Naples Bluff coastal section based on diatoms and benthic Foraminifera, in R. E. Casey, and J. A. Barron, eds., Siliceous microfossil and microplankton of the Monterey Formation and modern analogs: Pacific Section, SEPM, p. 121-136.

Atwater, T., 1970, Implications of plate tectonics for the Cenozoic tectonic evolution of western North America: GSA Bulletin, v. 81, p. 3513-3536.

Baldauf, J., and J. A. Barron, 1982, Diatom stratigraphy and paleoecology of the type section of the Luisian stage, central California: Micropaleontology, v. 28, p. 58-84.

Barron, J. A., 1975, Marine diatom biostratigraphy of the upper Miocene-lower Pliocene strata of southern California: Journal of Paleontology, v. 49, p. 619-632.

Barron, J. A., 1976, Revised Miocene and Pliocene diatom biostratigraphy of upper Newport Bay, Newport Beach, California: Marine Micropaleontology, v. 1, p. 27-63.

Barron, J. A., 1981, Late Cenozoic diatom biostratigraphy and paleoceanography of the middle-latitude eastern North Pacific: Initial Reports of the Deep Sea Drilling Project, Leg 63, p. 507-538.

Barron, J. A., 1986a, Paleoceanographic and tectonic controls on deposition of the Monterey Formation and related siliceous rocks in California: Paleoceanography, Paleoclimatology, Paleoecology, v. 53, p. 27-45.

Barron, J. A., 1986b, Updated diatom biostratigraphy for the Monterey Formation of California, in R. E. Casey, and J. A. Barron, eds., Siliceous microfossil and microplankton of the Monterey Formation and modern analogs: Pacific Section, SEPM, p. 105-120.

Barron, J. A., and G. Keller, 1983, Paleotemperature oscillations in the middle and late Miocene of the northeastern Pacific: Micropaleontology, v. 29, p. 150-181.

Barron, J. A., G. Keller, and D. A. Dunn, 1985, A multiple microfossil biochronology for the Miocene, in J. P. Kennett, ed., The Miocene Ocean: paleoceanography and biogeography: GSA Memoir 163, p. 131-177.

Berggren, W. A., D. V. Kent, J. J. Flynn, and J. A. Van Couvering, 1985, Cenozoic geochronology: GSA Bulletin, v. 96, p. 1407-1418.

Billman, H. G., and A. A. Hopkins, Jr., 1980, The stratigraphic distribution of the Foraminifera at the type locality of the Luisian Stage, in G. H. Blake, ed., Neogene biostratigraphy of the northern La Panza Range, San Luis Obispo County, California, Guidebook: Pacific Section, SEPM, p. 1-9.

Blake, G. H., 1985, The faunal response of California continental margin benthic Foraminifera to the oceanographic and depositional events of the Neogene [unpublished Ph.D. dissertation]: University of Southern California, 316 p.

Blake, G. H., 1987, Biostratigraphic correlations in California marginal basins, in D. S. Gorsline, ed., Depositional systems in active margin basins: Pacific Section, SEPM, p. 53-80.

Blow, W. H., 1969, Late middle Eocene to Recent planktonic foraminiferal biostratigraphy: Proceedings of the First International conference on planktonic microfossils, Geneva 1967, Leiden, E. J. Brill, v. 1, p. 199-422.

Bottjer, D. J., S. P. Buck, and E. J. Enzweiler, 1982, Trace fossils from Upper Cretaceous strata in Silverado Canyon, Santa Ana Mountains, California, in D. J. Bottjer, I. P. Colburn, and J. D. Cooper, eds., Late Cretaceous depositional environments and paleogeography, Santa Ana Mountains, Southern California: Pacific Section, SEPM, p. 81-88.

Bukry, D., 1973, Coccolith and silicoflagellate stratigraphy: Initial Reports of the Deep Sea Drilling Project, Leg 18: p. 817-831.

Bukry, D., 1981, Pacific Coast coccolith stratigraphy between Point Conception and Cabo Corrientes: Initial Reports of the Deep Sea Drilling Project, Leg 63, p. 445-472.

Carlos, A., 1985, Comparative study of paleoenvironmental factors on the preservation of calcareous microfossils in the Miocene Monterey Formation, upper Newport Bay, Newport Beach, California [unpublished Master's thesis]: University of Southern California, 164 p.

Casey, R. E., 1972, Neogene radiolarian biostratigraphy and paleotemperatures: southern California, the experimental Mohole, and Atlantic core 14-8: Paleogeography, Paleoclimatology, Paleoecology, v. 12, p. 115-130.

Casey, R. E., A. B. Price, and C. A. Swift, 1972, Radiolarian definition and paleoecology of the late Miocene to early Pliocene in southern California, in E. H. Stinemeyer, ed., The proceeding of the Pacific Coast Miocene biostratigraphic symposium: Pacific Section, SEPM, p. 226-238.

Casey, R. E., and A. R. Price, 1973, A tentative radiolarian zonation and paleoceanographic interpretation from Newport Bay, California, in Miocene sedimentary environments and biofacies, southeastern Los Angeles basin, SEPM Field Trip #1 Guidebook: SEPM, p. 67-70.

Conrad, C. L., and P. L. Ehlig, 1983, The Monterey Formation of the Palos Verdes Peninsula, California—an example of sedimentation in a tectonically active basin within the California continental borderland, in D. K. Larue, and R. J. Steel, eds., Cenozoic marine sedimentation, Pacific margin: Pacific Section, SEPM, p. 103-116.

Conrad, C. L., and P. L. Ehling, 1987, The Monterey Formation of the Palos Verdes Peninsula, California— an example of sedimentation in a tectonically active basin within the California continental borderland, in P. J. Fischer, ed., Geology of the Palos Verdes Peninsula and San Pedro Bay: Pacific Section, SEPM, p. 17-30.

Conrey, B. L., 1967, Early Pliocene sedimentary history of the Los Angeles basin, California: California Division of Mines Special Report 93: California Division of Mines, 63 p.

Cooper, J. D., I. P. Colburn, and F. A. Sundberg, 1982, Upper Cretaceous environmental stratigraphy and field trip stops, Silverado Canyon area, in D. J. Bottjer, I. P. Colburn, and J. D. Cooper, eds., Late Cretaceous depositional environments and paleogeography, Santa Ana Mountains, southern California: Pacific Section, SEPM, p. 11-24.

Crouch, J. K., 1979, Neogene tectonic evolution of the California continental borderland and western Transverse Ranges: GSA Bulletin, v. 90, p. 338-345.

Daviess, S. N., and A. O. Woodford, 1949, Geology of the northwestern Puente Hills, Los Angeles County, California: USGS Oil and Gas Inv. Prelim. Map 83.

Doyle, L. J., and D. S. Gorsline, 1977, Marine geology of Baja California continental borderland, Mexico: AAPG Bulletin, v. 61, p. 903-917.

Durham, D. L., and R. F. Yerkes, 1964, Geology and oil resources of the eastern Puente Hills area, southern California: USGS Professional Paper 420-B, 62 p.

Eldridge, G. H., and R. Arnold, 1907, The Santa Clara Valley, Puente Hills, and Los Angeles oil districts, southern California: USGS Bulletin, v. 309, 266 p.

English, W. A., 1926, Geology and oil resources of the Puente Hills region, southern California: USGS Bulletin, v. 768, 110 p.

Ford, D. W., 1972, A re-examination of the Mohnian type section, in E. H. Stinemeyer, ed., The proceeding of Pacific Coast Miocene biostratigraphic symposium: Pacific Section, SEPM, p. 192-198.

Haq, B. U., J. Hardenbol, and P. R. Vail, 1987, Chronology of fluctuating sea levels since the Triassic: Science, v. 235, p. 1156-1166.

Hoots, H. W., 1931, Geology of the eastern part of the Santa Monica Mountains, Los Angeles County, California: USGS Professional Paper 165-C: p. 83-134.

Hoskins, E. G., and J. R. Griffiths, 1971, Hydrocarbon potential of northern and central California offshore, in I. H. Cram, ed., Future petroleum provinces of the United States—their geology and potential: AAPG Memoir 15, p. 212-228.

Howard, J. L., 1987, Paleoenvironments, provenance and tectonic implications of the Sespe Formation, southern California [unpublished Ph.D. dissertation]: University of California at Santa Barbara, 306 p.

Imlay, R. W., 1963, Jurassic fossils from southern California: Journal of Paleontology, v. 37, p. 97-107.

Ingle, J. C., Jr., 1967, Foraminiferal biofacies variation and the Miocene-Pliocene boundary in southern California: Bulletin of American Paleontology, v. 52, p. 210-394.

Ingle, J. C., Jr., 1971, Paleoecologic and paleobathymetric history of the late Miocene-Pliocene Capistrano Formation, Dana Point area, California, in Fall Field Trip Guidebook: Pacific Section, SEPM, p. 71-88.

Ingle, J. C., Jr., 1972, Biostratigraphy and paleoecology of early Miocene through early Pleistocene benthonic and planktonic Foraminifera, San Joaquin Hills-Newport Bay, Orange County, California, in E. Steinmeyer, ed., Proceedings of the Pacific Coast Miocene biostratigraphic symposium: Pacific Section, SEPM, p. 255-283.

Ingle, J. C., Jr., 1973, Summary comments on Neogene biostratigraphy, physical stratigraphy and paleoceanography in the marginal northeastern Pacific Ocean, in

Initial Reports of the Deep Sea Drilling Project, Leg 18: U.S. Government Printing Office, v. 18, p. 949-960.

Ingle, J. C., Jr., 1979, Biostratigraphy and paleoecology of early Miocene through early Pliocene benthonic and planktonic Foraminifera, San Joaquin Hills-Newport Bay-Dana Point area, Orange County, California, in C. J. Stuart, ed., A guidebook to Miocene lithofacies and depositional environments, coastal southern California and northwestern Baja California: Pacific Section, SEPM, p. 53-77.

Ingle, J. C., Jr., 1980, Cenozoic paleobathymetry and depositional history of selected sequences within the southern California continental borderland: Cushman Foundation Foraminiferal Research Publication 18, p. 163-195.

Ingle, J. C., Jr., and J. A. Barron, 1978, Neogene biostratigraphy and paleoenvironments of the San Joaquin Hills and Newport Bay areas, California, in W. O. Addicott, ed., USGS Open-File Report 78-446, p. 3-28.

Jackson, E. D., H. R. Shaw, and K. Bargar, 1975, Calculated geochronology and stress field orientations along the Hawaiian Chain: Earth Planetary Science Letters, v. 26, p. 145-155.

Keller, G., and J. A. Barron, 1982, Paleoceanographic implications of Miocene deep-sea hiatuses: GSA Bulletin, v. 94, p. 590-613.

Kew, W. S. W., 1924, Geology and oil resources of a part of Los Angeles and Ventura counties: USGS Bulletin 753, 202 p.

Kleinpell, R. M., 1938, Miocene stratigraphy of California: AAPG, 450 p.

Kleinpell, R. M., 1980, Miocene stratigraphy of California revisited: AAPG Studies in Geology 11, 182 p.

Lagoe, M. B., and P. R. Thompson, 1988, Chronostratigraphic significance of late Cenozoic planktonic foraminifera from the Ventura basin, California: potential for improving tectonic and depositional interpretation: Journal of Foraminiferal Research, v. 18, p. 250-266.

Lipps, J. H., 1964, Miocene planktonic Foraminifera from Newport Bay, California: Tulane Studies in Geology, v. 2, p. 109-133.

Lipps, J. H., 1968, Mid-Cenozoic calcareous nannoplankton from western North America: Nature, v. 218, p. 1151-1152.

Natland, M. L., 1952, Pleistocene and Pliocene stratigraphy of southern California [unpublished Ph.D. dissertation]: University of California at Los Angeles, 350 p.

Natland, M. L., and W. T. Rothwell, 1954, Fossil Foraminifera of the Los Angeles and Ventura regions, California, in R. H. Jahns, ed., Geology of southern California: California Division of Mines Bulletin, n. 170, p. 33-42.

Obradovich, J. D., and C. W. Naeser, 1981, Geochronology bearing of the age of the Monterey Formation and siliceous rocks of California, in R. E. Garrison, and R. G. Douglas, eds., The Monterey Formation and related siliceous rocks of California: Pacific Section, SEPM, p. 87-96.

Obradovich, J. D., C. W. Naeser, and G. A. Izett, 1978, Geochronology of late Neogene marine strata in California—correlation of tropical through high latitude marine Neogene deposits of the Pacific basin [abstract]: Intl. Geol. Correlation Prog., p. 40-41.

Okada, H., and D. Bukry, 1980, Supplementary modification and introduction of code numbers to the "Low-latitude coccolith biostratigraphic zonation" (Bukry, 1973, 1975): Marine Micropaleontology, v. 5, p. 321-325.

Pierce, R. L., 1956, Upper Miocene Foraminifera and fish from the Los Angeles area, California: Journal of Paleontology, v. 30, p. 1288-1314.

Pierce, R. L., 1972, Re-evaluation of the late Miocene biostratigraphy of California—summary of evidence, in E. H. Stinemeyer, ed., The proceedings of the Pacific Coast Miocene biostratigraphic symposium: Pacific Section, SEPM, p. 334-340.

Poore, R. Z., K. McDougall, J. A. Barron, E. E. Brabb, and S. A. Kling, 1981, Microfossil biostratigraphy and biochronology of the type Relizian and Luisian stages of California, in R. E. Garrison, and R. G. Douglas, eds., The Monterey Formation and related siliceous rocks of California: Pacific Section, SEPM, p. 15-41.

Redin, T., 1991, Oil and gas production from submarine fans of the Los Angeles basin, in K. T. Biddle, ed., Active margin basins: AAPG Memoir 52, p. 239-259.

Reed, R. D., 1932, Section from the Repetto Hills to the Long Beach oil field, in H. R. Gale, ed., Guidebook 15, Excursion C-1, southern California: Sixteenth Intern. Geol. Congress, p. 1-31.

Riedel, W. R., and A. Sanfilippo, 1978, Stratigraphy and evolution of tropical Cenozoic radiolarians: Micropaleontology, v. 24, p. 61-96.

Rowell, H. C., 1981, Diatom biostratigraphy of the Monterey Formation, Palos Verdes Hills, California, in R. E. Garrison, and R. G. Douglas, eds., The Monterey Formation and related siliceous rocks of California: Pacific Section, SEPM, p. 55-70.

Schoellhamer, J. E., D. M. Kinney, R. F. Yerkes, and J. G. Vedder, 1954, Geologic map of the northern Santa Ana Mountains, Orange and Riverside counties, California: USGS Oil and Gas Inv. Map OM-154.

Schoellhamer, J. E., J. G. Vedder, R. F. Yerkes, and D. M. Kinney, 1981, Geology of the northern Santa Ana Mountains, California: USGS Professional Paper 420-D, p. D1-D109.

Schwartz, D. E., and I. P. Colburn, 1987, Late Tertiary to recent chronology of the Los Angeles basin, southern California, in P. J. Fischer, ed., Geology of the Palos Verdes Peninsula and San Pedro Bay: Pacific Section, SEPM, p. 5-16.

Sloan, J. R., 1987, Age and paleoceanography of the Point Fermin fan complex, in P. J. Fischer, ed., Geology of the Palos Verdes Peninsula and San Pedro Bay: Pacific Section, SEPM, p. 47-51.

Stuart, C. J., 1979, Middle Miocene paleogeography of coastal southern California and the California borderland—evidence from schist-bearing sedimentary rocks, in J. M. Armentrout, M. R. Cole, and H. TerBest, Jr., eds., Cenozoic paleogeography of the western United States: Pacific Coast Paleogeography Symposium 3, Pacific Section, SEPM, p. 29-44.

Turner, D. L., 1970, Potassium-argon dating of Pacific Coast Miocene foraminiferal stages: GSA Special Paper 124, p. 91-129.

Vedder, J. G., 1979, The Topanga Formation in the San Joaquin Hills, Orange County, California, in C. J. Stuart, ed., Miocene lithofacies and depositional environments, coastal southern California and northwestern Baja

California: Pacific Section, SEPM, p. 19-24.

Vedder, J. G., and D. G. Howell, 1976, Review of the distribution and tectonic implications of Miocene debris from the Catalina Schist, California continental borderland and adjacent coastal areas, in D. G. Howell, ed., Aspects of the geologic history of the California continental borderland: Pacific Section, AAPG Miscellaneous Publication 24, p. 326-340.

Warren, A. D., 1972, Luisian and Mohnian biostratigraphy of the Monterey Shale at Newport Lagoon, Orange County, California, in E. H. Stinemeyer, ed., The proceedings of the Pacific Coast Miocene biostratigraphic symposium: Pacific Section, SEPM, p. 27-36.

Warren, A. D., 1980, Calcareous nannoplankton biostratigraphy of Cenozoic marine stages in California, in R. M. Kleinpell, ed., The Miocene stratigraphy of California revisited: AAPG Studies in Geology 11, p. 60-69.

White, W. R., 1956, Pliocene and Miocene Foraminifera from the Capistrano Formation, Orange County, California: Journal of Paleontology, v. 30, p. 237-270.

Wilcoxon, J. A., 1969, Tropical planktonic zones and calcareous nannoplankton correlations in part of the California Miocene: Nature, v. 221, p. 950-951.

Wissler, S. G., 1943, Stratigraphic relations of the producing zone of the Los Angeles basin oil fields: California Division of Mines, Bulletin, n. 170, Part II, p. 209-234.

Wissler, S. G., 1958, Correlation chart of producing zones of Los Angeles basin oil fields, in J. W. Higgins, ed., A guide to the geology and oil fields of the Los Angeles and Ventura regions: Pacific Section, AAPG, p. 59-61.

Woodring, W. P., and M. N. Bramlette, 1950, Geology and paleontology of the Santa Maria district California: USGS Professional Paper 222, 185 p.

Woodring, W. P., M. N. Bramlette, and W. S. Kew, 1946, Geology and paleontology of Palos Verdes Hills, California: USGS Professional Paper 207, 145 p.

Woodring, W. P., M. N. Bramlette, and R. M. Kleinpell, 1936, Miocene stratigraphy and paleontology of Palos Verdes Hills, California: AAPG Bulletin, v. 20, p. 125-149.

Woodruff, F., 1985, Changes in Miocene deep-sea benthic foraminiferal distribution in the Pacific Ocean: relationship to paleoceanography, in J. P. Kennett, ed., The Miocene Ocean; paleoceanography and biogeography: GSA Memoir 163, p.131-177.

Wright, T., 1987, Geologic summary of the Los Angeles basin, in T. Wright, and R. Heck, eds., Petroleum geology of coastal southern California: Pacific Section, AAPG, p. 21-31.

Yeats, R. S., and J. M. Beall, 1991, Stratigraphic controls of oil fields in the Los Angeles basin, in K. T. Biddle, ed., Active margin basins: AAPG Memoir 52, p. 221-237.

Yerkes, R. F., 1972, Geology and oil resources of the western Puente Hills area, southern California: USGS Professional Paper 420-C, p. 63.

Yerkes, R. F., and R. H. Campbell, 1979, Stratigraphic nomenclature of the central Santa Monica Mountains, Los Angeles County, California: USGS Contrib. to Strat., Bulletin 1457-E, p. E1-E31.

Yerkes, R. F., T. H. McCulloh, J. E. Schoellhamer, and J. G. Vedder, 1965, Geology of the Los Angeles basin, California—an introduction: USGS Professional Paper 420-A, p. A1-A57.

Chapter 5

Central Los Angeles Basin
Subsidence and Thermal Implications for Tectonic Evolution

Larry Mayer

*Miami University
Oxford, Ohio, U.S.A.*

ABSTRACT

The central Los Angeles basin represents the deepest part of a basin that apparently resulted from rapid and prolonged lithospheric thinning owing to extension between rotating blocks. Subsidence in this tectonic setting began about 18 Ma and presumably reflects isostatic adjustment to the thinning of the buoyant crust. Sediment starving in the period immediately following the initiation of rapid subsidence resulted in a deep water-filled basin that reached water depths in excess of 2 km during Pliocene time. Sedimentation accelerated immediately following the widespread extrusion of andesitic and basaltic volcanics about 16 Ma. Maximum tectonic subsidence, which may require 50% to 75% of lithospheric thinning under the central deep, is about 3 km depending on assumptions. This amount of thinning can be used to estimate the maximum time-temperature history of basin sediments. The pattern of subsidence is best explained by a model of crustal rotation between right-slip faults that results in both extension in the early development of the basin and compression in the later phase of basin development.

INTRODUCTION

The Los Angeles basin is known as a great oil-producing area and also for its interesting and complex geologic history. In this paper, I examine a generalized subsidence history of the Los Angeles basin and try to relate subsidence mechanisms to the basin's subsidence and thermal evolution. The presence of oil is geologically important because it places constraints on the thermal evolution of the basin; therefore, any proposed basin model must attempt to satisfy the empirical constraint imposed by the presence of mature hydrocarbons. The purpose of this paper is to apply subsidence analysis to the Los Angeles basin, to discuss whether this technique can be appropriately applied to relatively small rift basins and related uncertainties, and to propose a simple model of the basin relating faulting, extension, and subsidence.

Subsidence analysis permits the partitioning of total subsidence into sediment-loading induced subsidence and tectonic subsidence; it is generally reported as a water-loaded depth (Steckler and Watts, 1978). Tectonic subsi-

Date Submitted: 6/25/87
Date Accepted: 6/8/88

dence is simply defined as that part of the total subsidence that cannot be explained by sediment loading. Once tectonic subsidence is determined, mechanisms that explain the magnitude of tectonic subsidence and that are also consistent with other geologic constraints can be proposed.

Subsidence analysis has been applied to Atlantic-type passive margins (Steckler and Watts, 1978), thrust-loaded foreland basins (Beaumont, 1981), as well as small transtensional basins (Dickinson et al., 1987). Ideally, the pattern of subsidence revealed through subsidence analysis provides some insight into the processes that cause subsidence. Subsidence analysis can proceed in two ways. One way is first to determine the tectonic subsidence history and then, based on the pattern of subsidence, decide on a model of basin formation that is consistent with or that explains the subsidence history. The second method is to select a preliminary basin model—a causal mechanism for the subsidence based on the geologic setting—then to proceed with the subsidence analysis and finally determine whether all pieces fit together in a consistent fashion. One reason for selecting the second method is that, for basins with a relatively short history (less than the period of time for which, for example, thermal decay is recognizable on a subsidence curve), the pattern of tectonic subsidence is not always distinct between models and, therefore, cannot be used as the sole criterion of model selection. In other words, in the early stages of basin formation, many causal mechanisms may result in similar subsidence histories.

GEOLOGY

Major Structures

The Los Angeles basin (Figure 1) can be divided into four distinct fault-bounded blocks (Yerkes et al., 1965) and perhaps into several less distinct geologic terranes (Howell et al., 1987). Yerkes et al. (1965) made a distinction between its "western basement" and "eastern basement," which are roughly separated from each other by the Newport-Inglewood fault zone. These "basements" may correlate with the Catalina and Santa Ana terranes of Howell et al. (1987). The central block is bounded by the Whittier-Elsinore fault on the northwest and by the Newport-Inglewood fault on the southeast. The Whittier-Elsinore fault may have had 5 km of right slip since Miocene time and 1.7 km of right slip since Pleistocene time (Yerkes et al., 1965; Sharp in Lamar, 1972). The Newport-Inglewood fault is known to have had up to 760 m of right slip since the early Pliocene (Hill, 1971; Harding, 1973); however, there is little constraint on any possible earlier slip history. Speculation requires the Newport-Inglewood fault to be a key player in post–mid-Miocene tectonics of the region.

Regionally, the most important faults, based on total possible slip and the requirement of kinematic linkages for palinspastic reconstructions (cf. Hornafius et al., 1986), appear to be the Hosgri fault and the San Andreas fault. The Hosgri fault is believed to have had about 115 km of right slip during the interval from 15 to 5 Ma (Graham and Dickinson, 1976, 1978a, b; Graham, 1978), and the San Andreas has had about 200 km of right slip since Pliocene

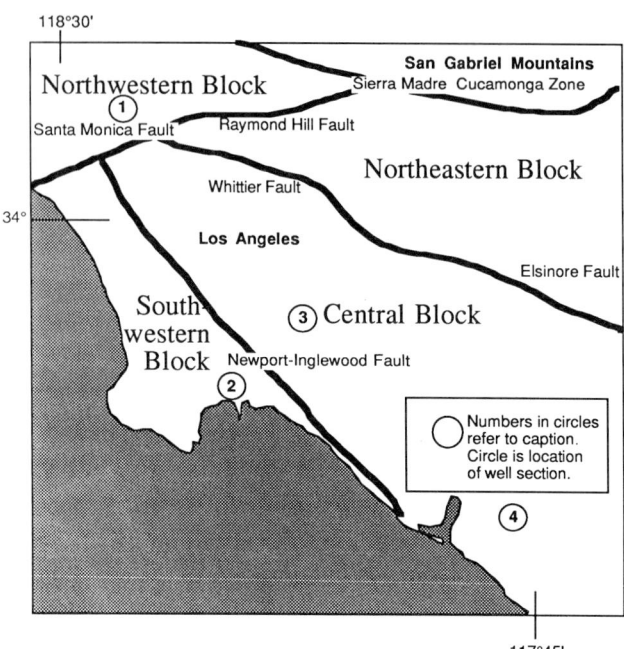

Figure 1. Map of the Los Angeles basin showing location of faults, names of blocks, and location of composite well sections that are used as the basis for tectonostratigraphic and subsidence analysis (from Yerkes et al., 1965). Location 1, eastern Santa Monica Mountains; location 2, Wilmington oil field; location 3, central deep; and location 4, San Joaquin Hills. Locations 1 through 4 correspond to section 1, section 4, section 6, and section 10 of Yerkes et al. (1965), respectively.

time (Crowell, 1962, 1968, 1971; Ross et al., 1973; Nilsen and Clarke, 1975).

Two models have been proposed to explain the post–18 Ma history of the Los Angeles basin. Both models involve substantial amounts of crustal thinning. Crowell (1974) suggested that the Los Angeles basin was a pull-apart basin formed by a releasing bend in a strike-slip fault. The particular fault or faults related to the original pull-apart style of basin formation are not known. Luyendyk et al. (1980) proposed a different pull-apart mechanism for formation of the Los Angeles basin. Based on paleomagnetic data in the Transverse Ranges that suggested significant clockwise rotations during the period from 16 to 5 Ma [with most of the rotation possibly occurring in the 16 to 10 Ma time window (Hornafius et al., 1986)], they proposed a model of block rotations that results in crustal thinning and the formation of deep sphenochasms. In their model, the sphenochasms are bounded on three sides by faults. The faults that could be invoked to satisfy the rotational model as the cause of subsidence in the central block of the Los Angeles basin are the Newport-Inglewood fault, the Whittier-Elsinore fault, and the Santa Monica fault.

The Santa Monica–Raymond Hill system of faults was active in the post–middle Miocene and accounts for up to 13 km of left slip (Lamar, 1961; Wright, *Structural Geology and Tectonic Evolution of the Los Angeles Basin*, this volume). Sixty km of left slip between the Santa Ana and Santa Monica Mountains (Yeats, 1968; Campbell and Yerkes, 1976) may be reasonable. Isopachs of Mohnian strata, which are thickest in the Puente Hills, (the thickness of Mohnian strata in the central trough is not documented) indicate that both the Whittier fault and the Newport-Inglewood fault were active dip-slip faults during Mohnian time and that the Santa Monica fault must also have had a significant dip-slip component as well (Yeats and Beall, *Stratigraphic Controls of Oil Fields in the Los Angeles Basin*, this volume).

Tectonostratigraphy

Mayer (1987) divided the stratigraphy of the Los Angeles basin into three tectonostratigraphic units based on significant unconformities and changes in sedimentation rates throughout the basin. The oldest tectonostratigraphic unit, which rests on the Los Angeles erosion surface (Woodford and Gander, 1977), consists of nonmarine conglomerate of the Trabuco Formation and the overlying shallow-marine sandstone, siltstone, and pebble conglomerate of the Ladd, Chico, and Williams formations. The next younger tectonostratigraphic unit is defined by an unconformity of unknown duration that occurs near the Paleocene-Cretaceous boundary and consists of the Silverado, Santiago, Vaqueros, and Sespe formations. These consist of marine and nonmarine sandstone and arkosic sandstone. The next younger tectonostratigraphic boundary is considered to be at the base of the Topanga Formation.

The Topanga Formation together with all younger units in the basin are grouped into the youngest tectonostratigraphic unit. During deposition of the Topanga Formation, there was significant extrusion of andesitic and basaltic volcanics that may be related to extension under the basin and that may therefore represent rift-related volcanism. The youngest tectonostratigraphic unit represents the part of the basin evolution that is of interest here. The southwestern block, which lies west of the Newport-Inglewood fault zone, shares only the youngest tectonostratigraphic unit in common with the rest of the basin. This fact may permit substantial mobility of blocks west of the Newport-Inglewood fault prior to mid-Miocene time.

Oil Fields

The occurrence of oil is relevant because it requires a time-temperature history for the source rocks that is sufficient to produce mature hydrocarbons. Source rocks in the basin are generally confined to those deposited during the time interval between the middle Miocene and the late Pliocene. Producing oil fields are distributed around the central block like a bathtub ring. The distribution of oil fields is apparently related to the location of suitable trapping structures and migration pathways like those found along the Newport-Inglewood fault.

SUBSIDENCE ANALYSIS

Tectonic forces that are related to the formation and evolution of a sedimentary basin as a physiographic entity are commonly reflected in the subsidence history of that basin. Deep basins that are preserved as thick accumulations of sediment imply that tectonic processes which operate at the scale of the lithosphere are the primary mechanism for basin formation. Sedimentation can also be modulated by changes in sea level or some other variable. If we can determine a basin's subsidence history, then presumably we will know or will be able to infer something about the tectonic mechanisms forming the basin. In addition it may be possible to relate sedimentation and thermal history. Because we cannot directly determine basin subsidence, it is necessary to calculate a subsidence history from chronostratigraphic, sedimentary thickness, and paleobathymetric data that are derived from the sedimentary record. The procedure of deriving a basin's subsidence history is loosely referred to as *subsidence analysis*. Subsidence analysis has been applied most successfully to Atlantic-type passive margins where it was found that tectonic subsidence followed a pattern of cooling and subsidence analogous to ocean floors spreading away from oceanic rifts (Steckler and Watts, 1978; Royden et al., 1980).

There are several possible tectonic mechanisms that result in deep sedimentary basins. Given the assumption, largely based on observations of oceanic basins, of isostatic equilibrium throughout basin development, mechanisms for basin formation are limited to those that change the relative thicknesses of crustal and subcrustal lithosphere and/or those that change the density structure of the lithosphere by way of thermal expansion or contraction. These tectonic processes include rifting, which results in crustal extension and subsidence; thrusting, which loads the elastic lithosphere and results in a flexural basin; and back-arc thermal perturbations, which cause a rise in geotherms, thermally thinned lithosphere; and eventual subsidence. In the case of rifting, subsidence occurs as a result of both lithospheric thinning and thermal contraction. However, if the rifting is considered to occur instantaneously, the isostatic adjustment to the thinning or thickening of the crust or lithosphere occurs rapidly in comparison to the adjustments related to thermal perturbations and contraction.

The steps involved in subsidence analysis are schematically shown in Figure 2 where only the steps used for a rifted basin and an Airy isostasy are shown. Since there is evidence for an extensional origin of the Los Angeles basin, as discussed above, a loading model must now be selected.

Loading Mechanisms

Sediments weigh down the elastic lithosphere and cause subsidence. In order to distinguish between the cause of subsidence and the weight of the sediments and the tectonic subsidence that originally formed the basin, we need first to choose a model that best represents the loading

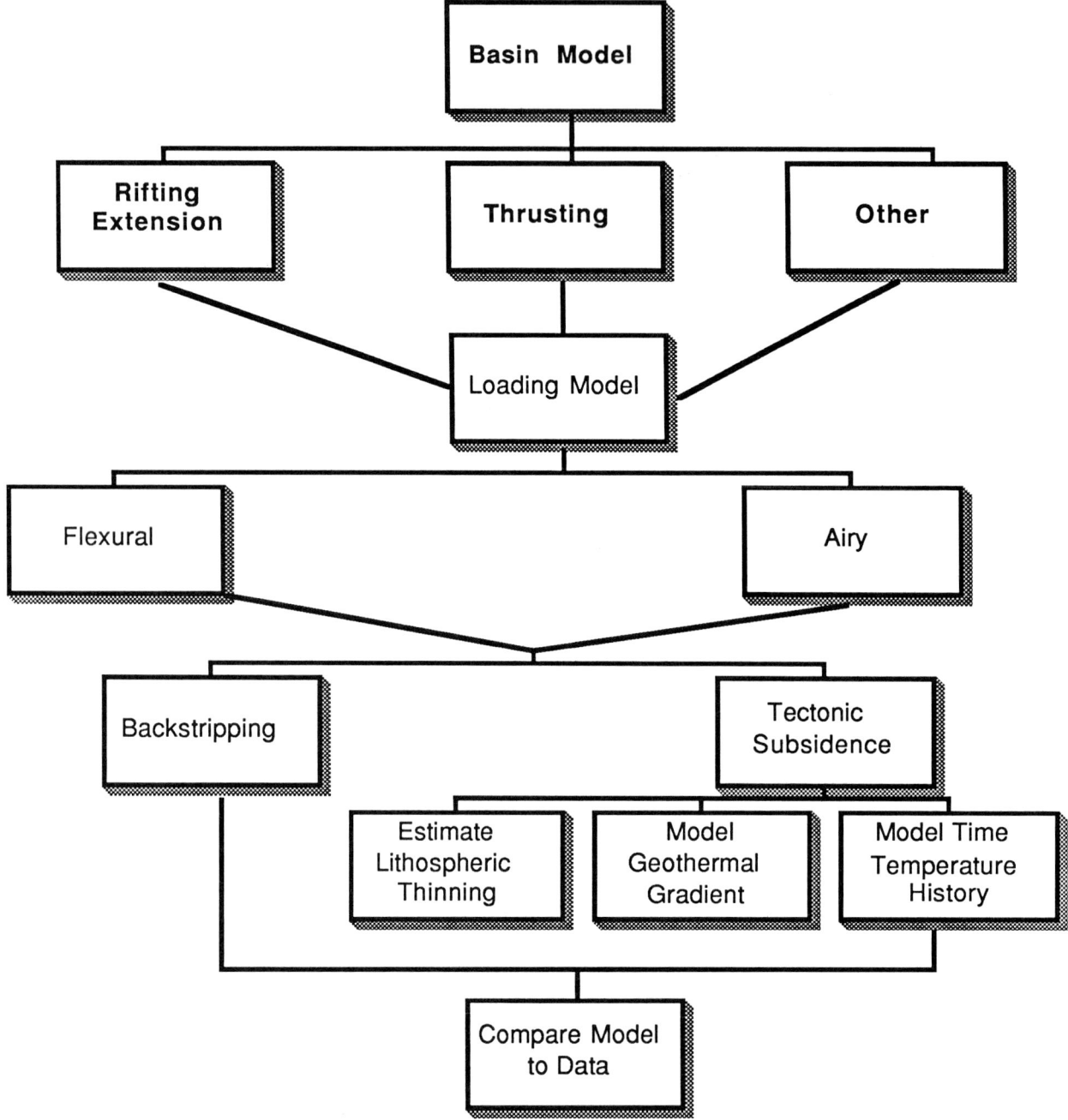

Figure 2. Organization of subsidence analysis studies.

mechanism. There are two common models of sediment, load-induced subsidence (Figure 3). The first model is that of Airy isostasy, which represents a block in hydrostatic equilibrium. The second model is flexural isostasy. The difference between the two models is that in the latter model, the lithosphere has strength and thus distributes the load across some distance. Foreland basins that result from the loads of thrust sheets have been modeled using flexural isostasy as have been Atlantic-type rifted margins. The Airy model assumes the loaded block moves vertically in response to loading and does not transmit loads laterally (i.e., it assumes no lateral strength). Because the Los Angeles basin is conveniently sliced into blocks by major strike-slip faults that probably would not permit the effective transmission of loads across them during extension, an Airy model seems most appropriate. The depth of a basin when the load of a sediment column is removed (tectonic subsidence) is given by

$$Z = WD + S\left(\frac{\rho_m - \rho_s}{\rho_m - \rho_w}\right)$$

where Z is the depth to the basin floor, WD is water depth, S is the thickness of the sediment column, ρ_w is the density of sea water (1.03 g/cc), ρ_m is the density of mantle (3.3

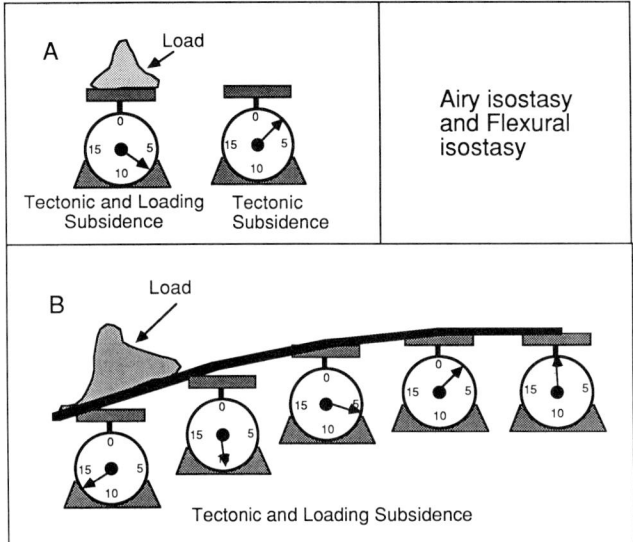

Figure 3. Isostatic mechanisms for loading of the lithosphere. (A) Airy mechanism, assumes no lateral strength. (B) Flexural isostasy, assumes lateral strength. Shape of deflection profile in B depends on flexural rigidity.

g/cc), and ρ_s is the average density of the sediment column. This equation is the *mechanism* of the scale shown in Figure 3A, and it explains why the scale does not return to zero after the sediment load is removed from a rift-formed basin. The average density of the sediment column is calculated using *backstripping* (Steckler and Watts, 1978).

Backstripping

Backstripping, or geohistory analysis (Van Hinte, 1978), a procedure that reconstructs the density structure of a sedimentary column through time, permits the calculation of tectonic subsidence. In backstripping, the average density of the entire sediment column is calculated; then this density is used to calculate how deep the basin would be at that time if the sediments were removed using Airy isostasy (above). The average density is

$$\rho_s = \frac{\sum_{\text{for all } i}[\phi_i \rho_w + (1 - \phi_i)\rho_g]T_i}{S}$$

where ϕ_i is porosity of the ith layer, ρ_g (2.65 g/cc) is the grain density, and T_i is the thickness of the ith sedimentary layer. In general, T is also recalculated for each layer as the sediments are progressively decompacted. T_i is calculated using this equation:

$$T_i = \frac{T_{net}}{1 - \phi(z)}$$

where T_{net} is the thickness of a layer at zero porosity and $\phi(z)$ is a porosity depth coefficient based on lithology or well measurements. For the sandstones within the basin, the functional relation between compaction with increasing burial and diagenetic reduction of porosity is highly uncertain (cf. McCulloh et al., 1978). I believe this correction is minor in the case of the Los Angeles basin when compared with the large changes in water depth.

This procedure is repeated for each stratigraphic unit for which age and paleobathymetric data are available. The result is a curve of depth versus time that shows the subsidence of the basin as if it were subsiding without sediments (though water filled). The resulting tectonic subsidence curve is as good (or as bad) as the data from which it is calculated.

The application of these techniques is shown in Figure 4. The data for this analysis represent a composite well section for the central block of the Los Angeles basin [locations 3 and 4 are based on descriptions given in Yerkes et al., 1965, for sections 6 (central deep) and 10 (San Joaquin Hills)]. There are several reasons for using a composite section. It is useful to show the long-term history of the basin and thus permit the study of the basin's recent history within the context of its long-term development. Another reason is the lack of penetration of the deepest part of the central block by wells. The only way to account for the unpenetrated sediments is to assume some stratigraphic analog from the margins of the basin. The data were chosen to reflect the general evolution of the central block and do not assume that the unpenetrated section in the central block is all Mohnian (cf. Yeats and Beall, this volume, for discussion).

The uppermost curve (Figure 4) represents the paleobathymetry of the basin (Woodring et al., 1946; Natland and Rothwell, 1954; Sullwold, 1960; Ingle, 1980). The next lower curve is the tectonic subsidence curve where the depth of the basin floor would lie without sediment loading. Note that the strongest part of the tectonic subsidence-curve *signal* is derived from the paleobathymetry, and, lacking that data, the tectonic subsidence curve might more closely represent a monotonically increasing function from 18 Ma to about 3 Ma.

An interesting characteristic of the tectonic subsidence curve is its steepness starting around 18 Ma: it gradually decreases at the point of maximum tectonic subsidence around 3 Ma and then reverses direction after 3 Ma. This means that, were it not for sediment loading pushing the basin floor down at a high rate, the basin floor would have gone up instead. This feature is found in several southern California basins (Dickinson et al., 1987). The bottom curve shows the subsidence of the basin floor and displays a rapid subsidence following 20 Ma. The total amount of tectonic subsidence prior to the reversal in tectonic subsidence is approximately 3 km.

Sawyer et al. (1987) also studied the subsidence history of the central block by backstripping. A comparison of the results presented here and their results is shown in Figure 5. They (Sawyer et al., 1987) used a cross section to obtain stratigraphic thickness for the central deep (from California Division of Oil and Gas, 1974) and apparently assumed that the entire section is Mohnian and younger (cf. Yeats and Beall, this volume, for an alternative interpretation),

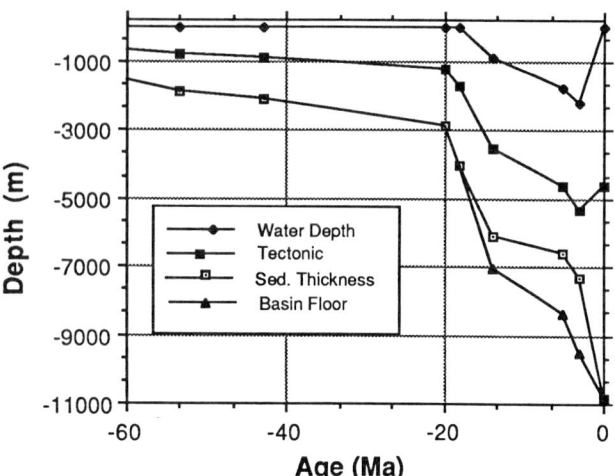

Figure 4. Geohistory plot for the composite well section in the central block showing the subsidence history of the Los Angeles basin since 60 Ma.

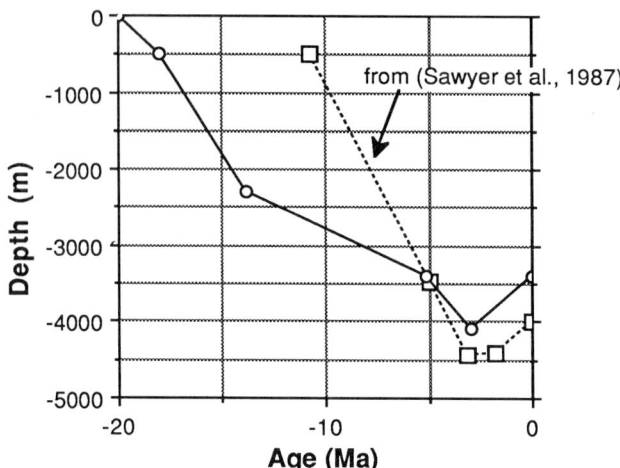

Figure 5. Tectonic subsidence curves for the central deep of the Los Angeles basin. The major difference between these curves stems from stratigraphic assumptions in the deep part of the basin (see text for discussion). The curve from Sawyer et al. (1987) shows their "maximum" tectonic subsidence.

which accounts for the differences in the early part of the two tectonic subsidence curves. In a sense, these two curves bracket the subsidence history of the central deep. However, in both cases, the curves change from subsidence to uplift at about 3 Ma. The subsidence pattern shown on the tectonic subsidence curves does not uniquely describe a tectonic mechanism. However, given that there are significant amounts of volcanics and extensional structures, it seems reasonable to assume that lithospheric thinning caused by extension may be involved in producing the basin. The relation between thinning or extension and tectonic subsidence is important for developing further constraints on the evolution of the Los Angeles basin.

Lithospheric Thinning

McKenzie (1978) proposed a model of simple stretching in which he related initial subsidence and postrift tectonic subsidence caused by the thermal contraction of the lithosphere to the magnitude of stretching. His model provided a way to estimate the factor of lithospheric stretching, β, or thinning, γ, where $\gamma = 1 - (1/\beta)$, by matching two components of subsidence: synrift, or initial subsidence, and thermal subsidence. Initial subsidence occurs rapidly following rifting and is caused by isostatic adjustment to a thinned lithosphere. The thermal subsidence occurs subsequently to the initial subsidence at a rate described by cooling of a lithospheric plate. The subsidence history of the Los Angeles basin is so short, relative to the time needed for thermal decay, that we do not see any evidence of separate initial and thermal subsidence. We can assume that both thermal and isostatic subsidence are occurring at the same time, or we can assume that all the subsidence is likely to be initial subsidence, if the rifting is not instantaneous. Given the uncertainties regarding how rapidly extension occurred and how a narrow basin will cool, it is simpler to use only the initial subsidence to estimate the magnitude of lithospheric thinning. This initial subsidence, assuming all the subsidence that has occurred is initial subsidence, is about 3 km. Initial subsidence is a linear function of the thinning factor, γ, and can be approximated by

$$\gamma = (1 - 1/\beta) = 0.2342\, U_i$$

where U_i is the amount of initial subsidence in kilometers. This relation assumes a prerift 35-km-thick crust, a 120-km-thick lithosphere, and a coefficient of thermal expansion of $3 \times 10^{-5}\,°C^{-1}$. Thus, an estimate of lithospheric thinning for the Los Angeles basin is 75%, if all the subsidence is initial, or less than 75%, depending on how much of the subsidence is really thermal and on the assumptions of initial crustal and lithospheric thickness.

Thermal Implications

According to the simple stretching model (McKenzie, 1978), stretching by a factor of β results in an increase in the geothermal gradient by a factor of β. Royden et al. (1980) discussed how the McKenzie model could be used to predict hydrocarbon maturation. In the case of the Los Angeles basin, we would like to check thermal evolution predicted from the simple stretching model against maturation criteria inferred from the occurrence of oil. Each thinning factor can be used to predict Q, the near-surface heat flux (Figure 6). The greater the thinning, the greater the heat flux during the early postrift phase of basin evolution. After about 150 m.y., heat flux tends to converge to equilibrium heat flux

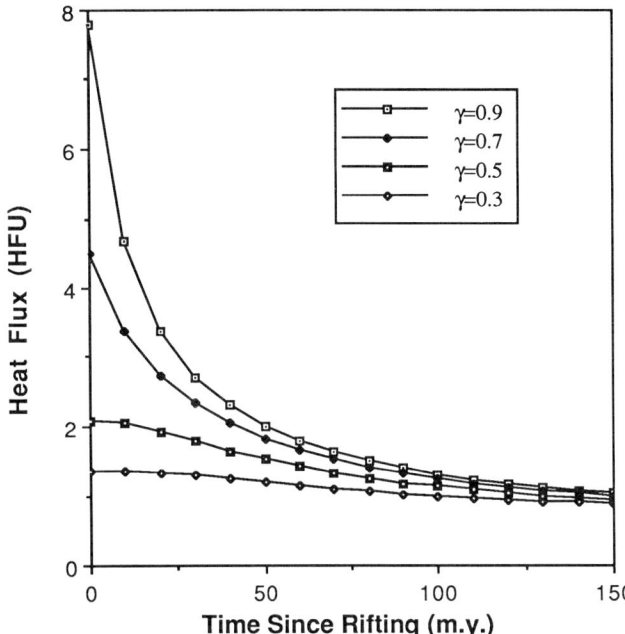

Figure 6. Change in heat flux with time for different amounts of lithospheric thinning, according to simple stretching model (Royden et al., 1980).

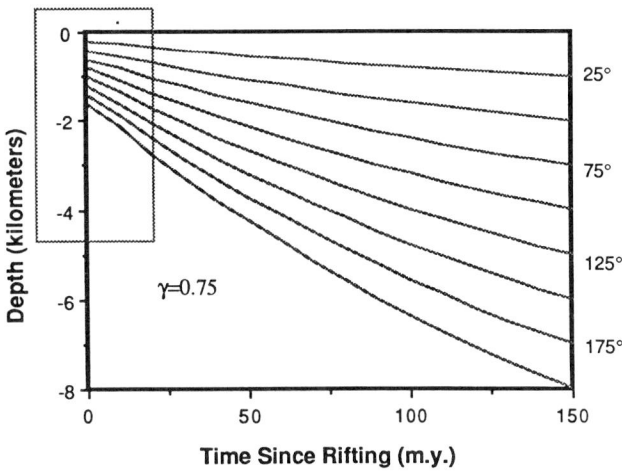

Figure 7. Changes in geothermal gradient over time for a basin rifted at time zero and thinned by a factor of 75% (assuming a thermal conductivity for sediments, k, of 4×10^{-3} cal cm^{-1} s^{-1} °C^{-1}). The small rectangle encloses that portion of the cooling history that applies to the Los Angeles basin. Note that these temperatures are much higher than found in the basin at present and suggest an alternative history.

levels (Figure 6). Once heat flux is known, it can be used to predict the thermal history of the basin (Figure 7).

There are several problems with the direct application of simple one-dimensional cooling scenarios of the thinned lithosphere under the Los Angeles basin. First, the basin is so narrow that it is likely for heat to flow laterally in addition to vertically (Sawyer et al., 1987); thus, the basin can cool more rapidly. Another complication results from the rapid pulse of sedimentation during the late Pliocene and Pleistocene, which can depress the near-surface temperature gradient. Finally, the migration of diagenetic fluids and oil clearly indicates that heat can be transferred by processes other than conduction.

To understand how these complicating factors can affect the thermal history of the basin and, therefore, levels of organic maturation, it is useful to consider two possible cooling scenarios. In the first scenario, simple one-dimensional cooling is assumed until about 3 Ma when an instantaneous 1-km depression of the geothermal gradients occurs. In the second scenario, simple cooling occurs for 1 m.y., at which point the effective cooling of the basin takes place at twice the rate normally expected, given the thermal conductivity of the sediments.

Using the 75% estimate of thinning by rounding the stretching factor to $\beta = 4$, the time-temperature history of the basin under the central deep can be described. Figures 8 and 9 show the time-temperature history of the basin assuming that instantaneous rifting occurred at 18 Ma for each cooling scenario. This age is chosen because it is the approximate age of the base of the Topanga and the youngest tectonostratigraphic unit. However, there is ample evidence, from the tectonic subsidence curve and given the time of structural displacements (Wright, this volume), to suggest that stretching might have continued to occur during the period 18 to 4 Ma. Thus, assuming that all the stretching is concentrated at 18 Ma may result in a maximum bracket of the basin's thermal history. Using the Lopatin method (Waples, 1980) and a very rough estimate of maturation, the time-temperature index (TTI) was calculated by

$$TTI = \sum_{n\,min}^{n\,max} (\Delta T_n)(r^n)$$

where ΔT is the length of time spent in the temperature interval, r is the reaction factor and is taken to equal two, and n is the index value that is equal to zero for the temperature interval 100–110°C and increases or decreases by one for each ten-degree change in temperature (up to a maximum of five, beyond which there are no data) (Waples, 1980). For the stepped cooling scenario, sediments deposited 3 Ma and buried to a depth of 3 km at a constant rate would spend 1 m.y. at the 25°C interval, 33°C interval, and 75°C interval. The resulting TTI for the stepped cooling case is 0.14. For the rapid cooling case, the temperatures for 1 m.y. intervals would be 25°C, 70°C, and 100°C, resulting in TTI = 0.63. Comparing organic maturation within sediments deposited at 10 Ma and buried at a uniform rate to 5-km depth, for the rapid-cooling model, TTI = 87; for the stepped cooling model, TTI = 132. Both of these TTI values are within the peak oil generation window. Thus for sediments deposited after 3 Ma, the rapid cooling model

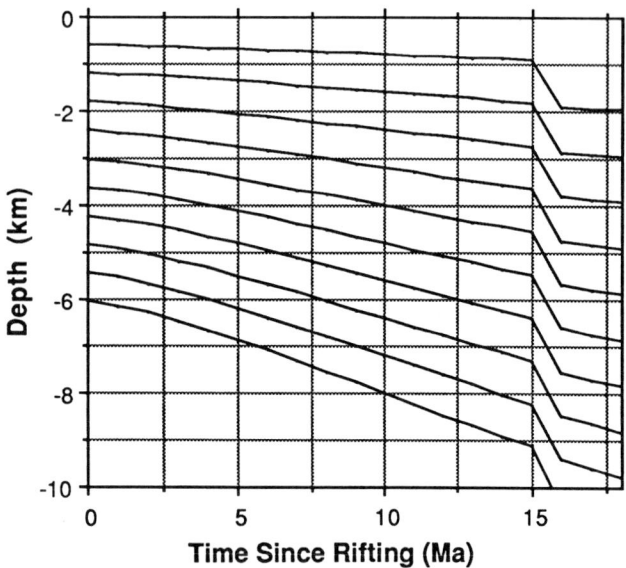

Figure 8. Evolution of geotherms in a basin that cools, according to the simple stretching model, after instantaneous thinning by 75% for the first 15 m.y., and then that experiences a 1-km depression of the gradient meant to reflect rapid sediment influx.

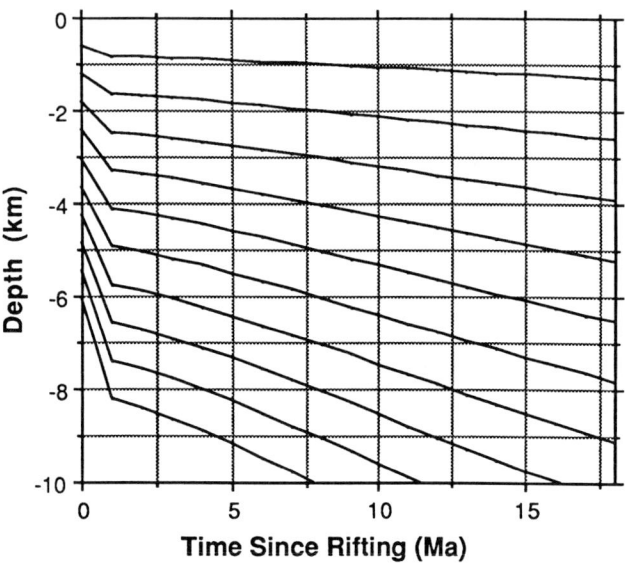

Figure 9. Evolution of geotherms in a basin that initially cools as in Figure 7, but then cools at twice its previous rate (i.e., twice the effective conductivity). Figure is meant to reflect heat loss by lateral heat flow and other mechanisms besides conduction.

results in greater maturity. For sediments deposited before the stepped depression of near-surface temperature gradient, the stepped cooling model results in greater maturity. The present temperature gradient (cf. Bostick et al., 1978, for discussion on reliability of temperature measurements) and the level of organic maturation as indicated by vitrinite reflectance for the central basin is shown in Figure 10. Both of these scenarios predict higher temperatures than presently observed at depths greater than 4 km; whereas the two-dimensional heat-flow model of Sawyer et al. (1987) predicts lower temperatures than observed in the central block.

A SYNTHESIS

The subsidence history of the Los Angeles basin does not appear to follow a model of instantaneous simple extension, nor is the thermal history of the basin easy to extract from its subsidence history. However, it seems apparent from both the subsidence and thermal history of the basin that some kind of extensional model is appropriate. Any explanation of the subsidence history must satisfy several key constraints on the physical evolution of the basin and its immediate boundaries. In addition to the thermal constraints discussed above, these constraints include the following:

1. significant (about 90°) clockwise rotations of the basin's bounding blocks (Luyendyk et al., 1980)

2. 60 km of left slip across the north-south extent of the basin (Santa Monica Mountains and Santa Ana Mountains), of which 13 km of left slip occurs across the northern boundary of the basin (Santa Monica fault)

3. a requirement for substantial amounts of right slip across the basin, although no significant right slip is permitted across the Newport-Inglewood or Whittier faults individually

4. extension that permits rates to change and ultimately to shift to compression

5. initial basin that must form in an extremely narrow zone (Wright, this volume)

6. tectonic processes that must allow for complex deformation, thrusting, and alignment of structures with the general orientation of (right-slip faults bounding) the basin

Mayer (1987) noted that uplift seemed to be common as the ultimate phase of many borderland basins' subsidence histories and speculated that there was a relation between rotating crustal slivers and the kinematics of the extending basins. One way to satisfy all the constraints given above is to propose a simple shear model of rotation and crustal extension (Figure 11). The simple rotational shear model requires that rigid block rotation occur between right-slip faults. The rotating blocks do not

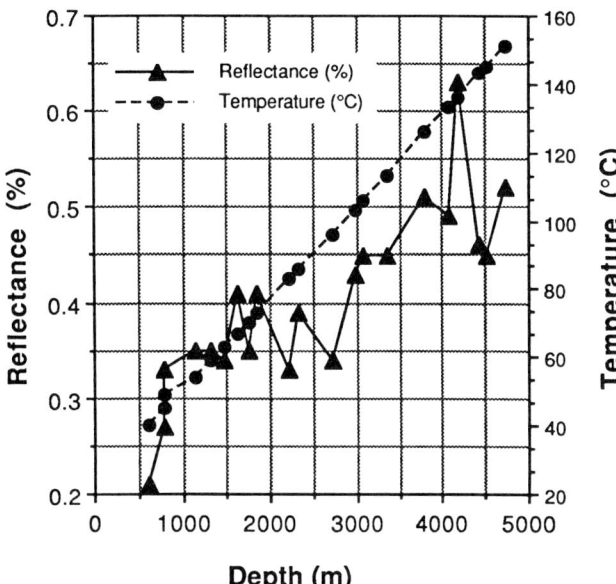

Figure 10. Vitrinite reflectance and temperature data from the central block of the Los Angeles basin (data from Bostick et al., 1978). Note the intercept of the temperature-depth line does not intersect the y-axis near 0°C.

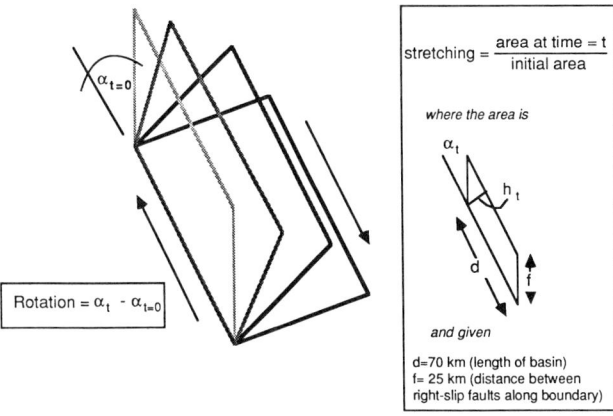

Figure 12. Relation between amount of extension and magnitude of clockwise rotations of basin-bounding blocks. α at time $= 0$ represents the initial angle between the left- and right-slip fault. With progressive dextral shear across the basin, the area of the basin changes. The boxed diagram shows the relevant line lengths and angles for computing the amount of stretching that results from rotation.

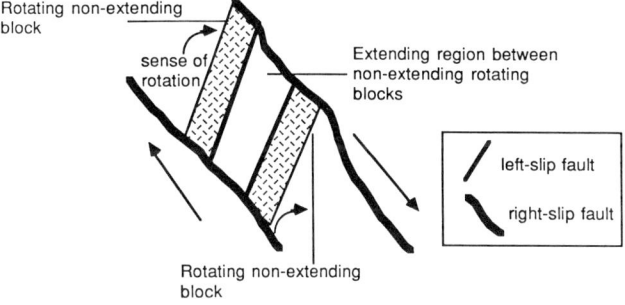

Figure 11. Model of how rotation of rigid blocks forms an extensional basin. This model may well apply to the Los Angeles basin. The right-slip faults are the Newport-Inglewood and Whittier-Elsinore faults. The left-slip fault bounding the northern margin of the basin is the Santa Monica fault. The southern structure is undetermined.

extend and, therefore, as they rotate they form a basin. The boundary between the nonextending rotation block and the basin is a left-slip fault. The basin forms in response to rotation and not in response to any regional extension.

The amount of extension or stretching is a function of the amount of rotation that has occurred (Figure 12). As rotation proceeds, the basin opens and deepens, and the left-slip faults rotate clockwise relative to the right-slip faults. Extension proceeds along with rotation, but if rotation proceeds beyond the point where the angle between the left-slip and right-slip basin-bounding faults is 90°, the basin begins to compress (Figure 13).

If we define the boundaries of the Los Angeles basin to be the Whittier-Elsinore fault on the east, the Newport-Inglewood fault on the west, the Santa Monica fault on the north, and an undetermined structure bounding the San Joaquin–Santa Ana Mountains on the south, then we have the structural elements needed to apply this simple model. For the initial angles between faults and the dimensions of the Los Angeles basin central block given in Figure 12, the stretching of the lithosphere continues up to a maximum of about $\beta = 3$ before the basin begins to compress. If we assume that the rotation occurs over time, then the resultant extension also occurs over time. Perhaps maximum extension in the rotation model occurred when the tectonic subsidence was greatest and, hence, was immediately followed by compression, as evidenced by a reversal in the tectonic subsidence curve after 3 Ma.

We can also calculate the amount of slip on each of the basin-bounding faults required by the model, using the scheme shown in Figure 14. The total amount of slip across the left-slip faults is directly determinable. However, the right slip is distributed immediately across neither the Newport-Inglewood fault nor the Whittier-Elsinore fault, but across the entire basin. Using the geometry shown in Figure 14, the amount of right slip distributed the entire central block and the amount of left slip across the Santa Monica fault can be determined (Figure 15). The model predicts a maximum of about 16 km of left slip on the

194 Mayer

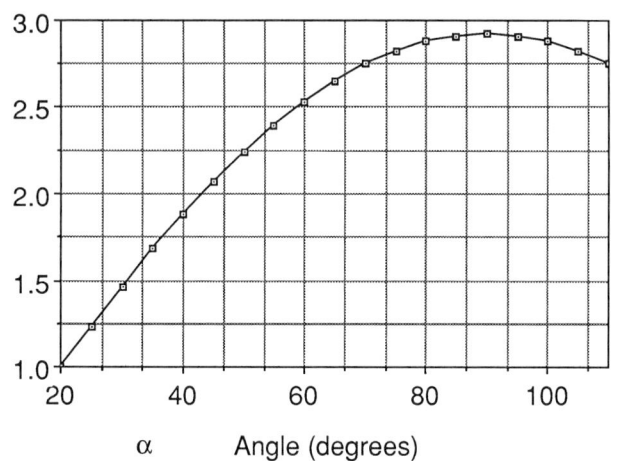

Figure 13. Relation of the angle between the left- and right-slip faults, α to the amount of extension calculated according to Figure 12. In this case, after 70° of clockwise rotation, where $\alpha = 90°$, the basin begins to compress.

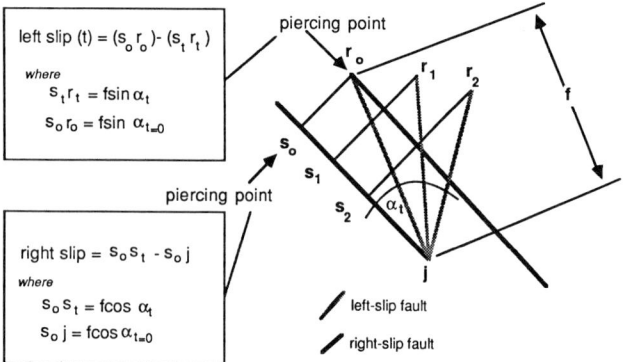

Figure 14. Geometry of slip resulting from clockwise rigid block rotation.

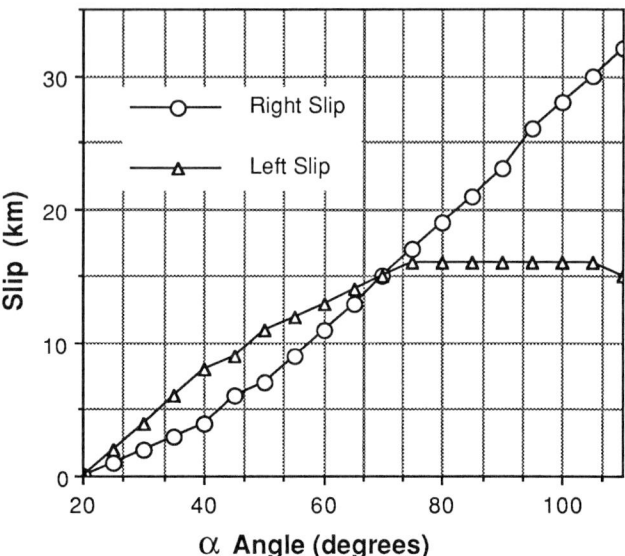

Figure 15. Relation between the angle α and the amount of right and left slip. The rate of left slip is initially rapid, reaching a maximum after $\alpha = 90°$. However, the amount of right slip monotonically increases with increased rotation.

Santa Monica fault, which is in good agreement with the 13 km of slip given by Lamar (1961). The total amount of left slip across the entire basin is much larger, about six times, or 96 km, larger than that for the Santa Monica fault alone. The total right slip distributed across the basin is predicted to be about 32 km.

That the Los Angeles basin formed in an extensional setting is supported by the subsidence analysis, thermal history, and apparent convergence of several kinematic lines of evidence from paleomagnetic and geologic data. If the rotational model presented here is correct, then the next step in unraveling the tectonic history of the Los Angeles basin is the linkage of rotation and extension through the study of new data to provide better timing constraints for tectonic and thermal modeling.

ACKNOWLEDGMENTS

Earlier drafts of this paper were reviewed by K. T. Biddle, G. H. Blake, D. W. Phelps, and C. A. Tapscott.

REFERENCES CITED

Beaumont, C., 1981, Foreland basins: Geophysical Journal of the Royal Society, v. 65, p. 291–329.
Bostick, N. H., S. M. Cashman, T. H. McCulloh, and C. T. Waddell, 1978, Gradients of vitrinite reflectance and present temperature in the Los Angeles and Ventura basins, California, in D. F. Oltz, ed., A symposium in geochemistry: low temperature metamorphism of kerogen and clay minerals: Pacific Section, SEPM, p. 65–96.
California Division of Oil and Gas, 1974, California oil and gas fields, south-central coastal and offshore California, v. 2: Sacramento, California Division of Oil and Gas.
Campbell, R. H., and R. F. Yerkes, 1976, Cenozoic evolution of the Los Angeles basin area—relation to plate tectonics, in D. G. Howell, ed., Aspects of the geological history of the California borderland: Pacific Section, AAPG Miscellaneous Publication 24, p. 541–558.
Crowell, J. C., 1962, Displacement along the San Andreas fault, California: GSA Special Paper 71, 61 p.
Crowell, J. C., 1968, Movement histories of faults in the Transverse Ranges and speculations on the tectonic history of California, in W. R. Dickinson and A. Grantz, eds., Proceedings of conference on geologic problems of the San Andreas fault system: Stanford University Publications, Geological Sciences, v. 11, p. 323–341.
Crowell, J. C., 1971, Tectonic problems of the Transverse

Ranges, California: GSA Abstracts with Programs, v. 3, p. 106.

Crowell, J. C., 1974, Origin of late Cenozoic basins in southern California, in W. R. Dickinson, ed., Tectonics and sedimentation: SEPM Special Publication 22, p. 190–204.

Dickinson, W. R., et al., 1987, Geohistory analysis of rates of sediment accumulation and subsidence for selected California basins, in R. V. Ingersoll, and W. G. Ernst, eds., Cenozoic development of coastal California: Englewood Cliffs, New Jersey, Prentice-Hall, Inc., Rubey v. VI, p. 2–23.

Graham, S. A., 1978, Role of the Salinian block in evolution of the San Andreas fault system, California: AAPG Bulletin, v. 62, p. 2214–2231.

Graham, S. A., and W. R. Dickinson, 1976, San Gregorio fault as a major right-slip fault of the San Andreas fault system: GSA Abstracts with Programs, v. 8, p. 890.

Graham, S. A., and W. R. Dickinson, 1978a, Evidence for 115 kilometers of right slip on the San Gregorio-Hosgri trend: Science, v. 199, p. 179–181.

Graham, S. A., and Dickinson, W. R., 1978b, Apparent offsets of on-land geologic features across the San Gregorio-Hosgri fault trend: California Division of Mines and Geology Special Report 137, p. 13–23.

Harding, T. P., 1973, Newport-Inglewood trend, California—an example of wrenching style of deformation: AAPG Bulletin, v. 57, p. 97–116.

Hill, M. L., 1971, Newport-Inglewood zone and Mesozoic subduction, California: GSA Bulletin, v. 82, p. 2957–2962.

Hornafius, J. S., B. P. Luyendyk, R. R. Terres, and M. J. Kamerling, 1986, Timing and extent of Neogene tectonic rotation in the western Transverse Ranges, California: GSA Bulletin, v. 97, p. 1476–1487.

Howell, D. G., D. E. Champion, and J. G. Vedder, 1987, Terrane accretion, crustal kinematics and basin evolution, southern California: in R. V. Ingersoll and W. G. Ernst, eds., Cenozoic development of coastal California: Englewood Cliffs, New Jersey, Prentice-Hall, Inc. Rubey v. VI, p. 243–258.

Ingle, J. C., Jr., Cenozoic paleobathymetry and depositional history of selected sequences within the southern California continental borderland: Cushman Foundation for Foraminiferal Research Special Publication 19, p. 163–195.

Lamar, D. L., 1961, Structural evolution of the northern margin of the Los Angeles basin [unpublished Ph.D. dissertation]: University of California at Los Angeles, 162 p.

Lamar, D. L., 1972, Microseismicity and recent tectonic activity in the Whittier fault area, California: USGS Final Technical Report, 44 p.

Luyendyk, B. P., M. J. Kamerling, and R. Terres, 1980, Geometric model for Neogene crustal rotations in southern California: GSA Bulletin, v. 91, pt. 1, p. 211–217.

Mayer, L., 1987, Subsidence analysis of the Los Angeles basin, in R. V. Ingersoll and W. G. Ernst, eds., Cenozoic development of coastal California: Englewood Cliffs, New Jersey, Prentice-Hall, Inc., Rubey v. VI, p. 300–320.

McCulloh, T. H., S. H. Cashman, and R. J. Stewart, 1978, Diagenetic baselines for interpretive reconstructions of maximum burial depths and paleotemperatures in clastic sedimentary rocks, in D. F. Oltz, ed., A symposium in geochemistry: low temperature metamorphism of kerogen and clay minerals: Pacific Section, SEPM, p. 18–46.

McKenzie, D. P., 1978, Some remarks on the development of sedimentary basins: Earth and Planetary Science Letters, v. 40, p. 25–32.

Natland, M. L., and W. T. Rothwell, Jr., 1954, Fossil foraminifera of the Los Angeles and Ventura regions, California: California Division of Mines Bulletin 170, p. 33–42.

Nilsen, T. H., and S. H. Clarke, Jr., 1975, Sedimentation and tectonics in the early Tertiary continental borderland of central California: USGS Professional Paper 925, 64 p.

Ross, D. C., C. M. Wentworth, and E. H. McKee, 1973, Cretaceous mafic conglomerate near Gualala offset 350 miles by San Andreas fault from oceanic sources near Eagle Rest Peak, California, USGS Journal of Research, v. 1, p. 45–52.

Royden, L., J. G. Sclater, and R. P. von Herzen, 1980, Continental margin subsidence and heat flow: important parameters in formation of petroleum hydrocarbons: AAPG Bulletin, v. 64, p. 173–187.

Sawyer, D. S., A. T. Hsui, M. N. Toksöz, 1987, Extension, subsidence, and thermal evolution of the Los Angeles basin: a two-dimensional model: Tectonophysics, v. 133, p. 15–32.

Steckler, M. S., and A. B. Watts, 1978, Subsidence of the Atlantic-type continental margin off New York: Earth and Planetary Science Letters, v. 41, p. 1–13.

Sullwold, H. H., Jr., 1960, Tarzana fan, deep submarine fan of late Miocene age, Los Angeles County, California: AAPG Bulletin, v. 10, p. 502–512.

Van Hinte, J. E., 1978, Geohistory analysis—application of micropaleontology in exploration geology: AAPG Bulletin, v. 62, p. 201–222.

Waples, D. W., 1980, Time and temperature in petroleum formation—application of Lopatin's method to petroleum exploration: AAPG Bulletin, v. 64, p. 916–926.

Woodford, A. O., and C. Gander, 1977, Los Angeles erosion surface of middle Cretaceous age: AAPG Bulletin, v. 61, p. 1979–1990.

Woodring, W. P., M. N. Bramlette, and W. S. W. Kew, 1946, Geology and paleontology of the Palos Verdes Hills, California: USGS Professional Paper 207, 145 p.

Wright, T. L., 1991, Structural geology and tectonic history of the Los Angeles basin, California, in K. T. Biddle, ed., Active margin basins: AAPG Memoir 52, p. 35–134.

Yeats, R. S., 1968, Rifting and rafting in the southern California borderland, in W. R. Dickinson and A. Grantz, eds., Proceedings of conference on geologic problems of the San Andreas fault system: Stanford University Publications, Geological Sciences, v. 11, p. 307–322.

Yeats, R. S., and J. M. Beall, 1991, Stratigraphic controls of oil fields in the Los Angeles basin: a guide to migration history, in K. T. Biddle, ed., Active margin basins: AAPG Memoir 52, p. 221–237.

Yerkes, R. F., T. H. McCulloh, J. E. Schoellhamer, and J. G. Vedder, 1965, Geology of the Los Angeles basin—an introduction: USGS Professional Paper 420A, 57 p.

Chapter 6

Geochemistry of Los Angeles Basin Oil and Gas Systems

Alan W. A. Jeffrey

Global Geochemistry Corporation
Canoga Park, California, U.S.A.

Hossein M. Alimi

Global Geochemistry Corporation
Canoga Park, California, U.S.A.

Peter D. Jenden

Global Geochemistry Corporation
Canoga Park, California, U.S.A.

ABSTRACT

The Los Angeles basin is one of the most prolific petroleum-producing provinces in the world. The basin has produced in excess of 6 billion barrels of oil (GBO) and over 7 trillion cubic feet (tcf) of gas and includes one of the world's largest single accumulations, the Wilmington field. Oil gravities are highly variable, ranging from less than 10° API in shallow producing zones to condensate (>50° API) in a few deep fields. However, much of the oil produced in the basin is rather heavy (<25° API) with an appreciable sulfur content (>1%).

The Los Angeles basin contains abundant organic-rich source rocks containing kerogen rich in sapropelic material. Maturity estimates of the source rocks based on vitrinite reflectance values are low. This appears to be related to the sapropel-rich kerogen, which may generate oil at lower maturities than is conventionally accepted or may cause suppression of vitrinite reflectivity in the kerogen. Maturity estimates based on bitumen production indicate that upper Miocene rocks in deeper parts of the basin are the source of the oil and gas accumulations. Vertical migration of this oil and gas into shallower reservoirs is a feature of many fields, especially the giant Wilmington and Huntington Beach fields.

Oil quality within the basin varies geographically, with higher quality (high API gravity, low sulfur content) oils in the northeast of the basin and lower quality oils in the west and south of the basin. The variation in oil quality does not appear to be caused by differences in kerogen type, because the sterane biomarker ratios and carbon isotope ratios of the oils are very similar and are consistent with a marine-derived kerogen as the source. Ratios of biomarkers and lower molecular weight hydrocarbons suggest that the superior quality of the oil in the northeast of the basin may be caused by higher maturity, greater migration, a more oxidizing depositional environment, or a combination of all three. Oils appear to be genetically related to the associated gases in a reservoir, except for some cases of localized mixing of gases with microbial methane.

Depth-related variations in oil quality and in carbon dioxide content in the associated gas appear to be caused by biodegradation. A suite of oils and associated gases from the Salt Lake area in the northwestern basin shows that a variety of interrelated physical and chemical changes in many shallow oil and gas deposits results from microbial oxidation of liquid and gaseous hydrocarbons to carbon dioxide. This is an important process in the giant Wilmington and Huntington Beach fields, where deep reservoirs contain a medium quality oil; whereas shallower reservoirs contain a genetically similar oil that has been microbially transformed to a lower gravity, higher sulfur crude.

INTRODUCTION

Since the inception of its exploration in the late nineteenth century, the Los Angeles basin (Figures 1, 2) has proven to be one of the most prolific petroleum-producing provinces in the world. It has produced in excess of 6 billion barrels of oil (GBO) and over 7 trillion cubic feet (tcf) of gas and includes one of the world's largest single accumulations, the Wilmington field. Fifteen of the Los Angeles basin fields are considered giant fields,[1] one of which (Beta field) was added to the list as recently as 1976. Virtually all of the production is from depths shallower than 10,000 ft (3000 m), and in the Wilmington and Huntington Beach fields, average production is from less than 5000 ft (1500 m). In common with other California Tertiary basins, the major product is oil and associated gas, with no significant nonassociated gas fields. Oil gravities are highly variable, ranging from less than 10° API in shallow producing zones to condensate (>50° API) in a few deep fields. However, much of the oil produced in the basin is rather heavy (<25° API) with an appreciable sulfur content (>1%). Multiple stacked pay zones are common, and oil quality varies considerably both vertically and horizontally from fault block to fault block.

Geochemistry of the oil and gas systems of the Los Angeles basin and of the sedimentary rocks believed to be the source of the hydrocarbons will be discussed in this chapter. Most of the source rock data has been taken from previously published material, and this chapter provides a review of both the hydrocarbon potential of the sedimentary rocks of the basin and the maturity of these rocks, including the question of oil generation from rocks regarded as immature by conventional estimates. Much of the geochemical data on oils and gases of the Los Angeles basin has been taken from unpublished reports of Global Geochemistry Corporation (GGC). The variety of oil types in the basin will be illustrated by a small number of representative oils, and the influence of source material, maturation, migration, and biodegradation on the oil chemistry will be discussed. Chemical and stable isotope analyses of a suite of gases from the basin will be used to identify effects of source and maturation on the gases, using conventional hydrocarbon gas genetic characterization schemes. Finally, there will be a discussion of the effects of biodegradation—an important process in the Los Angeles basin—on reservoired hydrocarbons.

---

[1] Ultimate production exceeds 100 million barrels of oil (MBO).
Date Submitted: 12/28/87
Date Accepted: 8/8/88

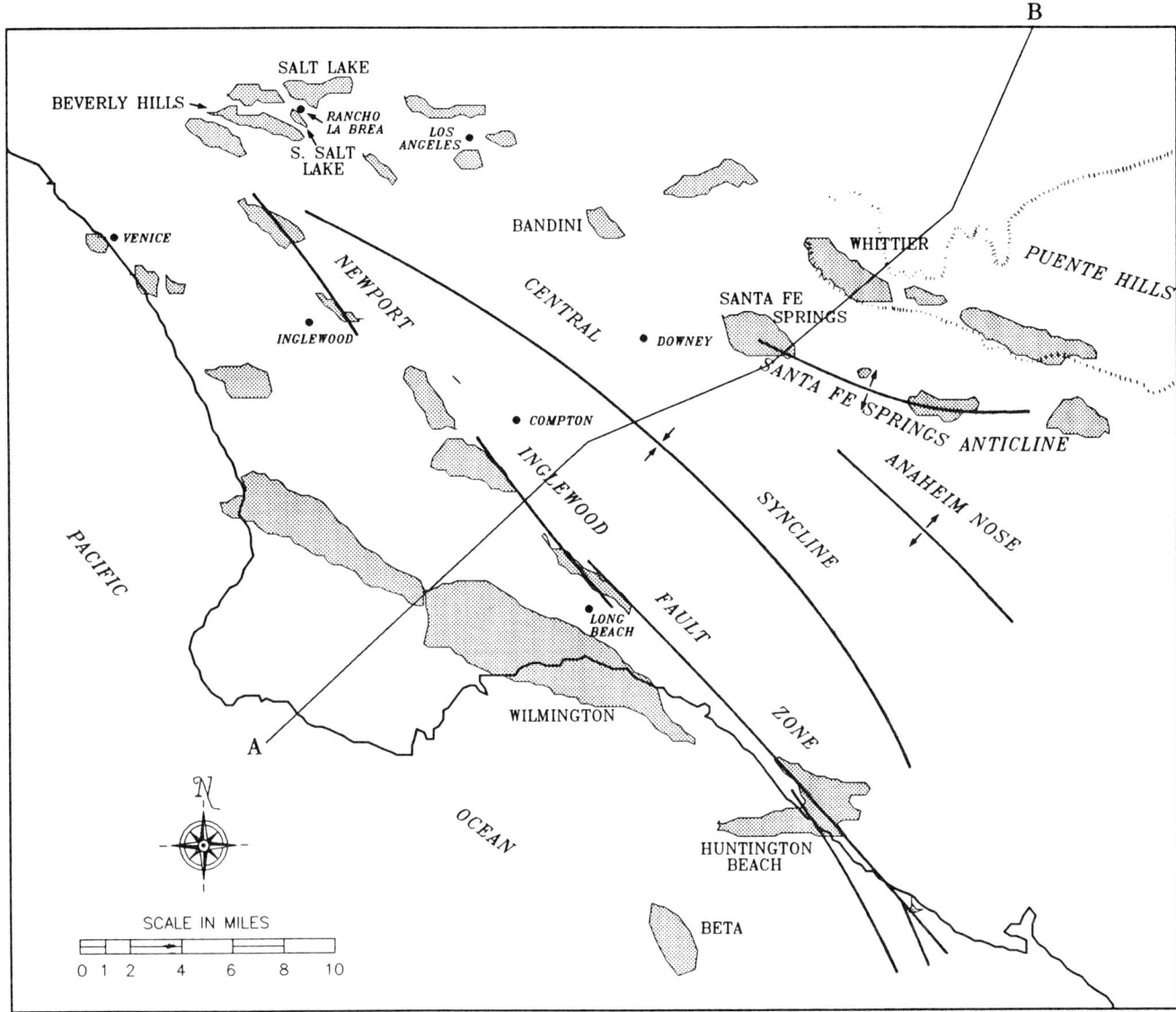

Figure 1. Map of the Los Angeles basin showing oil fields, major geological features, and topographical features discussed in the text (after Yerkes et al., 1965). Cross section A-B is shown in Figure 2.

SOURCE ROCKS

Maturity

In the Los Angeles basin, significant oil generation occurs in the deeply buried mature strata only. Philippi (1965) presented some of the essential changes that have taken place in the organic matter of shales in the Los Angeles and Ventura basins. The composition of shale bitumen changed gradually with depth of burial and age, and a pronounced increase in the hydrocarbon content of the bitumen marked the onset of oil generation. In the Ventura basin, the upper Miocene shales at around 15,000 ft (4.5 km) have not reached optimum maturity; whereas in the Los Angeles basin, the bulk of the oil generation takes place deeper than 8000 ft (2.5 km). The difference in the depth-maturity relationship between these two basins was ascribed by Philippi (1965) to differences in the geothermal gradient. The present-day geothermal gradient in the Los Angeles basin is 3.91°C/100 m (2.15°F/100 ft). The Ventura basin appears to have a lower geothermal gradient of 2.66°C/100 m (1.46°F/100 ft). McCulloh et al. (1978) indicated that geothermal gradients in the Los Angeles basin are highest along the southwest margin, the northern margin, and the uplifted area around the Santa Fe Springs anticline (Figure 3); they are lowest in the central synclinal trough. Bostick et al. (1978) believe that, in the central synclinal portion of the Los Angeles basin, the present rock temperatures are the historical maxima reached in the lithologic strata.

Vitrinite reflectance measurements (R_o) were reported by Bostick (1979) to be, in general, a good indicator of thermal

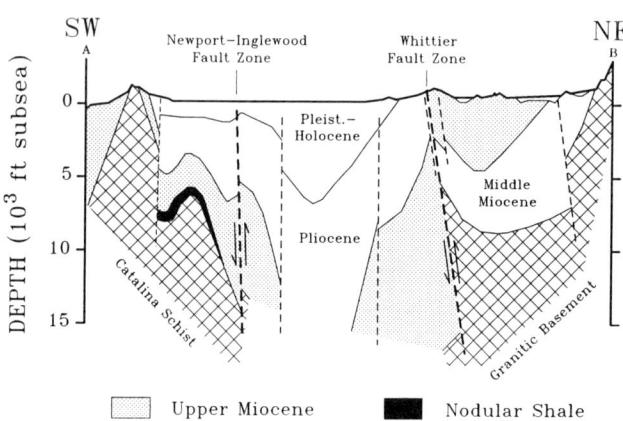

Figure 2. Subsurface cross section of the Los Angeles basin along line A–B in Figure 1 (from California Division of Oil and Gas, 1974).

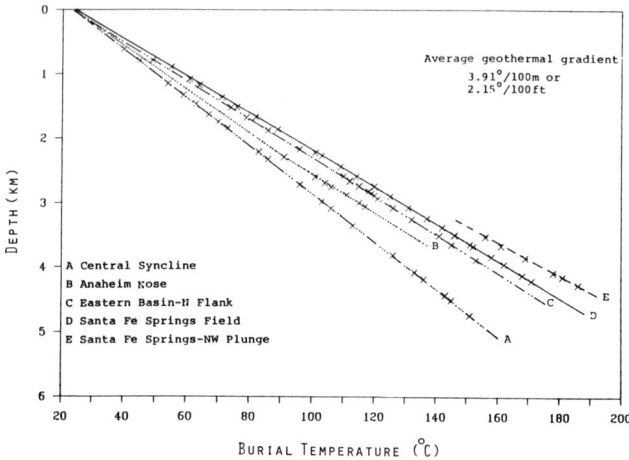

Figure 3. Geothermal gradients for five locations within the Los Angeles basin. Gradients range from lowest in the central syncline to highest at Santa Fe Springs (from Bostick et al., 1978).

maturity. Hutton and Cook (1980), Kalkreuth (1982), and Walker et al. (1983), however, measured anomalously low R_o values in source rocks that had experienced temperatures above 150°C and had, in some cases, generated oil. Walker et al. (1983) calculated time-temperature index (TTI) values (Lopatin, 1971; Waples, 1980) for sedimentary rocks in the northwestern Los Angeles basin and compared the calculated maturity estimates from the TTI values with the measured R_o values. The TTI values were in the range of 2 to 44, which Walker et al. (1983) calculated to be equivalent to vitrinite reflectances of 0.40% to 0.88%. The measured values, in contrast, ranged from 0.13% to 0.41%, with only two deeply buried samples showing reflectance values exceeding 0.35% R_o. The calculated values suggest that the organic matter in these deeply buried sedimentary rocks should be sufficiently mature for hydrocarbon generation.

Anomalously low R_o values have been reported for other oil-prone basins (Bostick et al., 1978; Hutton and Cook, 1980; Snowdon and Powell, 1982). Anomalously low R_o values may be related to the predominant kerogen type in the source rock: Jones and Edison (1978), Hutton and Cook (1980), Hutton et al. (1980), Kalkreuth (1982), Snowdon and Powell (1982), Walker (1982), Walker et al. (1983), and Price and Barker (1985) have demonstrated that, at the same level of maturity, the R_o value of a kerogen composed of largely herbaceous material is higher than that of a kerogen containing primarily alginite (Figure 4). Snowdon and Powell (1982) suggested that alginite (an exinite maceral) matures more rapidly than humic material, thus producing oil at lower vitrinite reflectivities. The trend shown in Figure 4 has been interpreted by Walker et al. (1983) and by Price and Barker (1985) as being due to the early maturation of alginite inhibiting maturation in vitrinite, resulting in low (suppressed) R_o values.

A mechanism for early generation of oil in sulfur-rich Monterey Formation kerogens of the Santa Maria basin has been reported by Orr (1986). The predominance of weak sulfur bonds in these kerogens leads to the generation of sulfur-rich oil at significantly lower maturities than similar kerogens with low sulfur content. This mechanism may account for the low maturity estimates of some apparently productive source rocks in the Los Angeles basin. However, this process is probably not as important volumetrically in the Los Angeles basin compared to the Santa Maria basin, because high-sulfur oils are much less prevalent in the Los Angeles basin (Global Geochemistry Corporation, unpublished data).

Hydrocarbon Source Potential

Of the Tertiary sediments deposited in the Los Angeles basin, those of Miocene age are considered to have the best oil source potential. Philippi (1965) concluded that oil-like shale hydrocarbons are not observed in sections above the upper Miocene Divisions D and E (Mohnian and Luisian, respectively) in the Los Angeles basin. Sedimentary rocks of late Miocene age are widely represented throughout the Los Angeles basin and are lithologically assigned to one of three formations: the Puente, Modelo, or Monterey formations.

Puente and Modelo Formations

Rocks of the Puente Formation, with a composite maximum thickness of about 13,000 ft (4 km) locally, consist of shales, siltstones, and sandstones. Most of the shale rock types of the Puente Formation are fairly rich in organic matter, with a wide range of total organic carbon (TOC) contents up to around 16%, with an average of about 4% (Global Geochemistry Corporation, unpublished data). The kerogen in these source rocks consists largely of marine sapropel (a mixture of type I and II kerogens) with excellent hydrocarbon generating potential (Global Geochemistry Corporation, unpublished data). Rough calculations [based on an average TOC of 4%, a kerogen composition of around

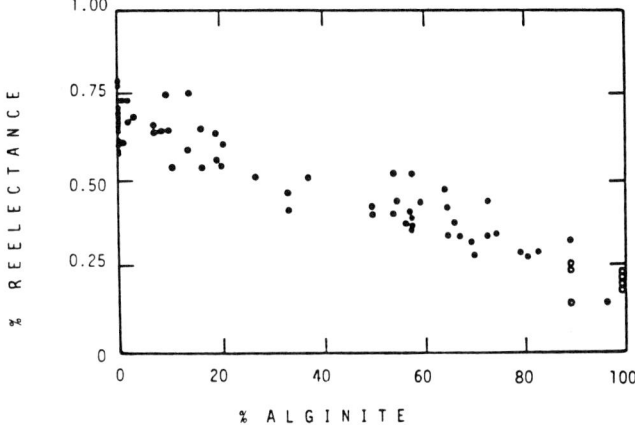

Figure 4. Plot of vitrinite reflectance versus percentage of alginite in kerogens with similar time-temperature histories (from Hutton et al., 1980; Walker et al., 1983).

80% sapropelic particles, and an aggregate shale thickness of about 1000 ft (300m)] provide an estimate that shales of the Puente Formation have the potential to generate around 515,000 barrels of oil/acre at optimum maturity. Thermal maturity data (Bostick et al., 1978), however, indicate that these potential source rocks are immature to marginally mature by conventional estimates and would require greater burial depth before any significant oil generation could occur.

In the northwestern and northeastern part of the Los Angeles basin, the rocks of late Miocene age have been assigned to the Modelo Formation. This formation consists of brown to brownish gray diatomaceous shales interbedded with sandstones. The "nodular shale" is a significant, phosphatic-rich, potential source unit at the base of the Modelo Formation and contains up to 10% total organic carbon content (Walker et al., 1983). Microscopic examination of organic matter in this source unit (Walker et al., 1983) shows that it consists of 80% to 90% sapropelic material with minor amounts of terrigenous particles, deposited in a marine environment under slightly reducing conditions. This distinctive unit is considered to have significant oil source potential in the western portion of the Los Angeles basin (Hoots et al., 1935; Metzner, 1935; Wissler, 1943; Reed, 1951; Philippi, 1965). "Nodular shales" with an aggregate thickness of around 500 ft (150 m), an average TOC of 5.6%, and a kerogen composition of about 85% sapropelic particles would have an estimated hydrocarbon potential of about 386,000 barrels of oil/acre at optimum maturity. These rocks appear to be only marginally mature by conventional estimates (Walker et al., 1983). However, in view of the high sapropel content of "nodular shales," these rocks may have generated significant quantities of liquid hydrocarbons, as discussed earlier.

Monterey Formation

The third formation to which Miocene rocks in the Los Angeles basin have been assigned is the Monterey Formation. The organic geochemistry of this formation has been discussed by Philippi (1965, 1975), Giger and Schaffner (1979, 1981), Claypool et al. (1979), King and Claypool (1983), Kruge (1983), Curiale et al. (1985), and Orr (1986).

The Miocene Monterey shales in the Los Angeles basin, especially those from Divisions D and E, were identified by Philippi (1965) as the major source of the oils in the basin. The Monterey shales are relatively organic rich in the Los Angeles basin, containing between 2.0% and 18.0% TOC, with an average of about 4.0% TOC (Philippi, 1975). This is relatively high compared to the worldwide average of 2.16% for shale petroleum source rocks (Tissot and Welte, 1978). Published geochemical data on the Los Angeles basin Monterey Formation source rocks (Philippi, 1965, 1975; Bostick et al., 1978; Price, 1983; Price and Barker, 1985) show that the rocks resemble, in many respects, Monterey Formation rocks from other southern California basins (Claypool et al., 1979; McEvoy et al., 1981; Rullkotter et al., 1981; Simoneit and Mazurek, 1981; Summerhayes, 1981; Katz and Elrod, 1983; King and Claypool, 1983).

Monterey shales contain a high proportion of marine sapropelic kerogen (Katz and Elrod, 1983) with subordinate quantities of terrigenous material. The Monterey shales were mainly deposited in a marine environment under anoxic conditions (Simoneit and Mazurek, 1981). Analysis of the lipids, n-alkanes, and aromatic hydrocarbons by various authors confirms that both marine and terrigenous organic matter are present. Gilbert and Summerhayes (1981) estimated the relative content of amorphous (marine) organic matter to be between 55% and 95%, with the median around 75% to 80%. These authors also reported the abundant occurrence of phosphorite in the Monterey Formation. Katz and Elrod (1983) reported that the majority of Monterey shales have relatively high hydrogen indices (mostly above 300 mg/g org C; Figure 5) indicative of type I and type II kerogen with excellent hydrocarbon-generating potential.

Assuming, in the Los Angeles basin, an average TOC content of 4.0%, a sapropelic kerogen content of about 75%, and a thickness of about 1000 ft (300 m), the diatomaceous Monterey shales have an oil-generating potential of about 455,000 barrels of oil/acre. At their present level of maturity, (mostly marginally mature; Bostick et al., 1978), a fair quantity of this potential will already have been generated.

Although most of the available data tend to point to in-situ production of the bitumen in the sedimentary rocks of the Monterey Formation, there is also evidence of migration. This evidence comes, in part, from the presence of mature sterane and hopane isomers in shallow, immature samples. In an early paper, Claypool et al. (1979) concluded that the composition of bitumens they studied in Miocene shales did not support contamination by migrating petroleum. They indicated that the bitumen present in the source rocks analyzed may have been generated at an early stage of thermal maturity. However, in a later study, King and Claypool (1983) extracted and fractionated bitumen from six upper Neogene shales and compared them with one oil from the Santa Maria Valley field (14° API). The

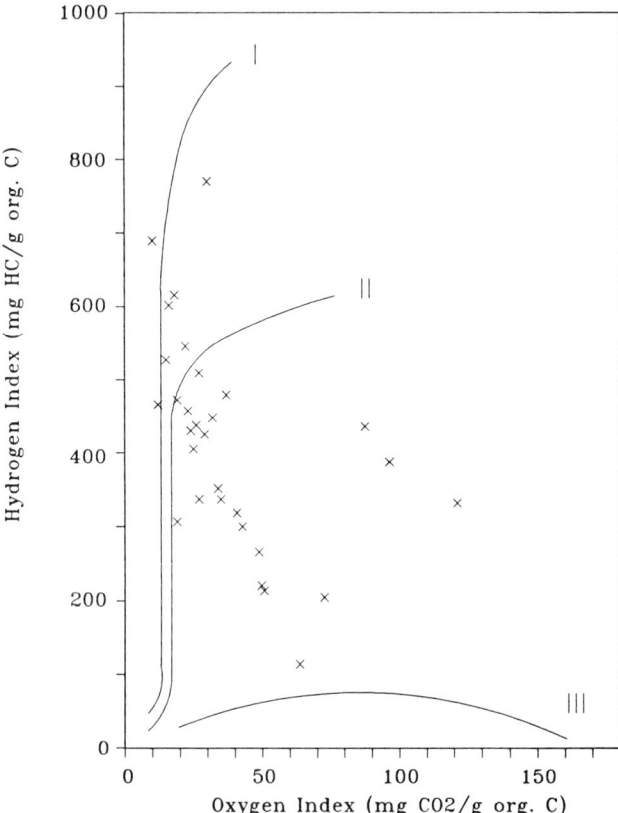

Figure 5. Rock-Eval data for Monterey Shale from the Lost Hills field plotted on a modified Van Krevelen diagram (from Kruge, 1983).

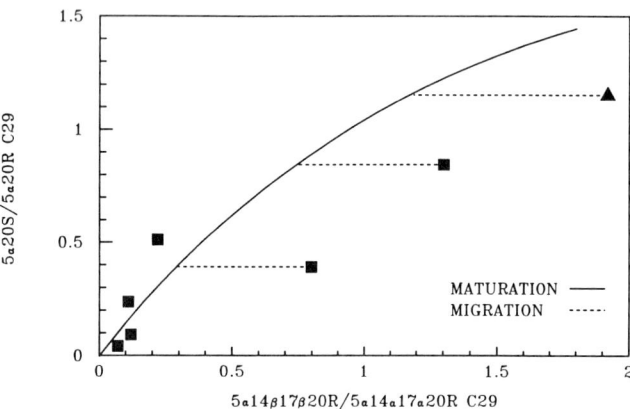

Figure 6. Maturation-migration pathways based on a plot of C_{29} sterane isomer ratios (Seifert and Moldowan, 1981), with bitumens from the Point Conception deep test well (squares) and oil from the Santa Maria basin (triangle) superimposed (from King and Claypool, 1983). The diagonal curve illustrates the normal maturity trend of increasing $\alpha\alpha\alpha$20S and $\alpha\alpha\beta$20R C_{29} steranes with respect to $\alpha\alpha\alpha$20R C_{29} sterane. Horizontal lines indicate enrichment in $\alpha\beta\beta$20R C_{29} sterane, which was attributed by Seifert and Moldowan (1981) to migration. Four bitumens near the maturity line appear to contain indigenous hydrocarbons, whereas two bitumens, shifted to the right of the line, are enriched in migrated hydrocarbons, similar to the oil sample from the Santa Maria basin.

main purpose was to investigate biomarker compounds in these shales to determine whether the relatively heavy bitumens were indigenous or had migrated from rocks at different stratigraphic horizons. Based on vitrinite reflectance data of these samples (R_o = 0.27% to 0.30%), the geothermal gradient, the age of the sedimentary rocks, and a Waples-type (1980) TTI value, King and Claypool concluded that the suite of samples contained immature kerogen. However, by superimposing the biomarker data of bitumens on other geochemical data, they suggested that two bitumen samples were more mature than the associated shale kerogen and had migrated from a deeper horizon. As further evidence, they plotted two other parameters (Figure 6): one indicative of maturation—the ratio of the 20S/20R 5α(H)14α(H)17α(H) isomers of the C_{29} sterane—and the other a migration indicator—the ratio of the 5α(H)14β(H)17β(H)20R/ 5α(H)14α(H)17α(H)20R isomers of the C_{29} sterane. This figure confirmed that the two bitumens migrated from depths greater than those at which they were found.

Conclusions

Most of the Miocene source rocks from the Los Angeles basin are organic rich and contain mainly sapropelic kerogens with good to excellent oil-generating potential. They are low in maturity by conventional estimates, although in deeper parts of the basin these potential source rocks would be able to generate significant quantities of oil and gas. The source rocks may also be capable of generating oil at a lower maturity than is conventionally accepted. The presence of heavy bitumens and/or low to medium gravity oils in fractured Monterey shale reservoirs may be the result of either fractionation during migration of initially heavy oil or a mixture of more mature oil from deeper parts of the depocenter with immature, in-situ bitumens.

OILS

Introduction

Oil quality is a major consideration in petroleum exploration in the Tertiary basins of California. This is especially true of Miocene targets where oils of Miocene Monterey shale origin can be extremely heavy and have very high sulfur contents. Low-quality oils such as these can cause production problems, especially in offshore fields.

Oil quality varies considerably throughout the Los Angeles basin and can be used to identify the following four major oil groups (with representative oils):

1. Light, low-sulfur oils (e.g., Bandini)

2. Medium gravity, medium-sulfur oils (e.g., Wilmington/237)

3. Heavy, medium-sulfur oils (e.g., Wilmington/Ranger)

4. Heavy, high-sulfur oils (e.g., Beta)

High-quality oils in group 1 are found mainly in fields to the north and east of the central syncline (see Figure 1). Oils in group 2 are present throughout the northwestern portion of the basin and in deeper reservoirs in the giant fields to the south (Wilmington and Huntington Beach). Oils in group 3 are found in shallower reservoirs in the southern fields. Farther south, the low-quality oils of group 4 are characteristic of the offshore Beta field.

There is a wide variation in the composition of the four representative oils (Table 1). Oils in the northeastern fields, illustrated by the Bandini oil, are enriched in saturated hydrocarbons; whereas oils in the southern fields, such as the Wilmington/Ranger sample, are enriched in polar compounds. The majority of oils in the Los Angeles basin have nearly equivalent amounts of saturated, aromatic, and polar constituents, as typified by the Wilmington/237 sample, and fall within the category of aromatic oils as classified by Tissot and Welte (1978).

The variation in oil quality does not appear to be related to significant differences in source rock type. Carbon isotope ratios of saturate and aromatic fractions of the oils are similar (Figure 7) and fall within the field of marine oils as defined by Sofer (1984). The two Wilmington oils and the Beta oil fall in the same isotopic range as extracts from Miocene source rocks in the Los Angeles basin (Global Geochemistry Corporation, unpublished data). The somewhat heavier (enriched in ^{13}C) values for the Bandini oil are similar to those for several mature, high-quality oils in the Santa Maria basin (Global Geochemistry Corporation, unpublished data).

Another source indicator, proposed by Huang and Meinshein (1979), is given by the relative amounts of C_{27}, C_{28}, and C_{29} $\alpha\alpha\alpha$(R)steranes. The four representative oils are very similar in the relative proportions of the three steranes, as shown in a ternary diagram (Figure 8), and the relative enrichment in the C_{27} sterane is consistent with a marine source. The progressive increase in C_{27} content from the Beta to the Bandini oil may be maturity related, because Mackenzie (1984), studying genetically similar oils, reported relatively higher C_{27} steranes in more mature oils.

Group 1 Oils

Several parameters (Table 2) indicate that the group 1 Bandini oil is more mature than the lower quality oils, with the group 4 Beta oil being the least mature. Pristane/n-C_{17} and phytane/n-C_{18} ratios from saturate gas chromatography (GC) analysis are commonly used as maturity indices, with lower ratios indicating more mature samples. The ratios can be altered by biodegradation with the normal alkanes being preferentially consumed. This process would result in higher pristane/n-C_{17} and phytane/n-C_{18} ratios leading to low maturity estimates. Extensive biodegradation appears to be significant only for the Wilmington/Ranger oil (Figure 9). Higher molecular weight biomarkers are less influenced by biodegradation, and isomerization at chiral centers, such as the C-20 position in steranes, have been extensively used to assess maturity (e.g., Mackenzie et al., 1980; Seifert and Moldowan, 1981). The biomarker ratios in Table 2 confirm the maturity trends based on GC data. Isomerization at C-14 and C-17 in steranes, as measured by the $\alpha\beta\beta/\alpha\alpha\alpha$ C_{29} sterane ratio (Table 2), may be influenced by migration as well as maturity (Seifert and Moldowan, 1981). The relationship of oil quality with maturity is also evident in a general trend of increasing API gravity and decreasing sulfur content with decreasing pristane/n-C_{17} ratios in a larger suite of nonbiodegraded oils throughout the basin (Global Geochemistry Corporation, unpublished data).

Migration parameters based on sterane ratios (Seifert and Moldowan, 1978) and terpane ratios (Seifert et al., 1980) are listed in Table 3. The Bandini oil appears to be somewhat more affected by migration than the lower quality oils, based on the relative enrichment in extended diterpanes compared to hopane.

Pristane/phytane ratios and two biomarker ratios found to be useful in distinguishing the depositional environment of source material of oils in other Tertiary basins of southern California indicate that the high-quality oils appear to have been generated from source rocks laid down in a more oxidizing environment; whereas the lower quality oils are associated with a more reducing environment (Table 4).

Table 1. Geochemical characteristics of four Los Angeles basin oils.

Field	Age/Formation	Zone/Member	Gravity (° API)	Saturate (% rec)	Aromatic (% rec)	Polar (% rec)	Asphalt (wt %)	Sulfur (wt %)	Nitrogen (wt %)
Bandini	Pliocene L. Repetto		35.6	65.4	24.0	10.6	0.5	0.06	0.08
Wilmington	Miocene Puente	237	27.6	38.2	34.1	27.8	5.3	1.22	0.48
Wilmington	Miocene Puente	Ranger	19.0	19.3	32.8	48.0	4.7	1.24	0.64
Beta	L. Miocene Delmontian		17.3	18.4	44.8	36.8	12.8	2.94	0.77

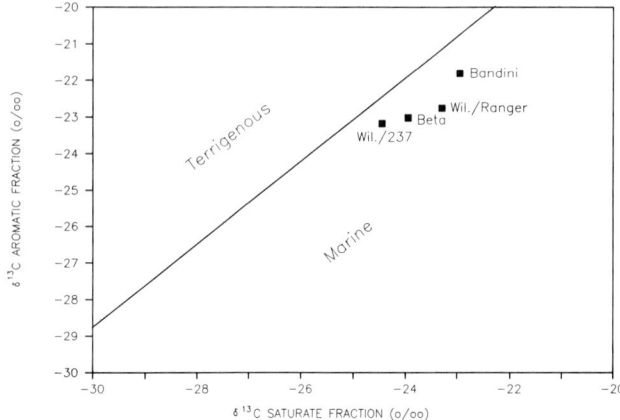

Figure 7. Plot of carbon isotope ratio of saturate fraction versus aromatic fraction of Los Angeles basin oils from the Bandini and Beta fields and the Ranger and 237 zones of the Wilmington field. Source environments are defined by Sofer (1984).

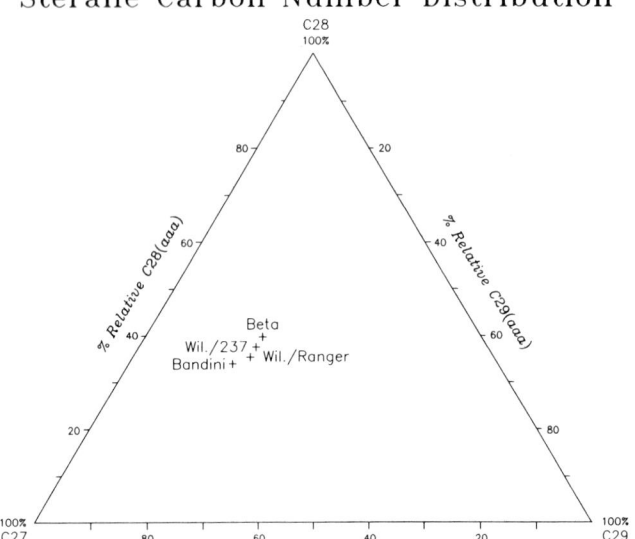

Figure 8. Ternary plot of the C_{27}, C_{28}, and C_{29} $\alpha\alpha\alpha$R sterane distribution in Los Angeles basin oils from the Bandini and Beta fields and the Ranger and 237 zones of the Wilmington field.

Table 2. Maturation parameters and the values for four Los Angeles basin oils.

Field/Zone	Pristane/ n-C_{17}[1]	Phytane/ n-C_{18}[1]	$\alpha\beta\beta$R/$\alpha\alpha\alpha$R C_{29} steranes [2,A]	20S/20R$\alpha\alpha\alpha$ C_{29} sterane [2,B]
Bandini	0.83	0.40	1.05	0.58
Wilmington/237	1.03	0.95	0.66	0.50
Wilmington/Ranger	—	—	0.80	0.53
Beta	1.99	1.57	0.59	0.28

[1]Parameter decreases with increasing maturity
[2]Parameter increases with increasing maturity
[A]Seifert and Moldowan (1979)
[B]Discussed in Mackenzie (1984)

The ratio of tri/mono + tri aromatic steroids in Table 4 has also been related to maturity by Mackenzie et al. (1981), but in the Santa Maria and Ventura basins, this ratio appears to be strongly influenced by the clastic content of the source rock and the redox conditions at deposition (Global Geochemistry Corporation, unpublished data). The relationship between oil quality and pristane/phytane ratio, illustrated by the four representative oils in Table 4, is typical of Los Angeles basin oils. The higher ratio in group 1 oils suggests a trend towards more oxidizing conditions of deposition from southwest to northeast in the basin.

Group 2 and Group 3 Oils

The lower quality oils of groups 2 and 3 are represented by the 237 zone and Ranger zone oils, respectively, of the Wilmington field. The almost complete lack of saturate hydrocarbons in the Ranger sample (Figure 9) prevents comparison of the oils by GC. However, comparison can be made by gas chromatography–mass spectrometry (GC–MS) analysis of biomarkers and aromatic compounds. The triterpane mass chromatograms are very similar (Figure 10). Also, the distributions of C_{27}, C_{28}, and C_{29} steranes (Figure 8) and of phenanthrenes, benzothiophenes and dibenzothiophenes (Figure 11) are very similar in both oils from the Wilmington field. These close correlations indicate that the oils are genetically similar and suggest that biodegradation of oils similar to the Wilmington/237 sample has resulted in a marked loss of saturate hydrocarbons and a decrease in API gravity, producing the group 3 oils characteristic of shallow accumulations in the Wilmington and Huntington Beach fields. This conclusion is confirmed by the depth distribution of API gravity within the

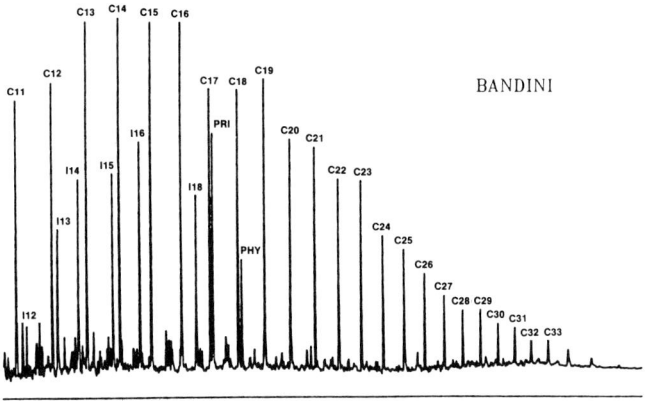
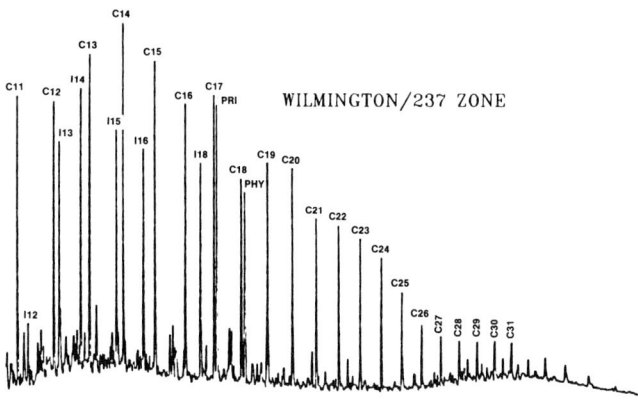
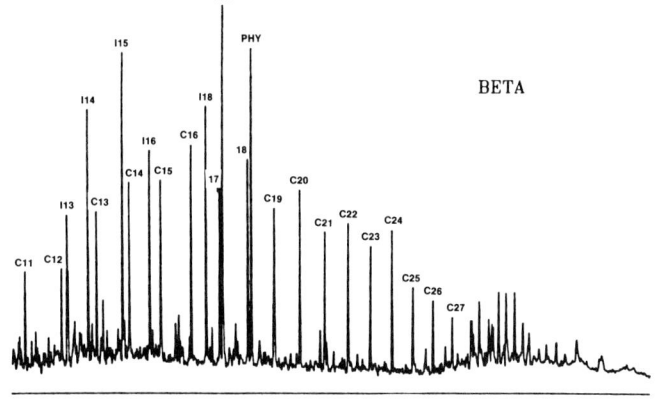
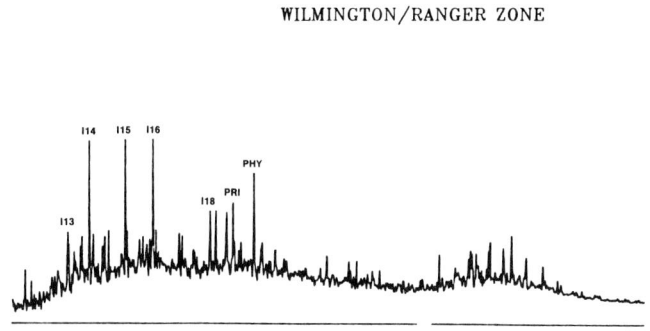

Figure 9. Gas chromatograms of Los Angeles basin oils from the Bandini and Beta fields and the Ranger and 237 zones of the Wilmington field. Absence of normal alkanes in oil from the Ranger zone of the Wilmington field indicates biodegradation.

Table 3. Migration parameters and the values for four Los Angeles basin oils.

Field/Zone	$\beta\alpha\alpha S/\alpha\alpha\alpha R$ $C_{27} + C_{28} + C_{29}$ steranes [1,A]	Extended diterpanes/ hopane [1,B]
Bandini	0.27	0.455
Wilmington/237	0.26	0.24
Wilmington/Ranger	0.17	0.265
Beta	0.11	0.30

[1]Parameter increases with increasing migration
[A]Seifert and Moldowan (1978)
[B]Seifert et al. (1980)

Wilmington field (Figure 12), showing progressively heavier oils nearer the surface. Biodegradation also correlates with a small increase in sulfur content (Figure 12), perhaps by concentration in the unaltered fraction. Biodegradation is an important influence on oil quality in the Los Angeles basin, and its effects on oil and natural gas are illustrated in more detail below in a suite of samples from the Salt Lake region in the northwest of the basin.

In the Wilmington field, the shallow Ranger oil appears to be at least as mature as the deeper 237 oil and is probably not derived from neighboring shallow sedimentary rocks. This inference is consistent with the conclusions of Philippi (1965) and with the more recent studies outlined in the previous section and points to a source in deeper, more mature rocks.

Group 4 Oils

Oils in group 4 have suffered only minor alteration, as illustrated by the saturate GC of the Beta sample (Figure 9), so the low API gravity and higher sulfur content do not appear to be due to biodegradation. The representative oil is relatively immature, as illustrated in particular by the sterane biomarker ratios in Table 2. In the aromatic fraction of the oil, the ratio of dibenzothiophene to phenanthrene is high; while this ratio in the high-quality oil from the Bandini field is low (Figure 11). The relative amounts of the two aromatic compounds is believed to be controlled by the depositional environment and redox conditions, as depicted in Figure 11. This scheme indicates

Table 4. Depositional environment parameters and the values for four Los Angeles basin oils.

Field/Zone	Pristane/ phytane [1,A]	Tri/mono + tri aromatic steroids [1,B]	Hopane/bisnorhopane + hopane[1,B]
Bandini	2.10	0.76	0.68
Wilmington/237	1.42	0.60	0.67
Wilmington/Ranger	0.92	0.57	0.69
Beta	1.09	0.45	0.53

[1]Parameter increases with more oxidizing environment
[A]Tissot and Welte (1978)
[B]Global Geochemistry Corporation, unpublished report

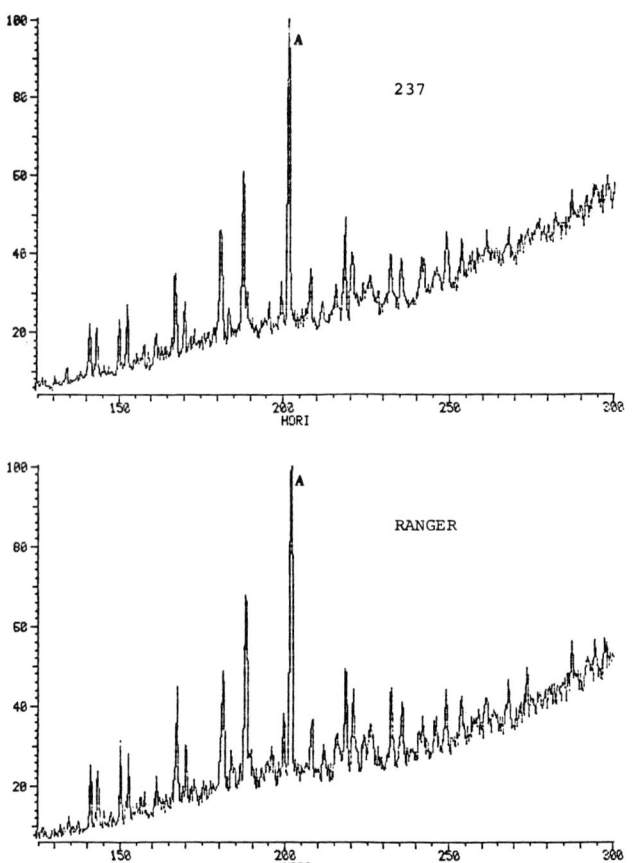

Figure 10. Triterpane mass chromatograms of oils from the Ranger and 237 zones of the Wilmington field. The major peak, designated as A, is 17α21β hopane.

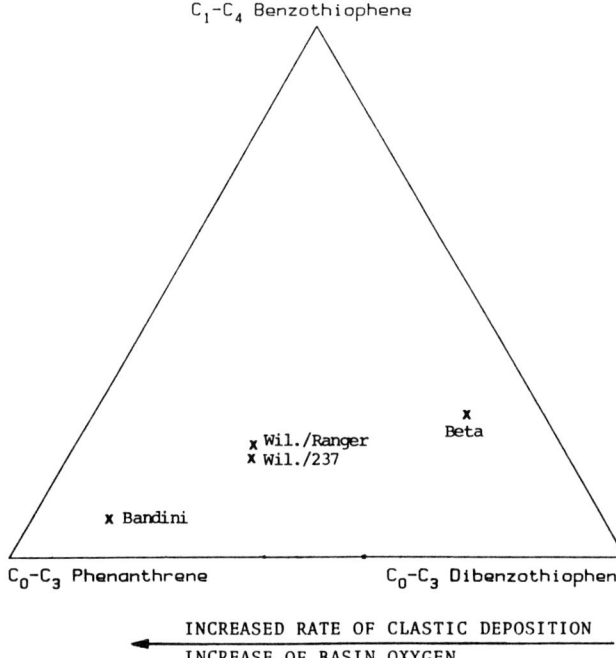

Figure 11. Ternary plot of the distribution of benzothiophenes, phenanthrenes, and dibenzothiophenes in Los Angeles basin oils from the Bandini and Beta fields and the Ranger and 237 zones of the Wilmington field. Interpretation of depositional environment derived from unpublished data of Global Geochemistry Corporation for the Santa Maria and Ventura basins. Application to the Los Angeles basin is discussed in the text.

a clastic-poor, reducing environment for the source of the Beta oil. The Beta oil is also rich in bisnorhopane, the structure of which was first reported by Seifert et al. (1978). Bisnorhopane is a characteristic biomarker in Monterey source rocks. The Beta oil is similar in several respects (e.g., enrichment in bisnorhopane and dibenzothiophene) to the heavy, high-sulfur oils of the Santa Maria basin that have been closely correlated with the siliceous Monterey Shale (Milner et al., 1977; Magoon and Isaacs, 1983; Global Geochemistry Corporation, unpublished report). It appears likely that the Beta oil and group 4 oils in general are derived from source rocks closely analogous to the Monterey.

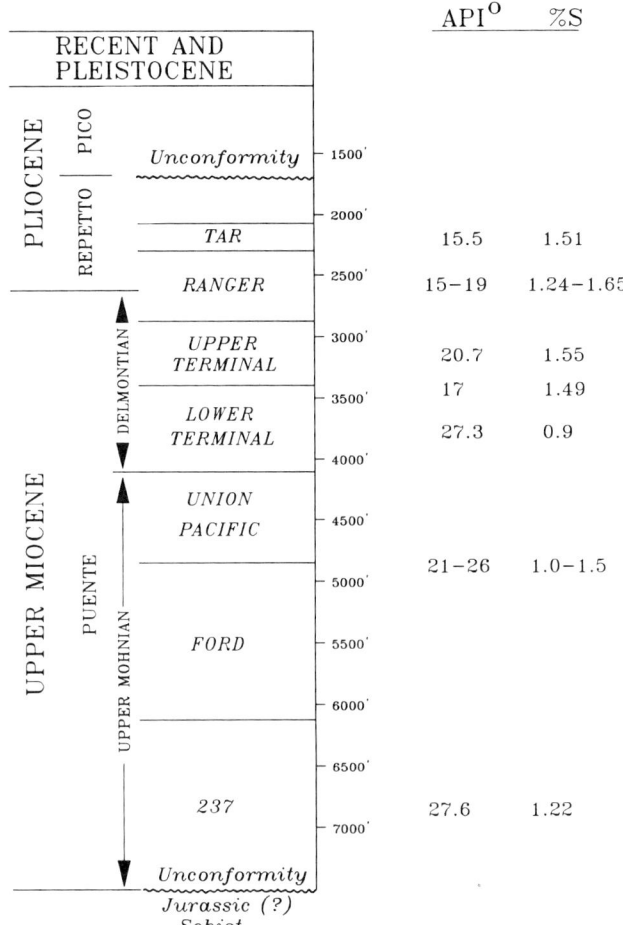

Figure 12. Depth distribution of API gravity and sulfur content of nine oils from several wells in the Wilmington field. Stratigraphy and oil-producing zones are from Mayuga (1970).

Discussion

Maturation, migration, and depositional environment may all contribute to the high quality of the oils in the reservoirs of the northeastern basin. These oils may have migrated from deeper, more mature source rocks in the northern arm of the central syncline, which is located closer to the presumed source of clastic sedimentary input to the forming basin. Oils in the southern part of the basin, in contrast, appear to be derived from sediments laid down in a more distal environment, farther from the source of clastic input. Less clastic input in a more reducing environment would enhance preservation of organic matter in sediments. This is consistent with higher TOC contents in Miocene source rocks west of the Newport-Inglewood fault zone (1.2% to 16.3%; mean 5.0%) compared to those east of the fault zone (1.6% to 3.9%; mean 2.6%) (Global Geochemistry Corporation, unpublished data). In addition, the "nodular shale," a black, oily phosphatic shale, considered to be the major source of the oil in the northwestern part of the Los Angeles basin by Walker et al. (1983), appears to be confined to the area west of the fault zone (Figure 2). East of the fault zone, time-equivalent rocks are present but are no longer lithologically identifiable as "nodular shale" (E. B. Edwards, personal communication, 1986). The organic richness of the source rocks to the west and south of the central syncline is the most probable reason for the greater quantities of oil in this part of the basin, and we propose that the depositional environment in which the rocks were laid down is, in part, responsible for the generally lower quality of the oils in this part of the basin.

Conclusions

Los Angeles basin oils, although markedly variable in physical and chemical properties, are nonetheless genetically similar and appear to be sourced in upper Miocene sedimentary rocks containing kerogen of marine origin. The high quality of oils in the northeast of the basin may be due to a combination of high maturity, migration, and a more oxidizing depositional environment of the source rocks. The lower quality of oils, reservoired in the northwest and southern parts of the basin, may be due to a more reducing depositional environment that has enhanced organic preservation in sediments. This difference probably accounts for the larger volumes of oil in reservoirs in the southern parts of the basin. Biodegradation of migrated oils in shallow reservoirs of the Wilmington and Huntington Beach fields has resulted in even heavier oils.

NATURAL GASES

Through 1985, 7.2 trillion cubic feet (tcf) of natural gas have been produced in the Los Angeles basin (California Division of Oil and Gas, 1986). Essentially all of this gas is derived from oil fields and is classified as associated.

Basic models for natural-gas geochemistry have long been established (Fuex, 1977; Bernard, 1978; Schoell, 1980; Rice and Claypool, 1981; Schoell, 1983), but many of the factors that control the geochemistry of associated gases are still poorly understood. It is well known, for example, that methane $\delta^{13}C$ values in nonassociated gases are strongly correlated with source rock maturity for rocks of similar source type (Stahl, 1977; Schoell, 1983). In associated gases, however, methane $\delta^{13}C$ values can be altered by the secondary cracking of oil, and a clear relationship with source rock maturity is not as apparent (Schoell, 1984). Genetic models based on isotopic differences among the light hydrocarbon gases have been proposed by James (1983) and Sundberg and Bennett (1983), but these models may not take adequate account of variations in organic facies and the timing of hydrocarbon generation versus trap formation. Another potential factor controlling the geochemistry of associated natural gases is production history. Oil production typically continues long after gas pressures have been exhausted and may produce isotopic as well as chemical fractionations. This problem has not been explicitly addressed in the literature, but it could be pertinent in the Los Angeles basin where many fields have been under production for over 50 years.

Despite such problems, natural-gas studies have made important contributions to our understanding of the origin

Table 5. Los Angeles basin gas samples.

Sample no.	Field	Well	Location	Formation	Formation age	Average depth (ft)
1360	Beta	C-1	OCS-P 0301	A through F sands	late Miocene	3,200
1361	Beta	C-34	OCS-P 0301	B, D, E, and F sands	late Miocene	3,900
1362	Beta	C-42	OCS-P 0301	A, B, D, E, and F sands	late Miocene	3,200
1185	Huntington Beach	North Bolsa Lease #111-G	Sec. 28 T5S R11W	Lower Ashton	late Miocene	4,915
1183	Huntington Beach Offshore	State PRC 425.1 #504-A	Sec. 4 T6S R11W	Lower Main	late Miocene	4,200
1184	Huntington Beach Offshore	State PRC 426.1 #4148	Sec. 4 T6S R11W	Upper Main	late Miocene	4,000
1092	Salt Lake, South	P-75	Sec. 29 T1S R14W	Clifton-Dunsmuir	late Miocene/ early Pliocene	2,180
1091	Salt Lake, South	P-80	Sec. 29 T1S R14W	Clifton	early Pliocene	2,930
1093	Salt Lake, West	S-4	Sec. 20 T1S R14W	Puente	late Miocene	1,960
1359	Santa Fe Springs	Sante Fe #757-S	Sec. 5 T3S R11W	Lower Santa Fe-Bell 100	late Miocene	8,500
1173	Whittier	Central Fee #129	Sec. 23 T2S R11W	Puente	late Miocene	2,500
1175	Whittier	Murphy Whittier #228	Sec. 26 T2S R11W	Repetto (2nd & 3rd zone)	early Pliocene	2,880
1174	Whittier	Murphy Whittier #257	Sec. 26 T2S R11W	Puente (6th zone)	late Miocene	3,950
1171	Wilmington Offshore	Island Chaffee C-416	Sec. 16 T5S R12W	Puente (237 zone)	late Miocene	9,610
1057	Seep	L.A.C. Well, Ogden & 3rd	Sec. 21 T1S R14W	Alluvium	late Pleistocene	40
1058	Seep	L.A.C. Well, Ogden & 3rd	Sec. 21 T1S R14W	Alluvium	late Pleistocene	40
1059	Seep	L.A.C. Well, Ogden & 3rd	Sec. 21 T1S R14W	Alluvium	late Pleistocene	40
1053	Seep	Fairfax District	Sec. 21 T1S R14W	Surface	---	---
1054	Seep	Fairfax District	Sec. 21 T1S R14W	Surface	---	---
1055	Seep	Fairfax District	Sec. 21 T1S R14W	Surface	---	---
1056	Seep	Fairfax District	Sec. 21 T1S R14W	Surface	---	---
1060	Seep	Lake Pit, Rancho La Brea	Sec. 21 T1S R14W	Surface	---	---
1061	Seep	Lake Pit, Rancho La Brea	Sec. 21 T1S R14W	Surface	---	---

and fate of oil, the genetic relationships between different oil fields, and the secondary processes that have influenced their geochemistry. In the Los Angeles basin, natural gas data suggest the presence of regional differences in gas maturity, of localized mixing with microbial gas, and of widespread biodegradation in reservoirs at depths less than 5000 ft (1500 m).

Table 5 lists 23 gases from the Los Angeles basin that were analyzed at Global Geochemistry Corporation between 1985 and 1987. Fourteen of the samples were collected from producing wells in the Beta, Huntington Beach, Salt Lake, Santa Fe Springs, Whittier, and Wilmington oil fields. Nine gas samples were collected from surface seeps. Natural oil and gas seeps, including the famous Rancho La Brea tar pits, are found throughout the basin and have played an important role in the discovery of many fields during the late 1800s and early 1900s (Stock, 1956; Brown, 1968).

Gas compositions and stable isotope data are listed in Tables 6 and 7. Compositions were determined by GC and are uncorrected for small amounts of air contamination introduced during sampling. Major components were measured with a relative precision of plus or minus 3% (1 σ). The average analytical total of 97.3% reflects the fact that water vapor and C_{6+} hydrocarbons were not analyzed. $\delta^{13}C$ measurements were made with a 1 σ precision of plus or minus 0.2‰ or better; δD measurements were made with a 1 σ precision of plus or minus 5‰.

Not surprisingly, the oil field gases contain high concentrations of C_{2+} hydrocarbons. Wetness, the percentage of C_{2+} hydrocarbons relative to total hydrocarbon gases, ranges from 2% to 30%. The seep gases are relatively dry. Despite their association with biodegraded oil, the tar pit samples have less than 2% wetness. Carbon dioxide is a significant component in both types of gas and varies from less than 1% to as much as 37%. Many of the seep gases contain significant amounts of air.

Hydrocarbon Gases

With the exception of the Santa Fe Springs sample, methane $\delta^{13}C$ values vary from −51‰ to −40‰ and are within the range expected for associated thermogenic gases (Figure

Table 6. Chemical composition of Los Angeles basin gas samples.

Sample no.	C_1 (%)	C_2 (%)	C_3 (%)	i-C_4 (%)	n-C_4 (%)	i-C_5 (%)	n-C_5 (%)	Wetness (%)	CO_2 (%)	N_2 (%)	O_2 (%)	Ar (%)	He (%)	Total (%)
1360	84.2	1.85	2.67	0.808	1.45	0.510	0.352	8.32	3.99	0.866	0.148	0.009	n.d.	96.9
1361	85.0	2.01	2.28	0.569	1.05	0.386	0.299	7.20	2.53	1.32	0.132	0.009	0.003	95.6
1362	88.0	0.642	0.499	0.235	0.266	0.102	0.068	2.02	4.82	0.534	0.106	tr.	n.d.	95.2
1185	69.5	6.64	9.31	1.83	4.16	1.14	0.973	25.7	1.20	1.50	0.100	0.020	0.001	96.3
1183	69.3	3.77	5.77	2.22	2.42	1.17	0.286	18.4	8.34	2.19	0.881	0.056	n.d.	96.4
1184	85.6	1.32	1.92	0.846	1.47	1.36	1.12	8.59	0.257	2.43	0.122	0.039	n.d.	96.4
1092	82.5	2.35	1.53	0.384	0.643	0.269	0.145	6.06	10.0	0.096	0.020	tr.	n.d.	97.9
1091	78.8	4.78	5.29	1.18	2.87	1.23	0.526	16.8	1.55	0.081	0.022	0.001	n.d.	96.3
1093	80.5	1.59	0.598	0.230	0.378	0.171	0.068	3.63	14.4	0.071	0.014	0.001	n.d.	98.0
1359	76.8	6.85	5.23	1.21	2.56	1.03	0.975	18.9	0.269	0.575	0.154	tr.	n.d.	96.0
1173	71.4	3.19	1.94	0.495	0.653	0.366	0.180	8.72	18.7	1.04	0.178	0.013	n.d.	98.2
1175	58.9	1.37	0.205	0.117	0.038	0.089	0.015	3.02	37.2	0.188	0.047	0.003	n.d.	98.2
1174	86.5	6.08	3.93	0.596	1.27	0.374	0.337	12.7	0.477	0.425	0.014	0.005	0.002	100
1171	66.2	7.93	10.5	1.35	4.70	1.33	1.44	29.2	2.58	1.24	0.005	0.006	0.001	97.2
1057	82.6	0.734	0.238	0.034	0.035	0.073	0.038	1.38	2.12	10.5	1.13	0.146	tr.	97.6
1058	87.1	0.998	0.373	0.050	0.059	0.108	0.060	1.86	6.18	2.27	0.324	0.034	n.d.	97.6
1059	91.0	0.990	0.383	0.046	0.055	0.113	0.062	1.78	6.26	2.04			0.002	101
1053	82.5	1.09	0.381	0.048	0.073	0.163	0.080	2.18	1.36	8.80	2.88	0.172	n.d.	97.5
1054	10.6	0.794	0.257	0.019	0.040	0.006	0.004	9.60	0.389	68.0	16.3	0.845	0.003	97.2
1055	14.3	0.158	0.034	0.004	0.004	0.010	0.009	1.51	1.78	62.0	16.4	0.841	n.d.	95.5
1056	78.1	0.992	0.371	0.043	0.053	0.090	0.050	2.10	8.93	6.33	2.33	0.145	n.d.	97.4
1060	66.7	0.604	0.149	0.019	0.002	0.003	tr.	1.15	13.0	12.9	2.67	0.159	n.d.	96.2
1061	82.7	0.675	0.163	0.021	n.d.	n.d.	n.d.	1.03	14.5	1.28	0.405	0.029	n.d.	99.7

Table 7. Stable isotopic composition of Los Angeles basin gas samples.

Sample no.	$\delta^{13}C(C_1)$ (‰)	$\delta D(C_1)$ (‰)	$\delta^{13}C(C_2)$ (‰)	$\delta^{13}C(C_3)$ (‰)	$\delta^{13}C(i\text{-}C_4)$ (‰)	$\delta^{13}C(n\text{-}C_4)$ (‰)	$\delta^{13}C(CO_2)$ (‰)
1360	−49.8	−208	−31.2	−31.2	−28.6	−27.3	5.0
1361	−50.4	−209	−31.6	−31.2	−28.9	−27.3	0.0
1362	−49.3	−202	−29.3	−30.6	−29.0	−27.6	4.2
1185	−51.4	−217	−30.4	−28.7	−28.1	−25.8	−10.9
1183	−45.9	−204	−30.7	−27.3			−3.4
1184	−48.6	−196	−25.8	−24.7	−28.4	−22.7	
1092	−43.1	−191	−28.3	−23.7			14.6
1091	−42.9	−205	−28.5	−26.0	−26.7	−23.8	0.3
1093	−42.5	−181	−29.4	−21.1			15.6
1359	−35.0	−165	−26.6	−25.1	−24.7	−23.6	
1173	−40.4	−181	−28.9	−23.9			21.7
1175	−42.3	−171	−28.5	−20.3			17.9
1174	−46.0	−223	−29.5	−27.3			
1171	−45.4	−260	−31.4	−27.1	−28.3	−24.6	−0.9
1057	−44.1	−199	−29.6	−23.6			−13.5
1058	−43.1	−189	−29.7	−23.7			3.0
1059	−43.3	−196	−29.6	−23.5			3.1
1053	−41.9	−189	−29.9	−24.1			−22.9
1054	−44.4	−186	−30.3	−26.1			−3.9
1055	−42.4	−191	−29.7				20.2
1056	−42.0	−189	−30.0	−22.7			−11.3
1060	−42.4	−178	−29.1	−19.6			25.9
1061	−42.4	−183	−29.3	−19.0			25.4

13) (Fuex, 1977; Schoell, 1983). The Santa Fe Springs gas ($\delta^{13}C = -35$‰) is enriched in ^{13}C relative to the others, which indicates that the Santa Fe Springs gas is derived from either a more mature source or a source enriched in terrestrially derived kerogen. A more mature source is more likely, based on the isotopic separation of methane, ethane, and propane in the gas (Figure 14). The Santa Fe Springs field is located on the same trend as the Bandini field in the northeast portion of the basin and the oils are compositionally similar and typical of group 1 oils, as discussed above. A more mature source for hydrocarbons in these fields compared to other Los Angeles basin fields is indicated by both the oil and the gas data.

All methane δD values but one fall between −225‰ and −165‰ and are consistent with the literature for associated thermogenic gas (Figure 15) (Schoell, 1980; 1984). A δD value of −260‰ was measured for the Wilmington/237 sample, which, together with the methane $\delta^{13}C$ value of −45.4‰, suggests a relatively early thermogenic origin. Source effects may explain why the Wilmington δD value is 40‰ to 60‰ more negative than gases with similar $\delta^{13}C$ values from Beta, Huntington Beach, and Whittier fields.

As expected from theoretical and empirical evidence (Silverman, 1967; Smith et al., 1971; Des Marais et al., 1981), the C_1 to C_4 hydrocarbons become less negative with increasing carbon number. Among the commercial gases, ethane $\delta^{13}C$ varies from −32‰ to −26‰ (average = −29.3‰); propane $\delta^{13}C$ varies from −31‰ to −20‰ (average = −26.3‰); and n-butane $\delta^{13}C$ varies from −28‰ to −23‰ (average = −25.3‰). Figure 16 shows that ethane $\delta^{13}C$ and methane $\delta^{13}C$ are correlated and that as maturity (and $\delta^{13}C$) increases, the difference between these parameters decreases. For most samples, the isotopic difference $\Delta^{13}C$(ethane − methane) falls between 5‰ and 18‰, bounding values for single-sourced thermogenic gases estimated from the work of James (1983) and Sundberg and Bennett (1983). The Beta gases and two Huntington Beach gases fall outside this range and suggest mixing with significant amounts of microbial methane. Microbial methane is very dry and generally has methane $\delta^{13}C < -60$‰. A thermogenic gas that has mixed with microbial methane will shift to the left in the direction of the arrow. Mixing has probably occurred in the nearby Ventura basin also. Unpublished data for 50 Ventura basin gases analyzed

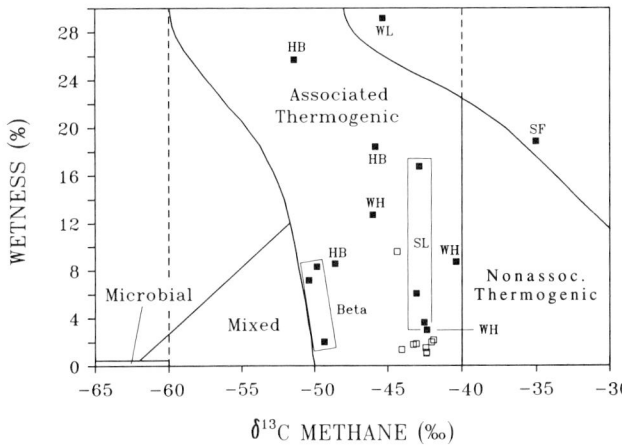

Figure 13. Plot of gas wetness (C_{2+} hydrocarbons/total hydrocarbons) versus methane $\delta^{13}C$. Genetic fields are generalized after Schoell (1983). Open symbols are seeps; solid symbols are oil-field gases. HB = Huntington Beach, SF = Santa Fe Springs, SL = Salt Lake, WH = Whittier, WL = Wilmington Offshore.

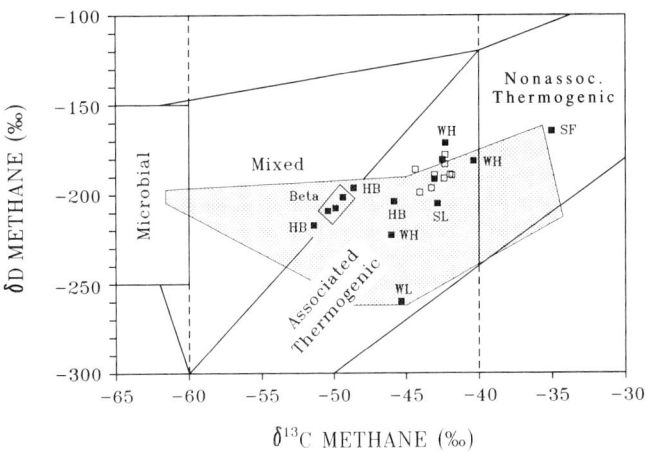

Figure 15. Plot of methane δD versus methane $\delta^{13}C$. Stippled area indicates the range of unpublished Global Geochemistry Corporation data for 50 Ventura basin gases. Genetic fields are generalized after Schoell (1983). Abbreviations as in Figure 13.

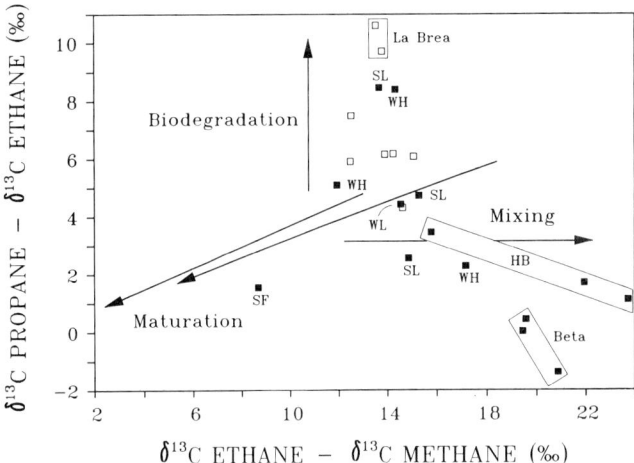

Figure 14. Plot of $\Delta^{13}C$(propane − ethane) versus $\delta^{13}C$(ethane − methane). Thermogenic gas maturation lines are taken from the models of James (1983) and Sundberg and Bennett (1983). Arrows indicate the secondary influence of biodegradation and mixing with microbial methane. Abbreviations as in Figure 13.

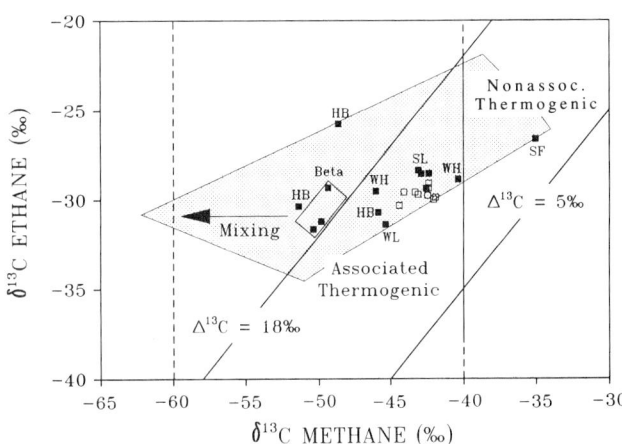

Figure 16. Plot of ethane $\delta^{13}C$ versus methane $\delta^{13}C$. Data for Beta and Huntington Beach gases together with data for many gases from the Ventura basin (stippled area) are consistent with a mixed thermogenic and microbial origin for methane. Genetic fields are modified from Schoell (1983) as described in the text. Abbreviations as in Figure 13.

at Global Geochemistry Corporation are indicated by the stippled area in Figure 16. The Ventura basin data plot well into the region occupied by mixed microbial and thermogenic gases. Further evidence for mixing in the Los Angeles and Ventura basins is provided by the plot of methane δD versus $\delta^{13}C$ in Figure 15.

Figure 14 is a plot of $\Delta^{13}C$(propane − ethane) versus $\Delta^{13}C$(ethane − methane). The illustrated maturation lines are adapted from the models of James (1983) and Sundberg and Bennett (1983) and indicate the general path followed by thermogenic gases derived from a single source rock. Apart from the possible influence of organic facies and trap timing, gases may deviate from the maturation path due to postgenerative effects, such as biodegradation and mixing. Biodegradation of wet natural gases appears to be accompanied by selective removal of propane (James and

Burns, 1984; Jenden and Kaplan, 1987). As propane is removed, the δ¹³C value of the residual propane becomes less negative and Δ¹³C(propane − ethane) increases. Biodegraded gases, therefore, plot above the thermogenic maturation lines. Figure 14 correctly indicates that the La Brea tar pits gases are biodegraded and suggests that several seep gases and two casinghead gases from the Salt Lake and Whittier oil fields have been altered to lesser degrees. As expected, the biodegraded gases have relatively low concentrations of propane (Figure 17).

The addition of microbial methane to a thermogenic gas will cause Δ¹³C(ethane − methane) to increase without changing Δ¹³C(propane − ethane). Thermogenic gases that have mixed with microbial methane will, therefore, plot to the right of the maturation lines in Figure 14 (e.g., Beta and Huntington Beach gases). It is not apparent why Δ¹³C(propane − ethane) is anomalously low for many of the mixed-origin gases. In the Beta field, low Δ¹³C(propane − ethane) is attributable to "anomalously" negative propane δ¹³C values. These may, in turn, be related to low maturity and/or source control (Figure 17), as suggested by the oil data in the previous section. More work is required to elucidate the factors that control the isotopic composition of the C$_{2+}$ hydrocarbon gases.

Carbon Dioxide

The origin of carbon dioxide in gases from the Los Angeles basin is intriguing. Biodegraded gases from the La Brea tar pits and the Salt Lake and Whittier oil fields have carbon dioxide concentrations from 13% to 37%, suggesting that the carbon dioxide is produced by hydrocarbon oxidation. Carbon dioxide derived from hydrocarbon oxidation ought to be isotopically light (Stahl, 1980; Coleman et al., 1981). Carbonate cements and tar-stained carbonate vein-filling minerals in Miocene sedimentary and volcanic rocks from the northeastern Puente Hills probably originate from the oxidation of oil and gas and have δ¹³C values from −45‰ to −19‰ (Table 8) (Shelton, 1965). Rather than having strongly negative carbon dioxide δ¹³C values, however, gases with the highest concentrations of carbon dioxide have δ¹³C values up to +26‰ (Figure 18).

Isotopically positive carbonate cements have been found in a number of organic-rich shales and have been attributed to intense early diagenetic bacterial activity (Friedman and Murata, 1979; Reitsema, 1980). Isotopically positive carbon dioxide could be formed by the reaction of these cements with organic acids released during subsequent thermal alteration. In order to test this possibility, several core samples of fine-grained rocks and a near-surface carbonate concretion were analyzed for carbonate δ¹³C (Table 9). The δ¹³C values range from −3.2‰ to +3.4‰ (similar to marine carbonates), suggesting that the high concentrations of carbon dioxide are not derived from carbonate dissolution.

Dimitrakopoulos and Muehlenbachs (1987) presented data showing that reservoir rocks bearing biodegraded oil may contain secondary carbonate minerals with δ¹³C values ranging from −21‰ to +15‰. The origin of the isotopically heavy secondary carbonates is not clear but probably involves a two-stage process. In the first stage, oxygen introduced by the intermittent invasion of meteoric water is utilized by hydrocarbon-oxidizing bacteria to form alcohols, ketones, and fatty acids. In the second stage, after the oxygen is consumed, the products of oxidation are fermented and part of the carbon dioxide is reduced to methane. It is well known that bacterial reduction of carbon dioxide involves a strong carbon isotope fractionation and can produce residual carbon dioxide δ¹³C values up to +25‰ (Claypool and Kaplan, 1974; Jenden and Kaplan, 1986). Depending upon the formation-water chemistry, the isotopically heavy carbon dioxide could either partition into

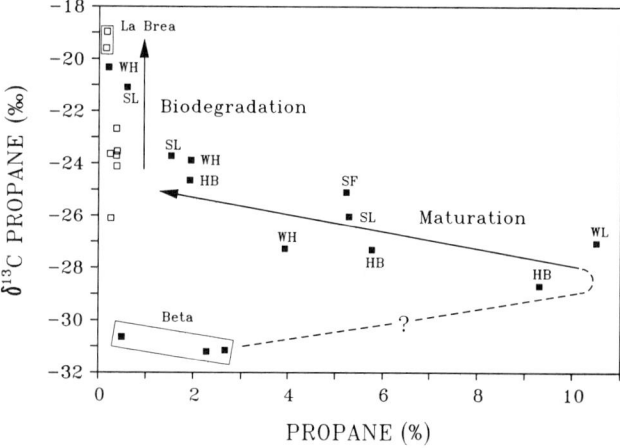

Figure 17. Plot of propane δ¹³C versus propane concentration showing the effects of maturation and biodegradation. Abbreviations as in Figure 13.

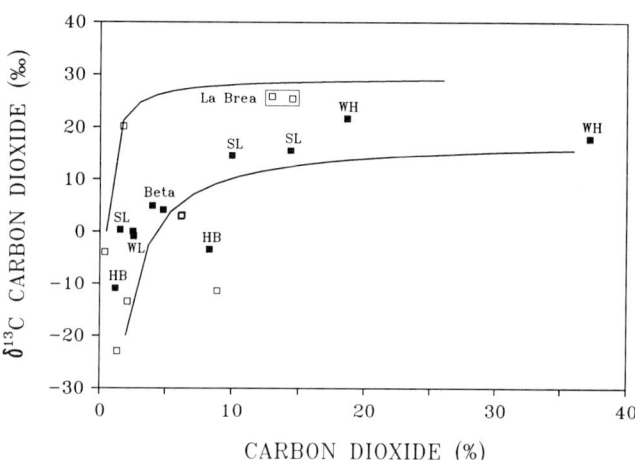

Figure 18. Plot of carbon dioxide δ¹³C versus concentration. Mixing lines are shown for two sets of hypothetical end-members. The low-CO$_2$ end-members represent unaltered thermogenic gases; the high-CO$_2$ end-members represent gases formed by the biodegradation of oil following the model of Dimitrakopoulos and Muehlenbachs (1987). Abbreviations as in Figure 13.

Table 8. $\delta^{13}C$ values of carbonate minerals in Miocene rocks outcropping in the northeastern Puente Hills.

Carbonate cements and veins in the Soquel Member of the upper Miocene Puente Formation

Fine-grained dolomite cement, basal dacite breccia	−18.7‰
Finely laminated, tan-colored sandy dolomite vein, dacite breccia	−36.4
Fine-grained calcite cement, sandy conglomerate	−20.2
Fine-grained calcite cement, massive wacke	−26.1
	−32.8
Very coarse grained amber calcite vein	−25.8

Carbonate veins in the middle Miocene Glendora Volcanics

Very finely laminated milky dolomite	−36.1‰
Finely laminated light gray calcite	−28.3
Fine- to medium-grained gray-and-white-banded calcite	−33.4
	−34.5
Coarse-grained light gray calcite	−22.1
Very coarse-grained amber calcite with hydrocarbon inclusions	−45.6

Data from Spira (1986)
Analyses were run at Global Geochemistry Corporation and the University of California, Los Angeles

Table 9. $\delta^{13}C$ values of carbonate cements in core samples from the Salt Lake area.

Occidental "West Pico" #10, Beverly Hills field, Miocene Modelo Formation

5747–5755 ft Massive gray mudstone	−1.2‰
5747–5755 ft Thin-bedded shale	−0.8
6049–6058 ft Microfractured mudstone	3.4
6256–6265 ft Gray siltstone	−3.2
6283–6287 ft Mudstone with slump structure	2.3

Occidental "West Pico" #18, Beverly Hills field, Pliocene "Repetto" Formation

4455–4464 ft Fractured mudstone	−1.1

Occidental "West Pico" #18 R/D, Beverly Hills field, Pliocene "Repetto" Formation

3900–3908 ft Massive gray siltstone	−0.6

Rancho La Brea Tar Pits, Pit #91, Quaternary alluvium

10 ft Tar-impregnated carbonate nodule	1.5
	1.9

a free-gas phase or be precipitated as a secondary carbonate mineral.

A number of observations suggest that Dimitrakopoulos and Muehlenbachs's model for isotopically positive carbon dioxide can be applied to the Los Angeles basin. Figure 19 shows that carbon dioxide $\delta^{13}C$ values are highest closest to the surface, where the availability of oxygen and the presence of moderate temperatures ought to provide optimum conditions for biodegradation. As reservoir depths increase, $\delta^{13}C$ values decrease. Figure 20 shows that a crude correlation exists between carbon dioxide $\delta^{13}C$ and propane $\delta^{13}C$. With the exception of the Beta gases and a group of seep gases that may have been oxidized in the near-surface, isotopically light carbon dioxide values are associated with unaltered (isotopically light) propane and isotopically heavy carbon dioxide values are associated with altered propane. Finally, the carbon dioxide concentration of gases in the Salt Lake field is closely related to the degree of alteration of the associated oils as measured by n-alkane content, sulfur content, and API gravity (see discussion below).

The relation between carbon dioxide concentration and $\delta^{13}C$ in Figure 18 can best be explained by the mixing of carbon-dioxide-rich gas produced by fermentation of hydrocarbon oxidation products (Dimitrakopoulos and Muehlenbachs, 1987) with unaltered low–carbon dioxide gas of thermogenic origin. The bounding curves illustrate mixing lines calculated for two sets of hypothetical end-members. The fermentation end-members were selected at 36% carbon dioxide with $\delta^{13}C = 15.7‰$ and at 26% carbon dioxide with $\delta^{13}C = 29‰$, values that are consistent with mathematical models for carbon dioxide reduction based on closed-system Rayleigh fractionation. The thermogenic end-members are taken at 0.5% carbon dioxide with $\delta^{13}C = 0‰$ and at 2.0% carbon dioxide with $\delta^{13}C = -20‰$. Most data fall between the two curves and appear compatible with the mixing model. The seep gas sample falling outside

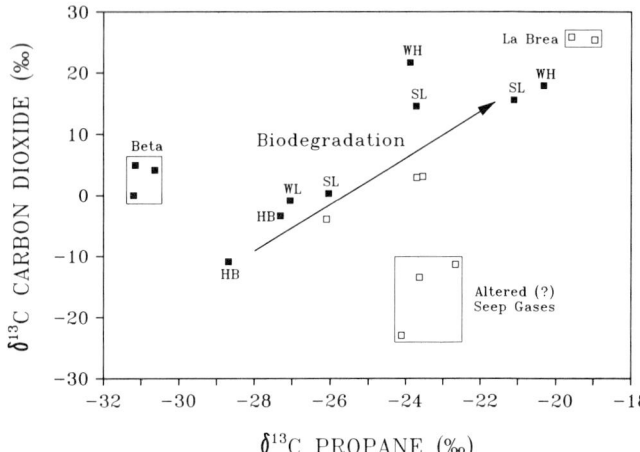

Figure 20. Plot of carbon dioxide $\delta^{13}C$ versus propane $\delta^{13}C$ showing the influence of biodegradation. Abbreviations as in Figure 13.

the mixing range has probably been altered by near-surface oxidation (Figures 19 and 20).

BIODEGRADATION

Biodegradation is a major influence on oil quality in the Los Angeles basin and appears to be largely responsible for the heavier oils in the shallow reservoirs of the giant Wilmington and Huntington Beach fields. The effects of progressive biodegradation on reservoired hydrocarbons can be clearly seen in a suite of oils and associated gases from the Salt Lake area in the northwest of the Los Angeles basin.

Three oil and associated gas samples were obtained from reservoirs in the South Salt Lake and West Salt Lake fields, and one oil and two gas samples were taken from the La Brea tar pits in the same vicinity (Table 5, samples 1091, 1092, 1093, 1060, and 1061). The saturate GC chromatograms (Figure 21) indicate a progressive loss of normal hydrocarbons with decreasing reservoir depth, culminating in loss of all saturated hydrocarbons in the tar pits surface seep. Due to the severe alteration of the oils, correlation rests on biomarker ratios (Table 10). The tar pits oil is so severely altered that even the regular steranes are absent. However, the triterpane ratios and diasterane ratios indicate that the four oils are genetically very similar, with the oil from well P-75 being somewhat more mature than the others.

Many of the physical and chemical differences between the oils and between the gases are directly related to reservoir depth (Table 11). The following trends can be identified with decreasing reservoir depth:

1. decrease in API gravity

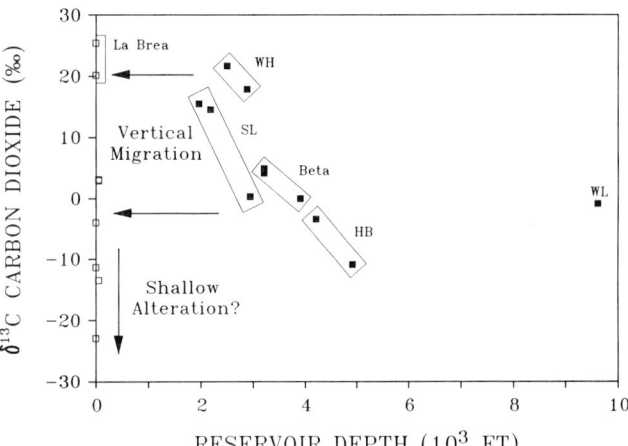

Figure 19. Plot of carbon dioxide $\delta^{13}C$ versus subsurface reservoir depth. Abbreviations as in Figure 13.

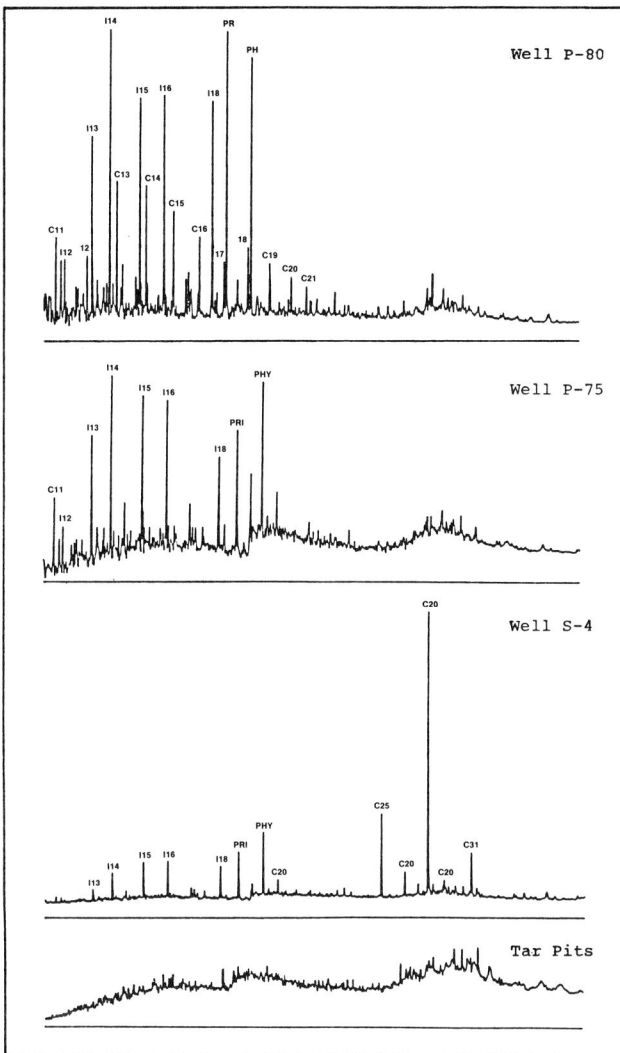

Figure 21. Gas chromatograms of Salt Lake area oils. Increasing degree of biodegradation is illustrated by progressive loss of saturated hydrocarbons.

shallow sediments, as observed by Coleman (1976) and Schoell (1983).

The other changes appear to be related to biodegradation, in which microbes preferentially oxidize saturated hydrocarbons to carbon dioxide that accumulates in the gas phase. The carbon dioxide produced in this process is isotopically heavy ($\leq +25.9$‰), and a possible mechanism for its formation has been discussed in the previous section. Of the gaseous hydrocarbons, propane is preferentially consumed, as indicated by the trend towards progressively heavier propane with severity of biodegradation. Methane and ethane, however, appear to be unaffected by microbial activity, because their carbon isotope ratios remain unchanged in the suite of gases (Table 11).

Methane oxidation has been inferred by Bernard (1978) from the progressive depletion of methane, along with its progressively heavier carbon isotope ratio, towards the surface in shallow sediments of the Gulf of Mexico. This system appears to mirror, on a much smaller scale, the progressive microbial oxidation observed in the Salt Lake area. The major difference arises in the source of the hydrocarbons in the two systems: thermogenic hydrocarbons in the Salt Lake area are rich in C_{2+} hydrocarbons, whereas microbial hydrocarbons in near-surface sediments consist almost entirely of methane. Consequently, the sole hydrocarbon available for oxidation in surface sediments is methane, in contrast to the availability of a whole suite of higher hydrocarbons in oil reservoirs. Although methane oxidation may represent the ultimate stage in destruction of a hydrocarbon reservoir by microbial activity, it is not observed in the Salt Lake reservoirs or even in the extensively altered surface seep.

This suite of samples from the Salt Lake area illustrates that biodegradation causes a number of interrelated physical and chemical changes in oil and its associated gas. From a commercial point of view, these changes are deleterious; a relatively light, sweet crude is transformed to a heavy, sour oil, and a relatively high BTU gas is transformed to a low BTU, carbon dioxide-rich gas. Shallow reservoirs in the Los Angeles basin are likely to contain the heavy biodegraded product, whereas deeper reservoirs in the same field may have the lighter, unaltered oil.

2. increase in sulfur concentration

3. loss of saturated hydrocarbons

4. increase in carbon dioxide concentration in gas

5. isotopically heavier carbon in carbon dioxide

6. decrease in concentrations of C_{2+} hydrocarbons in gas

7. isotopically heavier carbon in propane

Of these depth-related changes, decreased C_{2+} concentrations may be explained by migration, involving preferential adsorption of C_{2+} hydrocarbons compared to methane in

CONCLUSIONS

The Los Angeles basin contains abundant organic-rich source rocks containing kerogen rich in sapropelic material. Maturity estimates of the source rocks based on vitrinite reflectance values are low. This appears to be related to the sapropel-rich kerogen, which may generate oil at lower maturities than is conventionally accepted or may cause suppression of vitrinite reflectivity in the kerogen. Maturity estimates based on bitumen production indicate that upper Miocene rocks in deeper parts of the basin are the source of the oil and gas accumulations. Vertical migration of this oil and gas into shallower reservoirs is a feature of many fields, especially the giant Wilmington and Huntington Beach fields.

Table 10. Selected biomarker ratios for oil samples from the Salt Lake area.

Well no.	A	B	C	D	E	F	G	H	I
P-80	0.53	0.32	0.61	54	34.0	0.73	0.41	36.5	36.7
P-75	0.68	0.49	1.13	58	40.0	0.68	0.35	39.5	33.8
S-4	0.58	0.37	0.84	54	37.0	0.69	0.45	37.5	36.9
Tar Pits	0.44	1.21	0.71	60	31.0	0.92	0.36	34.5	34.2

Well no.	J	K	L	M	N	O	P	Q	R
P-80	26.8	0.81	2.80	0.13	0.13	1.53	1.47	0.05	0.40
P-75	26.7	0.91	3.20	0.13	0.14	1.55	1.52	0.06	0.40
S-4	25.8	0.81	2.90	0.13	0.14	1.50	1.52	0.05	0.60
Tar Pits	31.5	6.22	2.05	0.15	0.14	1.60	1.58	0.06	0.51

(A) C_{29} $\alpha\alpha\alpha$ (20S)-sterane/C_{29} $\alpha\alpha\alpha$ (20R)-sterane
(B) C_{29} $\alpha\beta\beta$ (20S)-sterane/C_{29} $\alpha\alpha\alpha$ (20R)-sterane
(C) C_{29} $\alpha\beta\beta$ (20R)-sterane/C_{29} $\alpha\alpha\alpha$ (20R)-sterane
(D) C_{27} (20S) diasterane/[same + C_{27} (20R) diasterane] $\times$ 100
(E) C_{29} $\alpha\alpha\alpha$ (20S)-sterane/[same + C_{29} $\alpha\alpha\alpha$ (20R)-sterane] $\times$ 100
(F) C_{29} $\alpha\alpha\alpha$ (20R)-sterane/C_{27} $\alpha\alpha\alpha$ (20R)-sterane
(G) C_{27} (20R) diasterane/C_{29} (20R) diasterane
(H) C_{27} $\alpha\alpha\alpha$ (20R)-sterane/[(C_{27} + C_{28} + C_{29}), 20R-sterane] $\times$ 100
(I) C_{28} $\alpha\alpha\alpha$ (20R)-sterane/[(C_{27} + C_{28} + C_{29}), 20R-sterane] $\times$ 100
(J) C_{29} $\alpha\alpha\alpha$ (20R)-sterane/[(C_{27} + C_{28} + C_{29}), 20R-sterane] $\times$ 100
(K) C30-hopane/C_{29} (20R + 20S) $\alpha\alpha\alpha$-steranes + C_{29} (20R + 20S) $\alpha\beta\beta$-steranes
(L) 17 α (H)-trisnorhopane (Tm)/18 α (H)-trisnorhopane (Ts)
(M) C_{29}-normoretane/C_{29}-norhopane
(N) C_{30}-moretane/C_{30}-hopane
(O) 22S/22R-C_{31}-hopane
(P) 22S/22R-C_{32}-hopane
(Q) 18 α (H) trisnorhopane (Ts)/C_{30}-hopane
(R) 28,30-bisnorhopane (C_{28})/C_{30}-hopane

Oil in the Los Angeles basin varies considerably in gravity and sulfur content, with higher quality (high API gravity, low sulfur content) oils in the northeast of the basin and lower quality oils in the west and south of the basin. The variation in oil quality does not appear to be caused by differences in kerogen type, because the sterane biomarker ratios and carbon isotope ratios of the oils are very similar and are consistent with a marine-derived kerogen as the source. The superior quality of the oils in the northeast of the basin may be caused by higher maturity, greater migration, a more oxidizing depositional environment, or a combination of all three.

Gases analyzed from the Los Angeles basin can be classified according to conventional genetic characterization schemes. Most gases appear to be genetically related to the oils in a reservoir, except in some cases where localized mixing of gases with microbial methane has occurred. Many shallow gases contain significant quantities of carbon dioxide that is commonly enriched in ^{13}C and appears to be derived from microbial degradation of oil in the reservoir.

Biodegradation causes important changes in both the oil and gas components in a reservoir, including a decrease in the API gravity and an increase in sulfur content of the oil and an increase in the carbon dioxide content of the associated gas. From an economic standpoint, these changes result in a lower-quality product, exemplified by the heavy oils in shallow reservoirs of the Wilmington and Huntington Beach fields.

ACKNOWLEDGMENTS

We thank Chevron U.S.A., Mobil Oil, Phillips Petroleum, Shell Oil, and THUMS Long Beach companies for gas samples and well data. We are grateful for valuable discussions we had with Ian Kaplan. This paper benefited from careful reviews by Kevin Biddle, Neely Bostick, Caroline Isaacs, and Rick Weber. Partial support for this study was provided by the Gas Research Institute (contract no. 5081-360-0533).

Table 11. Chemical and isotopic characteristics of oil and gas samples from the Salt Lake area.

Well	Depth (ft)	Oil °API	%S	Gas %CO$_2$	δ^{13}C-CO$_2$ (‰)	%C$_{2+}$	δ^{13}C-C$_1$ (‰)	δ^{13}C-C$_2$	δ^{13}C-C$_3$ (‰)
P-80	2930	——	0.01	1.6	+0.3	16.5	−42.9	−28.5	−26.0
P-75	2180	20.7	0.88	10.2	+14.6	5.4	−43.1	−28.3	−23.7
S-4	1960	14.4	1.03	14.7	+15.6	3.1	−42.5	−29.4	−21.1
Tar Pits	Surface	7.9	2.83	14.8, 15.6	+25.4, +25.9	0.88, 0.93	−42.4, −42.4	−29.1, −29.3	−19.0, −19.6

REFERENCES

Bernard, B. B., 1978, Light hydrocarbons in marine sediments [unpublished Ph.D. dissertation]: College Station, Texas, Texas A&M University, 144 p.

Bostick, N. H., 1979, Microscopic measurement of the level of catagenesis of solid organic matter in sedimentary rocks to aid exploration for petroleum and to determine former burial temperatures: a review: SEPM Special Publication 26, p. 17-23.

Bostick, N. H., S. M. Cashman, T. H. McCulloh, and C. T. Waddell, 1978, Gradients of vitrinite reflectance and present temperature in the Los Angeles and Ventura Basins, California, in D. F. Oltz, ed., Symposium in geochemistry: low temperature metamorphism of kerogen and clay minerals: Pacific Section, SEPM, p. 65-96.

Brown, J. B., 1968, Gas in the Los Angeles basin, California, in B. W. Beebe and B. F. Curtis, eds., Natural gases of North America: AAPG Memoir 9, v. 1, p. 149-163.

California Division of Oil and Gas, 1986, 71st annual report of the state oil and gas supervisor, 1985: California Department of Conservation, Division of Oil and Gas, Publication PR06, 157 p.

California Division of Oil and Gas, 1974, California oil and gas fields, v. 2, south, central coast and offshore California: California Department of Conservation, Division of Oil and Gas, Publication TR12, 350 p.

Claypool, G. E., and I. R. Kaplan, 1974, The origin and distribution of methane in marine sediments, in I. R. Kaplan, ed., Natural gases in marine sediments: New York, New York, Plenum Press, p. 99-139.

Claypool, G. E., J. P. Baysinger, C. M. Zubeck, and A. H. Love, 1979, Organic geochemistry, in H. E. Cook, ed., Geologic studies of the Point Conception deep stratigraphic test well OCS-CAL 78-164 No. 1, outer continental shelf, southern California, United States: U.S. Geol. Survey Open-File Rept. 79-1218, p. 109-124.

Coleman, D. D., 1976, Isotopic characterization of Illinois natural gas [unpublished Ph.D. thesis]: Champaign, Illinois, University of Illinois at Urbana-Champaign, 175 p.

Coleman, D. D., J. B. Risatti, and M. Schoell, 1981, Fractionation of carbon and hydrogen isotopes by methane-oxidizing bacteria: Geochimica et Cosmochimica Acta, v. 45, p. 1033-1037.

Curiale, J. A., D. Cameron, and D. V. Davis, 1985, Biological marker distribution and significance in oils and rocks of the Monterey Formation, California: Geochimica et Cosmochimica Acta, v. 49, p. 271-288.

Des Marais, D. J., J. H. Donchin, N. N. Nehring, and A. H. Truesdell, 1981, Molecular carbon isotopic evidence for the origin of geothermal hydrocarbons: Nature, v. 292, p. 826-828.

Dimitrakopoulos, R., and K. Muehlenbachs, 1987, Biodegradation of petroleum as a source of ^{13}C-enriched carbon dioxide in the formation of carbonate cement: Chemical Geology (Isotope Geoscience Section), v. 65, p. 283-291.

Friedman, I., and K. J. Murata, 1979, Origin of dolomite in Miocene Monterey Shale and related formations in the Temblor Range, California: Geochimica et Cosmochimica Acta, v. 43, p. 1357-1365.

Fuex, A. N., 1977, The use of stable carbon isotopes in hydrocarbon exploration: Journal of Geochemical Exploration, v. 7, p. 155-188.

Giger, W., and C. Schaffner, 1981, Unsaturated steroid hydrocarbons as indicators of diagenesis in immature Monterey shales: Naturwissenschaften, v. 68, S37, p. 819-820.

Giger, W., and C. Schaffner, 1979, Hydrocarbon indicators of diagenesis in Monterey Shale, California [abstract]: Geological Society of America Abstracts with Programs, v. 11, n. 7, p. 432.

Gilbert, J., and C. P. Summerhayes, 1981, Distribution of organic matter in sediments along the California continental margin, in R. S. Yeats, B. U. Haq et al., Initial Reports DSDP: Washington, D.C., U.S. Government Printing Office, p. 757-761.

Hoots, H. W., A. L. Blount, and P. H. Jones, 1935, Marine oil shale, source of oil in Playa del Rey field, California: AAPG Bulletin, v. 19, p. 172-205.

Huang, W. Y., and W. G. Meinschein, 1979, Sterols as ecological indicators: Geochimica et Cosmochimica Acta, v. 43, p. 739-746.

Hutton, A. C., and A. J. Cook, 1980, Influence of alginite on the reflectance of vitrinite from Joadja, NSW, and some other coals and oil shales containing alginite: Fuel, v. 59, p. 711-714.

Hutton, A. C., A. J. Kantsler, A. C. Cook, and D. M. McKirdy, 1980, Organic matter in shales: Journal of the Australian Petroleum Exploration Association, v. 20, p. 44-63.

James, A. T., 1983, Correlation of natural gas by use of carbon isotopic distribution between hydrocarbon components: AAPG Bulletin, v. 67, p. 1176-1191.

James, A. T., and B. J. Burns, 1984, Microbial alteration of subsurface natural gas accumulations: AAPG Bulletin, v. 68, p. 957-960.

Jenden, P. D., and I. R. Kaplan, 1987, The origin of natural gas in the Sacramento basin and its relation to the tectonic evolution of northern California: submitted to AAPG Bulletin.

Jenden, P. D., and I. R. Kaplan, 1986, Comparison of microbial gases from the Middle America Trench and Scripps Submarine Canyon: implications for the origin of natural gas: Applied Geochemistry, v. 1, p. 631-646.

Jones, R., and T. Edison, 1978, Microscopic observation of kerogen related to geochemical parameters with emphasis on thermal maturation, in D. F. Oltz, ed., Symposium in geochemistry: low temperature metamorphism of kerogen and clay minerals, Pacific Section, SEPM, p. 1-12.

Kalkreuth, W., 1982, Rank and petrographic composition of selected Jurassic-Lower Cretaceous coals of British Columbia, Canada: Canadian Petroleum Geology Bulletin, v. 30, p. 112-139.

Katz, B. J., and L. W. Elrod, 1983, Organic geochemistry of DSDP site 467, offshore California middle Miocene to lower Pliocene strata: Geochimica et Cosmochimica Acta, v. 47, p. 389-396.

King, J. D., and G. E. Claypool, 1983, Biological marker compounds and implication for generation and migration of petroleum in rocks of the Point Conception deep stratigraphic test well, OCS-CAL 78-164 No. 1, offshore California, in C. M. Isaacs and R. E. Garrison, eds., Petroleum generation and occurrence in the Miocene Monterey Formation, California: Pacific Section, SEPM, p. 191-200.

Kruge, M. A., 1983, Diagenesis of Miocene biogenic sediments in Lost Hills field, San Joaquin Basin, California, in C. M. Isaacs and R. E. Garrison, eds., Petroleum generation and occurrence in the Miocene Monterey Formation, California: Pacific Section, SEPM, p. 39-51.

Lopatin, N. V., 1971, Temperature and geologic time as factors in coalification (in Russian): Moscow, Acad. Nauk. SSSR, Izv., Ser. Geol., n. 3, p. 95-106.

Mackenzie, A. S., 1984, Applications of biological markers in petroleum geochemistry, in J. Brooks and D. H. Welte, eds., Advances in petroleum geochemistry: London, Academic Press, v. 1, p. 115-214.

Mackenzie, A. S., C. F. Hoffman, and J. R. Maxwell, 1981, Molecular parameters of maturation in the Toarcian shales, Paris basin, France: III: Changes in aromatic steroid hydrocarbons: Geochimica et Cosmochimica Acta, v. 45, p. 1345-1355.

Mackenzie, A. S., R. L. Patience, J. R. Maxwell, M. Vandenbroucke, and B. Durand, 1980, Molecular parameters of maturation in the Toarcian shales, Paris basin, France: I: Changes in configuration of acyclic isoprenoid alkanes, steranes and triterpanes: Geochimica et Cosmochimica Acta, v. 44, p. 1709-1721.

Magoon, L. B., and C. M. Isaacs, 1983, Chemical characteristics of some crude oils from the Santa Maria basin, California, in C. M. Isaacs and R. E. Garrison, eds., Petroleum generation and occurrence in the Miocene Monterey Formation, California: Pacific Section, SEPM, p. 201-211.

Mayuga, M. N., 1970, Geology and development of California's giant Wilmington field, in M. T. Halbouty, ed., Geology of giant petroleum fields: AAPG Memoir 14, p. 158-184.

McCulloh, T. H., 1979, Implication for petroleum appraisal, in H. E. Cook, ed., Geologic studies of the Point Conception deep stratigraphic test well OCS-CAL 78-164 No. 1 outer continental shelf southern California, United States: USGS Open-File Report 79-1218, p. 26-42.

McCulloh, T. H., S. M. Cashman, and R. J. Stewart, 1978, Diagenetic baseline for interpretive reconstructions of maximum burial depth and paleotemperatures in clastic sedimentary rocks, in D. F. Oltz, ed., Symposium in geochemistry: low temperature metamorphism of kerogen and clay minerals: Pacific Section, SEPM, p. 18-46.

McEvoy, J., G. Eglinton, and J. R. Maxwell, 1981, Preliminary lipid analyses of sediments from section 467-3-3 and 467-97-2, in R. S. Yeats, B. U. Haq et al., Initial Reports DSDP: Washington, D.C., U.S. Government Printing Office, p. 763-774.

Metzner, L. H., 1935, The del Rey Hills area of the Playa del Rey oil field: California Division of Natural Resources, Summary of Operations, v. 21, n. 2, p. 5-26.

Milner, C. W. D., M. A. Rogers, and C. R. Evans, 1977, Petroleum transformations in reservoirs: Journal of Geochemical Exploration, v. 7, p. 101-153.

Orr, W. L., 1986, Kerogen/asphaltene/sulfur relationships in sulfur-rich Monterey oils, in D. Leythauser and J. Rullkotter, eds., Advances in organic geochemistry, 1985: Organic Geochemistry, v. 10, p. 499-516.

Philippi, G. T. 1965, On the depth, time and mechanism of petroleum generation: Geochimica et Cosmochimica Acta, v. 29, p. 1021-1049.

Philippi, G. T., 1975, The deep subsurface temperature controlled origin of the gaseous and gasoline - range hydrocarbons of petroleum: Geochimica et Cosmochimica Acta, v. 39, p. 1359-1373.

Price, L. C., 1983, Geologic time as a parameter in organic metamorphism and vitrinite reflectance as an absolute paleogeothermometer: Journal of Petroleum Geology, v. 6, p. 5-38.

Price, L. C., and C. E. Barker, 1985, Suppression of vitrinite reflectance in amorphous rich kerogen—a major unrecognized problem: Journal of Petroleum Geology, v. 8, p. 59-84.

Reed, R. D., 1951, Geology of California: AAPG, 355 p.

Reitsema, R. H., 1980, Dolomite and nahcolite formation in organic-rich sediments; isotopically-heavy carbonates: Geochimica et Cosmochimica Acta, v. 44, p. 2045-2049.

Rice, D. D., and G. E. Claypool, 1981, Generation, accumulation and resource potential of biogenic gas: AAPG Bulletin, v. 65, p. 5-25.

Rullkotter, G., H. Von der Dick, and D. H. Welte, 1981, Organic petrography and extractable hydrocarbons of sediments from the Eastern North Pacific Ocean: in R. S. Yeats, B. U. Haq et al., Initial Reports DSDP: Washington, D.C., U.S. Government Printing Office, p. 819-836.

Schoell, M., 1980, The hydrogen and carbon isotopic

composition of methane from natural gases of various origins: Geochimica et Cosmochimica Acta, v. 44, p. 649-661.

Schoell, M., 1983, Genetic characterization of natural gases: AAPG Bulletin, v. 67, p. 2225-2238.

Schoell, M., 1984, Wasserstoff und Kohlenstoffisotope in Organischen Substanzen, Erdolen und Erdgasen: Geologisches Jahrbuch, v. D67, p. 161.

Seifert, W. K., and J. M. Moldowan, 1981, Paleoconstruction by biological markers: Geochimica et Cosmochimica Acta, v. 45, p. 783-794.

Seifert, W. K., and J. M. Moldowan, 1979, The effect of biodegradation on steranes and terpanes in crude oils: Geochimica et Cosmochimica Acta, v. 43, p. 111-126.

Seifert, W. K., and J. M. Moldowan, 1978, Application of steranes, terpanes and monoaromatics to the maturation, migration and source of crude oils: Geochimica et Cosmochimica Acta, v. 42, p. 77-95.

Seifert, W. K., J. M. Moldowan, and R. W. Jones, 1980, Application of biological marker chemistry to petroleum exploration: Proceedings of the 10th World Petroleum Congress, v. 2, p. 425-440.

Seifert, W. K., J. M. Moldowan, G. W. Smith, and E. V. Whitehead, 1978, First proof of structure of a C_{28}-pentacyclic triterpane in petroleum: Nature, v. 271, p. 436-437.

Shelton, J. S., 1965, Glendora Volcanic Rocks, Los Angeles Basin, California: GSA Bulletin, v. 66, p. 45-90.

Silverman, S. R., 1967, Carbon isotopic evidence for the role of lipids in petroleum formation: Journal of the American Oil Chemists' Society, v. 44, p. 691-695.

Simoneit, B. R. T., and M. A. Mazurek, 1981, Organic geochemistry of sediments from the southern California borderland, DSDP project Leg 63, in R. S. Yeats, B. U. Haq et al., Initial Reports DSDP: Washington, D.C., U.S. Government Printing Office, p. 837-853.

Smith, J. E., J. G. Erdman, and D. A. Morris, 1971, Migration, accumulation and retention of petroleum in the earth: Proceedings of the 8th World Petroleum Congress: London, Applied Science Publishers, v. 2, p. 13-26.

Snowdon, L. R., and T. G. Powell, 1982, Immature oil and condensates-modification of hydrocarbon generation model for terrestrial organic matter: AAPG Bulletin, v. 66, p. 775-788.

Sofer, Z., 1984, Stable carbon isotope compositions of crude oils: application to source depositional environments and petroleum alteration: AAPG Bulletin, v. 68, p. 31-49.

Spira, M. O., 1986, Calcite veins in the northeastern Puente Hills [unpublished B.S. thesis]: Pomona, California, California State Polytechnic University, 25 p.

Stahl, W. J., 1977, Carbon and nitrogen isotopes in hydrocarbon research and exploration: Chemical Geology, v. 20, p. 121-149.

Stahl, W. J., 1980, Compositional changes and $^{13}C/^{12}C$ fractionations during the degradation of hydrocarbons by bacteria: Geochimica et Cosmochimica Acta, v. 44, p. 1903-1907.

Stock, C., 1956, Rancho La Brea: a record of Pleistocene life in California: Los Angeles County Museum of Natural History, Science Series 20, Paleontology 11, 81 p.

Summerhayes, C. P., 1981, Oceanographic controls on organic matter in the Miocene Monterey Formation, offshore California, in R. E. Garrison, R. G. Douglas, K. E. Pisciotto, C. M. Isaacs, and J. C. Ingle, eds., The Monterey Formation and related siliceous rocks of California: Pacific Section, SEPM, p. 213-219.

Sundberg, K. R., and C. R. Bennett, 1983, Carbon isotope paleothermometry of natural gas, in M. Bjoroy, ed., Advances in organic geochemistry 1981: New York, New York, John Wiley & Sons, p. 769-774.

Tissot, B. P., and D. H. Welte, 1978, Petroleum formation and occurrence: a new approach to oil and gas exploration: New York, New York, Springer-Verlag, 538 p.

Walker, A. L., 1982, Comparison of anomalously low vitrinite reflectance values with other thermal maturation indices in and near the Playa del Rey oil field, California [unpublished M.S. thesis]: Seattle, Washington, University of Washington, 190 p.

Walker, A. L., T. H. McCulloh, N. F. Petersen, and R. J. Stewart, 1983, Anomalously low reflectance of vitrinite in comparison with other petroleum source-rock maturation indices from the Miocene Modelo Formation in the Los Angeles Basin, California, in C. M. Isaacs and R. E. Garrison, eds., Petroleum generation and occurrence in the Miocene Monterey Formation, California: Pacific Section, SEPM, p. 185-190.

Waples, D. W., 1980, Time and temperature in petroleum formation: application of Lopatin's method to petroleum exploration: AAPG Bulletin, v. 64, p. 919-926.

Wissler, S. G., 1943, Stratigraphic formations of the producing zones of the Los Angeles basin oilfields: California Division of Mines Bulletin, v. 118, p. 209-234.

Yerkes, R. F., T. H. McCulloh, J. E. Schoellhamer, and J. G. Vedder, 1965, Geology of the Los Angeles basin, California—an introduction: USGS Professional Paper 420-A, 57 p.

Chapter 7

Stratigraphic Controls of Oil Fields in the Los Angeles Basin
A Guide to Migration History

Robert S. Yeats

*Oregon State University
Corvallis, Oregon, U.S.A.*

John M. Beall

Tomball, Texas, U.S.A.

ABSTRACT

Unmetamorphosed strata of Turonian to middle Miocene age exposed around the margins of the Los Angeles basin (LAB) predate the formation of the basin and respond to a different structural framework. Highly organic middle and late Miocene (Luisian and Mohnian) strata also predate the present LAB framework: the Puente Hills received a thicker sequence than did the area southwest of the Whittier fault, and the Santa Monica Mountains were at the distal end of Mohnian turbidites derived from farther north. The central trough became a major depocenter during deposition of the upper Mohnian (after 8 Ma), and it achieved its present northwest-southeast trend about 4 Ma. The central trough filled with "Delmontian" and Repettian turbidites shallowing upsection to Pliocene and Pleistocene deposits. Miocene and early Pliocene source rocks were buried beneath the oil-generating thermal threshold so that oil migrated to stratigraphic and broad structural traps formed during deposition. Many oil fields contain a thick stack of reservoir turbidites; the boundary between highly productive turbidites and overlying water-bearing turbidites, also in trapping position, is abrupt. Late Quaternary deformation after basin filling distorted rather than enhanced oil traps, and some oil accumulations were breached by erosion. New oil prospects could result from a better understanding of (1) the Mohnian basin framework in contrast to the far different post-Mohnian framework and (2) the post-Mohnian fold and thrust belt tectonically loaded by the southern margin of the Transverse Ranges onshore and offshore.

INTRODUCTION

Los Angeles is the site of the most prolific petroleum basin known in the world, based on hydrocarbon productivity with respect to total sediment volume (Biddle, *The Los Angeles Basin: An Overview*, this volume). T. H. McCulloh (*in* Bostick et al., 1978) calculated that more than 340,000 m^3 of oil plus gas equivalent on an energy-content basis have been discovered for each cubic kilometer of total basin fill, a world record. This paper is concerned with the reasons for such high productivity, and it focuses on stratigraphic controls and on the timing of migration routes and trap formation with respect to maturation of source beds in the central trough.

The surface geology around the Los Angeles basin is relatively well known through publications by Hoots (1931) on the eastern Santa Monica Mountains, by Lamar (1970) on the Repetto Hills and nearby areas, by Durham and Yerkes (1964) and Yerkes (1972) on the Puente Hills, by Schoellhamer et al. (1981) on the northern Santa Ana Mountains, by Vedder et al. (1957) on the San Joaquin Hills, and by Woodring et al. (1946) on the Palos Verdes Hills. In addition, regional subsurface cross sections such as the Newport-Inglewood trend cross section by Traxler et al. (1962) and the northeast-southwest cross section across the basin by Knapp et al. (1962); structure-contour maps by McMurdie et al. (1973); subsurface stratigraphic correlations by Wissler (1943); reports on individual oil fields summarized by California Division of Oil and Gas (1974); and detailed studies of individual oil fields, such as the report on Wilmington oil field by Truex (1974), have given much insight into the subsurface. Yet most subsurface well data have not been integrated into any published regional geologic framework. We constructed stratigraphic sections hung on the Repettian-"Delmontian" boundary that include all major producing trends (located on Figure 1). Simplified versions of representative stratigraphic sections are shown in Figure 2. These sections show lithofacies as well as the distribution and API gravities of major oil accumulations. Isopach maps were constructed from these sections to show major changes in basin geometry after the end of the middle Miocene (Figures 3 through 6).

Here it is necessary to restrict what is meant by the term *Los Angeles basin*, and this restriction affects how old we say it is. If all unmetamorphosed strata in the surrounding hills are included, then the basin is as old as Turonian. But pre-middle Miocene strata of these uplands reflect tectonic frameworks far different than that controlling the modern Los Angeles basin. We follow Barbat (1958) in deciding to date the Los Angeles basin from the time that it assumed its modern structural configuration: a northwest-trending central trough flanked on the west by a structural shelf controlled by right-lateral strike-slip faulting, on the north by an east-trending fold-thrust belt at the southern margin of the Santa Monica Mountains, and on the southeast by the Santa Ana Mountains and San Joaquin Hills (pre-Pliocene highlands that were overlapped on their northwest margins by Pliocene and Pleistocene sediments of the basin). We show that this structural configuration was not fully achieved until close to the beginning of Repetto deposition, about 4 Ma, although the central trough first appeared during the late Mohnian, about 8 Ma.

REGIONAL GEOLOGIC SETTING

The Peninsular Ranges, including the Santa Ana Mountains at their northwestern end, tilted westward in Late Cretaceous and early Tertiary time. Upper Cretaceous strata were derived largely from the Peninsular Ranges, and they occur in a westward-deepening and westward-thickening wedge, including proximal turbidite fans. The Cretaceous is overlain locally by Eocene strata, but Paleocene strata intervene in some places (Popenoe and Woodring, 1945; Sage, 1973; Schoellhamer et al., 1981). Unlike the Cretaceous, the Paleocene and Eocene conglomerates are characterized by exotic clasts, possibly derived from northwest Mexico or southern Arizona (Minch, 1972; Yeats et al., 1973). Although Paleogene sequences of the Santa Ana Mountains can be correlated with those of the California continental borderland, Paleogene and Upper Cretaceous strata are absent in the western Los Angeles basin, intermediate between the two regions, and Miocene rocks rest on a metamorphic basement not represented in clasts in Paleogene conglomerate (Yeats, 1968). Cretaceous and Paleogene strata show no tendency to thicken toward the center of the Los Angeles basin.

In the Santa Monica Mountains, San Joaquin Hills, northwestern Santa Ana Mountains, and eastern Puente Hills, the marine Eocene sequence grades upward into nonmarine Sespe Formation of Oligocene to early Miocene age. The Sespe is interbedded with and overlain by shallow-marine Vaqueros Formation. A similar facies gradation characterizes the overlying Topanga Formation of middle Miocene age. These strata, like the older Paleogene, show no tendency to thicken toward the center of the Los Angeles basin.

The Topanga Formation is locally interbedded with and overlain by volcanic rocks and is cut by dikes and sills, also of middle Miocene age (Woodring et al., 1946; Vedder et al., 1957; Eaton, 1958; Yerkes et al., 1965), radiometrically dated as 14 to 17 Ma (Turner, 1970; Weigand, 1982). This sequence also shows no consistent tendency to thicken toward the basin, and it is locally very thick in adjacent areas that are now highlands, particularly the western Santa Monica Mountains. Middle Miocene volcanic rocks and associated strata also occur in the western Los Angeles basin, including the Palos Verdes Hills and the northern part of the Newport-Inglewood fault zone (Wright, 1987). The San Onofre Breccia, a western facies of the middle Miocene and upper part of the lower Miocene sequence, is characterized by locally derived clasts of basement rocks of the inner California continental borderland (Woodford, 1925; Stuart, 1976).

The middle Miocene was characterized by crustal extension (Campbell and Yerkes, 1976), and during this time, the Santa Monica Mountains may have moved 90 km westward with respect to the Santa Ana Mountains along a left-lateral fault zone near the northern margin of the

---

Date Submitted: 8/25/87
Date Accepted: 9/1/88

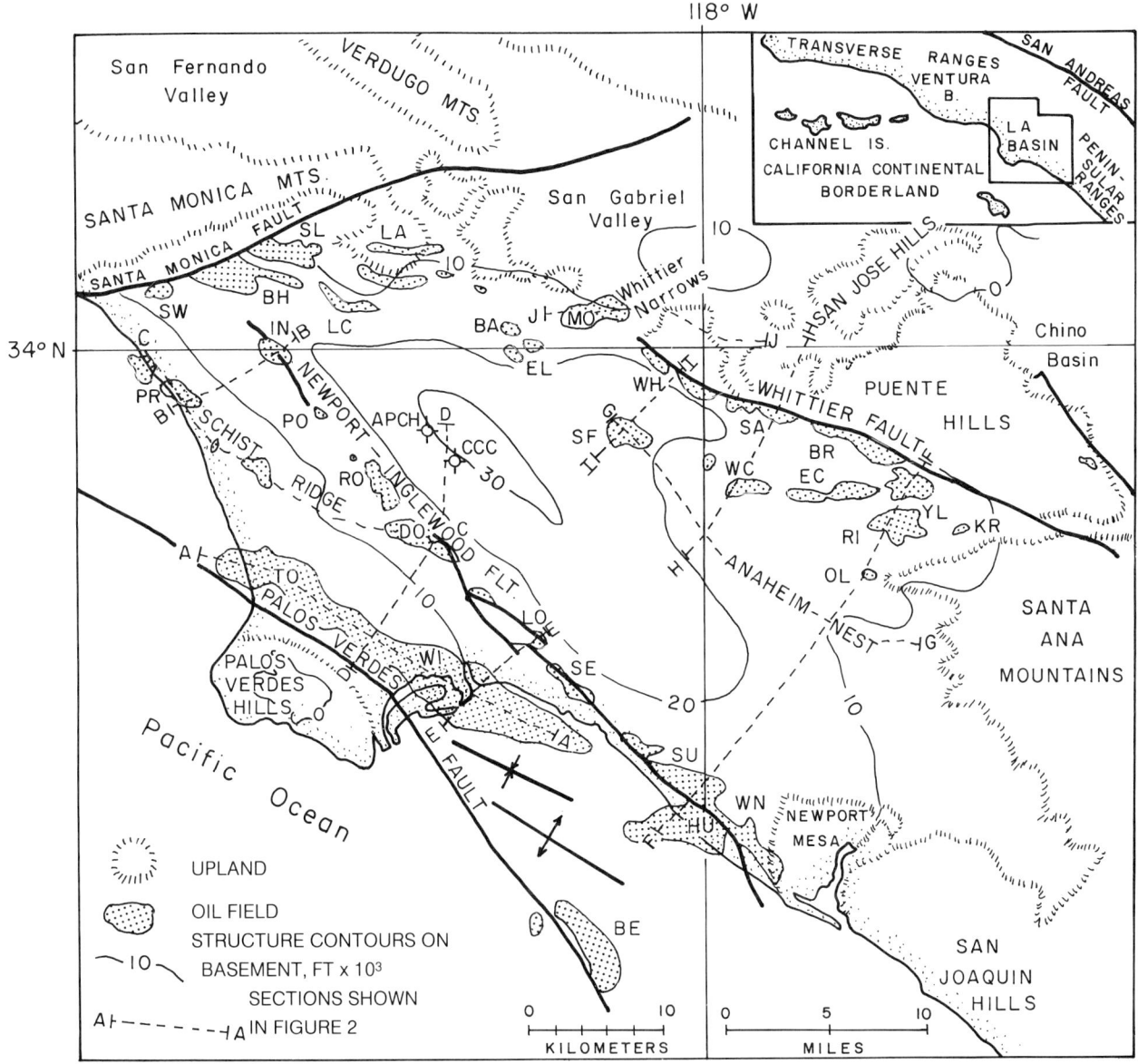

Figure 1. Index map of Los Angeles basin showing major faults, uplands, and oil fields (shaded pattern) modified from Yerkes et al. (1965). Basement contours (contour interval = 10,000 ft) from McCulloh (1960). Oil fields identified as follows: BA, Bandini; BE, Beta; BH, Beverly Hills; BR, Brea-Olinda; DO, Dominguez; EC, East Coyote; EL, East Los Angeles; HU, Huntington Beach; IN, Inglewood; KR, Kraemer; LA, Los Angeles Downtown; LC, Las Cienegas; LO, Long Beach; MO, Montebello; OL, Olive; PO, Potrero; PR, Playa del Rey; RI, Richfield; RO, Rosecrans; SA, Sansinena; SE, Seal Beach; SF, Santa Fe Springs; SL, Salt Lake; SU, Sunset Beach; SW, Sawtelle; TO, Torrance; WC, West Coyote; WH, Whittier; WI, Wilmington; WN, West Newport; YL, Yorba Linda. Deep wells in central trough are APCH (American Petrofina Central Core Hole 1) and CCC (Chevron Carlin Community 1). A–A through J–J locate stratigraphic sections of Figures 2A through 2J.

Los Angeles basin (Yeats, 1968; Campbell and Yerkes, 1976). In addition, much of the California continental borderland moved northwest with respect to the Peninsular Ranges along a right-lateral fault, probably at the same time (Abbott et al., 1983).

West of the Newport-Inglewood fault, the oldest strata to be preserved atop Catalina Schist basement consist of a middle Miocene basal conglomerate deposited on a surface of moderate relief. In the Palos Verdes Hills (Woodring et al., 1946) and along the coast to the north, the middle and lower upper Miocene (Luisian and Mohnian Stages of Kleinpell, 1938) consists of cherty, phosphatic, and diatomaceous shale and mudstone similar to the Monterey Shale elsewhere. Throughout most of the basin and in the Santa Monica Mountains and the Puente Hills, similar fine-grained strata are interbedded with sandstone turbidites and are called the Puente Formation. The Puente Formation is thicker in the Puente Hills than in known subsurface sequences in the Los Angeles basin, indicating that the basin had not achieved its present form at the time of Puente deposition.

Upper Miocene and Pliocene strata are conformable with Mohnian beds except on basin margins, but in contrast to the Mohnian, these strata thicken toward the center of the basin (Conrey, 1967). Microfossils are assigned to the upper part of the Mohnian Stage and the Delmontian Stage of Kleinpell (1938) and to Repettian and younger stages of Natland and Rothwell (1954), and in most of the basin, these microfossils are similar to assemblages found today in deep water. Shallower water equivalents are found near the eastern edge of the basin. The Pleistocene is best known around the flanks of the Palos Verdes Hills and includes the Timms Point Silt, Lomita Marl, and San Pedro Formation, the latter composed mainly of shallow-marine sand and gravel. These formations also thicken toward the center of the basin (Poland and Piper, 1956; California Department of Water Resources, 1961; Yerkes et al., 1965; Johnson and Duke, 1973).

Many of the oil fields in the Los Angeles basin occur on anticlines and upturned stratigraphic traps aligned over northwest-trending basement faults that have a component of right slip (Harding, 1974). The Whittier fault is the northwestern extension of the right-lateral Elsinore fault, and it also has a reverse-slip component with the Puente Hills upthrown with respect to the Coyote and Richfield trend to the south (Durham and Yerkes, 1964; Yerkes, 1972; Harding, 1974; Truex, 1975). Farther west, the Newport-Inglewood fault zone is expressed at the surface mainly as aligned anticlines housing oil fields (Harding, 1973; Yeats, 1973; Barrows, 1974); total right-lateral displacement is small. Farther west, the Wilmington and Torrance oil fields occur on broad, northwest-trending folds that began to form in the late Miocene (Truex, 1974), possibly related to movement on the Palos Verdes fault farther southwest. On the continental shelf south of the Wilmington field, the Beta field is east of and adjacent to the Palos Verdes fault. Oil fields along the northwestern margin of the Los Angeles basin (Lamar, 1970; Jacobson and Lindblom, 1987) are related to the fold-and-thrust belt accompanying north-south convergence between the Santa Monica Mountains and the Los Angeles basin.

ORGANIZATION OF PAPER

We correlated exploratory and oil-field wells throughout the Los Angeles basin on the basis of electric-log comparisons and on the microfaunal zones of Kleinpell (1938), Wissler (1943; 1958), and Natland and Rothwell (1954) as applied to the Los Angeles basin (for discussion, see Blake, *Review of the Neogene Biostratigraphy of the Los Angeles Basin*, this volume). These correlations were used to construct a series of stratigraphic correlation sections along and across producing trends (Figure 2). Surface locations of wells are given by Munger (1987). These sections show lithofacies (interbedded turbidity-current sandstone and mudstone, hemipelagic mudstone, proximal turbidite-fan conglomerate, shallow-marine to nonmarine sandstone, mudstone and conglomerate) and benthic microfaunal zone boundaries, assuming that these mark time lines. We recognize that these zones are actually time-transgressive because benthic zones are strongly influenced by water depth (Ingle, 1980; Blake, 1976, this volume), but this should not significantly influence their use in this analysis because most of the critical correlations are within the same environment, i.e., the basin plain. Subsurface correlations are made independently of the published literature and, thus, may not agree in detail with cross sections published by others. The correlation sections are hung on the Repettian–"Delmontian" boundary, which is intra-Pliocene in parts of coastal California (Blake, 1987), but was formerly assumed to be the boundary between the Pliocene (Repettian) and Miocene (Division A of Wissler, 1943). The "Delmontian" as used here and in Blake (1987, this volume) is considerably younger than the type Delmontian of Kleinpell (1938) in the central Coast Ranges.

In addition to the correlation sections (Figure 2), we present isopachs of Divisions D and E of Wissler (1943), consisting of most of the Mohnian (Figure 3); Divisions A, B, and C of Wissler (1943), consisting of the "Delmontian" and uppermost Mohnian (Figure 4); Repettian (Figure 5); and post-Repettian (Figure 6) layers. The post-Repettian isopach is taken at maximum burial, restoring sediments removed by erosion in late Pleistocene time. For this reason, the post-Repettian isopach is limited to areas at or near maximum burial today; original depositional thicknesses are not restored for surrounding uplands, where only older rocks are exposed.

The paper discusses these four layers as based on the correlation sections and resultant isopachs, then it considers the thermal history of the oil-generating center of the basin as a guide to the present distribution of oil fields. Finally, the tectonic significance of isopach trends in the Los Angeles basin is analyzed.

DIVISIONS D AND E (MOHNIAN)

Divisions D and E were correlated to the Mohnian and Division C to the lower Delmontian by Wissler (1943), but later, Wissler (1958) correlated Division C to the upper Mohnian, an interpretation accepted by Blake (this volume). Divisions D and E span the time interval 13.8 to 8 Ma, with Division D occupying only the last million years of this time (Blake, this volume). Division D and E deep-marine strata are widely exposed in the Santa Monica Mountains and Puente Hills and are less widely exposed in the Palos Verdes Hills and the western margin of the northwest Santa Ana Mountains, indicating that these uplands were sites of deposition during that time. The Mohnian is conformable with the Luisian (late middle Miocene) in the Puente Hills, Santa Fe Springs–Coyote trend, and Newport-Inglewood trend, but in the Santa Monica Mountains, the Mohnian rests unconformably on older rocks, including Mesozoic basement. The thickest Division D and E sequence is found in the Puente Hills (Figure 3), and thick sequences also occur in the subsurface in the Richfield area and in downtown Los Angeles north of the Las Cienegas fault. The Mohnian largely postdates volcanism, but at one locality in the Puente Hills, Mohnian strata are intruded by diabase (well 16, Figure 2J).

Division D and E strata show basin eversion or reversal in a section between West Coyote field across Brea-Olinda field into the Puente Hills (Figure 2H). In the Puente Hills, the Soquel Member of the Puente Formation comprises a thick sequence of turbidites that thins and lenses out to mudstone in the direction of the Whittier fault. Division D and E strata at West Coyote are fine grained and thin relative to the Puente Hills sequence. The isopach map shows the greatest thickness of Division D and E strata in the Puente Hills close to the Whittier fault. Thicknesses on the southwest side are much less and decrease to zero in the vicinity of the Anaheim nose.

Division D and E thicknesses in the upthrown block northeast of the Las Cienegas fault are much less than those in the Inglewood field to the southwest; both these areas were high in the Pliocene-Pleistocene, although Division D and E thicknesses in the intervening deep trough are not known. The Mohnian is much thicker north of the Santa Monica-Hollywood fault zone than it is to the south (Wright, 1987, his figure 9), but this may be explained in part by 10 km to 14 km of post-Mohnian left slip on this fault system (Lamar, 1970; Wright, 1987).

The isopachs of Figure 3 show that it is not straightforward to correlate Division D and E sequences across the Newport-Inglewood, Las Cienegas, or Whittier faults. The Tarzana fan of the Santa Monica Mountains (Sullwold, 1960) can be correlated into the northeastern Los Angeles basin, and most of the fans clearly have a northeastern source (Redin, *Oil and Gas Production from Submarine Fans of the Los Angeles Basin*, this volume); but aside from the piercing-point offset described by Lamar and Wright across

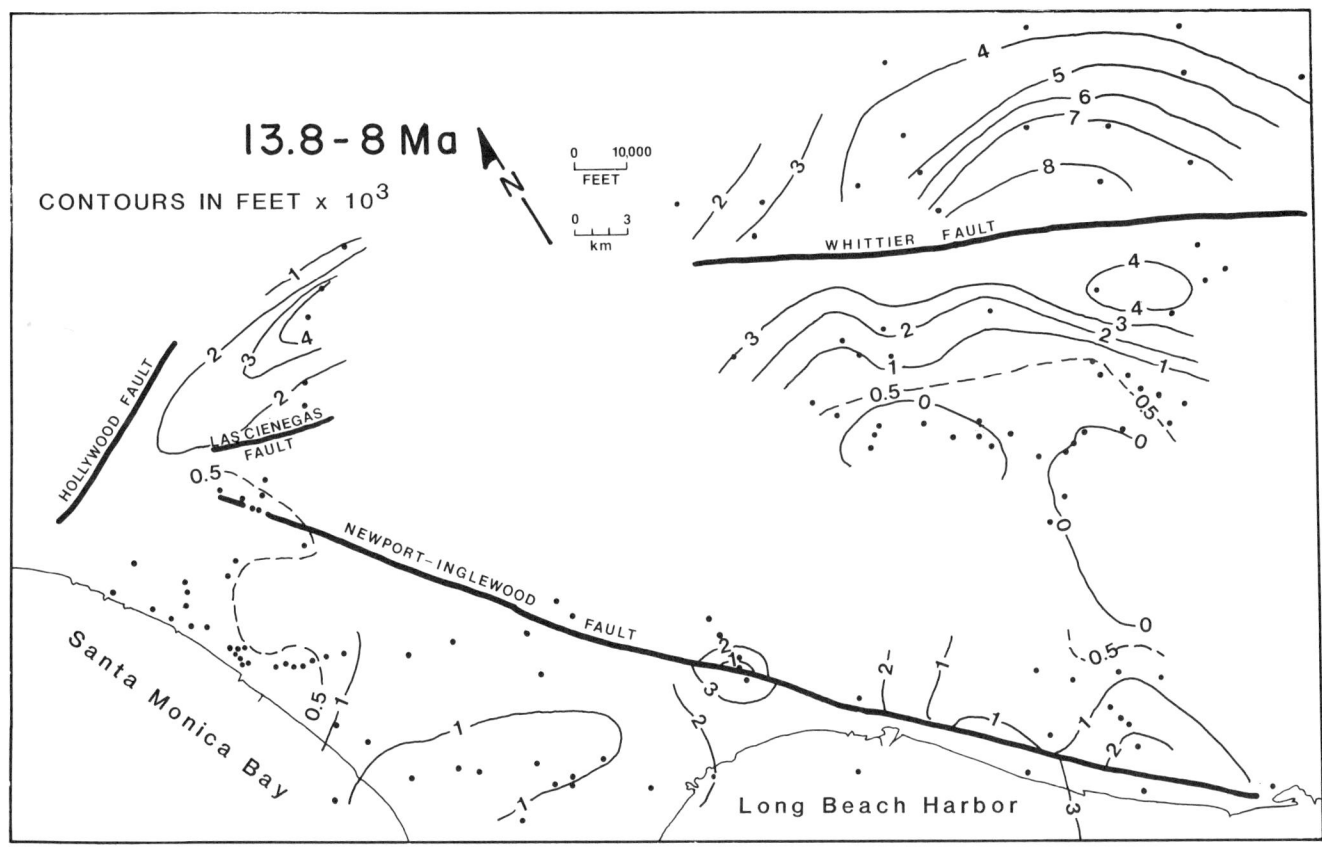

Figure 3. Isopachs of Mohnian Division D and E (upper Miocene) strata; isopach interval = 1000 ft (300 m). Dots show well control. Thickest Mohnian strata are in Puente Hills near the Whittier fault; other thick sections occur south of the Whittier fault and north of Las Cienegas fault. Contrasts in thickness across Newport-Inglewood, Las Cienegas, and Whittier faults are not explained by assuming strike slip on any of these faults; more likely, the thickness differences are due to dip slip during deposition. Mohnian strata do not consistently increase in thickness toward the present central trough.

the Santa Monica fault zone, it is not possible to obtain unambiguous piercing-point offsets across the Whittier or Newport-Inglewood faults, even though both are recognized as right-lateral strike-slip faults. Most workers believe that total right slip on these faults is small; hence, the abrupt thickness variations may be caused by local dip slip on ancestors of these faults. Campbell and Yerkes (1976) showed that the middle Miocene of the Los Angeles basin was dominated by extension, and extension may have characterized the late Miocene as well (Crowell, 1987).

The Division D and E isopachs show no tendency to thicken toward the present central trough, nor do they show the pronounced northwest-southeast orientation of the present basin. It is not possible to document this in wells in and adjacent to the central trough because the Mohnian there is not penetrated by wells. However, the contrast between the Mohnian and the post-Repetto can be documented to the southeast in a cross section between Huntington Beach field and Anaheim nose (Figure 2F). In this cross section, the up-plunge extension of the trough is expressed by thicker post-Repetto strata and, to a lesser extent, thicker Repetto strata. The Mohnian is thinner from the High School (Newport-Inglewood) fault northeast and is fine grained, whereas at Huntington Beach field, the Mohnian is thicker and in trough-turbidite facies.

DIVISIONS A, B, AND C ("DELMONTIAN" AND UPPER MOHNIAN)

The "Delmontian" of the Los Angeles basin comprises Divisions A and B of Wissler (1943; 1958), and the upper Mohnian comprises Wissler's Division C. The age of Divisions A, B, and C ranges from 8.0 Ma ($\pm$ 0.9 m.y.) to 4.3 Ma ($\pm$ 0.9 m.y.) (Blake, this volume). Most deep-water sequences of Divisions A, B, and C are found in the subsurface of the Los Angeles basin; few are exposed in the hills surrounding the basin, indicating that the basin *sensu stricto* originated as this sequence was deposited. However, the isopach map (Figure 4) shows that the depocenter is oriented nearly north-south, in contrast to younger depocenters that follow a northwest-southeast trend. The north-south trend is established by well control northeast and southwest of the central trough (Figure 4). Thickest sections, rich in sand, occur in the East Los Angeles-Montebello area (Figure 2J), in the path of the main entry point for sand transported from mountainous areas farther northeast. The eastern edge of this thick sequence is exposed in the western Puente Hills as the Sycamore Canyon Conglomerate Member of the Puente Formation (Figure 2J; Yerkes, 1972). Division A, B, and C strata are thick and sand rich in the central trough, as documented by the American Petrofina Central Core Hole 1 well; but throughout most of the central trough, this sequence is too deep to be reached by most wells. The axis of the sand-rich Division A, B, and C trough crosses the Newport-Inglewood trend in the vicinity of Long Beach and Seal Beach fields, and sandstones lens out southeastward at West Newport field, northwestward between Potrero and Inglewood fields, and southwestward across the Schist Ridge structural shelf. Playa del Rey, Torrance, Wilmington, and Inglewood fields became positive during the Delmontian; and at Inglewood, Division A (youngest "Delmontian") rests directly on Division E (oldest Mohnian) strata. The isopachs in the western Los Angeles basin trend northwest, parallel to younger isopachs and to the Newport-Inglewood fault.

A second Division A, B, and C trough is found south of and parallel to the Whittier fault; it is thickest in the Yorba Linda field (Figure 2F). Sandstone in this trough lenses out to mudstone in the direction of the Whittier fault and the Anaheim nose (Figure 2H). The Anaheim nose was structurally positive during this time (Figure 2G) because sandstone at West Newport and Huntington Beach fields on the Newport-Inglewood trend also lenses out toward the Anaheim nose. During the deposition of Division A strata, the Anaheim nose ceased being structurally positive. Youngest Division A sandstones (II Clark and Hathaway zones of Santa Fe Springs field) are buttressed against the Anaheim nose, trapping oil at La Mirada field.

REPETTO

The Repettian Stage of Natland and Rothwell (1954) lasted from about 4.5 to 3 Ma (Blake, this volume). By the beginning of the Repettian, the Los Angeles basin had acquired its present structural configuration: northwest-southeast elongation parallel to the Whittier, Newport-Inglewood, and Palos Verdes faults and an east-trending northern margin controlled by the southern edge of the Transverse Ranges (Figure 5). The Anaheim nose, a submarine ridge that was gradually covered by basin-plain turbidites that buttressed against it, was on the southeast. This paper confirms the Repetto basin framework described by Conrey (1967), adding information only in the Los Angeles downtown area and in the central trough.

As in the "Delmontian," the main sediment source was from the northeast through the Whittier Narrows, west of the Puente Hills, but a subsidiary source was from the east, parallel to and south of the Whittier fault. At this time, structures housing the Santa Fe Springs, Bandini-East Los Angeles, Inglewood, and Playa del Rey oil fields began to grow. Most of these growing structures were located far enough within the submarine fan that turbidite sandstone extended over the structures even as they grew. Toward the margins of the fan (e.g., close to the Palos Verdes fault, near West Newport field, and even along the Whittier fault), sandstone gave way to mudstone.

An up-plunge cross section across the central trough between Huntington Beach field and the Anaheim nose (Figure 2F) shows that the new trough axis formed between the Newport-Inglewood fault and the now-inactive Anaheim nose. The Repetto is at least 8000 ft (2.5 km) thick in the central trough, based on the two deep wells located on the southwest flank (American Petrofina Central Core Hole 1 and Chevron Carlin Community 1). Farther southwest, a syncline did not develop between the Newport-Inglewood fault and the Schist Ridge and Wilmington trends until after Repetto time: these areas were in updip drainage position with respect to the central trough at the end of Repetto deposition. (A reviewer, T. H. McCulloh, disagreed with this conclusion based on his isopach map of the Mohnian "nodular shale" that shows

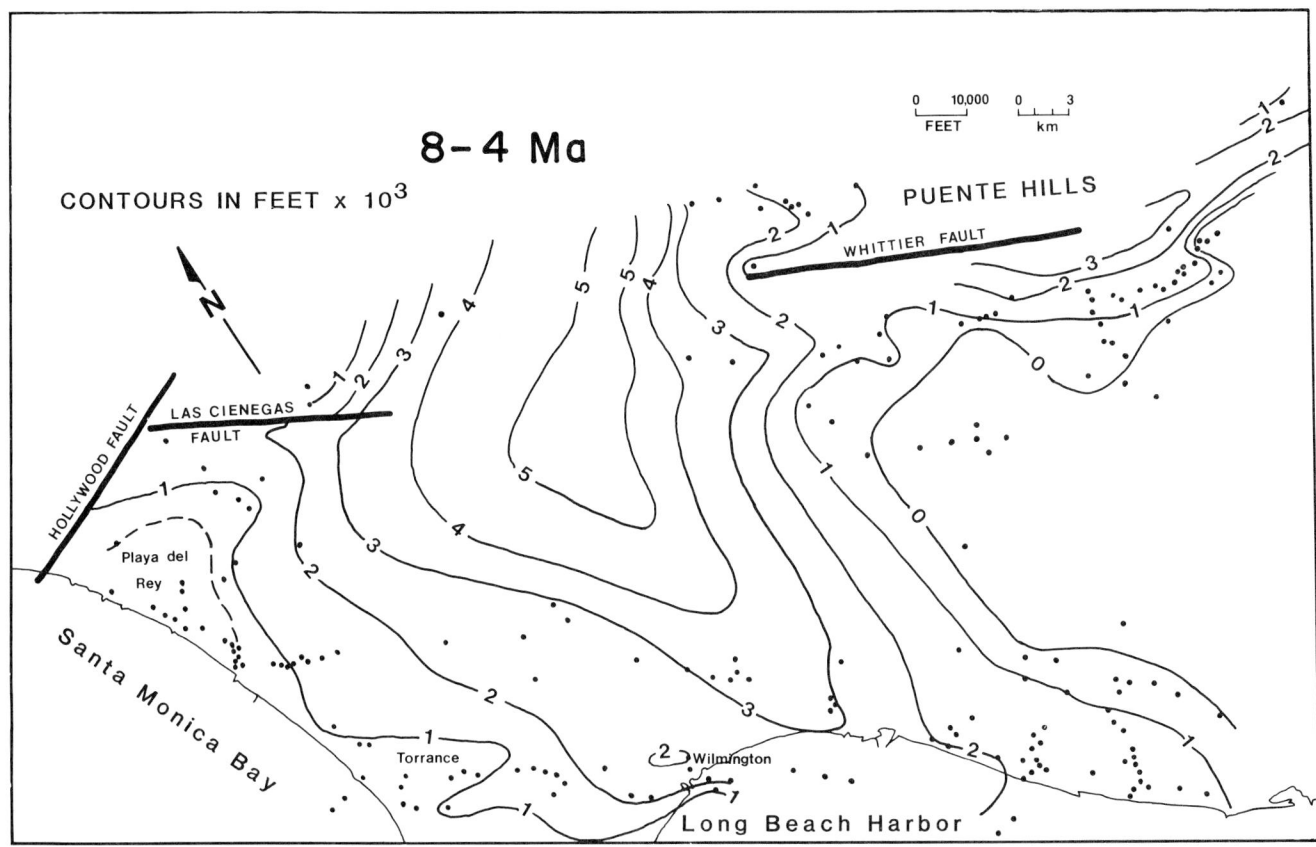

Figure 4. Isopachs of upper Mohnian Division C and "Delmontian" Division A and B (upper Miocene–lower Pliocene) strata; see Figure 3 for explanation. The central trough received a thick sequence of trough turbidites for the first time, but the basin was oriented north-south, not northwest-southeast. On the other hand, isopachs west of the Newport-Inglewood fault trend northwest, parallel to that fault. Playa del Rey, Torrance, and Wilmington fields were positive, as was the Anaheim nose.

a thickening of this unit west of Long Beach field. Further discussion of this controversy must await publication of more detailed stratigraphic studies of the Miocene strata west of the Newport-Inglewood fault.)

POST-REPETTO STRATA

Included in this sequence are the Pico Formation (Upper Member of the Fernando Formation of some authors; cf. Yerkes et al., 1965), in part trough turbidites like the underlying Repetto, and the San Pedro and La Habra formations (shallow-marine and nonmarine, in part coarse-grained clastic strata that make up the youngest part of the Los Angeles basin fill). On the coast, this includes the Lomita Marl, Timms Point Silt, and San Pedro Formation. Amino-acid racemization and magnetostratigraphic data show that the Timms Point Silt is 300 to 600 ka (ka = 1000 years). The San Pedro Sandstone and overlying Palos Verdes Sandstone are virtually the same age (95 to 120 ka) (Lajoie, 1986). Marine sedimentation in the southwestern Los Angeles basin was continuous to about 300 ka and episodic thereafter. It was controlled by eustatic fluctuations; marine deposits younger than about 200–300 ka do not appear east of the Newport-Inglewood fault (Lajoie, 1985). However, nonmarine deposition probably continued in the central trough in the last few hundred thousand years because late Quaternary aquifers are thickest in the axis of the trough, and the trough probably continued to trap floodplain sediments.

The structural grain begun in the Repettian is accentuated and exaggerated in the younger sequence (Figure 6). In the central trough, the American Petrofina Central Core Hole 1 showed that post-Repetto sediments are slightly more than 10,500 ft (3 km) thick, even though the hole is on the lower southwest flank of the trough (Figure 2D). Up plunge to the southeast, where there is well control, the axis shifted northeast from its position in the Repetto toward the Anaheim nose (Figure 2F). A syncline developed between the Newport-Inglewood fault and the Schist Ridge, isolating structures on the oil fields on that trend from Dominguez field northward (Figure 2B). The Santa Fe Springs structure continued to grow (Figure 2I). Turbidite sandstone fans were not quite as widespread as they were

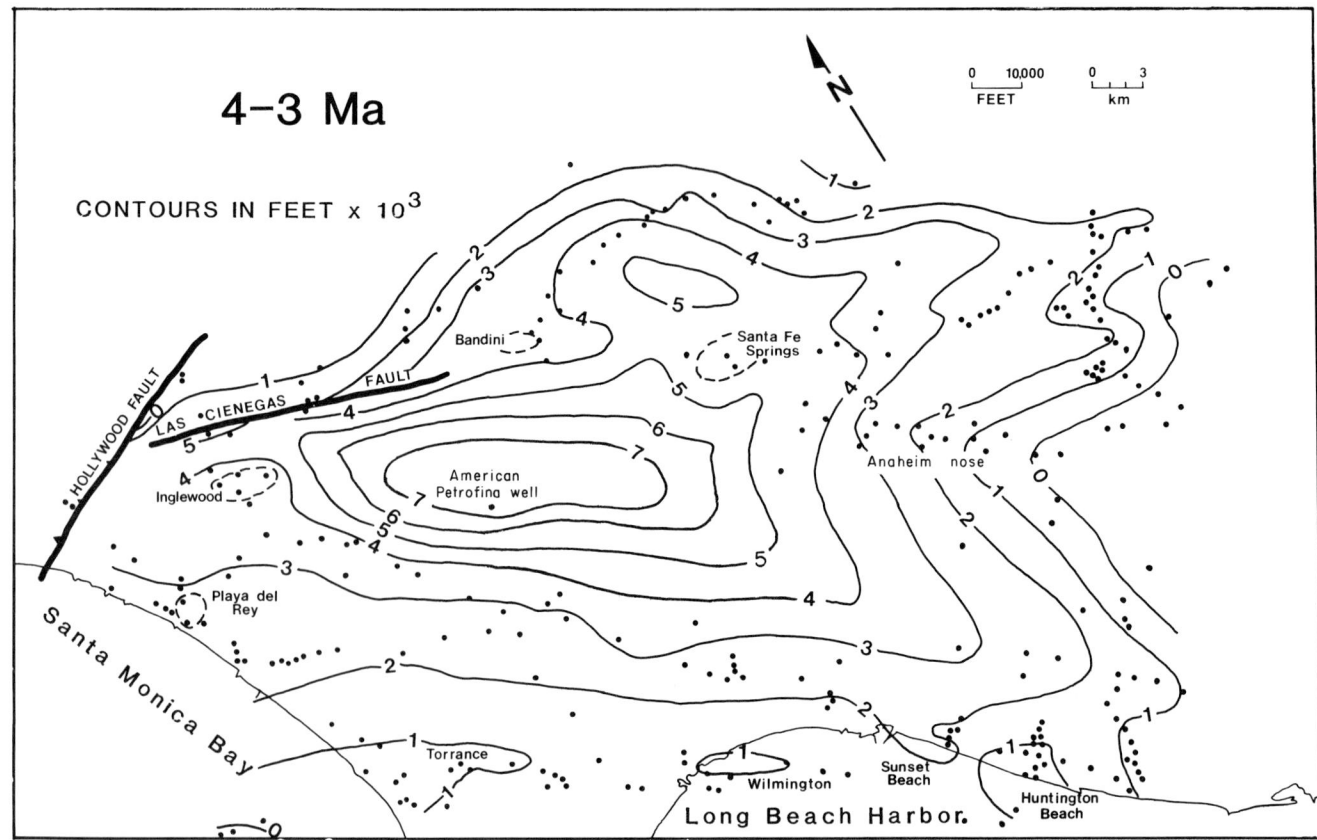

Figure 5. Isopachs of Repetto (lower Pliocene) strata; see Figure 3 for explanation. The central trough assumed its northwest-southeast configuration for first time; control point near 8000-ft (2.5-km) isopach is American Petrofina well. Isopachs around Anaheim nose reflect buttressing of Repetto turbidites against a northwest-sloping ridge. Sites of several oil fields flanking the central trough were positive during Repetto deposition.

in the Repetto. The Pico turbidite sandstone gives way upsection to a mudstone slope facies, then to the shallow-marine to nonmarine, coarse-grained clastic strata that compose the top of the basin sequence.

QUATERNARY STRUCTURE

Modern structures began to grow during deposition of the Repetto, and growth accelerated during and after Pico deposition (Yeats, 1978). However, in contrast to the onshore part of the Ventura basin, deformation was not accompanied by extensive uplift. Pliocene and younger deposits are exposed in the Repetto Hills and vicinity (Lamar, 1970) and on the flanks of the Palos Verdes Hills (Figure 2D; Woodring et al., 1946), but for the most part, these exposures are of basin-marginal facies. Most of these deposits are still preserved in the basin. Minor uplifts within the basin, such as those at Inglewood, Dominguez, and Long Beach fields, are large-scale pressure ridges associated with right slip along the Newport-Inglewood fault; other uplifts along the Whittier fault and along the Coyote trend may have a similar origin.

DISTRIBUTION OF MAJOR OIL ACCUMULATIONS

Most of the oil in the Los Angeles basin is found in Mohnian, "Delmontian," and lower and middle Repettian turbidites. Long Beach and Inglewood fields are two notable exceptions where the oil occurs in the Pico as well as older turbidites (Figures 2B, 2E). Fields in which the top of the oil column is within the Repetto include Huntington Beach, Seal Beach, Dominguez, and Potrero on the Newport-Inglewood trend; Wilmington, Torrance, and Playa del Rey on the structural shelf to the west; and Santa Fe Springs, West Coyote, East Coyote, Yorba Linda, Brea-Olinda, Whittier, and Montebello fields in the eastern part of the basin. At Yorba Linda (Figure 2F), younger traps have probably been breached by erosion, and much of the oil accumulation may be sealed by tar. At East Coyote, Brea-Olinda, and Whittier, the top of the oil column is also the top of the sandstone sequence. But in most of the other fields, the oil-bearing sandstones are overlain by water-bearing sandstones, also in trapping position. Why are these sandstones not oil-bearing?

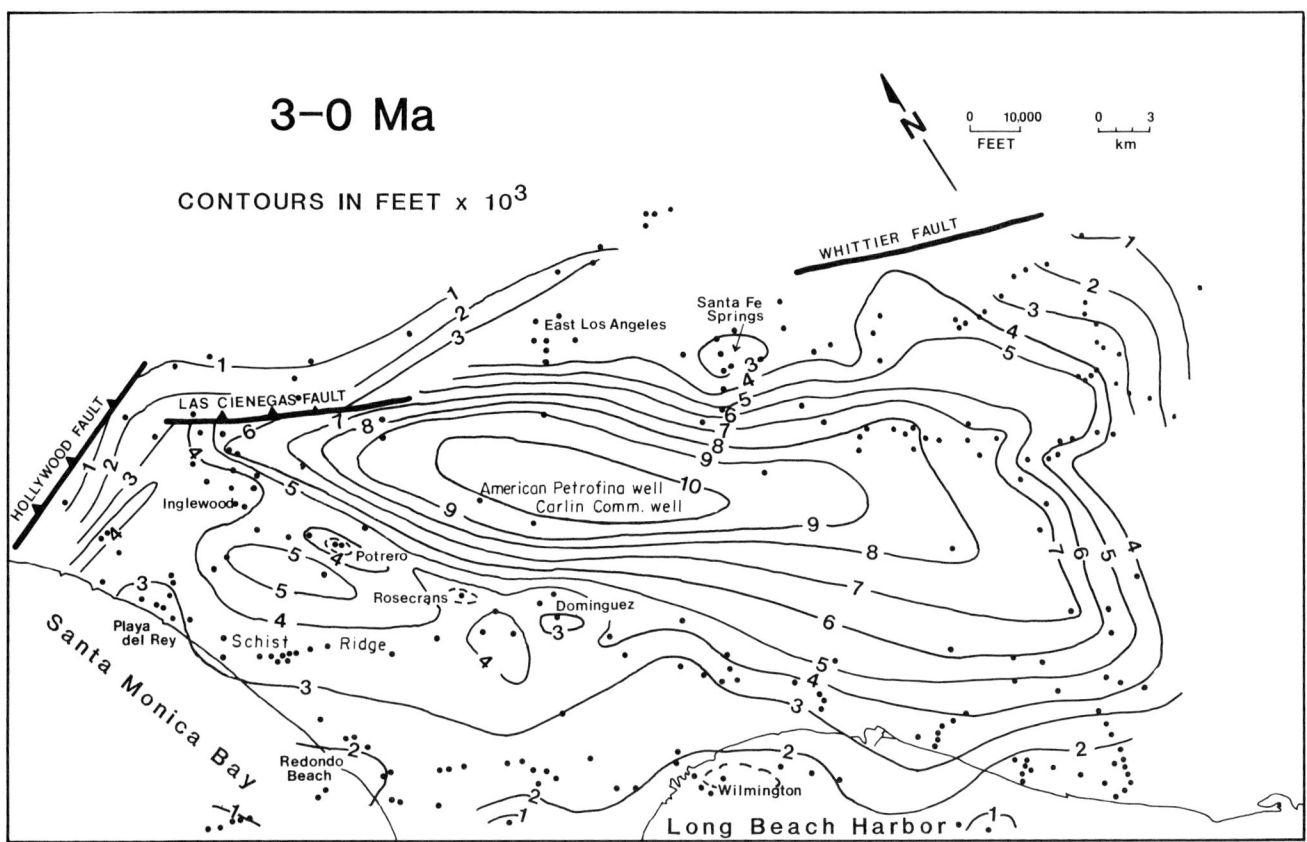

Figure 6. Isopachs of post-Repetto strata; see Figure 3 for explanation. Sequence includes Pico, San Pedro, and upper Quaternary deposits; where these sediments have been removed by late Quaternary erosion, sediment thickness has been restored so that isopachs represent thicknesses at maximum burial. Central trough axis shifts northeast from Repetto position in southeast Los Angeles basin. Syncline develops between the Inglewood and Potrero fields and the Schist Ridge; a less prominent syncline south of Schist Ridge only is locally expressed in isopachs. Most major oil fields flanking central trough were structurally high during deposition.

The West Newport, Sunset Beach, and Beta fields produce from Miocene turbidites, and there is no Pliocene production (Redin, this volume). At Sunset Beach and West Newport, the producing sandstones are overlain by water-bearing turbidite sandstones. On the Schist Ridge (except for Playa del Rey), only the basal transgressive Miocene sandstone and conglomerate, together with weathered and fractured Catalina Schist basement (Figure 2C), produce. Overlying turbidite sandstones at El Segundo and Santa Fe Springs fields locally produce gas, but not oil.

DISCUSSION

Burial and Thermal History of the Central Trough

We now consider the central trough, because it is the most likely source of most of the oil of the Los Angeles basin. We conclude, as do Philippi (1965) and Jeffrey et al. (*Geochemistry of Los Angeles Basin Oil and Gas Systems*, this volume), that a major source is the upper Miocene sequence. We also consider the possibility that younger fine-grained sediments, including the Repetto, may also be important sources. To evaluate this possibility, it is necessary to consider the thermal history of the central trough and the timing of its burial with respect to the formation of traps in adjacent upstructural oil fields.

The deepest well in the basin is the American Petrofina Central Core Hole 1, drilled to 21,215 ft (6466 m) and bottoming in the "Delmontian" (Figure 2D). This well and the nearby Chevron Carlin Community 1 well were drilled downstructure from the Dominguez field, close to, but not at the bottom of, the trough. The American Petrofina well was directionally drilled, and our figures as well as those of Bostick et al. (1978) are based upon vertical depth, not well depth.

Geothermal gradients in the Los Angeles basin have been collected by Carlson (1930), French (1940), and most recently by Bostick et al. (1978) who summarize criteria for determining in situ temperatures based on well data. Some previous analyses of thermal history (for example,

Turcotte and McAdoo, 1979; Mayer, 1987) are based on geothermal gradients from wells on the flanks of the basin, but these gradients are generally higher because the more highly conductive basement is closer to the surface. In basins where subsidence and sedimentation are rapid and fairly recent, the blanketing effect of added sediment cover displaces isotherms downward (Bostick et al., 1978), resulting in a lower geothermal gradient. For the central Los Angeles basin, Bostick et al. (1978) used a gradient of 23.3°C + (0.0082 × depth in feet), which is in good agreement with a temperature of 193.3°C at 6320 m vertical depth in Central Core Hole 1 based on downhole logs. The temperature in sedimentary rocks at any given point in the basin must be at its maximum if it is now at maximum burial, provided conduction is the only significant heat-transfer mechanism. However, the geothermal gradient was probably higher in the past for three reasons: (1) the thermally blanketing sediment cover was thinner in the past, (2) depositional rates were slower in the Miocene, and (3) the basement underneath the entire basin is still cooling from its thermal peak in the middle Miocene (from about 17 to 14 Ma), the last episode of widespread volcanism (Turner, 1970; Weigand, 1982) and crustal extension (Campbell and Yerkes, 1976) in this part of southern California.

Figure 7 is a stratigraphic column for the central deep, based on thicknesses in the American Petrofina well. The total thickness of basin fill in the trough was calculated by McCulloh (1960) on the basis of gravity data. The thickness of Mohnian Division D and E sediments is based on thicknesses in wells flanking the central trough. We do not assume, as others have, that the Mohnian becomes very thick in the central trough; instead we use the average thickness measured on the flanks of the trough. If the total thickness estimate of McCulloh (1960) is correct, then the pre–Division C section is not much more than 2800 m thick, and the Mohnian Division D and E makes up about half of that. The pre-Mohnian section, which could extend from middle Miocene to Upper Cretaceous, may be little more than 1300 m thick. Thus, subsidence rates immediately following middle Miocene rifting and volcanism must have been slow, and the thermal model of isostatic subsidence of Turcotte and McAdoo (1979) is inadequate to account for the central trough. Thermal isostatic subsidence may, however, explain the thick sequences of late Miocene strata in the Puente Hills, Santa Monica Mountains, and elsewhere.

Following the general recognition that both time and temperature are important controls in the generation of petroleum, Lopatin (1971) described a method in which the effects of time and temperature can be incorporated in a calculation of thermal maturity of organic matter in sediments. Lopatin recognized that the dependence of maturity on time is linear, but the dependence of maturity on temperature, based on chemical reaction-rate theory, must be exponential. Waples (1980) tried to define better the relation of reaction rate to temperature based on vitrinite reflectance, ratio of bitumen to organic carbon, and other thermally controlled factors. We do not estimate the position of the oil-generation window for the Los Angeles basin because of uncertainties in (1) the reaction rate, (2) the geothermal gradient for the last 8 m.y., and (3) possible

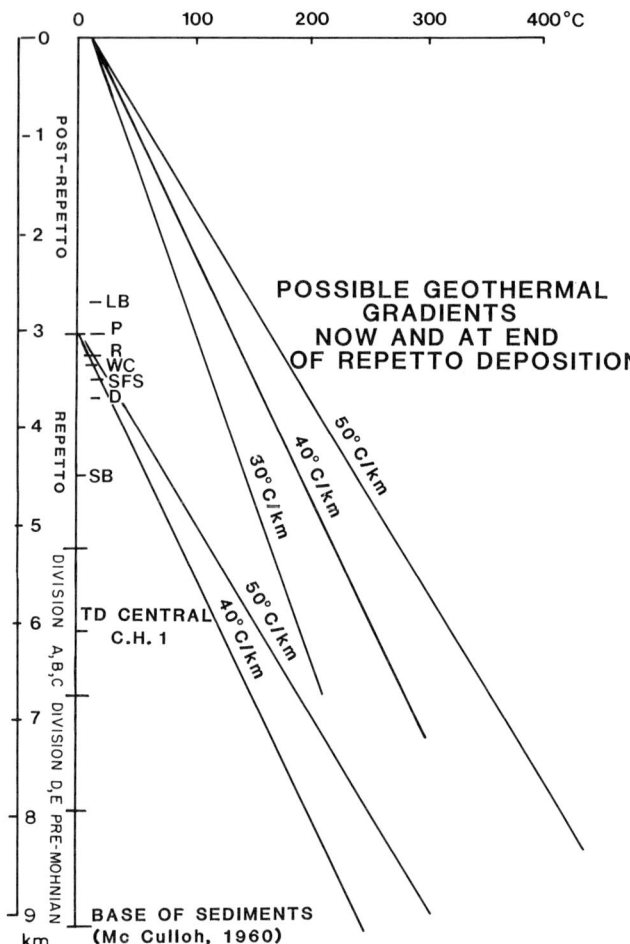

Figure 7. Estimated sediment thickness and geothermal gradients in central trough, based on American Petrofina Central Core Hole 1. Stratigraphic position of bottom of hole shown in figure. Present geothermal gradient from Bostick et al. (1978); geothermal gradients at end of Repetto, "Delmontian," and Division D deposition are assumed to be progressively higher because the sediment blanket would be thinner, and the time following the middle Miocene thermal peak would be shorter. Surface temperature now is 23°C (Bostick et al., 1978); earlier surface temperatures assumed to be 5°C on the sea floor, based on analogy with the present California continental borderland (Douglas, 1987).

different thermal response when subsidence is very rapid and very young. However, estimates of thicknesses in the central trough and of ages of critical horizons permit a qualitative estimate of the thermal history of individual horizons (Figure 8). Note that by the end of the Repettian, the base of the Mohnian was already at 200°C and the

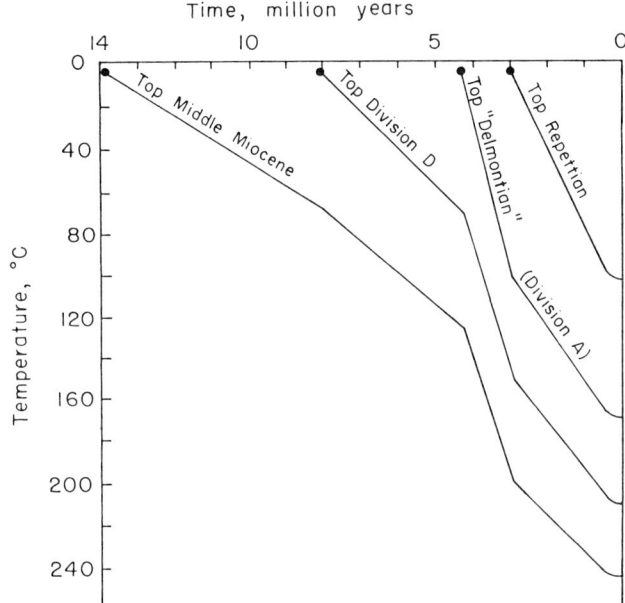

Figure 8. Time-temperature plots for four horizons in central trough, based on assumed geothermal gradients in Figure 7. See text for discussion.

top was at 150°C. At the present time, a horizon in the central trough corresponding stratigraphically to the top of the producing zone of many Los Angeles basin fields (upper Repetto) is 120–140°C.

Figure 9 is an east-west cross section across the central trough at the end of the Pleistocene and at the end of the Repettian. The cross section at the end of the Repettian and the isopachs of the Division A, B, and C and Repettian sequences show that the oil fields of the structural shelf west of the Newport-Inglewood fault were, at the end of the Repettian, in a position to receive oil migrating upward from the central trough parallel to bedding. Later, the structural shelf was cut off from the central trough by development of a syncline west of the Newport-Inglewood fault, implying that the oil of the Schist Ridge (except for the Repetto oil of Playa del Rey field) could have migrated into place prior to development of the syncline in the Pico.

Oil at Wilmington field occurs as high stratigraphically as the Repetto, indicating that this field continued to receive oil from the central trough much later than the Schist Ridge. The post-Repetto isopach map, together with the stratigraphic section between Wilmington and Long Beach fields (Figure 2E), shows that the syncline between Wilmington and Long Beach may not have been well developed enough to prevent oil from migrating across the Newport-Inglewood fault at Long Beach. Oil fields at the southern end of the basin (West Newport, Sunset Beach, and Beta) produce from Miocene reservoirs but not from Pliocene, suggesting that the top of the oil-generating window in the up-plunge extension of the central trough adjacent to these fields is in older strata than it is in adjacent Dominguez and Santa Fe Springs fields. However, this explanation does not account for Pliocene pools (Bolsa, Ashton) at Huntington Beach field (Figure 2F).

Tectonic Implications

The width of Mohnian Division D and E troughs (Figure 3) is about 7 km at Las Cienegas and on the Coyote-Richfield trend and about 12 km in the Puente Hills, considerably less than lithospheric thickness. These troughs thus are not controlled by lithospheric flexure accompanying tectonic loading of adjacent areas, although the timing of subsidence would be appropriate for thermal isostatic cooling. The troughs may accompany lithospheric stretching or may be pull-apart basins controlled by en echelon strike-slip faults, although faults with the appropriate geometry and coeval displacements have not been documented for the Mohnian. The "Delmontian" Division A, B, and C trough, too, may have a stretching or pull-apart origin. The trough trends northeast-southwest, but bounding normal faults of that age and trend have not been identified.

The modern structural configuration of the basin was adopted at the beginning of the Repettian, 4 Ma, at the time of the opening of the Gulf of California and the time of acceleration of subsidence of basins in southern California (Yeats, 1978). Unlike the deepest part of the Ventura basin, the rate of Repetto deposition in the central trough was greater than that for the post-Repettian. However, in the up-plunge extension of the central trough to the southeast, Repetto deposition was much slower, probably slower than that for the post-Repettian. The northwest-southeast orientation of the trough appears to be related to right-lateral shear along bounding strike-slip faults (Newport-Inglewood, Whittier), but the northwest plunge of the trough may be related to flexural loading by thrust faults bringing the Santa Monica Mountains southward over the Los Angeles basin. Oil-field development in the Los Angeles Downtown area (Jacobson and Lindblom, 1987; Wright, *Structural Geology and Tectonic Evolution of the Los Angeles Basin*, this volume) shows that strong folding and reverse faulting occurred south of the Santa Monica fault zone in latest Miocene and early Pliocene time and that tectonic loading of the Los Angeles basin on these faults and on the main Santa Monica fault to the north would coincide with the time of rapid deepening of the trough.

Future Oil Prospects

Many deep tests have been drilled on the belief that the "Delmontian" and Mohnian sequences responded to the same structural and stratigraphic controls as the post-"Delmontian." The recognition that these older sequences were controlled by a different basin configuration and tectonic regime should lead to a search for Mohnian and "Delmontian" stratigraphic traps in heretofore unexplored areas where the post-"Delmontian" sandstones would not be expected to lens out. There is so little well control on the downdip flanks of the central trough that new seismic lines will probably be required to develop these leads.

If the plunge of the central trough is controlled by structural loading from the north, the northwest end of

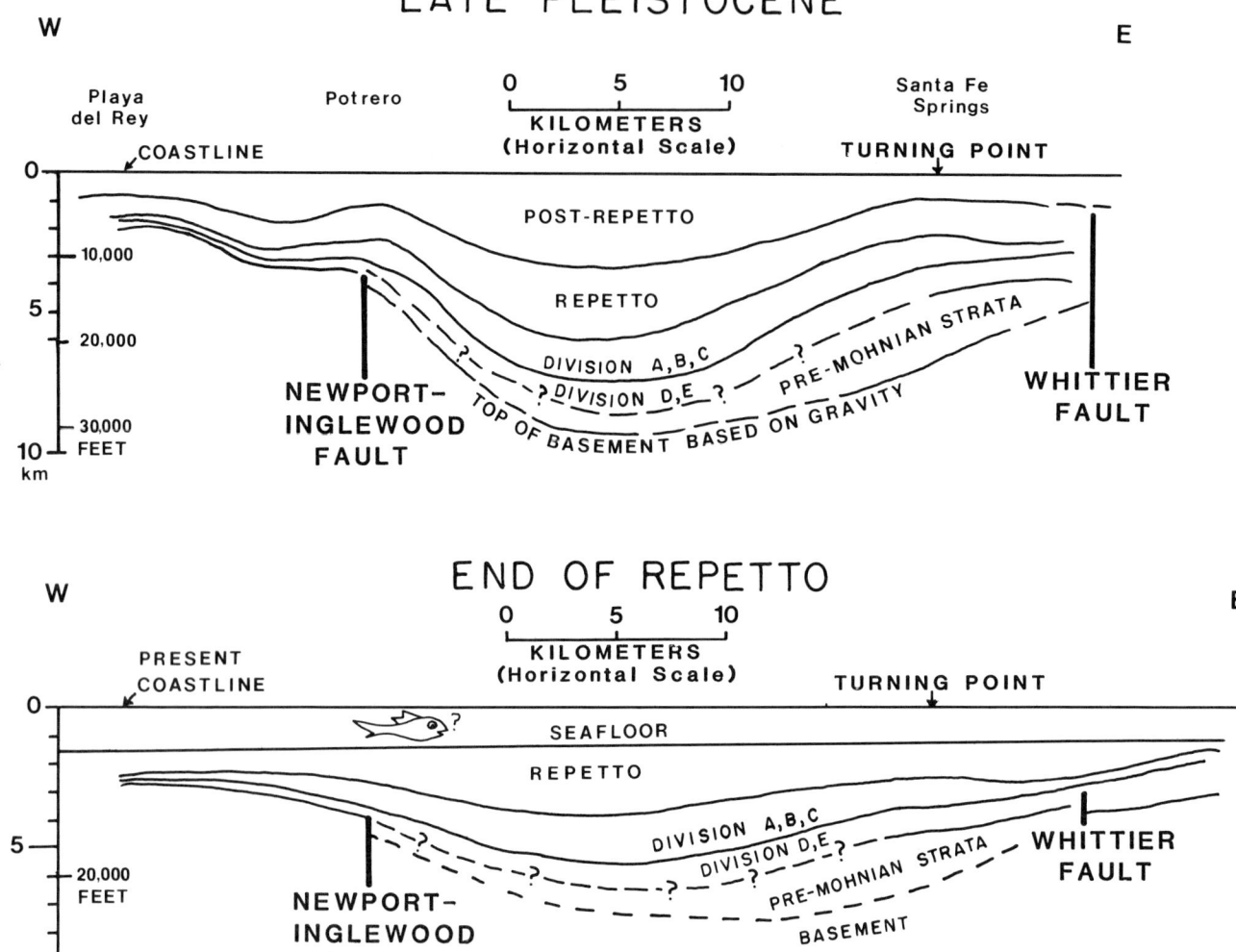

Figure 9. Diagrammatic east-west cross section across Los Angeles basin at end of Repetto and late Pleistocene time. Section not structurally balanced. Top of basement from McCulloh (1960). Note that the western structural shelf was in updip drainage position with respect to central trough at end of Repetto, but became isolated later due to development of syncline west of Newport-Inglewood fault.

this trough, together with the syncline north of Inglewood and Playa del Rey fields, may be deeper than previously believed. In addition, a deep trough may be localized immediately south of the northern Channel Islands, where southward-verging thrusting of the Transverse Ranges over the California continental borderland may also occur.

CONCLUSIONS

Current thermal-isostatic models for the Los Angeles basin do not account for its stratigraphic complexity, in particular the contrast between the Mohnian, "Delmontian," and post-"Delmontian" sequences. Mohnian strata may have responded to thermal-isostatic subsidence, but the central trough of the Los Angeles basin did not begin to receive thick sediments until at least 5 m.y. after the end of the thermal peak in the middle Miocene. Pre-Repetto depocenters may respond in part to lithospheric stretching and pull-apart on faults not clearly identified as yet. For the last 4 m.y., the central trough has formed parallel to flanking right-slip faults (Newport-Inglewood, Whittier), and its northwest plunge may be influenced by flexural loading from thrust faults bringing Santa Monica Mountains basement southward over the basin. The stratigraphic top of the oil column in many oil fields in the upper Repetto may be related to an upper Repetto age of the top of the oil-generating window in the central trough.

ACKNOWLEDGMENTS

This project was supported by industry contributions to the Southern California Fault Studies Project at Oregon State University. Tom Wright permitted us to use his unpublished cross sections across the Los Angeles basin, and he obtained data for the American Petrofina Central Core Hole 1 and Chevron Carlin Community wells. Reviews by K. T. Biddle, T. H. McCulloh, and P. D. Snavely III significantly improved the paper.

REFERENCES CITED

Abbott, P. L., R. P. Kies, W. R. Bachman, and C. J. Natenstedt, 1983, A tectonic slice of Eocene strata, northern part of California continental borderland, in D. K. Larue and R. J. Steel, eds., Cenozoic marine sedimentation Pacific margin, U.S.A.: Pacific Section, SEPM, p. 151-168.

Barrows, A. G., 1974, A review of the geology and earthquake history of the Newport-Inglewood structural zone, southern California: California Division of Mines and Geology Special Report 114, 115 p.

Barbat, W. F., 1958, The Los Angeles basin area, California, in L. G. Weeks, ed., Habitat of oil—a symposium: AAPG, p. 62-77.

Biddle, K. T., 1991, The Los Angeles basin—an overview, in K. T. Biddle, ed., Active margin basins: AAPG Memoir 52, p. 5-24.

Blake, G. H., 1976, The distribution of benthic foraminifera in the outer borderland and its relationship to Pleistocene facies [unpublished Ph.D. thesis]: Los Angeles, University of Southern California, 143 p.

Blake, G. H., 1987, Biostratigraphic correlations in California marginal basins, in D. S. Gorsline, ed., Depositional systems in active margin basins: Pacific Section, SEPM v. 54, p. 53-80.

Blake, G. H., 1991, Review of the Neogene biostratigraphy and stratigraphy of the Los Angeles basin and implications for basin evolution, in K. T. Biddle, ed., Active margin basins: AAPG Memoir 52, p. 135-184.

Bostick, N. H., S. M. Cashman, T. H. McCulloh, and C. T. Waddell, 1978, Gradients of vitrinite reflectance and present temperature in the Los Angeles and Ventura basins, California, in D. F. Oltz, ed., A symposium in geochemistry: low temperature metamorphism of kerogen and clay minerals: Pacific Section, SEPM, p. 65-96.

California Department of Water Resources, 1961, Planned utilization of the ground water basins of the coastal plain of Los Angeles County; Appendix A, Ground water geology: California Department of Water Resources Bulletin 104, 181 p.

California Division of Oil and Gas, 1974, California oil and gas fields, v. 2, South, central coastal and offshore California: California Division of Oil and Gas.

Campbell, R. H., and R. F. Yerkes, 1976, Cenozoic evolution of the Los Angeles basin area—relation to plate tectonics, in D. G. Howell, ed., Aspects of the geologic history of the California continental borderland: Los Angeles, California, Pacific Section, AAPG Miscellaneous Publication 24, p. 541-558.

Carlson, A. J., 1930, Geothermal variations in oil fields of the Los Angeles basin, California: AAPG Bulletin, v. 14, p. 997-1011.

Conrey, B. L., 1967, Early Pliocene sedimentary history of the Los Angeles basin, California: California Division of Mines and Geology Special Report 93, 63 p.

Crowell, J. C., 1987, Late Cenozoic basins of onshore southern California: complexity is the hallmark of their tectonic history, in R. V. Ingersoll and W. G. Ernst, eds., Cenozoic basin development of coastal California: Englewood Cliffs, New Jersey, Prentice-Hall, Inc., Rubey v. VI, p. 207-241.

Douglas, R. G., 1987, Paleoecology of continental margin basins: A modern case history from the borderland of southern California, in D. S. Gorsline, ed., Depositional systems in active margin basins: Pacific Section, SEPM, p. 81-117.

Durham, D. L., and R. F. Yerkes, 1964, Geology and oil resources of the eastern Puente Hills area, southern California: USGS Professional Paper 420-B, 62 p.

Eaton, G. P., 1958, Miocene volcanic activity in the Los Angeles basin, in J. W. Higgins, ed., A guide to the geology and oil fields of the Los Angeles and Ventura regions: Pacific Section, AAPG, p. 55-58.

French, R. W., Jr., 1940, Geothermal gradients in California oil wells: American Petroleum Institute, Drilling and Production Practice, 1939, p. 653-658.

Harding, T. P., 1973, Newport-Inglewood trend, California—an example of wrenching style of deformation: AAPG Bulletin, v. 57, p. 97-116.

Harding, T. P., 1974, Petroleum traps associated with wrench faults: AAPG Bulletin, v. 58, p. 1290-1304.

Henry, M. J., 1987, Los Angeles basin—an overview, in D. Clarke and C. Henderson, eds., Oil producing areas in Long Beach: Pacific Section, AAPG Publication GB 58, p. 1-29.

Hoots, H. W., 1931, Geology of the eastern part of the Santa Monica Mountains, Los Angeles County, California: USGS Professional Paper 165-C, p. 83-134.

Ingle, J. C., 1980, Cenozoic paleobathymetry and depositional history of selected sequences within the southern California continental borderland: Cushman Foundation Foraminiferal Research Publication 18, p. 163-195.

Jacobson, J. B., and R. G. Lindblom, 1987, The East Beverly Hills, South Salt Lake and San Vicente fields—examples of urban oil development, Los Angeles, California, in T. Wright and R. Heck, eds., Petroleum geology of coastal southern California: Pacific Section, AAPG Book 60, p. 32-39.

Jeffrey, A. W. A., H. M. Alimi, and P. O. Jenden, 1991, Geochemistry of Los Angeles basin oil and gas systems, in K. T. Biddle, ed., Active margin basins: AAPG Memoir 52, p. 197-219.

Johnson, J. A., and Duke, C. M., 1973, Subsurface geology of portions of San Fernando Valley and Los Angeles basin, in L. M. Murphy, coordinator, San Fernando, California earthquake of February 9, 1971: U.S. Department of Commerce, v. III, p. 155-164.

Kleinpell, R. M., 1938, Miocene stratigraphy of California: AAPG, 450 p.

Knapp, R. R., J. D. Traxler, T. J. Newbill, D. J. Laughlin, R. D. Stewart, E. G. Heath, H. E. Stark, S. G. Wissler, and W. H. Holman, 1962, Cenozoic correlation section across Los Angeles basin from Beverly Hills to Newport, California: Pacific Section, AAPG.

Lajoie, K. R., 1985, Coastal tectonics, western U.S.: USGS Open-File Report 86-31, p. 132-133.

Lajoie, K. R., 1986, Coastal tectonics, western U.S.: USGS Open-File Report 86-383, p. 139-140.

Lamar, D. L., 1970, Geology of the Elysian Park-Repetto Hills area, Los Angeles County, California: California Division of Mines and Geology Special Report 101, 45 p.

Lopatin, N. V., 1971, Temperature and geologic time as factors in coalification (in Russian): Moscow, Akad. Nauk SSSR, Izv. Ser. Geol. n. 3, p. 95-106.

Mayer, L., 1987, Subsidence analysis of the Los Angeles basin, in R. V. Ingersoll and W. G. Ernst, eds., Cenozoic basin development of coastal California: Englewood Cliffs, New Jersey, Prentice-Hall, Inc., Rubey v. VI, p. 299-320.

McCulloh, T. H., 1960, Gravity variations and the geology of the Los Angeles basin of California: USGS Professional Paper 400-B, p. B320-B325.

McMurdie, D. S., J. C. Taylor, and J. N. Truex, 1973, Southern Los Angeles basin [structure] contours on top Miocene: AAPG, map.

Minch, J. A., 1972, The late Mesozoic-early Tertiary framework of continental sedimentation, northern Peninsular Ranges, Baja California, Mexico [unpublished Ph.D. thesis]: Riverside, California, University of California, 192 p.

Munger, A. H., ed., 1987, Munger map book: California-Alaska oil and gas fields: Los Angeles, California, Munger Oil Information Services, 311 p.

Natland, M. L., and W. T. Rothwell, Jr., 1954, Fossil foraminifera of the Los Angeles and Ventura regions, California: California Division of Mines Bulletin 170, chapter 1, p. 33-42.

Philippi, G. T., 1965, On the depth, time, and mechanism of petroleum generation: Geochimica et Cosmochimica Acta, v. 29, p. 1021-1049.

Poland, J. F., and A. M. Piper, 1956, Ground-water geology of the coastal zone, Long Beach-Santa Ana area, California: USGS Water-Supply Paper 1109, 162 p.

Popenoe, W. P., and W. P. Woodring, 1945, Paleocene and Eocene stratigraphy of northwestern Santa Ana Mountains, Orange County, California: USGS Oil and Gas Investigations, Preliminary Chart 12.

Redin, T., 1990, Oil and gas production for submarine fans of the Los Angeles basin, in K. T. Biddle, ed., Active margin basins: AAPG Memoir 52, p. 239-259.

Sage, O. G., Jr., 1973, Paleocene geography of southern California [unpublished Ph.D. thesis]: Santa Barbara, California, University of California, Santa Barbara, 250 p.

Schoellhamer, J. E., J. G. Vedder, R. F. Yerkes, and D. M. Kinney, 1981, Geology of the northern Santa Ana Mountains, California: USGS Professional Paper 420-D, 109 p.

Stuart, C. J., 1976, Source terrane of the San Onofre Breccia—preliminary notes, in D. G. Howell, ed., Aspects of the geologic history of the California continental borderland: Pacific Section, AAPG Miscellaneous Publication 24, p. 309-325.

Sullwold, H. H., Jr., 1960, Tarzana fan, deep submarine fan of late Miocene age, Los Angeles County, California: AAPG Bulletin, v. 44, p. 433-457.

Traxler, J. D., T. J. Newbill, D. J. Laughlin, R. D. Stewart, E. G. Heath, H. E. Stark, S. G. Wissler, and W. H. Holman, 1962, Cenozoic correlation section across Los Angeles basin from Beverly Hills to Newport, California: Correlation Section, Pacific Section, AAPG.

Truex, J. N., 1974, Structural evolution of Wilmington, California, anticline: AAPG Bulletin, v. 58, p. 2398-2410.

Truex, J. N., ed., 1975, A tour of the oil fields of the Whittier fault zone, Los Angeles basin, California: Pacific Sections, AAPG-Society of Geophysical Explorationists-SEPM, Joint Annual Field Trip, Long Beach, California, April 26, 1975, 76 p.

Turcotte, D. L., and D. C. McAdoo, 1979, Thermal subsidence and petroleum generation in the southwestern block of the Los Angeles basin, California: Journal of Geophysical Research, v. 84, p. 3460-3464.

Turner, D. L., 1970, Potassium-argon dating of Pacific Coast Miocene foraminiferal stages, in O. L. Bandy, ed., Radiometric dating and paleontological zonation: GSA Special Paper 124, p. 91-129.

Vedder, J. G., R. F. Yerkes, and J. E. Schoellhamer, 1957, Geologic map of the San Joaquin Hills-San Juan Capistrano area, Orange County, California: USGS Oil and Gas Investigations Map OM 193.

Waples, D. W., 1980, Time and temperature in petroleum formation: application of Lopatin's method to petroleum exploration: AAPG Bulletin, v. 64, p. 916-926.

Weigand, P. W., 1982, Middle Cenozoic volcanism of the western Transverse Ranges, in D. L. Fife and J. A. Minch, eds., Geology and mineral wealth of the California Transverse Ranges (Mason Hill volume): Santa Ana, California South Coast Geological Society, p. 170-188.

Wissler, S. G., 1943, Stratigraphic formations of the producing zones of the Los Angeles basin oil fields: California Division of Mines Bulletin 118, p. 209-234.

Wissler, S. G., 1958, Correlation chart of producing zones of Los Angeles basin oil fields, in J. W. Higgins, ed., A guide to the geology and oil fields of the Los Angeles and Ventura regions: Pacific Section, AAPG, p. 59-61.

Woodford, A. O., 1925, The San Onofre Breccia: California University Publications, Bulletin of the Department of Geological Sciences, v. 15, p. 159-280.

Woodring, W. P., M. N. Bramlette, and W. S. W. Kew, 1946, Geology and paleontology of the Palos Verdes Hills, California: USGS Professional Paper 207, 145 p.

Wright, T., 1987, Geologic summary of the Los Angeles basin, in T. Wright and R. Heck, eds., Petroleum geology of coastal southern California: Pacific Section, AAPG Book 60, p. 21-31.

Wright, T. L., 1991, Structural geology and tectonic evolution of the Los Angeles basin, California, in K. T. Biddle, ed., Active margin basins: AAPG Memoir 52, p. 35-134.

Yeats, R. S., 1968, Rifting and rafting in the southern California borderland, in W. R. Dickinson and A. Grantz, eds., Proceedings of conference on geologic problems of San Andreas fault system: Stanford University

Publications, Geological Sciences, v. 11, p. 307–322.

Yeats, R. S., 1973, Newport-Inglewood fault zone, Los Angeles basin, California: AAPG Bulletin, v. 57, p. 117–135.

Yeats, R. S., M. R. Cole, W. R. Merschat, and R. M. Parsley, 1973, Poway fan and submarine cone and rifting of the inner southern California borderland: GSA Bulletin, v. 85, p. 293–302.

Yeats, R. S., 1978, Neogene acceleration of subsidence rates in southern California: Geology, v. 6, p. 456–460.

Yerkes, R. F., 1972, Geology and oil resources of the western Puente Hills area, southern California: USGS, Professional Paper 420-C, 63 p.

Yerkes, R. F., T. H. McCulloh, J. E. Schoellhamer, and J. G. Vedder, 1965, Geology of the Los Angeles basin—an introduction: USGS Professional Paper 420-A, 57 p.

Appendix 1

Wells Used in Figure 2

Figure 2A: Torrance-Wilmington

1. C. H. 9-27 (offshore Redondo Beach)
2. C. H. 9-28
3. C. H. 9-29
4. C. H. 9-30
5. C. H. 9-31, 9-32
6. Signal CORB 100
7. Signal CORB 103
8. Signal CORB 1
9. Jergins-Torrance 1
10. CCMO Torrance 94
11. Calif. Southern WP 119
12. Superior Torrance 66
13. Chevron-Joughin 42, 43
14. Texaco-Oakley 2
15. Zephyr-Banning 8-1, 8-5
16. Mobil-Patton 1
17. Mobil-Terminal 1, Ford 3
18. LBHD Gurnsey 1
19. LBOD J-50
20. City of Long Beach C. H. 7
21. City of Long Beach C. H. 1
22. City of Long Beach C. H. 2
23. City of Long Beach C. H. 3
24. MTS 3

Figure 2B: Playa del Rey-Inglewood

1. Union-Vidor 18
2. Union Del Rey 16
3. Vulcan Del Rey 1
4. Superior Inglewood Ext. 1
5. Chevron LAI Three-1, Mobil Sentous 1
6. Chevron Baldwin Cienega 105, Shell Rindge 24
7. Shell Baldwin Hills Comm. 1

Figure 2C: Schist Ridge

1. Mobil Venice 1
2. Ohio RGC 1, 2
3. Ohio RGC 11
4. Union Del Rey 3
5. Union Vidor 18
6. Texaco Inglewood Ext. 1

7. Chevron Six Companies 3
8. Chevron Six Companies 1
9. Chevron Six Companies 2
10. Wilshire El Segundo 1
11. Apex Elsie 6
12. Seaboard Johnson 1
13. B. A. Texaco Bodger 1
14. Ohio Gardena 6-1
15. Texaco Gardena 5-1
16. Union Callendar 79
17. Shell Reyes 135

Figure 2D: Palos Verdes-Central Trough

1. Tidelands-Chandler-McBurney 1
2. O'Donnell Dawn 2
3. Superior Torrance 66
4. Selbar Watson 1
5. Chevron Del Amo 1
6. Shell Reyes 135
7. Union Hellman 58
8. Union Hellman 67
9. Chevron Carlin Comm. 1
10. American Petrofina Central C. H. 1

Figure 2E: Wilmington-Long Beach

1. LBOD Z-92
2. LBOD J-50
3. LBHD Gurnsey 1
4. Van Trees 1
5. Shell S. R. 1
6. Shell Alamitos 48A
7. Shell Dolley 2
8. Texaco Bixby (NCT-1)-2

Figure 2F: Huntington Beach-Yorba Linda

1. Signal State 426.1, C. H. 3
2. Signal H. S. 392.1-65
3. Chevron Huntington B-92
4. Chevron Huntington B-111
5. Chevron Huntington A-74
6. Chevron Huntington A-113

7. Union Copeland 11
8. Seguro Seguro 1
9. Easley Patten 2
10. Exxon Isenor 1
11. Shell Anaheim Sugar 55-23
12. Hillman Westminster 1
13. Seaboard Bauer 44-17
14. Chevron Stanley Comm. 1
15. Superior Garden Grove Unit One-1
16. McVicar Rood Hall Helsinger 1
17. McVicar Rood Hall Foiles 1
18. Shell Mathis 1
19. Br. American Mauerhan 1
20. Texaco Hodges 1
21. Texaco Olive Unit Two 1
22. Texaco Olive Unit One 2
23. Chevron Kraemer 2-27
24. Union Chapman 29
25. Union Chapman 51
26. Hathaway A. H. A. 5
27. Union N. Richfield Unit 1, 1-3
28. Union Thompson 1
29. Shell Olinda Fee 4 305
30. Shell Olinda Fee 4 67-16
31. C. H. "J"
32. C. H. "K"
33. Union Del Giorgio 1

Figure 2G: Santa Fe Springs–Anaheim Nose

1. Union Bell 100
2. Mobil La Mirada Comm. 46-1
3. Mobil Librown 1
4. Mobil Heath 1
5. Exxon BPU 1
6. Texaco BPU 4E-1
7. Texaco BPU 4F-1
8. Texaco Stern 1
9. Texaco Spencer 1
10. Texaco BPU 1-1
11. Texaco A-13-1
12. Amerada-Anaheim Comm. 48-8
13. Shell Harbeson 1
14. Shell Weisel 1
15. McVicar Rood Hall Helsinger 1
16. McVicar Rood Hall Foiles 1
17. Shell Mathis 1
18. British American Fluor 1
19. Tidewater Olive Orange 1

20. Orange Comm. 1
21. Chevron Tustin Comm. 1

Figure 2H: West Coyote–Puente Hills

1. Mobil Heath 2
2. Mobil Heath 1
3. Chevron Pacific Comm. 1
4. Chevron Emery 92
5. Chevron Stern. Comm. 2-1
6. Dunlap Apex 1
7. Erin-Freedman North La Habra 1
8. Shell Puente A-3
9. Rucker Grazide 1
10. Honolulu Butler 1
11. Dietzel Rowland 1

Figure 2I: Santa Fe Springs–Whittier

1. Mobil Allen 1
2. Mobil Comm. 14-1
3. Chevron Woodhead 1
4. Mobil Santa Fe 190, Chevron Walker Comm. 25
5. Union Bell 100, 107
6. CCMO 165, Albercalif Arcarius 1
7. Ryan & Morrow 1
8. CCMO E. Whittier Comm. 1-1
9. Oakes & Combs Whittier 1, Shell Hillside 1
10. Partridge Hillside 1

Figure 2J: Montebello–Puente Hills

1. Ridge-Seacliff 1
2. Union Howard & Smith 2, 3
3. Chevron Baldwin 82
4. Chevron Baldwin 129
5. Chevron Baldwin 75
6. Texaco Baldwin 16
7. Texaco Baldwin 18
8. Texaco-City of Whittier 1
9. Shell Pellissier 1
10. Hilo-Bishop Hilo-Pellissier 1
11. Conoco Baldwin 1
12. Barnsdall Baldwin 1
13. Killingsworth Tandburg 1
14. Conoco Turnbull 5
15. Conoco Turnbull 3
16. Morton & Sons Rowland 3-1

Chapter 8

Oil and Gas Production from Submarine Fans of the Los Angeles Basin[1]

Tom Redin

Consultant
Ventura, California, U.S.A.

ABSTRACT

The Los Angeles basin is a small, deep Neogene basin located in the northeast portion of the southern California continental borderland. It was formed along a transform margin during early to middle Miocene time, as were numerous basins within the continental borderland south of the Transverse Ranges.

A series of submarine fans were deposited in the Los Angeles basin during the middle to late Miocene, Pliocene, and Pleistocene. The configuration of these fans appears to fit several basin-floor fan models, but the fan morphologies were greatly influenced by local paleobathymetry. The primary sediment transport mechanism was turbidity flows from submarine canyons, but other mass sediment transport mechanisms, such as debris flows, fluidized sediment flows, and grain flows, were also significant.

Three primary, commonly coalescing, submarine fans have been recognized: the Tarzana fan in the northwestern Los Angeles basin, the San Gabriel fan in the north central portion, and the Santa Ana fan in the eastern portion of the basin. Most of the oil produced in the Los Angeles basin comes from upper Mohnian, Delmontian, and Repettian sandstone and conglomerate reservoirs of these submarine fans.

The Los Angeles basin was a probable silled basin that intersected the oxygen-minimum oceanographic zone during the late Miocene and Pliocene. A combination of the rich biogenic sedimentation along with rapid burial by coarse- to fine-grained clastics and moderate paleo–heat flow provided almost perfect conditions for the generation and migration of oil and gas. Structural deformation was intermittent throughout the Neogene but reached a culmination in the late Pleistocene to Recent.

---

[1]This paper was originally presented to the 1984 AAPG convention symposium on submarine fans held in San Antonio, Texas.

INTRODUCTION

The Los Angeles basin is a Neogene depositional basin presently bounded on the north by the Santa Monica and San Gabriel Mountains of the Transverse Ranges. The basin in bounded on the east and southeast by the San Joaquin Hills, Santa Ana Mountains, and the east side of the Puente Hills. The southwest and western edge of the basin is marked by the northwesterly trending Lausen Seamount, the Palos Verdes Hills, and an unnamed basement high in Santa Monica Bay (Figures 1, 2, 4).

The purposes of this paper are (1) to review the regional setting and depositional history of the Los Angeles basin as related to the development of Neogene submarine fans; (2) to outline the submarine-fan morphology and mechanisms of deposition and show the relationship of major Los Angeles basin oil fields to the fan morphologies; and (3) to suggest why the basin, per unit volume, is one of the world's most prolific oil- and gas-producing areas.

Over 520 wells were used in this study (Figure 2). Each well was correlated by either electric-log character or paleontological data. All wells were analyzed for gross sand penetrated within the upper Mohnian, Delmontian, and Repettian stages. To obtain true stratigraphic thickness and total sandstone present, all wells were corrected for well deviation and structural dip. Paleobathymetry was determined by benthic foraminifera (See Blake, *Review of Neogene Biostratigraphy and Stratigraphy of the Los Angeles Basin and Implications for Basin Evolution,* this volume). Well locations were digitized with structural dip, stratigraphic points, and well courses compiled into a computer data set. Net unit sandstone isopachs, total unit isopachs, and unit paleobathymetric maps were generated upon which the interpretive geology was drawn.

REGIONAL GEOLOGIC SETTING AND FORMATION OF THE LOS ANGELES BASIN

The Los Angeles basin is one of many active margin Neogene basins that make up portions of onshore southern California and the adjacent continental borderland. Figure 1 shows the location of the Los Angeles basin relative to other nearby Neogene basins. The Los Angeles basin is located within the Peninsular Ranges province, just south of and adjacent to the central Transverse Ranges province.

Figure 3 shows the location of the Los Angeles basin relative to the remnants of disrupted Late Cretaceous-Paleogene fore-arc basins. By comparing Figure 1 with Figure 3, one can see that Neogene basins are commonly superimposed on or contain remnants of older Late Cretaceous-Paleogene sedimentary sections. The Los Angeles basin contains both Late Cretaceous-Paleogene and Neogene sedimentary rocks. Other southern California basins that include both Neogene and Upper Cretaceous-Paleogene rocks are the Ventura-Santa Barbara basin, the Santa Cruz basin, the San Nicolas basin, and most of the Oceanside-Capistrano embayment basin.

Date Submitted: 6/10/87
Date Accepted: 5/24/88

The Los Angeles basin was formed during the early part of the middle Miocene, and there are still basic questions remaining about its origin (Atwater, 1970; Campbell and Yerkes, 1976; Yeats, 1976; Howell et al., 1980; Dickinson, 1981; Crowell, 1984). The Los Angeles basin, along with other Neogene basins of the southern California continental borderland, appears to have been formed, in part, by extensional and compressional tectonics associated with right-lateral strike-slip faulting (Harding, 1974; Crowell, 1976; Dibblee, 1977; Turcotte and McAdoo, 1979; Ballance and Reading, 1980; Rodgers, 1980; Wright, *Structural Geology and Tectonic Evolution of the Los Angeles Basin,* this volume).

STRATIGRAPHIC SETTING

Basement Complex

The floor of the Los Angeles basin is a heterogeneous group of metamorphic rocks intruded by igneous plutonic rocks and middle Miocene volcanic rocks (Hoots, 1931; Schoellhamer and Woodford, 1951; Yerkes et al., 1965; Jones et al., 1976; Platt, 1976; Sorenson, 1985). Figure 4 shows four basement types based on various basement terranes exposed or penetrated by wells within and around the basin. The distribution of basement terranes is based on a compilation of data published by Yerkes et al. (1965). The rifted zone of middle Miocene volcanics, diagrammatically shown in the center of the basin, has been inferred from outcrop data in the San Joaquin Hills and the western Santa Monica Mountains. Middle Miocene volcanics have been penetrated on the east side of the Newport-Inglewood fault, within the questionable basement terrane, as shown on Figure 4, but it is uncertain if these volcanics extend into the center of the basin.

Upper Cretaceous and Paleogene Formations

Upper Cretaceous and Paleogene sedimentary rocks are exposed in the Santa Monica and Santa Ana Mountains and in the San Joaquin Hills (Figure 3) and have been penetrated by wells on the Anaheim nose and along the East and West Coyote anticlinal trend (Figures 5, 6). The Upper Cretaceous and Paleogene rocks are generally unconformably overlain or in fault contact with lower and middle Miocene sedimentary rocks or volcanic rocks. The northwestern extent of Late Cretaceous and Paleogene rocks from the Anaheim nose into the deep Los Angeles basin is unknown. To date this sequence has not been found productive of oil and gas.

Middle Miocene Topanga and San Onofre Breccia Formations

By the early middle Miocene Relizian Stage, the Los Angeles basin was in the process of being formed. The first sediments deposited in the newly formed basin are included in the Topanga Formation, which contains conglomerates, sandstones, and shales with intercalated middle Miocene volcanic and intrusive rocks. At about the same time and

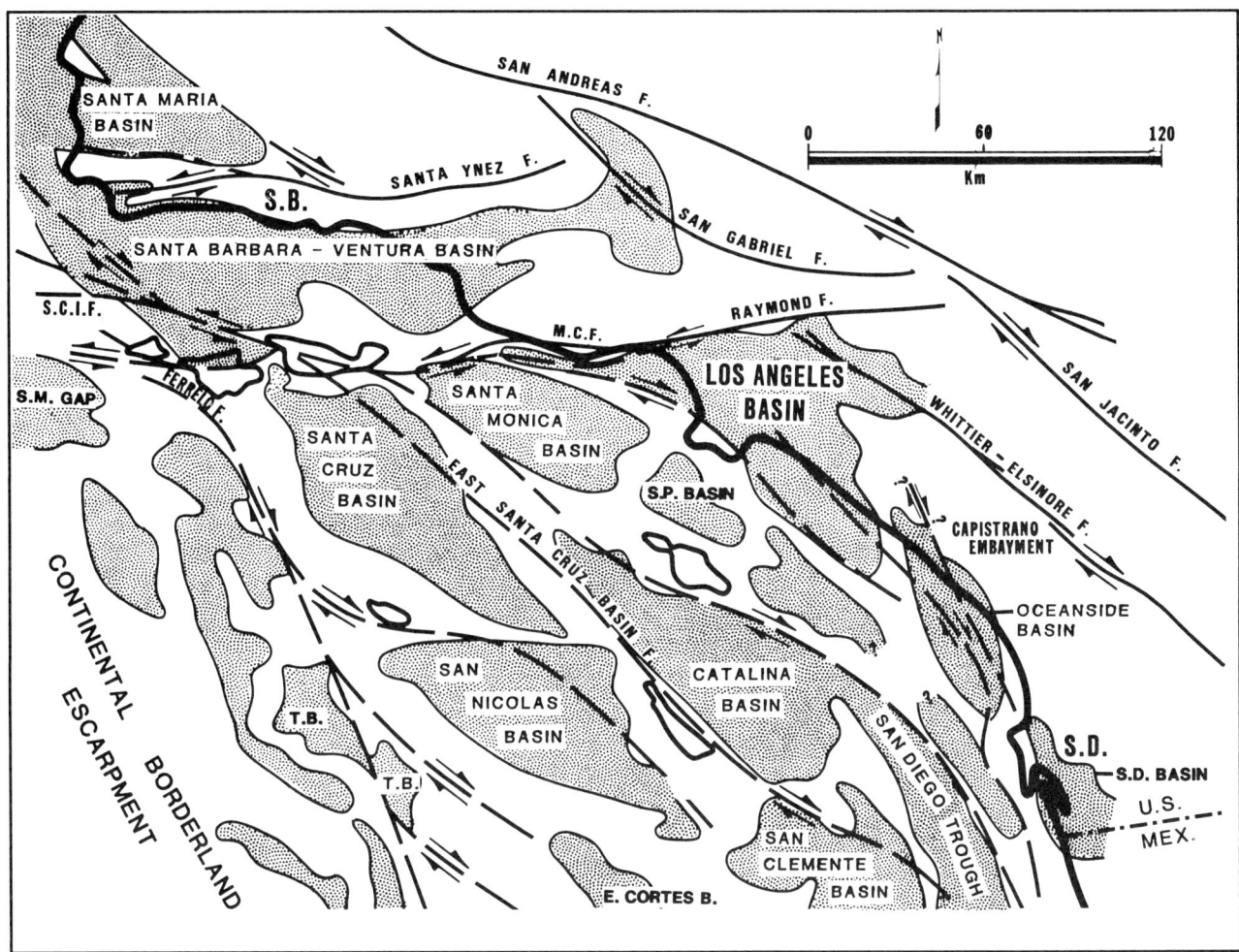

Figure 1. Neogene basins within the southern California borderland region. S.C.I.F. = Santa Cruz Island fault; S.M. GAP = San Miguel gap, T.B. = Tanner basin; S.P. BASIN = San Pedro basin; M.C.F. = Malibu Coast fault; S.B. = Santa Barbara; S.D. BASIN = San Diego basin; S.D. = San Diego.

continuing into the Luisian Stage, the westerly sourced San Onofre Breccia was being deposited.

Topanga Formation

The Topanga Formation may have been deposited as a shallow- to moderately deep-water submarine-fan channel system in which the submarine canyon or canyons were landward from active right- and left-lateral wrench fault systems. As the coarse clastics were deposited in the newly formed basin, some may have been translated to the northwest or west by strike-slip faulting; thus, only a small portion of the total Topanga stratigraphic sequence is present at any given location. During this time, there was active volcanism along the edge and possibly out in the middle of the newly formed basin. In the Santa Monica Mountains, the Topanga Formation is now referred to as the Topanga Group. The Topanga Group includes the Calabasas Formation, the Conejo Volcanics, and the Topanga Canyon Formation (Campbell and Yerkes, 1980).

The source terrane for the Topanga Formation is unknown but was probably located to the north and/or east of the present Los Angeles basin.

San Onofre Breccia

Within and on the southwest side of the Los Angeles basin, the San Onofre Breccia contains a clast assemblage derived from metamorphic rocks found within the Catalina Schist. During the Relizian Stage, much of the inner California borderland had a series of Catalina Schist topographic highs that were subject to subaerial erosion. The adjoining marine basins were at bathyal depths. The San Onofre Breccia was commonly deposited as a thick wedge of sedimentary breccia resembling a marine talus-slope deposit that thinned basinward. By contrast, in the east-central Santa Barbara basin, the San Onofre Breccia has a southeasterly source and is greater than 2000 ft (600 m) thick near the center of the basin. In the Los Angeles basin, the San Onofre Breccia is commonly productive of oil and is locally known by various operators as the "Schist Conglomerate."

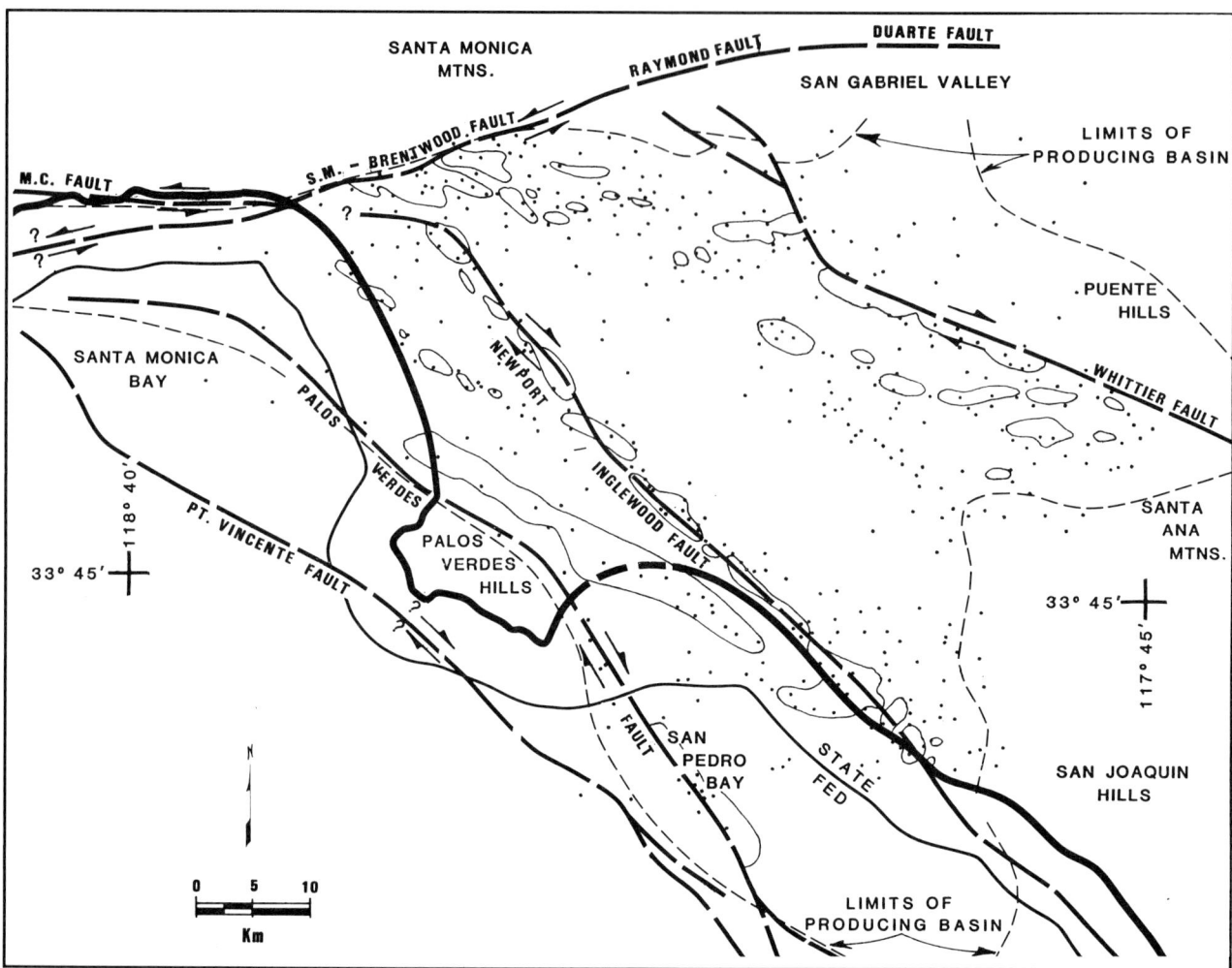

Figure 2. Los Angeles basin map showing the locations of all wells used in this study. Each small black dot represents either a field well or an exploratory well. An outline of the major oil fields is also shown. For the names of all oil fields see Figure 5. M.C. FAULT = Malibu Coast fault; S.M. = Santa Monica-Brentwood fault; FED. = Federal.

Middle and Upper Miocene Luisian and Lower Mohnian Monterey Formation

The Monterey Formation is primarily a biogenic and fine-grained clastic lithologic unit where the lower and upper stratigraphic boundaries are time transgressive. In the central and eastern Los Angeles basin, the Monterey Formation ranges in age from upper Luisian to early Mohnian. However, in many areas Luisian Stage rocks are thin or absent. By late Mohnian time, an influx of coarse clastics formed a series of prograding submarine fans, leaving only the basin margins available for pelagic and hemipelagic Monterey-type sedimentation without too much clastic contamination. An example of basin-margin sedimentation may be found in the Palos Verdes Hills. Here the Monterey Formation ranges in age from Luisian to early Pliocene (Conrad and Ehlig, 1983; Blake, this volume). Elsewhere in the subsurface of the Los Angeles basin the Monterey is commonly referred to as the Phosphatic nodular shale or as the lower Mohnian part of the Puente Formation (Figure 6). In the Puente Hills and northern Santa Ana Mountains, the lower Mohnian La Vida Member of the Puente Formation is lithologically similar to the Monterey Formation (Figure 6 and Blake, this volume).

The Monterey Formation commonly shows high resistivity on an electric log, a feature typical of the Monterey Formation in other basins. Because of fine-grained clastic contamination, thick sequences of fairly pure biogenic, carbonate, or siliceous sediments are generally absent throughout the basin.

Upper Miocene-Upper Mohnian and Delmontian Stages—Monterey, Modelo, and Puente Formations

By late Mohnian time, the future oil-producing submarine fans of the Los Angeles basin were well developed. Upper

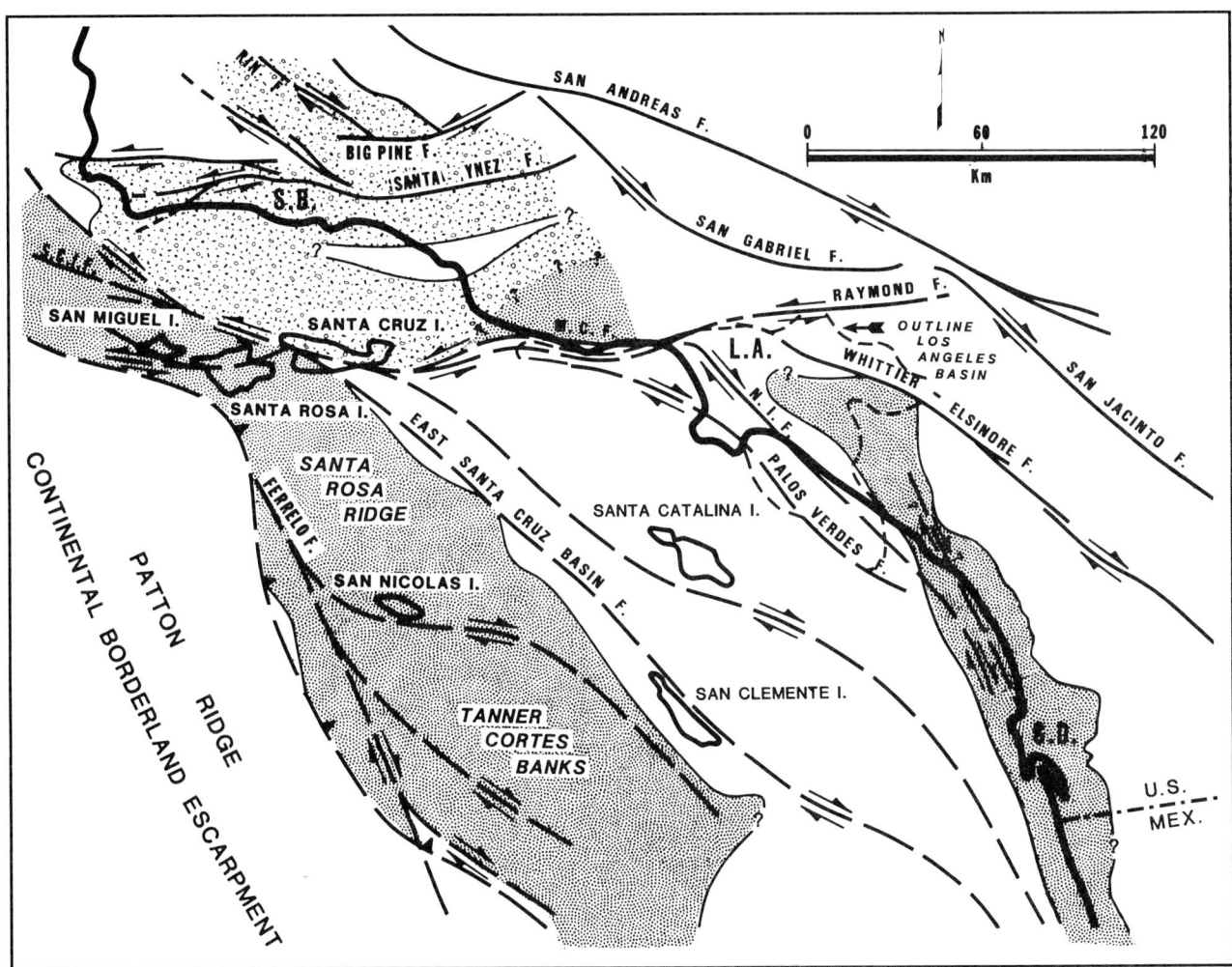

Figure 3. Distribution of Upper Cretaceous and Paleogene sedimentary rocks within the southern California borderland region. Shaded patterns show two different depositional terranes that are in probable fault contact in the western Channel Islands area and the western Santa Barbara Channel. The term "depositional terrane" or "stratigraphic terrane" is used here for remnants of a disrupted fore-arc basin that have been translated to the northwest by a combination of right-lateral strike-slip faulting and extension in the southern California borderland. A dashed outline of the Neogene Los Angeles basin is also shown. S.B. = Santa Barbara; L.A. = Los Angeles; S.D. = San Diego; N.I.F. = Newport-Inglewood fault; M.C.F. = Malibu Coast fault; S.C.I.F. = Santa Cruz Island fault; RIN. F. = Rinconada fault.

Miocene sedimentary rocks within the basin are known by a variety of formation names. Some of the most common formation names are the Modelo Formation of the Santa Monica Mountains, the Monterey Formation of the Palos Verdes Hills, and the Puente Formation of the Puente Hills and the northern Santa Ana Mountains (Blake, this volume). The Puente Formation has four members: the La Vida Member of early Mohnian age, the Soquel and Yorba members of late Mohnian age, and the Sycamore Canyon Member of Delmontian age (Durham and Yerkes, 1964).

Lower Pliocene-Repettian Stage—Repetto and Fernando Formations

The future oil-producing submarine fans of the Los Angeles basin continued to expand during Repettian time. In the eastern part of the basin, Repettian sediments were deposited unconformably on upper or middle Miocene formations. Repettian Stage sediments exposed near the edge of the Los Angeles basin and within the subsurface are known either as the Repetto Formation or the lower

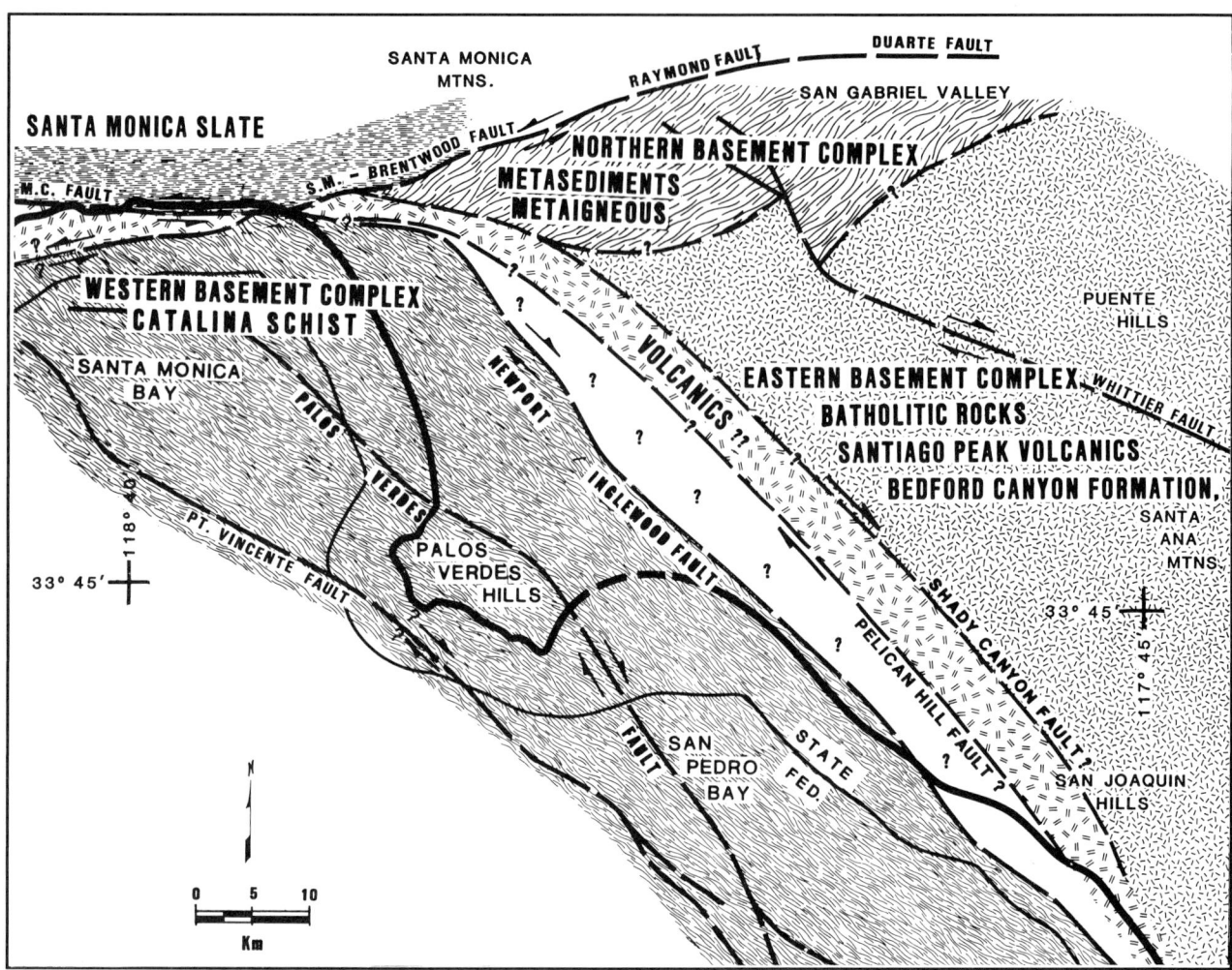

Figure 4. Approximate distribution of basement rocks for the Los Angeles basin. The suggested rifted zone, shown in the center of the basin as volcanic rocks and question marks, may consist of middle Miocene volcanic rocks. M.C. Fault = Malibu Coast fault; S.M. = Santa Monica-Brentwood fault. The eastern basement complex includes the Jurassic Bedford Canyon Formation overlain unconformably by the Upper Jurassic to Lower Cretaceous Santiago Peak Volcanics. These formations have been intruded by plutonic rocks of the Southern California Batholith (Yerkes et al., 1965). The western basement complex consists of glaucophane-lawsonite bearing schist of unknown age and stratigraphic position (Yerkes et al., 1965; Platt, 1976). The northern basement complex consists of variety of metamorphic rocks described in well cores and cuttings. The Santa Monica Slate consists of "fine- to medium-grained metagraywacke, black metashale, metasiltstone and meta-argillite with minor amounts of impure quartzose and quartzofeldspathic metasandstone" (Sorensen, 1985).

Fernando Formation (Lamar, 1961; Yerkes et al., 1965; Blake, this volume). The name Repetto Formation is in more common usage now (Figure 6).

Upper Pliocene–Pleistocene Venturian, Wheelerian, and Hallian Stages—Pico and San Pedro Formations

The Venturian, Wheelerian, and Hallian stages are benthic foraminiferal stages used in the Ventura basin and generally not used by the oil industry in the Los Angeles basin. The U.S. Geological Survey does not show a stage for early or late Pliocene or the Pleistocene (Yerkes et al., 1965). Blake (this volume) shows that the Pico Formation is late Pliocene–Pleistocene in age and contains benthic foraminifera representing the Venturian, Wheelerian, and Hallian stages. The overlying San Pedro Formation is Pleistocene in age and contains a Hallian Stage fauna.

The submarine fans of the Pico Formation continued to fill the Los Angeles basin for the remainder of the Pliocene and most of the Pleistocene. The San Pedro Formation was deposited unconformably on older formations and

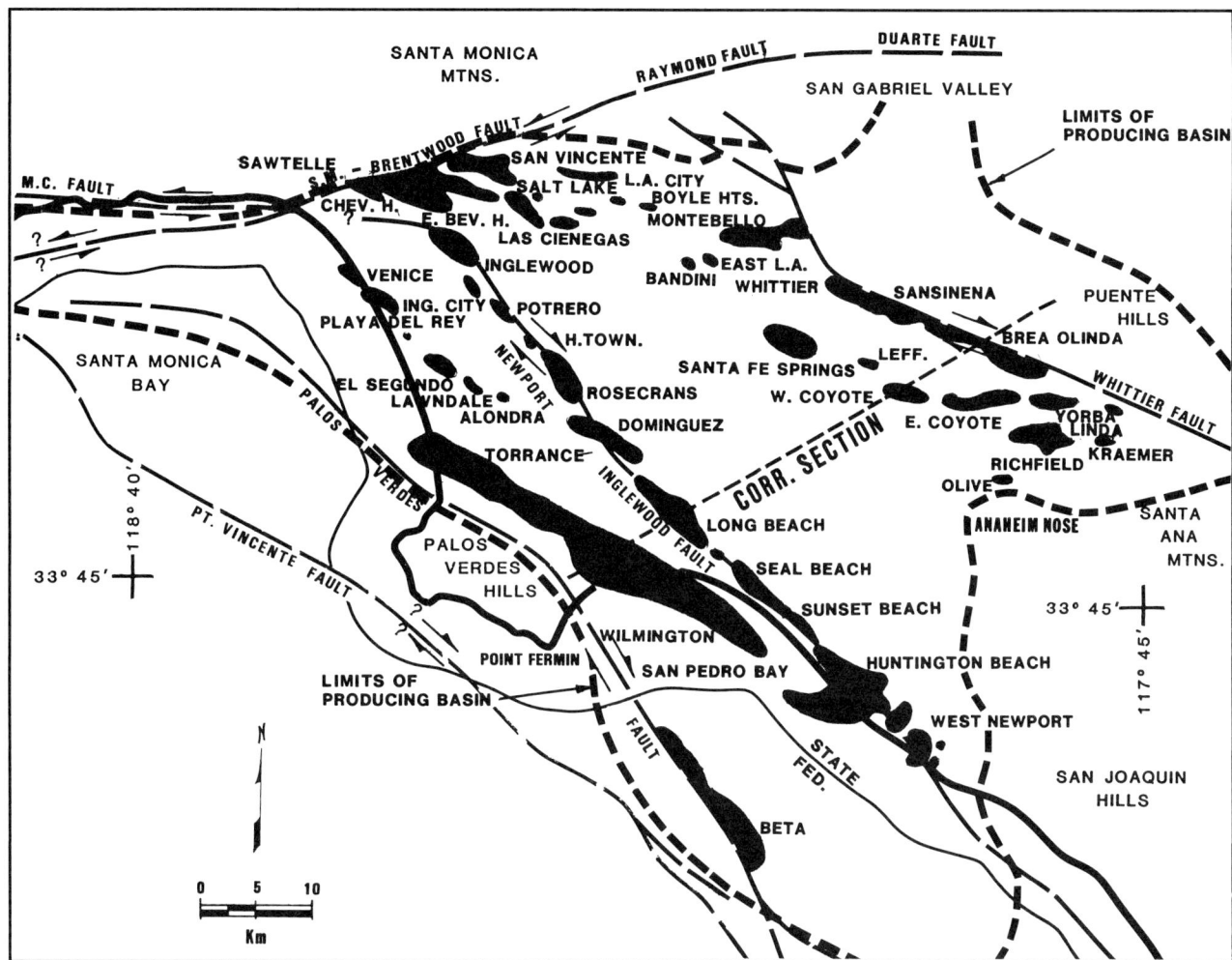

Figure 5. Oil-producing areas of the Los Angeles basin showing all major oil fields. Location of Figure 6 correlation section is shown as a dashed line. M.C. FAULT = Malibu Coast fault; S.M. = Santa Monica–Brentwood fault; CHEV. H. = Cheviot Hills; E. BEV. H. = East Beverly Hills; ING. CITY = Inglewood City; H.TOWN. = Howard Townsite; LEFF = Leffingwell; FED. = Federal. Cumulative production through 1986 = 7.6 billion barrels of oil and 5.9 trillion cubic feet of gas.

represents the final withdrawal of the ocean from the Los Angeles basin. The San Pedro Formation consists of interbedded marine and nonmarine siltstones, sandstones, and gravels.

NEOGENE SUBMARINE FANS

Fan Model

Figure 7 is a modern submarine-fan model proposed by Normark (1978). Normark's model has been modified slightly to fit the paleobathymetry of the Los Angeles basin during the late Miocene and early Pliocene. Some of the difficulties in applying a simplistic Normark-type model to the Los Angeles basin are (1) the Los Angeles basin has at least three major points of sediment input; (2) the fan margins are commonly destroyed or have coalesced and, therefore, are difficult to locate; and (3) the basin geometry is more like an elongated trough rather than an unrestricted plain, thus distorting a true fan configuration. Applying modern morphological submarine-fan model divisions to an ancient fan is very tenuous at best. However, apparent absence of detached sand lobes (Mutti, 1977) and highly channelized mid-fan areas with outer fan depositional lobes (Ricci Lucchi, 1975; Walker, 1978) leave Normark's model the most useable.

Hsü (1977) suggested that within the small, tectonically active transform margin basins of California the basin geometry has so greatly modified the way that fans were built into a basin that no simple fan model is applicable. Furthermore, these basins were filled with a thick sequence of turbidites showing no consistent proximal to distal change along the trough axis in a downslope direction. Seismic coverage of modern sedimentary basins within the California borderland shows, for the most part, that this is not the case. These basins are relatively flat bottomed where the water is deepest, and the late Neogene sedimentary section is thick. These basins, including the

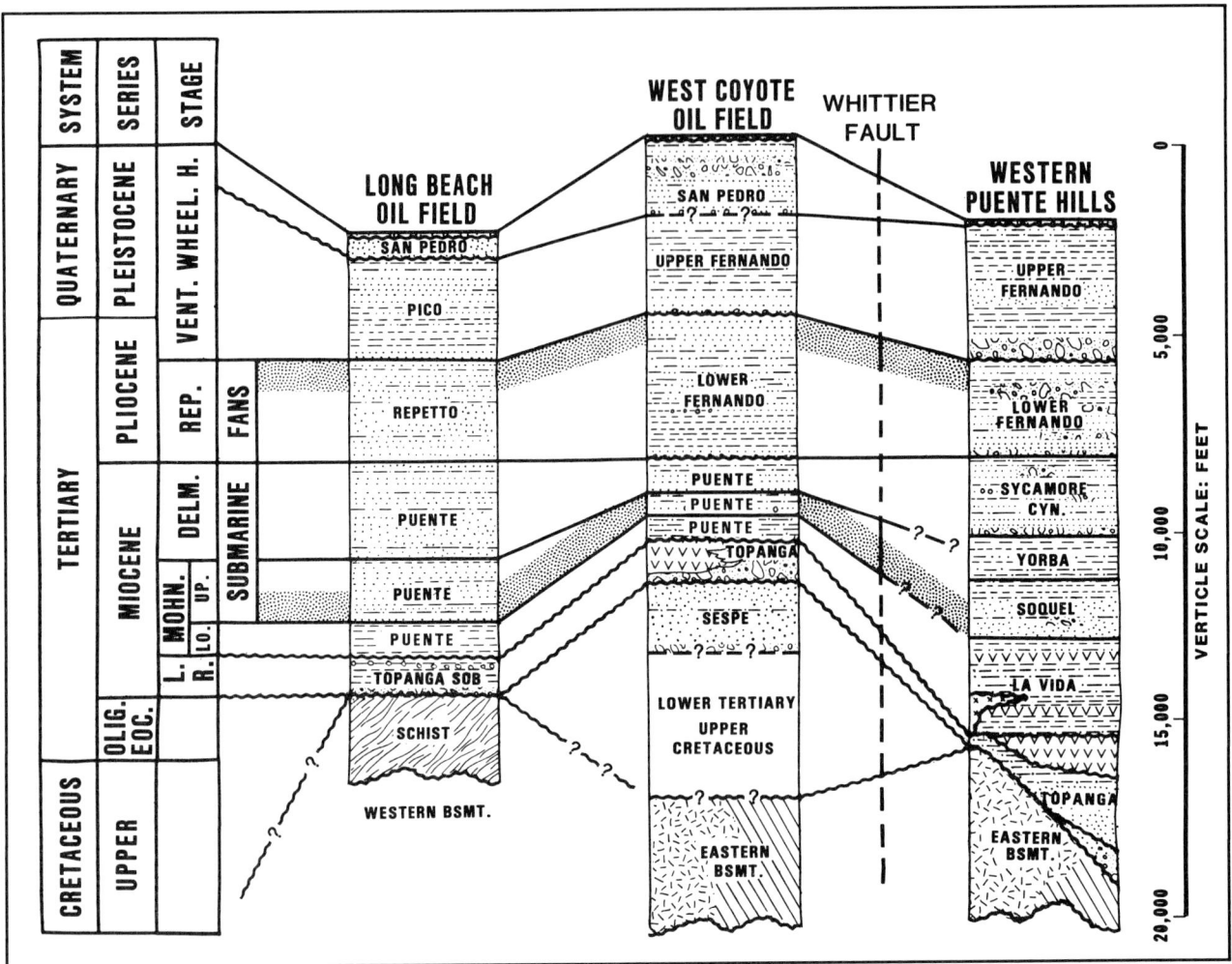

Figure 6. SW–NE correlation section across the Los Angeles basin. See Figure 5 for section location. Shaded contacts show primary producing submarine fans across the basin. The La Vida, Soquel, Yorba, and Sycamore Canyon are members of the Puente Formation. For all stratigraphic equivalents to the formation names shown above see Blake, this volume, his Figure 2. In the western Puente Hills, the volcanic unit above the Topanga Formation has not been named, and the wedge of sedimentary rocks below the Topanga Formation is generally called Vaqueros-Sespe undifferentiated. The lithologic patterns shown for the eastern basement complex represent plutonic rocks that intruded metasediments and metavolcanics. SOB = San Onofre Breccia; L. = Luisian; R. = Relizian; LO. = lower; UP. = upper; H. = Hallian; WHEEL. = Wheelerian; VENT. = Venturian; REP. = Repettian; DELM. = Delmontian; MOHN. = Mohnian; OLIG. = Oligocene; EOC. = Eocene. Columnar sections are modified slightly from Yerkes et al., 1965.

Los Angeles basin, were filled by a series of submarine fans interbedded with pelagic, hemipelagic, and mass sediment gravity slides. This is not to imply that the floor of the Los Angeles basin was simple and flat. The basin configuration and consequently its deepest part varied greatly throughout the Neogene (Blake, Wright, and Yeats and Beall, *Stratigraphic Controls of Oil Fields in the Los Angeles Basin*, this volume). For example, during the late Mohnian, the Puente Hills, the Salt Lake oil field area, and parts of San Pedro Bay received a thicker sequence of sediments than many other areas of the Los Angeles basin. To be sure, the fan shape is greatly modified by basin geometry, but the depositional processes remain similar to many ancient submarine fans [see Normark and Piper (1984) for a description of the relatively modern Navy submarine fan located in a complex tectonic setting]. As petroleum exploration geologists we need to predict, before drilling, the presence of adequate source and reservoir rock. By developing a submarine-fan depositional model, the stratigraphic risk is more fully understood when compared with using just formation and net sand isopachs.

A good knowledge of modern fan morphology is definitely needed when working with ancient fans. Most submarine-fan morphology is based on modern fans as seen

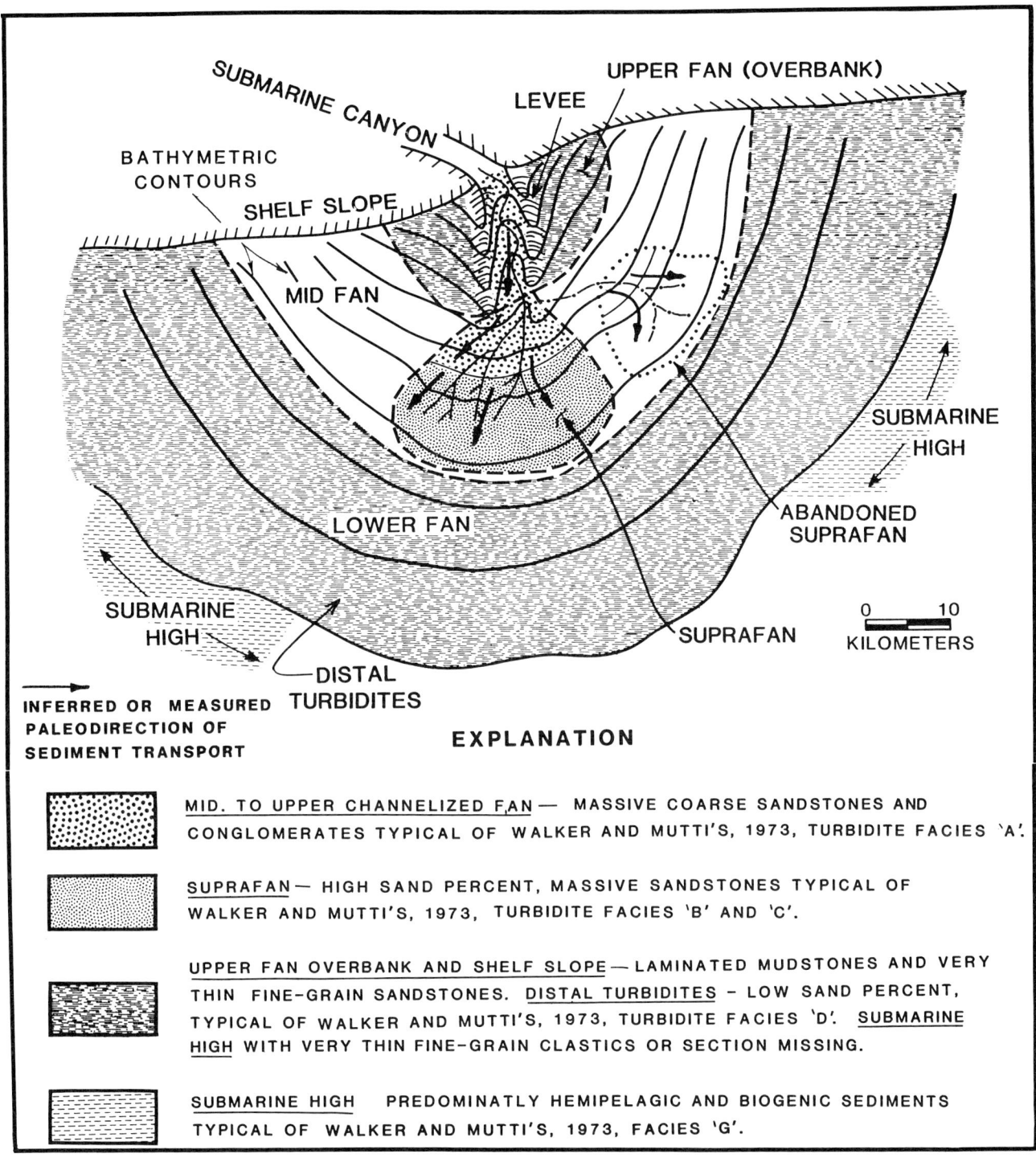

Figure 7. Submarine-fan model from Normark (1978) modified to apply to ancient submarine fans of the Los Angeles basin. The modifications include the additions of submarine highs, thus distorting the fan shape configuration, adding distal turbidites to the lower fan, changing continental slope to shelf slope, adding overbank and levee to the upper fan area, and adding arrows showing the inferred or measured paleodirection of sediment transport. Note: Bathymetric contours only partially apply to the Los Angeles basin. The lithofacies patterns, shown in the explanation above, apply to the upper Mohnian, Delmontian, and Repettian submarine fans (Figures 8, 9, 11, 14).

by long-range and deep-towed side-scan sonar, high-resolution subbottom seismic profiles, and detailed bathymetric data. In contrast, most ancient fans have been mapped based on vertical sedimentary facies succession analysis, including paleocurrent directions and lateral distribution of lithofacies. Simply put, turbidite facies have been used as an indication of depositional environments. Shanmugam et al. (1985) suggested that turbidite facies associations for interpreting ancient submarine-fan environments should be considered tenuous until confirmed by coring of modern fans. Mutti and Normark (1987) presented a good discussion on the difficulties and complications arising when comparing modern and ancient turbidite systems.

Side-scan sonar of modern submarine fans shows a modern time-correlative horizon. For the Los Angeles basin, the fan surface morphology has been inferred by correlating over 500 wells but not on a single horizon. Figure 9 represents what the average upper Mohnian submarine fan may have looked like. Figures 8, 11, and 14 are maps that represent what the average corresponding submarine fan looks like today in the subsurface. The arrows indicating inferred or measured direction of sediment transport are based on paleobathymetry, net sand and gross interval isopachs, field reports (Bennett, 1967), and, to a very minor extent, seismic data. Due to the irregular basin shape, dividing the Neogene submarine fans morphologically into strictly upper, middle, and lower fan divisions did not seem appropriate (Figure 7). The fan divisions used are based on sandstone and conglomerate isopachs, sand-shale ratios, and sandstone bed thickness. This relates to Normark's model (Figure 7) in the middle fan–suprafan area. The

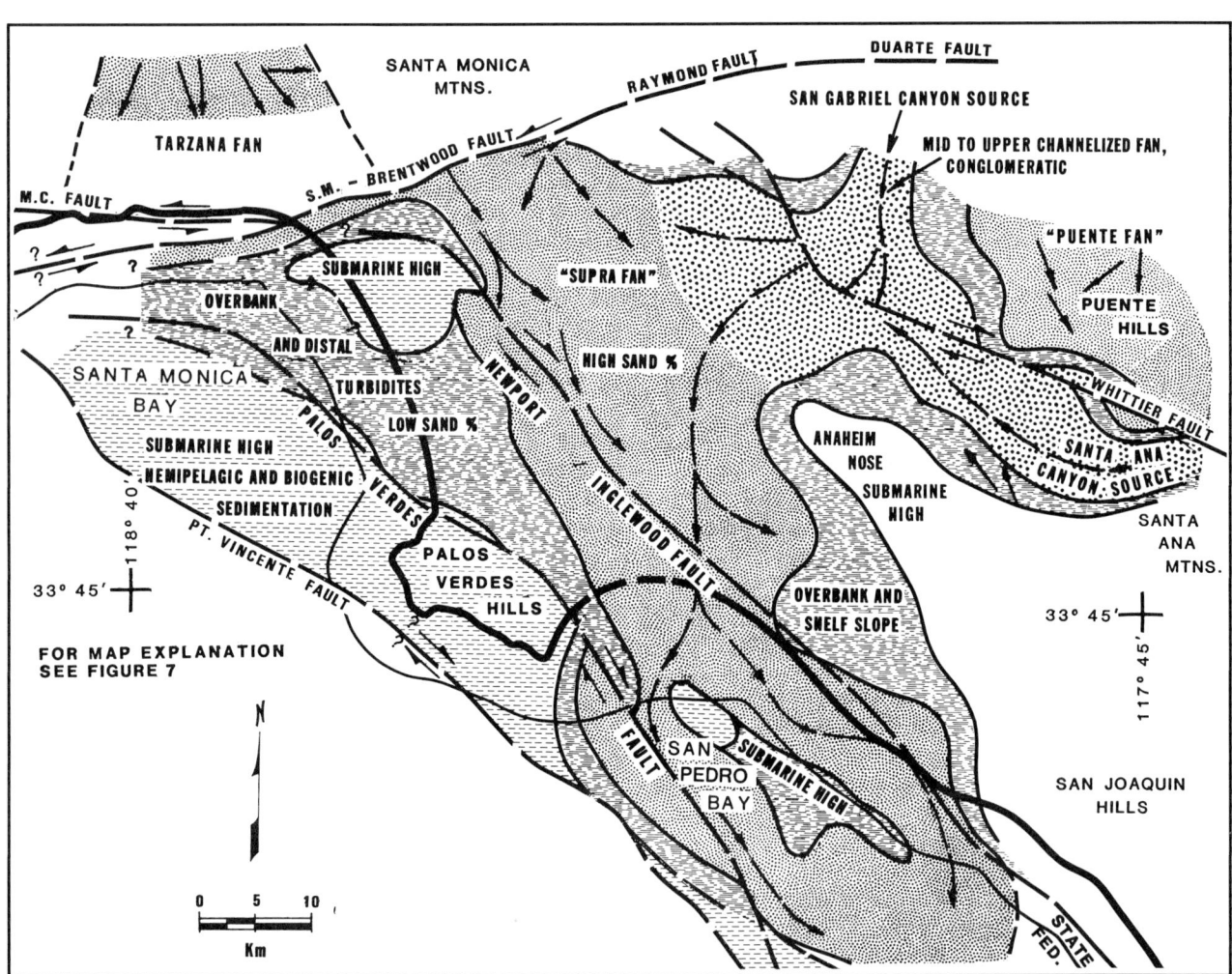

Figure 8. Present-day upper Mohnian submarine-fan facies map, Los Angeles basin, represented by the Soquel and Yorba members of the Puente Formation, a portion of the Modelo Formation in the Santa Monica Mountains, and a portion of the Monterey Formation in the Palos Verdes Hills. For an explanation of the lithofacies patterns see Figure 7. M.C. FAULT = Malibu Coast fault; S.M. = Santa Monica–Brentwood fault.

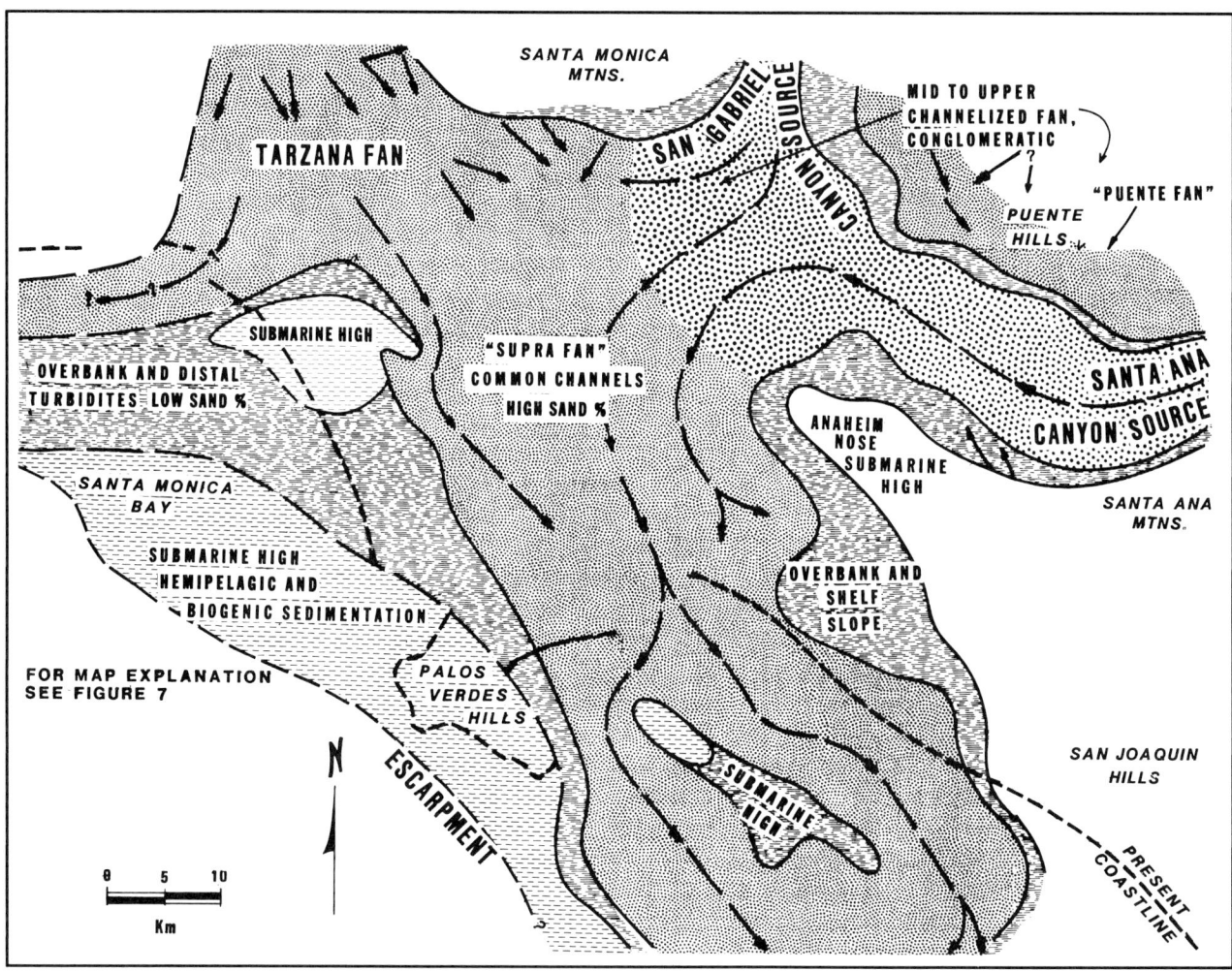

Figure 9. Palinspastic restoration of the upper Mohnian submarine-fan facies, Los Angeles basin. This restoration assumes approximately 2 mi (3 km) of north-south shortening; 4 mi (6 km) of northeast-southwest shortening; 5 mi (8 km) of right-lateral displacement on the Palos Verdes fault; 10 mi (16 km) of left-lateral displacement on the Malibu Coast, Santa Monica, Brentwood, Raymond, and Duarte fault systems; 6 to 7 mi (10 to 11 km) of right-lateral displacement on the Whittier fault; and 1 to 2 mi (2 to 3 km) of right-lateral displacement on parts of the Newport-Inglewood fault.

coarsely stippled pattern represents middle to upper channelized fan with conglomerates. The remaining portion of the mid-fan area has been labeled suprafan.

Lower Mohnian Stage

Practically the entire lower Mohnian subsurface section of the Los Angeles basin is composed of fine-grained clastic rocks that are commonly called the "Phosphatic nodular shale." The Phosphatic nodular shale is commonly interbedded with thin, fine-grained sandstones and black organic shales with rare cherts, limestones, dolomites, and bentonites. One exception to this may be the La Vida Member of the Puente Formation in the Puente Hills, eastern Los Angeles basin. Gourley (1971) referred to portions of the Puente Formation as the "Puente Fan" with a northeasterly source. The lower Mohnian Stage nodular shale represents the closest encounter the Los Angeles basin had with Monterey-type deposition. In general, there was just enough clastic contamination to prevent deposition of a thick section of pure pelagic and hemipelagic sediments that usually make up the Monterey Formation. The nodular shale should be a good exploratory objective where its lithology is similar to the Monterey Formation.

The lower Mohnian part of the Monterey Formation was deposited on the Palos Verdes Hills and their northwest and southeast offshore extensions. The lower Mohnian Nodular Shale of the Los Angeles basin is paleontologically and in part lithologically equivalent to the uppermost Altamira Shale of the Monterey Formation in the Palos Verdes Hills (Driver et al., 1952). Of interest here are the Catalina Schist debris lithic sandstones and conglomerates exposed at Point Fermin as part of the Altamira Shale.

Conrad and Ehlig (1983) believed these sandstones constituted a small submarine-fan channel system with a source area from an uplifted Catalina Schist block to the northeast in the western Los Angeles basin. It is suggested here that the eroded schist source area is presently in San Pedro Bay, about 10 to 15 km to the southeast. Right-lateral offset on the Palos Verdes fault may account for all or for a good portion of this separation. Figure 8 shows the eroded schist area as submarine high.

Upper Mohnian Submarine Fans

Fan Morphology

Throughout the discussion of the Los Angeles basin submarine fans, the morphological features will refer to Normark's 1978 model (Figure 7). During the late Mohnian, there were three primary sources of sediment input into the Los Angeles basin (Figure 8).

The most westerly source of sediment input was an inferred submarine canyon, buried somewhere beneath the San Fernando Valley, that formed the Tarzana fan (Sullwold, 1960). During the late Mohnian, the Tarzana fan was a part of the Los Angeles basin (Figure 9). Since late Mohnian time, the northern portion of the Tarzana fan has been offset left laterally about 10 to 15 km along the Malibu Coast and other faults. Sullwold (1960) speculated that the Tarzana fan had a crystalline-basement-complex sediment source in the Pacific Mountain area of the San Gabriel Mountains. He also suggested that the submarine canyon for the Tarzana fan is now buried somewhere beneath the San Fernando Valley. The Tarzana fan was mapped as the Modelo Formation in the Santa Monica Mountains (Hoots, 1931) and is called the Rancho sands in the subsurface of the northwest Los Angeles basin. In turn the Rancho sands are part of the Puente Formation. The Tarzana fan is difficult to identify in the north-central Los Angeles basin where it coalesces with suprafan sediments derived from the San Gabriel and Santa Ana submarine canyons.

Sediment input to the north-central portion of the basin was through the San Gabriel submarine canyon. It should be noted that the San Gabriel submarine canyon is an inferred Neogene submarine canyon and should not be confused with the present-day San Gabriel Canyon. Sediments derived from the San Gabriel submarine canyon lose their identity in the central Los Angeles basin where they coalesce with suprafan sediments of the Tarzana fan and middle to upper channelized fan sediments derived from the Santa Ana submarine canyon. The San Gabriel and Santa Ana submarine fans were mapped in the Puente Hills and vicinity as the Soquel and Yorba members of the Puente Formation. Gourly (1971) referred to these members as part of the Puente fan.

The easterly sediment source was the Santa Ana submarine canyon. Again, this refers to an inferred Neogene submarine canyon and not the present-day Santa Ana Canyon. The Santa Ana submarine canyon fan-channel system loses its identity in the Whittier Narrows area where it coalesces with the middle to upper channelized fan from the San Gabriel submarine canyon source.

In the area between the Anaheim nose and the north central margin of the Los Angeles basin, the division between Normark's upper and middle fan was difficult to establish from well logs. Therefore, an arbitrary division was made (Figure 8) for the division between the mid to upper channelized fan and the suprafan. The suprafan occupies the remainder of the mid-fan area. The lower fan appears to be almost nonexistent; however, lower fan, fine-grained sedimentary rocks are present. This is due to the basin shape that restricted lateral migration in the middle fan region. Fine-grained sediments may be deposited at higher submarine elevations while coarse-grained material will seek low areas and follow the basin deep.

Upper Mohnian suprafan sediments were confined to an elongate trough extending south to a present offshore position near Dana Point. The east side was bounded by the Anaheim nose and the San Joaquin Hills. The west side margin was formed by an irregular submarine high or ridge extending from Santa Monica Bay in a southeast direction to the Lausen Seamount in San Pedro Bay (Nardin and Henyey, 1978).

For a less complicated and more realistic look at the upper Mohnian fan, an upper Mohnian submarine-fan palinspastic map has been included (Figure 9). This map restores relative motion on all wrench faults back to the beginning of late Mohnian time and removes all post–early Mohnian west- and northwest-trending compressional folds. The map implies that any clockwise rotation of the basin that may have taken place did so, for the most part, before late Mohnian time.

Mechanisms of Sedimentation

During the late Mohnian, the Los Angeles basin received sediments by three primary mechanisms: hemipelagic and pelagic sedimentation, mass sediment gravity slides and slumps, and mass sediment gravity or density flows. Field's (1981) classification of sediment mass-transport in basins is used here. For most submarine fans, the third mechanism is the most important volumetrically. When dealing with hydrocarbon source rock, the first two mechanisms are much more significant.

The most common type of mass sediment density flows are turbidity flows, but certainly grain flows and debris flows are important contributors. Classical Bouma turbidite sequences are commonly seen in cores and outcrops but are difficult or impossible to recognize on logs. Massive coarse-grained sandstones that may be grain-flow deposits are easily recognized on logs, but the typical dish structures associated with grain-flow deposits are usually only recognized in outcrop. Debris flows are generally very lenticular and are commonly made up of shell debris transported *en masse* from shallow water to deep water. The arrows on Figures 8 and 9 indicate the paleodirection of mass sediment density flows and, in some instances, may parallel or subparallel many of the presumed fan channels. The arrows are also drawn to approximate the deepest parts of the basin.

Within the Los Angeles basin, as within most active margin basins around the northern Pacific rim, the middle Miocene was marked by a sudden increase in nutrient-rich upwelling with a corresponding increase in planktonic blooms and the deposition of organic-rich sediments (Garrison, 1981). Thus, Neogene hemipelagic and biogenic sedimentation took place in the Los Angeles basin and was particularly evident during the early Mohnian. During the late Mohnian, similar deposition continued but was overwhelmed by an influx of coarse-grained terrigenous clastics. Considerable amounts of hemipelagic sediments are recorded as Bouma "E" zones in a classical turbidite sequence; however, thick [50+ ft (15 m)] organic-rich shales commonly lie between massive sandstones. It is suggested here that some of these thick shales are not Bouma "E" zones but are the results of mass sediment gravity slides. Applying Walker and Mutti's (1973) submarine-fan lithofacies classification, many of the thick organic shales may be facies F or G. It appears that while coarse- to fine-grained clastics were being deposited within the deep-water portions of the basin, hemipelagic and biogenic deposition was continuing on the basin slope as well as the basin plain. As the basin slope hemipelagic and biogenic sediments thickened they became very unstable, particularly when deposited under anoxic conditions (Field, 1981). This unstable condition plus local seismic events could cause the entire sedimentary unit to move *en masse* downslope and out onto the basin floor. The next mass sediment density flow would commonly bury the pelagic sediments, thus producing thick organic-rich shales interbedded with coarse-grained clastics.

Oil Fields That Produce from the Upper Mohnian Submarine Fan

Most of the upper Mohnian oil-producing sandstones are part of the middle to upper channelized portion of the fan or the suprafan (Figure 10). Notable exceptions to this are the lower Main zone and Del Amo zone at Torrance. The thin sandstones and fractured shales at Torrance are shown as distal turbidites in Figure 8 and are lithologically similar to Normark's lower fan (Figure 7).

The upper Mohnian Rancho sandstones, productive at Sawtelle, Cheviot Hills, East Beverly Hills, and Inglewood City are believed to be part of the Tarzana fan. At Las Cienegas and farther east, the Rancho sandstones coalesce with the San Gabriel submarine-canyon sourced sandstones and thus become more difficult to recognize.

Delmontian Submarine Fans

Fan Morphology

During the Delmontian Stage, the San Gabriel and Santa Ana submarine-canyon sources were still quite active but, by the end of Delmontian time, deposition on the Tarzana fan was waning. By comparing the Delmontian fans (Figure 11) to the upper Mohnian fans (Figure 8) we see very little if any progradation or retrogradation suggesting basin subsidence and sedimentation rates were about equal. The Delmontian suprafan is very similar to the late Mohnian suprafan with two minor exceptions: (1) the late Mohnian suprafan sandstones are more widespread in San Pedro Bay, and (2) the Delmontian fan did not extend into the Salt Lake–San Vincente oil field area.

Mechanisms of Sedimentation

During Delmontian time, the Los Angeles basin continued to receive sediments by the three primary mechanisms described for the late Mohnian: hemipelagic and biogenic sedimentation, mass sediment gravity slides, and mass sediment density flows. Volumetrically, the ratio of the three mechanisms is about the same as the late Mohnian; thus, the Delmontian Stage was essentially a continuation of late Mohnian deposition with the same mechanisms applying.

Paleobathymetry

Mutti and Normark (1987) discussed the pitfalls of interpreting ancient submarine fan morphology from vertical sequences such as well logs and outcrop exposures. The concentration of work has been on submarine-fan lithofacies, including sandstone-shale ratios, bed thickness and continuity, fining or coarsening upward, and sedimentary structures. One factor, paleobathymetry, has been somewhat overlooked. Good paleobathymetric control within an active margin basin refines the interpretation of the corresponding submarine-fan morphology. Figure 12 is a paleobathymetric map on a benthic foraminiferal horizon near the top of the Delmontian Stage. This map was constructed using all the wells shown in Figure 2 that penetrated Delmontian sedimentary rocks. Gross similarities and many dissimilar features can be seen in a comparison of Figure 12 with Figure 11. What should be remembered here is that Figure 11 represents an approximation for the average Delmontian submarine-fan morphology while Figure 12 represents the paleobathymetry for a time horizon during the Delmontian Stage.

Good paleobathymetric information based on benthic foraminifera (see Blake, this volume) from near the top of the Delmontian Stage within the Los Angeles basin indicates the following:

1. The submarine-fan shape was greatly modified by tectonics. Note the series of bumps along the Newport-Inglewood fault trend and the abrupt changes near the Palos Verdes fault and the Whittier fault.

2. The suprafan, in general, occupies the deepest water in the basin.

3. The amount of sand, either coarse or fine grained, has nothing to do with the distance from its entry point into the basin but is controlled, at least in part, by paleobathymetry. This last point, of course, assumes a rather effective sediment transport system with a fairly sandy sediment supply.

Oil Fields That Produce from the Delmontian Submarine Fans

Practically every field along the Newport-Inglewood fault trend is productive from Delmontian sandstone reservoirs

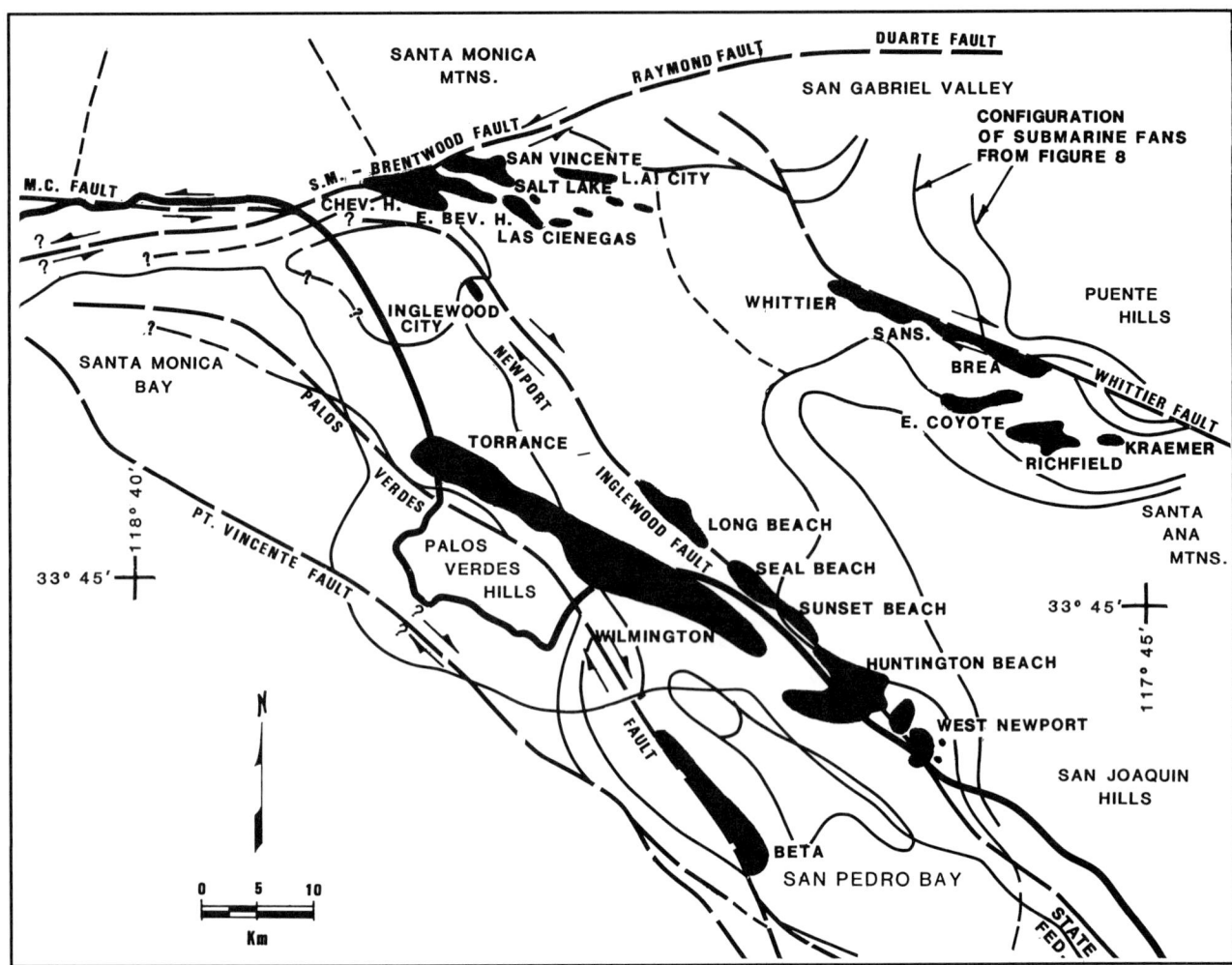

Figure 10. Oil fields that produce from the upper Mohnian submarine fans, Los Angeles basin. The oil fields (black) are shown upon the upper Mohnian submarine-fan configuration from Figure 8.

(Figure 13). Most of the Newport-Inglewood producing trend lies within the suprafan portion of the Delmontian submarine fans, while all the eastern Los Angeles basin oil fields produce from the middle to upper channelized portion of the fan (Weser, 1977). There is a possibility that some oil is produced from the Delmontian Tarzana fan sandstones in the East Beverly Hills–Las Cienegas area, but, for the most part, Delmontian sand deposition on the Tarzana fan was very minor.

Repettian Submarine Fans

Fan Morphology

The Repettian submarine-fan morphology is markedly different from the upper Mohnian-Delmontian fans. By Repettian time, sediment deposition on the Tarzana fan had ceased within the Los Angeles basin. The Tarzana fan may have been cut off from the basin by left-lateral strike-slip displacement along the Malibu Coast fault system. In addition, the basin deepened to lower bathyal and abyssal depths from middle to upper bathyal depths. On the east side of the basin, tectonic activity at the close of the Miocene caused the Repetto Formation to be deposited unconformably on upper Miocene rocks. The San Gabriel and Santa Ana submarine-canyon systems were still very active; so much so, that the uppermost fan areas prograded basinward several kilometers. The Repettian middle to upper channelized fan region is very conglomeratic.

Comparison of the Repettian submarine fan (Figure 14) with the upper Mohnian and Delmontian fans (Figures 8, 11) illustrates one remarkable difference: the Repettian fan more closely resembles Normark's 1978 submarine-fan model than do the upper Mohnian and Delmontian fans. If it were not for the Palos Verdes Hills–Santa Monica Bay bathymetric high, the Repettian fan would have almost an unrestricted submarine-fan radial shape. The San Gabriel and Santa Ana submarine-canyon systems were close enough together that the suprafan and lower fan

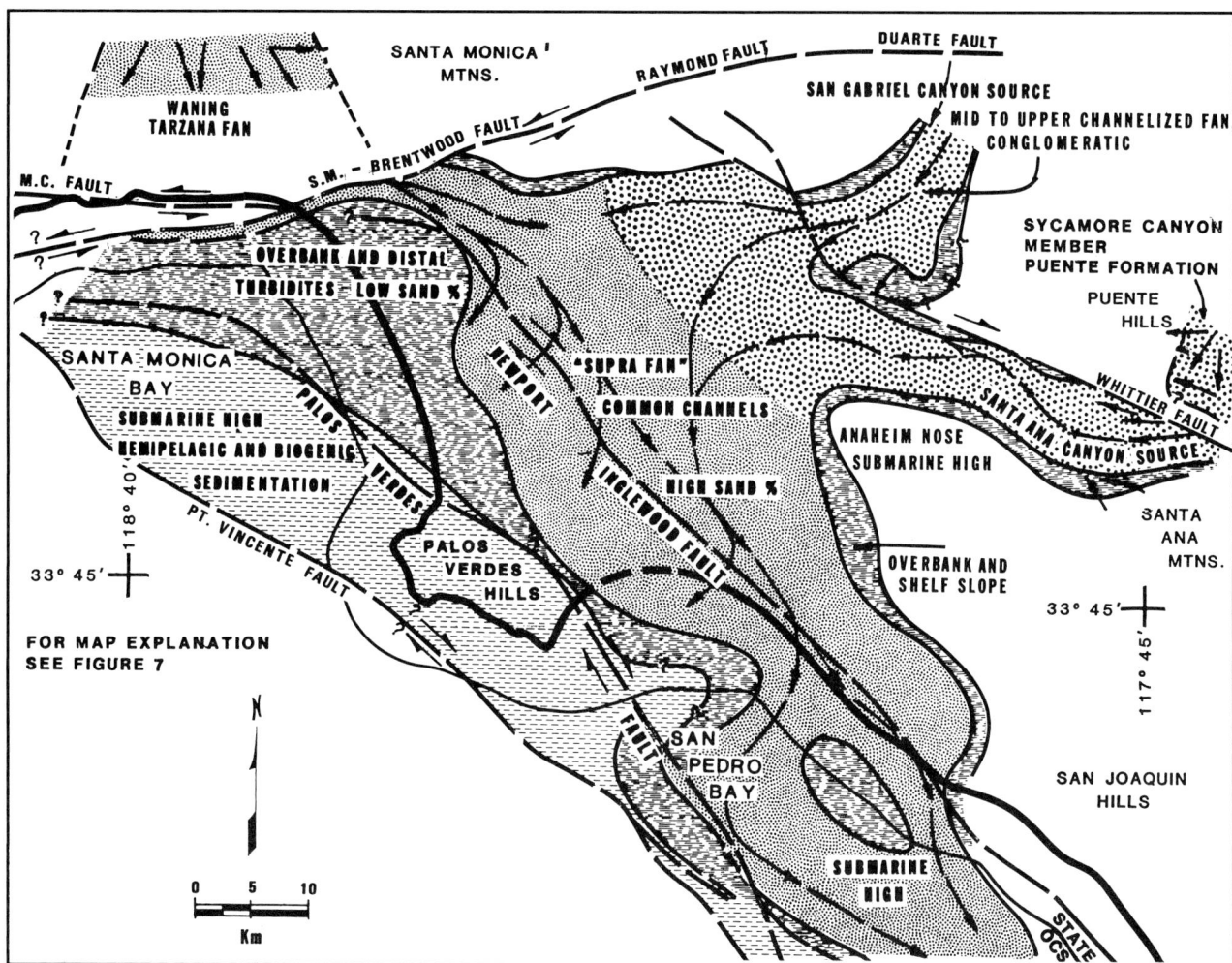

Figure 11. Present-day Delmontian submarine-fan facies map, Los Angeles basin, represented by the Sycamore Canyon Member of the Puente Formation, part of the upper Modelo Formation in the Santa Monica Mountains, and part of the Monterey Formation in the Palos Verdes Hills. For an explanation of the lithofacies patterns see Figure 7. M.C. FAULT = Malibu Coast fault; S.M. = Santa Monica–Brentwood fault.

equilavents appear to have a single source. The only other major difference between the Repettian fan and the upper Mohnian–Delmontian fans is that the Repettian fan covered much more of the Anaheim nose area and the suprafan extended west into Santa Monica Bay.

Mechanisms of Sedimentation

During Repettian time, the Los Angeles basin continued to receive sediments by the three primary mechanisms described for the upper Mohnian and Delmontian stages. There was, however, one noticeable difference. The higher influx rate of coarse-grained clastics caused the pelagic and hemipelagic sediments to be of less importance volumetrically. Consequently, the Repetto Formation is commonly very sandy or conglomeratic. Mass sediment density flows were the predominant mechanisms of sediment transport. Turbidity flows and channelized grain flows were quite common.

Oil Fields That Produce from the Repettian Submarine Fans

Most of the major oil fields in the Los Angeles basin produce from sandstones and conglomerates of the Repettian submarine fans (Figure 15). About one-half the cumulative production to date has come from the Repetto Formation. The most important Repetto oil production is found at the Inglewood, Rosecrans, Dominguez, Long Beach, and Huntington Beach oil fields on the west side of the basin. On the east side, the Montebello, Santa Fe Springs, West Coyote, and East Coyote fields are the most significant.

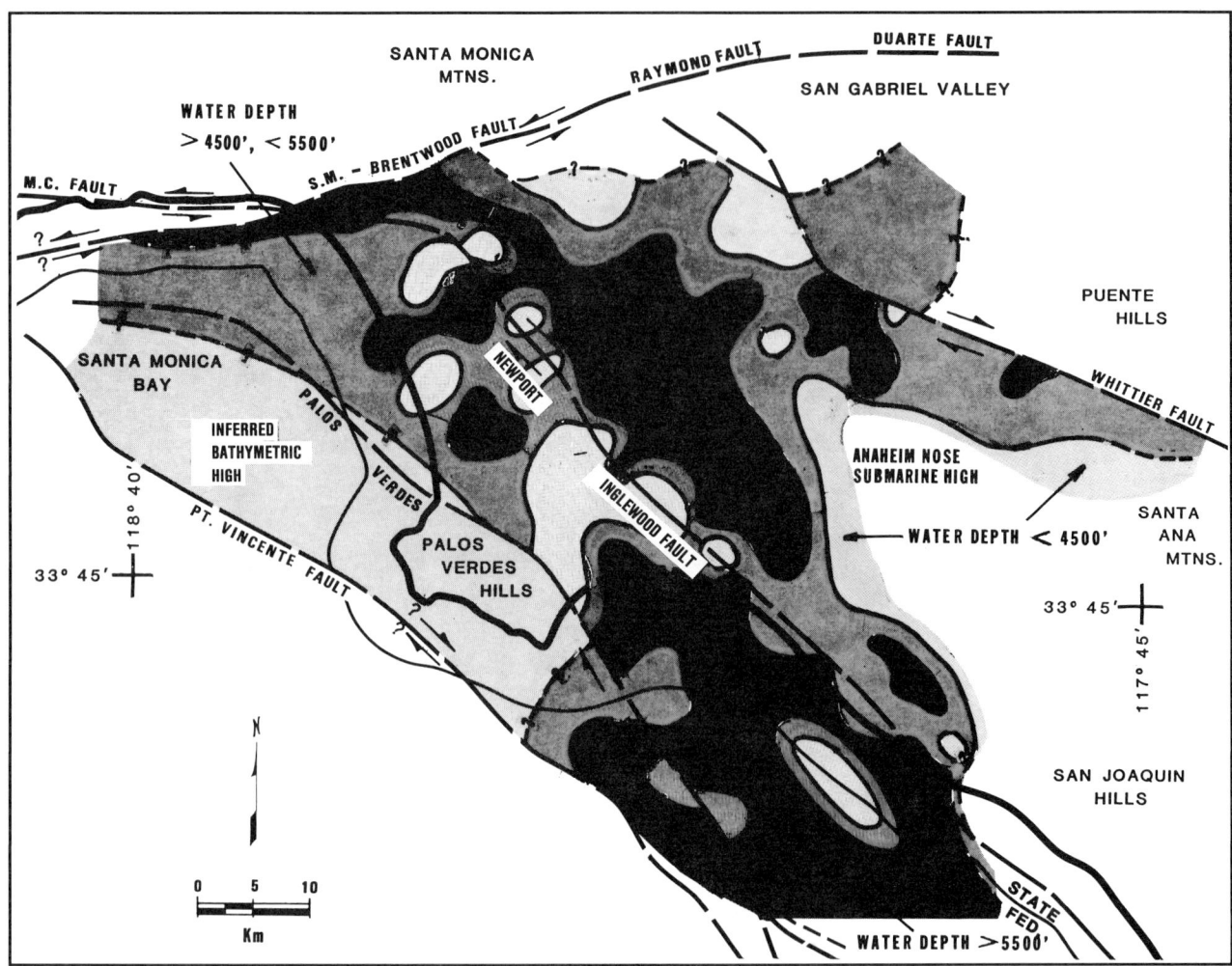

Figure 12. Present-day paleobathymetry near top Delmontian Stage, Los Angeles basin. Darkest shade = water depths greater than 5500 ft (1700 m). Medium shade = water depths less than 5500 ft (1700 m) but greater than 4500 ft (1400 m). Lightest shade = water depths less than 4500 ft (1400 m). M.C. FAULT = Malibu Coast fault, S.M. = Santa Monica–Brentwood fault.

FACTORS THAT MADE THE LOS ANGELES BASIN A PROLIFIC OIL PRODUCER

On a gross sediment volumetric basis, the Los Angeles basin is one of the world's richest oil-producing basins. The following factors or conditions have made this basin such a prolific oil producer.

Sedimentation

The following points, when combined, contributed to the hydrocarbon source and reservoir rocks of the Los Angeles basin.

1. Nutrient-rich Miocene and early Pliocene upwelling gave rise to high planktonic productivity along the northeast Pacific coast producing rich biogenic sedimentation (Garrison, 1981; Ingle, 1981).

2. Potential source-rock deposition occurred in a probable silled basin, under anoxic or close to anoxic conditions with negligible bioturbation, thus preserving the hydrocarbon precursors.

3. Almost immediate burial of pelagic and hemipelagic sediments by mass sediment density flows under anoxic or close to anoxic conditions helped preserve the organic matter.

4. Unstable slope deposition caused mass sediment gravity slides of organic-rich sediments into the basin deep where they were soon buried by coarse-grained mass sediment gravity flows (Field, 1981).

5. The source sediments had a kerogen type that is oil prone with a high (>1%) total organic carbon content (Jeffrey et al., *Geochemistry of Los Angeles Basin Oil and Gas Systems*, this volume).

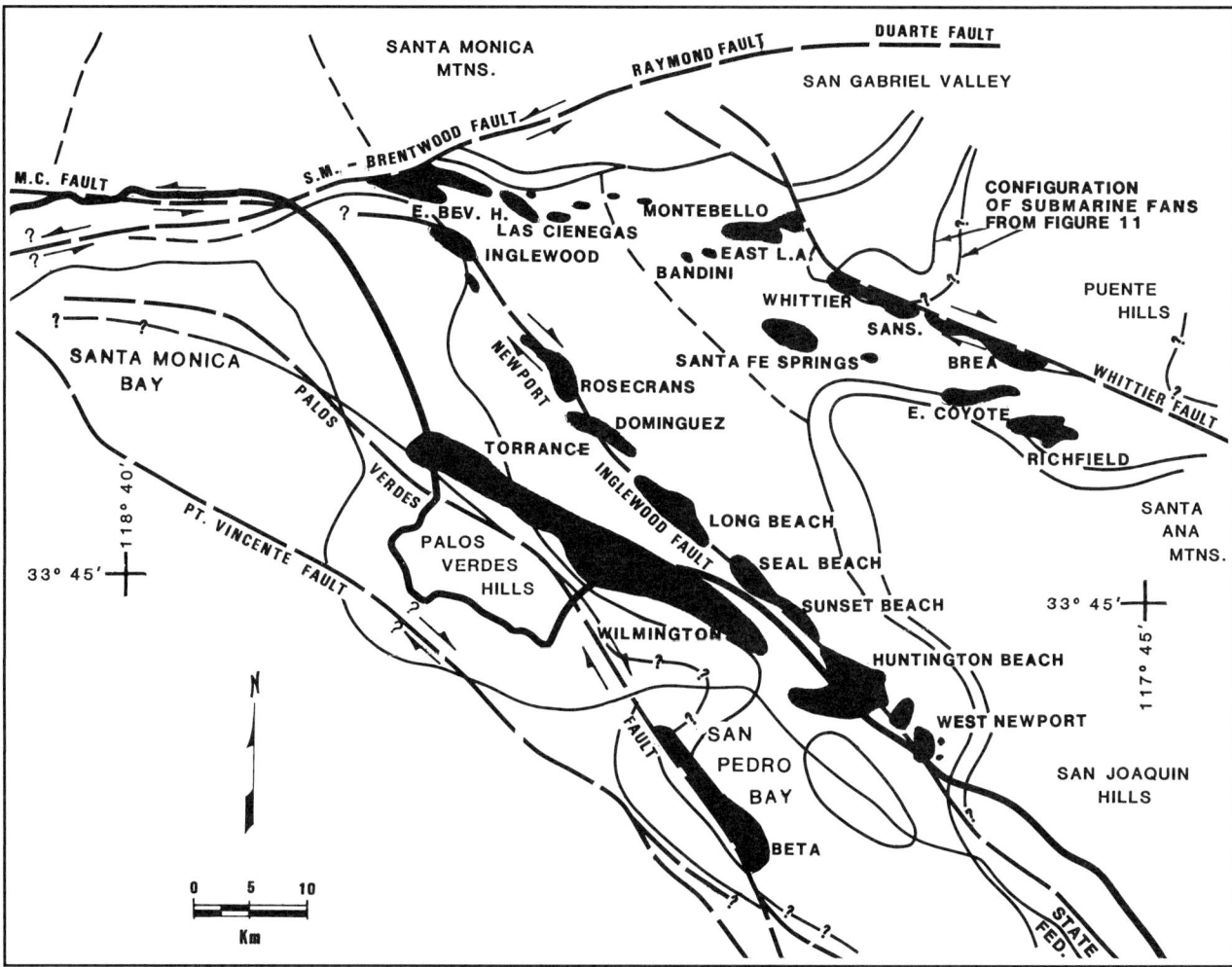

Figure 13. Oil fields that produce from the Delmontian submarine fans, Los Angeles basin. The oil fields (black) are shown upon the Delmontian submarine-fan configuration from Figure 11.

6. Thousands of feet of potential reservoir sands with good lateral continuity were deposited over areas of future oil fields.

These conditions existed from the Luisian Stage to at least the end of the Repettian Stage. Jeffrey et al. (this volume) consider the upper Miocene, Divisions D and E (upper and lower Mohnian), to be the major source of the oil discovered to date in the Los Angeles basin. This includes the Phosphatic nodular shale and its stratigraphic equivalents. They also concluded that the only significant oil generation occurred in the "deeply buried mature strata." Since about one-half of the oil produced to date comes from the Repetto Formation and at least another one-third comes from the upper Miocene Divisions A, B, and C (upper Mohnian and Delmontian), the question arises: could the Repetto Formation and the upper Miocene Divisions A, B, and C have generated their own oil? In other words, is the oil found in these rocks indigenous to their respective formations? It is suggested here that this is still a possibility. Vertical oil migration, whether along faults, along unconformities, or around impermeable beds, still presents a problem. Migrating oil into overpressured lenticular sandstones has not been adequately explained.

Other Factors

1. Incipient Neogene structural growth helped early hydrocarbon migration. Referring back to the Delmontian paleobathymetric map (Figure 13), we see marked changes in submarine topography along the Newport-Inglewood fault trend. It appears that this is due to incipient structural growth. Some Neogene oil may have migrated into these "early structures" prior to major Pleistocene folding and faulting (see Wright, this volume, for a fuller discussion of oil migration).

2. Obviously the paleogeothermal gradients were high enough to generate oil. For a discussion of Los Angeles basin subsidence and thermal implications of paleo-

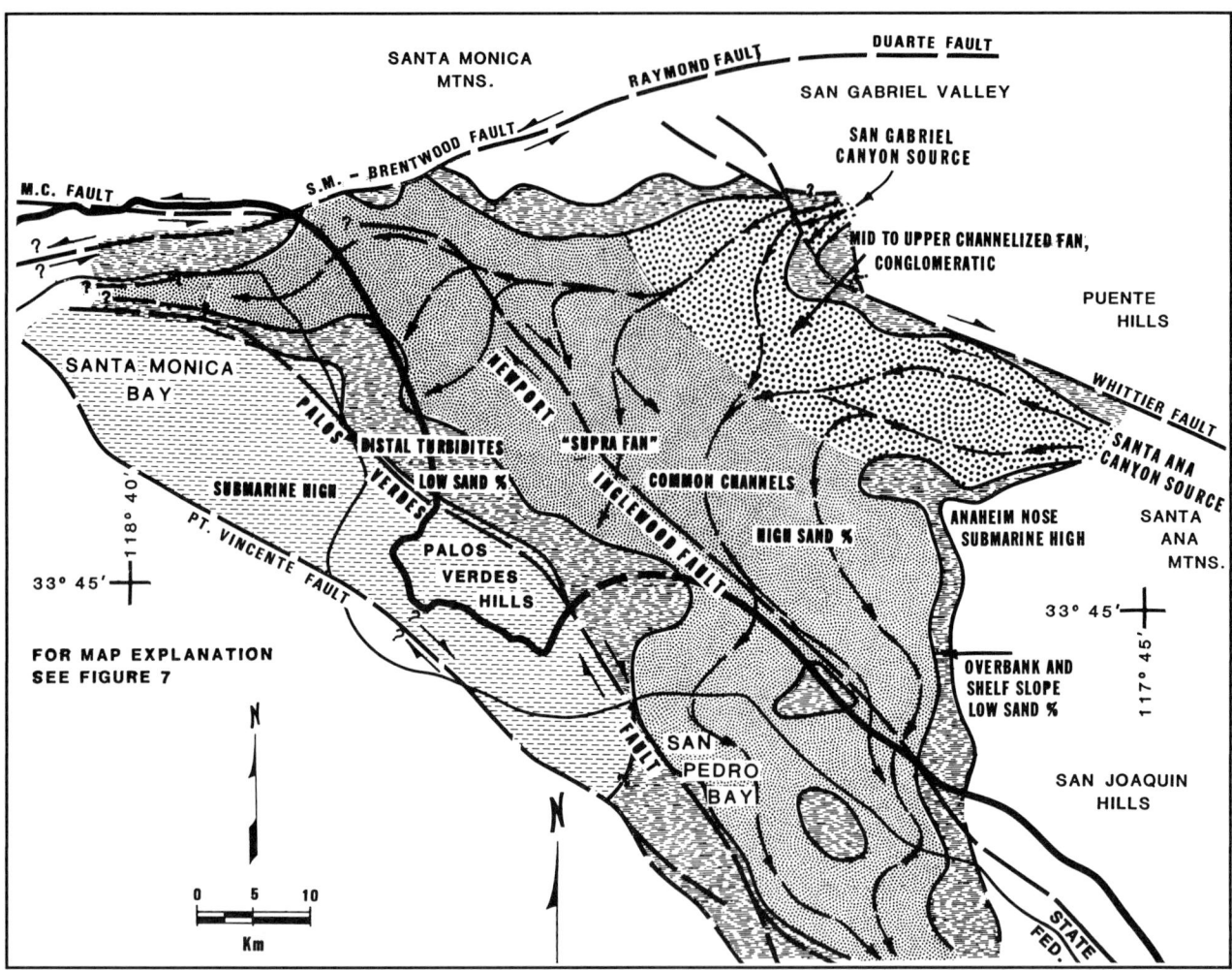

Figure 14. Present-day Repettian submarine-fan facies map, Los Angeles basin, represented by the Repetto Formation throughout most of the Los Angeles basin. Yerkes et al. (1965) referred to the lower Pliocene sedimentary rocks as the "lower member of the Fernando Formation." Lamar (1961) referred to the lower Pliocene sedimentary rocks as "part of the Fernando Group." For an explanation of the lithofacies patterns see Figure 7. M.C. FAULT = Malibu Coast fault; S.M. = Santa Monica–Brentwood fault.

heat flow see Mayer, *Central Los Angeles Basin: Subsidence and Thermal Implications for Tectonic Evolution*, and Yeats and Beall, this volume. Jeffrey et al., this volume, discuss oil quality and API gravity.

CONCLUSIONS

The Los Angeles basin formed during the early and middle Miocene within the North American–Pacific plate transform margin prior to major late Neogene offset on the San Andreas fault. The basin is the result of primarily dextral offset on major wrench faults that led to extension in the region, thus forming a depositional trough that may be floored by volcanic rocks. Regional east-west extension, accompanied by "tectonic block rotations," may also have aided basin formation.

The probably silled Los Angeles basin was filled by a series of Neogene submarine fans interbedded with pelagic, hemipelagic, and mass sediment gravity slides. Coarse-grained clastic sediment input was derived mainly from the inferred submarine canyon that formed the Tarzana fan, the San Gabriel submarine canyon, and the Santa Ana submarine canyon. Some of the thick organic shales found interbedded with mass sediment density flows may be due to mass sediment gravity slides of unstable pelagic and hemipelagic deposits from the basin slope to the basin plain. The morphology of the submarine fans was extensively controlled by paleobathymetry, but the mechanisms of sedimentation were typical of many active margin Neogene basins. The suprafan would often occupy the deepest water in the basin, and the amount of sand was controlled by paleobathymetry and the efficiency of the mass sediment transport system.

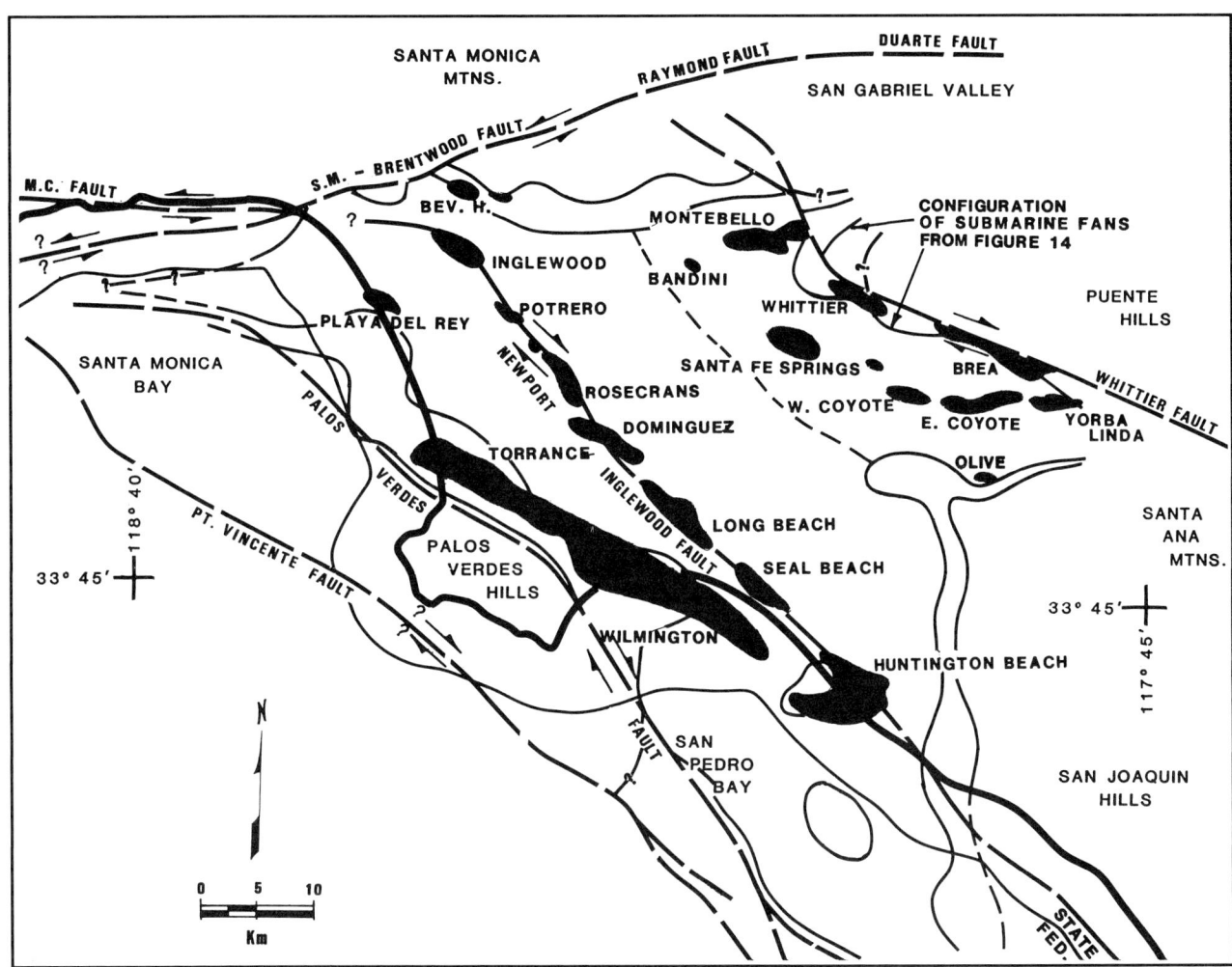

Figure 15. Oil fields that produce from Repettian submarine fans, Los Angeles basin. The oil fields (black) are shown upon the Repettian submarine-fan configuration from Figure 14.

Most of the oil produced in the basin to date comes from the middle to upper channelized fans or suprafan sands of late Mohnian to Repettian age. In general, the paleobathymetry was controlled in large part by regional tectonics.

From the middle Miocene Luisian Stage until at least the end of the Pliocene Repettian Stage, the Los Angeles basin was situated within an almost ideal environment for the deposition of organic-rich source rock and potential reservoir rock. Continued subsidence and deposition with some early structural growth and adequate paleo–heat flow provided the physical conditions needed for the efficient generation and migration of hydrocarbons.

ACKNOWLEDGMENTS

Much of the basic data used in this interpretation was assembled by Jan Vargo, presently employed by Tenneco. The paleobathymetric data was provided by Gregg Blake, presently employed by Unocal in Ventura, CA. The facilities, drafting (in part), and reproduction were provided by Unocal's office in Ventura, CA. A special thank you to Nancy Craigie and Susan Lewis, also of Unocal, for helping me prepare this paper for publication. The manuscript was reviewed by K. T. Biddle, R. G. Stanley, and an anonymous reviewer.

REFERENCES

Atwater, T., 1970, Implications of plate tectonics for the Cenozoic tectonic evolution of western North America: Geological Society of America Bulletin, v. 81, p. 3513-3536.

Ballance, P. F., and H. G. Reading, 1980, Sedimentation in oblique-slip mobile zones: an introduction, in P. F. Ballance and H. G. Reading, eds., Sedimentation in oblique-slip mobil zones: International Association of Sedimentologists Special Publication 4, p. 1-5.

Bennett, J. N., 1967, Paleocurrent analysis of the upper Miocene formations, Los Angeles basin, California [unpublished M.S. thesis]: Tucson, Arizona, University of Arizona, 80 p.

Blake, G. H., 1991, Review of the Neogene biostratigraphy and stratigraphy of the Los Angeles basin and implications for basin evolution, in K. T. Biddle, ed., Active margin basins: AAPG Memoir 52, p. 135-184.

Campbell, R. H., and R. F. Yerkes, 1976, Cenozoic evolution of the Los Angeles basin area—relation to plate tectonics, in D. G. Howell, ed., Aspects of the geologic history of the California continental borderland: Pacific Section, AAPG Miscellaneous Publication 24, p. 541-558.

Campbell, R. H., and R. F. Yerkes, 1980, Geologic guide to the stratigraphy and structure of the Topanga Group, central Santa Monica Mountains, southern California: Guidebook 49, Los Angeles Basin Geological Society Publication, 19 p.

Conrad, C. L., and P. L. Ehlig, 1983, The Monterey formations of the Palos Verdes Peninsula, California—an example of sedimentation in a tectonically active basin within the California continental borderland, in D. K. LaRue and R. S. Steel, eds., Cenozoic marine sedimentation Pacific Margin, U.S.A.: Pacific Section, SEPM Special Publication, p. 103-116.

Conrey, F. L., 1967, Early Pliocene sedimentary history of the Los Angeles basin, California: California Division of Mines and Geology Special Report 93, 63 p.

Crowell, J. C., 1976, Implications of crustal stretching and shortening of coastal Ventura basin, California, in D. G. Howell, ed., Aspects of the geologic history of the California continental borderland: Pacific Section, AAPG Miscellaneous Publication 24, p. 365-382.

Crowell, J. C., 1984, Origin of late Cenozoic basins in southern California, in A. G. Sylvester, ed., Wrench fault tectonics, selected papers reprinted from the AAPG Bulletin and other geological journals: AAPG Reprint Series, n. 28, p. 195-209.

Dibblee, T. W., 1977, Strike-slip tectonics of the San Andreas fault and its role in Cenozoic basin evolvement, in T. H. Nilsen, ed., Late Mesozoic and Cenozoic sedimentation and tectonics in California: San Joaquin Geological Society Short Course, p. 26-38.

Dickinson, W. R., 1981, Plate tectonics and the continental margin of California, in W. G. Ernst, ed., The geotectonic development of California: Englewood Cliffs, New Jersey, Prentice-Hall, Inc., Rubey v. 1, p. 1-28.

Driver, H. L., E. C. Edwards, D. T. Graves, C. W. Johnson, D. I. Johnstone, and J. S. Shelton, 1952, Cenozoic correlation section across Los Angeles basin from Palos Verdes Hills to San Gabriel Mountains: Pacific Section, AAPG.

Durham, D. L., and R. F. Yerkes, 1964, Geology and oil resources of the eastern Puente Hills area, southern California: USGS Professional Paper 420-B, 61 p.

Field, M. E., 1981, Sediment mass-transport in basins, in R. G. Douglas, I. P. Colburn, and D. S. Gorsline, eds., Depositional systems of active continental margin basins: short course notes: Pacific Section, SEPM, 165 p.

Garrison, R. E., 1981, Pelagic and hemipelagic sedimentation in active margin basins, in R. G. Douglas, I. P. Colburn, and D. S. Gorsline, eds., Depositional systems of active continental margin basins: short course notes: Pacific Section, SEPM, 165 p.

Gourley, J. W., 1971, Stratigraphy and sedimentation patterns upper Miocene Puente Formation (Mohnian-Delmontian) northwestern Puente Hills, Los Angeles County, California [unpublished M.S. thesis]: Los Angeles, California, University of Southern California, 138 p.

Harding, T. P., 1974, Petroleum traps associated with wrench faults: AAPG Bulletin, v. 58, p. 1290-1304.

Hoots, H. W., 1931, Geology of the eastern part of the Santa Monica Mountains, Los Angeles County, California: USGS Professional Paper 165-C, p. 83-134.

Howell, D. G., J. K. Crouch, H. G. Green, D. S. McCulloch, and J. G. Vedder, 1980, Basin development along the late Mesozoic and Cenozoic California margin: a plate tectonic margin of subduction, oblique subduction and transform tectonics: International Association of Sedimentologists, Special Publication 4, p. 43-62.

Hsü, K. J., 1977, Studies of Ventura field, California, 1: facies geometry and genesis of lower Pliocene turbidites: AAPG Bulletin, v. 61, p. 137-168.

Ingle, J. C., 1981, Origin of Neogene diatomites around the north Pacific rim, in R. E. Garrison et al., eds., The Monterey Formation and related siliceous rocks of California: Special Publication of the Pacific Section SEPM, p. 159-179.

Jeffrey, A. W. A., H. M. Alimi, and P. O. Jenden, 1991, Geochemistry of Los Angeles basin oil and gas systems, in K. T. Biddle, ed., Active margin basins: AAPG Memoir 52, p. 197-219.

Jones, D. L., M. C. Blake, and Claude Rangin, 1976, the four Jurassic belts of northern California and their significance to the geology of the southern California borderland, in D. G. Howell, ed., Aspects of the geologic history of the California continental borderland: Pacific Section AAPG Miscellaneous Publication 24, p. 343-362.

Lamar, D. L., 1961, Structural evolution of the northern margin of the Los Angeles basin: unpublished Doctor of Philosophy dissertation, University of California at Los Angeles, 142 p.

Mayer, L., 1991, Central Los Angeles basin: subsidence and thermal implication for tectonic evolution, in K. T. Biddle, ed., Active margin basins: AAPG Memoir 52, p. 185-195.

Mutti, E., 1977, Distinctive thin bedded turbidite facies and related depositional environments in the Eocene Hecho Group (south-central Pyrenees, Spain): Sedimentology, v. 24, p. 107-131.

Mutti, E., and W. R. Normark, 1987, Comparing examples of modern and ancient turbidite systems: problems and concepts, in J. K. Leggett and G. G. Zuffa, eds., Modern clastic sedimentology: concepts and case studies: Graham and Trotman, p. 1-38.

Nardin, T. R., and T. L. Henyey, 1978, Pliocene-Pleistocene diastrophism of Santa Monica and San Pedro shelves, California continental borderland: AAPG Bulletin, v. 62, p. 247-272.

Normark, W. R., 1978, Fan valleys, channels, and depositional lobes on modern submarine fans: characters for recognition of sandy turbidite environments: AAPG Bulletin, v. 62, p. 912-931.

Normark, W. R., and D. J. W. Piper, 1984, Navy fan, California borderland: growth pattern and depositional processes:

Geo-Marine Letters, v. 3, p. 101–108.

Platt, J. P., 1976, The significance of the Catalina Schist in the history of the southern California borderland, in D. G. Howell, ed., Aspects of the geologic history of the California continental borderland: Pacific Section, AAPG Miscellaneous Publication 24, p. 47–52.

Ricci Lucchi, F., 1975, Depositional cycles in two turbidity formations of northern Apennines (Italy): Journal of Sedimentary Petrology, v. 45, p. 3–43.

Rodgers, D. A., 1980, Analysis of pull-apart basin development produced by en echelon strike-slip faults, in P. F. Ballance and H. G. Reading, eds., Sedimentation in oblique-slip mobile zones: International Association of Sedimentologists Special Publication 4, p. 27–41.

Schoellhamer, J. E., and A. O. Woodford, 1951, The floor of the Los Angeles basin, Los Angeles, Orange and San Bernadino counties, California: USGS Oil and Gas Investigation Map OM117, 2 sheets, scale 1 in. = 2 mi.

Shanmugam, G., J. E. Damuth, and R. J. Moiola, 1985, Is the turbidite facies association scheme valid for interpreting ancient submarine fan environments?: Geology, v. 13, p. 234–237.

Sorensen, S., 1985, Petrologic evidence for Jurassic, island-arc–like basement rocks in the southwestern Transverse Ranges and California continental borderland: GSA Bulletin, v. 96, p. 997–1006.

Sullwold, H. H., 1960, Tarzana fan, deep submarine fan of late Miocene age, Los Angeles County, California: AAPG Bulletin, v. 44, p. 433–457.

Turcotte, D. L., and D. C. McAdoo, 1979, Thermal subsidence and petroleum generation in the southwestern block of the Los Angeles basin, California: Journal of Geophysical Research, v. 84, n. B7, p. 3460–3464.

Walker, R. G., 1978, Deep-water sandstone facies and ancient submarine fans: models for exploration for stratigraphic traps: AAPG Bulletin, v. 62, p. 932–966.

Walker, R. G., and E. Mutti, 1973, Turbidite facies and facies associations, Part IV, in A. H. Bouma and G. V. Middleton, eds., Turbidites and deep water sedimentation: Pacific Section, SEPM Short Course, p. 119–158.

Weser, O. E., 1977, Deep-water oil sand reservoirs—ancient case histories and modern concepts: Pacific Section, AAPG, Coast Geological Society Publication Short Course, Note Series 6, 170 p.

Woodford, A. O., T. G. Moran, and J. S. Shelton, 1946, Miocene conglomerates of Puente and San Jose Hills, California: AAPG Bulletin, v. 30, p. 514–560.

Wright, T. L., 1991, Structural geology and tectonic evolution of the Los Angeles basin, in K. T. Biddle, ed., Active margin basins: AAPG Memoir, 52, p. 35–134

Yeats, R. S., 1976, Extension versus strike-slip origin of the southern California borderland, in D. G. Howell, ed., Aspects of the geologic history of the California continental borderland: Pacific Section, AAPG Miscellaneous Publication 24, p. 455–485.

Yeats, R. S., and J. M. Beall, 1991, Stratigraphic controls of oil fields in the Los Angeles basin: a guide to migration history, in K. T. Biddle, ed., Active margin basins: AAPG Memoir 52, p. 221–239.

Yerkes, R. F., T. H. McCulloh, J. E. Schoellhamer, and J. G. Vedder, 1965, Geology of the Los Angeles basin, California—an introduction: USGS Professional Paper 420-A, 57 p.

Chapter 9

Taranaki Basin, New Zealand

Julie Palmer

Petrocorp Exploration Limited
Wellington, New Zealand

Geoff Bulte

Petrocorp Exploration Indonesia Limited
Jakarta, Indonesia

ABSTRACT

The Taranaki basin is a Cretaceous and Tertiary sedimentary basin located along the western side of the North Island, New Zealand. Initiated during the Cretaceous, the Taranaki basin lies at the southern end of a rift that developed subparallel to the Tasman Sea rift, which now separates Australia and New Zealand.

Structure of the basin has been controlled by movement along the Taranaki and Cape Egmont fault zones. Subsidence commenced in the Cretaceous and continued until the Pliocene. The predominant tectonic regime in the Taranaki basin changed from one of extension to one of compression in the early Miocene. Late Tertiary tectonics formed three primary structural types: faulted anticlines, high-angle overthrust structures, and tilted fault blocks. Kapuni and Maui gas-condensate fields are faulted anticlines; the McKee (oil) and the Ahuroa and Tariki (gas-condensate) fields are overthrust structures.

Sandstones within the Eocene Kapuni Group and the Oligocene Otaraoa Formation are the only producing reservoirs. The gas-condensate and oil are sourced from nonmarine to paralic coals and carbonaceous shales of the Late Cretaceous–Eocene Pakawau and Kapuni groups. The overlying marine sequences are organically lean and have negligible source potential.

A large proportion of the sedimentary succession was deposited during the late Tertiary. Consequently, the geothermal gradient in the Taranaki basin is moderately low ($<3°C/100$ m). Maturation studies show that only those source rocks buried between 4000 and 4950 m are in the present oil generation and gas expulsion window.

INTRODUCTION

The Taranaki basin (Figure 1) is one of New Zealand's largest Cretaceous-Tertiary sedimentary basins and is the country's only hydrocarbon-producing basin. Production is from the Eocene Kapuni Group and Oligocene Otaraoa Formation. This paper is an overview of exploration in the Taranaki basin and, in particular, presents the current ideas on the stratigraphy and the tectonic and structural evolution of the basin.

REGIONAL SETTING

The Taranaki basin lies along the western side of North Island, New Zealand (Figure 1). It extends westward from the Taranaki fault zone, underlies the onshore Taranaki Peninsula, and continues offshore beyond the edge of the continental shelf (McBeath, 1977) (Figure 2). The basin extends northward from Nelson and the northwest end of the South Island to about the latitude of Aukland (Pilaar and Wakefield, 1978) (Figure 1).

During the Paleozoic and early middle Mesozoic the microcontinent of New Zealand was located along the eastern margin of the Australian continent (Carter and Norris, 1976). About 75 Ma sea-floor spreading commenced, separating the Australian and New Zealand continental masses and forming the Tasman Sea (Doutch et al., 1981). At about the same time a "subparallel" rift system developed between the Lord Howe Rise and the present New Caledonia-New Zealand landmasses, forming the Reinga basin-New Caledonia trough and Fairway trough systems (Eade, 1988) and their southward extension, the Taranaki basin. Spreading of the rift system had ceased by the early Eocene, approximately 53 Ma (Eade, 1988), but regional subsidence and deposition has continued, with local, tectonically controlled interruptions to the present.

The Taranaki basin contains a thick succession of Upper Cretaceous to Recent sedimentary rocks overlying Paleozoic and Mesozoic basement rocks (Figure 3). Isolated occurrences of Upper Cretaceous and Tertiary sedimentary rocks crop out in the northwest of the South Island, at the southern end of the basin, and upper Tertiary sedimentary rocks crop out in eastern Taranaki basin. Elsewhere, the older sequences lie beneath a cover of Pliocene-Pleistocene marine sediments and subaerial volcanics. Most information on the Upper Cretaceous-Tertiary rocks comes from wells.

The first significant well, Kapuni-1 (Figure 2), was drilled in 1959 and discovered New Zealand's first gas-condensate field. Production is from upper Eocene sandstones of the Kapuni Group. Since 1959 a further 50 wildcat wells have drilled in the Kapuni Group. From these, there were four major discoveries: the offshore Maui gas-condensate field, the onshore McKee (oil), and Tariki and Ahuroa (gas-condensate) fields, in the Tarata thrust zone (Figure 2). Subcommercial oil discoveries were made in the offshore Maui-4 and Moki-1 wells and the commerciality of the Kupe South-1 oil and gas discovery is currently being assessed.

All production in Taranaki basin is from the Eocene part of the Kapuni Group (Kapuni, Maui, and McKee fields) (Figure 4) and the lower Oligocene Otaraoa Formation (Tariki and Ahuroa fields) (Figure 5).

The Taranaki basin is subdivided into two main structural units: the Western platform and the Taranaki graben (Figure 2). The Western platform extends west from the Cape Egmont fault zone to at least the continental shelf edge (McBeath, 1977) and is characterized by broad, simple structure and 2000 to 5000 m of Late Cretaceous to Recent sediments (Figure 6). The Western platform was affected by Late Cretaceous to Eocene block faulting (Figure 7) and, except for its southeastern margin, has been relatively stable throughout the remainder of the Cenozoic (Pilaar and Wakefield, 1978).

The north- to northeast-trending Cape Egmont fault zone (Figure 2), which separates the Western platform and Taranaki graben, consists of a series of steep, subparallel, normal and reverse faults (McBeath, 1977). Maximum throw along the fault zone is about 2100 m.

The Taranaki graben extends from the downthrown eastern side of the Cape Egmont fault zone east to the Taranaki fault zone. The eastern side of the graben is dominated by compressional features, including the Tarata thrust zone and the large, fault-bounded Kapuni structure (Figure 8).

The Taranaki fault zone trends north-south, and vertical throw along it is estimated to be 7000 m (McBeath, 1977), placing the top of the Kapuni Group, adjacent to the fault, at least 5000 m below mean sea level. East of the Taranaki fault lies the Patea-Tongaporutu High, an upthrust basement block that separates the Taranaki basin from the late Cenozoic Wanganui basin to the east (Figure 2).

STRATIGRAPHY

The surface geology of Taranaki Peninsula is dominated by Quaternary andesitic volcanoes and associated lava, ash, and laharic deposits (Neall, 1979) that mask the underlying Upper Cretaceous-Recent sedimentary succession. The nature and extent of the Upper Cretaceous and lower Tertiary rocks were unknown prior to drilling. The oldest sedimentary rocks known in the basin are altered breccias, probably of Early Cretaceous (Albian-Aptian) age, drilled by the offshore Tasman-1 well. In the north, Tangaroa-1 bottomed in Cretaceous basalt with a radiometric age of 128 ($\pm$ 4 Ma) (Shell, BP and Todd Oil Services Ltd., 1981). The Lower Cretaceous sedimentary rocks, probably lateral equivalents of the Hawks Crag Breccia, that crop out in Buller Gorge (Figure 1), South Island (Nathan et al., 1986), are believed to be unprospective and are considered economic basement.

The oldest prospective sedimentary rocks known in Taranaki basin belong to the Pakawau Group (Figure 3), a Cretaceous-Paleocene sequence of nonmarine conglomerates, sandstones, and coal measures that crops out in the northwest of the South Island and has been drilled in several offshore wells. During the Late Cretaceous-early Tertiary, the Taranaki basin was an extensional basin dominated by normal faults and fault-angle depressions, or half-

Date Submitted: 2/17/88
Date Accepted: 5/4/88

Text continues on p. 271.

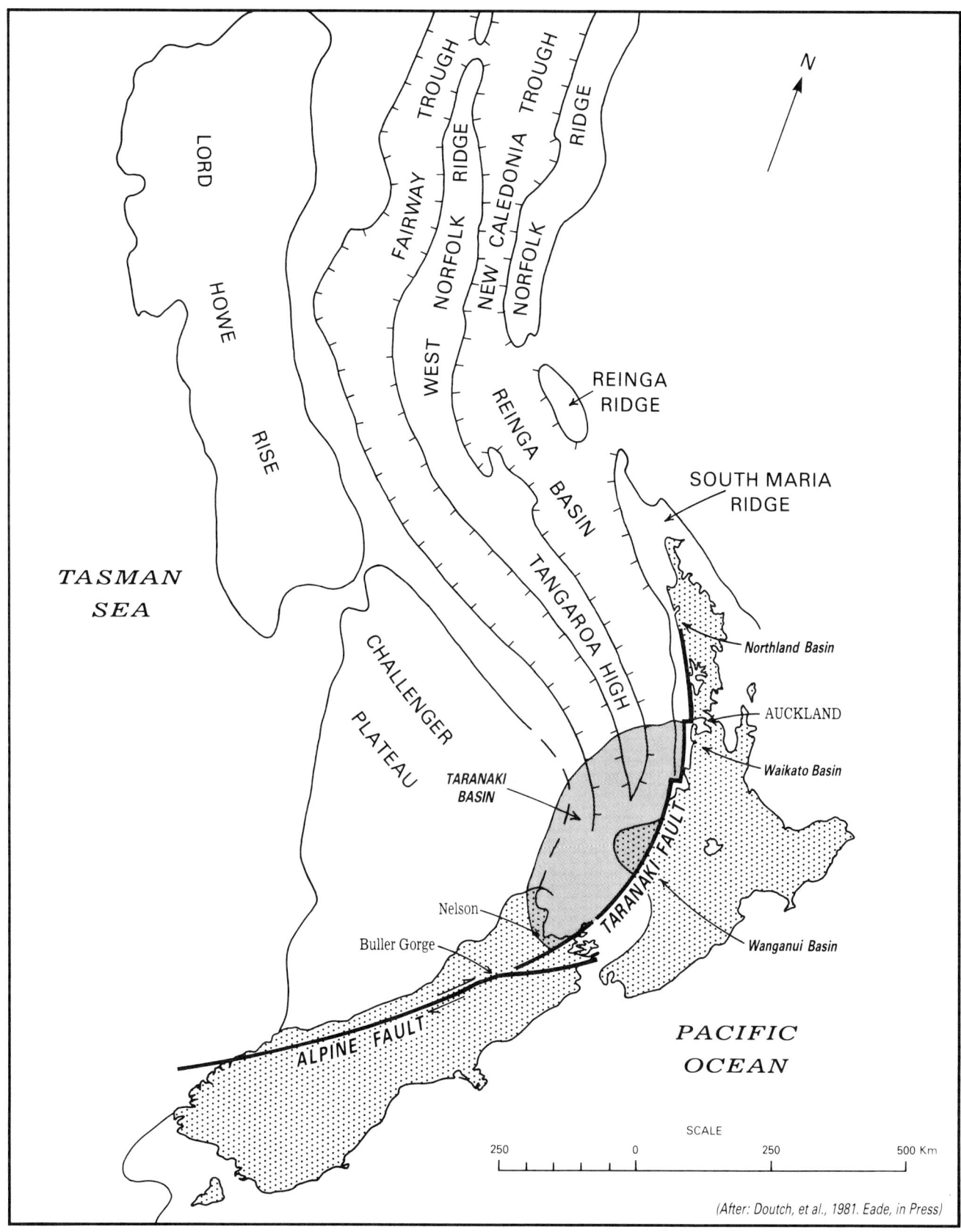

Figure 1. Map showing the main morphological units of the continental shelf west of North Island, New Zealand, and the regional setting of the Taranaki basin.

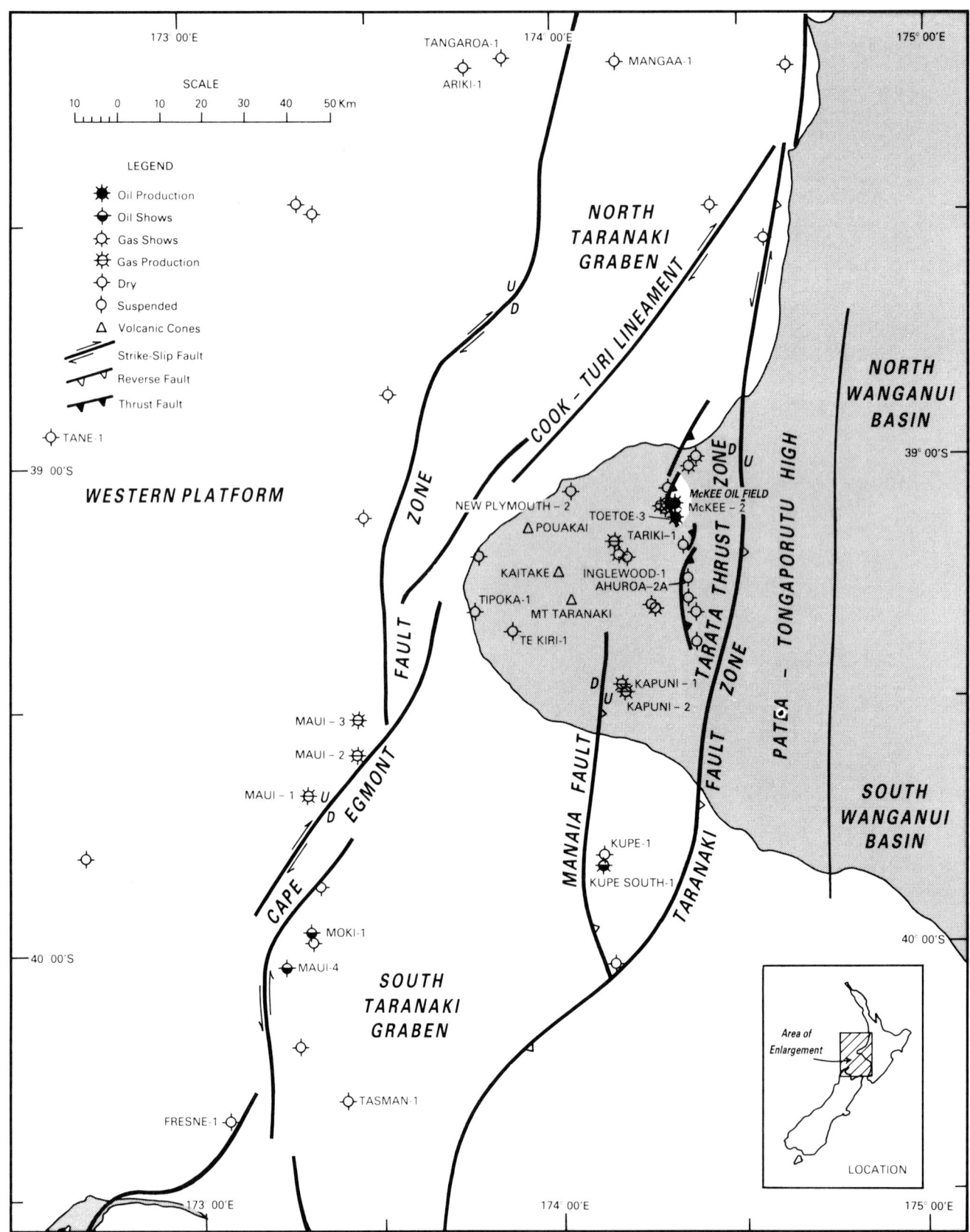

Figure 2. Map of the major structural elements of the Taranaki basin showing the major faults and the location of exploration wells. Wells mentioned in the text are named.

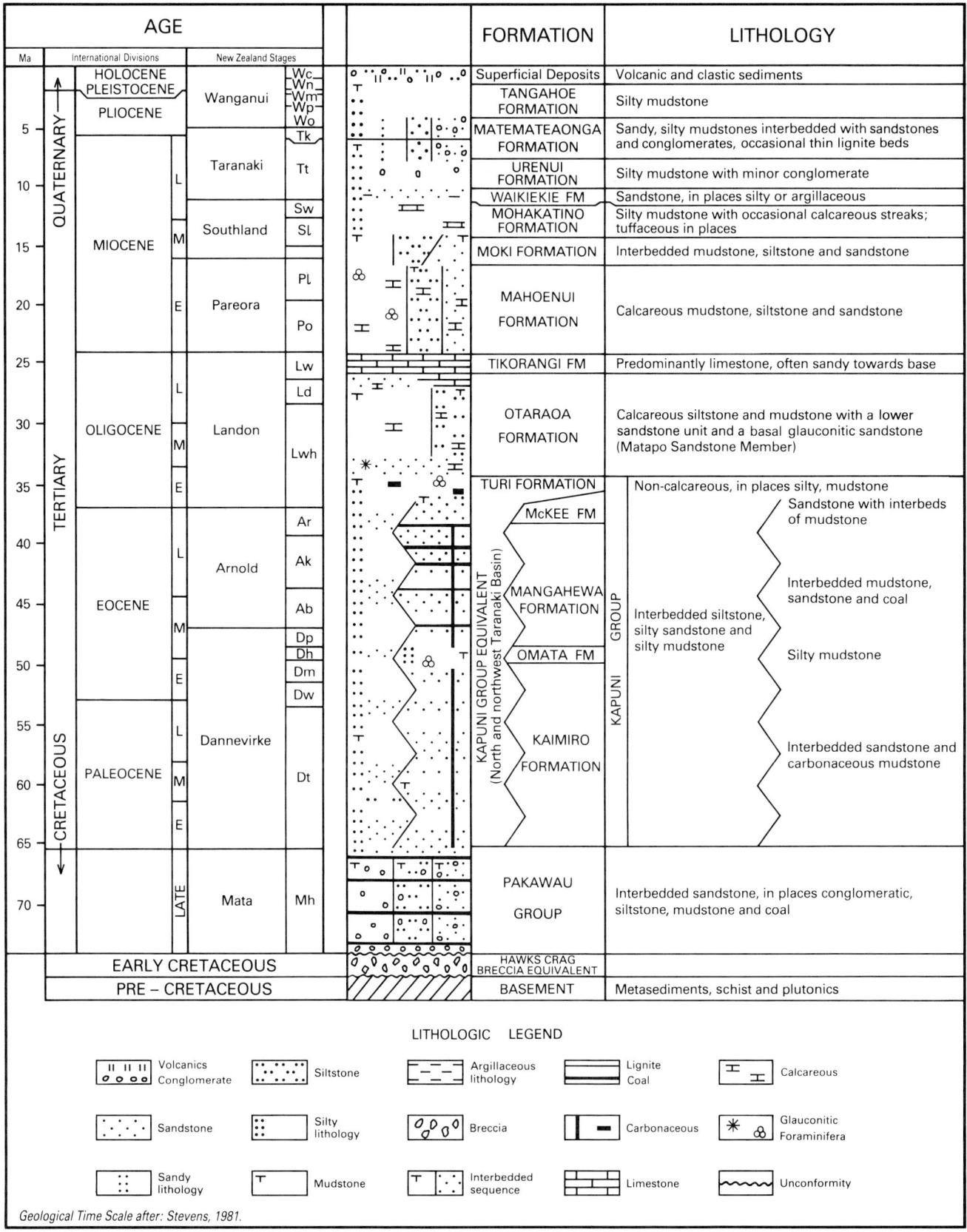

Figure 3. A generalized stratigraphy for the Taranaki basin. The time scale shows international divisions and New Zealand stages (after Stevens, 1981).

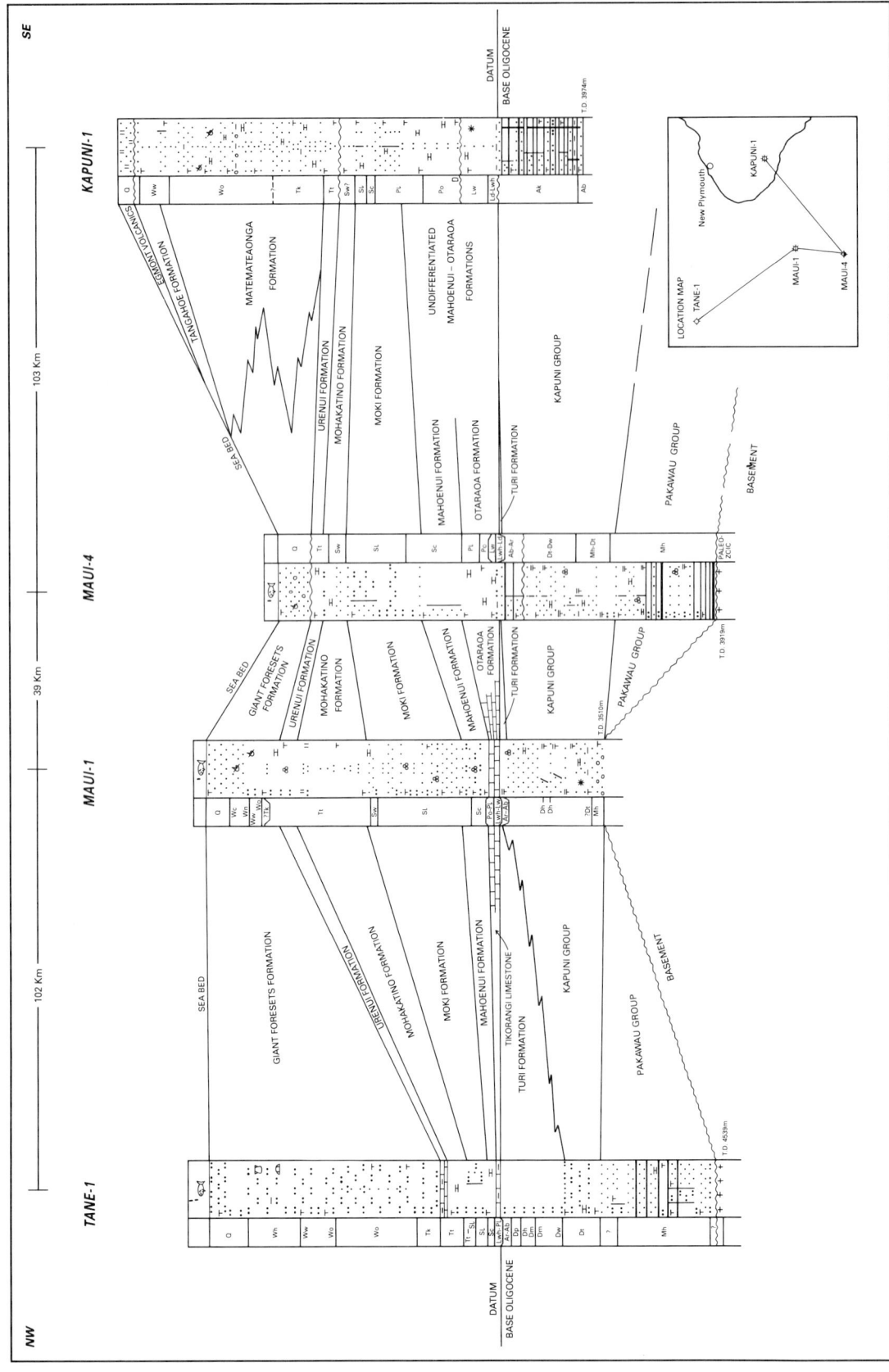

Figure 4. Stratigraphic correlation from Tane-1 (on Western platform) to Maui-1, to Maui-4, and to Kapuni-1 (in Taranaki graben). Datum is the base of the Oligocene.

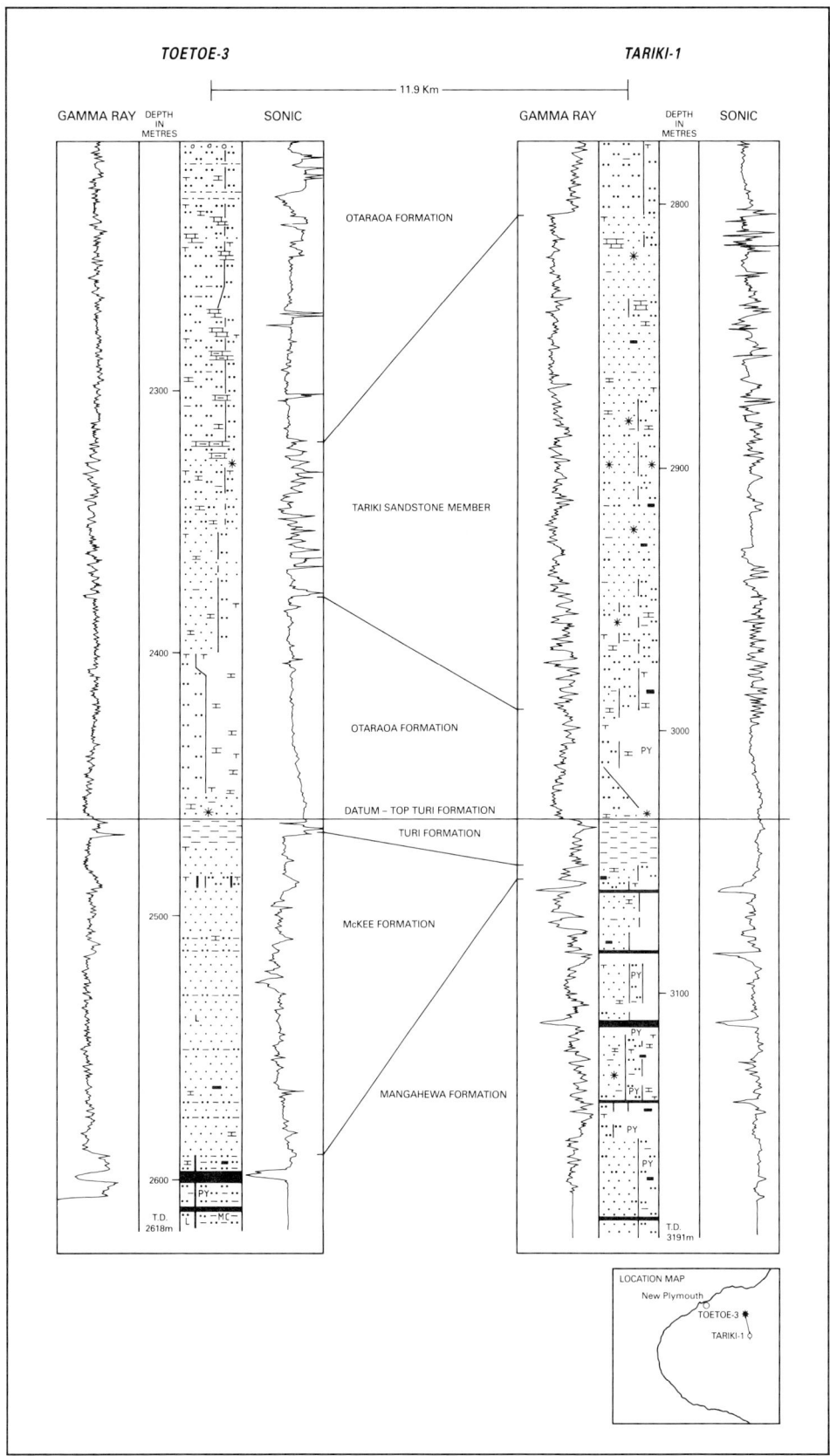

Figure 5. Log correlation from Toetoe-3 to Tariki-1 showing the well-developed Tariki Sandstone Member at Tariki-1 and the rapid thinning of the McKee Formation from Toetoe-3 to Tariki-1.

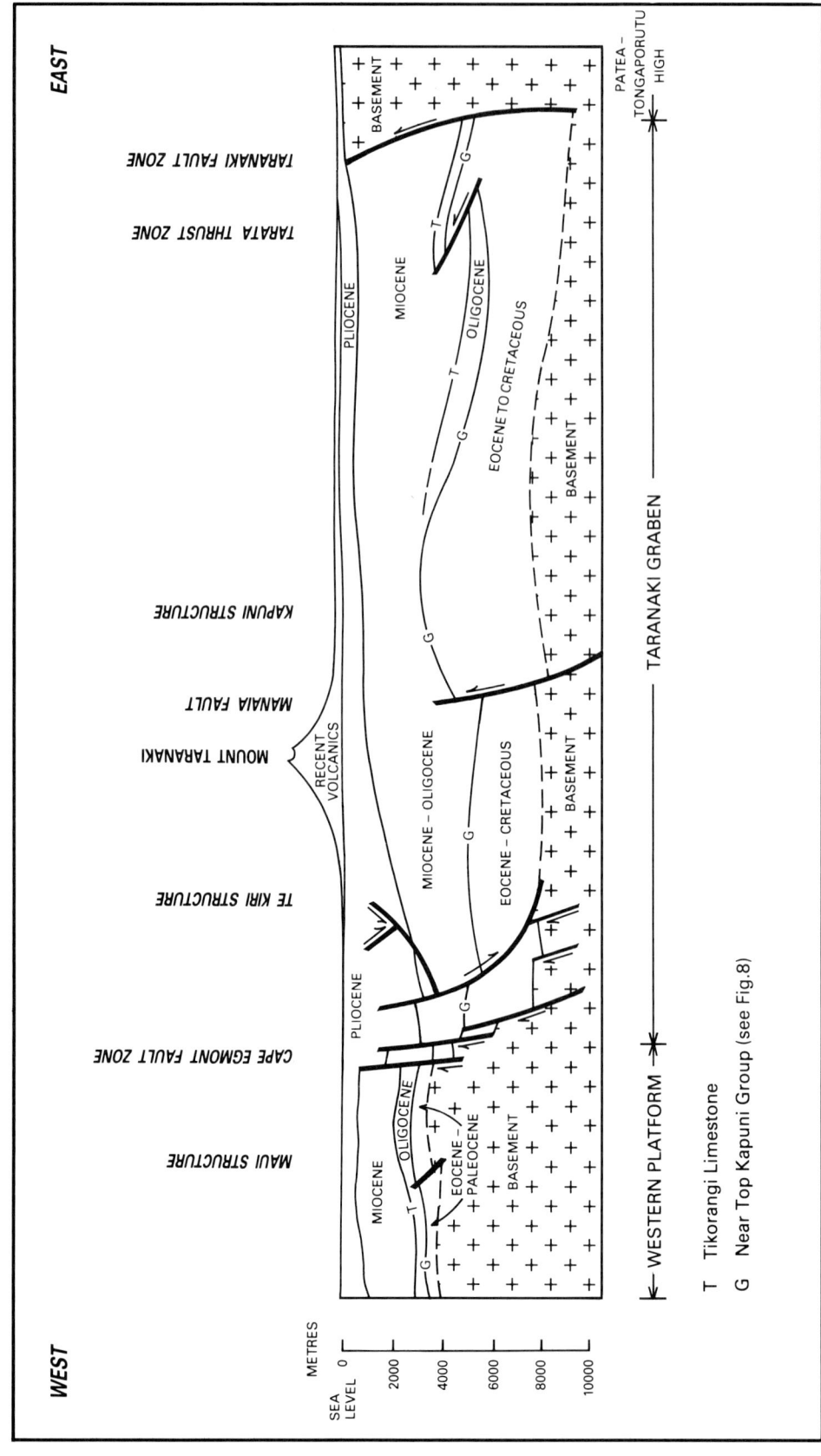

Figure 6. Schematic cross section through Taranaki basin showing the structural style of the traps. The major types of structures are (1) faulted anticlines, represented by the Kapuni and Maui structures; (2) overthrust structures along the Tarata thrust zone, including the McKee structure; and (3) tilted fault blocks, represented by the Te Kiri structure.

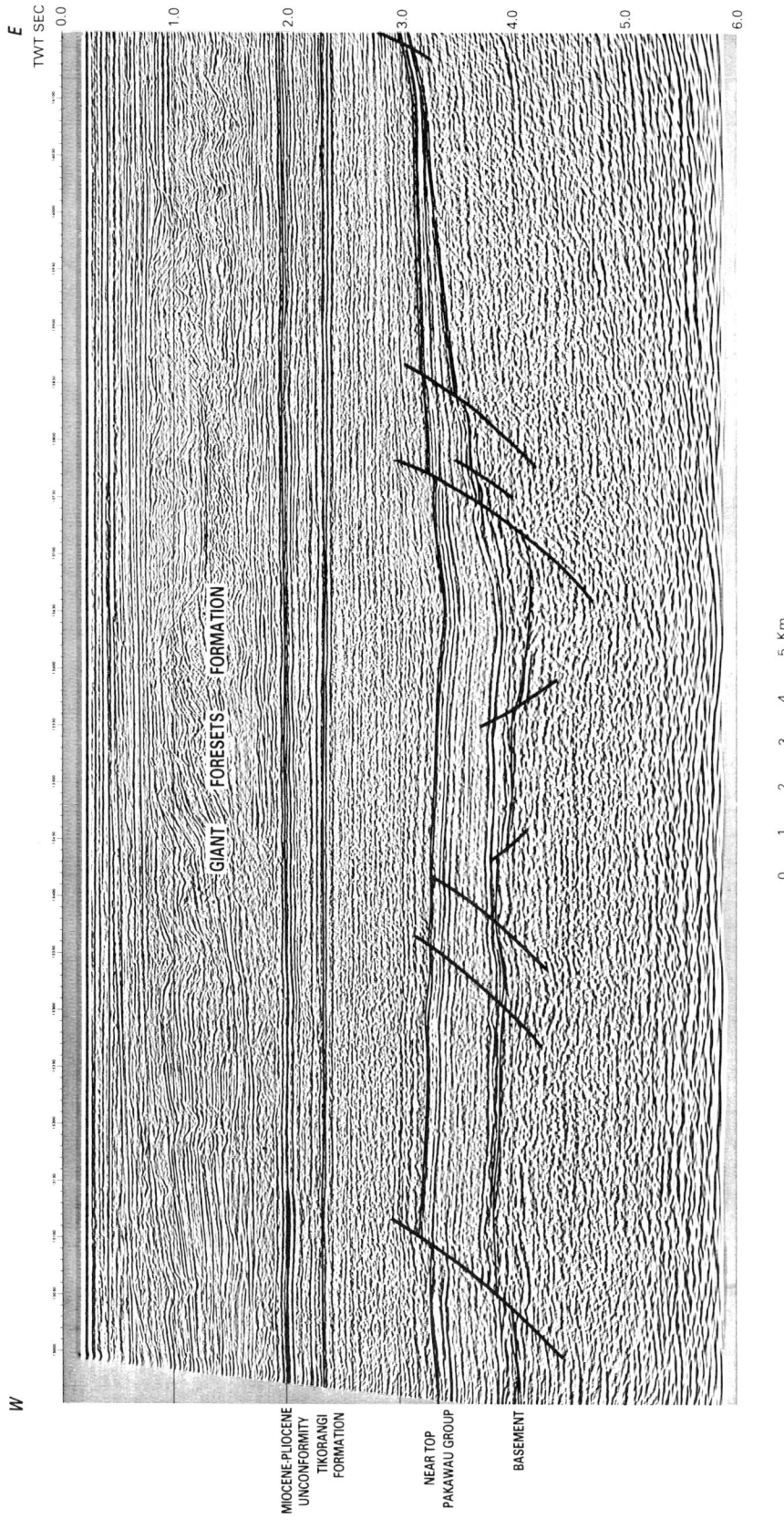

Figure 7. Seismic line across the Western platform showing how the accumulation of the Upper Cretaceous sequence was controlled by the block faulting of basement. Westward progradation of the Giant Foresets Formation is clearly seen in the upper part of the section.

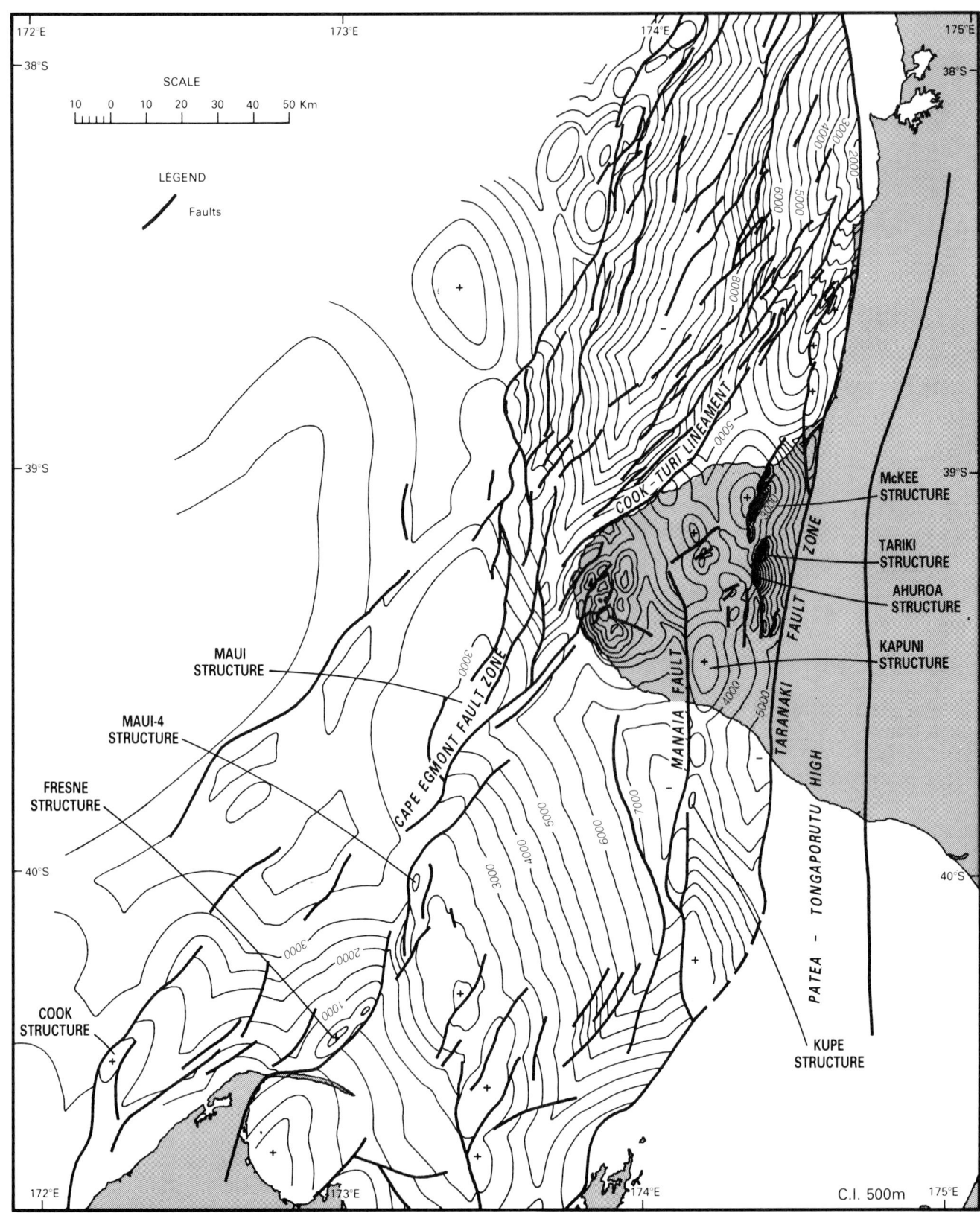

Figure 8. Structural contour map on seismic reflection near top of Kapuni Group, Taranaki basin. The Kapuni Group is the primary drilling objective in Taranaki. The contour interval is 500 m.

grabens (Pilaar and Wakefield, 1978). The half-grabens were the principle depocenters and, in places, more than 2000 m of Late Cretaceous and early Tertiary sediments accumulated (Figure 6). In southwestern Taranaki basin, the Pakawau Group consists of beds of conglomerate and coarse sandstone alternating with beds of sandstone, mudstone, and coal. Northward the succession is more uniform, consisting of fine- to coarse-grained sandstone with thin beds of mudstone and coal. Marine influences are evident in the Pakawau Group in the northernmost wells, and presumably (beyond available well control) the Pakawau Group grades north into an open-marine equivalent.

The Pakawau Group is overlain by the Kapuni Group, a sequence of Paleocene to lower Oligocene sandstone, mudstone, and coal. This group, which is widely distributed across the Taranaki basin, is at least 2000 m thick and accumulated in terrestrial and paralic environments of deposition. Toward the end of the Cretaceous, the sea transgressed southeastward across Taranaki basin (Palmer, 1985). The Kapuni Group is the lateral age equivalent of those transgressive marine siltstones and shales drilled in offshore wells to the north and west. In the northernmost wells, Ariki-1 and Tangaroa-1 (Figure 2), nonmarine Kapuni Group is absent. The top of the nonmarine sequence becomes progressively younger in a southeasterly direction (Figure 4).

Palmer (1985) subdivided the Kapuni Group into four formations. In ascending order, they are Kaimiro Formation, Omata Formation, Mangahewa Formation, and McKee Formation (Figure 3). Subsequent work supports inclusion of the overlying Turi Formation in the Kapuni Group. The Kaimiro Formation is a coastal-plain sandstone and mudstone deposit. It is interrupted by a thin unit of early-middle Eocene marine siltstone and shale (Omata Formation) in the Maui field and Kaimiro-Inglewood areas. The Omata Formation may thicken to the north and northwest and is particularly important in the Maui field where it seals the Kaimiro Formation reservoir, known informally as the D1 Sand (McBeath, 1977) (Figure 4). The overlying Mangahewa Formation is a coal measure sequence that occurs widely across the southeastern part of the Taranaki basin. Deposited in lowland swamps on the coastal plain or in back-beach lagoons, the Mangahewa Formation deposits are coeval with the sandstones and mudstones of the Kaimiro Formation. The two formations interfinger in places. The McKee Formation, which is now known to be restricted to northeast onshore Taranaki, consists of sandstone with minor, wide-spaced interbeds of siltstone, mudstone, and coal. It is interpreted as a coastal shallow-marine deposit.

When conformable with the Pakawau Group (for example, in Maui-4), the base of the Kapuni Group is difficult to define because it has been drilled in few wells and does not crop out. No lithological characteristics are known to distinguish the Kapuni and Pakawau groups; the top of the Pakawau Group is taken arbitrarily as the top of the youngest coal measure sequence of indisputable Late Cretaceous age (Figure 4). In other locations (for example, in Maui-1), the Kapuni Group unconformably overlies basement.

The paralic and nonmarine Kapuni Group is overlain by (and interfingers to the north and west with) the upper Eocene–lower Oligocene Turi Formation, a sequence of marine siltstones and mudstones distinguished by their medium to dark brownish gray color. The marine Turi Formation is the lateral time equivalent of the upper part of the nonmarine Kapuni Group. Inclusion of the Turi Formation in the Kapuni Group means the top of the Kapuni Group means the top of the Kapuni Group is approximately synchronous and is overlain everywhere by a calcareous sequence (Figure 4).

A decrease in clastic deposition during the Oligocene resulted in an increase in calcareous sedimentation and the deposition of the Otaraoa, Tikorangi, and Mahoenui formations. The Otaraoa Formation, which overlies the Turi Formation, probably disconformably, is a sequence of calcareous mudstones, siltstones, and sandstones with a basal glauconitic sandstone, the Matapo Sandstone Member. It is an outer shelf to upper bathyal deposit and is restricted to the Taranaki graben, where it is overlain by the bioclastic bathyal limestone, the Tikorangi Formation. Farther west, on the Western platform, the Otaraoa Formation is absent. Here the Tikorangi Formation conformably overlies the Turi Formation, and the lower part of the limestone is age equivalent to the Otaraoa Formation (Figure 4).

In eastern onshore Taranaki basin, a sequence of quartzose sandstones, informally named the Tariki sandstone (de Bock, personal communication, 1986), occurs within the lower part of the Otaraoa Formation (Figure 5). These recently discovered sands are interpreted as submarine-fan deposits; they constitute the producing horizons in the Tariki and Ahuroa gas-condensate fields.

The Tikorangi Formation is widespread in the Taranaki basin and is a regional seismic marker. The Tikorangi Limestone, which accumulated in a tectonically stable setting, is laterally equivalent to the shallow-water Takaka Limestone of northwest South Island and the Te Kuiti Limestone of Waikato basin (Figure 1). By late Oligocene the transgression had reached its maximum extent. Limestone deposition ceased and was followed by the accumulation of a thick, fine-grained detrital marine sequence.

The Miocene was an epoch of relative sea-level fluctuation. The return to detrital sedimentation, which was predominantly tectonically controlled, commenced with deposition of deep-water calcareous siltstones and mudstones of the Mahoenui Formation. The regression, which began in the early Miocene, continued into the middle Miocene and submarine-fan sediments, collectively known as Moki Formation (Lock, 1985) (Figure 4), were deposited widely across the Taranaki basin.

The succeeding transgressive episode coincided with the initiation of andesitic volcanic activity, which commenced in the northern part of the Taranaki basin during the middle–late Miocene. The Mohakatino Formation, which is tuffaceous in places, overlies the Moki Formation (Figure 3). Mangaa-1 (Figure 2) bottomed in an andesitic volcanic mass, one of several such volcanic masses evident on seismic sections across offshore North Taranaki basin.

In northeast onshore Taranaki basin, the Waikiekie Formation (Pilaar and Wakefield, 1978), a sandy sequence of late Miocene outer shelf and slope sediments, was deposited. Robinson et al. (1986), who correlated the subsurface Waikiekie Formation with the outcropping Mt. Messenger Sandstone, suggested that the influx of coarse sediment occurred in response to a rapid 150–200 m fall in global sea level about 10 Ma. Massive mudstones of the upper Miocene Urenui Formation, which overlie the Waikiekie Formation, mark the last transgressive phase within the Taranaki basin (Pilaar and Wakefield, 1978). By the end of the Miocene, regression ensued; the Urenui Formation was succeeded by a series of prograding strata, the inner to middle shelf siltstones, sandstones, and conglomerates of the Matemateaonga Formation in the east, and the deeper water siltstones and mudstones of the Giant Foresets Formation on the Western platform (Figure 7). Marine deposition continued through the Pliocene until a regional middle Pleistocene unconformity developed, possibly as a result of the combined effects of the volcanism that had migrated as far south as Taranaki Peninsula and eustatic sea-level fall. At about that time, Taranaki Peninsula began to emerge; the Holocene in Taranaki Peninsula has been dominated by emplacement of the volcanic cones Pouakai, Kaitake, and Taranaki (Figure 2).

STRUCTURE

The Taranaki basin is the net result of a complex structural history related to the plate tectonics of the Southwest Pacific. The main structural features of the basin are shown in Figures 2 and 8.

The original structural framework of the Taranaki basin was formed in the Late Cretaceous in response to rifting that opened the Tasman Sea and the New Caledonia and Fairway trough systems. The emergent landmass was dissected by numerous north-trending grabens and half-grabens in which were deposited thick sequences (up to 4000 m) of mainly continental sediments (Pakawau Group).

Sea-floor spreading in the Tasman Sea ceased at 58 Ma (Doutch et al., 1981). Soon after, the Pacific and Austro-Indian plate boundary was propagated through the New Zealand microcontinent (Eade, 1988). From at least the Eocene (45 Ma), the pole of rotation of the Pacific plate relative to the Austro-Indian plate migrated from a point east of New Zealand progressively to the southeast (Figure 9) (Stock and Molnar, 1982; Walcott, 1987). In response, the orientation of the principal horizontal stress (PHS) across the plate boundary changed from approximately north-south (45 Ma) to east-west (present day). The change in PHS has resulted in the progressive modification of the plate boundary from a north-south–oriented transform boundary (45 Ma) to a northeast-southwest–oriented convergent boundary (present day) (Walcott, 1987).

In the Taranaki basin, the change in orientation of PHS from parallel to, to oblique to the preexisting structural grain, caused old normal faults to become reactivated as compressional wrench faults. This reactivation occurred in two phases, with the first phase occurring along the eastern margin of the basin and the second forming the western margin of the Taranaki graben.

The first phase occurred in the early Miocene, when crustal stresses were being propagated into the basin via the Taranaki fault (Bulte, 1987). This caused reversal of movement on the Taranaki fault and several of its splay

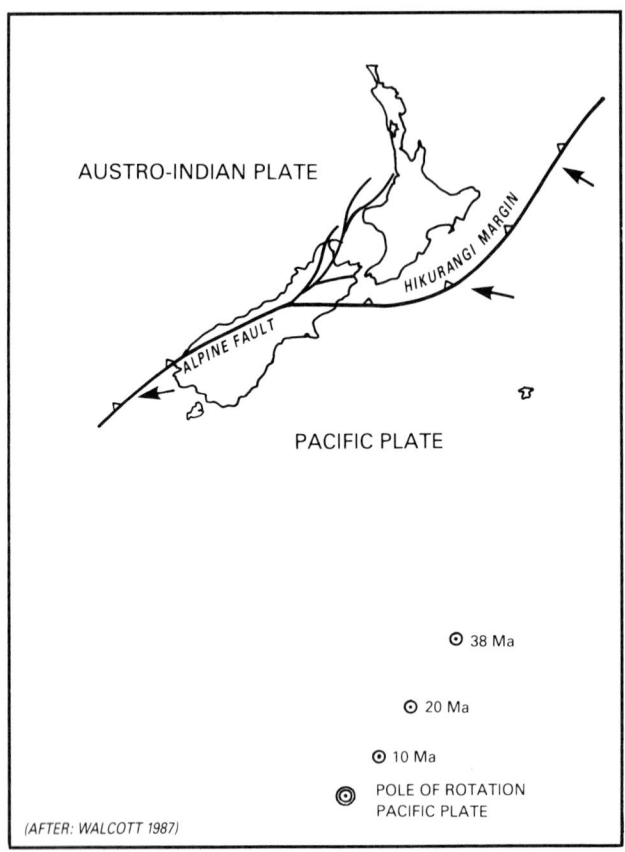

Figure 9. Present-day New Zealand showing the position of the subduction zone and trench (Hikurangi margin), the Alpine fault, and the southeasterly migrating position of the pole of rotation of the Pacific plate.

faults, in particular the Manaia fault and Tarata thrust zone (Figure 8). Reversal of movement on the Manaia fault formed the Kapuni and Kupe structures (Figure 10). Wrench movement, with a large component of compression, along the Tarata thrust zone formed the McKee, Tariki, and Ahuroa structures (Figure 11). These structures are similar in style to "flower structures," but because of the intense compression, the upper portion of the structure has been detached and thrusted, up to 2000 m, over the underlying rocks.

By the late Miocene, the plate boundary had migrated eastward away from the lower North Island, and movement on the Taranaki fault ceased. The same crustal stresses were then transferred to the western side of the basin via faults in and to the west of the South Island, and the second phase of compressional wrenching began. Reversals of throw of up to 4000 m occurred at the restraining bends along the Cape Egmont fault zone. This caused the inversion of Late Cretaceous half-grabens and the formation of the Cook, Fresne, and Maui-4 structures (Figure 12).

By the late Miocene a new structural grain was being superimposed on the North Island; the Cook-Turi lineament was initiated and several northeast-southwest–oriented

Text continues on p. 276.

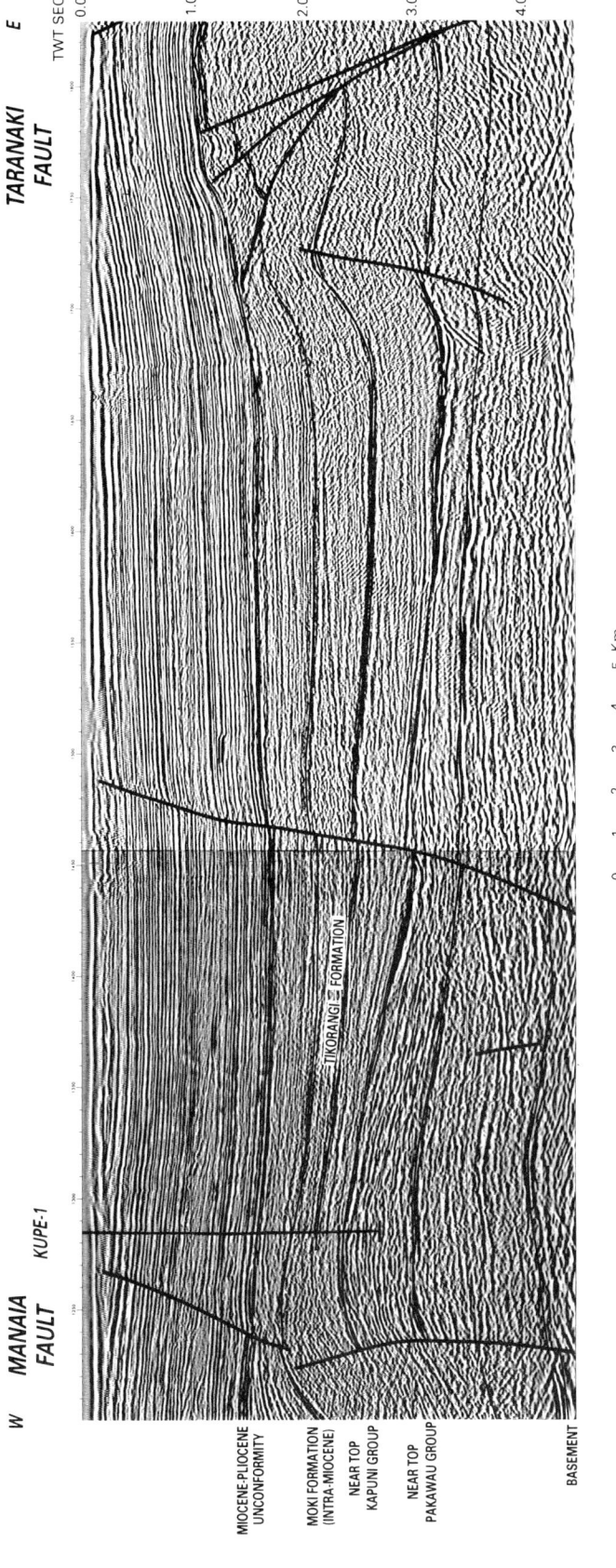

Figure 10. Seismic line through the Kupe structure showing the structural style associated with the high-angle reverse Manaia and Taranaki faults.

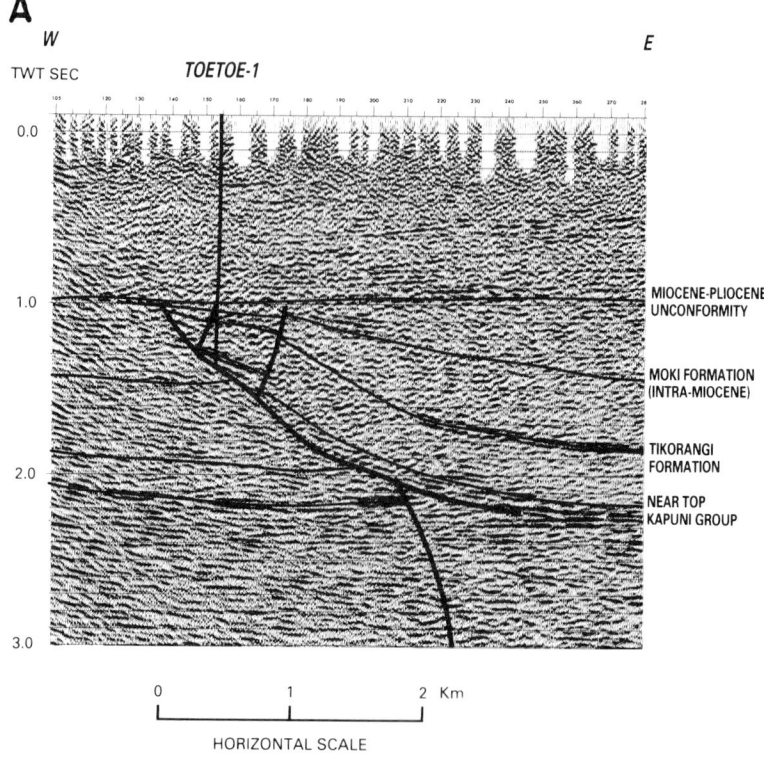

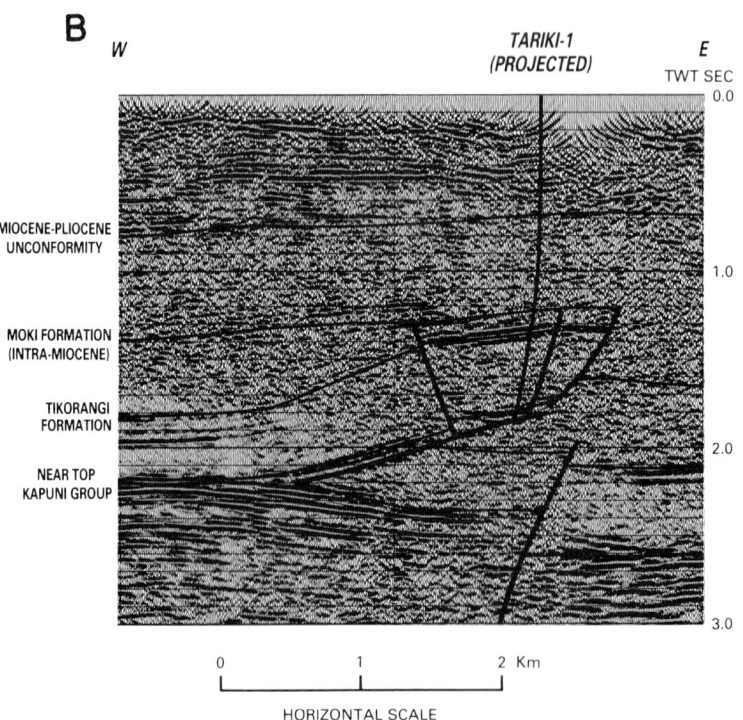

Figure 11. (A) Seismic line through the Toetoe structure (southern part of the McKee field), an overthrust structure that has been thrust to the west. (B) Seismic line through the Tariki structure showing the eastward thrusting of that overthrust structure.

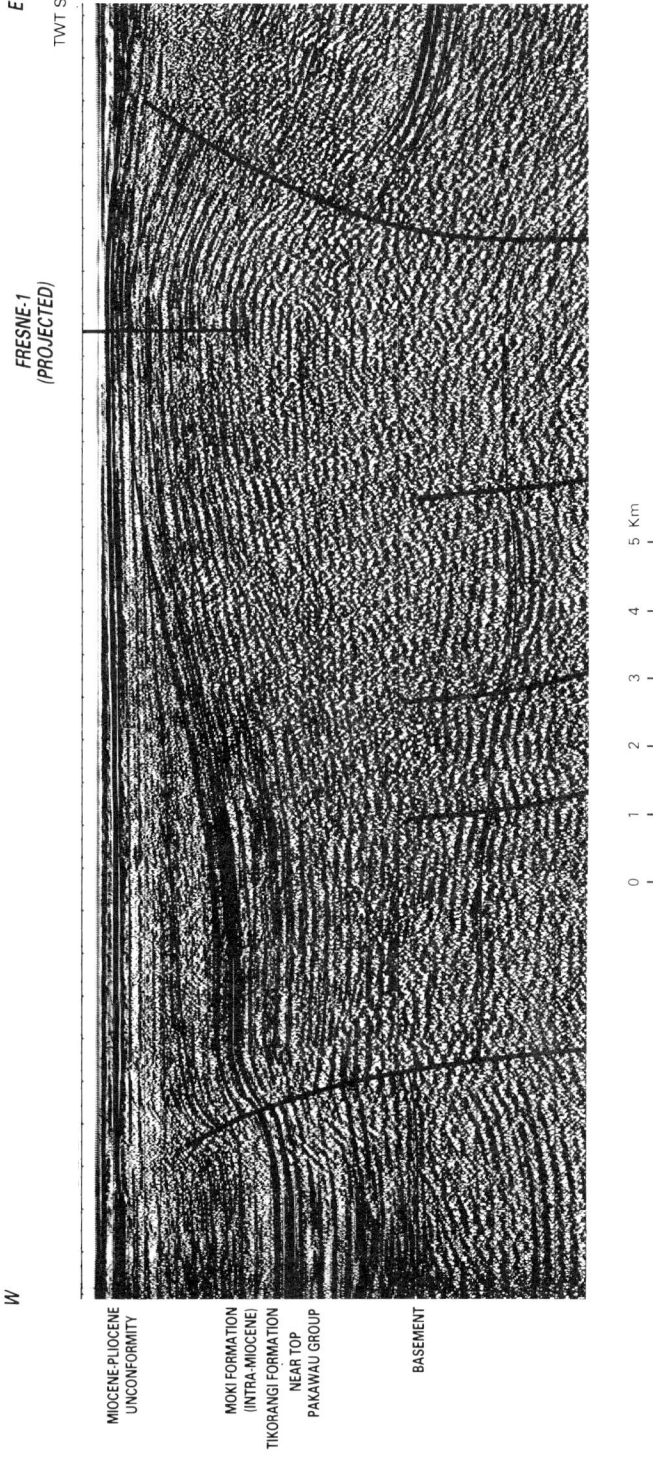

Figure 12. Seismic line through the Fresne structure showing inversion of a Late Cretaceous half-graben.

extensional basins were formed. These basins include the North Wanganui basin and the North Taranaki graben (essentially a subbasin). Andesitic volcanism commenced in these extensional basins during the middle-late Miocene and migrated southward, ending with the Recent cone of Mount Taranaki (Figure 2). The Maui structure was formed in the late Miocene-early Pliocene at the southwest end of the Cook-Turi lineament where it meets the Cape Egmont fault zone (Figure 8).

Late Miocene-Pliocene deformation in the Taranaki graben is characterized by down-to-the-basin, listric, normal faults and associated tilted fault blocks. This structural style dominates onshore western Taranaki (Figure 6), where two such fault blocks were drilled by the Te Kiri-1 and Tipoka-1 wells in 1986 (Figure 2). To date, there is no production from this type of structure.

RESERVOIRS

Sandstones are the only proven reservoirs in the Taranaki basin. Sandstone intervals occur throughout the stratigraphic sequence, but the only producing sequences, and therefore the primary exploration objectives, are the Kapuni Group and the Tariki Sandstone of the Otaraoa Formation. However, the Upper Cretaceous terrestrial Pakawau Group, the Miocene Moki Formation (submarine-fan sands), the Waikiekie Formation (marine sands), and the Miocene-Pliocene regressive Matemateaonga Formation all possess reservoir potential. The Oligocene Tikorangi Limestone is generally very fine grained and, apart from sparse secondary vugs, has no proven reservoir potential.

Core data are relatively sparse and reservoir quality is commonly inferred from drillstem testing. The available data, from both core and testing, indicate that reservoir quality is highly variable, ranging from very poor to very good. Porosities of up to 26% and permeabilities of more than 6 d have been measured. These are maximum values. For the Kapuni Group, average porosity is in the 8-12% range and permeabilities are commonly less than 20 md.

Reservoir quality is primarily controlled by present depth of burial, but it is also related to depositional facies type. For example, core data from Tane-1, Maui field and McKee field indicate that the most porous and permeable reservoirs are well-sorted and coarse-grained sands deposited in high-energy coastal environments (for example, beach and shoreface sands); the nearby terrestrial and lower energy shallow-marine sediments are less porous and permeable.

Postdepositional alteration has modified the sands to varying degrees. To date the Kapuni Group is the only sequence in which reservoir quality has been studied. Hill and Collen (1978) and Hogan (1979) analyzed core data from the onshore Inglewood-1 and Kapuni wells. Besides compaction, they noted that the quartzo-feldspathic sandstones of the Kapuni Group have been considerably modified by diagenesis. The diagenetic features noted include feldspar alteration, carbonate cement, pressure solution, extensive intergranular authigenic clays, and quartz overgrowths. In wells where the Kapuni Group is now more than 3000 m deep, diagenesis has severely reduced primary porosity, and subsequent secondary porosity, which was moderately to poorly developed, has been largely obliterated. Since 1979 several wells have drilled the Kapuni Group in shallower overthrust structures (1800-2400 m deep). In cores of overthrust McKee Formation, authigenic kaolinite is the dominant diagenetic mineral (Figure 13A). Syntaxial quartz overgrowths are rare. Pressure solution of quartz grains is common and carbonate cement is present in varying amounts. Secondary porosity is evident, generally as dissolution of feldspar and quartz grains (Figures 13B, 13C), but it has not greatly enhanced reservoir porosity.

Some core material is available from the Tariki, Waikiekie, and Moki sands, but to date no detailed reservoir studies have been undertaken. The sands are of submarine-fan origin and are commonly very fine grained, argillaceous, and interbedded with mudstone. The limited test data available for the Moki Formation indicate that the sandstones have extremely variable permeability and may be of limited lateral continuity.

SOURCE ROCKS

Geochemical studies have identified the stratigraphic intervals that have yielded the hydrocarbons in the Taranaki basin. Terrestrial source material is indicated by biomarkers (Cook, 1987). The McKee field oil has a characteristically high wax content. It is sourced from coals and carbonaceous shales of the Upper Cretaceous to Eocene Pakawau and Kapuni groups, in which marine influence was negligible. Geochemical studies (Oyen, 1971; Archer et al., 1980; Robertson Research, 1984) show that the upper Eocene Mangahewa Formation coal measures and the Upper Cretaceous Pakawau Group are rich in oil-prone organic material, whereas the Eocene Kaimiro Formation contains a predominance of gas-prone kerogens. The marine shales overlying the coal measures are organically lean and have negligible source potential.

In contrast to Paleozoic coal measures in Europe that have yielded huge amounts of gas but no significant oil, the Cretaceous-Tertiary coal measures of Australia and New Zealand are source rocks for both oil and gas (Tissot, 1984; Shanmugam, 1985). The vegetation dominant at that time included waxy conifers, such as *Agathis*. *Agathis* was present only in Australia and New Zealand during the Cretaceous and Tertiary and is still represented today in New Zealand by *Agathis Australis* (the Kauri) (Shanmugam, 1985). This coniferous vegetation was the source of organic matter rich in exinite macerals, which, when mature, are capable of generating liquid hydrocarbons. The resins, derived from the cuticles of leaves, stems, cones, and fruits, give the terrestrially sourced oil its characteristically high wax content and high pour point (Shanmugam, 1985).

SUBSIDENCE AND THERMAL HISTORY

The Taranaki basin has been subsiding, at variable rates, since the Late Cretaceous. Hayward (1985, 1986a, b) generated uncorrected subsidence curves for New Plymouth-2, Kapuni-2, and Tane-1 by plotting paleo-bathymetry and lithologic thickness against geologic time (Figures 14, 15, 16). The curves show that the Taranaki basin

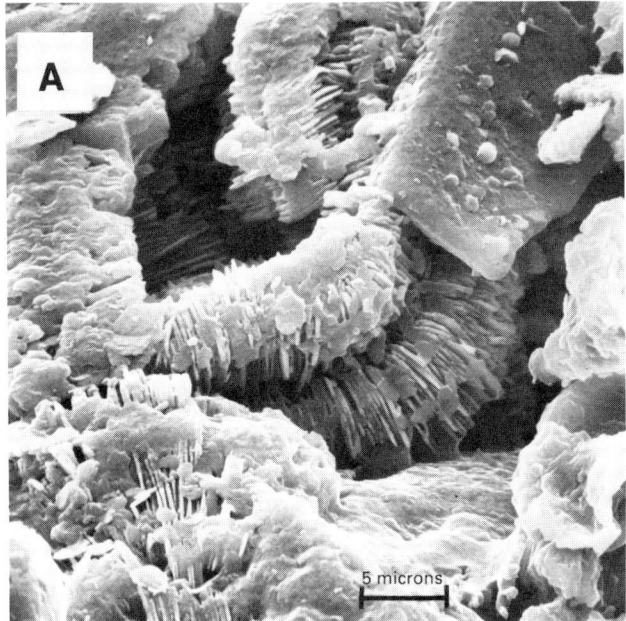
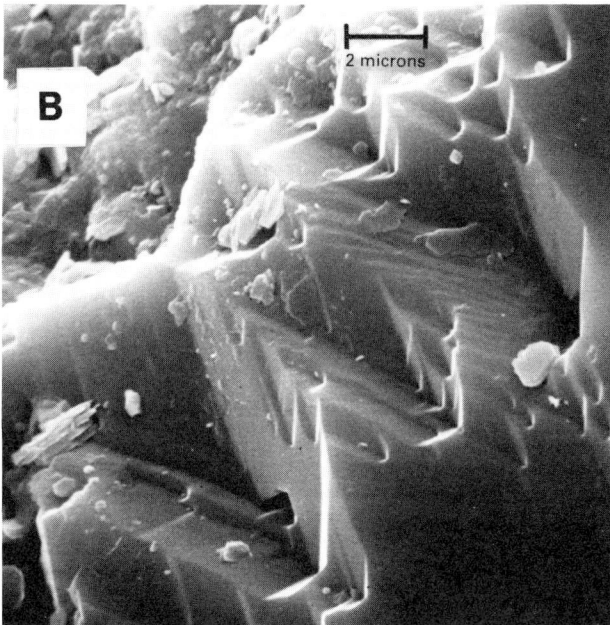
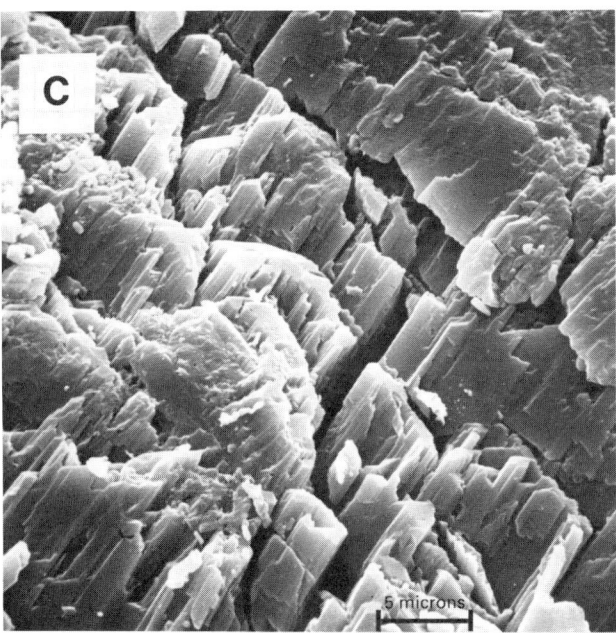

Figure 13. (A) An example of the authigenic kaolinite that typically occurs in overthrust McKee Formation. Sample from 2349 m, McKee-2. Scale represents 5 μm. (B) Solution pits on the surface of a detrital quartz grain in the McKee Formation. Sample from 2356 m, McKee-2. Scale represents 2 μm. (C) Dissolution of potassium feldspar grains occurs preferentially along cleavage planes. Sample from 2359 m, McKee-2. Scale represents 5 μm.

underwent steady subsidence from Late Cretaceous through Eocene time and rapid subsidence from late Oligocene to early Miocene. Subsidence waned through most of the Miocene and resumed again, at a slow rate, in the Pliocene. During the Pleistocene, the basin experienced minor uplift, evidence for which can be observed in shallow-water sequences but not in the deeper marine environments.

The present-day geothermal gradients, calculated from the bottomhole temperature (BHT) measured in wells, are low (<3°C/100 m) (Pilaar and Wakefield, 1984) despite proximity to Miocene–Recent volcanic activity. The

Text continues on p. 281.

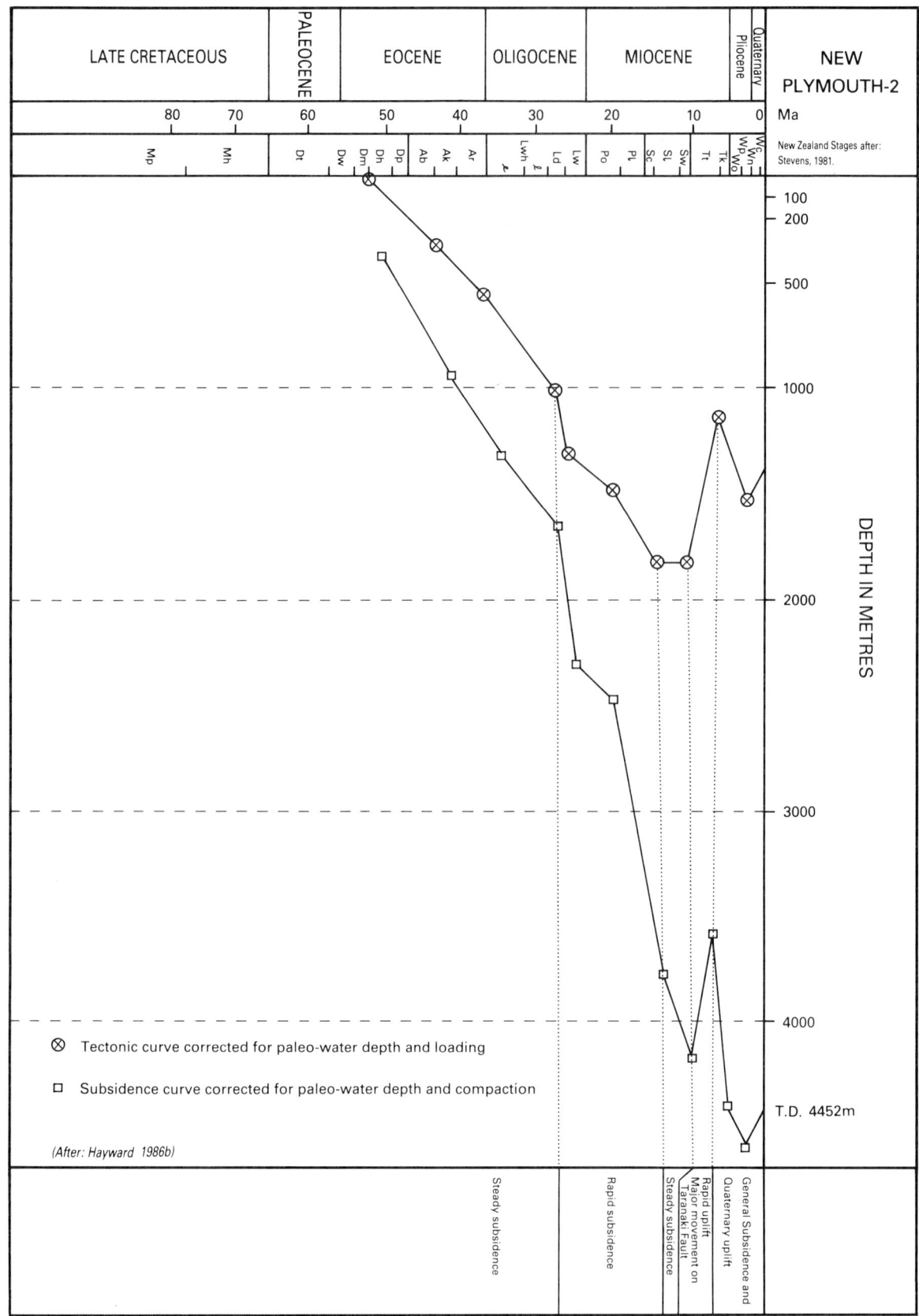

Figure 14. Subsidence and tectonic curves for New Plymouth-2, corrected for paleo-water depths and compaction and, in the latter case, loading. The tectonic curve shows steady to rapid subsidence from early Eocene to late Miocene followed by an uplift event that commenced about 10 Ma.

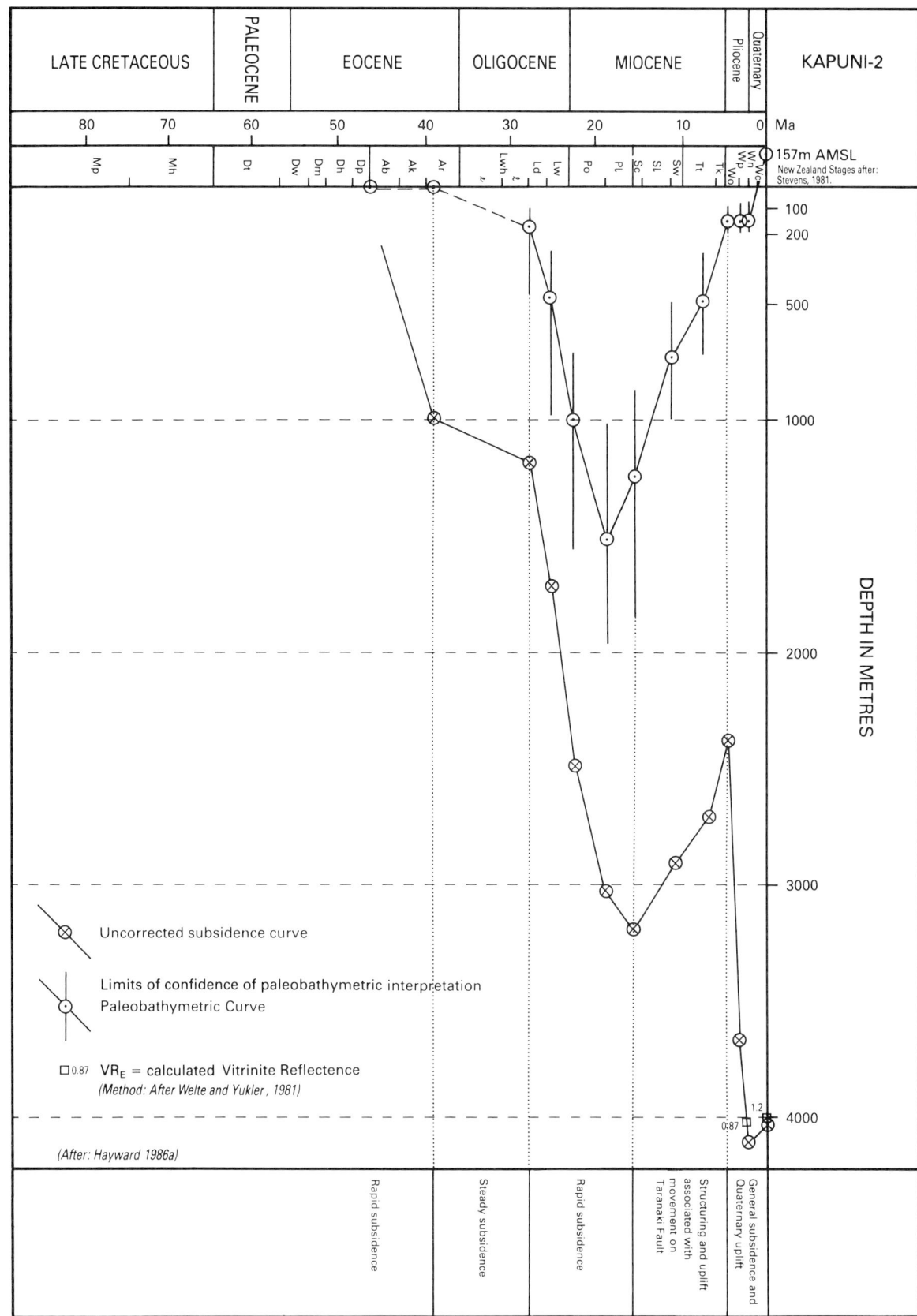

Figure 15. Uncorrected subsidence curve and paleodepth curve for Kapuni-2. The subsidence curve shows steady to rapid subsidence from middle Eocene to middle Miocene (about 16 Ma) when the main deformation and uplift event occurred. Subsidence resumed during the Pliocene.

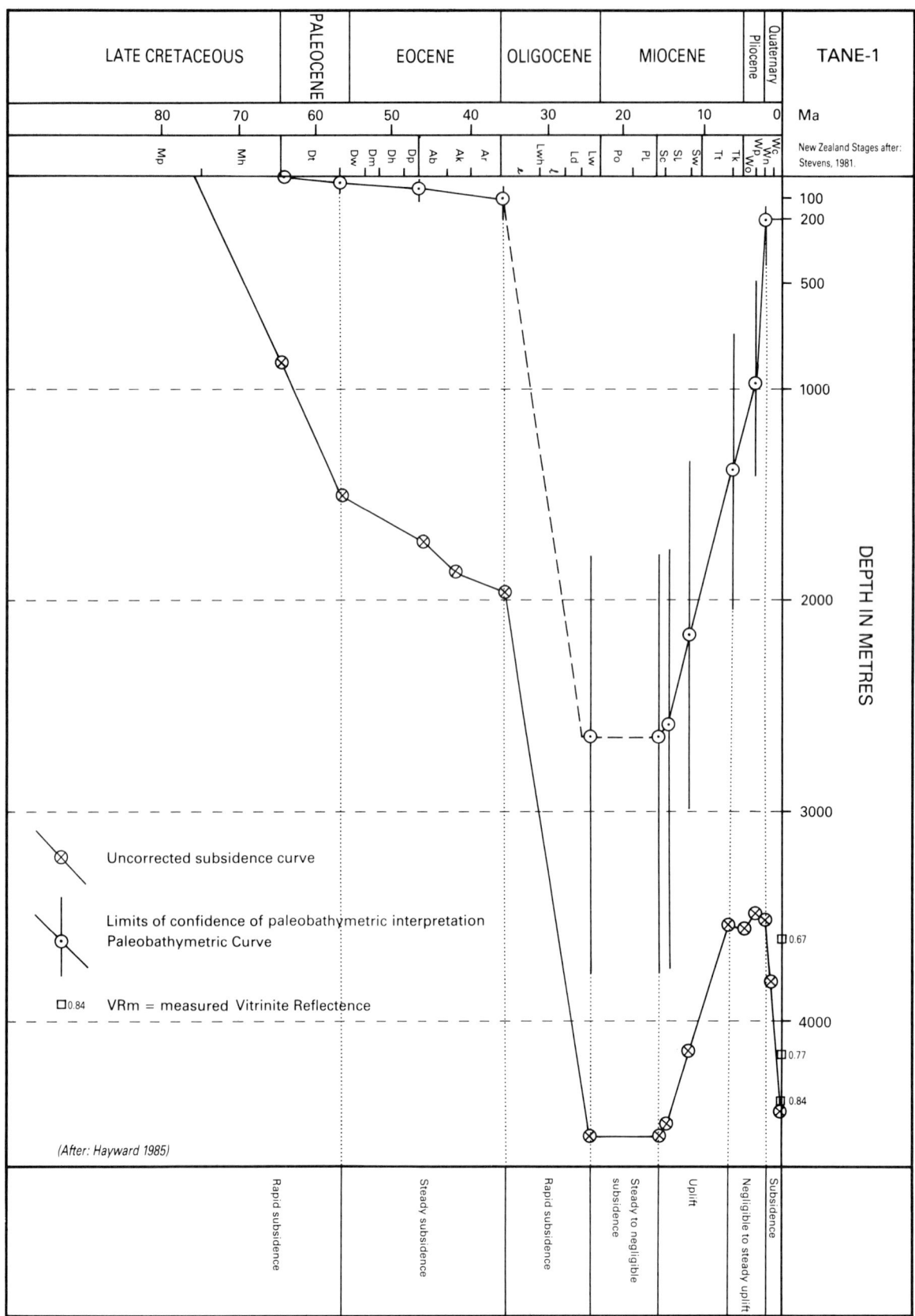

Figure 16. Uncorrected subsidence curve and paleodepth curve for Tane-1. The subsidence curve shows negligible to rapid subsidence from Late Cretaceous to middle Miocene (about 16 Ma) when an uplift event occurred. Subsidence resumed during the late Pliocene.

geothermal gradients calculated from well data are confirmed by heat-flow measurements (Pandey, 1981). The thick, late Tertiary basin-fill sequence has been the major influence on the thermal history of Taranaki, and the volcanism appears to have had little effect.

MATURATION

Geochemical studies of immature potential source material show that the threshold for generation and expulsion of significant quantities of hydrocarbons occurs at vitrinite reflectance (VR) = 1.0% (Saxby, 1982). Saxby's studies show that the oil expulsion window coincides with VR values of 1.0 to 2.0%, gas-condensate with 1.5 to 3.0%, and dry gas with 3.0 to 5.0%. In their study of hydrocarbon generation in Taranaki basin, Pilaar and Wakefield (1984) discussed the importance of time and geothermal gradient to maturation of source material. The subsidence curves for Taranaki basin (Figures 14, 15, 16) show that a large proportion of the sequence was deposited during the late Tertiary. The VR values plotted on the curves show that neither Kapuni-2 nor Tane-1 drilled any significant thickness of mature source rocks. Source material at the base of the drilled Kapuni-2 sequence entered the oil generation and expulsion window at 1.5 Ma. The complete Tane-1 sequence is immature. Applying Lopatin's method (Waples, 1980) to the present-day Taranaki basin indicates that source material in the graben attains maturity at about 4000 m, whereas on the Western platform source rocks are not mature above 4500 m. Pilaar and Wakefield (1984) noted that earliest generation and expulsion of hydrocarbons probably started in the late Miocene in the Taranaki Graben. At every drilled location, the top of Kapuni Group is immature and only those source rocks buried between 4000 m and 4950 m are in the oil generation and expulsion window. Over much of the Western platform generation has only recently commenced (Pilaar and Wakefield, 1984).

MIGRATION

The movement of hydrocarbons through rock is perhaps one of the least understood aspects of petroleum geology, and nothing is known of the mechanisms controlling migration in the Taranaki basin. Source materials drilled in the McKee, Maui, and Kapuni fields are immature. Hence these accumulations have resulted from migration, presumably from a more deeply buried sequence of nonmarine carbonaceous sediments. The migration conduits can only be assumed and probably include faults and bedding planes. The areas adjacent to both the McKee and Maui fields are highly faulted (Figure 8), hence migration along fault planes is highly likely. The Kapuni field is situated updip from a large potential kitchen (Figure 8). Migration into that field may not involve faults.

CONCLUSIONS

The Taranaki basin, a Late Cretaceous–early Tertiary rift basin, was subjected to intense compressional deformation, associated with the propagation of the Alpine fault through New Zealand, during the Miocene. Structures formed during this compressional episode have been major exploration targets since the late 1950s. Fields discovered include the giant offshore Maui gas-condensate field, the onshore Kapuni gas-condensate field, and McKee oil field.

Hydrocarbon production is primarily from the paralic to terrestrial Kapuni Group. In 1986, the Tariki and Ahuroa wells discovered gas-condensate in Oligocene sandstones deposited on submarine fans (Tariki Sandstone of the Otaraoa Formation). This latest discovery highlights the reservoir potential of sandstones within the predominantly marine post-Eocene sequence.

Perhaps the greatest contribution to exploration in Taranaki basin this decade was the discovery of the McKee oil field in 1980. Previously considered a gas-condensate province, the discovery of oil was a major breakthrough. The production of hydrocarbons from high-angle overthrust structures was equally significant, for it identified an additional style of structural trap and confirmed the late generation and expulsion of the hydrocarbons.

The Taranaki basin is not a prolific hydrocarbon producer. Production is hampered by problems associated with reservoir quality and low geothermal gradient, hence restricted kitchen areas. Compressional tectonics, associated with the propagation of the Austro-Indian–Pacific plate boundary through New Zealand during the late Tertiary, were the main structural events. Coincident late Tertiary maturation of much of the source material has greatly enhanced the prospectivity of Taranaki basin.

ACKNOWLEDGMENTS

The writers are indebted to the Petrocorp Exploration for permission to publish this manuscript. Many thanks go to our fellow workers for their valuable and often lively discussion, with particular thanks to Bernice Solomon, Mark Webster, Peter Andrews, and Gail Brooks. The manuscript benefitted from constructive reviews by Michael Fitzgerald, Tom Loutit, and Kevin Biddle.

REFERENCES

Archer, R., S. Thompson, K. H. Ng, 1980, A petroleum geochemical evaluation of the 1,553 to 3,890 metres interval of the McKee-1 well: Robertson Research (Singapore) Private Ltd., Report 849, 33 p.

Bulte, G. A., 1987, Hydrocarbon bearing overthrust structures, eastern Taranaki Basin [abstract]: New Zealand Oil Exploration Conference 1987, New Zealand Ministry of Energy and Pacific Seismic Group, 49 p.

Carter, R. M., and R. J. Norris, 1976, Cainozoic history of southern New Zealand: an accord between geological observations and plate-tectonic predictions: Earth and Planetary Science Letters, v. 31, p. 85–94.

Cook, R. A., 1987, The geochemistry of oils of Taranaki and West Coast region [abstract]: New Zealand Oil Exploration Conference 1987, New Zealand Ministry of Energy and Pacific Seismic Group, 49 p.

Doutch, H. F., G. H. Packham, W. A. Rinehart, T. Simkin, L. Siebert, G. W. Moore, X. Golovchenko, R. L. Larson, W. C. Pitmann III, 1981, Plate-tectonic map of circum-Pacific region southwest quadrant: AAPG Circum-Pacific Council for Energy and Mineral Resources.

Eade, J. V., 1988, The Norfolk Ridge system and its margins, in A. E. M. Nairn, F. G. Stehli, and S. Uyeda, eds., The ocean basins and margins, The Pacific Ocean, v. 7B: New York, New York, Plenum Press, p. 303-324.

Hayward, B. W., 1985, Foraminiferal biostratigraphy and paleobathymetry of Tane-1 offshore well, west Taranaki: New Zealand Geological Survey Report PAL 100, 51 p.

Hayward, B. W., 1986a, Foraminiferal biostratigraphy and paleobathymetry of Kapuni-2 onshore well, south Taranaki, and formation of Kapuni dome: New Zealand Geological Survey Report PAL 108, 39 p.

Hayward, B. W., 1986b, Foraminiferal biostratigraphy, paleobathymetry and subsidence history of New Plymouth-2 onshore well, Taranaki: New Zealand Geological Survey Report PAL 115, 39 p.

Hill, P. J., and J. D. Collen, 1978, The Kapuni sandstones from Inglewood-1 well, Taranaki—petrology and the effect of diagenesis on reservoir characteristics: New Zealand Journal of Geology and Geophysics, v. 21, p. 215-228.

Hogan, J. A., 1979, Stratigraphy and sedimentology of the Kapuni Formation, Taranaki, New Zealand [unpublished MSc thesis]: Wellington, New Zealand, Victoria University, 189 p.

Lock, R. 1985, The distribution, sedimentology and petroleum prospects of the Moki Formation, Taranaki, New Zealand [unpublished]: Petroleum Corporation of New Zealand Limited Report 0296, 37 p.

McBeath, D. M., 1977, Gas-condensate fields of the Taranaki basin, New Zealand: New Zealand Journal of Geology and Geophysics, v. 20, p. 99-129.

Nathan, S., H. J. Anderson, R. A. Cook, R. H. Herzer, R. H. Hoskins, J. I. Raine, and D. Smale, 1986, Cretaceous and Cenozoic sedimentary basins of the West Coast region, South Island, New Zealand: New Zealand Geological Survey Basin Studies 1, 90 p.

Neall, V. E., 1979, Geological map of New Zealand—New Plymouth, Egmont, and Manaia: Wellington, New Zealand, Department of Scientific and Industrial Research, sheets P19, P20, and P21 scale 1:50,000.

Oyen, van, F. H., 1971, Geochemical reports on Tasman-1, Maui-1, Maui-4 and Cook-1: New Zealand Geological Survey Open-File Petroleum Report 966, 33 p.

Palmer, J., 1985, Pre-Miocene lithostratigraphy of Taranaki basin, New Zealand: New Zealand Journal of Geology and Geophysics, v. 28, p. 197-216.

Pandey, P., 1981, Terrestrial heatflow in New Zealand [unpublished Ph.D. thesis]: Wellington, New Zealand, Victoria University, 194 p.

Pilaar, W. F. H., and L. L. Wakefield, 1978, Structural and stratigraphic evolution of the Taranaki basin, offshore North Island, New Zealand: Australian Petroleum Exploration Association Journal, p. 93-101.

Pilaar, W. F. H., and L. L. Wakefield, 1984, Hydrocarbon generation in the Taranaki basin, New Zealand, in G. Demaison and J. M. Roelef, eds., Petroleum geochemistry and basin evaluation: AAPG Memoir 35, p. 405-423.

Robertson Research (U.S.) Inc., 1984, Petroleum geochemical evaluation of the Taranaki basin, New Zealand: New Zealand Geological Survey Open-File Petroleum Report 1022, 252 p.

Robinson, P., A. Sherwood, and B. Morris, 1986, Neogene sediments of the Taranaki region: Geological Society of New Zealand 16th Annual Conference Field Trip Guides, Geological Society of New Zealand Miscellaneous Publication 35B, 158 p.

Saxby, J. D., 1982, Reassessment of the range of kerogen maturities in which hydrocarbons are generated: Journal of Petroleum Geology, v. 5, p. 117-128.

Shanmugam, G., 1985, Significance of coniferous rain forests and related organic matter in generating commercial quantities of oil, Gippsland basin, Australia: AAPG Bulletin, v. 69, p. 1241-1254.

Shell, BP and Todd Oil Services Ltd., 1981, Well resume Tangaroa-1 (offshore): New Zealand Geological Survey Open-File Petroleum Report 793, 82 p.

Stevens, G., 1981, Geological time scale: Geological Society of New Zealand Miscellaneous Publication.

Stock, J., and P. Molnar, 1982, Uncertainties in the relative positions of the Australia, Antarctica, Lord Howe Rise, and Pacific plates since the Late Cretaceous: Journal of Geophysical Research, v. 87, p. 4697-4714.

Tissot, B. P., 1984, Recent advances in petroleum geochemistry applied to hydrocarbon exploration: AAPG Bulletin, v. 68, p. 545-563.

Walcott, R. I., 1987, Geodectic strain and the deformational history of the North Island of New Zealand during the late Cainozoic: Philosophical Transactions of Royal Society of London, v. A321, p. 163-181.

Waples, D. W., 1980, Time and temperature in petroleum formation: application of Lopatin's method to petroleum exploration: AAPG Bulletin, v. 64, p. 916-926.

Welte, D. H., and M. A. Yukler, 1981, Petroleum origin and accumulation in basin evolution—a quantitative model: AAPG Bulletin, v. 65, p. 1387-1396.

Chapter 10

Middle and Upper Magdalena Basins, Colombia

Steven Schamel

Earth Sciences and Resources Institute
University of South Carolina
Columbia, South Carolina, U.S.A.

ABSTRACT

The Magdalena River flows northward across the Colombian Andes traversing a series of en echelon, sediment-filled structural depressions. The Magdalena basins resist easy classification in that, until the late Miocene, they have been parts of much more extensive basins: an extensional, back-arc basin during the Triassic–Jurassic; a pericratonic trough during the Cretaceous and early Tertiary; the inner margin of a broad, east-facing foreland trough during the mid-Tertiary; and more recently an array of intermontaine or "successor" basins. The geologic character of the Magdalena basins is tied intimately to that of the bordering Central and Eastern Cordilleras. Since 1918, there has been nearly continuous exploration activity in the Magdalena basins resulting in the discovery of more than 2.6 billion barrels of oil (GBO) and 2.7 trillion cubic feet (tcf) of gas—more than half of the total oil and about a third of the total gas reserves of the country. As of the end of 1989, the daily production from the basins averaged 143,432 bbl of oil and 182.8 mcf of gas.

The abundant hydrocarbon resources of the Magdalena basins are based on the presence of a thick, organic-rich limestone and shale succession (La Luna or Villeta) deposited in an extensive pericratonic trough along the northwest margin of the Guyana shield during the Cretaceous. In the south, nearer the paleogeographic margin of the trough, shallow-marine sandstones (Caballos and Monserrate) bounding the Cretaceous marine megacycle are the prime reservoirs. To the north, nearer the axis of the trough, Cretaceous sandstone reservoirs are absent and production is almost exclusively from mid-Tertiary molasse deposits. The Magdalena basins contain a wide variety of structural and stratigraphic traps, most developed during or prior to peak of maturation of the Cretaceous source beds. Recent discoveries of giant oil accumulations, such as the San Francisco field, were made in large, hanging-wall anticlines previously considered breached and unproductive. The testing of deeper reservoirs and new structural concepts during the 1980s have resulted in many important discoveries. From the standpoint of hydrocarbon exploration and exploitation, the Magdalena basins are not yet "mature." The potential for additional major discoveries is excellent and it is certain that with improved production techniques current estimates of remaining ultimately recoverable reserves in the producing fields will be revised upward.

INTRODUCTION

The Magdalena River flows from south to north through the easternmost of a series of linear valleys that divide the Colombian Andes into four prominent mountain ranges or cordilleras. Between the lat. 2.0°N and 9.0°N, the river wends its way northward through three en echelon, sediment-filled structural depressions: the relatively small Neiva and Girardot subbasins of the Upper Magdalena Valley and the considerably larger Middle Magdalena basin (Figure 1). Each segment of the Magdalena intermontane trough is separated from the next by a structurally controlled "choke point," located where a major structural trend (Figure 2) obliquely crosses the axis of the trough.

Oil seeps are common features in the Magdalena Valley. Their presence had been recorded by the earliest Spanish explorers. It was the presence of seeps along the Colorado River in the Middle Magdalena Valley that led in 1918 to the discovery of Colombia's first oil field, the Infantas field, in the rich De Mares Concession (Taborda, 1965). Since that time, there has been nearly continuous exploration activity in the Magdalena basins resulting in the discovery of more than 2.6 billion barrels of oil (GBO) and 2.7 trillion cubic feet (tcf) of gas (Ecopetrol, 1990), more than half of the total oil and about a third of the total gas reserves of the country. As of the end of 1989, the daily production from the Magdalena basins averaged 143,432 bbl of oil and 182.8 mcf of gas.

The abundant hydrocarbon resources of the Magdalena basins are based on the presence of a thick, organic-rich limestone and shale succession that was deposited in an extensive pericratonic trough along the northwest margin of the Guyana shield during the Cretaceous. This same Cretaceous succession provides the source for the even more productive Maracaibo and Oriente basins to the north and south, respectively. In the Tertiary, the pericratonic trough was deformed and elevated to form the external ranges of the northern Andes.

The Magdalena basins resist easy classification in that, until the late Miocene, they have been parts of much more extensive basins: (1) an extensional, back-arc? basin during the Triassic–Jurassic; (2) a pericratonic trough during the Cretaceous and earliest Tertiary; (3) the inner margin of a broad, east-facing foreland trough during the mid-Tertiary; and more recently (4) intermontaine or successor basins. The geologic character of the Magdalena basins is tied intimately to that of the bordering Central and Eastern Cordilleras. It is from this point of view that the basins and their hydrocarbon resources will be discussed.

REGIONAL SETTING

The Colombian Andes (Figure 1), which are less than 300 km wide in the south near the border with Ecuador, broaden northward to a maximum width in excess of 600 km at about lat. 7°N. In the north, the belt of late Neogene deformation extends from the Sinú basin, where the Magdalena and North Panama accretionary complexes join, eastward to the foothills of the Eastern Cordillera bordering the Llanos foredeep basin. Within the Colombian Andes are four principal mountain ranges, each separated from the next by long intermontaine troughs through which flow the major rivers of Colombia. The ranges are

1. *Serrania de Baudó or Baudó terrane*, the southward extension of the Panama volcanic arc that developed on a remnant of the Farallon plate and was sutured to South America in the late Miocene (Kellogg et al., 1985; Burke, 1988).
2. *Western Cordillera*, fragments of oceanic crust and deformed deep-marine sediments, possibly the Caribbean oceanic platform (Burke, 1988), accreted to South America during the mid- and Late Cretaceous (Aspden and McCourt, 1986). The process of plate collision leading to suturing involved large-scale dextral strike-slip motion along the Romeral fault, the reactivated continental-oceanic suture.

Date Submitted: 6/5/89
Date Accepted: 10/10/89

Text continues on p. 287.

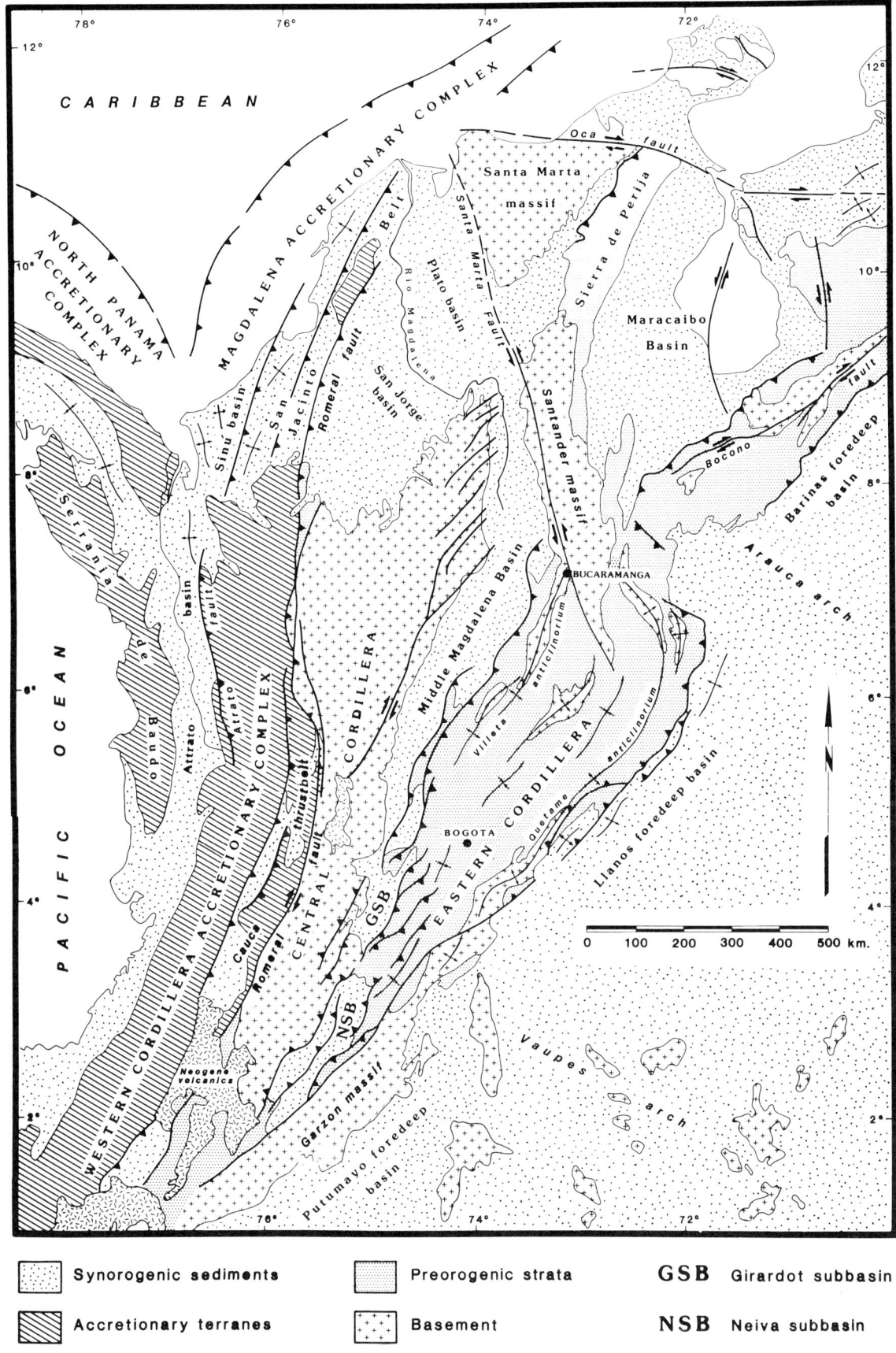

Figure 1. Tectonic map of the northern Andes of Colombia and western Venezuela.

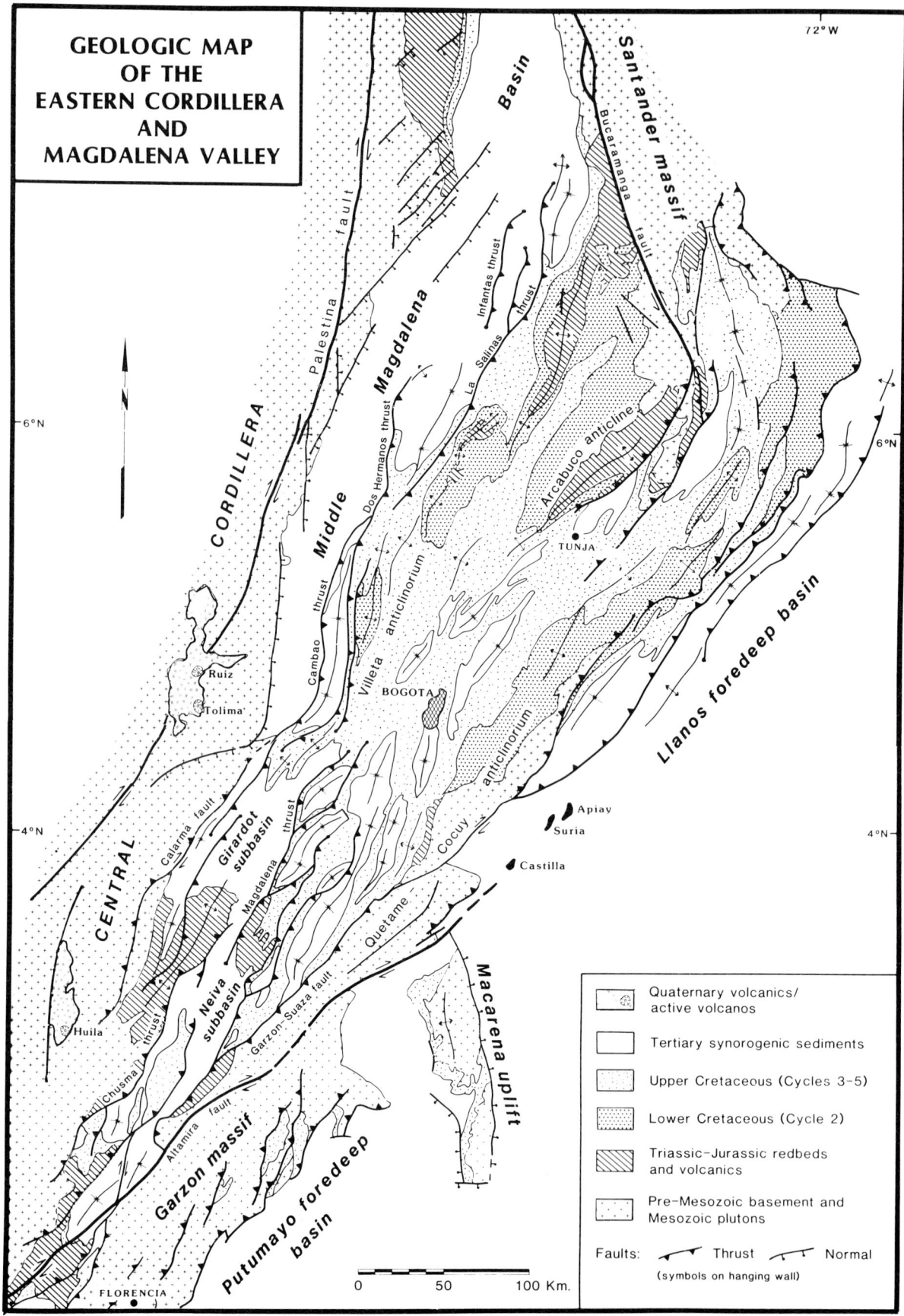

Figure 2. Generalized geologic map of the Eastern Cordillera and Magdalena Valley. Modified after Cediel and Caceres (1988) and Gonzáles et al. (1988).

3. *Central Cordillera*, the outer margin of the South American continental crust: late Proterozoic rocks of the Guyana shield pervasively injected by granitic plutons of Paleozoic (Gonzáles et al., 1988), Jurassic (Mojica and Dorado, 1987), and Late Cretaceous (Aspden et al., 1987) age. From at least the Late Cretaceous, the Central Cordillera has been a positive tectonic feature; it has been the source of coarse clastics since the middle Eocene.
4. *Eastern Cordillera*, a now elevated, thick Early Jurassic-Cretaceous sedimentary basin situated between the Vaupes arch (Garzón massif) on the south and the Arauca arch (Santander massif) on the north (Figure 1). The basin was buried beneath the eastern Central Cordillera foredeep during the Eocene through late Oligocene, but was elevated in the late Miocene-Pliocene (Campbell and Burgl, 1965; Van der Hammen et al., 1973; Lundberg et al., 1986). The Eastern Cordillera is a broad foreland inversion structure thrust both to the east and to the west along the approximate limits of the Cretaceous basin.

The Magdalena basins, which occupy positions between the Central and the Eastern Cordillera (Figure 2), share structures and geologic history with both mountain ranges.

STRATIGRAPHIC SUMMARY

The oldest rocks exposed in the Eastern and Central Cordillera are late Proterozoic polymetamorphosed gneiss, amphibolite, and metasedimentary rocks that represent the northwesternmost limit of the Guyana shield. In the Garzón massif, these rocks have been dated by Rb-Sr whole-rock methods as 1180 Ma (Alvarez, 1981). Distributed across the basement complexes are a variety of Paleozoic rocks, including Cambro-Ordovician metaclastics (quartzite, phyllite, and schist) and Middle Devonian and younger shallow-marine and paralic sedimentary rocks (Mojica et al., 1985; Gonzalez et al., 1988). The younger Paleozoic sedimentary rocks are preserved only as localized remnants; originally they may have been deposited in fault-bounded basins.

Resting with sharp angular unconformity on the underlying cratonic margin rocks is a thick, heterogeneous succession of Triassic-Jurassic continental red beds and silicic to intermediate volcanics (Figure 3). These rocks record (1) the rift-drift separation of northwest South America from continental blocks now repositioned in southern Mexico and Central America (Pindell, 1985; Rowley and Pindell, 1989) and (2) the transition of the region into a Pacific-type marginal basin (Maze, 1984). Packages of sedimentary and volcanic rocks up to 5000 m thick fill an array of northeast-trending rift basins distributed across a broad area between the eastern margin of the Central Cordillera and the Llanos foreland (Mojica and Dorado, 1987). In the Girardot basin, a 700-m-thick, shallow-marine limestone sequence (Payandé Formation) of Late Triassic age (Cediel et al., 1980) precedes deposition of a volcanic succession more than 2000 m thick in which marine strata are interbedded with continental welded tuffs, flows, and agglomerates (Saldaña Formation). The Saldaña and equivalent volcanics are merely the effusive expression of pervasive plutonism affecting large segments of the region during the Jurassic. As a consequence of the widespread plutonism, the Triassic-Jurassic succession is part of the mechanical basement; from the standpoint of hydrocarbon resources, it is also the economic basement.

Everywhere along the Magdalena River and in the Eastern Cordillera, the Cretaceous succession (Figure 3) begins with a transgressive, basal, fining-upward quartzarenite and conglomerate unit—the Yavi and Caballos sandstones in the Upper Magdalena Valley (Beltrán and Gallo, 1968; Mojica and Macia, 1983) and the Tambor or Arcabuco Sandstone along the Middle Magdalena Valley (Taborda, 1965; Galvis and Rubiano, 1985; Renzoni, 1985). Generally, the basal sandstones rest unconformably on the underlying rocks. The sandstones pass upward into an open-marine limestone and black shale sequence that is about 800 m thick in the south and over 4000 m thick along the eastern margin of the Middle Magdalena Valley. The Cretaceous succession ends in the Upper Magdalena Valley with the renewed deposition of shallow-marine quartzarenites (Monserrate/Guadalupe Formation), followed by variegated paralic siltstones and continental red beds of the Maastrichtian-Paleocene Guaduala Group. In the Middle Magdalena Valley, the comparable units (Bueno, 1971) are gray shales (Umir Shale) that pass upward into variegated mudstones and fluvial sandstones (Lisama Formation).

The Cretaceous sequence represents a marine megacycle (Gonzáles et al., 1988) that extended through the entire Cretaceous and into the Paleocene. Within the megacycle, Macelleri (1988) recognized five separate transgressive-regressive cycles that can be correlated broadly to the Vail sea-level cycles (Haq et al., 1987). The younger four cycles are represented in the stratigraphy of the Upper and Middle Magdalena valleys (Figure 3). Beginning in the earliest Cretaceous, the transgressive basal sandstone gradually spread outward from a basin center located in the central and northeast portion of the Eastern Cordillera (Fabre, 1985) to reach maximum flooding during the Turonian-Coniacian (Figure 4). During the Turonian-Coniacian (cycle 4), a marine seaway extended along the entire length of the Magdalena Valley and Eastern Cordillera, partially restricted on the west by the newly emergent Central Cordillera. To the east, a sandy coastal plain developed on the edge of the low-relief Guyana shield. As eustatic sea level dropped at the end of the Cretaceous (cycle 5), the sandy and muddy coastal facies prograded westward and northward across the Cretaceous seaway, eventually pushing marine waters north of the Middle Magdalena Valley by Paleocene time.

The marine Cretaceous succession is overlain by three separate molasse sequences (Figure 3), each of which is related to specific tectonic events in the Central and Eastern cordilleras. These sequences were deposited in similar continental settings adjacent to rising mountain ranges. All contain a variety of alluvial-fan, braided-stream, and lacustrine facies. Clast composition and color of associated mudstones are the principal features distinguishing the molasse sequences.

The Gualanday and Chorro/Chuspas groups (Sequence I) consist of coarse conglomerate and sand packages

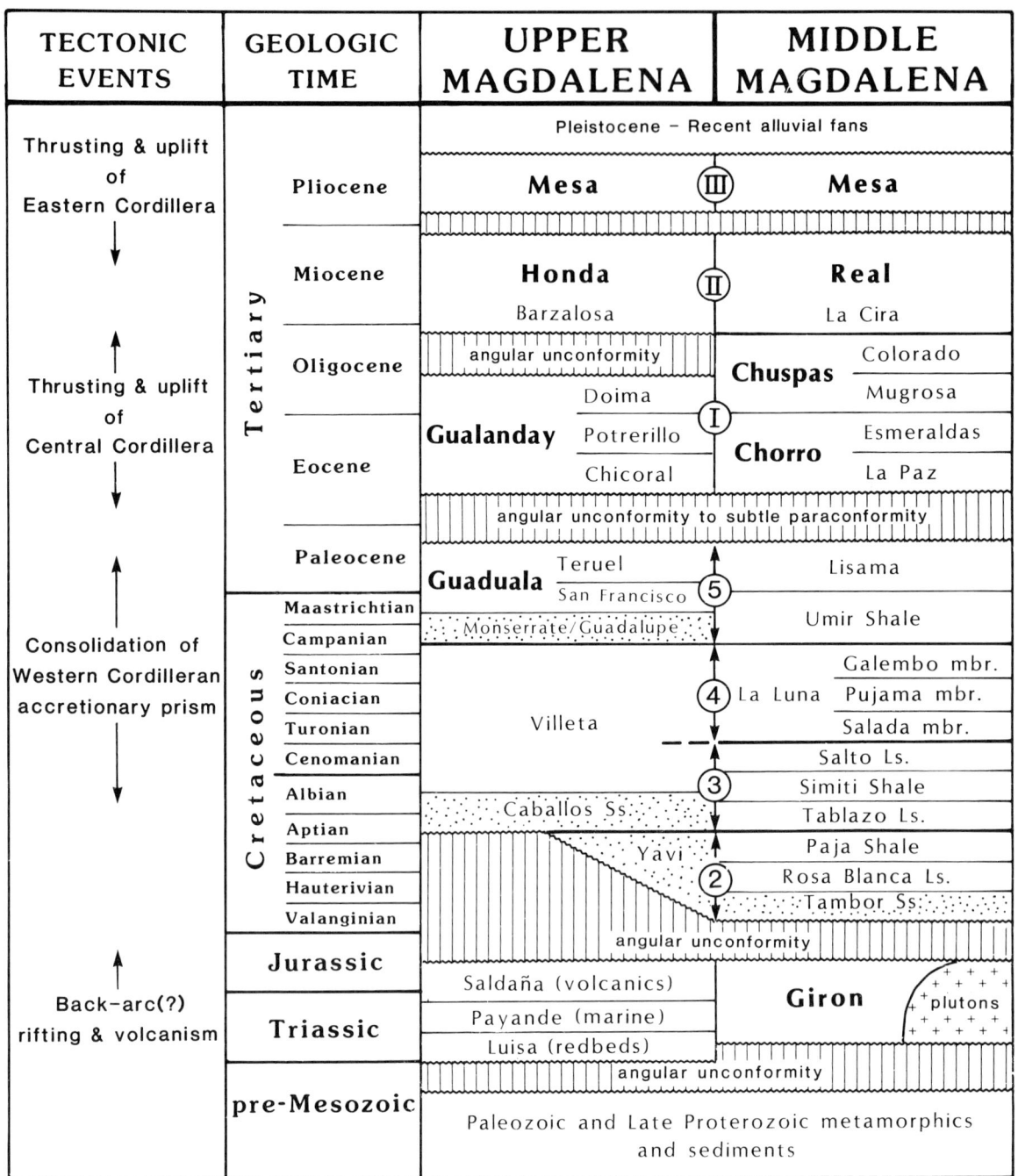

Figure 3. Generalized stratigraphy of the Upper and Middle Magdalena valleys, Colombia. Arabic numerals identify the Cretaceous–Paleocene marine cycles in the Magdalena basins delineated by Macellari (1988); roman numerals identify the various Tertiary molassic sequences.

alternating with red mudstone intervals. The conglomerates in this earliest molasse sequence record the progressive erosional unroofing of the Central Cordillera that occurred as a result of the Eocene–Oligocene deformation and uplift of the range. In the Magdalena Valley, the sequence is a proximal foredeep succession dominated by coalescing alluvial-fan deposits. These rocks originally extended eastward across the Eastern Cordillera as a broad alluvial plain passing into coalescing deltaic complexes (Mirador Formation) near the northwestern edge of the Llanos basin (Bogota, 1988). In the southern Upper Magdalena Valley, the basal Gualanday unit appears to be paraconformable with underlying Paleocene red beds. However, in the western Girardot subbasin (Cediel et al., 1980) and to the north (Morales and the Colombian Petroleum Industry, 1956), the Gualanday/Chorro Group rests with pronounced

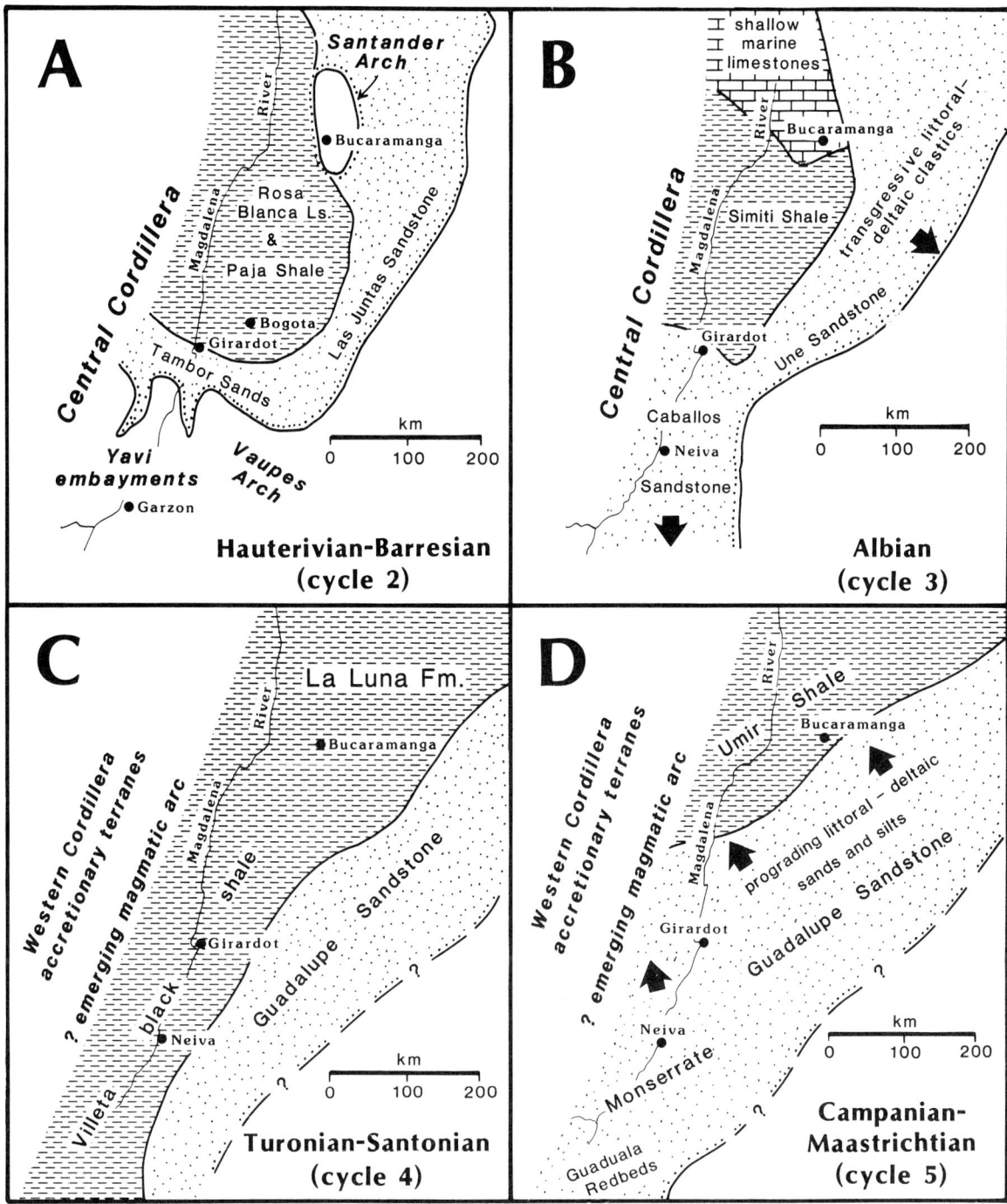

Figure 4. Distribution of sedimentary facies in the Magdalena Valley and Eastern Cordillera during Cretaceous cycles 2 through 5. Compiled from Julivert (1970), Fabre (1985), Macellari and DeVries (1987), and Macellari (1988).

angular unconformity on a variety of older rocks. The Gualanday Group in the Neiva subbasin is a westward thickening succession 1100 to 3000 m thick (Anderson, 1972; Butler, 1983); in the Middle Magdalena Valley, the comparable sequence (Chorro/Chuspas) is well in excess of 4000 m thick (Morales and the Colombian Petroleum Industry, 1956).

The second molasse sequence, the Miocene Honda and Real groups, is dominated by volcanoclastics derived from the Central Cordillera. The sequence records a period of andesitic volcanism that began along the axis of the Central Cordillera in the early Miocene (Van Houten, 1976) and, although less intense, continues today. The Honda Group contains a series of fining-upward cycles in which polymictic debris flows and stream-channel sandstones are succeeded upward by buff-colored floodplain mudstones. This sequence, like that of the Gualanday, was deposited in the proximal portions of a foredeep basin on the eastern flank of the Central Cordillera.

The third molasse sequence, the Pliocene–Pleistocene Mesa Group, developed in response to the uplift of the Eastern Cordillera. It is a varicolored sandstone and conglomerate unit formed by the filling of the Magdalena Valley with coarse debris brought into the basin from both the east and the west across broad alluvial fans and braided-stream complexes. The sequence is very varied and reflects the character of the rocks exposed in the immediately adjacent Cordillera. The unit is rarely more than a few hundred meters thick.

TECTONICS AND STRUCTURAL GEOLOGY

The Magdalena basins (Figure 2) do not constitute a single structural feature separating the Central and Eastern cordilleras. Rather, each basin is a separate structural entity in the broader architecture of the region. The basins are related to structures in both the Central Cordillera, which forms their substrate, and the Eastern Cordillera, which upon uplift transformed them into narrow, successor basins. Following the basins southward (Figure 2), they can be seen to cross obliquely the structural axis of the Eastern Cordillera. The Middle Magdalena basin is located to northwest of the Villeta anticlinorium. Where that anticlinorium dies out to the south, the Girardot subbasin steps in an en echelon fashion southeastward to a position along strike with the broad northeast-trending synclinorium that underlies the high plateau surrounding Bogotá. In turn, the Neiva subbasin is situated farther to the southeast across the general trend of structures in the Eastern Cordillera and is adjacent to the Quetame-Cocuy anticlinorium, the eastern flank of the Eastern Cordillera.

The prominent northeast-trending structural grain (Figure 2) appears to be inherited from the Triassic–Jurassic rifting. The northeast-striking thrusts along the Llanos foredeep arc developed on the Cretaceous hinge zone, a structurally controlled feature coincident with the southeastern margin of the Late Jurassic–Early Cretaceous, fault-bounded Cocuy trough (Fabre, 1987). The Arcabuco anticline (Figure 2), which also parallels the northeast structural grain, appears to be an inverted half-graben. The Lower Cretaceous section within the anticline is considerably thicker and more complete than the comparable section exposed immediately to the east of the bounding, east-vergent thrust.

The dominant style of the structures flanking the Magdalena basins is that of large-scale, paired asymmetric synclines and broad, arched, basement-cored anticlines (Figure 5A). This style of structure is observed along the eastern margins of the Middle Magdalena and the Girardot basins and along the western margin of the Neiva subbasin. It is expressed in structures developed during both the Eocene–Oligocene and the late Miocene–Pliocene phases of Andean deformation. That this structural style is not restricted either to a specific place or phase of deformation suggests that it is related to the "mechanical stratigraphy" of the Eastern Cordillera. The pre-Cretaceous mechanical basement is overlain by a 1- to 6-km-thick Cretaceous succession that is dominated by ductile shales. The thick ductile interval is overlain by a several-kilometer-thick molasse succession of interbedded conglomerates, sandstones, and mudstones that appear to respond to thrusting as a single rigid beam unit. Basement-rooted thrusts ramping upward out of the deep brittle-ductile transition zone in the midcrust follow a trajectory (Figure 5B) that passes parallel to bedding within the Cretaceous shales for a distance of as much as 10–20 km before continuing to ramp through the Tertiary beam unit to the surface. The resulting structure is a paired syncline and basement-rooted, fault-bend anticline.

Modifications to this simple model are required where one or several rigid beam units are situated between the mechanical basement and the thick Tertiary beam. This is the situation in the Upper Magdalena Valley, where the approximately 100-m-thick Monserrate/Guadalupe Sandstone is a mechanically rigid unit. A portion of the displacement on the main, basement-rooted thrust is transferred downward into thrust splays in the footwall block that carry Villeta shales out over Guaduala red beds (Butler, 1983). Although duplexes would be equally likely in this setting, they are virtually absent along the Magdalena Valley.

The Middle Magdalena basin (Figure 6) is distinctly asymmetric. The partially eroded rocks of the Central Cordillera dip gently eastward beneath the west-vergent thrusts of post-Real (Miocene) age that form the western boundary of the Eastern Cordillera (Figure 7). The floor to the basin, the Central Cordillera, had been deformed, uplifted, and eroded down to basement prior to deposition of the middle or late Eocene molasse sequence (Figure 8). The down-to-the-east normal faults on the west side of the basin have lead to its designation as a "half-graben" (Taborda, 1965). However, the basin is not the product of extension. Rather, the normal faults are an expression of the interplay of crustal loading by the Eastern Cordilleran thrust sheets and post-Miocene dextral movement along the Palestina fault system (Feininger, 1970) that splays out into the central and northern Middle Magdalena basin (Figure 6). The anticline-syncline pairs relay one another along the east margin of the basin, stepping en echelon eastward to the north. Minor thrusts between the paired folds increase in throw to the north as the outer (westerly) bounding thrust dies out to take up the position as the next outer thrust to the north.

The Miocene-age Girardot subbasin (Figure 9) is developed on the eroded crest of a broad Oligocene-age basement arch, the Pata arch (Figures 10, Section B–B′, 11). This structure is the fault-bend fold carried on the hanging wall of the Chusma thrust, the structure that bounds the Neiva subbasin on the west (Figure 10, Section C–C′). The Neiva subbasin is bordered on the west and/or underlain by a series of pre-Miocene, east-vergent paired synclines–fault-bend anticlines. With respect to the late Eocene–early Oligocene generation of east-vergent thrusts, the Girardot subbasin is in a structural position above the Neiva subbasin (Figure 11). The Middle Magdalena basin, in turn, lies structurally above the Girardot subbasin. The irregular trace of the late Miocene, west-vergent thrusts that bound

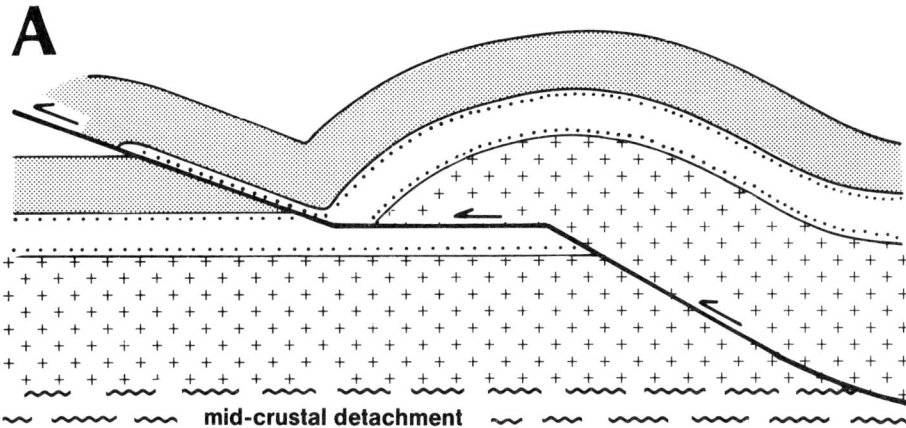

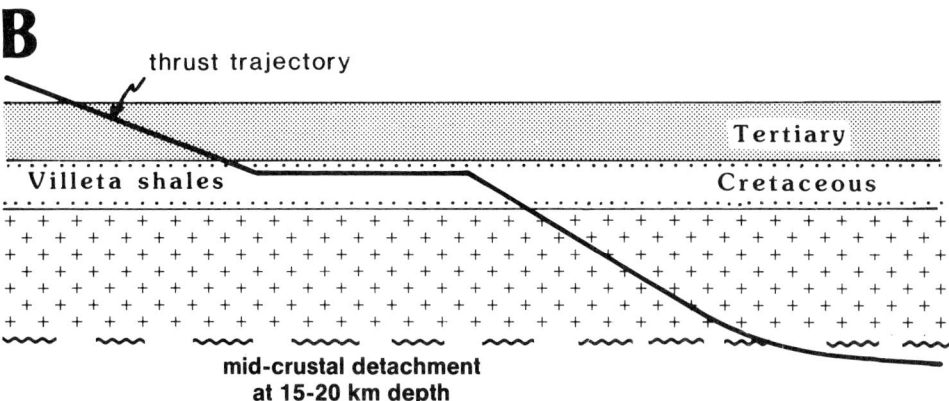

Figure 5. Model for deep basement-rooted thrusting or intra-crustal delamination (Balkwill et al., 1988) resulting in asymmetric paired synclines and basement-cored anticlines, the prominent structural style bordering the Magdalena basins. The mechanically weak Cretaceous section serves as a detachment horizon between the thick, competent Tertiary "beam" unit and the rigid, delaminated basement slab. (A) Generalized cross-section. (B) Restored section showing thrust trajectory.

the Magdalena basins on the east is the result of buttressing by the north-plunging stack of basement-rooted Central Cordilleran thrust sheets. The Neiva subbasin is terminated to the south by the Altamira fault, a post-Miocene dextral fault that obliquely crosses the axis of the Garzón massif and, in part, is responsible for the structural closure in the Castilla, Suria, and Apiay fields in the southwest Llanos basin (Figure 2).

Timing of Deformation

Three major structural events involving the Magdalena basins are recognized in the Tertiary:

1. *Eastward-vergent thrusting along the eastern margin of the Central Cordillera during the middle Eocene–early Oligocene.*

This phase of deformation occurred earliest in the north and closest to the axis of the Central Cordillera. Although not observed directly, this generation of thrusts is inferred in the Middle Magdalena Valley from the pre-Chorro subcrop pattern that indicates that the Central Cordillera was deformed and uplifted prior to the late Eocene (Morales and the Colombian Petroleum Industry, 1956). There the Eocene thrust front may have been overridden by younger, west-vergent thrusts and now may lie buried beneath the western flank of the Eastern Cordillera.

2. *Westward-vergent thrusting along the western margin of the Eastern Cordillera in the late Miocene–Pliocene.* In part, these thrusts extend west across the Magdalena Valley overprinting older Eocene–Oligocene thrusts in the northwest Girardot subbasin (Figure 9). Also, early Oligocene thrusts in the Neiva subbasin are reactivated

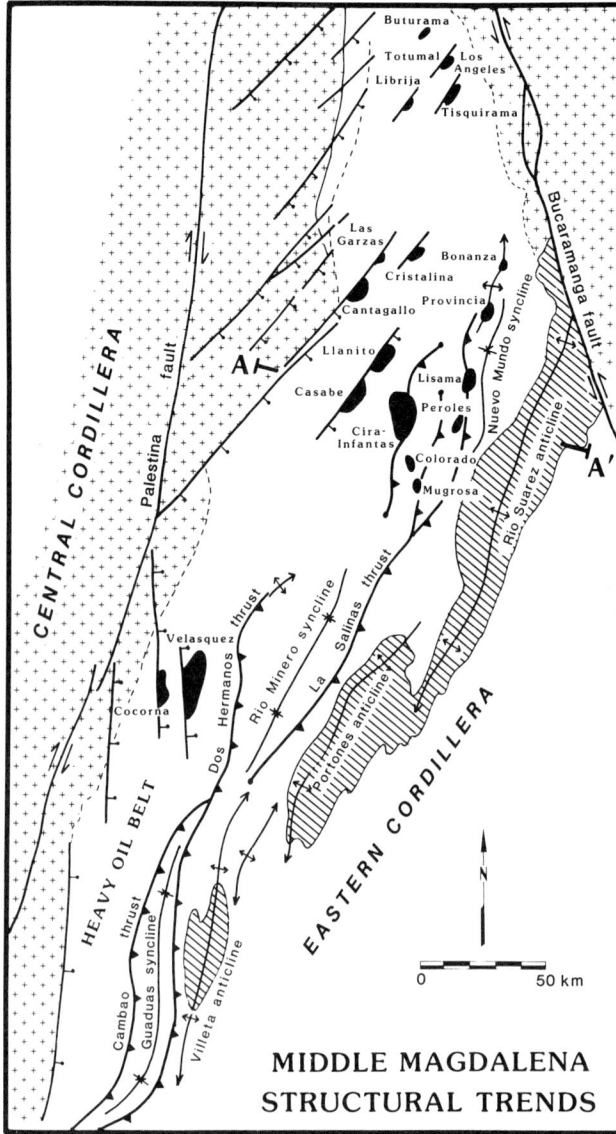

Figure 6. Principal structures and producing fields in the Middle Magdalena Valley. Pre-Cretaceous basement is shown in the cross pattern; Lower Cretaceous (and older) sediments coring the broad anticlines Villeta anticlinorium are shown in the diagonal hatchure. Oil/gas fields are shown in black.

by this thrusting event (Butler and Schamel, 1989; Figure 10, Section C-C').

3. *Probable dextral strike-slip movement during the late Pliocene on a few steep, northeast-striking faults, such as the Altamira fault (Figure 2), that appear to cut and offset the late Miocene thrusts.*

BASIN EVOLUTION

The continental crust underlying the Magdalena basins is late Proterozoic (Guyana shield) in age, but it has been substantially modified at various times in the Paleozoic by thermal, structural, and plutonic events. The pre-Mesozoic history of the Colombian Andes is still poorly known.

The Mesozoic began with an episode of rifting that was related initially to the opening of the westernmost Tethys and the breakup of Pangea. Continued rifting in the Early Jurassic? occurred in a back-arc basin setting that also was the locus for important granitic and granodioritic plutonism and volcanism through the end of the Jurassic (Mojica and Dorado, 1987). The northeast trend of the early Mesozoic rifts provides a mechanical grain in the pre-Cretaceous basement that was reactivated at various times during the Tertiary deformation of the Colombian Andes.

During the Cretaceous, the locus of plutonism shifted from the area of the Magdalena Valley and Eastern Cordillera to the axis of the Central Cordillera (Aspden et al., 1987). A basin developed in the present region of the Eastern Cordillera that filled ultimately with up to 8000 m of organic-rich shale and limestone. A broad belt of shallow-marine and paralic sands bordered the basin on the east and south; a comparable belt of sands may have developed across much of the elevated Central Cordillera during the Early Cretaceous (Rodriguez and Rojas, 1985), but subsequent uplift and erosion has destroyed most of that record. During the Late Cretaceous eustatic sea-level highstand, the basin expanded southward and eastward so that a broad seaway lay immediately east of the Central Cordillera in the area now occupied by the Magdalena basins and the Eastern Cordillera. It was during this time that the important source beds were deposited. This episode closed in the Paleocene as sea level dropped and the clastic shoreline facies prograded northwestward out across the Cretaceous marine basin.

The tectonic evolution of the Colombian Andes has taken place symmetrically about the Central Cordillera. The late Eocene–early Oligocene, east-vergent thrusting along the eastern margin of the Central Cordillera is mirrored by west-vergent thrusting and uplift at about this time in the San Jacinto belt (Duque-Caro, 1984) and initial deformation of the west-vergent Cauca thrust belt (Alfonso et al., 1989). Likewise, the late Miocene uplift (inversion) of the Eastern Cordillera is mirrored by west-vergent thrusting in both the Cauca and Atrato (Sinú) basins (Figure 1).

The Magdalena basins evolved during the Tertiary first by accumulation of 1500 to 4000 m of westward-thickening, proximal foredeep sediments immediately east of the rising Central Cordillera (late Eocene) and subsequently by entrapment of a series of narrow successor basins between the western margin of the Eastern Cordillera and the already elevated Central Cordillera. The Eocene–Oligocene age, en echelon thrust margin of the Central Cordillera may have controlled the location of late Miocene thrust ramps along the eastern side of the Magdalena basins.

The Magdalena basins are relatively cold. The mean geothermal gradient for the Middle Magdalena basin is approximately 1.6°C/100 m, with a range from 1.3–2.2°C/100 m. The mean gradient in the Upper Magdalena basin

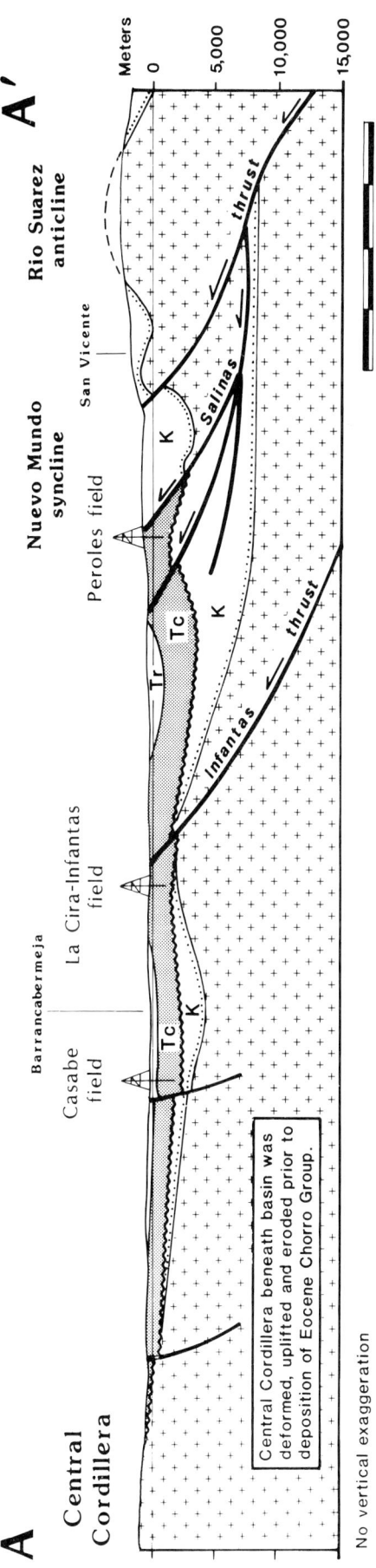

Figure 7. Generalized structural cross section through the northern Middle Magdalena basin illustrating the style of thrusts bordering the basin on the east and the pre-middle Eocene deformation and erosion of the Central Cordillera beneath the basin. See Figure 6 for the location of the Section A-A'. Key to units: Tr = Real Group, Tc = Chorro-Chuspas Group, K = Cretaceous and Paleocene rocks; pre-Cretaceous basement is shown in the cross pattern.

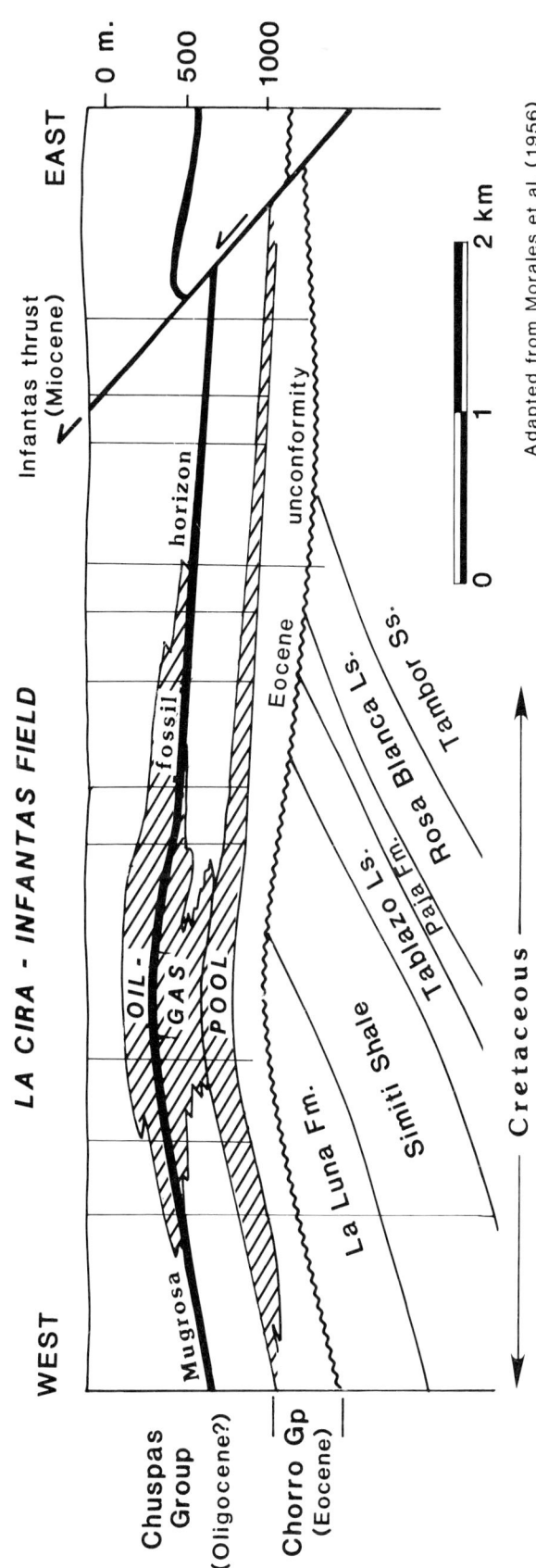

Figure 8. East-west cross section through the giant La Cira-Infantas field, Middle Magdalena Valley. Note the strong angular unconformity beneath the Eocene Chorro Group and the distribution of the producing reservoirs within the Eocene-Oligocene molasse sequence.

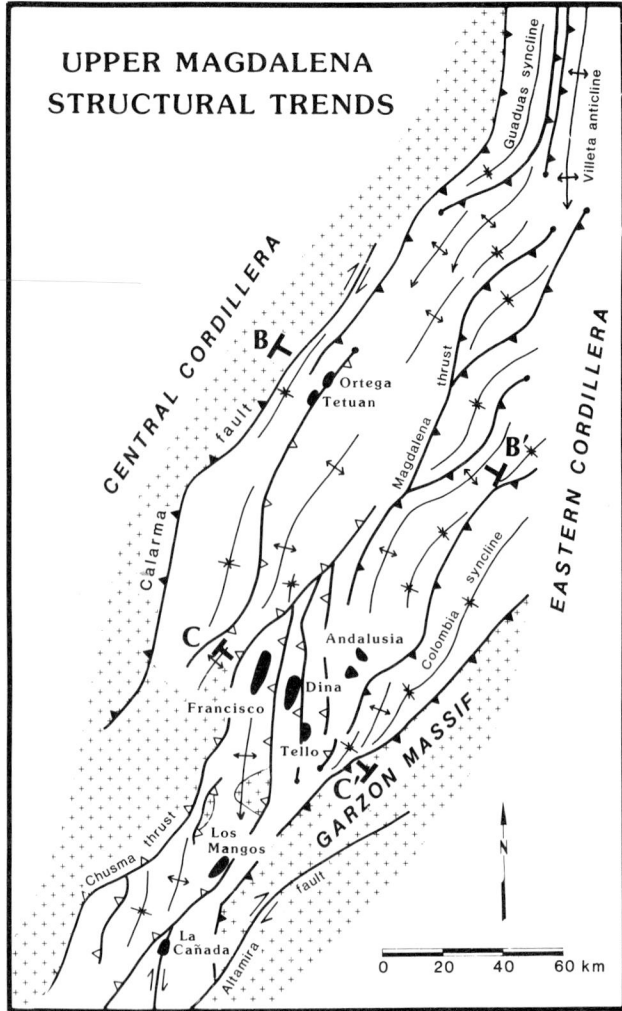

Figure 9. Principal structures and producing fields in the Upper Magdalena Valley. Exposure of pre-Cretaceous basement is shown in the cross pattern. Oil/gas fields are indicated in black.

is 1.7°C/100 m, with a range from 1.1–2.2°C/100 m. These low modern gradients may reflect the rapid infilling of Neogene molassic sediments into the basins.

OIL AND GAS SYSTEMS

As of the end of 1989, there were 45 fields in the Middle Magdalena Valley together producing a daily average of 86,372 bbl of oil and 160.6 mcf of gas. In terms of both discovered recoverable reserves and cumulative production (Table 1), this basin is Colombia's richest petroleum province. Estimates of remaining reserves do not reflect the great probability of additional discoveries nor the certainty of increases due to the eventual adoption of secondary recovery methods. The oldest field in the basin, the La Cira-Infantas field (Figure 8), is also the largest with initial recoverable reserves of 737.4 million bbl of oil and 447.9 billion cubic feet (bcf) of gas (Table 2). Four other fields are comparable in size: Provincia-Payoa, Casabe-Galán, Velasquez, and Cantagallo-Yarigui. Many of the fields in the basin have virtually insignificant reserves.

The Upper Magdalena Valley is a much younger petroleum province; it is still in its early development. The first field, Ortega (Figure 9), was discovered in 1951, and at least a third of the fields have been discovered since 1984. The newly discovered fields have more than doubled the estimates for recoverable reserves for the basin, which were officially 486.7 million bbl of oil and 68.2 bcf of gas as of the end of 1988 (Table 3; Ecopetrol, 1990). Very recent discoveries and further evaluation of the new fields may increase these estimates by at least 100 million bbl. As of the end of 1990, 20 fields in the Upper Magdalena Valley were together producing a daily average of 57,060 bbl of oil and 22.2 mcf of gas. An additional field in the Girardot subbasin and two fields in the Neiva subbasin are in evaluation and will soon come on line.

Reservoirs

In the Middle Magdalena Valley, the principal reservoirs are fluvial channel sandstones and conglomerates in the Chorro and Chuspas groups (Morales and the Colombian Petroleum Industry, 1956; Taborda, 1965). These Eocene-Oligocene molasse beds are excellent reservoirs, having average porosities in the range 20–25% and permeabilities of 0.5 to 1.0 d (Taborda, 1965). The channel sandstones are sealed by interbedded mudstone and shale. Four fields in the basin also produce from sandstones in the Paleocene Lisama Formation: Provincia, Cantagallo, Cristalina, and Las Monas (Govea and Aguilera, 1985). Although the Miocene Real Group contains voluminous porous sandstones and conglomerates, only minor production occurs from this succession (Bonanza field) owing to the lack of effective seals. The basal Cretaceous sandstone units, such as the Tambor and Arcabuco, have not yet been tested successfully.

In the Upper Magdalena Valley, the principal reservoir units are Cretaceous nearshore and paralic sands of the Campanian-Maastrichtian Monserrate/Guadalupe Formation and the Aptian-Albian Caballos Formation. Porosities as great as 30% are measured in the Monserrate Sandstone; associated permeabilities are in the range of 20–1000 md (Waddell, 1985). The porosity is secondary intergranular, the result of partial dissolution of secondary calcite and silica cements. Sandstone and conglomerate lenses within the Miocene Honda Group serve as an additional reservoir in the basins, but due to poor seals their importance is restricted to the Dina field and a few very small associated fields in the central Neiva subbasin.

Source Rock and Maturation

Since the earliest years of exploration in the northern Andes (Hedberg, 1931), geologists have been so certain that the source of hydrocarbons in the region is the thick succession of Cretaceous marine shales and limestones that there have

Text continues on p. 299.

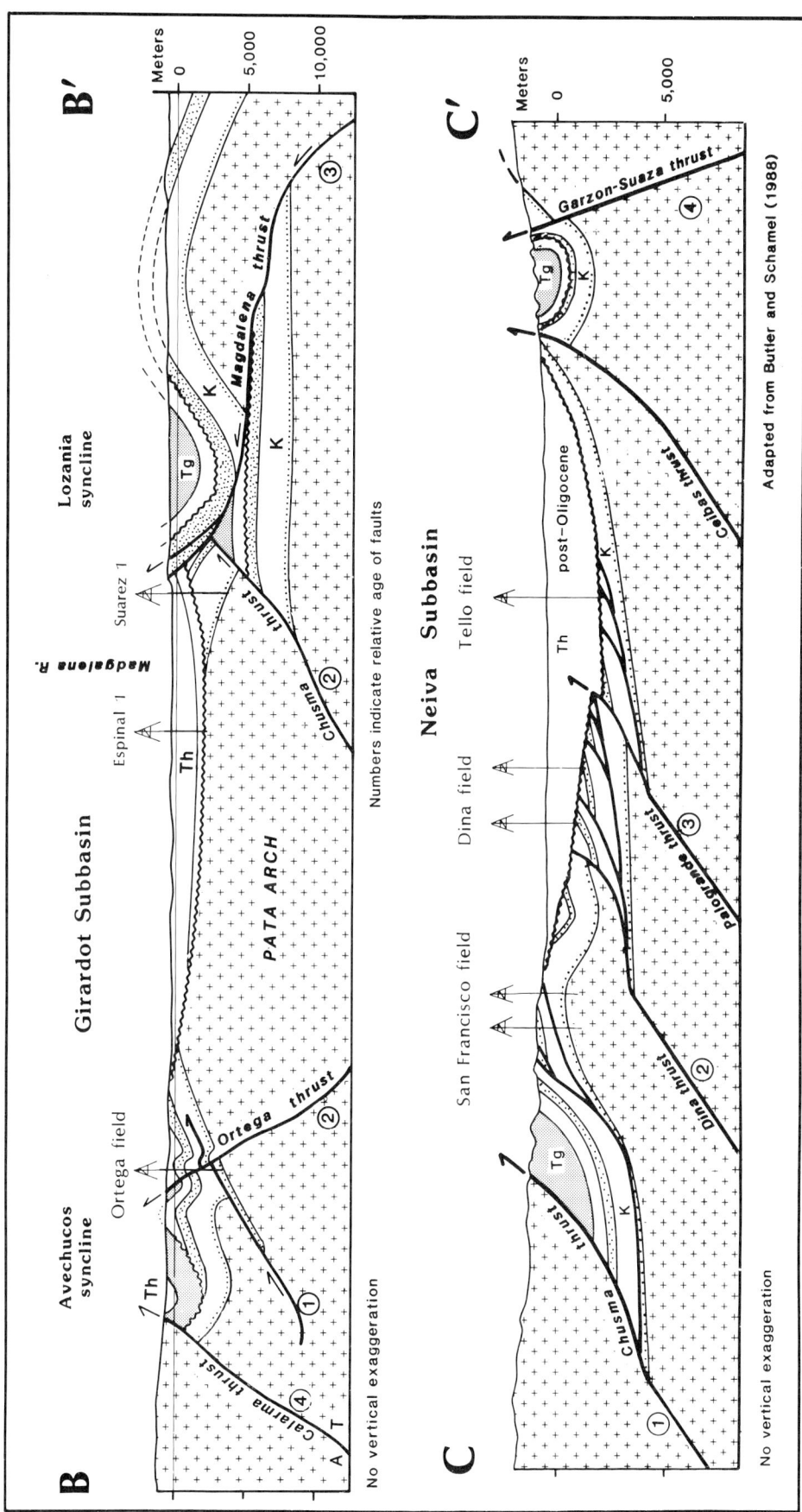

Figure 10. Generalized structural cross sections for the Girardot (B-B') and the Neiva (C-C') subbasins of the Upper Magdalena Valley. The numbers next to the faults indicate the inferred relative sequence of movement on the various faults.

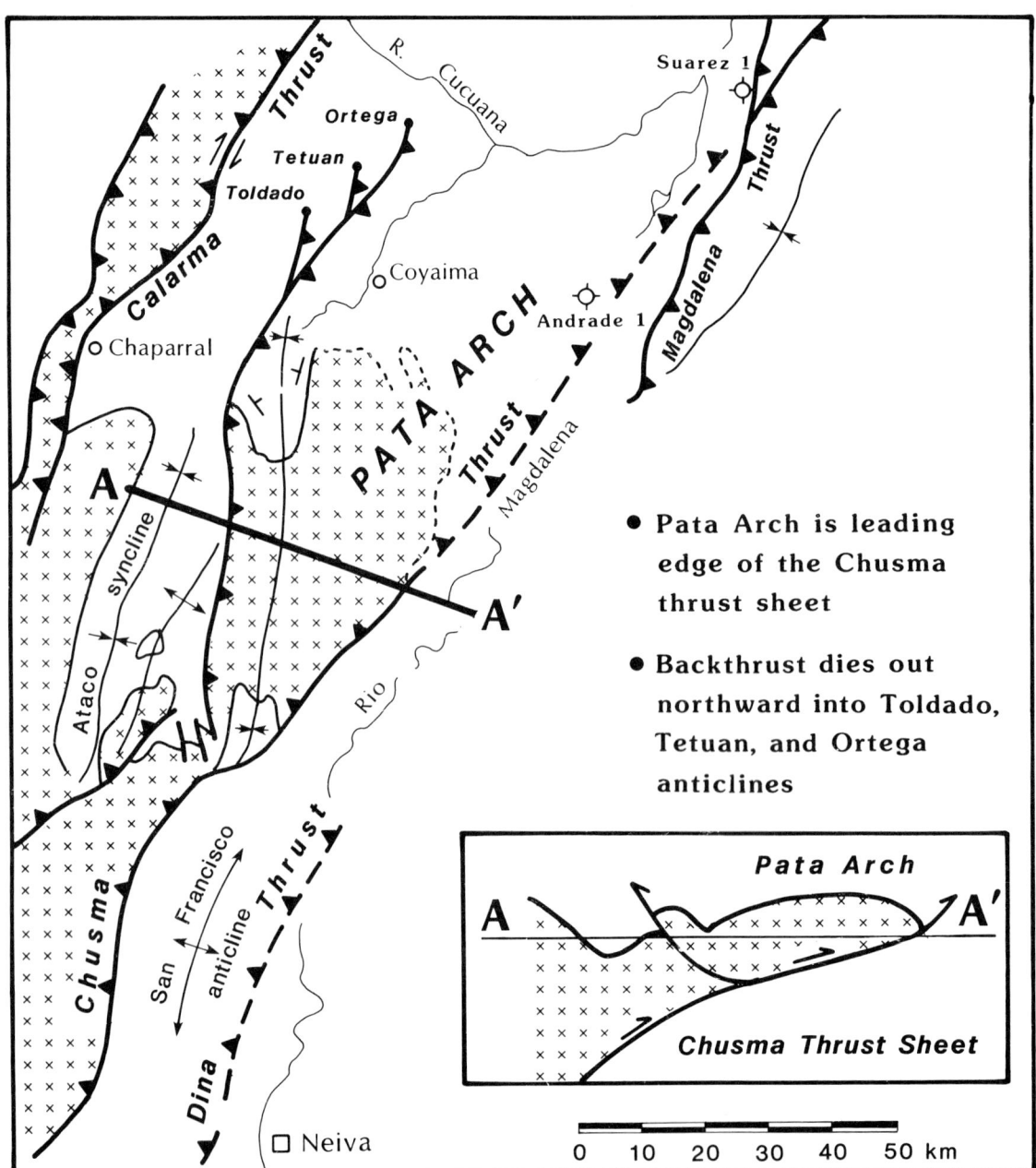

Figure 11. Sketch map showing the structural relationships across the Pata arch, the structure underlying the central portion of the Girardot subbasin. All of the current oil and gas production in the Girardot subbasin is situated along the western back-thrusted margin of the structural arch, a broad fault-bend anticline carried on the early Oligocene, eastward-verging Chusma thrust.

Table 1. Hydrocarbon reserves (December 31, 1988).

	Discovered		Cumulative production		Remaining	
	Oil (M bbls)	Gas (bcf)	Oil (M bbls)	Gas (bcf)	Oil (M bbls)	Gas (bcf)
Middle Magdalena Valley	2,115.6	2,644.2	1,756.6	2,134.4	359.0	509.8
Upper Magdalena Valley	486.7	68.2	188.0	45.5	298.7	22.8
All Colombian basins	4,966.7	8,580.2	2,989.2	4,604.9	1,977.5	3,975.3

Data from Ecopetrol, 1990.

Table 2. Principal fields in the Middle Magdalena Valley.

Field	Discovery date	Depth (m)	API gravity	Recoverable reserves		Cumulative production (December 1989)	
				Oil (M bbls)	Gas (bcf)	Oil (M bbls)	Gas (bcf)
Tertiary production within thrust belt							
Bonanza	1963	1,220	30.0	N.A.	N.A.	13.2	13.8
Infantas	1918	975	25.8	234.4	143.2	226.1	134.7
La Cira	1925	990	24.0	503.0	304.7	468.7	287.1
Las Monas	1938	—	—	93.2	642.8	87.0	543.4
Lisama-Peroles	1957	2,895	31.0	99.6	287.0	42.0	65.0
Payoa	1962	2,440	27.0	93.2	642.8	70.0	524.9
Provincia	1962	2,440	33.0	169.6	762.6	166.9	643.7
Tertiary production associated with normal fault traps							
Cantagallo-Yarigui	1941	2,073	19.7	160.6	78.5	135.0	71.1
Casabe-Galan	1941	1,158	20.7	331.5	203.7	253.8	191.5
Cocorna	1963	640	12.0–13.0	70.9	1.5	44.5	6.2
Llanito	1960	2,225	21.0	33.4	26.9	25.9	12.0
Velasquez	1946	2,286	19.0–24.0	196.1	88.4	167.2	78.6
Cretaceous production associated with normal fault traps							
Los Angeles	1983	1,951	14.0–15.0	N.A.	N.A.	0.4	—
Tisquirama	1983	2,438	16.0–28.0	2.6	—	5.8	1.1

Data from Ecopetrol, 1990; Asociacion Colombiana de Ingenieros de Petroleos monthly summary, December, 1989; *Oil and Gas Journal* worldwide production tables.

Table 3. Principal fields in the Upper Magdalena Valley.

Field	Discovery date	Depth (m)	API gravity	Recoverable reserves		Cumulative production (December 1989)	
				Oil (M bbls)	Gas (bcf)	Oil (M bbls)	Gas (bcf)
Dina-K	1969	1,981	22.0	201.0	49.2	97.7	23.9
Dina-T	1962	1,097	20.0	(see Dina-K)	(see Dina-K)	(see Dina-K)	(see Dina-K)
La Canada	1966	213	22.0	N.A.	N.A.	1.2	0.1
Los Mangos	1987	949	24.0	64.0	—	In evaluation	In evaluation
Ortega-Tetuan	1951	1,768	28.0	15.1	7.0	12.4	5.1
Pacande	1989	2,037	28.4	10.5	1.1	In evaluation	In evaluation
San Francisco	1985	853	27.0	134.9	1.0	25.9	8.3
Santa Clara	1987	640	17.0	11.8	—	0.9	0.3
Tello	1972	2,286	22.0	(see Dina-K)	(see Dina-K)	41.1	7.4
Toldado	1987	1,967	20.5	40.7	4.1	1.0	0.2
Toy	1987	1,232	18.0	19.8	—	0.01	0.0

Data from Ecopetrol, 1990; Asociacion Colombiana de Ingenieros de Petroleos monthly summary, December, 1989; *Oil and Gas Journal* worldwide production tables.

been few published geochemical studies of this unit. However, it is established that along the Magdalena Valley there are intervals at least many hundreds of meters thick having average TOC in the range of 3.0–4.0 wt. %. The La Luna Formation in the Middle Magdalena Valley and the middle Villeta Formation (cycle 4) in the Upper Magdalena Valley are especially organic rich. Kerogen petrography, H/C and O/C values, and biomarker analysis indicate that the kerogen is dominantly marine or type II (Zumberge, 1984). Regionally, the richness of organic matter in the Cretaceous succession and the relative proportions of marine versus terrestrial kerogen varies both vertically and laterally (Talukdar et al., 1989), related to position within individual transgression-regression cycles (Macellari, 1988) and proximity to the paralic margins of the Cretaceous pericratonic basin. Fortuitously, the present sites of the Magdalena basins are close to the center of the pericratonic basin, where during the early Late Cretaceous, thick, exceptionally organic-rich sediments accumulated. Macellari and DeVries (1987) attributed the generation and preservation of organic-rich sediments during cycle 4 to coastal upwelling within a partially restricted marginal seaway. However, the late Cenomanian to Santonian was a time of maximum sea-level stand (Haq et al., 1987) and exceptionally anoxic conditions in the world's oceans (Schlanger and Jenkyns, 1976; Jenkyns, 1980). It was during this period of time that approximately 29% of the world's source rocks were deposited (Klemme and Ulmishek, 1989).

Virtually no data on the degree of maturation have been published for the Magdalena basins and the Eastern Cordillera. However, numerical maturation modeling and field observations by the author suggest that the lower third of the 6000+-m-thick Cretaceous section in the Villeta anticline is overmature, but that the Cretaceous source rocks higher in the section and virtually the entire Villeta Formation in the Upper Magdalena basins is in the hydrocarbon generative window. Probably none of the Cretaceous source rock in the Magdalena basins is undermature.

Pratsch and Lawrence (1982) proposed that the oils in the Middle Magdalena fields were generated to the east within the deep Bogotá Cretaceous basin and had migrated updip to the west just prior to uplift of the Eastern Cordillera beginning in the late Miocene. Generation and migration from beneath the Eastern Cordillera is plausible and also has been proposed as the origin of the oil in the supergiant Caño Limon field in the northern Llanos basin (Fabre, 1987; Sutherland et al., 1989). The very large volumes of oil trapped in the basin would require a proportionally large kitchen. However, it is not necessary in the Middle Magdalena basin to appeal to such a distant and now inoperative source for the hydrocarbons. The Cretaceous rocks beneath the three major synclines (Figure 6) bordering the basin on the east—the Nuevo Mundo, the Rio Mineras, and the Guaduas synclines—are at depths in excess of 4000 m and are well within the oil window. If this thick Cretaceous section had not already been "cooked out" prior to late Miocene thrusting, the synclines may constitute at least an additional source of hydrocarbons for charging the young traps within the basin.

In the Upper Magdalena basins, where the Cretaceous source bed interval is generally less than 1000 m thick, there have been at least two periods of hydrocarbon generation (Butler and Schamel, 1989). The earliest phase of generation coincided with the latest Eocene–Oligocene east-vergent thrusting along the eastern margin of the Central Cordillera. The second phase is related to the late Miocene to Recent infilling of the successor basin, particularly in the Neiva subbasin. Oil is currently being generated in the vicinity of most of the major fields in the Neiva basin.

Trapping Mechanisms

Three types of traps have been recognized in the Middle Magdalena Valley (Taborda, 1965). Anticlines associated with late Miocene–Pliocene west-vergent thrusting are the most productive (Table 2). Fields in this setting are located along and to the east of the La Cira–Provincia trend in the north-central portion of the basin (see Figures 7, 8). Of almost equal importance are subtle anticlines that developed on the hanging wall of down-to-the-east normal faults along the western side of the basin (Figure 6). The Casabe-Galán, Velasquez, and Cantagallo-Yarigui fields are the most important examples of this type of trap. Production in both types of traps is from porous Tertiary sandstones. A third type of trap is exploited only in very small fields in the northernmost part of the basin (Figure 6). Here intensely fractured La Luna limestones adjacent to northeast-striking normal faults are partially sealed by overlying shales. Production from these fields is virtually insignificant (Table 2).

In the Upper Magdalena Valley, there are two important types of traps, one purely structural and the other structural/stratigraphic. Fields along the western margin of the Neiva and Girardot subbasins (Figure 9), such as San Francisco (Figure 10, Section C–C'), Los Mangos, La Cañada, and Ortega-Tetuan fields, produce from conventional fault-bend anticlines. Fields in the interior of the Neiva subbasin, principally Dina-K and Tello (Figure 10, Section C–C'), are developed in west-dipping thrust splays that were erosionally truncated prior to the Miocene, but subsequently sealed by the lower Miocene Barzalosa shales. Smaller fields in the basin that produce from the Miocene Honda Group, such as the Dina-T and Andalusia, are localized by small anticlines resulting from late Miocene–Pliocene reactivation of older, east-vergent thrusts (Butler and Schamel, 1989).

Future Opportunities

From the standpoint of hydrocarbon exploration and exploitation, the Magdalena basins are not yet "mature." The potential for additional major discoveries is excellent, and it is certain that current estimates of remaining ultimately recoverable reserves (Table 1) in the producing fields will be revised upward. Exploration in the basins is still guided by surface seeps and surface anticlines, albeit backed up with seismic-reflection data. Subsurface stratigraphic models for reservoir intervals have aided in the discovery of only two fields, the Peroles and Lisama fields (Figure 6), and even here surface structures served as the principal guide. The search for purely stratigraphic traps in the fluvial reservoirs of the Middle Magdalena Valley has hardly begun. The heavy oil belt in the southern Middle Magdalena Valley (Figure 6) immediately updip from the Guaduas syncline is known to contain large quantities of hydrocarbons; however, evaluation and exploitation of this resource is just beginning. Secondary recovery efforts are underway at only one field in the older Middle Magdalena basin (Casabe) (Figure 6) and are projected to result in the production of an additional 70 million bbl of oil from the field (Bueno, 1987).

In the Upper Magdalena Valley, several newly discovered major fields have just begun producing; none of the fields has reached the point where secondary recovery is required. Indeed, all along the length of the Magdalena Valley there remain prominent surface anticlines, as well as many buried anticlines, that are as yet untested.

ACKNOWLEDGMENTS

This review has benefited from the research and insightful criticism of many past and present colleagues in ESRI and Ecopetrol, most prominantly Kim Butler, Richardson Allen, Egon Castro, Camilo Hernandez, Carlos Macellari, Victor Perez, and Robert Ressetar. Careful reviews of the first draft of this paper by Kim Butler, Christine Rossen, and Kevin Biddle are gratefully acknowledged.

REFERENCES

Alfonso, C. A., P. E. Sacks, and D. T. Secor, Jr., 1989, Late Tertiary northwestward-vergent thrusting in Valle del Cauca, Colombian Andes [abstract]: AAPG Bulletin, v. 73, p. 327.

Alvarez, J., 1981, Determinacion de edad Rb/Sr en rocas del Macizo de Garzón, Cordillera Oriental de Colombia: Geologia Norandina, v. 4, p. 31-38.

Anderson, T. A., 1972, Paleogene non-marine Gualanday Group, Neiva basin, Colombia, and regional development of the Colombian Andes: GSA Bulletin, v. 83, p. 2423-2438.

Aspden, J. A., and W. J. McCourt, 1986, Mesozoic oceanic terrane in the central Andes of Colombia: Geology, v. 14, p. 415-418.

Aspden, J. A., W. J. McCourt, and M. Brook, 1987, Geometrical control of subduction-related magmatism: the Mesozoic and Cenozoic plutonic history of western Colombia: Journal of the Geological Society of London, v. 144, p. 893-905.

Balkwill, H. R., F. I. Paredes, and J. P. Almeida, 1988, Relationships of intra-crustal delamination, stratigraphy, and oil prospectivity, Oriente basin, Ecuador: Conexpo Arpel '88 (Rio de Janeiro, October 16-21), TT-72, 15 p.

Beltrán, N., and J. Gallo, 1968, The geology of the Neiva sub-basin, Upper Magdalena basin, southern portion: 9th Annual Field Conference, Colombian Association of Petroleum Geologists and Geophysicists, 44 p.

Bogota, J., 1988, Contribución al conocimiento estratigráfico de la Cuenca de los Llanos (Colombia): Caracas, III Simposio Bolivariano, Exploración Petrolera de las Cuencas Subandinas, v. 1, p. 309-346.

Bueno, R., 1971, A geologic section between Bucaramanga and La Uribe, Middle Magdalena Valley: 12th Annual Field Conference, Colombian Society of Petroleum Geologists and Geophysicists, 22 p.

Bueno, R., 1987, Presente y futuro de la exploración petrolera en Colombia: Bogotá, Ecopetrol (June 1987), 35 p.

Burke, K., 1988, Tectonic evolution of the Caribbean: Annual Review of Earth and Planetary Sciences, v. 16, p. 201-230.

Butler, K., 1983, Andean-type foreland deformation: structural development of the Neiva basin, Upper Magdalena Valley, Colombia [unpublished Ph.D. dissertation]: Columbia, South Carolina, University of South Carolina, 272 p.

Butler, K., and S. Schamel, 1988, Structure along the eastern margin of the Central Cordillera, Upper Magdalena Valley, Colombia: Journal of South American Earth Science, v. 1, p. 109-120.

Butler, K., and S. Schamel, 1989, Upper crustal control of deformation and hydrocarbon traps along the Upper Magdalena Valley, Colombia [abstract]: AAPG Bulletin, v. 73, p. 339.

Campbell, C. J., and H. Burgl, 1965, Section through the Eastern Cordillera of Colombia, South America: GSA Bulletin, v. 76, p. 567-590.

Cediel, F., J. Mojica, and C. Macia, 1980, Definición estratigráfica del Triásico en Colombia, Suramérica: Formaciones Luisa, Payandé y Saldaña: Newsletter Stratigr., v. 9, p. 73-104.

Cediel, F., and C. Caceres, 1988, Mapa Geologico de Colombia, Bogotá, Geotec Ltda., scale 1:200,000.

Duque-Caro, H., 1984, Structural style, diapirism, and accretionary episodes of the Sinú-San Jacinto terrane, southwestern Caribbean borderland, in W. E. Bonini, R. B. Hargraves, and R. Shagam, eds., The Caribbean-South American plate boundary and regional tectonics: GSA Memoir 162, p. 303-316.

Ecopetrol, 1990, Industria Petrolera, Estadisticas 1989: Bogota, Ministerio de Minas y Energia, 115 p.

Fabre, A., 1985, Dinamica de la sedimentacion cretacico en la region de la Sierra Nevada del Cocuy (Cordillera Oriental de Colombia), in F. Etayo-Serna and F. Laverde, eds., Proyecto Cretacico: Ingeominas (Bogotá), Publicacion Especial 16, p. XIX 1-20.

Fabre, A., 1987, Tectonique et génération d'hydrocarbures: un modèle de l'évolution de la cordillère orientale de Colombie et du bassin des Llanos pendant le Crétacé et le Tertiaire: Arch. Sc. Genève, v. 40, p. 145-190.

Feininger, T., 1970, The Palestina fault, Colombia: GSA Bulletin, v. 81, p. 1201-1216.

Galvis, J. N., and J. L. Rubiano, 1985, Redefinición estratigráfica de la Formación Arcabuco, con base en el análisis facial, in F. Etayo-Serna and F. Laverde, eds., Proyecto Cretacico: Ingeominas (Bogotá), Publicacion Especial 16, p. VII 1-16.

Gonzáles, H., A. Núñez, and G. Paris, 1988, Mapa geologico de Colombia: Memoria explicativa: Ingeominas, Bogota, 71 p.

Govea, C., and H. Aguilera, 1985, Cuencas sedimentarias

de Colombia, *in* II Simp. Explor. Petrol. Cuencas Subandinas: Colombian Society of Petroleum Geologists and Geophysicists, v. 2, 93 p.

Haq, B. V., J. Hardenbol, and P. R. Vail, 1987, Chronology of fluctuating sea levels since the Triassic: Science, v. 235, p. 1156-1166.

Hedberg, H. D., 1931, Cretaceous limestones as petroleum source rocks in northwestern Venezuela: AAPG Bulletin, v. 15, p. 229-244.

Jenkyns, H. C., 1980, Cretaceous anoxic events: from continents to oceans: Journal of the Geological Society of London, v. 137, p. 171-188.

Julivert, M., 1970, Cover and basement tectonics in the Cordillera Oriental of Colombia, South America, and a comparison with some other folded chains: GSA Bulletin, v. 81, p. 3623-3646.

Kellogg, J. N., I. J. Ogujiofor, and D. R. Kansakar, 1985, Cenozoic tectonics of the Panama and North Andes blocks: Memoirs of the 6th Latin American Congress on Geology, v. 1, p. 40-59.

Klemme, H. D. and G. F. Ulmishek, 1989, Depositional controls, distribution, and effectiveness of world's petroleum source rocks [abstract]: AAPG Bulletin, v. 73, p. 372-373.

Lundberg, J. G., A. Machado-Allison, and R. Kay, 1986, Miocene characid fish from Colombia: evolutionary stasis and extripation: Science, v. 234, p. 208-209.

Macellari, C. E., 1988, Cretaceous paleogeography and depositional cycles of western South America: Journal of South American Earth Science, v. 1, p. 373-418.

Macellari, C. E., and T. J. DeVries, 1987, Late Cretaceous upwelling and anoxic sedimentation in northwest South America: Palaeogeography, Palaeoclimatology, Palaeoecology, v. 59, p. 279-292.

Maze, W. B., 1984, Jurassic La Quinta Formation in the Sierra de Perija, northwestern Venezuela: geology and tectonic environment of red beds and volcanic rocks, *in* W. E. Bonini, R. B. Hargraves, and R. Shagam, eds., The Caribbean-South American plate boundary and regional tectonics: GSA Memoir 162, p. 263-282.

Mojica, J., and J. Dorado, 1987, El Jurasico anterior a las movimientos intermalmicos en las Andes colombianos: Bioestratigráfia de los Sistemas Regionales del Jurasico y Cretacico de America del Sur, Mendoza, p. 49-110.

Mojica, J., and C. Macia, 1983, Caracteristicas estratigraficas y edad de la Formación Yavi, Mesozoico de la region entre Prado y Dolores, Tolima, Colombia: Geologia Colombiana, v. 12, p. 7-32.

Mojica, J., C. Villarroel, F. Colmenares, and K. Bayer, 1985, Avances en conocimiento del Paleozoico superior del Macizo de Garzón, Tercio meridional de la Cordillera Oriental de Colombia: Abstracts of Annual Meeting, IGCP Project 211, Bogotá, October 9-11, 1985, 7 p.

Morales, L. G., and the Colombian Petroleum Industry, 1956, General geology and oil occurrences of Middle Magdalena Valley, Colombia, *in* L. G. Weeks, ed., Habitat of oil—a symposium: AAPG, p. 641-695.

Pindell, J. L., 1985, Alleghenian reconstruction and subsequent evolution of the Gulf of Mexico, Bahamas, and proto-Caribbean: Tectonics, v. 4, p. 1-39.

Pratsch, J. C., and P. L. Lawrence, 1982, Hydrocarbon concentration through preferred migration—Middle Magdalena basin, Colombia, South America: Oil and Gas Journal, v. 80(1), p. 42-52.

Renzoni, G., 1985, Paleoambientes del la Formación Tambor en la Quebrada Pujamanes, *in* F. Etayo-Serna and F. Laverde, eds., Proyecto Cretacico, Ingeominas (Bogotá), Publicacion Especial 16, p. III 1-18.

Rodriquez, C., and R. Rojas, 1985, Estratigráfia y tectonica de la Serie Infracretacica en los alrededores de San Felix, Cordillera Central de Colombia, *in* F. Etayo-Serna and F. Laverde, eds., Proyecto Cretacico, Ingeominas, (Bogotá), Publicacion Especial 16, p. XXI-1-XXI-21.

Rowley, D. B., and J. L. Pindell, 1989, End Paleozoic-early Mesozoic western Pangean reconstruction and its implications for the distribution of Precambrian and Paleozoic rocks around Meso-America: Precambrian Research, v. 42, p. 411-444.

Schlanger, S. O., and H. C. Jenkyns, 1976, Cretaceous oceanic anoxic events: causes and consequences: Geol. Mijnbouw, v. 55, p. 179-184.

Sutherland, J. A., L. E. Pena, F. Munoz, and J. H. Cristancho, 1989, Petroleum geology of Llanos basin of Colombia [abstract]: AAPG Bulletin, v. 73, p. 417.

Talukdar, S., O. Gallango, F. Cassani, and C. Vallejos, 1989, Upper Cretaceous source rocks of northern and northwestern South America [abstract]: AAPG Bulletin, v. 73, p. 418.

Taborda, B., 1965, The geology of the de Mares Concession, 1965, *in* Geological Field Trips, Colombia, 1958-1978: Colombian Society of Petroleum Geologists and Geophysicists, p. 119-159.

Van der Hammen, J. H. Werner and H. van Dommelen, 1973, Palynological record of the upheaval of the Northern Andes: a study of the Pliocene and lower Quaternary of the Colombian Eastern Cordillera and the early evolution of its high-Andean biota: Review of Palaeobotany and Palynology, v. 16, p. 1-122.

Van Houten, F. B., 1976, Late Cenozoic volcaniclastic deposits, Andean foredeep, Colombia: GSA Bulletin, v. 87, p. 481-495.

Waddell, M. G., 1985, The relationship of porosity development and diagenesis in the Upper Cretaceous Guadalupe Formation, Neiva basin, Colombia, *in* II Simposio Bolivariano: Bogotá, Asociacion Colombiana de Geologos y Geofisicos del Petroleo, v. 2, 18 p.

Zumberge, J. E., 1984, Source rocks of the La Luna Formation (Upper Cretaceous) in the Middle Magdalena Valley, Colombia, *in* J. G. Palacas, ed., Petroleum geochemistry and source rock potential of carbonate rocks: AAPG Studies in Geology 18, p. 127-133.

Chapter 11

A New Geologic Model Related to the Distribution of Hydrocarbon Source Rocks in the Falcón Basin, Northwestern Venezuela

Tito Boesi

Maraven, S.A.
Caracas, Venezuela

Donald Goddard

Maraven, S.A.
Caracas, Venezuela

ABSTRACT

The Falcón basin, located in northwestern Venezuela, has been intermittently explored since 1912. Since 1912, 200 exploratory wells have been drilled and 12,000 km of seismic lines have been acquired. This exploration effort has resulted in the discovery of eight small producing fields in both onshore and offshore areas.

The geologic history of the basin began in the late Eocene, and deposition continued through the Pliocene to the Recent. Because the basin is located at the boundary between the Caribbean and South American plates, sedimentation was controlled primarily by tectonism as evidenced by seismic and well data.

Three structural systems developed as a result of east-west dextral crustal movement. The first, consisting of a set of normal faults and associated horsts and grabens, forms the northern extension of the Oligocene-Miocene basin. The second system, known as the Falcón anticlinorium, includes east-northeast-striking parallel folds located in the center of the basin. The third structural system encompasses the active east-striking right-lateral strike-slip faults of which the Oca fault is the most relevant, owing to its regional extent.

The stratigraphic discontinuity within the basin is one of its principal features. The two stratigraphic stages that have been recognized are the result of a late Eocene to early Miocene transgression and a middle Miocene to Recent regression.

The northern flank of the basin, including the offshore area, has generated hydrocarbons from Oligocene and lower Miocene marine source rocks. However, small quantities of crude oils of terrigenous origin have been generated from Eocene source rocks.

Based on the tectonic and stratigraphic framework of the Falcón basin, a new conceptual model is proposed that can be applied to future hydrocarbon exploration in the area.

INTRODUCTION

The Falcón basin includes the state of Falcón and the northern part of the state of Lara in northwestern Venezuela (Figure 1). The boundaries of the basin have undergone only slight changes during evolution of the basin. The limits of the basin on land are considered to be those areas that include Oligocene and earlier sedimentation.

The first regional geological studies took place between 1912 and 1920 with the beginning of petroleum exploration in the area. The geologists of the British Oil Field Company and Standard Oil Company discovered some oil seeps close to the villages of Dabajuro, Mene de Mauroa, and Mene de Acosta. They also reported the presence of kerogen-impregnated shales to the east of the basin. The spectacular structures that were present throughout the area eventually caught the attention of various groups within the oil industry. At the same time, the first discoveries were being made in the Maracaibo basin. These discoveries helped to encourage exploratory drilling activity in the adjacent Falcón basin. The location of the wells was based solely on surface geologic information, and these wells were drilled downdip from the structures where the more important oil seeps were found.

The first commercial oil field in the Falcón basin, the Mene de Mauroa field, was discovered in 1921. Afterwards, a series of discoveries were made: Mamón and Hombre Pintado fields in 1926, Mene de Acosta field in 1927, Campo Media in 1928, Cumarebo in 1931 (Payne, 1951), and finally the Tiguaje field in 1953, when gravity and seismic methods began to be used in the area. By this time, 126 exploratory wells had been drilled in an area that covered 20,000 km². Reserves were estimated to be 1 billion bbl of oil in place. The results of this costly exploration effort did not satisfy the expectations of the oil industry.

Following a prolonged standstill in the systematic exploration, the Corporación Venezolana del Petróleo (CVP) began an exploration campaign in offshore La Vela Bay in 1970. The company acquired 4000 km of seismic lines and drilled 26 wells that resulted in the discovery of an additional 200 million bbl of oil in place. The discovery of this commercial amount of hydrocarbons was the second made in the offshore areas of Venezuela after the Gulf of Paria was discovered in 1958, and it led to a better understanding of the hydrocarbon habitat in the Falcón basin. Prior to this, however, Zambrano et al. (1971) produced a synthesis of the geological history of the area that provided a stepping stone for future work.

At the beginning of 1980, the exploration effort was renewed in western and eastern Falcón by Maraven and Corpoven oil companies. Since then, new seismic data has been obtained and 50 additional wells have been drilled, bringing totals to 12,000 km of seismic lines, 200 exploratory wells, and some 800 development wells in the basin, including La Vela Bay. One of the most important aspects of this campaign has been the results of the geochemical analyses that were aimed at recognizing the source rocks and locating the areas of hydrocarbon generation.

Since 1912, the oil industry has explored the Falcón basin with considerable effort. This has resulted in an enormous amount of unpublished technical and geological reports. Muessig (1984) has thoroughly documented the more important geological studies that were undertaken by universities, government institutions, and industry. This paper shall deal primarily with the tectonic setting and sedimentary aspects that provided the conditions for hydrocarbon accumulation.

The major elements of the geologic history of the Falcón basin, such as its depositional sequences and basin-forming tectonics, are somewhat comparable with those of the Los Angeles basin (Barbat, 1956). However, where the minor production of the Falcón basin in no way compares to the extraordinary production of the Los Angeles basin. Therefore, in light of the similarities between the two basins, we shall describe the more relevant features of the Falcón basin.

GEOLOGIC HISTORY

Although the geologic history of the Falcón basin begins during the Tertiary, earlier geologic conditions played an important role in basin formation and must be taken into account. Based on the classification system of Kingston et al. (1983), the basin-forming tectonic style is related to plate convergence. With regard to the depositional cycles, two can be described. The first is considered to be pre–late Eocene time and to have resulted in deposition of basal sediments. The second cycle occurred between late Eocene and Recent and consists of several depositional stages that were later deformed by various tectonic episodes. Little is known about the sediments that belong to the first cycle

Date Submitted: 5/4/87
Date Accepted: 4/1/88

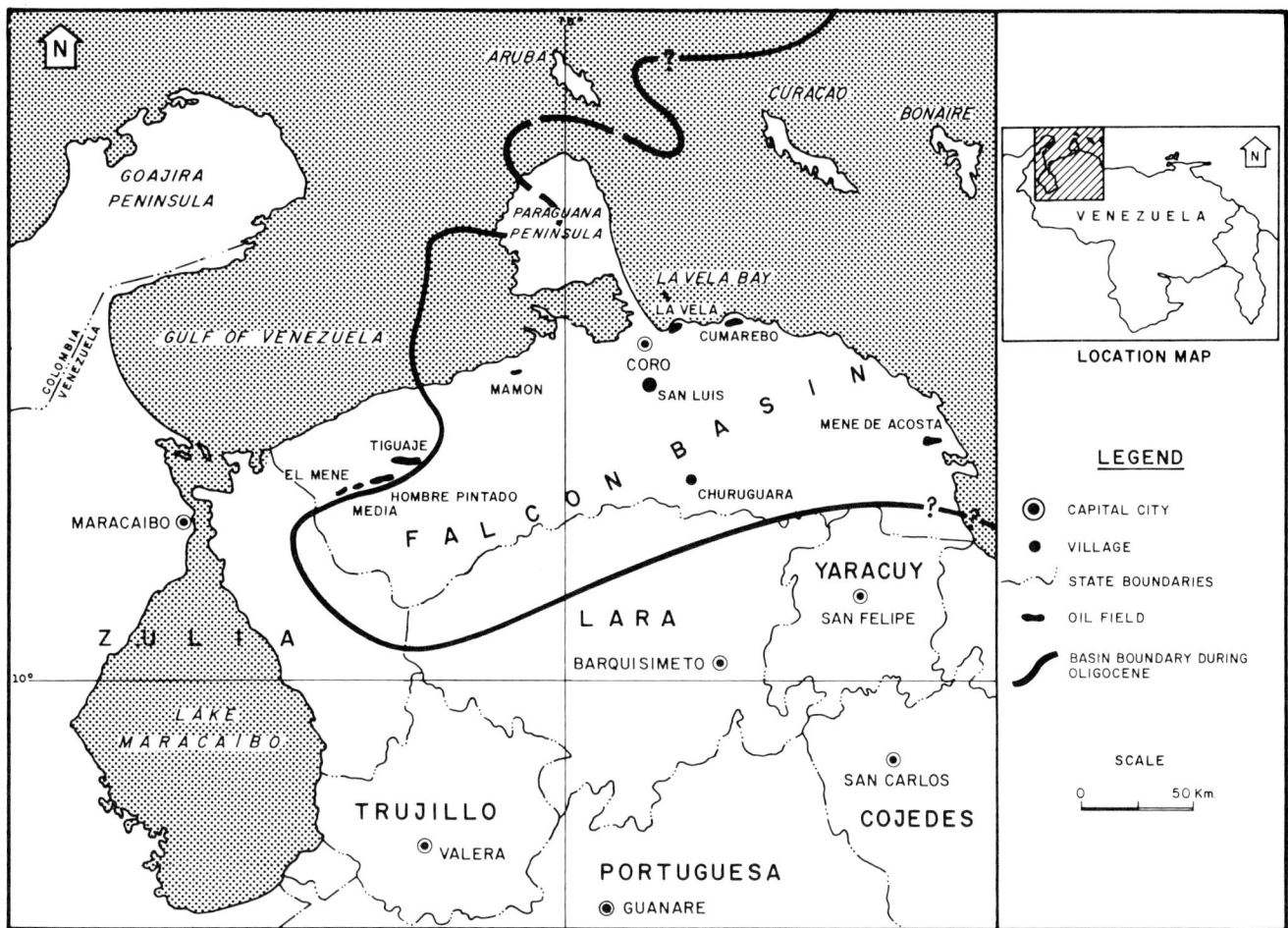

Figure 1. Location map of the Falcón basin with its northern offshore extension. Outline shows the basin boundary during the Oligocene.

because outcrops toward the northwest and southeast of the basin are of only limited extent. A few wells have penetrated Paleocene–Eocene sediments and have reached what may be an Upper Cretaceous section.

The sedimentary rocks that correspond to the second cycle provide the better known sections in the Falcón basin. This can be attributed to the widespread nature of the outcrops and to the large number of wells drilled into these sedimentary rocks, particularly on the north flank of the basin.

Tectonic Evolution and Principal Structures

The Falcón basin is located within the area of interaction between the South American and Caribbean plates. It originated as a pull-apart basin, beginning in the early Tertiary as a result of right-lateral, east-west strike-slip movement along the plate boundary (Muessig, 1984).

As a result of a single phase of crustal movement, three structural systems developed simultaneously within the basin. Each evolved with a maximum intensity at a different time. The first structural system consists of a set of northwest-striking normal faults located in the northern part of the basin. Their interpretation is based on Bouguer anomaly gravity maps (Bonini et al., 1977) (Figure 2). This set of faults forms a series of horst and graben structures. Among these are the Urumaco trough, the Coro-Paraguaná High, the La Vela Bay structure (trough with local southern horst), the Aruba High, and similar horsts and grabens to the east (Figure 3A).

With the integration of seismic and well data, the tectonic evolution of the areas of major exploration activity, such as the Urumaco trough, can be dated. Seismic lines 80-DU-G and 80-OC-I were joined to form one section that crosses the trough in a transverse, east-west direction (Figure 4). The western and eastern boundaries of the Urumaco trough are defined by the Lagarto fault and the Sabaneta fault. Gradual subsidence with an eastern tilt started in the Oligocene and continued until the early Miocene. A more homogeneous subsidence after this time is evident from the section.

The second structural system consists of several east-northeast-striking parallel folds of great length that are situated in the center of the basin. These folds are collectively known as the Falcón anticlinorium and are the result of a northwest compressive component (Figure 3B). In areas of maximum stress, thrusting developed parallel

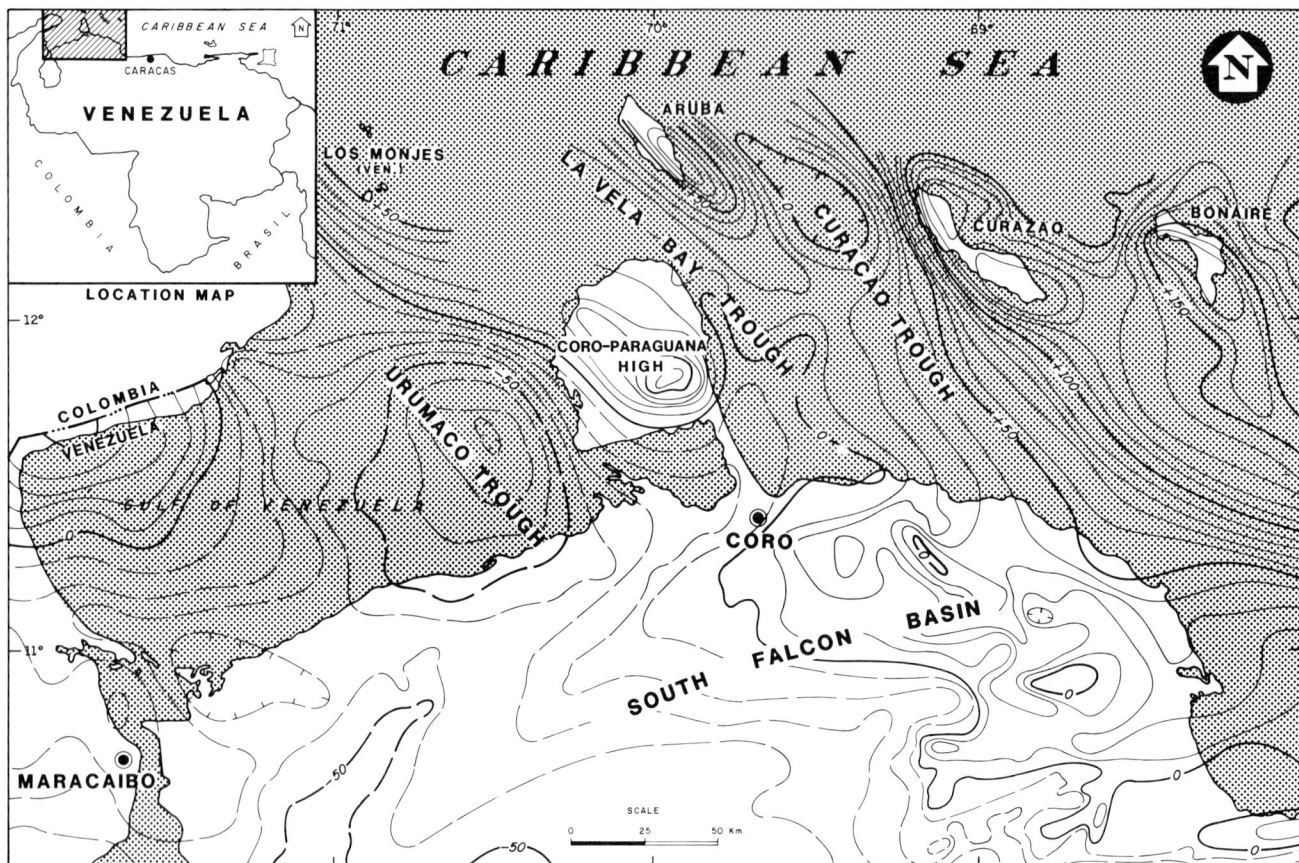

Figure 2. Bouguer gravity anomalies map of the Falcón basin and surrounding areas (after Bonini et al., 1977). Main gravity lows are interpreted as pull-apart grabens trending northwest-southeast.

to the folds. The most notable thrust is the Guadalupe thrust near the town of La Vela. This deformational system grew in intensity with time, ending in late Miocene with basin inversion.

It is believed that dextral movement of blocks towards the northeast along the Boconó fault (Schubert, 1980) has played an important role in the folding of the Oligocene-Miocene formations in the basin. This movement is also responsible for overthrusting of the Cretaceous metamorphics on top of Tertiary turbiditic sediments in the Barquisimeto-Carora area after basin formation (Figure 3B).

A third structural system of east-striking right-lateral strike-slip faults started developing within the basin in late Eocene and continued through the Tertiary up to the Recent. These faults resulted from shifting and regional shearing of the entire crustal block in the basin. As the basin was gradually filled with sediments, the zone of shear was reduced, thus concentrating the intensity of the strike-slip movement along the Oca fault system (Feo Codecido, 1972; Vásquez and Dickey, 1972) (Figure 3C).

The more recent seismic-reflection studies undertaken in the western part of Falcón State, along the Oca fault system, have detected three right-lateral strike-slip faults forming flower structures that can be observed along parts of the fault zone and that are separated from one another by synclinal blocks. The names of these three faults are Oca, El Mayal and Ancón de Iturre (Figure 5). A late Eocene to early Oligocene age was determined for the infill sediments of these synclinal blocks from sparse well data, thus indicating an approximate time for the beginning of the Oca fault system. To the west, the faults come together in the El Tablazo depression. To the east they penetrate the basin, where recent movements along the faults cut across the large folds of the Falcón anticlinorium (Figure 3C).

Stratigraphy and Sedimentation Prior to Basin Formation (Cretaceous-Paleocene)

The subsurface Cretaceous sequence of the Falcón consists of carbonate sediments in the basal section and clastics above. However, owing to the scarcity of wells that penetrate the Cretaceous in this area, only speculative interpretations can be attempted. From three wells on the Dabajuro platform and one on the Coro-Paraguaná High, it was determined that the Cretaceous has undergone diagenetic alteration and metamorphism, with growing intensity towards the east and north. These changes can be attributed to a series of events that took place during the Cretaceous-Paleocene period and should be considered, in the regional sense, as part of Caribbean tectonism in

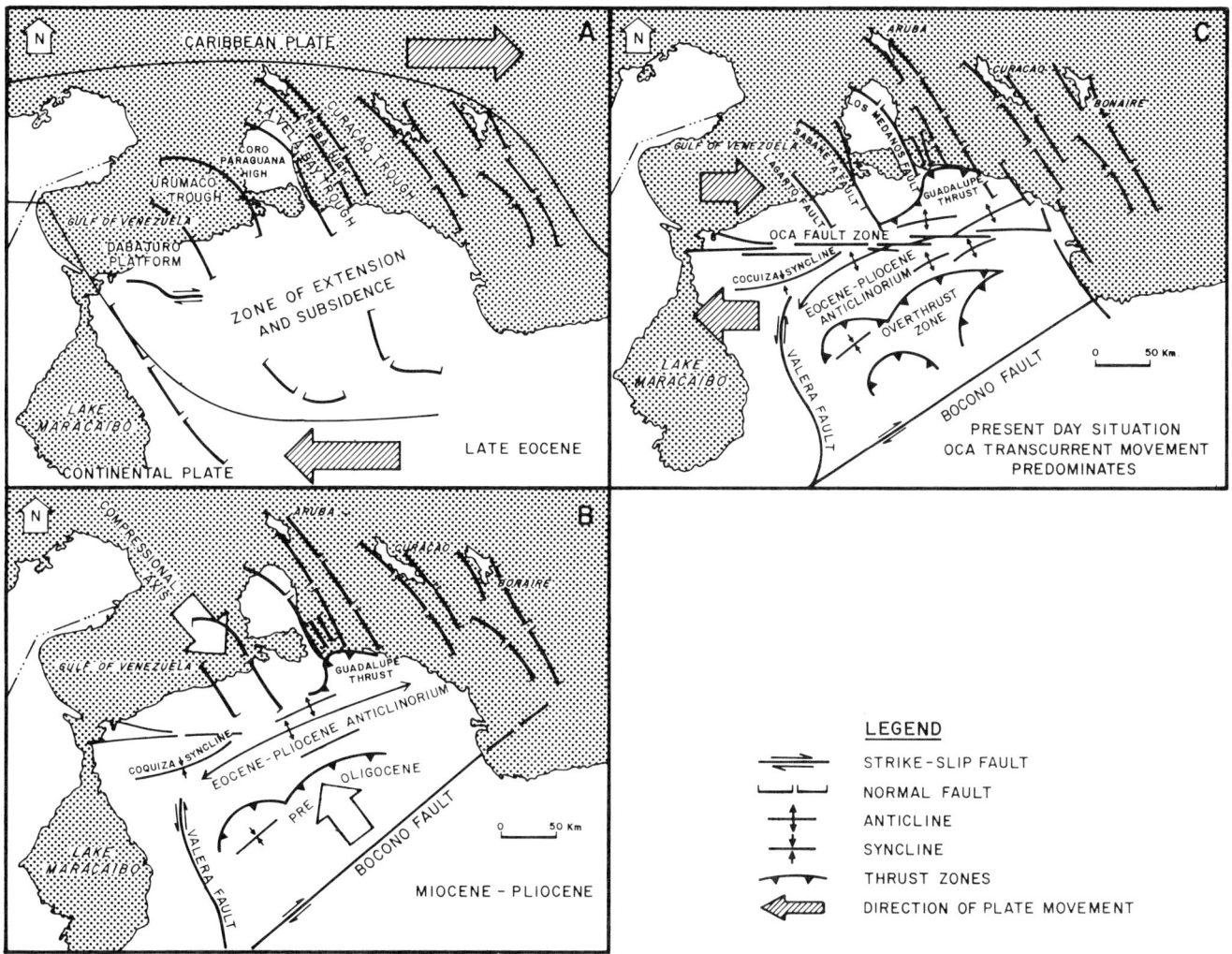

Figure 3. Tectonic history of the Falcón basin showing (A) late Eocene, (B) Miocene–Pliocene, and (C) present situation.

which volcanic activity played a major role (Stephan, 1977; Pumpin, 1978; James, 1984).

At this time, an east-trending volcanic island arc developed as a result of subduction of the Caribbean plate below the South American plate (Burke et al., 1984). This island arc was responsible for controlling the sedimentation into the Cretaceous-Paleocene seas and for setting up environments of deposition of which the La Luna Formation, of middle Late Cretaceous age is the most representative and is considered the source rock for most of the petroleum produced in the Maracaibo basin (Hedberg, 1931; Blaser, 1979).

The sedimentary processes that took place from Late Cretaceous to Paleocene were associated with volcanic and tectonic activity. The igneous intrusions, extrusives, and emplaced ophiolites (Compañía Shell de Venezuela, 1965; Coronel, 1970) occurred simultaneously with sedimentation in an unstable crustal zone that was gradually being deformed owing to subduction to the south (Bellizzia et al., 1972). At the end of the Paleocene and the beginning of the Eocene, the slope angle increased and huge masses of turbidites as well as volcanic and metamorphic material were transported toward Lara State to the South.

Early to Middle Eocene

Eocene sedimentation in western Venezuela is characterized by a predominance of clastics, particularly in Lara and Zulia states. During this period, regional crustal movements continued. These grew more intense, thus affecting the paleogeographic configuration of the region.

Two sedimentary provinces existed in northwestern Venezuela during the Paleocene and the Eocene. To the west, in the Zulia province, Paleocene carbonates of the Guasare Formation were deposited on top of a platform and were covered by deltaic sediments in the early to middle Eocene. To the east, in the Lara province, Matatere Formation turbidites from a Caribbean source gradually filled a large depression that included the Falcón region (Hunter, 1972).

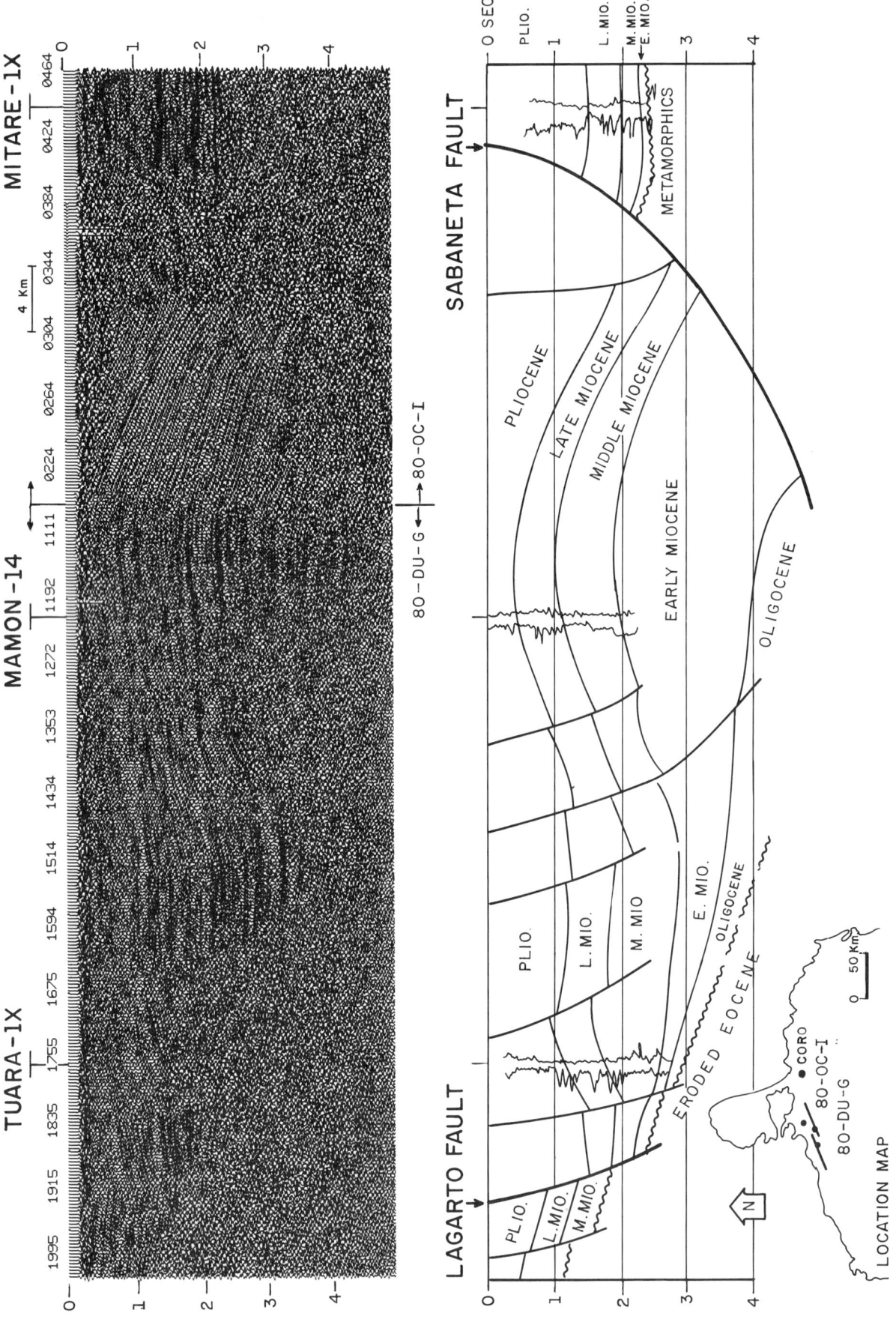

Figure 4. Combined seismic-reflection profiles 80-DU-G/80-OC-I across the Urumaco trough with transit time converted logs of wells Tuara-1X, Mamón 14, and Mitare-1X.

Distribution of Hydrocarbon Source Rocks in the Falcón Basin   309

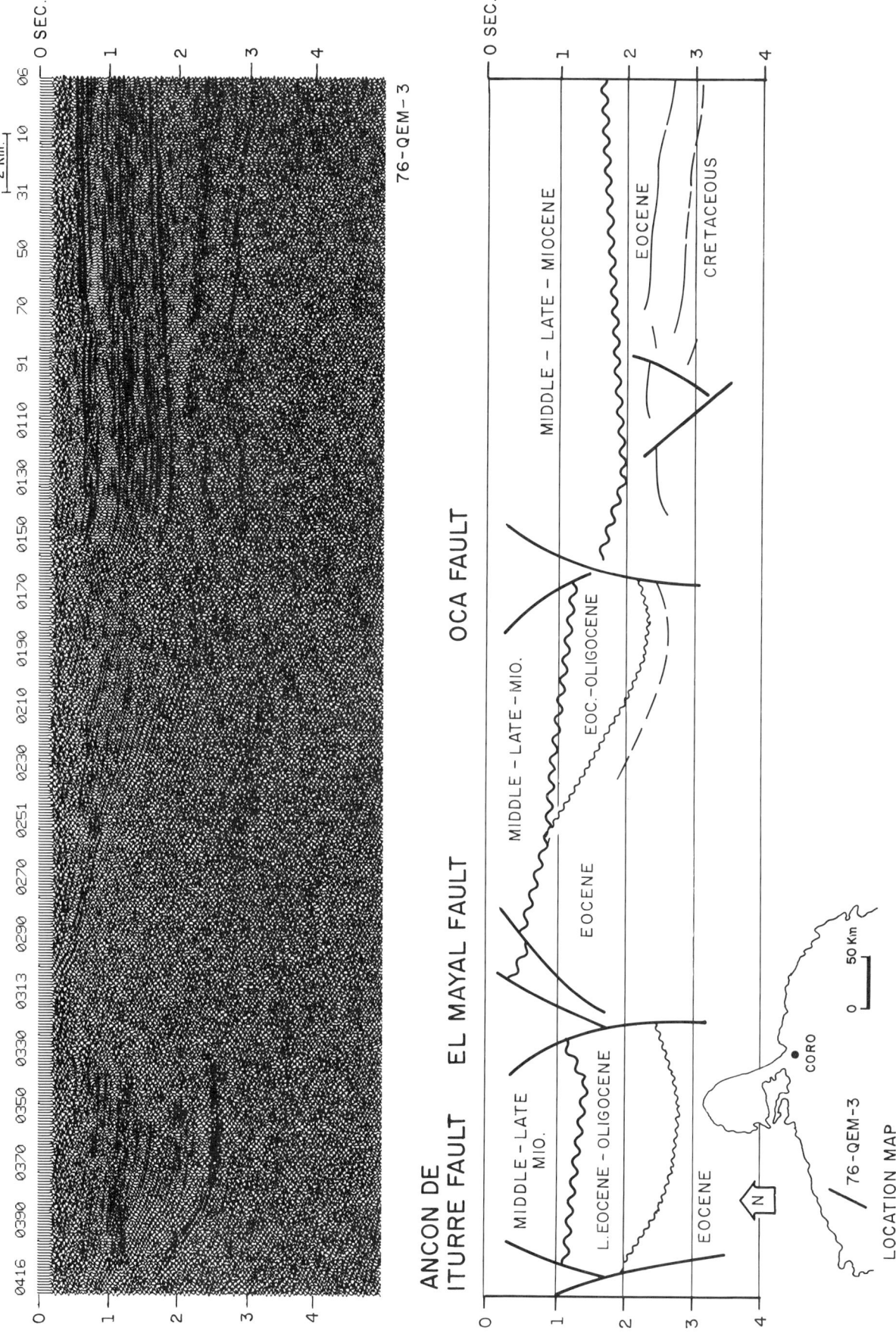

Figure 5. Seismic-reflection profile 76-QEM-3 across the Oca fault zone in the extreme west side of the Falcón basin.

The platform sediments of the Zulia province and the turbidites of the Falcón–Lara depression, of equivalent ages, meet in a transition zone commonly known as the "Hinge Line." This zone is located along the Bolívar coastal district of Zulia State along the northeastern shore of Lake Maracaibo. It contains a shaly sequence of sediments of mixed composition from the two depositional regimes and is known as the Trujillo Formation. The thick Paleocene-Eocene turbidite sequence of Lara State belonging to the Matatere Formation is believed to be an allocthonous deposit (Renz et al., 1955; Bellizzia and Rodríguez, 1968; Stephan, 1977). It is interpreted as an erosional remnant of an extensive system of nappes derived from the Caribbean area during the middle Eocene. These rocks of Cretaceous to middle Eocene age constitute the basal rocks on top of which the sedimentation in the Falcón basin began in the late Eocene.

Stratigraphy and Sedimentation during Basin Formation

Wheeler (1963) presented the first stratigraphic compilation of the Falcón basin that confirmed some conclusions arrived at earlier by several other geologists (González de Juana; 1938, González de Juana et al., 1980). These conclusions are based on the description of the boat shape of the basin with an east-west axis and a narrow central section that is located between the towns of Churuguara and San Luis (Figure 6).

According to this model, the shape of the basin determined the sedimentary distribution during the Oligocene-Miocene depositional history (Díaz de Gamero, 1977). The sandy and calcareous formations were deposited in shallow water in the western part of the basin. In the center and to the east, finer grained sediments were deposited in a less energetic marine environment. The sedimentary fill in the basin occurred in two main stages. The first was a transgressive stage that took place between the late Eocene and middle Miocene and is characterized by continued subsidence and progressive invasion of the sea from the northeast. The second stage is predominantly regressive, beginning in the middle late Miocene as a result of basin inversion that caused the sea to retreat toward the northeast.

The stratigraphy in the center of the basin has been determined from surface geology; no wells have been drilled in the area. The ages that were obtained are based on planktonic foraminiferal and macrofaunal studies (Renz, 1948; Wheeler, 1963; Díaz de Gamero, 1977). With the exploration wells drilled onshore before 1960 and offshore in the La Vela Bay in the 1970s, an attempt has been made to correlate this area with the known surface stratigraphy. Owing to rapid facies changes, correlation between wells is difficult, especially between wells located in troughs and on horst blocks.

Fortunately, in the western areas, in the Urumaco trough, palynological studies undertaken in 1950, together with more recent studies by Maraven palynologists E. Di. Giacomo and R. Pittelli have improved age determinations. These data have been integrated with abundant seismic information (Figure 4), and as a result, correlations could

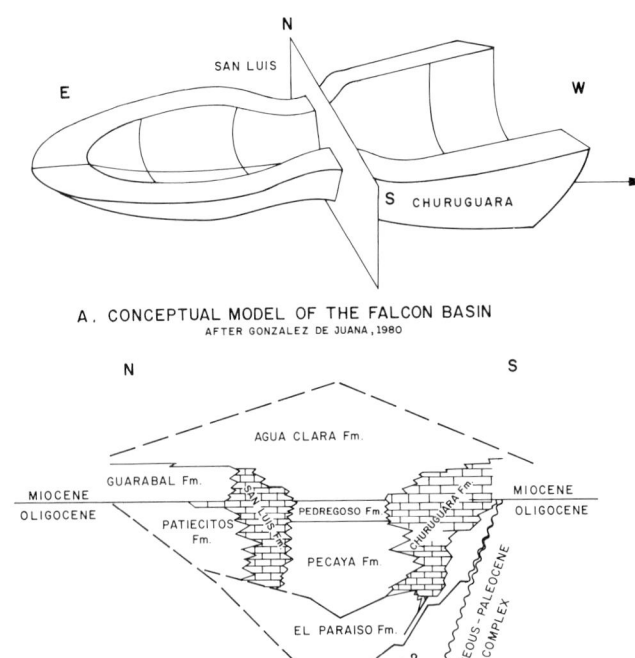

Figure 6. (A) The original conceptual model for the basin (after González de Juana, 1980) and (B) a stratigraphic compilation (after Wheeler, 1963) based on a transverse section through the center of the basin between the villages of San Luis and Churuguara whose geographic positions are indicated in Figure 1.

be made between subsurface and surface information thus providing a better understanding of the synsedimentary history. Figure 7 shows the stratigraphic relationship between the facies in the Urumaco trough (Wells Tuara-1X and Mamón-14) and those on the Coro-Paraguaná horst block that was penetrated by the Mitare-1X well. Also included here are the results of the La Vela-6 well, whose stratigraphy represents the complicated La Vela Bay trough and southern La Vela horst areas (Findlay, 1985).

The stratigraphic nomenclature (Figure 8), originally developed for the Falcón basin, combines the work of various authors (González de Juana, 1938; Wheeler, 1963; Díaz de Gamero, 1977). A few minor changes have been made to the existing nomenclature in order to adapt it to current tectonic and stratigraphic ideas (Figures 3, 6), based on the latest seismic and well data. Now, a different model for the basin is proposed (Figure 9). This model includes the northwest-trending horsts and troughs mentioned earlier that are responsible for the northern extension of the central basin sedimentary environments. Combining these areas to the north with the original basin has led to a better understanding of the hydrocarbon habitat in the Falcón basin.

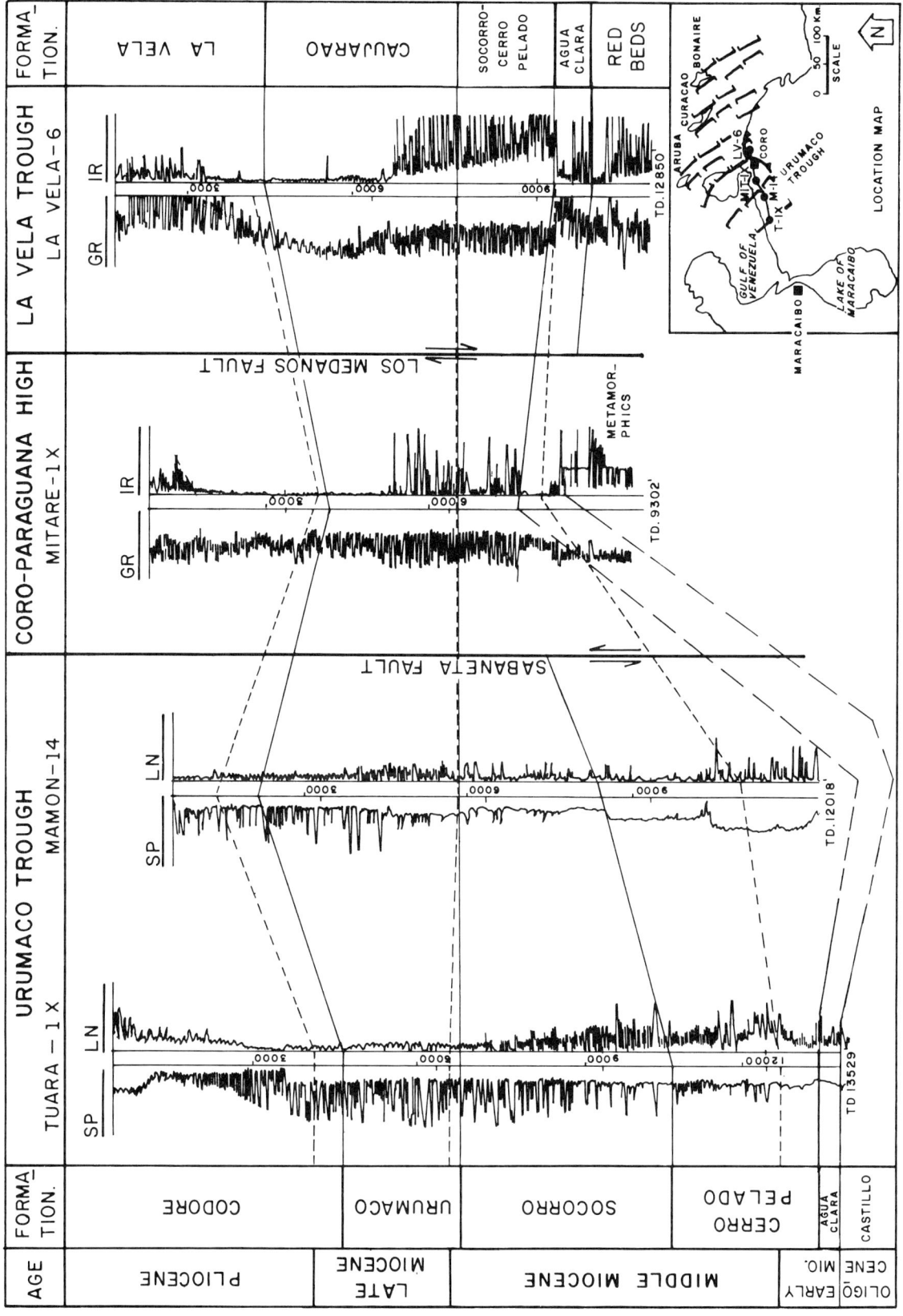

Figure 7. Lithostratigraphic and chronostratigraphic relationship between the Urumaco trough, Coro-Paraguaná High, and La Vela Bay sediments. Time units have been determined by palynological and micropaleontological integrated zonations.

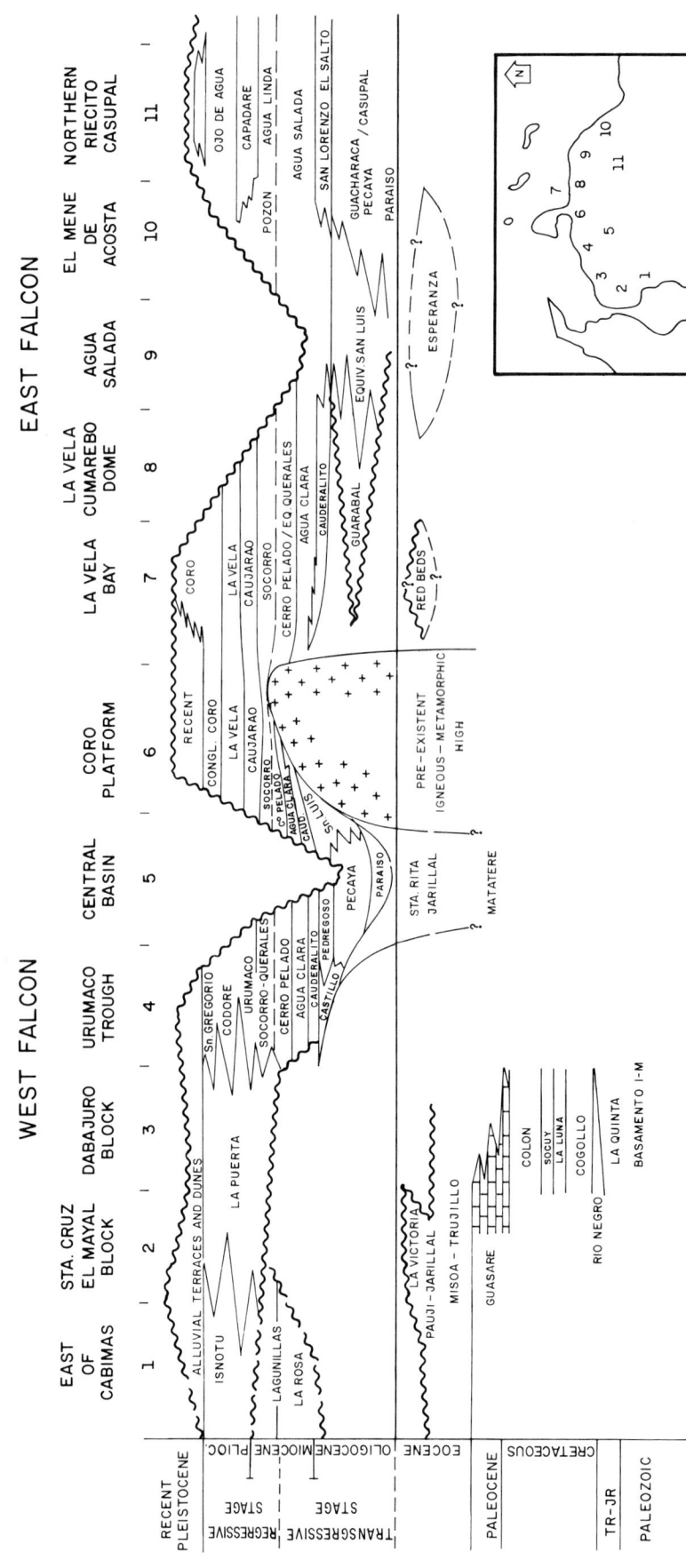

Figure 8. Stratigraphic nomenclature of the Falcón basin from west to east.

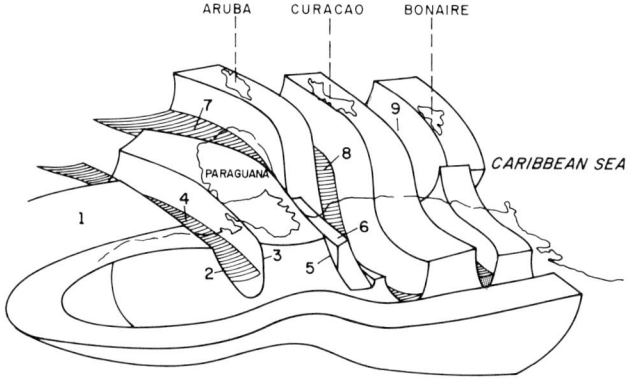

LEGEND

1 DABAJURO PLATFORM
2 LAGARTO FAULT
3 SABANETA FAULT
4 URUMACO TROUGH
5 LOS MEDANOS FAULT
6 LA VELA HORST
7 ARUBA TROUGH
8 CURACAO TROUGH
9 BONAIRE TROUGH

Figure 9. The proposed conceptual model for the Falcón basin.

Late Eocene

In contrast to other regions of Venezuela, the Eocene sedimentation continued uninterrupted from middle to late Eocene. At the time that the erosional period was starting in Zulia, Lara, and the Caribbean, the formation of the basin in Falcón was just beginning (Figure 10A). Also, orogenic movements in the southern Caribbean were at a peak. These movements were responsible for building a mountain range of which the islands of Aruba, Bonaire, and Curacao are the present-day remnants.

The orogenic period is also responsible for uplifting areas that later became the present-day Mérida and Perijá mountains. The extensive sea into which Cretaceous to middle Eocene sediments were being deposited was reduced to a smaller gulf. This gulf was bounded on the north by a locally uplifted area of the Caribbean, to the west in Zulia by an erosional platform, and to the south by the uplifted Cretaceous metamorphic blocks of Barquisimeto (Pumpin, 1978) (Figure 10A). These blocks were later thrust upon the Matatere flysh during the Andean orogeny.

Oligocene

With increasing subsidence, the basin took the shape of an elongated boat with steeply sloping sides. Northwesterly oriented synsedimentary trenches developed in the northern sector with a narrowing of the central area. This can be attributed to forces from the south that pushed the entire region against an uplifted region to the north, part of which is the Paraguaná High (Figure 10B).

Igneous intrusions and submarine basalt flows (Coronel, 1970; Muessig, 1978, 1984) are found in the deeper parts of the basin, in subsiding trenches, and are common in the northern sector.

Miocene

The tectonic stresses became increasingly complex during the Miocene and caused successive paleogeographic modifications that in turn were responsible for the complex facies distributions we find today. During the early Miocene, the northeastern marine transgression came to an end. As a result of this transgression, marine sediments also covered the entire Maracaibo basin, being represented in that basin by the La Rosa Formation. Well and seismic data show a passage termed the "Cabimas strait" that formed a channel between the La Rosa and the Agua Clara seas (Figure 11A). The sediments that were deposited in seas of the Falcón basin during the early Miocene transgression are represented by the Agua Clara Formation and the lower part of the Agua Salada Group (Figure 8).

In the central Falcón area, sediment deposition no longer took place. On the northern flank and in the Caribbean region, however, the synsedimentary trenches continued to fill up as subsidence continued, due to factors related to extensional tectonics. At the same time, the southern compressional stresses from the Andean region persisted.

Tectonic activity partly controlled the middle Miocene sedimentation. The basin inversion that began at this time was responsible for uplift and erosion of the sediments that were deposited in the channel. The position occupied by the axis of the basin in Oligocene and early Miocene now became the axis of the mountain range that is known as the Falcón anticlinorium. This was the initiation of the subbasins and the separation of the area into two distinct sedimentary provinces. To the south, shallow-marine and lagoonal facies and fluviatile sediments predominated, while to the north, a deeper marine to coastal transition facies developed.

At the end of the Miocene, tectonic movements related to lateral movements along the Caribbean–South American plate boundary were evident. The area of deposition was toward the Caribbean Sea and the sediment was derived mainly from the south as recycled deposits (Figure 11B). Toward the west, a progradational delta complex gave rise to the La Puerta Formation, and toward the east, the depositional environments were more marine. The carbonate rocks that belong to the Caujarao and La Vela formations of the eastern area were deposited on top of the horsts and platform highs. A large volume of sands, clays, and calcareous marls found their way into the troughs (Figure 11B).

Pliocene

Later tectonic activity took place in the Pliocene, affecting mainly the northern Falcón area, and was responsible for the uplift of the current mountain system. The sedimentary processes responsible for the alluvial terraces, beaches, and dunes along the piedmont zone, are believed to have been affected by the tectonic activity. An orogenic period also occurred during the Pleistocene, similar processes were repeated. This repetition resulted in a number of terraces of different ages but with the same environment of deposition. These can be observed today along the entire coastal plain (Graf, 1972).

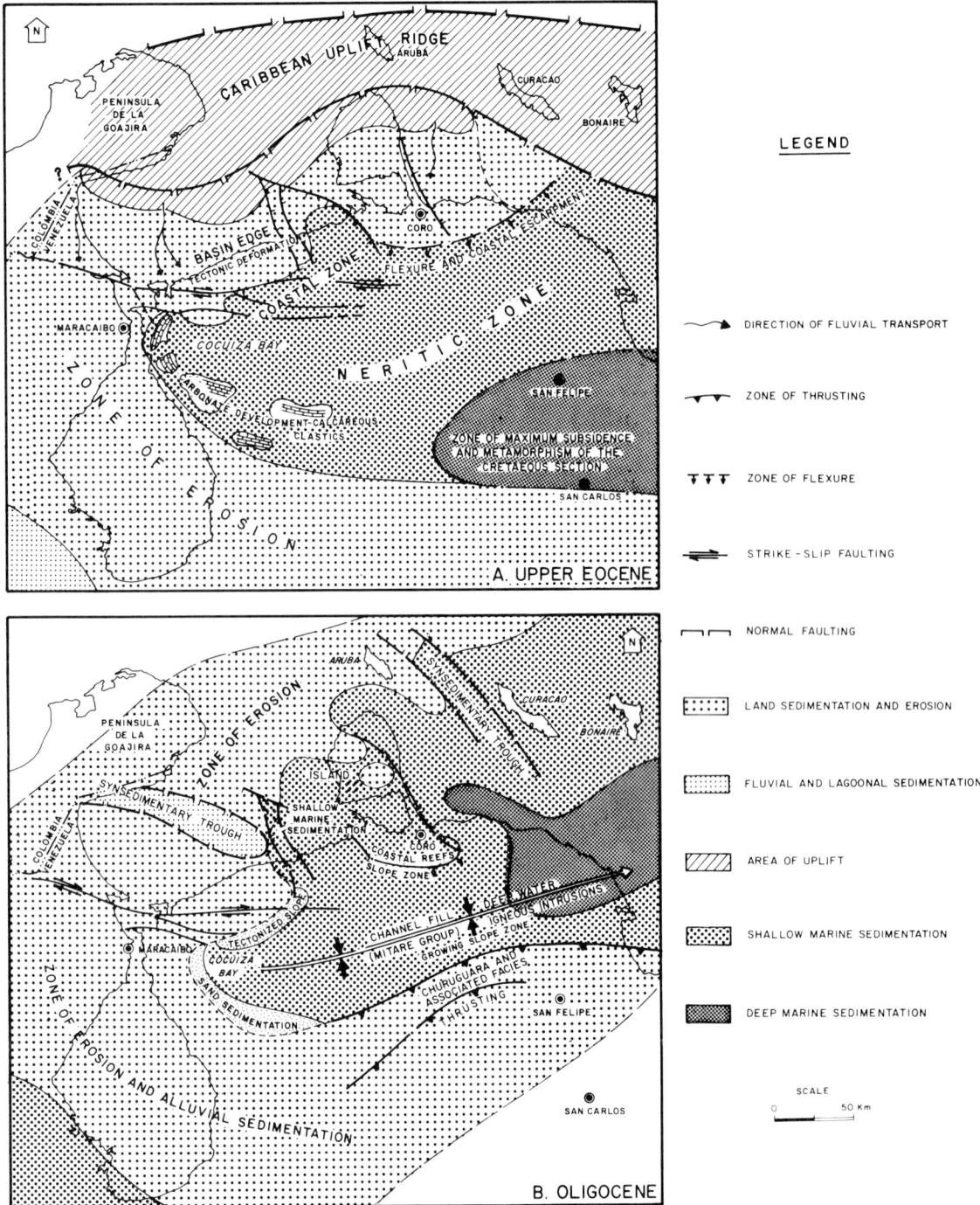

Figure 10. Paleogeographic situation (A) in the late Eocene and (B) in the Oligocene.

The tectonic and stratigraphic history of the basin is synthesized in the schematic cross sections shown in Figure 12. The first three cross sections (A-A', B-B', and C-C') show the effect of compression that occurred after the early Miocene when the basin inverted. An important compressional component related to these three sections resulted in the strike-slip movement of the Oca fault (Figures 3C, 5). The northern flank of the basin is represented by the longitudinal cross section in D-D', which shows the effect of extension that gave rise to the series of horsts and troughs where a thick sequence of Miocene–Pliocene sediments accumulated (Figure 4).

HYDROCARBONS

From earlier geochemical studies (Hartog and Syaps, 1973), it was observed that the C_7-saturate composition of the crude oils from the Falcón basin, in addition to their sulfur, nickel, and vanadium contents, was quite different when

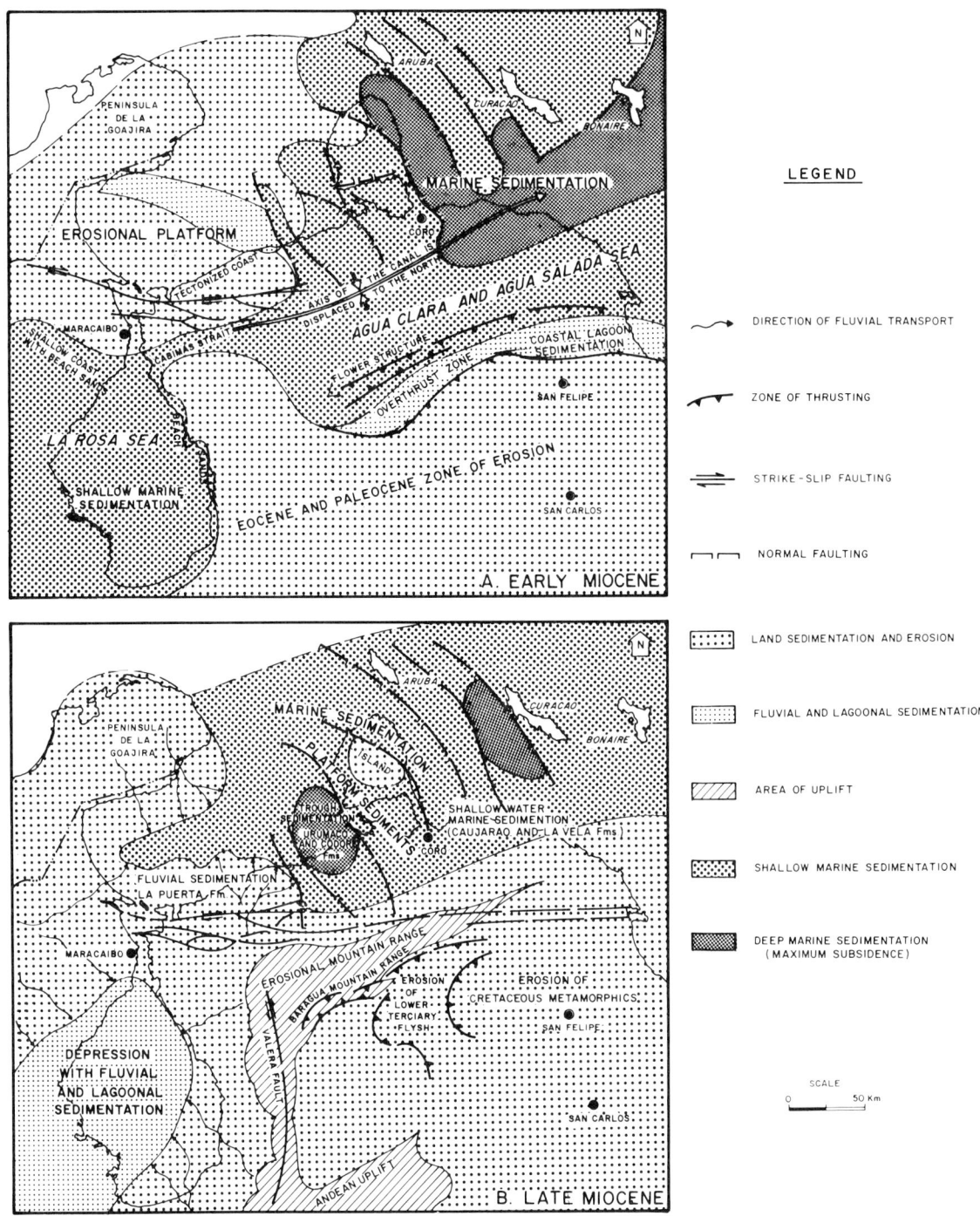

Figure 11. Paleogeographic situation (A) in early Miocene and (B) in the late Miocene.

compared to crudes from the Maracaibo basin. This has led geologists to consider the Falcón basin as a separate hydrocarbon province. Additional studies (Van der Veen and Posthuma, 1983; Buiskool and Van der Veen, 1984) were carried out using state-of-the-art techniques aimed at classifying the crudes and determining their origin.

The gas chromatograms of the saturated hydrocarbons and the C_7-alkane distribution of samples from different areas of the Falcón basin indicate various degrees of bacterial degradation. Most crudes in the western areas are derived from source rocks rich in land-plants deposited in terrestrial (swampy) environments. A typical chromatogram was obtained from samples of the Tiguaje field by Buiskool and Van der Veen (1984) (Figure 13A). Chromatograms of the eastern crudes from the Cumarebo field were obtained and interpreted by the same authors as crudes of marine origin (Figure 13B). Sofer's (1984) approach, using isotopic relationships, has been applied to the Cumarebo crudes (Lew, 1985), and the results indicate once more that these crudes were derived from a marine organic source.

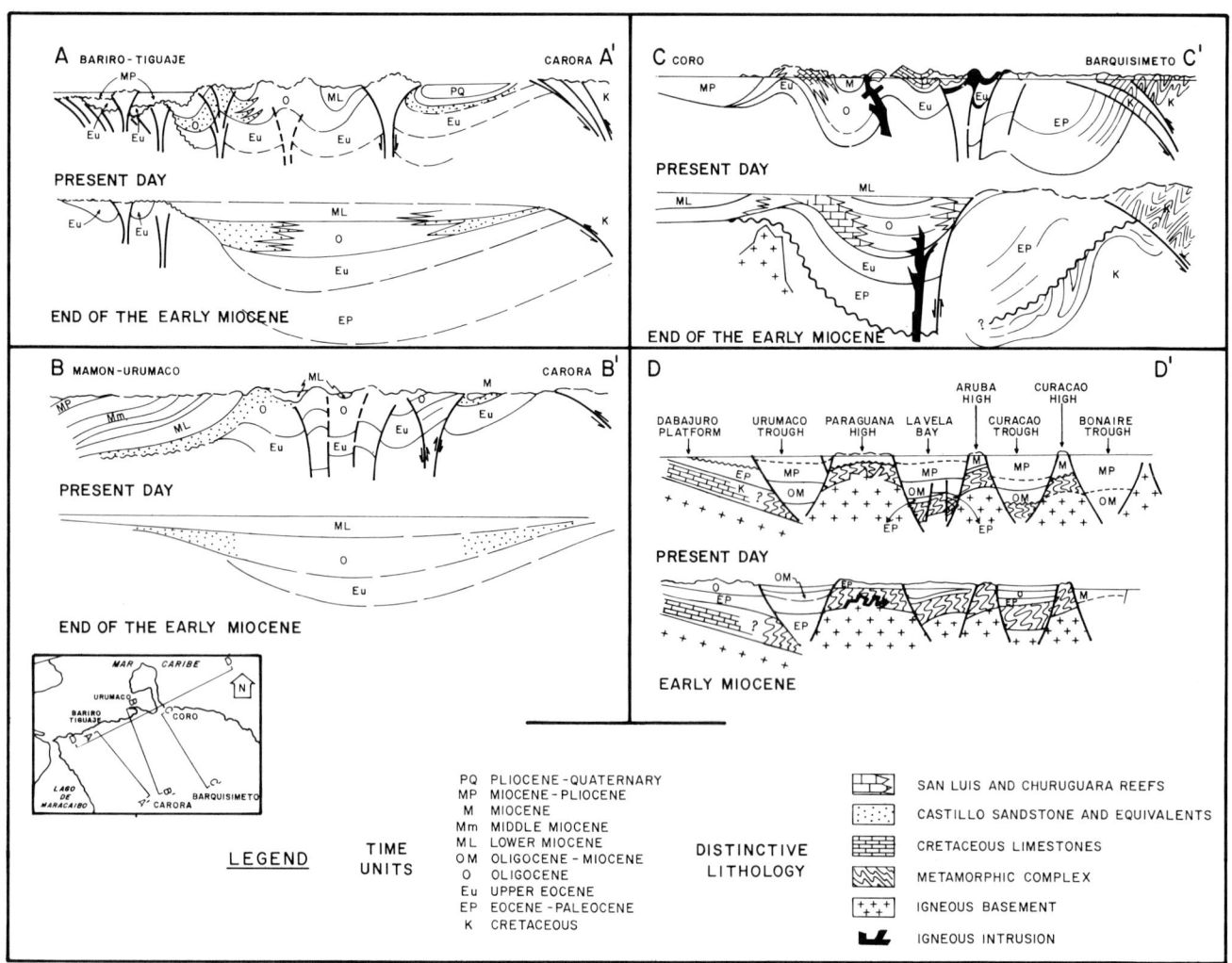

Figure 12. Schematic sections across the Falcón basin. Three northwest-southeast transverse sections (A–A', B–B', C–C') showing extension and compression and one northeast-southwest longitudinal section (D–D') showing continued extension.

The gravity of the crudes varies from west to east between 8° API in the Quiróz area and 48° API in the Cumarebo field. However, accumulations, both in the west and the east, range between 25° and 40° API.

Analyses were carried out on six gas samples from the Tiguaje field (well T-1-41x) and on three nearby wells (Van der Veen and Posthuma, 1983). By plotting carbon isotope values of these gases versus $C_1/\Sigma\, C_n$, the results show that the gas is associated with oil generation and is of thermal origin (Figure 13C).

SOURCE ROCKS

Owing to the complexity of the area, source-rock locations and their levels of maturity could only be determined by using a multidisciplinary approach. Source-rock analyses were carried out on both cuttings and core samples from several key wells using standard geochemical techniques. These included pyrolysis techniques for measuring the amount of organic matter, techniques for determining the type of organic matter, and finally analyses of total organic carbon and vitrinite reflectance. These data have been combined with stratigraphic, sedimentological, and geophysical information in order to identify the source rocks and to get a feel for their distribution within the Falcón basin. Documented source rocks in the basin are listed here in order of importance:

Formation	Age
Equivalent of San Luis	Oligocene–Miocene
Agua Clara	Miocene
Cerro Pelado	Miocene
Jarillal	Eocene
Misoa	Eocene

Based on all the above information and the new tectonic model (Figure 9), a map showing mature source-rock distribution has been produced (Figure 14). It is apparent

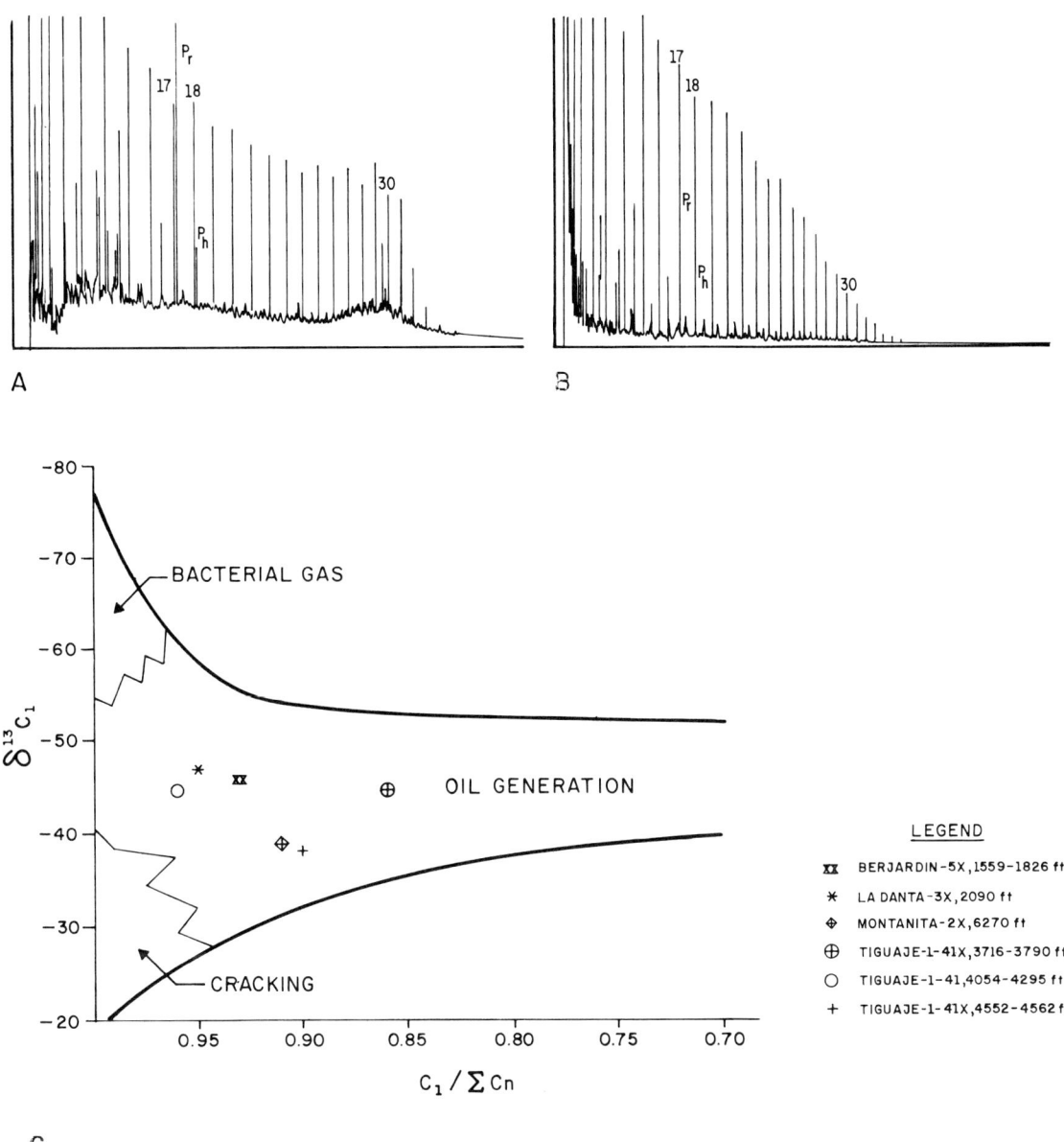

Figure 13. Gas chromatograms of saturated hydrocarbons of the Falcón basin (after Buiskool and Van der Veen, 1984). (A) An example of a waxy crude of terrestrial origin from the western Falcón Tiguaje field (lump in the C_{22}–C_{31} region), (B) crudes of marine origin from the eastern Falcón Cumarebo field, and (C) carbon isotopes versus $C_1/\Sigma C_n$ plot indicating the association of the gases with oil generation (after Van der Veen and Posthuma, 1983).

that the areas of hydrocarbon generation are limited primarily to the troughs where the migration paths are short. This concept, for the time being, is supported by the location of commercial fields in the northeastern part of the basin that are located in troughs (Boesi, 1985).

Source rocks containing marine organic matter exist mainly in the Oligocene and lower Miocene sequences in the northern part of the basin (Findlay, 1985). These rocks were eventually buried in the Urumaco trough (Lew, 1985) and La Vela Bay to depths greater than 10,000 ft (3050 m), depths at which favorable conditions existed for the maturation of the organic matter. Toward the center and western part of the basin, the Oligocene and Miocene rocks were exposed and eroded and are therefore of no interest with regard to hydrocarbon generation. However, the El Mene, Media, and Hombre Pintado fields (Figure 14) in the extreme western Falcón basin produce crude oils of terrigenous origin that were generated from source rocks of Eocene age (Lew, 1985). The possible extension of these Eocene source rocks toward the east could lead to the discovery of additional hydrocarbons in the region north of Carora (Figure 14).

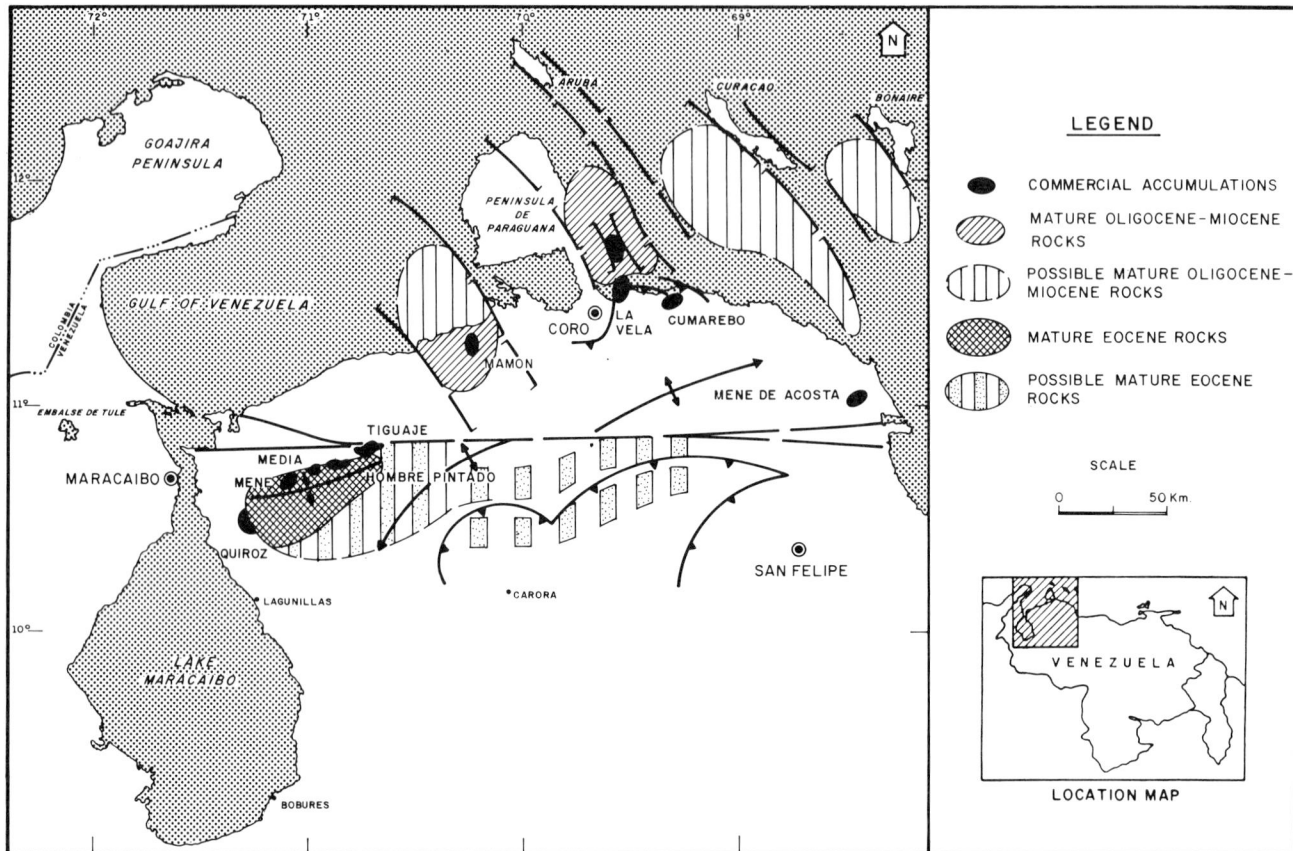

Figure 14. Mature areas and commercial hydrocarbon accumulations in the Falcón basin.

CONCLUSIONS

During more than 70 years of petroleum activity in the Falcón basin, some 12,000 km of seismic lines have been acquired, 1000 wells have been drilled, and numerous technical reports and publications have been produced. With all of this information, it has been possible to obtain a fairly complete picture of the tectonic regime and to establish the basic stratigraphic relationships in the basin.

Basin formation included subsidence and sedimentary filling of a large trench beginning in late Eocene and lasting until middle Miocene. Toward the end of the middle Miocene, the sediments in the basin were compacted, and during the late Miocene–Pliocene orogeny, the basin inverted. This tectonic inversion is responsible for the mountain building and reworking of a large mass of sediments. These sediments have been transported to the north, to the coastal areas, and into the Caribbean Sea.

As a consequence of the latest geological, geophysical, and geochemical findings, the northern flank of the Falcón basin, which encompasses offshore areas, suggests a somewhat different model than previously envisioned. This new model is based on the existence of northwest-southeast horsts and troughs that are now considered as the northern extension of the basin.

Geochemical analyses have shown that crudes from western Falcón basin were derived from source rocks containing terrestrial organic matter. These rocks generated only minor quantities of hydrocarbons. However, the source rocks in the troughs of the northern flank of the basin generated crudes of marine origin. On this flank are located the areas of main hydrocarbon interest. The gas found in the Falcón basin is associated with oil generation and is of thermal origin.

The important source rocks are found in the Miocene Agua Clara and Cerro Pelado formations in the Urumaco trough and in the Eocene Jarillal and Misoa formations located in the western part of the basin. The Agua Clara Formation has proven to be the richest in organic material, and its burial history indicates favorable conditions for the generation of hydrocarbons in the Urumaco trough. This area and other troughs located in the offshore should be looked at more carefully in the future.

ACKNOWLEDGMENTS

This paper is published with the kind permission of Maraven, S.A. and Petróleos de Venezuela, S.A. The authors acknowledge that reference has been made to internal company reports, and they thank Drs. C. Schubert, K. T. Biddle, and K. Burke for their constructive reviews.

REFERENCES

Barbat, W. F., 1956, The Los Angeles Basin area, California, in L. G. Weeks, ed., Habitat of oil: AAPG, p. 62-78.

Bellizzia, A., and D. Rodríguez, 1968, Consideraciones sobre la Estratigrafía de los estados Yaracuy, Lara, Cojedes y Carabobo: Caracas, Bol. de Geología, v. 9, n. 18, p. 515-563.

Bellizzia, A., D. Rodríguez, and M. Graterol, 1972, Ofiolitas de Siquisique y Río Tocuyo y sus Relaciones con la Falla de Oca: Transactions, VI Caribbean Geological Conference, Margarita, Venezuela, p. 182.

Blaser, R., 1979, Source rock and hydrocarbon generation in the Maracaibo basin, western Venezuela: Maraven Report EPC-5841.69.

Boesi, T., 1985, Estudio Geológico Regional de Falcón Occidental (Resumen): Maraven Report EPC-10063.

Bonini, W. E., C. Pimstein de Gaete, and V. Graterol, compilers, 1977, Mapas de Anomalías Gravimétricas de Bouger de la Parte Norte de Venezuela y Areas Vecinas: Venezuela, Ministerio de Energía y Minas, escala 1:1.000,000.

Buiskool, T., and F. Van der Veen, 1984, Geochemical analysis of ten crude oils and one oil seepage (Mene Las Palmas) from Venezuela: Maraven Report EPC-7678.

Burke, K., C. Cooper, J. F. Dewey, P. Mann, and J. L. Pindell, 1984, Caribbean tectonics and relative plate motions, in W. E. Bonini, R. B. Hargraves, and R. Shagam, eds., The Caribbean-South American plate boundary and regional tectonics: GSA Memoir 162, p. 31-63.

Compañía Shell de Venezuela, 1965, Igneous rocks of the Siquisique region, State of Lara: Asociación Venezolana de Geología, Minería y Petróleo, Boletín Informativo v. 8, p. 286-305.

Coronel, G., 1970, Igneous rocks of central Falcón: Asociación Venezolana de Geología, Minería y Petróleo, Boletín Informativo, v. 13, p. 155-162.

Díaz de Gamero, M. L., 1977, Estratigrafía y Micropaleontología del Oligoceno y Mioceno Inferior del Centro de la Cuenca de Falcón, Venezuela: Escuela de Geología y Minas, Universidad Central de Venezuela, Caracas, GEOS, n. 22, p. 3-60.

Feo Codecido, G., 1972, Breves Ideas sobre la Estructura de la Falla de Oca, Venezuela: Transactions, VI Caribbean Geological Conference, Margarita, Venezuela, p. 184-190.

Findlay, A. L., 1985, The Prospectivity of eastern Falcón: Corpoven Report CAIGEPET-CDG, 3658.

González de Juana, C., 1938, Contribución al Estudio de la Cuenca Sedimentaria Zulia-Falcón: Caracas, Bol. Geol. y Min., v. 2 (2-4), p. 123-138.

González de Juana, C., J. M. Iturralde, and X. Picard, 1980, Geología de Venezuela y sus Cuencas Petrolíferas: Caracas, Foninves.

Graf, C., 1972, Relaciones entre Tectonismo y Sedimentación en el Holoceno del Noroeste de Venezuela: IV Congreso Geol. Venez., Caracas, 1969, Memoria, Tomo II, p. 1125-1144.

Hartog, J. J., and R. J. Syaps, 1973, The use of C_7 hydrocarbons for crude oil typing compilation and evaluation of data on 500 non-USA crudes: Maraven Report EPC-5355.

Hedberg, H. D., 1931, Cretaceous limestones as petroleum source rocks in northwestern Venezuela: AAPG Bulletin, v. 15, n. 3, p. 229-244.

Hunter, V. F., 1972, A middle Eocene flysh from east Falcón, Venezuela: Transactions, VI Caribbean Geological Conference, Margarita, Venezuela, 1971, p. 126-130.

James, K., 1984, The Falcón area: a geological appraisal and analysis of hydrocarbon potential: Maraven Report EPC-7768.

Kingston, D. R., C. P. Dishroon, and P. A. Williams, 1983, Global basin classification system: AAPG Bulletin, v. 67, n. 12, p. 2175-2193.

Lew, M., 1985, Estudio Geológico Regional de Falcón Occidental, Estudio Geoquímico, Apéndice VII: Maraven Report EPC-10063.7.

Muessig, K. W., 1978, The central Falcón igneous suite, Venezuela: alkaline basaltic intrusions of Oligocene-Miocene age: Geologic en Mijnbouw, v. 57, p. 261-266.

Muessig, K. W., 1984, Structure and Cenozoic tectonics of the Falcón basin, Venezuela and adjacent areas, in W. E. Bonini, R. B. Hargraves, and R. Shagam, eds., The Caribbean-South American plate boundary and regional tectonics: GSA Memoir 162, p. 217-230.

Payne, A. L., 1951, Cumarebo oil field, Falcón, Venezuela: AAPG Bulletin, v. 35, p. 1850-1878.

Pumpin, V. F., 1978, The structural setting of northwestern Venezuela: Maraven Report EPC-6094.

Renz, H. H., 1948, Stratigraphy and fauna of the Agua Salada Group, State of Falcón, Venezuela: GSA Memoir 32, p. 199.

Renz, O., R. Lakeman, and E. Van der Muelen, 1955, Submarine sliding in western Venezuela: AAPG Bulletin, v. 39, p. 2053-2067.

Schubert, C., 1980, Late Cenozoic pull-apart basins, Boconó fault zone, Venezuelan Andes: J. Struct. Geol., n. 2, p. 463-468.

Sofer, Z., 1984, Stable carbon isotope compositions of crude oils: application to source depositional environments and petroleum alteration: AAPG Bulletin, v. 68, p. 31-49.

Stephan, J. F., 1977, El Contacto Cadena Caribe-Andes Merideños entre Carora y El Tocuyo (Edo. Lara): V Venezuelan Geologic Congress, v. 2, p. 789-810.

Van der Veen, F., and J. Posthuma, 1983, Geochemical analysis of 22 gas samples from Los Lanudos, Falcón, Mara-La Paz and Zuata wells, Venezuela: Maraven Report EPC-7610.

Vásquez, E., and P. A. Dickey, 1972, Major faulting in northwestern Venezuela and its relation to global tectonics: Transactions, VI Caribbean Geological Conference, Margarita, Venezuela, p. 191-202.

Wheeler, C. B., 1963, Oligocene and lower Miocene stratigraphy of western and northeastern Falcón Basin, Venezuela: AAPG Bulletin, v. 47, p. 35-68.

Zambrano, E., E. Vásquez, B. Duval, M. Latraille, and B. Coffinieres, 1971, Síntesis Paleogeográfica y Petrolera del Occidente de Venezuela: Memoria IV Congreso Geológico Venezolano, Caracas, 1969: Venezuela Ministerio de Minas e Hidrocarburos, Boletín de Geología, Publicación Especial, n. 5, v. 1, p. 481-552.

Index

Anaheim nose, 83
  cross section of, 87
Andes Mountains, tectonic map of, 285
Austro-Indian plate, 272

Backstripping, 189-190
Basin productivity, 8
Basins, location of, 3
Beta oil field, cross section of, 55
Brea Olinda oil field, cross section of, 54

Capistrano Formation, 146
Central Los Angeles basin
  geohistory plot of, 190
  geology of, 186-187
  major structures of, 186-187
  oil fields of, 187
  subsidence of, 185-194
  tectonic evolution of, 185-194
  tectonic subsidence curves for, 190
  tectonostratigraphy of, 187
  temperature data from, 193
  thermal implications for, 185-194
  vitrinite reflectance data from, 193
Coyote Hills
  subsurface structure map of, 85
  surface geology of, 85
  topography map of, 85

Dominguez oil field, subsurface structure map of, 69

East Beverly Hills oil field, cross section of, 61
Eastern Cordillera
  distribution of sedimentary facies in, 289
  geologic map of, 286
Extensional basin formation, model of, 193

Falcón basin, 303-319
  Bouguer gravity anomalies map of, 306
  chronostratigraphic relationships among several structures of, 311
  combined seismic-reflection profiles across, 308-309
  conceptual model of, 310
  gas chromatograms of saturated hydrocarbons in, 317
  geologic history of, 304-314
  hydrocarbons in, 314-316
  lithostratigraphic relationships among several structures of, 311
  location map of, 305
  map of mature areas and hydrocarbon accumulations in, 318
  map of tectonic evolution of, 307
  paleogeographic situation of
    during Eocene, 314
    during Miocene, 315
    during Oligocene, 314
  proposed conceptual model of, 313
  schematic cross sections across, 316
  sedimentation of
    during basin formation, 310
    prior to basin formation, 306-307
  source rocks in, 316-317
  stratigraphic nomenclature of, 312
  stratigraphy of
    during basin formation, 310
    prior to basin formation, 306-307
  tectonic evolution and principal structures of, 305-306
  transverse stratigraphic section through, 310
Fernando Formation, 149, 243-244
Fresne structure, seismic line through, 275
Fullerton embayment, 84-88

Geothermal gradient, changes in, 191
Geotherms, evolution of, 192
Girardot subbasin, map of structural relationships in, 296
Glendora Volcanics, 97

Huntington Beach oil field
  cross section of, 74
  subsurface structure map of, 73
Huntington Beach-Yorba Linda, cross section of, foldout following 224

Inglewood oil field
  cross section of, 68
  subsurface structure map of, 21

Kleinpell, biostratigraphic scheme of, 150-153
Kupe structure, seismic line through, 273

La Habra Formation, 150
Las Cienegas fault, 107
Lithospheric thinning, 190
  changes in heat flux and, 191
Loading mechanisms, 187-189
  isostatic, 189
Long Beach Airport oil field
  cross section of, 71
  subsurface structure map of, 70
Los Angeles basin. *See also* Central Los Angeles basin; individual formation names, Los Angeles basin oils; Los Angeles basin cross sections
  basement complex of, 240
  biodegradation and influence on oil quality in, 214-215
  biomarker ratios for Salt Lake area oils in, 215
  boundary structures of, 47-66
  central block of, 137
  central trough of, 82-83
    burial and thermal history of, 229-231
    geothermal gradients of, 230
    sediment thickness of, 230
    time-temperature plots for, 231
  chemical characteristics of Salt Lake area oils and gases in, 217
  chronology of tectonism of, 99-101
  chronostratigraphic framework of, 159
  correlation of benthic foraminiferal biostratigraphic zonations in, 151-152
  correlation of benthic stages to planktonic microfossil zonations in, 155-170
  correlation section across, 246
  $\delta^{13}C$ values of carbonate cements in, 213
  $\delta^{13}C$ values of carbonate minerals in, 213
  deposition of, 106
  discovery history of, 12
  distribution of basement rocks in, 41, 244
  Division A ("Delmontian") of, 226
  Division B ("Delmontian") of, 226
  Division C (upper Mohnian) of, 226
  Divisions D and E (Mohnian) of, 224-226
  during Pasadenan deformation of, 115
  dynamic setting of, 107-108
  early Miocene reconstruction of, 90-92
  early Pliocene reconstruction of, 105
  factors leading to oil productivity of, 254-256
  field-size distribution of, 16
  gas chromatograms of Salt Lake area oils in, 215
  generalized stratigraphic chart of, 139
  geologic/geomorphic provinces of, 39-43
  geothermal gradients within, 200
  growth of internal structures of, 106-107
  history of oil discoveries in urban setting of, 25-34
  Holocene deformation of, 109
  index map of, 48, 223
  isopachs of, 225, 228-229
  isotopic characteristics of Salt Lake area oils and gases in, 217
  Late Cretaceous Series in, 138
  late Miocene stability of, 101-102
  late-stage shortening of, 16
  location of oil fields in, 10, 15, 199
  location of surface faults, 10
  location of wells in, 242
  lower Miocene Series in, 139-143
  lower Mohnian formations of, 242-243
  lower Mohnian Stage of, 249-250
  lower Pliocene Stage of, 243-244
  major internal structures of, 66-89
  map of major faults in northwestern, 57
  map of major structures in, 199
  map of producing areas, 245
  map of stratigraphic sections of, 138
  middle Miocene deformation of, 92-95
  middle Miocene formations of, 240-242
  middle Miocene reconstruction map of, 93
  middle Miocene Series in, 138-139, 143-146
  Mio-Pliocene deformation of, 102-108
  northern margin of, 63
  northern shelf of, 83-84
  northwestern, subsurface stratigraphic section of, 104
  northwestern block of, 137
  oil and gas fields of, 13-14
    stratigraphic controls of, 221-232
  oil and gas systems of, 6, 197-217
  origin of, 7
  overview of, 5-24
  Paleocene Series in, 138-139
  Paleogene formations of, 240
  Pasadenan deformation of, 108
  peripheral basins of, 46
  plate-tectonic framework of, 43-46
  Pleistocene Series in, 146-150, 244-245
  Pliocene Series in, 146-150
  post-mid-Miocene displacements of, 104-106
  post-Repettian strata of, 226-227
  predecessor basins of, 43
  present-day paleobathymetry of, 254
  Quaternary structure of, 228
  questionable structures of, 89-90
  regional biostratigraphic correlations of, 170-180
  regional correlation to planktonic zonations of, 167-170
  regional setting of, 35-45, 137, 222-224, 240
  regional stratigraphy of, 137-150
  Repettian Stage of, 226-227, 243-244
  right-lateral displacements of, during Pasadenan deformation, 108-110
  sedimentation and, 254-255
  source rocks in, 199-202
    hydrocarbon potential of, 200-202
    maturity of, 199-200
  southeastern margin of, 63
  southwestern block of, 137
  stratigraphic setting of, 240-245
  structural geology of, 46-89
  structure-contour map of, 52, 53
  style of crustal extension of, 94
  submarine fans of. *See* Submarine fans, Los Angeles basin
  subsidence and uplift of, during Pasadenan deformation, 110

subsurface stratigraphy of, 147-148
tectonic evolution of, 90-116
top-of-basement surface of, 11
topography of, 9
Upper Cretaceous formations of, 240
upper Eocene Series in, 139-143
upper Miocene formations of, 242-243
upper Miocene Series in, 146
upper Pliocene of, 244-245
vitrinite reflectance versus alginite percentage in kerogens in, 201
western shelf of, 78-82
Los Angeles basin cross sections. See also names of individual oil fields
end of Repetto, 232
late Pleistocene, 232
regional, 50-51
restored, 20
retrodeformable, 113
simplified retrodeformable, 19
subsurface, 200
Los Angeles basin natural gases, 207-214
carbon dioxide in, 212-214
chemical composition of, 209
stable isotopic composition of, 210
Los Angeles basin oils, 202-207
accumulations of, 228-229
depositional environment parameters for, 206
depth distribution of API gravity and sulfur content in, 207
future prospects of, 231-232
gas chromatograms of, 205
geochemical characteristics of, 203
introduction to, 202-203
maturation parameters for, 204
migration parameters for, 205
ternary plot of, 206
ternary plot of sterane distribution in, 204
triterpane mass chromatograms of, 206

Magdalena basins, 283-300
evolution of, 292-294
future opportunities for hydrocarbon exploitation of, 299-300
model of basement-rooted thrusting bordering, 291
oil and gas systems of, 294-300
regional setting of, 284-287
reservoirs in, 294
source rock in, 294-299
stratigraphic summary of, 287-290
tectonics and structural geology of, 290-292
timing of deformation in, 291-292
trapping mechanisms in, 299
Magdalena Valley
cross section through, 293
distribution of sedimentary facies in, 289
generalized stratigraphy of, 288
geologic map of, 286
hydrocarbon reserves in, 296
principal fields in Middle, 297
principal fields in Upper, 298

principal structures and producing fields in Middle, 292
principal structures and producing fields in Upper, 294
structural cross sections of Upper, 295
structural cross section through, 293
Middle and Upper Magdalena basins, 283-300. See Magdalena basins
Mio-Pliocene deformation, onset of marginal uplifts during, 102-104
Modelo Formation, 146, 200-201, 242-243
Montebello oil field, subsurface structure map of, 89
Montebello-Puente Hills, cross section of, foldout following 224
Monterey Formation, 146, 200-202, 242-243
Monterey Shale, Van Krevelen diagram of, 202

Natland, biostratigraphic scheme of, 155
Neogene basins, California, location of, 7
Newport Bay area, stratigraphy of, 143-144
Newport Bay section, 163-166
Newport-Inglewood fault zone, 66-78
New Zealand, continental shelf of, 263
North Los Angeles thrust system, 111-115

Pacific-North American plate boundary, Cenozoic evolution of, 44
Pacific plate, 272
Palos Verdes-central trough, cross section of, foldout following 224
Palos Verdes fault, 55-56
Palos Verdes Hills, stratigraphy of, 141
Palos Verdes Hills section, 157-160
Palos Verdes uplift, 55-56
Pico Formation, 149, 244-245
Playa del Rey oil field, subsurface structure map of, 79
Playa del Rey-Inglewood, cross section of, foldout following 224
Puente fan, 101
Puente Formation, 146, 200-201, 242-243
Puente Hills, stratigraphy of, 144-145
Puente Hills composite section, 167-168
Pull-apart basin, map of idealized, 16

Repetto Formation, 149, 243-244

San Joaquin Hills, 96-97
geologic map of, 64
restored cross sections of, 65
stratigraphy of, 142
San Onofre Breccia Formation, 240-241
San Pedro Formation, 150, 244-245
Santa Ana Mountains, stratigraphy of, 144-145
Santa Fe Springs-Anaheim nose, cross section of, foldout following 224
Santa Fe Springs oil field, subsurface structure map of, 88
Santa Fe Springs-Whittier, cross section of, foldout following 224
Santa Monica fault system, 56-63, 107
cross sections of, 58

Neogene tectonic evolution of, 59
Santa Monica Mountains, 95–96
    stratigraphy of, 140
San Vicente oil field, cross section of, 61
Sawtelle oil field, cross section of, 60
Schist Ridge, cross section of, foldout following 224
Seal Beach oil field, topography and subsurface structure map of, 72
Sespe Formation, 139
Sespe/Vaqueros undifferentiated, 143
Slip, geometry of, 194
Southern California
    geometric model of, 18
    initial fracture pattern of, 17
    palinspastic reconstruction of, 17
Southern California borderland
    distribution of sedimentary rocks within, 243
    Neogene basins within, 241
Submarine fans, Los Angles basin, 239–259
    Delmontian, 251–252
        present-day facies map of, 253
        producing oil fields in, 255
    Mohnian, 248–249
        facies map of present-day, 248
        palinspastic restoration of facies of, 249
    Neogene, 245–254
        model of, 245–249
    Normack's model of, 247
    Repettian, 252–253
        present-day facies map of, 256
        producing oil fields in, 257
    upper Mohnian, 250–251
        producing oil fields in, 252
Subsidence analysis, organization of, 188
Sunset Beach oil field
    cross section of, 74
    subsurface structure map of, 73

Taranaki basin, 261–282
    authigenic kaolinite in, 277
    cross section through, 268
    log correlation of, 267
    major structural elements of, 264
    map of regional setting of, 263
    maturation of potential source material in, 281
    migration of hydrocarbons in, 281
    paleodepth curve for, 279–280
    regional setting of, 262
    sample of detrital quartz grain in, 277
    sample showing dissolution of potassium feldspar grains from, 277
    seismic line across, 269
    stratigraphic correlation of, 266
    stratigraphy of, 262–272, 265
    structural contour map of, 270
    structure of, 272–276
    subsidence and tectonic curves for, 278–280
    subsidence and thermal history of, 276–281
Taranaki basin reservoirs, 276
Taranaki basin source rocks, 276
Tariki structure, seismic line through, 274
Tarzana fan, 101
    cross sections of, 103
Toetoe structure, seismic line through, 274
Topanga Canyon section, 156–157
Topanga Formation, 240–241
Topanga Group, distribution and thickness of, 42
Torrance-Wilmington, cross section of, foldout following 224

Vaqueros Formation, 139–143
Venezuela, western, tectonic map of, 285
Venice oil field, subsurface structure map of, 79

West Coyote oil field, cross section of, 87
West Coyote–Puente Hills, cross section of, foldout following 224
Whittier fault, 107
Whittier oil field, cross section of, 54
Wilmington anticline
    cross sections of, 81
    subsurface structure map of, 80
Wilmington–Long Beach, cross section of, foldout following 224
Wissler, biostratigraphic scheme of, 153–155

QE 693.5 .A28 1991

Active margin basins